Nucleus and Nation

Nucleus and Nation

Scientists, International Networks, and Power in India

ROBERT S. ANDERSON

The University of Chicago Press
Chicago and London

Robert S. Anderson is professor in the School of Communication at
Simon Fraser University in Vancouver, British Columbia, Canada.

The University of Chicago Press, Chicago 60637
The University of Chicago Press, Ltd., London

Printed in the United States of America
19 18 17 16 15 14 13 12 11 10 1 2 3 4 5

ISBN-13: 978-0-226-01975-8 (cloth)
ISBN-10: 0-226-01975-6 (cloth)

Library of Congress Cataloging-in-Publication Data

Anderson, Robert S., 1942–
Nucleus and nation: scientists, international networks, and power in India / Robert S. Anderson.
p. cm.
Includes bibliographical references and index.
ISBN-13: 978-0-226-01975-8 (cloth: alk. paper)
ISBN-10: 0-226-01975-6 (cloth: alk. paper) 1. Science—India—History—20th century. 2. Nuclear industry—India—History—20th century. 3. Saha, Meghnad, 1893–1956. 4. Bhatnagar, Shanti Swarupa, Sir, 1894–1955. 5. Bhabha, Homi Jehangir, 1909–1966. I. Title.
Q127.I4A69 2010
509.54'0904—dc22 2009036012

♾ The paper used in this publication meets the minimum requirements of the American National Standard for Information Sciences—Permanence of Paper for Printed Library Materials, ANSI Z39.48-1992.

Dedicated to the scientists, technologists, observers, historians, and teachers who have given their time and insights so generously to me, and to this book

CONTENTS

A gallery of photographs appears following page 226.

PREFACE

Early in 1962, when I was a nineteen-year-old university student at Santiniketan in rural West Bengal, we all experienced a rare conjunction of eight planets called "astograha." This three-day period was accompanied by a popular idea that the world might end in catastrophe as a result of this conjunction. Experts in Hindu cosmological theory sounded the alarm, and I watched large numbers of people coming together to spend the chilly winter night, wrapped in shawls and huddled in blankets, under tents and awnings outside their houses in the nearby town of Bolpur, warming fires flickering in the moonless night. They expected an earthquake or some other disaster to befall them. This conjunction was observed in many parts of India; it lasted for a few nights, and then the blanket-bundled groups diminished and ultimately disappeared back into their houses. Though learned astrological pundits debated the consequences of the rare event, I heard a few of my university classmates joke about it. The paper of record in English at that time in Bengal, the *Statesman* of Calcutta, cautioned people not to be persuaded by "unscientific thinking" and quoted distinguished scientists urging people to reason about this conjunction, to understand that rare cosmological events did not necessarily have disastrous consequences. I had never before met or seen large numbers of people expecting the end of the world, and my feeble Bengali was just good enough to understand that these people really *expected* a disaster. Though I don't think I had yet met any Indian scientists, in an inchoate way I sensed that there was an important and interesting relationship between science and culture in India, and that "unscientific thinking" must have a very special meaning.

I returned from India in 1962 to study at the University of British Columbia and began to think about science and culture in India. Employed in 1963 as an assistant in a research project on the experiences of foreign students in Vancouver, I became aware, when talking with young Indian scientists and engineers, that some of them experienced a tension with their families, their cultural origins, and their professional futures. I did not invent this tension just for something to talk about—they volunteered it; they tended not to see science as fully situated in the India they were returning to. My meeting with older and more mature scientists from India gradually confirmed that very interesting questions could be asked about the culture of science and scientific institutions in India.

I formulated this subject for an honors essay in 1964, with the guidance of my teachers in sociology, religious studies, and anthropology. Focusing on the experience of young Indian scientists, I tried to address the questions raised by C. P. Snow in *The Two Cultures and the Scientific Revolution* and Thomas Kuhn in *The Structure of Scientific Revolutions.*[1] I then discovered the marvelous books of Joseph Needham, which had begun to appear in the late 1950s, and looked for something comparable about India. One of my teachers, Cyril Belshaw, then handed me Edward Shils's book on Indian intellectuals, and I audaciously wrote to Shils, asking questions about his contact with scientists and sending a draft of my honors essay in 1964.[2] At twenty-two, I hardly knew the context of what I was talking about. I had no formal training in science, little study in the history and philosophy of science, and one year's experience in India. But Shils suggested I apply for graduate studies at the University of Chicago, which I did. I remember first meeting members of the Committee on South Asia at the University of Chicago in 1965, to discuss my "ideas," such as they were, particularly with Milton Singer, who, it turned out, had been trained in the philosophy of science and had spent time in South India, thinking about these questions. Though I was generously supported by scholarships, as a poor graduate student at Chicago in the summer of 1966 I needed a part-time job, and got work as a scanner in the high-energy physics bubble chamber group at the Enrico Fermi Institute for Nuclear Studies; now I thought I could understand things from the inside. If my choice of this subject for a dissertation was seen as rather eccentric, this perception arose in an anthropological community quite full of eccentrics.

I proposed a masters thesis about the social organization and culture of experiments and theoretical work at the Fermi Institute at the university, and my objective was diplomatically introduced to the Fermi Institute by Edward Shils, whose book about the consequences of secrecy was well known among nuclear physicists at Chicago.[3] I realized in 1966–67 that while working among this historic community of scientists, I was stumbling through the echoes of a history of science and physics in America, and I had better begin at "the beginning" (an arbitrary point, nevertheless). In a private sense, many physicists were very interested in history, but the history of the Fermi Institute was in their "collective" mind and not on paper. I was, to my surprise, allowed to read Fermi's own papers and files, held in a giant steel filing cabinet. Near the spot where the first sustained chain reaction occurred in 1942, I read and talked with people who knew Fermi and his colleagues; these conversations and files took me right back to the organization of the Metallurgical Laboratory, which housed the Manhattan Project.[4]

I understood that I was now in touch (through memos, scribbled notes, etc.) with Fermi, one of the deities of modern science, viewed widely as a genius for his combination of theoretical sophistication and experimental ingenuity. I also saw how he raised money and defended his staff from the intrusions of the government bureaucracies that funded their work. I read the file on his notes for the defense of Robert Oppenheimer and then came across the extraordinary papers of Leo Szilard. I think there was then no modern "history of science" of these phenomena, or individuals, and no histories of the Manhattan Project or the postwar period, for it was all still seized in secrecy. But I was learning that there is a politics of science, without which scientists seem not to function. I had not yet met historians like Charles Weiner at the American Institute of Physics, who had begun to put together the vast archives needed to do this kind of institutional history. But probably more important, I thought I was becoming an anthropologist and not an historian, preparing to go to India to do fieldwork in laboratories for two years, not in archives; I was to be prepared in the same way some of my classmates prepared to go to rural villages.[5] But the work of the historian was just waiting around the corner for me.

When I looked around, there was only one complete ethnographic study of a scientific institution done in the manner I intended. Its author, Gerald Swatez, had just moved in 1965 from studying the Lawrence Radiation Laboratory at the University of California, Berkeley, to work at the University of Chicago.[6] Swatez gave me good advice and continuing friendly and wise support throughout my work toward the dissertation. I assembled everything written at the time about scientific institutions and communities, taking a sociological-anthropological-cum-historical approach.

From anthropology I walked over to physics. I began a series of conversations with S. Chandrasekhar about scientists in India and attended his lectures, understanding little except his strong emphasis on the aesthetics of mathematical choice. But our stimulating conversations persisted. I remained an innumerate kind of Alice in Wonderland among physicists, who, it turns out, had Alice as a cultural heroine among them. The Physics

Colloquium at Chicago was an invaluable experience; attending it with physics graduate students, particularly with my roommate G. Srinivasan, I came to understand the public culture of physics in America, physics onstage, so to speak. I recall even now the excitement of the presence in the colloquium of someone who had just won the Nobel Prize and could see people among whom I worked (as their research assistant) catching some stardust in his presence, all in the human hope that it would lift them up in that direction. As a "scientific worker" in a bubble chamber group, I learned a lot about the electron that spins off when a muon decays from a pion colliding with a hydrogen atom, at certain specific energy levels (in my case 21 million electron volts), when an accelerator beam passed through a chamber of extremely compressed liquid hydrogen. I looked for that electron in hundreds of thousands of photographs, and gasped "yes" with excitement when I found and recorded one of our few perfect specimens. Tired or bored co-workers gathered round to compare my pi-mu-e result with their best ones. Now I knew how an experiment had to be organized and managed, though little more.

Though I had prepared for research in India, by 1967 I had actually to go and do it. Introductory letters were written from the University of Chicago by S. Chandrasekhar, E. Shils, and space physicist John Simpson to Vikram Sarabhai, the new chairman of the Atomic Energy Commission. Sarabhai liked the idea of anthropological study of scientific institutions and had already encouraged them in his own organizations; he introduced me to the director and council of the Tata Institute of Fundamental Research (TIFR), and the director of the Saha Institute of Nuclear Physics (SINP). I was based in Bombay but traveled everywhere to other sites of the scientific community's work. After a year in Bombay, I arranged to work for a year in the Saha Institute in Calcutta (1968–69), for purposes of comparison with TIFR. Beginning in the late 1960s, I have enjoyed hundreds of conversations about the lives of scientists and the life of science in India, on buses and trains, at seminars and conferences, in airports and markets, waiting for rockets to launch or balloons to fly, building telescopes or culturing organisms in petri dishes, and returning back to the lab from their home village.

I saw the old musty archives of Kolkata, but I could barely touch the histories of the Indian Association for the Cultivation of Science, the Bose Institute, and the Indian Statistical Institute, even Science College. More important, I could learn only a little at that time about the lives of Meghnad Saha, Shanti Bhatnagar, and Homi Bhabha, three people who dominate this book. Everything written about them was too thin to be interesting and too exemplary. I privately asked questions about each of them like "Did he take short-cuts? Did he have disappointments? Did he reach dead ends in his thinking? Was he ever opposed? Did he have a sense of humor?" I was asking how different epochs perceive the acts of great scientists, and how the heroic narratives about them are created and developed: the cultural expectations seemed to be that these scientists must be both romantic intellectual virtuosi and cunning bureaucratic strategists (and in India, good "family men"), all at once. And, in these biographies, where were all the people who made the work of these heroes possible?

But there was simply too little information then to answer these questions. I did what I could but encountered a respectful and secure wall around Bhabha, a decorative fence of flowers around Saha, and almost nothing around Bhatnagar. I learned about (but did not see) Bhabha's office, untouched since his recent death, and observed a 1967 visit to TIFR by his lifelong companion Pipsi Wadia for the first time since the 1966 plane accident. I looked many times at Homi Bhabha's paintings, and strolled often in his garden at the institute. I picked up Bhabha stories and Saha stories, some of them very interesting, but I had mostly to rely on authorized and rather administrative memorials. Bhabha's story was more closed because of his knowledge of the major strategic questions recently facing nuclear India. I read the published biographies and speeches of Shanti Bhatnagar but found little more written about this influential person. When I moved to the Saha Institute in 1968, Meghnad Saha had been dead for more than twelve years, but he at least had authorized a frank and colorful biography for his sixtieth birthday, just before his death. I saw only Saha's published speeches on science, statements in Parliament, and editorials from *Science and Culture,* but none of his copious correspondence, available only thirty years later. Though most of his immediate family was still alive, and his son Ajit Saha was a physics professor in the institute, I learned only enough to make a sketch and now knew how big a hole I would have to fill if I were to understand properly the lives of these men and their historic contexts: who was I, a mere dissertation writer, to fill that hole anyway?

While working with postwar refugees in Bangladesh in 1972–73, in communities with numerous Saha families in them, I was inspired to visit Saha's native village about an hour by bus north of Dhaka. I began to prepare a manuscript from the first chapter of my dissertation, and this was published two years later as *Building Scientific Institutions in India: Meghnad Saha and Homi Bhabha.*[7] When that monograph went to press in early 1974, this subject was, in North America at least, regarded as very obscure indeed. Then a few months later the first Indian nuclear test in the Rajasthan Desert changed that, and suddenly though briefly it was important to understand science and scientists in India. But since my book contained little more about the nuclear weapon readiness of India than could be seen in the pages of the *New York Times* or the *Times of India,* it appealed more to those with a long-term view of how science and technology developed in India and other poor countries. Although not easily available, the book was serialized in India in three issues of *Science Today* (then published monthly in Bombay by the *Times of India*), and thus reached about 1 million readers in India during 1976.[8] Apart from an article in *Contributions to Indian Sociology* in 1977, I accepted the disinterest in this subject in the world around me and turned to other research on tropical forests and rice cultivation systems. But eight months before the second Indian nuclear tests in 1998, I was in Cambridge as a visiting fellow and began to realize how many new sources were available to fill out the earlier picture I had formed, including formerly secret documents. People were now more willing to speak about the past. So I began to rework my earlier book on Saha and Bhabha, decided to include the new material on Bhatnagar, and began to build what is now seen in your hands. I visited India in January 1998 and had the conversations described in chapter 1, unaware of the preparations for the second nuclear bomb test a few months later.

ACKNOWLEDGMENTS

This book has been so long in writing that many people may now be oblivious of the debt I owe them. I wish I had thanked them earlier. In some cases, unfortunately, my appreciation is expressed posthumously. Some named here engaged in occasional yet important conversations; others helped over many years, as friends, teachers, and guides. Throughout this work I have received the kindness of strangers, and though I don't really know how to reciprocate, I only trust that this result is of value to most of them. I recognize that I may omit from acknowledgment some important names, and apologize now if this occurred. My colleagues in India always persuaded me that this work was worth rethinking and writing, after twenty-five years away from it.

Following a lucky introduction to India as a young student at Santiniketan and Madras Christian College in 1961 and 1962, I stumbled upon this subject. I learned first from my teachers at the University of British Columbia, particularly social scientists Kaspar Naegele, Joseph Richardson, and Cyril Belshaw, physicist Myer Bloom and a host of Indian students of science and engineering, most of whom will have forgotten their Canadian encounter with me in the 1960s. Later, in Vancouver, I had continuing support from Barrie Morrison, Michael Ames, John Wood, Edwin Levy, Tirthankar and Mandakranta Bose, Tony Beck, and in Seattle from Paul Brass. The regular collegial meetings of scholars of South Asia in Vancouver and Seattle kept me informed about studies of India.

At the University of Chicago, I am indebted to Edward Shils, who very patiently supervised the original dissertation in his thoughtful and penetrating manner (though it terrified me at moments). To my teachers Milton Singer, Bernard Cohn, McKim Marriot, Nur Yalman, A. K. Ramanujan, M. N.

Srinivas, Claude Lévi-Strauss, David Schneider, Clifford Geertz, Saul Bellow, Louis Dumont, Hannah Arendt, and Mircea Eliade, I offer a long and belated gesture of thanks. I am always grateful also to astrophysicist S. Chandrasekhar, experimental physicist Roger Hildebrand, and space researcher John Simpson for many conversations about research and the steps they took to facilitate my work in India. And to fellow students G. Srinivasan, Akos Ostor, Lina Fruzetti, and Paul Rabinow—now all distinguished researchers—I owe the debt for challenging conversations and the buzz of graduate student life in Chicago. To David DeVorkin at the National Air and Space Museum, Washington, Indira Chowdhury at TIFR, and Abha Sur at MIT, I send thanks for an exchange of ideas about Meghnad Saha and Homi Bhabha.

At the Centre for Developing Area Studies at McGill University in Montreal in 1970–71, I began the first version of this book, called *Building Scientific Institutions in India: Meghnad Saha and Homi Bhabha* (1975). The anthropologist Richard Salisbury, who graciously wrote its preface, and Rosalind Boyd-Jeeroburkhan, who skillfully edited it, both gave steady and thoughtful encouragement. I am grateful for the centre's permission to include that material here.

Some of the work for this book was done while I was a visiting fellow at the National Institute of Science, Technology, and Development in Delhi in 1998, and I am grateful for the facilities and encouragement provided for me there. I recommenced this project while a visiting fellow at Corpus Christi College in Cambridge in 1997 and 1998, and I thank the Master Sir Tony Wrigley and Fellows for that extraordinary opportunity.

In Delhi in 1998, I was aided by Ashok Jain, director of the National Institute for Science, Technology, and Development Studies, and his colleagues S. Irfan Habib, Dinesh Obrol, Druv Raina, and Rajeswari Raina, and encouraged by Shiv Visvanathan at the Centre for the Study of Developing Societies, Deepak Kumar and V. V. Krishna at Jawaharlal Nehru University; I began invaluable conversations with A. Rahman, Ashok Parthasarathy, Upen Trivedi, Sukhamoy Chakravarty, Kamla Chowdhury, Ram Prasad, and a couple of other experts who asked not to be named.

I am very grateful for conversations with or support by individuals in Kolkata, Himani Banerji, Satyendranath Bose, Shantimoy Chatterjee, B. D. Nagchaudhuri, Sajni Kripalani, Indranil Chakraborty (for his assistance with the photographs), and particularly at the Saha Institute, Manoj Pal, Dipti Pal, Bikash Sinha, Binayak Dutta-Ray, and Atri Mukhopadhyay; in Mumbai, I am indebted to Vikram Sarabhai, Rustom D. Choksi, B. Choksi, J. J. Bhabha; and particularly at the Tata Institute, M. G. K. Menon, Virendra Singh, P. P. Divakaran, Yash Pal, Obaid Siddiqi, R. Narasimhan, B. M. Udgaonkar,

Govind Swarup, Oindrila Ray, and B. V. Sreekantan; in Ahmedabad, I was greatly helped by Padmanabh Joshi and writer Amrita Shah; in Pune, I was kindly assisted by S. Ananthakrishnan and G. S. Swarup; in Chennai, by S. C. Sathya, formerly of Indian Space Research Organization, and N. Ram, editor of the *Hindu;* in Bangalore, I am grateful for guidance from G. Srinivasan, Sir C. V. Raman, Satish Dhawan, B. V. Subbarayappa, A. Ratnakar, V. Nanjundiah, M. N. Srinivas, Girija Srinivasan, and Amulya Reddy. In Tiruvanthapuram, I am grateful to writer Gopal Raj; I recognize particularly the work and influence of physicist and historian G. Venkatraman (formerly of Bhabha Atomic Research Centre and now at Puttaparathi), a person who has made the history of science in India come alive through his books and whose insight and sharp memory are seldom equaled.

In Cambridge, during a number of visits as visiting fellow at both Corpus Christi College and Clare Hall since 1997, I was encouraged and assisted by Sir Christopher Bayley, librarians and the weekly seminar on South Asia, the late Raj Chandravarkar, James Crawford, Romila Thapar, Asiya Siddiqui, Haroon Ahmed, Benjamin Zachariah, Sulagna Roy, Christopher Andrew, Arne Wested, Ekhard Salje, Stephen Hugh-Jones, and Alan Macfarlane.

Archivists have been most helpful and courteous to me at the following archives: Churchill College, Cambridge; University Library, University of Cambridge; the Royal Society, London; National Archives, Kew, London; British Library, London; Saha Institute for Nuclear Physics, Kolkata; Tata Institute for Fundamental Research, Mumbai; National Institute for Science, Technology, and Development (CSIR), Delhi; National Archives of Canada, Ottawa.

The financial support from Canada Council for support for the fieldwork in India in 1967–69 is gratefully acknowledged, as is the University of Chicago's financial support for my original studies. I have visited India numerous times since 1961 and have written books about forestry and rice agriculture there. All the ensuing experience provided the context for interpreting the evidence in this book. I thank again those hundreds of Indian researchers who spoke to me and acknowledge here that their insights about the sciences and the larger scientific community have been invaluable: their generosity of ideas is like a grand sunrise, and I could not have succeeded without their clarity and candor over the years. My graduate students at Simon Fraser University have also kept me informed about India. A Simon Fraser University President's Research Grant in 1997–98 and a University Publication Grant in 2008–9 supported completion of this work; the expert administrative team at the School of Communication under Lucie Menkveld cheerfully enabled me to complete this work.

My editors Christie Henry, Jean Eckenfels, Erin DeWitt, and Dmitri Sandbeck at the University of Chicago Press and indexer Margaret Manery have been able to turn an average writer's prose into something finer, and I am very grateful to have had their guidance and support.

It is customary to acknowledge the support of family and friends when putting a book to bed. Over such a long period, our children have grown and my dear wife, Kathy Mezei, has generously accompanied and loyally encouraged me all along, including during separations caused by this work. And because friends have grown to wonder about my long study of a subject so far away, in reply, here it is.

Robert Anderson
Vancouver, 2009

NOTE ON SPELLING, PHOTOGRAPHS, AND CURRENCIES

Many of the quotations are from British and Indian sources and where possible those original spellings have been conserved. Changes in the names of cities like Kolkata (Calcutta), Mumbai (Bombay), Pune (Poona), and Chennai (Madras) occurred after most of the usage in this book, so the original spelling has been largely conserved. Note also that names were and are variously spelled and printed in India, so that the same person Asutosh Mookerjee is also sometimes Asutosh Mukherjee, and B. D. Nag Chaudhury is also sometimes B. D. Nagchaudhuri.

All photographs are used with prior permission, as credited. Photographers are unfortunately unknown, except where mentioned in captions. In many cases the captions had to be verified by circumstantial evidence as original notes were unavailable. Note that a large photographic archive in Delhi associated with Shanti Bhatnagar of the CSIR appears to have been destroyed recently by administrative mistake, so that there are few pictures of him available. There is also an overrepresentation of certain scientific leaders in the available photographic archives, and this constraint explains the limits on the range of pictures shown here.

The rupee was exchanged at a stable rate with the UK pound and US dollar between 1926 and 1966, at Rs 4.75 = $1.00. The rupee was devalued 36 percent in 1966, and it thereafter exchanged at Rs 7.5 = $1.00 officially until 1975, when it rose to Rs 8.4 = $1.00. Though in 1980 the exchange rate was Rs 7.8 = $1.00, by 1985 the rate had fallen to Rs 12.4 = $1.00. Through this period it was pegged to a basket of trading currencies. But these official values do not account for the unofficial exchange rate, which in 1970 was Rs 13.00 = $1.00 (compared with the official rate of Rs 7.5 = $1.00) and in 1975 was Rs 16.00 = $1.00. Between 1969 and 1979 the unofficial value of the rupee with the UK pound declined by half. As a rule of thumb, before 1966 to convert rupees to dollars, divide by five; after 1966 up to 1980, divide by 8.

Atomic energy, space, and defense research centers in 1974

Atomic Energy, Space, and Defense Research Centers in 1974

Ahmedabad:	Physical Research Laboratory
	Experimental Satellite Communications Earth Station
	Satellite Instructional Television Experiment
Alwaye:	Rare Earths plant
Bangalore:	Indian Space Research Organisation
Baroda:	Heavy Water Plant
Bombay:	Tata Institute of Fundamental Research
	Microwave Antenna Systems Engineering group
	Tata Memorial Medical Centre
Calcutta:	Saha Institute of Nuclear Physics
	Variable Energy Cyclotron
Chandigarh:	Terminal Ballistic Research Laboratory
Chavara:	Mineral Sands Extraction
Delhi:	Atomic Minerals Division
	Defence Science Laboratory
Gauribidanur:	Seismic Station
Gulmarg:	High-Altitude Research Centre
Hyderabad:	Nuclear Fuel Reprocessing Complex
	Electronics Corporation
	National Balloon Launching Facility
	Defence Metallurgy Research Laboratory
	Defence Research & Development Laboratory
Jaduguda:	Uranium Corporation Mine and Refinery
Kalpakkam:	Atomic Power Station
	Breeder Reactor Research Station
Manavalakuruchi:	Mineral Sands Extraction
Nangal:	Heavy Water Plant
Ootacamund:	Radio Astronomy Centre
Pokhran:	Nuclear Bomb Test Site
Poona	Explosive Development Research Laboratory
	Armament Research & Development Laboratory
Rana Pratap Sagar:	Atomic Power Station
	Heavy Water Plant
Sriharikota:	Rocket Launching Station
	Solid Propellant Plant
	Static Test Evaluation Complex
Tarapur:	Atomic Power Station

Thumba:	Sarabhai Space Science and Technology Centre
	Equatorial Rocket Launching Station
	Rocket Propellant Plant
	Rocket Fabrication Facility
Trombay:	Bhabha Atomic Research Centre
Tuticorin:	Heavy Water Plant

Note: Map place-name spellings are consistent with usage in 1974.

ABBREVIATIONS

AEC	Atomic Energy Commission (India)
BARC	Bhabha Atomic Research Centre, Trombay
BEL	Bharat Electronics Ltd.
BHEL	Bharat Heavy Electricals Ltd.
CANDU	Canada Deuterium Uranium (reactor)
CIRUS	Canada-India-US Reactor, Trombay
COST	Committee on Science and Technology
CSIR	Council for Scientific and Industrial Research
DAE	Department of Atomic Energy
EPW	*Economic and Political Weekly*
FRS	Fellow of the Royal Society, London
IACS	Indian Association for the Cultivation of Science, Calcutta
IIM	Indian Institute of Management
IISc	Indian Institute of Science, Bangalore
IIT	Indian Institute of Technology
ISC	Indian Science Congress
ISRO	Indian Space Research Organisation, Bangalore
NCL	National Chemical Laboratory, Poona
NCST	National Committee on Science and Technology
NPL	National Physical Laboratory, Delhi
PRL	Physical Research Laboratory, Ahmedabad

SACC	Scientific Advisory Committee to Cabinet
SINP	Saha Institute of Nuclear Physics, Calcutta
SNEPP	Study of Nuclear Explosions for Peaceful Purposes, Trombay
TIFR	Tata Institute for Fundamental Research, Bombay

ONE

Introduction

When important discoveries about fission occurred in 1939, a handful of scientists in India read the news, understood the physics, and realized its implications. They passed the war in relative isolation until 1944, when they began to plan nuclear research and went on a tour of atomic research facilities in Britain, Canada, and the United States. Revealing curiosity and awareness while visiting some of these facilities, they were questioned by American intelligence officers early in 1945 to uncover how much they really "knew." This was six months before the first atomic bomb tests in New Mexico. On the eve of India's Independence in 1947, more than a handful of her scientists understood the potentials and risks of nuclear weapons and nuclear power: an Indian physicist had already been sent by the government of India in 1945 to ground zero at Hiroshima and returned to Delhi to report on the destruction. Scientists established the Indian Atomic Energy Committee in 1946, obtained hard currency for research and development, and, away from the sight of the preoccupied British governors, forged a program that eventually built laboratories, started research, and sent people abroad for training. They were building upon international networks they had already established over many years, long before this nuclear opportunity presented itself.

At this stage "nuclear India" was hardly taken seriously outside the country except by a few foreign scientists who in 1947–48 appreciated the intellectual sophistication of India's scientists and by a few strategic planners in four or five countries who wished to create a new relationship with India for their benefit. But even if a "nuclear India" was not taken seriously in its early days, not everyone was indifferent to India's usefulness to *their* interests, interests that were surely among the stimulants of the Indian nuclear program. Well before its Independence in 1947, India had more than a handful of people who anticipated what a modern scientific community would look like and what its scientific institutions would require to survive. Based in scientific institutions already established, this nucleus of scientific founders sought the ways and means to build new institutions that would both resemble those they admired abroad and be effective in the Indian context. They believed science was worth pursuing for its own sake, and for their own sake too. Equally important, they understood the implications for India of many of the strategic developments of their day. They had longed for the opportunity to get to work, on their own terms, in their own country. Nuclear development was at the center of those ambitions, and it took on an allure, an attraction, and importance that overshadowed other scientific initiatives. They expected that with political independence could come industrial and energy independence. At this stage in 1947–48 only one country had built an atomic bomb, and there is no evidence that Indian scientists thought that India should or could do so. But they viewed nuclear power to be clearly superior as a source of cheap energy. This was to be the inspirational basis for India's nuclear program, seen in the context of "scientific development."

Moreover, this nucleus of scientists starting to experiment with nuclear development had already established relationships with individuals in other countries who were engaged in similar applications of science to strategic development. People in political power in India, right up to the prime minister, thus relied heavily on the opinion and judgment of these scientists in a rapidly changing technical field. These scientists had earned international reputations, and India did not yet have much of an international reputation that could be used in the inevitable negotiations and transactions which Independence brought. Collectively and individually, well before 1947, scientists had set in motion the efforts to build new institutions to do research at international levels and train new generations of competent scientists who would stay and work in India. Scientific leaders sought to insert these new institutions into the financial bloodstream of the new state, to install and manage them as securely as their political influence would allow, and to attract people to come from abroad and work in these laboratories, including people not born in India.

There are few countries as large and complex as India or with as long a history of scientific inquiry and engagement with the intellectual scientific traditions of other countries. This has been, to use a tired phrase, a "centuries-long encounter." The sciences in India, as elsewhere, were site-specific and context-sensitive, drawing on the special characteristics of the cultures and

societies in which they grew. The scientists who formed the nucleus of a scientific community were about to learn just how context-sensitive science would be and how dependent on India's special characteristics. The youngest among them were also about to learn that a skeptical critique of science and scientists would grow up around them and that a resistance to big and expensive scientific projects, even an antiscience attitude, would eventually assert itself in the 1970s. Their seniors did not live long enough to see this happen: Meghnad Saha, Shanti Bhatnagar, Homi Bhabha, whose lives form a nucleus of the story here, and their patron Jawaharlal Nehru died before that more critical attitude flowered. For the younger scientists, this resistance contradicted and shook their confident belief in the virtues of scientific thinking and research. But India had many other needs and wants, and a creative and innovative scientific community was only one of them. More important, some of those other needs were unmet for an unacceptably long time, even according to most Indians. But, like other modern societies, India addressed, accommodated, and integrated that skepticism about scientific projects—an accommodation that is itself a key strand of the evolution of the sciences in every country.[1]

It is not an exaggeration to say that the nucleus of scientists only learned about the revolutionary and confrontational quality of their work as they went along. Though confident, they were relying most on their imaginations. There was no textbook for the development of science and technology, no formula for managing its relationship with power, no theoretic or conceptual analysis of the process, no workshops to attend. And they did not have a broad grasp of what had gone before the mid-twentieth century in India because they could not obtain one: the synthesis had not been carefully done, much of the evidence about the history of modern science and technology was lost from mainstream understandings, and the role of science and technology had been inevitably tucked in to some other sweeping interpretation of history that (often) sought to break fiercely upon the bulwark of colonialism. When I began this study in the 1960s, there was almost no sustained writing, based on empirical sources, about science and technology in the twentieth century in India.[2] Therefore, not only scientific leaders but also the Indian public and its leaders had incoherent and incomplete understandings of the institutional and individual origins of the development of science and technology, when this building of nuclear India began. Their view of scientific policies and practices thus had an incomplete and ineffective quality for a long time, at least until a more holistic grasp of the situation was possible. Then those understandings rose

to address and contest one another, leading to a theme of this book concerning the disagreements about the path toward self-reliance in a high-technology economy.

From being slightly marginal and rather eccentric intellectuals at the beginning of the twentieth century, by its middle, scientists commanded great respect. That some of them concentrated on nuclear power and nuclear weapons is not surprising; they were acting in concert with their peers all over the world, and they knew it. This book focuses on the long institutional and individual preparations for the first nuclear test in 1974 and its immediate consequences. Though some abhorred it, the first bomb test was the sign of achievement some scientists were waiting for. The test's consequences confirmed to the prime minister and her advisors both the risk and the advantages that they suspected lay in the test. But along the way a large and complex scientific community was created, without whom none of this would have been possible. Secrecy began to segment the work of some individuals and groups away from mainstream discourse. But if one focuses too narrowly on the threads leading to the bomb, one misses the complex embeddedness of these people, their research groups, and their institutions in a wider and supportive community. That very community lifted their careers and projects on a rising tide.[3] Though important and interesting, this theme of the bomb should not mislead us into thinking that those threads, broken or interrupted in places, stand all on their own; this book is about a texture, not some threads. Nor should we think that physics and engineering stand in for all the sciences; each has its distinct history, all forming part of the texture.

Nucleus and Nation

Although great attention has been paid to why India developed an atomic bomb, much less attention has been paid to how this was done, by whom, and with what means and objectives. Building on primary research in archives, personal interviews, observations, and newspapers, combined with the research and thoughts of others, I explore the establishment of the necessary scientific institutions and how they evolved to maturity, supporting a substantial scientific community. There emerged a nucleus of individuals who knew the physics, chemistry, mathematics, and engineering, made the plans, or built the context and apparatus for the nuclear explosion in 1974. Meanwhile the values of this scientific community and the interests of its institutions were articulated by a few individuals who are the heart of this story. Though their objectives could never be reduced simply to the construction of a bomb, they had long imagined a nuclear India and were

dedicated to building a modern scientific state in which they would have a central place and key role.

The definition of "community" is stretched widely here, beyond physics, to include most of those people who considered themselves part of the scientific community. Thus, not just mathematicians and astronomers but also doctors, engineers, lab technicians, industrial and medical researchers, and technologists were included by definition in India's scientific community. Indeed, a central figure in this story was trained as a colloidal chemist. This inclusiveness is necessary because in India all these occupations and professions were thought (by others) to embody scientific thinking or the scientific spirit. In the popular imagination, because doctors know chemistry and engineers know physics (no matter how much), they were included as members of the scientific community, even though only a few doctors or engineers would attend the annual Indian Science Congress, where scientists gathered to meet the prime minister. Nevertheless, even though their professions differed greatly, science was a point of reference for all these people with respect to values and their fascination with their subjects. They saw science more as a movement than as an institution, thus enabling them to join together. All of these different kinds of people would have been among the million people who read India's *Science Today* in the 1960s and 1970s. All these people would have stood up and testified in court on behalf of science, so to speak, particularly if scientists were contradicted or confronted by unscientific ideas or unscientific people.[4] They would have pointed unhesitatingly to their own scientific values. Much of the concluding chapter addresses how they articulated and defended these values when some thought they were at risk.

In the title of this book, "nation" has three senses, including first the mythic entity imagined by people in the pre-1947 period; second, the gradual political unification of the society's heterogeneous composition (eventually twenty-four "official" languages and twenty-eight states); and third, the instruments of a sovereign national state that were used from 1947 onward to define and control national sovereignty and finance its institutions. "Nucleus" in the title has two senses: one refers to a small group of influential scientists and the other to a tiny invisible entity in physics that many of them studied. The term "power" also has two senses here—the political variety and the physical variety; the first is the capacity to influence others and mobilize resources toward desired outcomes ("power over" and "power to"), and the second is the power that generates; for example, electricity generated by steam power heated by burning coal or by "burning" uranium fuel rods. This electrical power was to be the key to a modern industrial economy, and rapid rural electrification in countries like the Soviet Union and the United States excited scientists in India and citizens alike.

"International networks" here describe a number of elastic coalitions, formal alliances, and professional friendships by means of which information was exchanged and influence flowed. Networks often ran through institutions and were molded by technical things (like certain types of nuclear reactor and their fuel cycles), so these networks were always a combination of personal and material things. The properties of technical things, such as their difficulty, their scarcity, their risk, their beauty, their cost, and their unexpected applications all shaped and changed networks. The personal and material became inseparable in these networks, operating at a barely conscious level. But it is the information about these technical things, and the flow of influence around them, that gave life to these networks. These networks also ran both secretly and openly, along parallel paths. The secrecy required by the state was balanced by the openness required by scientific communication; the friendship or familiarity that made many links in these networks viable was both instrumental and deeply personal. Thus the faces of secrecy, professional openness, and private friendships were inherent properties of the networks, and each actor presented all three of these faces during their work. Actor network theory has guided my approach; in the networks shown here actors mobilize their resources and allies through their networks, constantly building and adding to them while letting drop those inactive parts. Actors perform well or poorly, mobilize widely or narrowly, and as agents often get themselves into irreversible situations (a form of entrapment, allocating more resources when earlier similar commitments have borne no fruit). Big technical projects like reactors or rockets were absorbed in the consciousness of scientists and technologists, so that the networks acted through these material things but were by no means reduced to them. Even those big things like reactors were unthinkable and undoable without very small things like calculators, slide rules, probability theory, data books—all of which were essential to these networks too. This complexity means that speaking in shorthand about them here leaves much important detail out, which cannot be helped. In this volume part of the activity of these networks is missing, in that some of the international and multilateral connectivity is treated in a following volume on nuclear bargains and bargaining.[5]

My focus is less on the first Indian bomb itself than on the nucleus of people who made it possible, whether knowingly or not, and on their relation to the nation and its political leadership, right up to prime ministers. The bomb is set in a web of prerequisites and long-term entailments, like nuclear reactors and their physicists and engineers, heavy water and its chemists, missiles and their metallurgists, and the money, power, and financiers necessary to their assembly and deployment. The purpose of this book is to see these developments through Indian lenses, and largely as an extension of Indian interests, which were naturally very sensitive to international trends and influences. The bomb and the reactors are a focusing device to understand the scientific community in India. This volume is a companion to *Negotiating Nuclear Power: Bargains, Reactors, and the Bomb in India,* where the operation of these networks and their developments will be seen through the lenses of foreign powers and interests, largely as extensions of *their* nuclear power.

The Indian Scientific Community and the 1998 Nuclear Tests: Why Does the History of Science and Technology Matter?

In May 1998 the desert under Rajasthan heaved, melted, and echoed with the blast of a series of nuclear bomb tests, followed by a wave of public enthusiasm and private apprehension. Almost immediately Pakistan tested its own nuclear bombs, deforming a whole mountain near the capital Islamabad. These were nuclear test explosions, devices more than bombs, and definitely not military weapons. But everyone called them bombs. Since the word is no longer reserved for something delivered and dropped from the air, but also for something that can be left in a bag in a train to explode lethally in the middle of Mumbai or London, these Indian devices *were bombs* after all. And the Indian path has been followed by others too—Pakistan, North Korea, and eventually perhaps Iran. Curiously, many people outside India (and some inside) thought the 1998 test was the first Indian nuclear test, didn't know about the 1974 test, and knew little about the history of science and how India's scientific institutions had made this possible. Such a history is now being told, notably by Perkovich, Abraham, Chengappa, Parthasarathi, Srinivasan, and others, and this book contributes an essential part of that deeper understanding.[6] Far from putting other science and technology activities "on hold," the 1998 bomb tests were embedded in a vast matrix of personal, institutional, national, and international initiatives, some routine and others extraordinary. The *Economic and Political Weekly* immediately chastised the chairman of the AEC, physicist R. Chidambaram, for saying that "without assured national security, development falters."[7] In this section I lay out what part of that matrix looked like in 1998. This is the approach I have taken to a detailed understanding of the matrix of initiatives surrounding the first atomic bomb test in 1974.

Though there was jubilation, there was also criticism of the tests in 1998, and condemnation by some countries, silence among others. While the prime minister talked naively of carrying Pokhran's blasted earth around India for people to see, dissenting voices emerged, and within the scientific community a disagreement with nuclear policy percolated again to the surface. Six months later, in November 1998, new legislation created the National Security Council involving the prime minister's office, the minister of Defence, and chiefs of staff in nuclear weapons decisions: it was striking that scientists were no longer to be included in the inner circle of nuclear deliberations, a break with forty years' practice. Also in November of that year state elections were held, and the prime minister's party, the Bharatiya Janata Party (BJP), did not gain seats or states, but lost them. At this stage he decided to appoint Abdul Kalam as principal scientific advisor to the government, a role that held cabinet rank: Kalam, whose professional origins lay in the satellite and missile program, was widely seen as a leader in the second bomb test program of 1998. On the first anniversary of the 1998 tests (known as Pokhran II) the prime minister celebrated the triumph of India's emergence (again) as a nuclear power, but now with an accurate missile delivery system that did not exist in 1974. He called it "a day of pride." But he did this from an election campaign trail he was forced onto when his government lost a confidence vote in Parliament. Again, it is *unlikely* that nuclear weapons enhanced the political fortunes of the people (and parties) who ordered them, although the tests confirmed the importance of the institutions, individuals, companies, and international relationships that made them possible.

Shortly after the first anniversary of these 1998 tests, 50,000 Indian troops were locked in high-altitude combat with 50,000 members of the Pakistan army at Kargil on the border of Kashmir in 1999, combat in which almost 1,700 Indian soldiers and about 700 Pakistani soldiers were killed.[8] There is evidence to suggest that one effect of this military confrontation on a strategically insignificant glacier was the reelection four months later of Prime Minister Vajpayee and the BJP, and a consensus suggested this combat in Kashmir, rather than the bomb, was decisive for that BJP election victory. The enthusiasm surrounding the previous year's nuclear tests alone did not seem to have provided the BJP with a reelection success in 1999. This is the not the first government to enjoy electoral support following an armed conflict, but this particular conflict was played out for the first time in the shadow of military nuclear bombs on both sides.

During 1998 and 2005 I revisited a number of scientific institutions in India, discussing the subject of this book with people I have known for thirty years, as well as people I had never met before.[9] The themes of these conversations circled around self-reliance in science and technology, the contest over intellectual property rights for ancient knowledge in India, arresting the brain drain of well-trained people out of India, and resistance to interference from outside India. Their other preoccupations in 1998 included how scientists are treated in the news and how to find ways to make scientific institutions innovate, produce revenue, and be less dependent on government funding. Each of these themes is dramatically foreshadowed during the history of modern science and technology from 1920 to 1980; each has a lineage back through the twentieth century that this book uncovers and explains.

My first stop in January 1998 was for the Indian Science Congress at Hyderabad, where a full day was given to celebrating and questioning the achievements of the nuclear power program.[10] Nuclear leaders all spoke about the huge demand for electric power and the limited means in India to generate it, concurring that nuclear sources still account for only a very small percentage of the total electrical supply in India and admitting that most nuclear power plants still do not produce electricity up to their technical capacity. There was a strong mood of optimism about the future contribution of atomic energy to India's energy budget. By 2006 there were fourteen power reactors operating with an installed capacity of 6,600 MW, all running at an (official) average of about 85 percent capacity, on a par with reactors in other countries. The capacity thus had doubled since the beginning of the century. This electricity, however, still provided a very small percentage of the country's overall requirements, and some experts did not accept the official operating performance data.

Five months before the Pokhran II tests, experts at the Science Congress made numerous references in 1998 to the effects of a "technological embargo" imposed by the United States because India had not signed the Nuclear Non-Proliferation Treaty. The discussion revolved around the recent achievement of an operational fast breeder reactor that will use India's 300,000 tons of thorium deposits and allow conversion of spent natural-uranium fuel into plutonium for the fast breeder reactor. According to one press report, the imported component of the fast breeder, including payments for French design and know-how, was about 22 percent of the overall cost.[11] In 1998 India was also building a reactor with Russian design and assistance in South India, puzzling local experts who expected that, after forty years, India would insist on doing such projects on her own. The Indian refusal to sign the NPT had its origins in a number of factors, among them the perceived hypocrisy of the Nuclear Club accepting China as a member but excluding India: this was one reason for the 1998 atomic tests. Another reason for the tests was the mirror of the previous one—that China was acknowledged as a great power with a seat on the Security Council whereas India was not. So, after much debate and handwringing, the United States and Canada both agreed on nuclear cooperation with India in 2005 (in the American case as part of a huge proposed sale of F16 fighter jets and in the Canadian case on terms that were characteristically more vague). Though the embargo was lifted in 2005–6, it was not until 2008 that these relationships were operational.

Irritation with the technology embargo was also evident in the public media in 1998, loudly amplified by attempts of US firms to establish patents on a number of plants and substances traditionally used in India, like neem, mustard, basmati rice, Darjeeling tea, among others. US patents have been taken out on these and many other plants, perhaps as many as one hundred patents, according to media reports. A patent granted by the US Patent Office for the wound-healing properties of turmeric powder was challenged by the government of India and overturned in August 1997, to much popular satisfaction in India. Another patent was filed in the United States for a strain of basmati rice. This was portrayed in India's popular media as an American attempt to ignore India's historical genetic heritage and traditional knowledge and manipulate the market to American advantage (no other country was singled out for this disapproval). The Council of Scientific and Industrial Research was advised by the prime minister, the chair of its board, to copy the powerful companies and countries and to rewrite India's intellectual property rights law in order to protect its heritage.[12] This book provides an account of the politics of the intellectual property issue from the time in which India introduced its own systems of patents. Involved in "misappropriation" of computer software, some of it written in India, and its unauthorized resale, Indians were told that they would have to accept the rules of other nations if they were going to establish their own. But they had a 200-year-long experience with rules set and manipulated by others and have not forgotten it, so theirs was a contradictory spirit of defiance (what's in it for me?) and interdependence.

Caught up in their personal squabbles and volatile alliances, none of the political parties campaigning for power in February 1998 seemed to understand these technical issues or could position themselves in a credible scientific sense for the election. One person, however, caused a stir inside the scientific community because he joined the BJP on the eve of its election, saying that he did so to bring this party into the world of science and technology, the party likely to govern. M. G. K. Menon disturbed many of his colleagues by joining the BJP's Manifesto Drafting Committee, explaining confidently (in his defense to a colleague) that the BJP had no one with a scientific background and he was necessary in order to provide ideas for policies. What surprised others was that Menon (seventy years old in 1998) had held most of the top positions in the scientific elite (see Short Biographical Notes) and could have gracefully retired. Acquaintances argued that Menon overlooked the deeper reason why the Hindu-oriented and traditionalist BJP had no core members with a scientific background and had no positions on science and technology issues, touching on very old sensitivities about the scientific community's commitment to political secularism going back more than one hundred years. The following chapters examine the tension between scientists and the appropriation of ancient beliefs in India, beliefs embodied in the BJP's outlook, particularly the section in chapter 25 on the "scientific temper" movement that began in the 1950s. In the subsequent years of the BJP government until its defeat in 2004, M. G. K. Menon was not made a cabinet minister or an ambassador.

At the conclusion of an academic session on science and society at the 1998 Indian Science Congress at Hyderabad, a very large crowd gathered and the path was strewn with rose petals by beautiful young women, preparing for the arrival of Swami Ramachandran. The "Hindu consciousness movement," linked with international communities of Hindus, including temple-based money-raising activities among the diaspora in other countries, has long engaged in outreach to sympathetic non-Hindus and was able to mobilize large amounts of money, both outside and within India. This lecture by the swami was separated from the rest of the Science Congress session but held in the same buildings at the same time, and the swami spoke in English to an overflowing and attentive crowd on science and spirituality, stressing the high calling of each "profession" (scientist and swami) and their potentially deep affinities in the study of the complex and the mysterious, a message of no-conflict and synthesis between spirituality and science, all for the greater glory of India, what publicists were soon to call "India shining." A swami's presence was unthinkable in the first Science Congress I attended thirty years before (1968), but it was approved and acceptable now in 1998.

Many scientists took acute interest not just in the operatic dramas of the mighty ones, but also in the quiet, sometimes tragic lives of ordinary scientists: they directed me to a hot issue in early January 1998, namely, the suicide of a scientific worker and the court case brought by CSIR's Scientific Workers Association against the CSIR. In the archives I discovered that there was a record of suicide among scientific workers, beginning in 1960 with perhaps a total of thirty suicide deaths in the scientific community since then, some widely publicized, others not (see chap. 14).[13] It is not the first time scientists had taken their employers to court in order to release an official inquiry's report about their working conditions, the very conditions that are often blamed for amplifying suicidal tendencies. But this was apparently the first case of filing for public interest litigation.[14]

The old debate of the 1940s and 1950s about self-reliance and the brain-drain shifted in the late 1990s to focus on the export of scientific and intellectual skills, particularly in software design, in which the classic fear of a "loss" presented itself again. This time the brain drain appeared in the form of the career drift of twenty-two-year-old whiz kids, trained in the Indian Institutes of Technology and other universities and polished in private computer schools. Offered twice the salaries even of their professors, they were then snapped up by multinationals in India, while ignoring good positions in scientific institutions in India—positions their professors had struggled long and hard to create. Enticed by other countries, many moved abroad. By 1999 during the dot-com bubble, it was quite easy to get a green card to work in the United States, at least until the crash of 2001. But a deeper apprehension was that the whiz kids were no longer coming to study mathematics and physics and other natural sciences—their traditional destination. By 2005 this apprehension had become a tough reality in scientific recruitment, matched by competition from other countries for the opportunity to train these same young people.

Studying patent regulations in the United States, contesting inequity in Indian labs, finding the right people to recruit to work in research groups—all of these issues have very old roots, as this book demonstrates. Ultimately this work must take its place among the histories and sociologies of science and technology in India, but that placement now cannot be my purpose. An excellent historiographic map is provided by Dhruv Raina.[15] He charts the ascendance in India of the idea of science as a transcendent and culturally universal enterprise, noting the tension between scientism and romanticism, or more precisely an overcommitment to scientific thought and practice as the answer to almost every question (scientism) and the deep skepticism about "modern and/or western" scientists and science in favor of alternate sciences including indigenous knowledge. In the 1970s, says Raina, there was a decline in the commitment to scientism and to Nehruvian socialism. Raina examines the objectives underlying the creation of and writing in the *Indian Journal of the History of Science* and decodes the tendencies of its authors and editors, up to the 1992 special edition on science in India 1900–1980, written by subject-specialists (e.g., botany by a botanist) about the intellectual traditions and challenges of their subjects but without much thought about the institutional or policy environments in which this took place. Raina calls it "a lost opportunity for the history of science" and "a product of the institutional dependence of the history of science on the world of science" (pp. 127, 131).

The Scope of This Book

The units of analysis in this book have to be not only the individual scientist but also the working group and the whole lab—with particular attention to the scientific elite and decision-making processes in and for the scientific community. These individuals and groups, and not just impersonal forces, built the networks, fashioned the scientific community in India, and exercised (or criticized) the power. Although there were *other* candidates for these roles, it was these key individuals discussed in this book, with these lives, who turned out to be the important agents and the embodiment of these wider forces and processes. Nevertheless, all along there were impersonal forces running through all their opportunities, posing the possibilities or obstacles to which individuals responded. The scientific community was built by a good deal of competition, imitation, repetition, and some raw genius. These stories are thus a little more about their "institutional lives" than their strictly professional scientific lives, and it was these lives through which individuals contested the limits set by an imperial-colonial system, demonstrating that those excluded (for, say, being Indian) would have to be admitted to the international system for reasons of their high performance (as scientists). Kumar, Baber, and Arnold provide an excellent account of the late nineteenth-century interplay of "European" and Indian working relations in science in places like Calcutta, giving us clear pictures of an almost apartheid-like working arrangement that was gradually breaking down in some places around 1914. The compulsion among British and Indian scientists to communicate and collaborate was, when this story begins in 1920, gradually having an effect. When Independence was achieved and the exclusionary principles of socioeconomic class and status became more powerful among Indians, a superior performance in science and technology remained an open path by which to contest and transcend limits set by others; these individual pioneers blazed trails for many followers.

Frequent reference is made in India to the exemplary lives and abilities of the scientist-founders of its scientific community—often without knowledge of what their lives really were. In part this is so because realistic accounts of these scientists' lives have simply not been available and in part because such references were less about realism and more about praise and celebration. The appetite for information about the "institutional lives" of these famous people continued undiminished but largely unsatisfied. Moreover, very little was written about the culture and organization of the scientific institutions that enabled scientists to establish and refresh their reputations, recruit new scientists, and sustain their relations with the powers in India and the rest of the world. Such research requires sustained and observant presence "in the field," an assiduous reading of boring administrative reports (if available), conversations involving patience and diplomacy: in short, "being there."[16]

It is sometimes said in India that to examine these great scientists' lives carefully is to disrespect them. But this is like saying that the critical analysis of poetry demeans it. We need to understand how these scientists and technologists overcame the many obstacles they faced and to understand the lessons their lives teach us. If we make a critical examination, then their achievements stand out more clearly, along with their own critical assessment of national development. The lives of scientific leaders under historical analysis become exemplary in a new sense; no longer are they superhumans or saints. Their stories are very human and demonstrate what others too could accomplish if they tried; they would learn what has been tried, and the consequences. In addition to their professional scientific achievements, we are drawn to these remarkable people because of their complexity, tenacity, and imagination. In fact with so few models available to them at the time, much of their work was an exercise in the imaginary.[17]

This book examines the careers of three men beginning in the 1920s, their interaction with each other, and their relations with other significant individuals like Jawaharlal Nehru. Meghnad Saha, Shanti Bhatnagar, and Homi Bhabha are ideal illustrations of general patterns because of the vivid contrasts between them. The book portrays their lives in terms of the institutions they built, the policies they pursued, and the battles they fought. In addition, I discuss other scientists no less interesting: C. V. Raman, P. C. Mahalanobis, K. S. Krishnan, S. K. Mitra, S. N. Bose, D. S. Kothari, all of them physicists, some with outstanding international reputations. Each of these contemporaries deserves biographical attention in his own right, but only C. V. Raman and to a lesser extent Meghnad Saha have received it. Finally, this book examines the activities of new leaders like Vikram Sarabhai and his colleagues who reconstituted the scientific establishment from the mid-1960s onward. I consider Sarabhai and his successors in the 1970s (M. G. K. Menon, Raja Ramanna, Homi Sethna, Satish Dhawan) and the institutions they directed to be the "offspring" of Saha, Bhatnagar, and Bhabha and their contemporaries like Raman and Krishnan.[18] These are the two generations that led India toward its high-tech future. But the collective memories of Saha, Bhatnagar, and Bhabha were at risk of becoming overdetermined, making them distorted historic "figurines," in that they carried too much baggage, were used to answer too many questions, and (like Nehru) overshadowed others who also had a right to be known on the historical stage. I have tried to unwind those distortions here, using archival material when possible.[19]

These individuals formed a nucleus of the scientific community; they built the labs, sat on the committees, managed the funds, and persuaded the powerful to support them and their institutions. They achieved international recognition, and they achieved recognition inside India far beyond the scientific community. They advanced their students and younger colleagues. Through their international networks they articulated new ideas and programs of action, and these networks reminded the powerful in India that the planning and development of science was a major responsibility that nonscientists had to acknowledge and allow scientists a free hand. They were the nucleus of a powerful new elite capable of exercising new kinds of power in a new country, and keen to do so.[20] Most of these ambitious individuals realized that the structure of the "other" nucleus (in this case of uranium and other radioactive fissile materials like thorium) was a source of potential industrial and military power.

In 1947, India's loss of oil fields in Burma and Pakistan brought the sharp realization that the petroleum path to wealth, so easily followed by India's neighbors Iran and Iraq, was barred. But, if correctly handled, they reasoned that the fissionable nucleus could carry the scientific community, and perhaps the nation itself, onto the world stage where India rightly belonged. Moreover, the "nucleus" would accomplish this by making India into a self-reliant, self-sufficient nation, and never again would she be a dependent colony, particularly not in energy. The potential political and economic power of nuclear fission was increasingly clear to them, even if their path to its effective applications was not: indeed the path often looked blocked. And equally important in this entire process were Jawaharlal Nehru and his daughter Indira Gandhi, prime ministers who for thirty-two years between them held the cabinet portfolios directly related to atomic energy research and development. Others held the portfolio for five years in the interim, but the cumulative influence of the Nehrus was profound.

Power was available to this nucleus of scientists through the prime minister, but they had to find the ways *to create an environment* in which they too could be powerful. There was no recipe book of methods to achieve this, nor could they agree on the right strategies to follow. They were not seizing an old kind of power from someone else, but fashioning a new kind of power. Even while cooperating, they competed for national attention and resources and for the means to compel (or seduce) the political and economic powers in India to underwrite their strategies. They also had to adapt to changes in international relations and the power politics of leading nuclear nations and to do so in and through their own networks of influence. International recognition and reputation was an essential dimension of their adaptation. To hold the attention of more powerful people, they engaged in discourse about industrial power, dams, solar power, petrochemicals and refineries, fertilizers and heavy water, and atomic energy reactors. All these projects required large amounts of electricity, which India did not have, one reason why atomic power reactors were seen so widely as the key to India's future.

More ambiguous, they had to contemplate a military-industrial complex of their own, one that would contain armaments, rockets, satellites, jet fighters, and submarines, and the means to design and produce them. Such a complex would arise where nothing of the sort had previously existed and where part of the population was opposed to it. Many scientists did not really want this complex either.[21] This industrial project was not necessarily in the sphere of their scientific expertise, their original scientific passion, or their personal ethical values; nevertheless, they gradually engaged with this larger vision. By and large scientists had been told by activists in the nationalist movement to stay out of jail, and so their engagement with the state was different from those who went to prison. Not motivated by personal ambition alone, they also knew that only the larger vision would convince more powerful agencies to allow this "nucleus" to build the scientific community in its own way. Only this larger vision, repeatedly and relentlessly articulated by "political scientists," could persuade the commanding heights of the state to give them enough freedom and capital to increase their international participation and to address pressing national objectives. Only the realization of this larger vision would enable them to be left alone, to pursue research independently. All along they had to demonstrate that they merited these privileges and that they were pulling at the wheel of national development. When they succeeded in this persuasive effort, they held a new kind of power in their hands and induced generations of young people to think that science would be both a noble and rewarding life.

Though this book is not exclusively about nuclear power or weapons, these themes run through it because of their great symbolic and material importance. Like rockets and satellites, nuclear reactors and weapons are, for some at least, not just good to build, see, and hold, but, like all symbols, they are *good to think,* as Claude Lévi-Strauss reminded us forty years ago. I have been asked if this book is about science and politics in India in general or about atomic science and politics in particular? The book's thesis is that there would not have been a sustained atomic energy program without a coevolving relationship between science and politics, which resulted in a large scientific community. A single political party and four prime ministers—all closely related (two were Nehrus)—were responsible for the justification for and furtherance of "the scientific community" and the atomic energy program within it through to 1980, the end of the book. The chief scientific protagonists for the program were not North Korean or Iranian specialists working in protected isolation behind walls; they were world-renowned experts deeply embedded in the open politics of the broader scientific community and articulating the whole of science's relationship to Indian history and culture—nothing less. Their lives are crucial to our understanding of the war over self-reliance, a war that moved the atomic energy program and the bomb ahead rapidly during the 1970s. All that occurred in the context of the government's pursuit of the "commanding heights strategy." So things like steel or fertilizer, which *appeared* to have had little to do with "atomic science and politics," need to be understood in this wider context.

This cannot be a history of *everything* related to India's science and technology, and so it omits health, agriculture, and environment, among other interesting subjects; such a larger multidisciplinary project is beyond my scope here. There are other modern histories of other sciences and technologies running in parallel here and those are yet to be written. Nor can I fully audit the process of innovation and R&D for any particular technology: issues like heavy water for reactors, solid and liquid fuel for rockets, cosmic ray and cyclotron experiments are explored because they illustrate a broader picture. I follow the patterns of interplay between institutions, individuals, and technologies surrounded by socioeconomic planning and power dynamics in the scientific community, all in relation to the political systems of India. Therefore, this is not and cannot be a complete history in the actuarial sense of reactors, space, or any other facet of India's high-tech economy, and some will be disappointed that their part in the drama has been missed, or perhaps obscured by others; it is not my intention to do so. I emphasize that for some situations in this book I have had to rely on the recollection of two individuals or even only one. Sometimes these sources were the sole person still alive who was present at a significant event; in other cases, individuals have offered their recollections of a key actor or episode, which otherwise are difficult to verify. Through the Internet and e-mail I have received personal communications from individuals whom I have not met and who are writing to me about things which had not previously been written down in answer to my questions. They have given me permission to credit them as a source. This collection of information resembles personal conversations with informants I had in the 1960s and 1970s about things that had not previously been written down. Therefore some of what is written here will contradict the experience or views of others and may be contested and eventually rewritten. This process seems to me one of the best ways to arrive at better understandings of the recent past and draw conclusions from them.

But there is ample material here for those interested in postcolonial theory to trace the continuities across pre- and post-Independence frontiers and to witness the versatility of scientists who worked and lived in two or three environments at once, managing loyalties to each of them. There is a subaltern voice in the 1948 formation of the Association of Scientific Workers of India, because it came to represent laboratory staff and knowledge workers, at least in parts of the scientific community and vigorously promoted improvement of the working conditions with which Nehru was himself concerned; for this work its officers were sometimes disadvantaged or penalized. The book concludes with notes on the (indirect) role of women in these stories, the discourse of the 1980s on "scientific temper" initiated (in part) by Nehru, and discussion of the quiet yearning in India for another Nobel Prize, "soon." One of the encouraging initiatives was the focus on subaltern histories starting from 1981; such historians will find riches here, complete with ample references to boys and girls from low-caste households embracing both experimental and theoretical research; here too is reference to Brahmins working in high-tech projects along with a son of a Muslim fishing boat owner, or to elitist scientists committed to the scientific workers movement while cafeteria workers without toilets got "hot" (radioactive) on the nearby beach. On the whole, however, modern scientists and technologists have not interested global postcolonial and/or "cosmopolitan" scholars very much, whether or not in India.[22]

As you consider the evidence before you, watch how some influences that appear to flow from the top down don't actually occur unless there is a concurrent movement from the bottom up, or from the sides inward. Watch how these locations (top, bottom, etc.) are metaphors in a multilayered structure, how they don't make much sense without each other. And see how that appearance of grand flow (top to bottom) misleads us away from the intense interactivity of many levels, from the prime minister to the boys in the lab; watch how that appearance of grand design and flow obscures the density and complexity of the community being constructed, year by year, stage by stage, among groups, circles, and networks with fractures and fusions, quarrels and cooperation.

Outlook

In the context of large research institutes, networks affected how they recruited new members and trained them for the renewal of research groups and the scientific community. Most important for scientists, institutions must provide the optimal supporting conditions not only for a new idea but also for the sustained work of individuals who are its proponents. The new idea was, after all, the central commodity in the process of scientific production, around which other things and people revolve and into which they must be enrolled. In an economy of novelty, the new idea must be turned creatively into a technique or an answer to a question; it must be made available to the processes of science, and this availability can then reinforce the reputation of the individual(s) and the groups and the labs where they work on that idea. Technique revolved around equipment and other things as frequently as it did around concepts. In that sense, the new idea must be thoroughly consumed to be acknowledged; it had to be fully appropriated and consumed to have currency. It must be placed in the cyclic drama of scientific consumption and new production, whether to success or disappointment. The best new idea for Indian scientists was the one that became renowned, for example, the Raman spectra, the Saha equation, Bhabha-Heitler scattering, or the Bhatnagar-Mathur magnetic interference balance. Embodiment in a "thing with a name" contributed to the drama of consumption and new production, satisfying the expectation that the potentials of science could and therefore should be tangibly demonstrated. This animated the networks and gave them their kinetics. Without supportive conditions for creativity and innovation, most efforts to maintain research institutions would be, at best, confirming and repetitive, and, at worst, derivative and marginal. A laboratory in India that was derivative and did not create new ideas was open to question. In this sense some of the characteristics of actor-network theory appear in this story.[23]

I also ask here particularly whether the separate organizational culture that grew up around the "nucleus" had special "Indian" characteristics? Were its projects—such as the DAE's reactors—built and finished in a style and speed comparable with other mega projects of the post-Independence period, or did they stand out as models of greater efficiency, humanity, and planning? How did these projects articulate with the larger systems of power inside and outside the country?

Finally, these ideas and institutions developed in a "laboratory state," in two senses. The first sense is that scientists before 1947 imagined a future in which an independent state would be fully available for experimentation, or more precisely, it was to be *their state* in which, and through which, their experiments could be conducted. Planning was, in part, an experiment to them. The second sense of "laboratory state" implies the state's maintenance of a relatively powerful apparatus, which—directly and indirectly—applies scientific expertise to many, if not most, problems of social life (malaria, AIDS, famine and malnutrition, polluted drinking water, train accidents, etc.). In this particular role, of course, scientists were asked to provide opinions, judgments, and solutions for most of the difficult questions facing the state and society: should this rocket be built at the cost of 100 million rupees? Should this chemical be banned from agriculture? Are these dams safe to live beneath? Can India's part in international trade in human body parts be controlled? If a new nuclear reactor is built, what design should it have? Is the test that shows that a fetus is genetically deficient and the infant will have no feet really a reliable test? Visvanathan's remarkable essay about "the laboratory state" brings to light these issues, but I do not join in his conclusion that development is a scientific project that is inherently genocidal.[24]

I do, however, agree with many scientists in India, and other observers, that some projects of the developmental state (and private ones too) have been reckless in scale and indifferent to the rights of specific groups in the name of "the greater good." But most scientists in their time described here thought science in the state and for the state would be liberating, not genocidal. Indian scientists and their institutions were eventually confronted with decisions on all these kinds of questions and imagined these questions well before India was their "own country." Their approaches and solutions remain relevant to this day, though Indian science has not been able to resolve every task it addressed. Amid the sophisticated missiles and reactors, there was still malnutrition, preventable disease, and unsafe drinking water. This is a story about international scientific development with a special focus on constructing India's capacity to create, adopt, or adapt radically new technologies. This is also a story about scientists and political power with a special focus on reactors and their relationship with bombs, seen through Indian lenses. It shows that Indian scientists were acutely aware of the shortcomings of their institutions and the limitations of their role: not uncritical or unaware of their privilege, some expected much more; a few insisted on much more.

Finally a note about the tone and point of view I have adopted throughout. I was asked by an anonymous reviewer "where is the author's sense of irony about these pipe dreams, the lack of realism in all these big projects, and where is his criticism of the great-power dream of the Indian elites?" To answer this important question I have decided to use the voices of others to articulate the Indian sense of irony and the structural contradictions in these large technical and industrial projects. India is usually characterized as the land of contradiction, contradictions with which its citizens are quite familiar. Though not wrong, too much is made of that characterization, because the sense of unease and criticism of a lack of realism is a very Indian tradition too. Using this point of view, I want to convey the Indian sense of proportion, and the double world in which they would acquiesce to something in public, even in writing, yet make sharp critical judgments in private. My intent is that readers learn this Indian sense of ambiguity and irony through Indian voices, rather than simply mine. This book demonstrates that not all Indians had unrealistic great-power dreams and that there was a contest among people (not simply in the elites) over the kind of power India needed and how it should best be achieved.

TWO

Building Scientific Careers in the 1920s: Saha and Bhatnagar, from London to Allahabad and Lahore

Two young men sat drinking tea together in London in 1920, talking about the future of science in India and their careers, forming a friendship that would eventually be severely tested in their later years. Yet they were destined to work together, become founders of India's nuclear program, and quarrel about its directions. One was a physicist, Meghnad Saha, and the other, Shanti Bhatnagar, was a chemist. Both were twenty-six years old and had arrived without a clear place or supervisor for research. Saha wanted to be in Cambridge but could not afford to, and Bhatnagar, whose scholarship was to study in the United States, could find no space on board any ship. Both therefore stayed in London, not a first choice for either. Then the two were guided by friendly intermediaries to the labs in London where they eventually worked. Because of their records of excellent performance, they were both given an office and place in a lab by two leading British researchers and freedom to work. Although in different fields, they had friends in common, and thus they met almost daily. Living in London on low incomes, they both learned to cut corners there too, giving them insights into the structure of British society.

Their scholarships, moreover, allowed for travel, and so Bhatnagar, having first learned some French from his London landlady, went to Paris, where he met the Curies and many chemists. He then went to Berlin, where he joined Saha, who had studied German, again in 1921 and was waiting outside the lab of Walther Nernst, along with a dozen other University of London students, all bearing a letter of introduction from chemist F. G. Donnan, who was also Bhatnagar's thesis supervisor. Both of them benefited greatly by this stay abroad, meeting the stars of European science; they also encountered for the first time influential Indians, like Rabindranath Tagore and Muhammad Ali Jinnah, who were visiting Berlin, a revolutionary, artistic, and intellectual center at this time. Both men gained in maturity and, following their studies, returned to India in 1921, already appointed at age twenty-eight to full professorships in Calcutta and Benares. But they would not stay long in either place. Their international reputations were good, and they showed great promise; years later they would be at the heart of the politics of scientific development of India, planning for the independence of the nation and scientific community. In this project they were joined by Homi Bhabha, sixteen years younger but someone who would equal and then exceed them both in influence. During this crucial 1939–47 period, the paths of the three crossed frequently, and while they engaged in a common cause, they measured the distance and difference between them carefully.

India's most important early research institutions in the physical sciences were founded by strikingly different men. Meghnad Saha (1893–1956) was the son of a poor shopkeeper from East Bengal (now Bangladesh), Shanti Bhatnagar (1894–1955) was born in a village in Uttar Pradesh, and Homi Bhabha (1909–66), sixteen years younger than the other two, was from a wealthy Parsi family of Bombay. With the "plain-living and high-thinking" that Bengalis were supposed to love, Saha was perceived in Calcutta as a rustic. Bhatnagar negotiated a transition from a poor but educated family to a position of wealth and social influence, building on his reputation in Lahore. Both Saha and Bhatnagar lost property and social networks of influence in 1947 when Dacca and Lahore were included in Pakistan through Partition. On the other hand, Bhabha's power lay in Bombay: with a sophisticated internationalism only slightly touched by India and a worldliness of which the middle-class dreamed, Bhabha was like the bright lights of the city. As archetypes these three evoked heated discussions among Indians about the lives and work of scientists, the growth of scientific institutions, and their relations to society.

The research institutes that these men created were expected to compete in similar fields and to achieve the same standards as the centers of the international scientific community. Differences between them of background, style, and method were subdued by science, because science gave the appearance (and opportunity) of allowing everyone an equal chance to succeed. Saha, Bhatnagar, and Bhabha established strong international reputations, judged by the same standards: Saha in astrophysics, Bhatnagar in colloidal chemistry, and Bhabha in cosmic ray and elementary particle theory. All three became Fellows of the Royal Society in London, built international and national networks of influence and information, and had numerous students and colleagues who, in turn, carried out their projects. Their successes were achieved through their scientific institutions. At that time, as now, fellowships in the Royal Society were granted, by election among Fellows, to citizens of countries in the British Empire and then afterward in the Commonwealth, and those Fellows (e.g., from India or Canada) held the same status as British Fellows: any others could be elected as Foreign Members.[1]

These individuals struggled to change the conditions of science that surrounded them and aspired to build the kind of scientific community they wanted. Their tales reveal how scientific institutions are affected by the preferences and judgments of their founders. Their struggles did not transform the forces of political economy but rather created the arena within which Saha, Bhatnagar, and Bhabha could mobilize and organize. Concerning what they felt they *should* do, we must view their vision within the limits of what seemed to them to be possible, from their particular points of view, according to their understandings in their contexts. They built the institutions and science policy of the country, not simply by applying abstract criteria but by vigorous debate and constant political engagement. Well before there was any fashionable concern with the study of national science policy, these scientists understood two central problems in India's development: securing national resources and instruments of power, and increasing the economy's self-reliance, problems that remained unresolved throughout their lives.

Meghnad Saha's Formative Influences

Meghnad Saha was born in 1893, one of eight children of a poor shopkeeper in Seoratali, 45 km from Dacca in East Bengal. His oldest brother had failed in his school studies, and so their father seems to have made Meghnad, the fifth son, return to selling groceries like the elder brother. It was Meghnad's mother who, with an uncle, intervened with the father to allow him to go to high school.[2] Primary school teachers persuaded Meghnad's father to allow Meghnad to study in an English school because of his unusual abilities. By the beginning of the twentieth century, low-caste Sahas were engaged in farming, trading, and the manufacture of products that ranged from furniture to liquor (involvement in the latter was considered defiling by high-caste people). Limited by class and prejudice in their educational progress in the nineteenth century, there were among the Sahas some families (*gushti*, in Bangla, "lineages") of considerable wealth, who pursued advanced education aggressively in the early twentieth century. (For example, one of young

Meghnad's chemistry teachers at Dacca College was Haridas Saha, the other E. C. Watson.)

Meghnad appears to have changed his own name at some point in his childhood, away from the name "Meghnath" (cloud-lord, or beloved-of-the-clouds) given to him by his parents. He chose a name given by his grandmother, Meghnad, the name of a son of the demon Ravana in the Mahabharata. The Meghnad character was given a renewed voice in a long poem by a nineteenth-century (Christian) poet Michael Mudhusudan Dutt's "Meghnad Badh," wherein Meghnad is portrayed as proud, competitive, and technologically superior to the deity Rama. According to Abha Sur, "Saha's alteration of his first name from Meghnath to Meghnad can only be understood as a political gesture."[3] Changing and choosing their name is not something young Bengali boys could commonly do.

Meghnad first went to an English school 10 km from his home where his elder brother had persuaded a local medical practitioner, a *kobiraj* of the ayurvedic system, to give Meghnad free board and lodging during the week. At age twelve, he was awarded a scholarship to Dacca Collegiate School but was expelled soon after arrival when he and some other older students, Nil Ratan Dhar among them, disrespectfully took off their shoes and staged a boycott of the visit of the governor during the Partition of Bengal that year.[4] The governor represented the dividers of a united Bengal, a division that Muslim elites in the east favored and some Hindus opposed. Having lost his free tuition and stipend, Meghnad went to a small private school where he prepared for the examination and nevertheless ranked first among thousands of students of East Bengal schools. His science studies began at age sixteen in Dacca College in 1909 (not yet a university), where his teachers were Haridas Saha and E. C. Watson in chemistry, B. N. Das in physics, and N. C. Ghosh and K. P. Basu in mathematics.[5] Saha also attended Bible classes at the Dacca Baptist Mission and, while polishing his English, won an all-Bengal competitive Bible examination and a deluxe copy of the Bible. Like other Hindu students he studied Sanskrit. He also began the study of German with a chemist who had just returned from doctoral studies in Vienna.

Meghnad did consistently well in all subjects and left Dacca two years later, at age eighteen, to begin his BSc at Presidency College in Calcutta. He ate in Eden Hindu Hostel from 1911 to 1913 and for the following two years ate at another student's mess at 110 College Street. Student hostels and messes, where one bought cooked food at a fixed rate based on a fixed menu, were then commonly divided into Brahmin and non-Brahmin sections. There was anxiety among some students and objections from a few about eating at the hostel with Saha because of his low-caste origins. At Eden Hindu Hostel, he was prevented by some Brahmins from making an offering to the goddess of learning, Saraswati, on the special day on which she is commonly worshipped by all Bengalis (Saraswati Puja).

Meghnad now earned the nickname "Eigenschaften" (for his invincibility and because he was learning German). His classmates were Satyen Bose, Jnan Ghosh, N. R. Sen, and J. N. Mukherji, all of whom became active scientists; Bose was soon to be the co-discoverer of the Bose-Einstein statistics. Nil Ratan Dhar, who found an open position at the university and encouraged Meghnad to come to Allahabad in 1922, was two years his senior, and P. C. Mahalanobis, who later founded the Indian Statistical Institute, was one year his senior. They were all taught at Presidency College by chemist and industrialist Prafulla Chandra Ray and biophysicist Jagadish Chandra Bose, although Meghnad primarily studied applied mathematics and was only secondarily in physics. Besides the hostel comrades and these teenage classmates who were later to be so influential in the growth of the scientific community, Meghnad also knew budding political figures. Rajendra Prasad, later president of India, attended social functions at the hostel, and Subhas Chandra Bose, later president of the Indian National Congress, was only three years behind Meghnad at Presidency College. "Bagha" Jatin Mukherjee, who became leader of the Jugantar Party after 1908, frequently ate in the student mess with Saha and others, while planning armed struggle in Bengal using German weapons and finances; he was killed in a fight with police at Balasore in 1915. "Saha's revolutionary friend Bagha Jatin advised him to keep away from politics; the task of building the country was no less urgent. Saha agreed. But the revolutionary embers continued to glow."[6] East Bengal was home to a high proportion of revolutionaries: Meghnad also knew Pulin Das, a rival of Jatin Mukherjee's Jugantar Party, who organized the revolutionary group called Anusilan Samity at Dacca, 200 km from Calcutta, and Sailen Ghosh, who began his revolutionary political work near Calcutta at the same time.[7]

While a graduate student at Presidency College, he experienced the 1916 Oaten incident, which became famous in the Independence movement: undergraduate students challenged the behavior of history professor E. F. Oaten, who had confronted them about noise they were making in the corridor near his classroom. Though he was popular as a cricketer, Oaten's recent remark about the British having a mission to civilize "barbarian" Indians was known to the students. Rebuking them for the noise, he threatened to have a fine imposed on them by the college. When the students complained to the principal, they were told to work it out on their own with

Oaten. Dissatisfied with that response, there was a confrontation leading to a three-day strike, led by Subhas Bose and others, but the college authorities imposed a Rs 5 fine on each striking student, and Oaten turned eleven graduate students out of his class because they participated in the strike. There was another confrontation between students and Oaten a month later, and this resulted in students pushing Oaten down some stairs and beating him. The principal identified eighteen-year-old Subhas Bose as a rebellious leader in the beating (without evidence, though Bose was present) and expelled him from the prestigious Presidency College. Four years older, Saha was not in the younger group being disciplined but observed at close hand how students, many based in hostels, were challenging the status quo and how the college authorities were really unable to contain the consequences of Oaten's disdainful behavior.[8] Saha's treatment from some Brahmins, his being barred from writing the civil service exams, and his association with revolutionaries like Jatin Das and student leaders like Subhas Bose had long-lasting effects on him.

After four years of study in Calcutta, Saha applied to appear to write the Indian Finance Service examinations in 1915. Though he had stood second in the MSc examination of Calcutta University, one place behind his friend Satyen Bose, Meghnad was refused permission to write the Finance Service examination on the grounds of his associations with political revolutionaries. He was well aware of the economic expectations his family had of him, had refrained from political action after the loss of the tuition and stipend with his expulsion from Dacca Collegiate School, and now was prepared to work in government. But in 1916, "with all hopes of securing a Government service permanently gone, he now decided to carry on research in applied mathematics and physics."[9]

Like many of his classmates, he decided to enter a scientific career at an opportune time. Sir Asutosh Mookerjee was expanding the Science College of the University of Calcutta using the large grants given by two very wealthy lawyers, and Saha became a doctoral student in this encouraging and constantly expanding environment. Asutosh Mookerjee was a very competent mathematician who, as vice-chancellor of the university from 1906 to 1914, effected the change to a teaching university in addition to a mere examining university (see Biographical Notes). He had used philanthropic grants to build up the most powerful law school in India and employed the same technique with the Science College. Even though he was not officially the vice-chancellor between 1914 and 1921, he continued to chair important committees and, as a member of the Senate and the Calcutta University Commission (1917–19), was determined to strengthen teaching *and* research. For example, he concluded the appointment of C. V. Raman as professor of physics, without an advanced degree, in 1917, against the explicit wishes of the very influential Jagdish Chandra Bose, who wanted the position to be given to his nephew D. M. Bose.[10] D. M. Bose was, at this time, a PhD student in Germany and not able to come home. No greater friend to young scientists could have been in the vice-chancellor's office than Mookerjee, and this continued even when he left that office to become justice of the Calcutta High Court in 1920 until his death in 1924.

Meghnad's group of MSc students thought their training would be incomplete without the use of laboratory facilities in physics (chemistry labs already existed), and so they approached Mookerjee to ask him about master's degree classes in experimental physics. He told them to prepare themselves for this goal and secured funds for the purchase of journals and some equipment. Mookerjee invited Meghnad to be a lecturer in the Department of Mathematics in 1916, along with Satyen Bose. But neither of them could get along with Ganesh Prasad, the head of that department, so they asked to be transferred to the Department of Physics. S. K. Mitra, himself only twenty-six, was already a lecturer there. Mitra had ranked first in physics MSc exams in 1912 at Presidency College and would later be Saha's somewhat rivalrous colleague in the university. As N. R. Sen reminisced:

> In the Presidency College Saha developed a strong love for mathematical physics and that decided the course of his future life. Professor D. N. Mallick the senior Professor of Mathematics in the Presidency College at the time, took immense pains for the top students of his class, and Saha read for two years with him in the post-graduate classes. He also had one year with Sir J. C. Bose for physics. But his interest in mathematical physics was stronger than the lure of Sir J. C. Bose's laboratory.[11]

Throughout his career Saha considered himself a phenomenological theorist, standing in the middle between experimenters and pure theorists, but he put a lot of energy into building good facilities for experimenters. In the years between 1916 and 1919, there was a great deal of activity in the new Science College: Saha was asked to organize teaching of graduate students in thermodynamics. At twenty-four he published his first paper in 1917 in the *Philosophical Magazine* on the theory of Maxwell's electromagnetic stress energy-tensor. The only experimental paper he did at this time (on measurement of the pressure of light) was done with apparatus he constructed at the Science College. Being locally published in the *Journal of the Asiatic Society*, it did not receive much notice, according to Saha. The library at the Science

College was so poor in resources that Saha had to do his reading a mile away through crowded lanes at Presidency College. Saha was so poor that while studying he also had to work as a tutor to raise money for his family, particularly after they had married him to Radharani Saha in 1918.

Saha began to read the new German work on quantum theory in 1918, when it arrived at the end of the war in Europe. D. M. Bose, interned as an alien in Germany until the war's end, arrived in Calcutta with his doctorate, and joined with physicist C. V. Raman to become the senior colleagues of Meghnad Saha, Satyen Bose, and S. K. Mitra in 1919. But there were few books available.[12] One day the daily Calcutta newspaper the *Statesman* sent a reporter to the astronomical observatory at Science College to get an explanation of a cabled confirmation of Einstein's prediction of the deflection of starlight by the gravitational field of the sun. Saha immediately wrote a popular explanation of this effect and gave it to the reporter. He and Satyen Bose then translated Einstein's classical papers on special and general relativity, which the University of Calcutta Press published in 1919 as a book, *Principles of Relativity;* this was *the first* book publication of Einstein's work in English.[13] Saha and Bose collaborated in a paper on a new equation of state, published in the *Philosophical Magazine,* in 1918.

When he met them in 1918 on his return from Germany to Calcutta, D. M. Bose found that Satyen Bose had been attracted to the logical elegance of Planck's *Thermodynamik* and *Warmestrahlung,* which D. M. Bose gave him to read. Having done so, Satyen Bose questioned Planck's theorem on black-body radiation, and this led to a comparison of the two thinkers:

> This intellectual dissatisfaction with Planck's deduction of his radiation formula led, I believe, to Bose's deduction on a combinatorial basis of Planck's formula in 1925. Meghnad Saha's approach was more direct; he wanted to learn from me the latest advances on the frontiers of research in physics in relation to Quantum Physics and Thermodynamics. . . . From time to time he discussed with me the theory of thermal ionization of gases and its application to the interpretation of the stellar spectra. Next year in 1920, as an examiner of Griffith Memorial Prize essays, for which the candidates had to give a nom-de-plume, I came across amongst the other papers one by "Heliophilus" [i.e., Saha] on "Origins of Lines in Stellar Spectra." As the paper was so outstanding compared to other essays submitted for the prize, there was no hesitation in recommending it.[14]

Saha also developed theories on selective radiation pressures on stellar atmospheres and sent a paper in 1920 to the *Astrophysical Journal* at the

Yerkes Observatory of the University of Chicago. The major portion of this work was unfortunately kept in a drawer until it was discovered there during his visit to the observatory in 1936. E. A. Milne, of Cambridge, developed this field subsequently and acknowledged that Saha's original ideas had stimulated him.[15] Saha frequently met incredulity that his earliest work was done in India, not abroad, and he sometimes projected among foreign scientists an image of neglected and isolated genius ignored by Western scientists. In India, however, he was careful to show how much his work was celebrated abroad, for which there was ample evidence; about 200 journal citations referred to Saha's work by the end of the 1920s.[16]

It is for his theory of thermal ionization that Saha gained his reputation, and he later explained that this idea came largely out of his reading the German literature arriving in Calcutta after the four years' lapse of the war. In a long and widely circulated letter to Canadian astronomer H. H. Plaskett in 1946, Saha said, "I was a regular reader of German journals which had just started coming after four years of the First World War."[17] While there may have been a prohibition on subscription payments to Germany or extreme delay in mailing and censoring procedures, the evidence suggests that it is unlikely that these journals were completely banned in India.[18]

Saha did not know that F. A. Lindemann at Oxford had suggested an explanation for the high-temperature ionization of hydrogen that same year, and Lindemann, who eventually became one of the most powerful scientists in Britain, harbored a long grievance that his work was seen only as subsequent to Saha's. E. A. Milne at Cambridge was also working on stellar spectra. But Saha was more attracted to a paper by one of Nernst's students, Eggert, at Berlin. In his letter to Plaskett in 1946, Saha wrote:

> While reading Eggert's paper I saw at once the importance of introducing the value of ionization potential in the formula of Eggert, for calculating accurately the ionization, single or multiple, of any particular element under any combination of temperature and pressure. I thus arrived at the formula which now goes by my name. Owing to my previous acquaintance with chromospheric and stellar problems, I could at once see its application.[19]

This seminal paper, his eleventh, was sent in 1919 to the *Philosophical Magazine* in London and published in 1920. On the basis of these papers, Saha was awarded the DSc by the University of Calcutta. An external examiner of his dissertation was O. W. Richardson, who won the Nobel Prize for Physics in 1928 for the theory of thermionic emission. Saha had now gained international recognition and was awarded two scholarships that enabled him to have two years of postdoctoral research in England and Germany. He was teaching regularly, influencing students with his "sense of invincibility," and earning an income to support his family. But there were obstacles, because "virtually no astrophysicist accepted Saha's assumptions and the techniques he used from physical chemistry as the means of arriving at his conclusions."[20] Moreover few in the astrophysics establishment understood that he had done the essential work in Calcutta, because it was commonly thought, except among his friends, that this creative thinking and writing had occurred only after he went abroad. These two reactions of scientists overseas were to affect Saha most of his life.

Not all his energies went to physics. Saha took a personal interest in the students' living conditions, based on his own difficulties in hostels, and at age twenty-six, in a written questionnaire to the University of Calcutta's Sadler Commission in March 1919, boldly said, "Now the members of the democratic classes feel that, in at least the hostels which have been constructed at public expense, they have the same rights as members of other classes."[21] Saha was beginning a life of standing up to hierarchies and declaiming his analysis of current conditions and reasoning why his critique should be realized. His early voice on behalf of "democratic classes," against the restrictions of caste, was to echo down through his life, and popular perception of him incorporated this voice into its favorable appraisal of Saha the scientist.

Shanti Bhatnagar's Formative Influences

Shanti's father had studied at Forman Christian College in Lahore while living in the house of a senior court official, so it was not a surprise that his son became a student at the same American missionary college. In fact his father died when Shanti was born in 1894, and his mother had great difficulty in raising the children in the small town of Sikanderabad because her husband had insisted on becoming a member of the reformist nineteenth-century Hindu movement called Brahmo Samaj, and thus the rest of the more orthodox Bhatnagar lineage ostracized him and excluded his family financially. This independence of spirit of the dead father left the family "entirely without means."[22] The boy was raised by a grandfather, a railway construction engineer who grew progressively reclusive and thus restricted the boy's own sociability; "I did not have company in my childhood," Shanti said later. This Brahmo Samaj commitment of Shanti's father, however, had a silver lining, because Brahmo patrons had founded a high school and college in Lahore to which Shanti was eventually sent. The

Bhatnagar family was of the *kayastha* group of castes, which included the occupation of scribes, clerks, and readers, and were thus heavily involved in administration, courts, and law in the twentieth century.

His father's attachment to the reforming Brahmo philosophy had a precedent; one of Shanti's grandparents was Munshi Hargopal, who held the title of *khwaja-i-khwaja* (a reference to his learning) conferred by the Moghul court at Delhi. His poetic name was "Tafta," and he was befriended by the older renowned poet Mirza Ghalib. Tafta collected rare books and gave them to the University of the Punjab in Lahore. Accepting the name "Mirza" himself from Ghalib, Tafta (Munshi Hargopal) dressed like a Muslim, including his turban: "In fact he was as much Muslim as Hindu. The dividing line between the two communities had, for him, been obliterated."[23] This dividing line would reappear sharply and violently, for Shanti, in Lahore in 1947, but at a young age Shanti knew to value this kind of nonconformity: he was sustained all the way until 1921 on Brahmo Samaj support and scholarships. He understood the importance of crossing between Hindu and Muslim cultures: beginning at an Urdu primary school and then studying at an English and Hindi medium school, Shanti was moved to Lahore in 1908 when he was fourteen and was cut off from his mother except on holidays. Among other things he studied Sanskrit, Persian, and Urdu, and his mature poetry was written in Urdu. He entered a school and college run by the Brahmo Samaj, of which the principal was his father's friend.

He became keen to do scientific experiments and set up a small lab of his own in one of the neglected passageways of the school. In 1911 he won a government scholarship and moved to the Dayal Singh College, where he met his drama teacher and eventual lifelong friend Norah Richards, whose stage name was Norah Doyle, the Irish wife of English professor P. E. Richards. Under her guidance, Bhatnagar wrote a play satirizing traditional healing practices as quackery, but the college principal, Bhatnagar's father-in-law, banned staging the play because he thought it would offend Hindu sentiments in Lahore, a city full of traditional healers. Nevertheless, wrote Bhatnagar thirty-five years later, Norah Richards "electrified the University by her literary merit and keen interest in modern drama and Shakespearean plays . . . [and] she chose me for some important parts in plays which she staged."[24] Clearly Richards electrified Bhatnagar too. But the two friends, the principal and Shanti's father, had spoken about joining their children in marriage, and Shanti followed the wishes of his deceased father when, at the age of twenty-one, he married Lajjawati, the daughter of the principal of Dayal Singh College in 1915.[25] Within two years he moved to Forman Christian College, where in 1916 he finally got a BSc in physics, having failed the chemistry exam! He was then awarded a prize for solving a problem in ink and printing, a prize offered by a merchant in Lahore to someone who could find a local substitute for materials that were unavailable during the war. As result of his good performance and the prize, he was also appointed as a demonstrator in physics and chemistry courses, finally obtaining much-needed income for his family. Shanti prepared himself for the MSc exams as a private student, and with coaching from diligent and dedicated British and Indian science teachers, particularly in chemistry (like lecturer B. M. Jones), he prepared his MSc thesis on the effect of absorbed gases on the surface tension of water. He ranked first in the physical and inorganic chemistry exams, and his thesis won the Dayal Singh Five-Year Traveling Scholarship to study in America. Like Saha he arrived in London in 1920.

Saha and Bhatnagar in London and Berlin

In spite of his two scholarships, Saha still could not afford to attend Cambridge or Oxford as he wished, so he attached himself to Imperial College, London. His friends from the College Street Hostel in Calcutta Jnan Ghosh and J. N. Mukherji were already working there under F. G. Donnan in physical chemistry. His relationship with Ghosh would gradually become more and more important to Saha, as Ghosh's national profile increased too. It was here, at Imperial College, that Saha and Bhatnagar met. Away from wives and families, they worked hard and developed the inexpensive convivialties of poor students in London. Saha even nicknamed his friend "Steamship Bhatnagar" because of his energy and determination, and the nickname stuck. Soon Bhatnagar began working in Donnan's lab, but Saha took a little longer to find a place to work. One of Saha's former classmates from Presidency College was also in Imperial College: Sneyhamaya Dutta, who later became director of public instruction for Bengal, was completing his doctorate in physics. Dutta advised him to meet with Professor Alfred Fowler, because Saha had not been firmly located anywhere; Fowler, a pioneer in astrophysics, was working in stellar spectroscopy at Cambridge, and Dutta's introduction succeeded. While working with him, Saha's now famous paper on ionization appeared in the *Philosophical Magazine,* and on its basis Fowler and Saha then collaborated on rewriting a paper Saha had written on the physical theory of stellar spectra. Lindemann and Kramers had just been working on the same problem of solar chromosphere ionization independently suggested to them by Bohr and Tolman.[26] Saha was now in a heady intellectual milieu and proposed experiments to J. J. Thompson at the Cavendish Laboratory at Cambridge, experiments which he thought might verify his theory. But when Thompson told him that Cambridge was not equipped for high-temperature work at that stage, Fowler advised Saha to seek assistance from Walther Nernst at Berlin, which he soon did.

Saha's long interest in Germany and its technical development was now to be satisfied. His knowledge of the language and sympathy for its conflict with Britain made the study visit to Nernst's lab quite logical. In late 1920, Saha left for a year's research in Germany, where he set up an experiment in Nernst's laboratory even though, as he said later, he still could not conclusively verify his theory of thermal ionization.[27] Saha and Bhatnagar were part of a group of fourteen students visiting from the University College of London; Bhatnagar carried the crucial letter of introduction to Nernst from Bhatnagar's famous teacher, Donnan. Nernst had just won the Nobel Prize for Chemistry a few months before in late 1920 and held numerous patents, including one for nerve gas used during the First World War. Nernst was also known as the author of the heat theorem whose validity for gases was later proven by Einstein based on Satyen Bose's statistics, later called the Bose-Einstein statistics.

Nernst at first refused to invite any of the students from London into his laboratories on the grounds that the German researchers were still too sensitive about the outcome of the war. But "later on a note came addressed to Saha saying . . . [that] Nernst would allow him and his Indian colleague Bhatnagar to see the laboratories because the last blow to the British Empire would come from India."[28] Here "the image Saha wished to project in his letters from Berlin to people like Hale and later to Russell was of a poor transient whose fate was to return to a distant and ill-equipped outpost in his native land."[29] This became a projection with deep roots, long after he had improved his conditions. During this year Meghnad began acquaintances with Einstein, Planck, Sommerfeld, and other European physicists, relationships that he sustained all his life. Berlin was a political and cultural center in Europe, free of British influence, with many Germans genuinely interested in India; they were ready to meet its intellectuals. The city was also then the center of the international face of the Communist movement.

Through Arnold Sommerfeld he met Rabindranath Tagore for the first time and became connected with the international network of Bengali revolutionary parties that used Berlin as a base. He met fellow-Bengali M. N. Roy, whom Trotsky had asked to found a Communist party in Mexico in 1919; Saha knew that Roy owed allegiance to other groups but notwithstanding Roy's prominence and charisma, Saha appears to have become more friendly with members of the revolutionary Jugantar Party, some of whom he already knew, being from East Bengal, and remained in contact with after his return to Calcutta, particularly with Nalini Gupta, organizer of the Bengal Workers and Peasants Party. Saha's year in Berlin resulted in his first paper published in German in the *Zeitschrift für Physik* (1921), on a physical theory of stellar temperatures. He visited both Switzerland and England during this year. Then Sir Asutosh Mookerjee, who had just been reappointed as vice-chancellor of the University of Calcutta, telegraphed him an offer of a new chair in physics at Calcutta, which Saha, who had been awarded a DSc degree, accepted only after writing letters to Mookerjee asking for a higher salary and for £500 to purchase equipment in Europe for his new lab.[30] He returned to Calcutta in late 1921 not knowing whether Mookerjee could meet his expectations about a new lab.

In the year that Saha was in Berlin, Shanti Bhatnagar worked toward his DSc at the University of London, having been told not to bother with the somewhat less prestigious PhD degree. He chose colloidal chemistry and emulsions as a focus. Chemistry was surging in postwar Britain, where relations with German scientists were quickly reestablished by British chemists, envying the German successes in transferring university-based chemical innovations to industry. Through his visit to Nernst in Berlin, Bhatnagar made connections in the German scientific world too, particularly with Fritz Haber, who, along with Nernst, had invented nerve gas and won the Nobel Prize for Chemistry in 1918 for making ammonia from nitrogen in the air; Haber would have to leave Germany fifteen years later because he was Jewish, as did Bhatnagar's other supporter H. M. F. Freundlich, who specialized in colloids and became the director of the Kaiser Wilhelm Institute in Berlin before he too left in 1938 because his father was Jewish. All this gave Bhatnagar an early entrée into the field of colloid chemistry for which he developed his reputation in the powerful field of industrial chemistry, eventually leading to his election as a Fellow of the Royal Society.

The huge size of the world market for chemicals and the war's stimulus propelled the formation or expansion of large American, British, French, and German international chemical corporations, all of which had big offices in London. So the professional lives of young chemists became a combination of academic and industrial networks, something physicists did not experience until twenty years later. Bhatnagar published eight papers during the year 1921, all in reputable journals, well making up for a relatively small publishing record in India, particularly in comparison with his contemporary, Saha.

London proved to be a center for influential Indian scientists. At the time Bhatnagar defended his DSc dissertation, C. V. Raman, J. C. Bose, and P. C. Ray, all professors from Calcutta University, were visiting the city. The new vice-chancellor of Benares Hindu University asked these three individuals for nominations to a new chair in chemistry at Benares. P. C. Ray already knew of Bhatnagar's abilities and heard from Donnan of his good performance at the thesis defense. J. C. Bose had just been elected a Fellow of the Royal Society, after five previously unsuccessful annual attempts. Bose's institute was now receiving an annual Rs 75,000 grant from the Bengal and India governments for "pure science," and this was soon to be increased to Rs 100,000.[31] P. C. Ray had, like Bose, been nominated for the five previous years as an FRS, but even after eight years was never elected. Raman, much younger than the other two and on his first visit to Britain, was to be elected to the Royal Society in 1924, well before his 1927–28 work that led to the 1930 Nobel Prize. The trio of Raman, Ray, and Bose cabled from London their unanimous recommendation of Shanti Bhatnagar for the Benares chair in chemistry, and so, like Saha, he returned to India in 1921 to a new full professorship at age twenty-eight, with the blessing of its most prestigious figures. The conditions under which Mahendralal Sircar established the Indian Association for the Cultivation of Science (IACS) in the 1880s, P. C. Ray created a nationalist enterprise such as Bengal Chemical in 1900, and Ronald Ross, born in India in 1957, won the first Nobel Prize in Physiology and Medicine in 1902 were founded on a prejudice among many British that Indians would not be capable of good experimental science.[32] But twenty years later, after the war, there had been effective challenges to this prejudice from talented Indians, and things were evidently beginning to change, with some British scientists playing intermediary roles in those changes.

Saha in Academia: From Allahabad to Calcutta, via Bangalore

When Meghnad Saha reached Calcutta from Berlin in November 1921, he found Sir Asutosh Mookerjee, again the university's vice-chancellor, in conflict with Lord Ronaldshay, the governor of Bengal. Because the governor was also the chancellor of the University of Calcutta, this conflict was soon to affect Saha. Pressing upon these relationships was the underlying question of the university's position in the noncooperation movement; it was turning out people with science degrees who were also essential to administration in a colonial bureaucracy, and its alumni and patrons were conflicted about this ambiguity. The governor and the viceroy above him, on the other hand, expected more loyalty on the symbolic level and less criticism, in short a public commitment to the civilizing mission of the university. But on the practical level, charges of "thoughtless expansion" were made against the postgraduate teaching part of the university, specifically the sciences. The university's minutes of 11 July 1921 state that of 174 students in the Science College, 48 were in physics with 3 professors and 13 lecturers, many more students than in applied mathematics, chemistry, applied chemistry, physiology, geology, zoology, or botany.[33] The Khaira Chair that Saha occupied for one year had been created through Mookerjee as the result of settlement in the Calcutta High Court of a Rs 600,000 lawsuit brought against the rajah of Khaira, Kumar Guruprasad Singh, in 1919 by his wife Rani Bageswari Devi.[34] Asutosh Mookerjee, being both a justice of the High Court and vice-chancellor, knew how to make best use of the gift made to the university: he favored Saha for the appointment to this chair, and he got it—at age twenty-eight. As specified in the other gifts from wealthy lawyers Sir Taraknath Palit and Rashbehary Ghosh, occupants of new academic posts (special professorial chairs) were to be Indians. By 1921 there were eleven such privately endowed chairs at the University of Calcutta, two Palit chairs, six Ghosh chairs, and three Khaira chairs.[35] The government of Bengal was expected to build the laboratories and provide staff but did not support this process in proportion to the commitments of private individuals.

Asutosh Mookerjee's difficulties with the government were so severe that Saha could obtain neither a research assistant nor facilities to set up a proper laboratory. This might be acceptable for a theorist, but because Saha wanted to start an experimental lab, he requested an oil pump for the lab worth £12, but it was rejected in 1922 because the departmental budget had been exhausted.[36] He may also have encountered prejudice focused on his low social origins and country-style manners and language. However difficult it is to understand all the reasons that influenced Saha to leave a science center like Calcutta, given that he knew the conditions so well there, one has to conclude that the conditions he found upon his return were worse than expected.[37]

There was a working laboratory at the IACS, but C. V. Raman practically ran it, and Saha had already disagreed with Raman's insistence on teaching at the university during the 1919–20 noncooperation movement. Though it certainly improved later, the IACS did not look promising to Saha at this stage. Moreover, some very creative former colleagues were dispersing to other places. Satyen Bose had already accepted the post of reader at the new University of Dacca in 1921, from where he sent his work to Einstein in January 1924, now celebrated in the famous Bose-Einstein statistics.[38] Einstein translated Bose's paper into German, and for years many Europeans thought Bose was a German, as the name was common there. Bose's work pointed first to understanding photons as particles in a free photon gas and then to the statistical problem of assigning probability to indistinguishable particles, like photons. Saha's other colleague S. K. Mitra was already doing research in Paris with Charles Fabry. Saha began looking for work outside Calcutta and (fatefully, in my opinion) decided to leave the Khaira physics professorship.[39] The chair he left behind in Calcutta haunted Saha, and he spent a lot of energy ten years later trying to get it back.

The three senior and most influential scientists in Calcutta at this time were J. C. Bose, P. C. Ray, and C. V. Raman. Since they were considerably older than Saha, he probably did not look upon them as potential colleagues, but certainly as authority figures—something with which Saha already had an ambivalent relationship. They controlled laboratories and committees through which Saha needed to obtain resources; they also had influence over university positions of the type for which Saha would be competing, as he had already seen in Bhatnagar's chemistry appointment at Benares, and they could raise or lower international opinion about research done in India. Saha was about to discover that Raman could have this influence on his reputation.

Saha began to look for a new job at the age of twenty-eight. He refused offers of positions from Aligarh Muslim University and from Benares Hindu University; the latter offer was strongly supported by Shanti Bhatnagar, who wanted him there as a colleague.[40] Saha also refused an offer to work with astronomer John Evershed at the Kodaikanal Observatory in South India, though the reason for the refusal in the authorized biography is hardly convincing. "Working under a man of Evershed's genius and genial personality would have been a great honour and privilege, and it probably would have brought him greater recognition, but his heart was on physics, and he would not change it even for astrophysics."[41] This is a curious (retrospective) statement by Saha because Saha's reputation was already established in astrophysics. Evershed had built an effective spectroheliograph, with which he had discovered the effect of radial flow of gases in sunspots in 1909, a discovery which became known as "the Evershed effect." This research position had been offered previously to K. S. Krishnan, studying at Madras Christian College in 1920, but when the government redefined it as work at a nearby meteorological observatory, Krishan declined the offer and decided to go to study with Raman in Calcutta.[42] (It is a measure of Evershed's professional reach and India's tight scientific referral system that Evershed offered this moderately paid and remote job to two people under the age of thirty, each from opposite sides of the country, both of whom were exceptionally good scientists and eventually became FRS.)

While waiting for the right working opportunity, Saha suddenly turned his attention to the disastrous North Bengal floods of July 1923. Here he worked for the first time publicly with the highly successful, respected, and very ascetic chemist and businessman P. C. Ray, who chaired the Flood Relief Committee. With Saha was Subhas Bose, just returned from completing a degree at Cambridge; Bose led a group of relief workers to Santahar, and Saha focused on publicity and collecting gifts and donations. Meghnad's successful efforts to raise Rs 2.3 million for this committee launched a lifelong fund-raising skill.[43]

In the end, Meghnad accepted an offer from the University of Allahabad, encouraged by chemist Nil Ratan Dhar, who had gone to school with Saha in Dacca, was two years Saha's senior at Presidency College, and later had been P.C. Ray's student. Dhar had recently become professor of chemistry at Allahabad, part of the huge expansion of Calcutta-trained Bengali scientists throughout northern India. But Saha, working through a powerful physical chemistry network (like Bhatnagar and Dhar), was going where virtually no physics was being done.

No physics at Allahabad, save for the teaching of 120 undergraduates! The laboratory was adequate for demonstrations in lectures but not for research, despite proclamations to the contrary. Saha does not seem to have investigated this prior to taking the appointment, and perhaps in the end he felt he had little choice. Under Saha's direction the physics department was given one reader (who became a controller of the undergraduates), one lecturer, and two demonstrators, one of whom only did clerical work. Saha spent much of his time carefully preparing his lectures, which he usually wrote out in full on the blackboard for the students to copy, there being few modern textbooks. This was the origin of his popular and eventually profitable textbook *Treatise on Heat.* The retired High Court judge who was the university's registrar, and in charge of university finances, resisted Saha's attempts to improve the obsolete library, to reorganize the laboratory for experimental research, and to replace the hand-powered tools in the small workshop with electric ones. So it was to be for the next five years, struggling with these difficult conditions, until Saha was awarded the FRS in 1927. This was a far cry from working conditions in Calcutta, and even further from conditions in London. It was far also from the entanglements of Calcutta. This was a form of exile, and one cannot avoid the conclusion that it was, at first, partly self-imposed.

He did not neglect his friends, however, and kept in regular communication with Satyen Bose. In 1923, while Bose was teaching thermodynamics and electromagnetic theory to the MSc classes at Dacca University, he was simultaneously studying the theory of relativity and quantum theory.[44] Now, he said, he felt the need for a logically more satisfactory derivation of Planck's law. Saha visited Dacca from Allahabad toward the end of 1923 or in early 1924 and had discussions with his friend Bose about Planck's law.[45] They had already published papers together on a number of occasions. Saha drew Bose's attention to some recent papers by Wolfgang Pauli, Einstein, and Ehrenfest, published in *Zeitschrift für Physik.* Bose studied them and wrote his now famous paper, first in English (not published by the *Philosophical Magazine* as Bose hoped) and then in German, as "Plancks Gesetz und Lichtquantenhypothese" (Planck's law and the light quantum hypothesis).[46]

Bose sent this paper about Planck's law to the *Philosophical Magazine* from Dacca in 1923, and, not hearing from them for a long time, he sent a copy to Einstein for his opinion with the request to have it translated into German and published in the German journal *Zeitschrift für Physik.* Bose's first letter to Einstein, dated 4 June 1924 and sent with the paper, began as follows:

> Respected Sir, I have ventured to send you the accompanying article for your perusal. I am anxious to know what you think of it. You will see that I have ventured to deduce the coefficient (8Πv^2/c^3) in Planck's Law independent of classical electrodynamics, only assuming that the ultimate elementary regions in the phase-space has the content h^3.[47]

Einstein was presumably very pleased by Bose's paper and that is why he sent a letter of praise to Bose, calling his work "a beautiful step forward." This was the step that led Bose to Europe a year later, after which this work gradually became known as the Bose-Einstein statistics.

Taking stock of his situation, Saha began to look for funds to support his research in Allahabad. Using his network of colleagues, he learned in 1924 that Niels Bohr had received an £800 grant from the Rockefeller-supported International Education Board. Saha wrote to Henry Russell at the board to ask him about an application for a £2,000 grant. As DeVorkin shows, Russell consulted Caltech's Robert Millikan, who was on important granting committees, and Millikan replied that the grant to Bohr was an exception and that American physicists would have priority over international applicants as this was an American foundation. Though Rockefeller grants to foreign scientists were not common at this time, this was a polite way to reject Saha, without mentioning that Millikan had formed a poor opinion of Saha's abilities as an experimenter, based almost entirely on C. V. Raman's advice. Millikan and Raman had conversations about Saha at Caltech in 1924, and because Millikan and Russell both respected Raman's opinion, the application was rejected. Henry Russell also recalled conversations with Ralph Fowler in Cambridge, who also raised doubts about Saha's abilities as an experimenter.[48]

Saha was unaware how this opinion against him might have been formed, but he probably realized its consequences. It should be remembered that Raman did not think *anyone* in India approached his own experimental abilities. Meanwhile Saha was nominated as Fellow of the Royal Society by his teacher Alfred Fowler in London and by Gilbert Walker, mathematician and meteorologist who spent much of his life in India and was now retired in England. Citations of Saha's work were increasing, and there was already talk in the literature of "Saha's equation." James Jeans, the well-informed secretary of the Royal Society, asked the viceroy's office in 1925 "whether Saha's political record is likely to be embarrassing afterwards to the Royal Society." The official reply from the viceroy's office was that there were grounds for not recommending Saha for a fellowship, and this reply caused a further investigation by a British intelligence officer in India, who eventually reported that Saha was sympathetic to revolutionary aims and activities, though not an active participant in them.[49] The evidence available now confirms that Saha gave money to "political sufferers" (code word for those who were penalized, lost jobs, or were imprisoned because of political activity against the British administration). Moreover, he was a conduit for Indian revolutionaries in Germany and Switzerland; his contact in Berlin with M. N. Roy while at Nernst's lab in 1921 led to Saha's becoming keeper of the secret code for Jugantar, most of whose members merged with the Communist Party of India when it formed in 1921–22. Saha acted for the party to set up the visit to India of Roy's emissary Nalini Gupta in 1921.[50]

It is unlikely that Saha played any role as code keeper or other function once he moved to Allahabad in 1923; being a Bengali-speaking outsider he would not have the required invisibility for this role, nor was Allahabad particularly receptive to the new Communist Party. Though it is not known whether the viceroy's intelligence officer uncovered Saha's role as code keeper, the Royal Society was so alarmed by other evidence in the viceroy's report that James Jeans asked Fowler whether or not he wished to continue to nominate a "rabid revolutionary" and suggested that Saha's name be withdrawn. Provided with the intelligence report, says DeVorkin, "Gilbert Walker [who had previously offered Saha a job in India] remained solidly behind Saha" and Alfred Fowler replied that the nominators would sustain the recommendation for Saha.[51] By now, Saha was third overall in accumulated citations to his work among all physics candidates for FRS in 1925. But other factors were against his election that year; historically, few Fellows have been elected in the first or second round. Saha's name came forward again in 1926. The nomination now needed and acquired further backing; signatures were sought of F. A. Lindemann at Oxford, J. Evershed, F. W. Dyson, and others. If anyone had doubts about his experimental ability or if Lindemann harbored antagonism to Saha at this stage, no one applied it to the vote in the following year. With Lindemann's vote and the others, Saha was finally elected in 1927 for "development of the theory of high-temperature ionization."[52]

We do not know if or when Saha learned of the scrutiny to which his candidacy had been subjected. We do know that Saha thought his competitor E. A. Milne at Cambridge disregarded his work on selective radiation pressure and avoided giving him credit. Nevertheless, when Saha planned to go to Oxford ten years later, Milne himself wrote to warn him that Lindemann was jealous of his success and said, "he has never forgiven himself for not discovering the physical nature of the stellar spectral sequence himself, which we all recognize was due to you." Milne here seems to concede what Saha thought Milne himself withheld. Milne also warned Saha that Lindemann appeared to be "not very well disposed to Indians."[53] Lindemann, who later became Winston Churchill's chief scientific advisor and was deeply involved in policy toward science and industry in India, had no prior contact with India when he signed Saha's nomination in 1926. But two years later, he first visited India as member of a review committee of the Forest Institute at Dehra Dun, beginning a long career of influence in India.

Saha thus encountered a surface of hospitality and collegial assistance but a deeper reservation, if not prejudice, one that was used by some people in Britain to convince themselves that ruling India was both ethically and politically correct. This scrutiny also made clear to Saha a distinction between the scientific reputation and the politics of the person behind it. Not only was Lindemann "not well disposed to Indians," but E. A. Milne himself was specifically opposed to the nationalist movement in India, according to astrophysicist S. Chandrasekhar, because Milne did not think India could survive without British presence. While saying that British scientists had no reason to be ashamed of their positive role in helping India in science, Chandrasekhar said he experienced a different reaction about their politics, in Britain, in the mid-1930s: "It was quite obvious that the establishment at Cambridge was quite against the Indian movement. I mean, I could feel it very strongly as a Fellow. . . . I found it very difficult to talk politics with

E. A. Milne or R. H. Fowler because they had opposite views."[54] This is relevant because Lindemann, Milne, and Ralph Fowler were precisely the individuals in Britain with whom Saha interacted most before and after he became FRS in 1927 and from whom he wanted scientific recognition during his exile in Allahabad. They were pillars of the conservative British scientific establishment. Nevertheless, Fowler and Lindemann had both supported his election, and Fowler had courageously persisted in doing so in spite of adverse reports about Saha. Milne did eventually give due recognition to Saha's work, according to DeVorkin. This interplay of scientific reputation and Indian politics was infinitely complex.

Though his reputation rose when Saha was elected president of the Physics Section of the Indian Science Congress meeting in 1925, "things began to change only after 1927."[55] By that time Saha attracted a number of quite resourceful students, three of whom eventually rose to top positions in science organizations. P. K. Kitchlu, from Lahore, was a research scholar who first published his work on the arc spectrum of copper in the *Indian Journal of Physics,* the year the journal was founded. In 1927 and 1928 Kitchlu published joint papers with Saha, then went to Cambridge to study with none other than Ralph Fowler. R. C. Majumdar published a paper with Saha on new methods in statistical mechanics in 1930 and also went to Cambridge to study, arriving—it is said—on the doorstep of Lord Rutherford's home on a Sunday afternoon. Daulat Singh Kothari was also Saha's student and published a paper with him in 1934 on the beta-ray activity in radioactive bodies. Kothari "was prevented by Professor Saha from writing the all-India civil service examinations" and went to Cambridge in 1933 to do his dissertation in astrophysics.[56]

On his return from Cambridge, Kothari was offered only his old job as demonstrator in physics at Rs 180 per month, as Saha had tried but could provide nothing else, there being no new position for a reader in the department. Before Kothari's departure for Cambridge, Saha had asked him to tutor his friend Chief Justice Sir Shah Mohammed Suleiman of the Allahabad High Court, in physics. Now Suleiman remembered his "intellectual companion . . . and he really went out of his way to recommend Kothari to the Delhi University authorities."[57] Kothari moved to Delhi with Saha's encouragement and, soon after Independence, became the scientific advisor to the minister of Defence and then chairman of the University Grants Commission. Kitchlu was appointed professor of physics in the University of Punjab at Lahore and was named director of the National Physical Laboratory in Delhi in 1961, succeeding Sir K. S. Krishnan. Majumdar later led the Department of Physics at the University of Delhi. In addition to these scientists, most of Saha's students became influential in political and administrative life outside the scientific community, and he relied on that—as we shall see.

On becoming an FRS in 1927, Saha was congratulated by the governor of the United Provinces, who had himself been a physics classmate of Lord Rutherford of the Cavendish Laboratory; the governor awarded him an annual research grant of Rs 5,000—a substantial sum at the time. (Note, however, the much larger grant of Rs 100,000 received by J. C. Bose from the government of Bengal.) Saha was still trying to gain experimental confirmation of his theory of thermal ionization at Allahabad, but, in his words, "the laboratory facilities were too inadequate and grants insufficient."[58] The governor's grant changed that. The ionospheric research laboratory was established and enabled the work of Saha's student C. R. Toshniwal from Rajasthan, who, like Kothari, became demonstrator in the physics department. Research was done mostly during the nonteaching period (15 April to 15 July) when it was almost 35 degrees centigrade in Allahabad. Little wonder that Saha stuck to his theoretical work. Kothari later remarked that Saha had little real skill or aptitude for practical experimental work, in spite of his keen interest in its results. This would certainly seem to apply to his efforts to build an experimental lab in Calcutta in the 1940s, when the money available was much greater. Nevertheless, Abha Sur's visit to that lab in 1994 showed that a vacuum furnace built by Saha and his lab staff was *still being used* in student spectroscopy experiments at Allahabad sixty years later![59] Saha believed, however, that physics would not progress in Allahabad unless experimental research was done and students were trained in the lab. Experimental physics was becoming stronger and stronger with new instrumentation, but, though Saha constantly stressed the importance of getting his own ideas about ionization of the sun's plasma "experimentally verified," he could not build a well-functioning lab for himself, whether in Allahabad or Calcutta. And he did not spend enough time in other labs to ensure that these tests occurred, and so he felt that he worked always at a disadvantage.

In this period half of his papers were published abroad each year. In 1927 Saha attended the Volta Centenary conference at Lake Como and went to Berlin and Copenhagen, where he first met E. O. Lawrence, who eleven years later, at Berkeley, arranged the transferal of a cyclotron to Saha's laboratory in Calcutta in the 1940s. He also traveled to Oslo, Paris, Leiden, and Utrecht.[60] When Saha returned to Allahabad in 1928, he began setting up equipment for spectroscopic research, very similar to work for which C. V. Raman would receive the Nobel Prize in 1930. Despite his limited research funds he managed to save a few students from joining the civil service or technical branches of the railways, meteorology, finance, and police. These careers were open to them because physics was a "high-scoring subject," and so the grades of a well-trained student in physics exams were often superior, even though one might not have a passionate interest in the subject itself. High-scoring subjects meant high marks and therefore the best potential for employment, so Saha's top students got top jobs outside physics. On the other hand, Saha seems to have been proud of his students who succeeded in careers elsewhere, and he gradually made more and more use of those contacts in his political work.

In 1929 physicist Arnold Sommerfeld visited Saha at Allahabad, and in 1930 Saha was photographed sitting beside his teacher J. C. Bose and Sommerfeld, in Calcutta.[61] Saha's scientific friends, however, Satyen Bose, Jnan Ghosh, and D. M. Bose, Jagadish Chandra's nephew, all stood behind the three chairs. The leading role of Meghnad Saha had become clear after his FRS. In 1931, he received a £1,500 grant from the Royal Society, London, for purchase of equipment, but he failed in his attempts to get matching funds from the government of India, which had never given much support to scientific research outside its own laboratories. He gained notoriety at this time for his confrontation with C. V. Raman, now knighted as Sir C. V. Raman, over the governance and management of the IACS in Calcutta. Though residing far away in Allahabad, Saha came to Calcutta for the extraordinary meetings of the IACS, following which Raman left his university chair in physics for Bangalore (see chap. 3). In the same year, Saha first published his *Treatise on Heat,* which became widely used as a textbook and ran to many editions. He wrote two short papers in *Nature* with his student Daulat Kothari in 1934 on beta decay, a subject which was under dispute until Fermi resolved it that year; Saha was, it seems, unaware of the neutrino Pauli discovered in 1930, though he still had not published about it in 1933–34. Then Saha stopped work in this direction, according to S. P. Pandya: "Why did he not pursue nuclear theory—which would have been so much easier? There were so many new developments—Yukawa's theory of nuclear forces, Bohr's work on nuclear reactions, beginnings of a shell model etc. Somehow he remained aloof from all these developments."[62]

Bhatnagar Marries Academia, Industry, and Administration

Shanti Bhatnagar was not particularly satisfied with his situation at Benares Hindu University from 1922 onward, although it was a new and nationalist-oriented university with a politically well-connected vice-chancellor, M. M.

Malaviya. It could be that Bhatnagar's Brahmo background was unsuited for the more straightforward conservative Hindu outlook embodied in the university leadership and the conservative city itself, a center of pilgrimage and ritual. But he was quickly given top university recognition and went as the university's delegate in 1923 to the Liverpool meetings of the British Association for the Advancement of Science, where he renewed old connections. Though Benares was not far from Calcutta, the center of academic and industrial chemistry in India, he undoubtedly experienced family pressures to return to his "native" Lahore. In 1924, Bhatnagar saw an ad in *Nature* for the position of director of Chemical Laboratories at the University of the Punjab in Lahore. This was an opportunity he had been waiting for, and, according to Richards, he already had support there. There were two nominating committees for this search, one based in the UK and the other in India, composed of P. C. Ray, H. R. Watson, and John Simonsen—the latter a real power in science in India.[63] There were three short-listed candidates, Bhatnagar and two Englishmen, one of whom, Dr. Willsden, was already deputy director and acting director of the chemical laboratories at Lahore. Willsden had been Bhatnagar's teacher at Lahore, and this fact caused Shanti to think of withdrawing from the competition; he said he might withdraw in deference to Willsden. But later he decided not to withdraw his candidacy, bringing the question to a vote in the committee. The vote was a tie, the chair abstained, and Willsden withdrew his candidacy, thus leaving the position open for Bhatnagar.[64]

This was Bhatnagar's early confrontation with a system of privileged appointments in many Indian universities, and that he decided to force a vote with an older, British, acting director shows he anticipated a possible success. Something must have made him change his mind against withdrawing. The abstention of the chair and Willsden's withdrawal show that others anticipated the nasty consequences of a further confrontation and saw how best to make the decision, and/or at the same time, the chairman detected in Bhatnagar someone with whose cooperative attitude he could work; perhaps that idea of cooperation could be extended to a bridge from British industry in India to Indian expertise and capital? The result in Bhatnagar's favor may have arisen from a comparison of credentials and age too, in which a younger chemist with a DSc for work done in the lab of F. G. Donnan, one of the most influential British chemists of the time, was compared with an older man whose academic achievements were more modest.[65]

Readers familiar with India after 1947 should remember that Lahore was not a marginal location in undivided North India but was in fact more important in educational, cultural, and intellectual terms than Delhi, India's capital after 1922. With Partition Punjab lost its center in 1947, and India lost the magnificent city Lahore to Pakistan. It is sometimes said that Lahore nurtured three Nobel Prize winners, Hargobind Khorana, Abdus Salam, and S. Chandrasekhar, who was born and went to school there as a child. Bhatnagar moved in to well-known and supportive surroundings and had a nice house built for his family. Here, he set about building academic and political connections in the scientific community. These efforts were realized when, in 1927, Bhatnagar organized the annual meetings of the Indian Science Congress in direct consultation with its president J. C. Bose. He even asked his friend and former drama teacher Norah Richards to come from her retreat in the Kangra Valley to be a hostess at the 1927 Congress dinners and parties, because his wife, Lajjawati, was evidently not comfortable playing this role. Richards the actress made the best use of this unique part in the drama of presenting Lahore to the scientists, and vice versa.

About this time he was also host to physicist Arthur Compton, of the University of Chicago, who came to conduct cosmic ray studies at a high altitude in Gulmarg, Kashmir, including 250 feet down at the bottom of a lake. Bhatnagar himself made new equipment in his labs for Compton's experiment and also learned what the search for new high-energy particles meant to physicists.[66] Though he had not been to the United States, Bhatnagar established contacts through Compton, which he would renew fifteen years later during the war when Compton joined the inner group planning the Manhattan Project, and made the famous 1942 phone call from Fermi's lab at the University of Chicago to announce the successful chain reaction. Working in Compton's group in Kashmir, Bhatnagar met Nazir Ahmed, a physicist who knew a lot about instrumentation because he was completing his PhD in physics under Rutherford at Cambridge, a lab in which students were expected to build their instruments. This eclectic group was held together by Kashmir's cool breeze at 3,000 meters during the hot season, Compton's dynamism, and a relentless quest to photograph elusive cosmic rays.

Continuing in a style he adopted while a student in Lahore, Shanti was always interested in solving problems brought to him by people outside the university, and here he cast his net quite widely. He was in regular contact with the agricultural economy of Punjab and developed a process for rich industrialist Sir Ganga Ram by which to convert bagasse, the waste after sugar is extracted from cane, into cattle fodder. But his star rose even higher in 1933 when he became interested in a viscosity problem in oil drilling, thus exciting his interest in the relative stability of colloids (a state midway between suspensions and solutions). India was dependent on petroleum from Burma and Iraq, and, given the rise in use of the internal combustion engine, exploration for oil was very active. During the war in 1915 new wells were struck at Kaur in Attock district in northwest Punjab, and the large Burma-British firm Steel Brothers began drilling for oil in the Punjab. The firm began asking researchers at the university in Lahore for help with the technical problems in the drilling but without success. The problem was that mud used in the Punjab drilling came into contact with subsurface salts and formed a solid mass that prohibited further drilling. Bhatnagar thought in 1931 it was a problem in colloid chemistry and tested the drilling mud in his lab; in his words, "the problem was solved by the addition of an Indian gum which had the remarkable property of lowering the viscosity of the mud suspension and of increasing its stability against the flocculating action of electrolytes at the same time."[67] This basically transformed an oil-in-water emulsion to a water-in-oil emulsion and thus introduced the idea of "mud oils" in India. This was a genuinely interesting intellectual problem that had resisted solutions proposed by other people, and Bhatnagar's training, laboratory staff, and diligence paid off, lowering the viscosity and enabling the drilling. The petroleum subsidiary of Rangoon-based Steel Brothers, Attock Oils, decided in 1934 to offer Shanti a grant of Rs 150,000 over five years to study the problems of his choice in petroleum chemistry.[68] This was a very large sum of money for Indian scientists, particularly during the Depression, and began Bhatnagar's long association with the petrochemical industry.

Steel Brothers had already engaged Indian researchers on subjects that challenged them; for example, in 1925 they brought senior researchers from the School of Tropical Medicine in Calcutta to the oil fields in Burma to advise them on dealing with rampant malaria among workers.[69] This relationship between the School of Tropical Medicine and Steels' oil fields lasted until the spring of 1942, when the school's last weekly malaria studies and reports were sent to Steels' head office in Rangoon just before the arrival of Japanese troops. This was not the first time in India that private money had benefited a university's research program, but this instance and the example it set made a major impact in the scientific community, and on Bhatnagar's career. Bhatnagar decided that he should not personally accept this money but should channel it through the university. The 1934 offer, accepted by the syndicate of Punjab University, was to be the basis of a new Department of Petroleum Research, with salaries for research assistants, purchase of apparatus and supplies, and an honorarium for Bhatnagar of Rs 6,000, a sum almost half his annual salary. In addition, "the salary and expenses of a fully qualified British petroleum research chemist, who would, as required, act as assistant to Professor Bhatnagar and relieve him of the laboratory's routine and detailed work connected with the running of the proposed new department."[70] The company's estimate of this expense was Rs 30,000 over five years. This would also provide experience to a British chemist who might then serve the company in the Attock district, or elsewhere. The offer made to Punjab's governor concluded:

> A feature of the proposal is that any results which in our opinion justify the taking out of patents will be exploited by us and such patents would be jointly in the name of Messrs. Steel Brothers & Co. Ltd and Professor Bhatnagar and/or his chemists, and any profits would be shared equally between the Company on one hand and the parties concerned on the other. We understand Professor Bhatnagar proposes to spend a large portion of any profits which may accrue to him from any patents for the encouragement and development of industrial and chemical research in your Province under the auspices of the Punjab University.[71]

When the vice-chancellor and syndicate accepted this proposal and approved a plan in which half of the patent profits accruing to Bhatnagar would flow back into the university and half would go to him directly, the company then made the payments. At the heart of Shanti's plan were six scholarships, which he shrewdly had named not after Punjabi nationalist revolutionaries but after well-known British figures: one after Sir Herbert Emerson, then governor of the Punjab, one after his old Lahore teacher Dunncliff, one after his DSc supervisor in London, F. G. Donnan, and so on.

Just at this time he published his textbook called *Principles and Applications of Magneto-Chemistry* with Macmillan in London, and like authors everywhere he made an effort to obtain favorable reviews. One person whose book review he especially wanted was C. V. Raman, and Bhatnagar's own "Personal Records" (circa 1947) quotes obtaining this favorable review of his textbook. Within months of the book and its review, however, Bhatnagar was to come into Raman's orbit in Bangalore and Raman's opinion of Bhatnagar was going to change dramatically, as will be shown in chapter 3.

Bhatnagar's reputation now spread in many directions. Steel Brothers invited him to London, and he arrived in 1936 in time for the Empire Universities Conference at Cambridge. This was crowned by his acceptance of an Order of the British Empire award and then a tour of Steel Brothers' rice and food oil mills in Germany and Poland, as well as the UK. So happy were Steel Brothers with the Lahore arrangement that in 1937 they extended the agreement from five years to ten and added Rs 250,000 to make a total of Rs 400,000. Over 40 percent of Steel Brothers' Attock Oils shares were then owned by Indo-Burmah Petroleum, a Steels company on whose oils and refineries the Indian economy and Indian Army already depended. The paving of most of the roads of North India was done using bitumen from the refinery in Attock district. A new oil strike at Dhulian in 1937 and Joya Mair in 1944 served to cement the strategic importance of local researchers, more acutely after the loss to the Japanese of the Burma oil fields in 1942. This kind of industrialization was precisely what Saha believed necessary for India, if it were in Indian hands. There was a widespread prejudice that advanced education in India was simply hostile to technical education: in fact, however, in the face of this presumed hostility, quite practical experiments were being made, sometimes in fields radically unconnected to each other. For example, in comparison with petroleum chemistry in Lahore, the evidence from both Baroda and Bombay between 1890 and 1930 shows that applied textile chemistry was being inserted into Kala Bhavan in Baroda and the University of Bombay.[72]

News of the Steel Brothers money brought Bhatnagar to everyone's attention. His colleagues from his London university days responded very positively: Jnan Ghosh wrote, "Your generosity has no parallel in India," and Meghnad Saha wrote him to say:

> You have thereby raised the status of the university teachers in the estimation of the public, not to speak of the benefit conferred on your Alma Mater. India does not lack in men earning millions, but if a few of those millionaires were guided by the fine example set by a comparatively poor teacher like yourself, I think her scientific and moral progress would have been rapid. Nobody but a true researcher can feel how much our energies in this country are being wasted for lack of funds.[73]

In 1938, Bhatnagar became president of the Chemistry Section of the Indian Science Congress, meeting at Calcutta. With P. C. Ray sitting in the audience, Bhatnagar's speech very skillfully but respectfully compared his turning the Steel Brothers' money over to the university with the self-sacrifices of P. C. Ray, chemistry professor and successful industrialist. Ray was the ascetic founder of the successful firm Bengal Chemical.[74] At these meetings, Jnan Ghosh described "The Three Musketeers" from Donnan's lab in London, Bhatnagar, K. N. Mukherjee, and himself. "But," Ghosh said, "Bhatnagar has provided a bridge of communication between scientists and industrialists. . . . In pre-war days such close intimacy existed in Germany alone, with the result that she rapidly outstripped her rivals in industrial enterprise."[75] This brought him further into contact with the increasingly nationalist textile, oil, and chemical company owners (like Tata Oils), with the Indian Chamber of Commerce, and with the British business elite in India. Through his receipt of the Order of the British Empire, and work with Steel Brothers, he met Indians and Britishers whose profits were impressive and reinvestments were directed to a stronger Indian economy. The link with Indo-Burmah Petroleum took Bhatnagar to the top of the British agenda because of the nervous (and profitable) search for more and more oil. Bhatnagar's achievements would have caused a stir in most scientific and industrial communities, and so he was, for example, soon invited by industrialist Lala Sri Ram to be the director of Sri Ram's own planned Industrial Research Laboratory in Delhi.[76] He needed these high contacts because he had been involved in a mighty confrontation at the same time with C. V. Raman in Bangalore. A member of a prestigious international committee chaired by a British chemist, Bhatnagar found himself one of the Indian figures in an inquiry into Raman's performance as director of the Indian Institute of Science (IISc) (see chap. 3).

The 1939 Science Congress again met in Lahore in January, and Bhatnagar was in charge of all local arrangements, under the presidency of Calcutta-based chemist Jnan Ghosh. Again he asked his friend and biographer Norah Richards to help with stage-managing the important events. It was a great success, and many people toured his University Chemical Laboratories, and Bhatnagar's hospitality warmed many hearts. This social success led to a visit, the following year, from Sir Ramaswami Mudaliar, who sat on the Viceroy's Executive Council. So impressed was economist-businessman Mudaliar with Bhatnagar and his research institute that he wrote to the governor of the Punjab to ask for Bhatnagar's services as advisor in scientific and industrial research.

This was one objective of the nationalist-oriented Planning Committee, of which Saha was a member and Nehru the chair, namely, to have active scientists move into the decision-making circles around the government of India and to guide the government toward "liberating" the scientific and industrial side of the economy. With the declaration of war in 1939, visionaries saw that Britain would need all the help it could get, and the perception of this need would further accelerate the demand for independence. Saha and others presumably saw Bhatnagar as an ideal broker in this complex transaction. Bhatnagar set conditions on his acceptance of the government's offer to move to Delhi, including the movement with him of six Steel Scholars and provision of labs for them wherever he went. The conditions were agreed to, and Bhatnagar left Lahore for Delhi in March 1940.

Saha's Reasons for Leaving Allahabad

It would be inaccurate to conclude that Saha was unhappy in Allahabad. He created a kind of extended joint family house, sustaining relatives in need along with a couple of his students living-in, plus a family of seven children. He took his family for picnics and swimming, notably at the auspicious confluence of the Ganga and Yamuna rivers (called Prayag by Hindus).[77] He also went for very long walks, including the 120 km distance from Allahabad to Kanpur. But something was missing and Saha was restless. In a period of his increasing political activity, between 1932 and 1936, Saha published only two papers—one was a joint paper with D. S. Kothari (1934) and the other reviewed his ideas on the upper atmosphere and was published in the first proceedings of the National Institutes of Science of India in 1935. His students continued to work in Allahabad, beset by difficulties, while Saha spent much of his time traveling to other cities. Saha later described these difficulties:

> The work was laborious but not brilliant. The laboratory was handicapped on account of the absence of proper apparatus which could be collected only in the course of years. Saha came very close to the discovery of the origin of complex spectra, but was forestalled by Hund by a few months. This was due to difficulties of publication, for papers sent outside India took at least a year to publish. Stimulated by Franck's work on the absorption spectra of molecules, he instituted studies on the photochemical action of light on molecules. Unfortunately, these ideas could not be prosecuted to their logical ends because students who were trained for the purpose left him for other occupations.[78]

In 1936–37, Saha enjoyed the relief of a year abroad for the first time in nine years, on a Carnegie Fellowship to be held largely at Harvard. He traveled overland through Iran and Iraq with his wife, Radharani, and thirteen-year-old son Ajit, leaving him in Zurich at Paul Geheeb's school, run on lines similar to Tagore's school at Santiniketan, a balance of labor, art, and scholastic study. The countryside of Iraq and Iran rekindled his love of archaeology. He again met Sommerfeld in Germany and stayed at the institute in Copenhagen with Bohr's group, where he met Bhabha for the first time. He also met Paul Dirac and apparently tangled with him about physics in conversation in Copenhagen. Dirac soon bluntly replied "with regard to your criticism of my wave equation for a magnetic pole, the use of the vector potential can be justified on the grounds that the only field acting on the magnetic pole is that arising from the electric charge, and this field can be described by a vector potential."[79] Saha stayed not only at Cambridge but also Oxford. Saha went for the first time to the United States for two months at the Harvard Observatory, having closely followed their work on stellar spectra. He also stayed as Ernest Lawrence's guest at the University of California at Berkeley, where he made arrangements for his student B. D. Nagchaudhuri, who arrived two years later to do his doctorate. Saha's diary shows he made detailed notes on the cyclotron being constructed at Berkeley. He also met Arthur Compton at the University of Chicago, long a friend of India and of Saha and Bhatnagar. A few months later, Saha received a letter confirming that Compton's nomination of Saha for a Nobel Prize had been acknowledged from Stockholm.[80] This was Saha's second nomination for the Nobel Prize in Physics.

Saha returned to Allahabad from Harvard in 1937, longing for a more active scientific life in the political center of Calcutta. An opportunity presented itself in 1938 when Saha's former guide to the world of German physics D. M. Bose left the Palit Professorship in the University of Calcutta to become the director of the Bose Institute, following the death of his uncle Sir Jagadish Chandra Bose in 1937.[81] Saha was approved by the university to occupy the Palit Chair, from which he had helped to unseat Raman in 1931–32. He returned to Calcutta with a reputation as a very political force in science with an influential national network in politics.

Now here were Bhatnagar and Saha in 1939, the two old friends, nineteen years after they met in London, poised to undertake new ventures in institution-building, both dissatisfied with the context they had found themselves in—Lahore and Allahabad. They had both participated in the confrontations with Raman at Calcutta and the IISc in Bangalore, and now had very favorable relationships with the IISc's leadership, the most prestigious and richest scientific institution in the country. Its new director, chemist Jnan Ghosh, was their friend. Their personal scientific reputations were well established; Saha had already been elected and Bhatnagar was about to be elected as Fellow of the Royal Society. With their international reputations established, they were launched in a career as planners of scientific institutions, in collaboration with the leading industrialists, top government officials, and nationalist politicians. In the process the differences between them would be revealed and amplified, opening an irresolvable tension in their working lives. Soon a new shadow would fall across their paths, the shadow of the young Homi Bhabha.

THREE

The Bangalore Affair, 1935–38: Scientists and Conflict around C. V. Raman

Because a more *national* scientific community was developing during the 1930s, the alignment of regional Indian scientific institutions shifted. Researchers were meeting frequently on a regional and national basis, travel by train was slightly easier for scientists and more frequent, the postal and telegraph systems were improving, and opportunities arose for both status and power that were not just local in character. Although the great majority were given very little social recognition, a few talented and ambitious scientists no longer depended on local support alone; instead they had built significant international reputations that brought them to the attention of local elites and the office of the governor of the province as well as the capital in Delhi. There was an enlargement and shift in the intensity of communication within the scientific community, brought about by the creation of academies and journals and new magazines. With this enlargement came an opportunity for mobility among scientists, and there was more direct competition for status and resources. This mobility, desired by so many, could also lead to conflict, as a dramatic illustration in this chapter demonstrates, showing that Saha and Bhatnagar were involved in a complex conflict around Sir C. V. Raman in faraway Bangalore in southern India. The only living winner of the Nobel Prize among Indian scientists (Ronald Ross died in 1932), Raman attracted unprecedented media and political attention, and that attention in turn brought a very bright light shining onto disagreements in the scientific community. This conflict went all the way up to the viceroy's office and the Cavendish Laboratory at Cambridge, showing that international networks were brought into play to moderate and manage conflict among Indian scientists.

This all took place during a shift in national politics. Congress undertook a new campaign of civil disobedience in 1930, and there was a revival of revolutionary activity in eastern Bengal (particularly successful in Chittagong). In 1931, the Second Round Table Conference between Gandhi and Lord Irwin, Viceroy of India (1926–31), occurred over several weeks in London but failed to secure any commitment for independence. Civil disobedience and militant opposition continued in 1932–33, bringing the appointment of Sir John Anderson, a civil servant with counterinsurgency experience in Ireland, as governor of Bengal (1932–37). Seven battalions of British infantry were moved to Bengal and deployed to restore British rule, in part by hunting down revolutionary groups in the countryside. Though Bhatnagar and Saha were far from this scene, Saha's sympathies were with the hunted and not the hunters.[1]

During this period Saha influenced the establishment of academies, research institutions, and journals as a result of foreign recognition for his work done in 1919–21. Beyond building a viable research group in Allahabad, he took firm positions on national political issues. It seems that Saha himself was aware of a change in his outlook and his practice after 1930, saying, "Apart from my association with the political movement of my juvenile years, I lived in the Ivory Tower till 1930."[2] Saha was also very focused on the politics within institutions in Calcutta, although he lived a twenty-hour train journey away in Allahabad. There is no other clear explanation for his engagement in a conflict with Raman in the Indian Association for the Cultivation of Science (IACS), founded by the rich homeopathic doctor Mahendralal Sircar in 1876 in Calcutta.[3]

This IACS lab was constructed as a gift from the maharajah of Vizianagaram in 1892, when Sircar had cured the maharajah of a rare disease and asked for a laboratory instead of payment. Though it eventually received government support, the IACS was private and nationalist in intent, and most of the association's first members were Bengalis. Through provision of a good library, research facilities to members, and popular lecturers by active scientists, it sought to cultivate the growth of science within Indian society, not simply among scientists. C. V. Raman was elected an ordinary member and began to use the equipment for his research in 1907. He started work in the lab early in the morning, then went to the Finance Department to his employment, and returned again in the evening. This pattern continued until he left to join Calcutta University full-time in 1917. Raman was the first person to use the building consistently for research; otherwise it had been used like a scientific gentlemen's club. He worked there alone for ten years, publishing papers in physics, while advancing upward in the Finance Department. This is precisely what Sircar and others had imagined in the 1870s—intelligent and curious Indians would turn up to do experimental science and learn to do it, not just talk about it, on their own terms. After all, Jagadish Chandra Bose had already blazed a path in that experimental direction before 1900 and was soon successfully advocating government support for research in the university.

When Raman became the first Palit Professor of Physics at the university in 1917, he left the Finance Department, to the great surprise of those who expected him to advance to a high position in government; he was short-listed for very senior position. Raman taught at the university but continued to do all his experiments in the association building. This arrangement had the full support of his patron Sir Asutosh Mookerjee. Slowly gaining control over the operation of the institution, as honorary secretary of IACS, Raman attempted to limit new membership to those of whom he approved, to make it more professional and less like the gentlemen's club it originally was. This created tension with a group of older members in the association who eventually opposed Raman. Although the association gradually invited both Bengalis and people from all other parts of India, some members said Raman favored South Indians at the association, of whom there were now a small number. Certainly his main assistant in the lab was K. S. Krishnan, from Madras, and good students like Raman's nephew S. Chandrasekhar and others from Mysore, Kerala, and Andhra Pradesh were being attracted to come to the IACS after award of Raman's Nobel Prize; these disparate origins were lumped together by Bengalis as "south Indian."[4]

The Saha-Raman Confrontation in Calcutta, 1931

Though based in Allahabad, Saha became a leader of this opposition to Raman's plans and used this charge of "favouring South Indians" against him. It would be misleading to say that Saha caused the conflict, but he appears to have seized this opportunity to make a new kind of reputation in Calcutta. Doubtless, had Saha been abroad, someone else might have stood up against Raman, but it is not clear who that person would have been, because Raman was supremely confident in his belief in his own correctness and very tough in his own defense. Some people must have been afraid of him; after all he could make future careers or break them. It is possible that other members of the association used Saha's willingness to confront Raman for their own objective of promoting Bengali identity and autonomy, a sign that Bengali elites had not forgotten losing the capital of India to that dusty distant place called Delhi. Raman's reported plan (what it really was has never been made very clear) threatened Bengali prominence and perhaps cast doubt on the clublike atmosphere in an attempt to professionalize the IACS; instead they polarized it. But he was doing nothing covert or illegal.

In addition to their insatiable egos, Saha and Raman had a curious unconscious thing in common: Saha had wanted to sit the examination for employment in the Finance Department but was barred in 1916 because of his political commitments. In bitterness he rejected that secure way of life entirely, well knowing the poverty of educated people who could not get civil service jobs. He therefore had to ride his bicycle through the rain to houses of students who needed to be prepared for exams, to make money for his family. Raman had no such commitments and already excelled at that very same examination. He had good career prospects and a large civil service salary *before* he was offered the Palit professorship. Saha came to despise the Finance Department and those who worked for it, ultimately controlling as it did the very funds for the university where he studied and worked. Raman was a senior officer in that department and had been identified by the finance member of the viceroy's council for promotion to the top, just before Raman became a university professor in 1917, a job he accepted at half his finance officer's salary.

And so for science, Raman voluntarily stepped away from that job. Saha, on the other hand, was driven involuntarily away from finance. We do not know whether he would have been happy in a finance job, or whether he would have been able to keep it, but he wanted one in 1916. Moreover, to help further explain their difficult relationship, during the noncooperation movement of 1919–20, Raman broke through the cordon of students trying to deter teaching of regular university classes and brusquely insisted on holding classes.[5] Saha knew this because he was there in the University of Calcutta at the time, teaching in the same physics department. Fifteen years later they had a full confrontation.

Before the conflict with Raman took its larger shape, Saha's stature had changed by being elected FRS in 1927. When it was clear about 1930 that Saha expected to get the appointment to a new chair in physics at the University of Calcutta, the lines were clearly drawn. The intermittent but lengthy build-up to the confrontation occurred while Raman gained his knighthood in 1929 and won the Nobel Prize in 1930. K. S. Krishnan, who was present, reported that during the Sommerfeld visit of 1928, at the time of the discoveries of spectral ("combinational") scattering later called the Raman effect, Saha stood up after Raman's lecture, with Sommerfeld present, and said that the discovery was no more than a confirmation of what Smekal had predicted. Saha also published a letter in *Nature* that belittled Raman's work and suggested it was wrong.[6]

Saha's reference to Smekal was well informed—Raman always referred to his work as testing and proving "Smekal's surmise"—but Saha's attack had no effect on Raman's world scientific reputation. This subject was, after all, hardly Saha's field of greatest competence, and so his remarks in Sommerfeld's presence and the letter in *Nature* just before Raman was awarded the Nobel Prize may have had an adverse effect on Saha's scientific reputation.[7] Prior to his departure from Calcutta, Raman had already been offered an appointment as director of the IISc and proposed a leave of absence from Calcutta while he tried out the Bangalore job for a year or two. To achieve this Raman had already persuaded the wealthy Sircar family to establish the conditions for a new chair in physics at Calcutta University but one also attached to the IACS, to which he might return from Bangalore if necessary.[8] Saha heard about this new chair but probably did not know of Raman's intention to go to Bangalore, so he wrote from Allahabad in 1930 and asked that Raman nominate him to the chair. Raman replied that Saha's earlier achievements were notable, but he was at present not very active in research and the association, which Raman lived next door to and practically ran from his private house, needed a younger researcher who had not reached a plateau, as Raman thought Saha had. This correspondence occurred after Saha's critique of Raman's work in Sommerfeld's presence—the first of a pattern in which Saha asked for something from someone he had attacked. Incensed by Raman's refusal to support him, Saha mobilized the sons of Mahendralal Sircar and Asutosh Mookerjee to "save the association," at the same time undermining Raman's reputation with respect to the Sircars, donors of the proposed chair, and Sir Asutosh Mookerjee's family, also Raman's patrons. Sir Asutosh's son was Shyama Prasad Mookerjee, now vice-chancellor of the University of Calcutta, politically aligned with the Hindu Mahasabha Party, and later to play a major role in Bengal's politics. Using Bangla-language newspapers too, Saha mobilized people around a call to save the association, a personal opposition to Raman and his known arrogance, with a subtext of resisting South Indian dominance in science in Calcutta. A physicist like K. S. Krishnan, who worked closely with Raman on his experiments, was implicated in this resistance. New members of the IACS were introduced and given voting rights, preparing for a dramatic showdown in an extraordinary meeting, at which a majority voted against Raman as honorary secretary of the IACS. As a result Raman's position was intractable, and despite his years of work for it, he left the association in 1931.[9] He soon left Calcutta, where he had lived twenty-six years, to be the first Indian to become the director of the Indian Institute of Science (IISc) in Bangalore; he was recommended to the post by Sir Ernest Rutherford, director of the Cavendish Laboratory. But he had already maneuvered to have himself replaced in "his" IACS role in Calcutta by the co-discoverer of the Raman effect, K. S. Krishnan.[10] In the end the now-empty Palit Chair was soon filled by D. M. Bose, nephew of J. C. Bose, and no Sircar chair was ever created. Raman's own choice, K. S. Krishnan, moved to teach physics at the new department in the University of Dacca, under the new professor Satyen Bose. This stalled Saha's long desire for a chair at the university, ever since he mysteriously abandoned the prestigious Khaira Chair in 1922.

But after Raman left Calcutta for Bangalore in 1932, the council of the IACS was reconstituted with Saha as an influential member."[11] Saha's influence increased over the next ten years, so that, in his words, "he has virtually controlled the destinies of the Association since 1943, because Sir U. N. Brahmachari had such faith in Saha that he left all the management to his discretion."[12] Saha became president of the association in 1946 and was responsible for building a new laboratory in South Calcutta, near the new nationalist Jadavpur University; when he left the university in 1952, he became the director of the IACS laboratory. One cannot say that Saha, based in 1931 in Allahabad, confronted Raman *in order to take over* the IACS, but it was one of the eventual results of the confrontation. One of the other results was a boost to the scientific fortunes of the IISc in Bangalore, which had not yet fulfilled the expectations of its influential founders, the Tata family.

Tension around New Academies and Communication

The Saha-Raman conflict in the IACS also showed up in a tangle over the creation of scientific academies and news journals. The result was damaging but served to strengthen the regional tendencies in Indian science and delay the growth of a national consensus over how the scientific community should regulate itself. Until the 1930s, the Indian Science Congress served as both the forum for technical exchange in the different disciplines and for political negotiations in the general meetings and policy committee. It was the ideal way by which to invite (and lobby with) leading political figures. The congress was founded in 1911 by P. S. MacMahon and J. L. Simonsen, from Lucknow and Madras, respectively. It first met in Calcutta in 1914 in the rooms of the Asiatic Society, Calcutta, and continued to have its business managed by the offices of the Asiatic Society in 1935. It echoed some of the work and ambitions of the British and American associations for the advancement of science. But by 1930 something more was needed in the internal formation of a scientific community, something to regulate membership and demonstrate recognition for superior performance. Scientists based in Bengal dominated the Science Congress, and competition between British and Indian scientists for its highest offices was keen in the 1930s. Despite talk of a national science forum with some money-granting powers, nothing was done until Saha founded the United Provinces Academy of Sciences in 1930, with membership coming from the United Provinces (later called Uttar Pradesh) and Allahabad as its center, where Saha worked. There was a large network of Bengali-speaking and Calcutta-trained scientists holding appointments throughout North India, and so, though this academy solicited membership from all over India, its focus was really in North India and Bengal.

The Indian Science Congress (ISC) held its 1930 annual meeting at Bangalore. The news magazine in India for rapid transmission of research notes, *Current Science,* was founded by a general resolution of that meeting. Sir Richard Gregory, editor of *Nature,* had discussed the formation of a national academy with Raman just before the meeting. This was ten months before the award of the Nobel Prize to Raman, something of which the participants could hardly have been aware. Raman, however, had expected the prize in each of the two previous years, without success. Though there was discussion of an academy in *Current Science* from 1930 onward, it was not until 1933, when Raman had left Calcutta and was living in Bangalore, that *Current Science,* of which he was the editor, contained a questionnaire soliciting support for a national academy. Scientists in Calcutta discussed the questionnaire and formally proposed that an academy be on the agenda of the 1934 ISC meetings.

In January 1934, the ISC met in Bombay. Saha was president of the whole congress and was co-chairman with Raman of the physics and meteorology section. In his presidential address, Saha proposed the formation of an Indian Academy of Science and thanked the editor of *Current Science* (Raman) for circulating the questionnaire. Consequently the Academy Committee was formed with members L. L. Fermor, director, Geological Survey of India; M. N. Saha; and S. P. Agharkar; Saha wrote the committee's final report.[13] Saha said the academy should (1) be limited in membership as a mark of distinction, (2) be associated with the state at the highest level, (3) publish Comptes Rendues or Proceedings, (4) take over the organization of the ISC, (5) secure and manage funds for research, and (6) be the apex of a pyramid of special societies devoted to particular subjects, such as the Indian Physical Society.[14]

Then came, according to Saha, "the unfortunate complication . . . Raman . . . criticized adversely the work of the Academy Committee as revealed by the draft minutes of the first meeting."[15] Raman gave a presidential address to the Conference of South Indian Scientists in Bangalore early in April and publicly resigned from the ISC, of which he had held the top positions. Fifteen days later, the Academy Committee met in Calcutta and gave a press report to clarify the situation. The committee asked Raman to withdraw his resignation and to start again to build a single academy. Two weeks later, on 30 April 1934, the establishment of an Indian Academy of Science at Bangalore was announced by Raman, preempting the academy being planned by the ISC. He reorganized his journal *Current Science* to serve the new academy in Bangalore. Eventually Raman attended a meeting in Calcutta where some agreement was reached on the founding of a National Institutes of Science of India (NISI) on the model of the Institut Français.[16]

The first meetings of NISI were soon held in 1935 in the Senate Hall of Calcutta University and the rooms of the Asiatic Society, formally opened by the governor of Bengal, Sir John Anderson, also chancellor of the university, and chaired by J. H. Hutton, president of the ISC. L. L. Fermor was elected to head NISI until Saha became chairman in 1937; the council of NISI had 14 British and 19 Indian members (14 non-Bengalis and 5 Bengalis). People on the council resided in Lucknow, Lahore, Delhi, Dacca, Madras, Bombay, Bangalore, Hyderabad, and Allahabad, but 15 members were based in Calcutta. This was due to the policy of "allotting to the headquarters station sufficient membership to ensure the possibility of always being able to ensure quorum."[17] When Saha became president, he encouraged NISI's move to Delhi, though its publishing unit remained for many years in the offices in the Asiatic Society building in Calcutta. Clearly NISI was being established in competition with Raman's academy in Bangalore and had a national ambition.

This is why in 1935, drawing on British and American models, Saha started the popular science monthly journal *Science and Culture*. With P. C. Ray's patronage, Saha stated it would play a role that *Nature* had in Britain or *Science* in the United States, saying it would interpret science in nontechnical language and advocate a planned application of science to India's problems.[18] It was to be produced by the Indian Science News Association, housed in a small office in Science College. Edited by one of his Allahabad students, the journal was used frequently by Saha to expound his views as unsigned editorials. This was done in direct competition with the Bangalore science news publication *Current Science*. There were now two popular science journals catering mostly to scientists, one in Bangalore edited by Raman, the other in Calcutta edited by Saha, and three science academies, one in Allahabad, the second in Calcutta, and the third in Bangalore.

Raman and Saha were also competing for membership in "their" academies by exercising their networks to secure invitations to international visitors for lectures and to seek and accept nominations for election of scientists as fellows. Naturally they tried to incorporate nearby institutions in their competitive strategies, Raman using the IISc in Bangalore of which he was director, Saha (though living in Allahabad) using the IACS and the University of Calcutta.

Understanding the Situation in Bangalore, 1935–38

But it was *within* the Indian Institute of Science (IISc) that these two central figures clashed between 1935 and 1938, revealing the strength of regional and colonial government tensions within the scientific community, the depth of disagreement between some English and Indian scientific faculty members, the jealousy between members of different professions like chemistry and physics, and the impact of personality clashes. So volatile and risky was this public conflict about Raman that the viceroy's office and the Tata Trust reluctantly became involved.

Founded in 1909 jointly by Jamshetji Tata, the government of India, and the maharajah of Mysore, the IISc was still known to many people in Bangalore as the "Tata Institute" seventy years after its founding! Johns Hopkins University in Baltimore had served as the model, and Tata's friend Burjoji Padshah served as the intellectual catalyst for its plans.[19] It was the most prestigious and best-funded science training and research institution in India. But criticism of the IISc was circulated in journals such as the *Calcutta Review* and the *Modern Review* following publication of an earlier report of a Review Committee of the IISc in 1922.[20] Saha was appointed the only Indian member of the IISc's Review Committee in 1930. The other three members of the committee were the director of the Zoological Survey of India (chairman), the chief engineer of hydroelectric development in Madras, and the inspector of schools for Bangalore (H. J. Bhabha, grandfather of young physicist Homi Bhabha). The committee interviewed twenty-three people about the institute's structure and operations in 1930, of whom five were Indians; among these were P. C. Ray and C. V. Raman. Questionnaires were sent to nine other Indians, one of whom was chemist Shanti Bhatnagar.

Saha's "Minute of Dissent" in the review published in 1931 dealt with the high salary of the institute's director, which was then Rs 5,000 per month.[21] (Saha may have known that the director, chemist Martin Forster, FRS, was about to retire.) This salary was not only much higher than the salary for vice-chancellors of universities in India, but also three times as high as the salary of the director of the National Physical Laboratory at Teddington in England (the exchange rate of the rupee and pound was then fixed and stable, so the currencies were effectively convertible). Saha said that the junior staff and laboratory assistants at the IISc were grossly underpaid, in comparison. He not only stressed the negative effect of discrepancies of salary between some mid-rank academic posts but also contrasted the salaries of "European and Indian" faculty. Saha approved, however, of the opinion of Dr. H. J. Bhabha, nominee of the Tata family on the council, who had said "Indian Professors should be patriotic enough to serve at a lower scale of pay than their European colleagues. [Saha] agrees to this in principle, and in proposing the following scale takes into consideration also the fact that a man of first rate merit will prefer the freer atmosphere of a university to the restriction that will be imposed on him in the Institute." This implied restriction surfaced again and again too, though it is curious that anyone would think the atmosphere of public universities was particularly free in colonial India, or that the IISc would be less free in practice.[22]

Saha was also of the firm opinion that everything should be done to increase the work of IISc scientists for industry, to increase industry's investment in work at IISc, and to free IISc people to do contract work for government and industry. Saha's opinions gained extra publicity when they were quite widely quoted by the chemist Sir P. C. Ray, in his autobiography published in 1932. Ray quoted Saha's complaint that the IISc had not served its purpose because it was too far from either Bombay or Calcutta for the students to compete for jobs in industry. Ray agreed with Saha's opinion that IISc would be unable to attract good scientists so long as administrative personnel received higher salaries than academic personnel (Saha did not invent this complaint, which was used consistently in India). Ray even favorably quoted Raman as saying, in 1930, that IISc had showed "a failure to produce . . . despite colossal sums."[23] Ray eventually became a member of the IISc council and also voted to accept Raman's resignation as director in 1937. The gathering tension between South Indian and Bengali scientists was showing up in the institute's governance, but there were other complexities as well.

Just as tensions between Raman and some scientists in Calcutta mounted in the battle of the academies in1932 and 1933, Saha was appointed a member of the Governing Council of IISc in 1932. This was part of a move to build the field of physics in the institute. Chemist Martin Forster retired from the post of director in 1932, and Raman was appointed director soon afterward. (Forster, long a Fellow of the Royal Society, was soon knighted for his work in Bangalore, retired to Mysore, and died there twelve years later.) As council member, Saha must have known of and approved of Raman's appointment; after all, it removed Raman from Calcutta. In effect the two men were arriving at the institute at the same time. Saha was already proposing the formation of a national science academy, clashing with Raman's idea of a similar academy but based in Bangalore. Behind some of this activity is the feeling among people from Bombay and Calcutta that Bangalore would be a nice place for a holiday or retirement, but not a serious place to live and work. To people from the south, like Raman, this was a kind of arrogance and interference worth fighting.

There was a mutual north-south disdain at work here: many northerners saw the south as slow and traditional (not innovative), while southerners saw the north as disorganized and uncultured. These mutually reinforcing stereotypes found their way into north-south jokes, just as they have in Italy or the United States. Moreover, Raman, with the glow of the Nobel Prize over him, with an acute mind and sharp tongue, seemed to provoke resentment and tension around him. He had been placed in the midst of people he did not know, some of whom largely looked upon their roles in the IISc as a series of sinecure teaching positions, and Raman set about to change that. As Max Born perceptively observed to Lord Ernest Rutherford about Raman and Saha, "It seems to me as if every prominent Indian has a secret 'complex': he feels himself being envied and intrigued by his compatriots, and not being taken quite seriously by Englishmen."[24] Rutherford replied prophetically to Born, "I am myself afraid that Raman has got into such a devil of a mess that he may be driven to resign to save his face."[25] Rutherford took this issue seriously only after others pressed him to intervene: he felt intervention was either dangerous or futile, but his resistance was ultimately overcome. Just before he died suddenly in 1937, he had accepted an invitation to participate in the ISC and was planning to go in late 1937 to find out what could be done about the situation in Bangalore, because many people, including Max Born and intermediaries from the viceroy's office, had begged him to intervene.

Anticipating the need for potential allies in high places in 1936 and mindful of his wife's very strong interest in the Indian National Congress, Raman invited Mahatma Gandhi to visit the institute, in company of a key Congress Party leader, Sardar Patel. This was not Gandhi's first visit: he had already come in 1927, asking the students, "How will you infect the people of the villages with your scientific knowledge?"[26] According to the memory of a young witness, "Lady Raman was an ardent supporter of Gandhiji and his philosophy," and Gandhi toured the physics department labs in 1936, admitting at the end, "I feel I am an ignorant man."[27] In the evening about twenty-five physics students and Lady Raman had a private audience with Gandhi to give him donations they had collected for the Harijan Fund. The conversation ranged from Gandhi's ideas for women's education and women's role in social work, to the recent conversion of one of Gandhi's sons to Islam: "while discussing this . . . the Mahatma became visibly angry and sad and was choked with emotion," wrote B. V. Thosar, who later became a physicist and worked with Bhabha at the Tata Institute for Fundamental Research (TIFR) in Bombay. In the end, Raman turned more to people like Rutherford at Cambridge and not Gandhi when he was under attack as director of the IISc.

Lady Raman also provided some support to and enjoyed the friendship of female graduate students in her husband's labs. Though there is little writing about women in this period of science in India, some examples are available to historians. The first at Bangalore was Kamala Sohonie, challenging Raman to admit her when he expressed his doubts about women students in 1933. Following this challenge she was admitted and completed her master's and went to do her doctorate in physics at Cambridge in 1936. In 1935, Lalitha Chandrasekhar, whose father and sisters were doctors, left a teaching job in Delhi and came south to work in Raman's lab. According to Sur, she ended a promising career in research in physics when she met a former classmate from Madras, Subhramanyam Chandrasekhar, on a visit home from Cambridge; she soon agreed to a marriage and eventually decided that support for her husband's career was worth relinquishing her own. In this decision, though not one urged by Chandrasekhar himself, she followed Lady Lokasundari Raman's path.[28] In 1939 Sunanda Bai came to work in Raman's lab in Bangalore, and during her five years there published ten single-authored papers on depolarization of light scattering. Then, on the eve of a journey to Sweden for a postdoc in 1945, having submitted but not yet defended a dissertation, Sunanda Bai committed suicide. Sur's investigation showed that those close to Sunanda Bai at the time thought the suicide was unrelated to either the IISc or her laboratory work there. So this is perhaps *not the first* "scientist's suicide" in India.[29] Anna Mani arrived in Bangalore in 1940, when Raman was most despondent and considering leaving the IISc, and she published five single-authored papers before submitting her dissertation and leaving for advanced training in the UK in 1945. Nevertheless, the University of Madras refused to award her a PhD without a basic MSc.[30] Raman himself had no degree higher than a BSc until the University of Calcutta conferred a PhD on him in 1921, *honoris causa*, at age thirty-three! Raman himself had never sought an MSc or higher degree and may have been rather indifferent to this question concerning his students—particularly women. We do not know if or how Raman intervened in Anna Mani's situation.[31]

"Slay Raman"

About three years after Raman's appointment, the institute's council called for an extraordinary review committee to visit Bangalore in late 1936. Saha, a member of the council, was in Europe and the United States on a Carnegie Fellowship; he was neither a witness before the review nor did he send a written opinion. As evidence of his own expanding influence, S. S. Bhatnagar was appointed the only Indian member of the Review Committee; the others were its chair, chemist Sir James Irvine who was vice-chancellor of St. Andrews University, the vice-chancellor of Osmania University in Hyderabad, and the inspector of schools of Bangalore, H. J. Bhabha. The presence of two chemists (Irvine, Bhatnagar) on the committee was important to the outcome; no physicist was on the committee. The Irvine Report first debated the philosophical differences between "academic" and "technical" research and then came directly to the point: "There must be in the Institute a spirit of harmony and cooperation which we fear does not at present exist. . . . There can be no doubt that an atmosphere of insecurity and misery has been created." It continues, "Evidence quickly revealed that physics was in the process of becoming the dominant feature of the Institute, while the Departments of Chemistry remained understaffed and were in consequence losing ground."[32]

As an indication of the relative importance of scientific and technical fields at the IISc, the Department of Physics had been created only when Raman arrived and Saha joined the council in 1933. Raman developed it rapidly so it could then compete with other departments in India, particularly with Calcutta, and tried hard to bring prominent people to it. He argued that chemistry was already well established at IISc and was developing adequately in other universities. Furthermore, Raman said that physical chemistry should be done in the physics department, and then ordered its equipment moved to the new physics building, against the will of the physical chemists. He had a vision of a creative frontier between physics and physical chemistry later shared by many researchers, but it was not shared at this time in the IISc; this was a time for tighter boundary maintenance.

Having looked at all the evidence, Venkataraman concluded, "There can be little question that the Irvine Committee did what it was supposed to do—slay Raman."[33]

Max Born, a distinguished refugee physicist from Germany, for whom Raman had created the Chair of Mathematical Physics in 1935, was suddenly in the center of this controversy, not least because Born's salary (set by Raman) was much larger than for other professors. Lord Rutherford told Born that he had been on the selection committee that recommended Born for the job. Born was not supposed to be the only distinguished visitor: Budapest physicist Rudolph Ortvay actually spent three months in India in 1935 but did not stay longer. Raman also tried to bring other refugees to Bangalore, like German physicist Ernst Schroedinger, Hungarian chemists George von Hevesy and V. M. Goldschmidt, and Nobel Prize–winning Dutch physicist Pieter Zeeman to work in the institute, but for various reasons they found it difficult to come, and news from Born about the 1936–37 situation at the IISc deterred them. Had they come it might have been an even more stimulating environment. Von Hevesy agreed to a contract, though in the end he did not come but went to Copenhagen to work with Bohr; he had co-discovered "hafnium" in 1923 and won the Nobel Prize for Chemistry in 1943 for development of isotopic tracer techniques.[34] Paul Dirac acted as intermediary for a number of these refugees, particularly since Dirac's first wife was the daughter of Hungarian physicist Eugene Wigner. Born wrote from Bangalore to Dirac about Ortvay's desire to return to Bangalore, saying, "I am very sorry for Ortvay but I am afraid there is just now little interest in India for other things than their own quarrels."[35] Clearly the institute had called into question the cost of these visits, though it was well known that these scientists in exile had few alternatives. Neither the Tatas nor the viceroy could quite appreciate what a galaxy of scientists might have been brought to India.[36]

One biographer described Raman at the institute as "a bull in a china shop."[37] Another biographer, Venkataraman, frequently referred to Raman's overriding ego, which blocked his ability to see things from the perspective of others. Sur's paper describes the sharp mode of argument and condescending tone Raman used in his scientific communication, which could not endear him to his readers or listeners.[38] The review committee was asked again to consider the budget and heard strong opinions from within the IISc that the chair occupied by Max Born should be dissolved, though there was a consensus that other money should be found to keep Born at Bangalore. Many economizing moves were suggested, mostly by reducing the salaries of lower staff who had no bargaining power. It should be remembered that though this was near the end of the global economic depression, commodities traded by India were still greatly undervalued in the imperial economic system.

Born's letters to Lord Rutherford during 1936 illuminate the situation around Raman and Saha.[39] These letters should be understood in light of the fact that Born later said he was critical of Saha's recent research, which they had discussed in Copenhagen and Cambridge after Born left Bangalore. He also stated that he was more attracted to "the vigour and intensity" of Raman instead of to Saha. The two old friends, Born and Raman, would come into conflict in the late 1940s when Raman refuted Born's classic lattice dynamics as applied to frequencies in crystals, in part because Born's idea did not agree with Raman's experiments.[40] Though Born later dismissed his own scientific research in India with one sentence in his autobiography,[41] his letter of 26 January 1936 is more revealing.

> In Aligarh I had to give some lectures and I got the idea of the standard and work of a remote Indian university. . . . One of my tasks was to discuss his theory of relativity with High Court Judge Sir Shah Suleiman who is attacking Einstein (in *Current Science*). . . . As to my duties at the Institute, I have to give two lectures a week on quantum optics; Raman and some other professors are attending them and we have some pretty heavy discussions. I have plenty of time to pursue my research work with the assistance of an extremely gifted (and very shy and modest) young theoretical physicist, Nagendra Nath, whose name will soon be known elsewhere. Besides, there are 3 or 4 quite good students who work with me. I myself try to continue my work on a unitary field theory, and I have written two papers for Raman's Academy Proceedings. But I am still far from a deciding result. It may be foolish that I follow so obstinately this line of thought, but I cannot help it—it fascinates me. And I am still convinced that it will lead to something.[42]

Lord Rutherford was now asked by Raman (and others) to adjudicate the conflict between Raman and the IISc. Born offered Rutherford an insider's view of the situation and said that, because of it, he wanted to leave Bangalore soon. Rutherford urged Born to remain in Bangalore because of the political situation in Europe and the scarcity of positions in universities in Britain, particularly at Cambridge, where he wished to be. Born was torn because the IISc salary was greater than he could earn in Britain, and he had lost everything when he left Germany, so he knew he could easily stay; moreover he liked working at the institute and with Raman. Born told Rutherford of the resentment among some British faculty about working under Raman, an experience they had never had before—not simply that he was Raman, grown so confident with the Nobel Prize, but also they had not before taken orders from an Indian, and Raman certainly *did* give orders. Born noted these British faculty members had gained the ear of the colonial government, which agreed to put pressure on the Tata family as a result.[43]

This is the not the first time the director's role proved difficult: as a sign of the Tata family's earlier frustration with its leadership, Sir Dorabji Tata, Bhabha's uncle, proposed to the 1930 Review Committee that the post of director of IISc be abolished and the role simply be rotated between heads of departments. Though an interesting and creative idea, no one took it seriously; there are few "headless" institutions, and leadership was simply not managed by rotation in India, and not in the Tata Group either. "Abolishing the post," however, referred to Martin Forster, FRS, a chemical industry insider who had come to Bangalore from Britain as director in 1922 and who stayed on ten years until 1932. Rutherford and Born discussed press releases in Bangalore hostile to the institute and Raman and anonymous pamphlets against Raman that circulated while the committee sat for hearings like a court. An understanding of the official view of this situation is gained through a "Confidential Memorandum on Sir Venkata Raman's Administration as director of the Indian Institute of Science, Bangalore."[44] Born paid little attention to the physics versus chemistry part of the conflict, though it was not trivial. He wrote the following to Rutherford from Edinburgh after he left Bangalore, and after he had also just met Saha in Copenhagen:

> Nothing can be easier in India than to rouse discord and to stir it. Once you allow and encourage people to speak, they will never end because they want an outlet. There is always a latent jealousy and dissatisfaction which could be directed against almost anybody and anything. Take Professor Saha. I know the following from him. Saha is one of the greatest enemies of Raman. I do not know whether he also had hoped to become director, but he had hoped to become the successor of Raman in Calcutta, and Raman may not have helped him to get this post. A very clever pupil of Raman—Krishnan—got it. Since then Saha attacks Raman when he can. They have another object of quarrel; Saha intended to found an all-India academy, but things went too slowly for Raman's temperament and he founded his own academy (Indian Academy of Science) in Bangalore, with his own Proceedings. Now there are two academies in India, not too many for such an enormous country, but they are bitter adversaries. All the north Indians joined Saha's party, and the south Indians that of Raman. These parties have their delegates in the Council

of the Institute, but the north Indians, and particularly the Bengalis, have the majority.[45]

Rutherford never really entered the conflict through correspondence because he felt he could do harm to Raman and the institute by making a judgment without being there; after all, he was on his way to India and could see for himself. Then he suddenly died in late 1937. The pressures Born described continued until Raman resigned as director in late 1937. Saha and Justice Suleiman of Allahabad, who by now fancied himself as something of a physicist, as Born observed, were appointed members of the search committee for a new director, and they succeeded in having Jnan Ghosh appointed director. Chemistry and Bengal had won the day. One of the most vigorous internal antagonists against Raman had been P. C. Guha, a Bengali professor of organic chemistry at IISc. This should be understood in terms of both regionalism and disciplinary boundary maintenance in an institution which was establishing its national reputation.

Advised of the crisis around the directorship by the colonial government's representative resident at the Mysore court (named as the visitor to the IISc), the embarrassed viceroy's office sent a secret emissary to consult the Tatas, asking whether they wished to have a formal review of the Executive Council's decision to accept Raman's resignation and appoint Ghosh as director. Instead of a forced resignation, the viceroy hoped that a one-year probation for Raman as director under new financial guidelines and oversight might succeed. If not, Raman had already offered to retire as director without a demand of compensation, at a later stage.

The newspapers and the public in Bangalore were now divided into pro-Raman and anti-Raman factions. The Tata's final position, reached on the basis of discussions with IISc council members whom they trusted, for example, lawyer J. J. Ghandy and H. J. Bhabha (Homi Bhabha's grandfather, a member of the extraordinary review committee), was to accept the council's decision without review, thus sustaining the authority of the council. Raman was, however, offering his resignation on a condition; he said he would "make over charge of my office when the sanction for the payment of the sum of Rs 100,000 has been received."[46] This was a very large sum of money then, seen as compensation for early departure, almost two years' salary for the director. The council clearly had second thoughts and could not answer the viceroy's question as to why Raman should resign. The original offer and settlement were withdrawn, and Raman was summoned and told he was "unfit to continue any longer as director."[47] The payment may not have been made, but Raman resigned as director while remaining on staff in the IISc as professor of physics.

Despite his troubles and subsequent resignation as director, from 1938 Raman continued to try to build up the physics department and sought unsuccessfully to secure government and private research funds from either the Birlas or Tatas. These two family trusts (foundations) were to support Saha in building the cyclotron in Calcutta in 1941–42, and Tatas were receptive to Bhabha's scheme for Bombay in 1943–44. Raman tried to bring Erwin Schroedinger to Bangalore in 1940, and another plan to bring Nobel Prize chemist von Hevesy again, in 1939, almost succeeded. Prior to getting a professorship at Birmingham in 1937, refugee scientist Rudolph Peierls was again communicating with Raman about a position at Bangalore.[48] All three scientists found an alternative to Bangalore at the last minute, but the IISc's reputation was strong enough for Raman almost to succeed; all three had exceptional reputations. Raman eventually left IISc entirely and devoted all his energy to building up the Raman Research Institute (by using his private funds) a few kilometers away. He began to edit and publish his students' papers in *Proceedings of the Indian Academy of Sciences*. This journal, which, like his institute, Raman owned privately, became his exclusive domain, and he was reputed to rarely send a paper out for peer review but to decide himself what to publish.

A Search for Harmony?

Raman had also acquired a reputation for unpredictable or argumentative behavior, which extended around 1940 to what eventually became a very public disagreement with Max Born about lattice dynamics. Starting with publications in 1941, in the early phase of the controversy Raman insisted that Born's lattice theory could not be used to explain quantum behavior because it was semiclassical. Sur carefully described the evolution of the disagreement after 1943–45: "Although Raman wanted to contest Born's theory [of lattice dynamics] on theoretical grounds, he had little patience with theoretical formulations in which simple spectroscopic manifestations of lattice dynamics were buried deep in mathematical rigour and tedious equations."[49] Among the "underlying" issues, Sur lists Raman's wish to avoid being classified *simply* as an experimental physicist, Born's greater theoretical confidence in the evolving lattice theory, and the reciprocal arrogance of the two men, their constant needling, and baiting each other. She points out that Raman "was deeply committed to the idea that the perfect symmetry and harmony of the crystals must be reflected in their dynamical properties; indeed, he held that nature could not be otherwise." Raman had become fascinated with the spectra of diamonds and began to collect them at his own great expense. Physicist G. Venkataraman devotes a chapter to the Born-Raman controversy, noting that they quarreled about physics even when Max Born lived in Bangalore, but they later carried the disagreement to the pages of *Nature* and *Reviews of Modern Physics*.[50] The disagreement had deep aesthetic and methodological roots, he said.

> The incommensurability between theory and experiment was thus transformed into a contest between two different theoretical frameworks. . . . Neither Raman nor Born cared to look beyond his own empirical experience and theoretical constructs to acknowledge and accommodate the differences between them. While Raman strove to generalize from a few specific experimental observations, Born expected experimental reality to conform to his theory.[51]

Because Saha's position was analogous to Born's—looking for experimental confirmation of a good theoretical idea, confirmation which he felt was being denied him—we can now understand one small dimension of the disagreement that grew between Raman and Saha. Raman made a powerful case against Saha by saying he "was not an experimenter" when Saha sought money for experimental equipment for his lab, but Raman knew that Saha had not yet personally done the experimental work that would justify research funds. Raman's disagreement with Born—and such disagreements are at the core of physics—reappeared when Raman used a 1946 visit to the Commonwealth Science Conference in London to return to their very public disagreement about lattice dynamics. The drama lay in the fact that these men were widely known to have been friends and been of great mutual assistance to each other.

Raman never forgot his anger with Saha and transferred it to the new institute director, Jnan Ghosh, and to Bhatnagar, who was on the Irvine Review Committee whose report sealed his fate. Raman was asked by Nehru in 1947–48 to accept an appointment as National Professor, with a good sustaining salary, and Raman said he would think about it, but he was concerned ("angry," said one observer) that through it the government might try to control what was done with it. Scientists heard this and met to depute Sir K. S. Krishnan to speak to Nehru, who sent the nawab of Rampur to Raman to assure him that the prime minister believed in him in spite of what other people might say. He was urged again to accept the position and finally did.[52] Krishnan later brought a proposal to Raman by which the Raman Institute would have regular government funding, but Raman dramatically rejected funding for the institute at a ceremonial dinner. I think Nehru's moves were genuine efforts to open a new door for Raman in a newly independent country, to demonstrate national admiration for him. Notwithstanding his evident achievements and the strength of the institute and academy he built, he repeated his anger when I met him in 1968, saying his institute would not be co-opted. But in the 1950s he invited Nehru to his institute, in order to persuade him to attach the national professorship not to Raman the person but to the institute, so that his successor would have a continuing salary. Raman's biographer Venkataraman, speaking of the expulsion from the IISc and subsequent dealings about the institute, said it is "possible that deep inside he was very much shaken. He was never again the same man, increasingly prone to cloudy judgment where both persons and scientific matters were concerned."[53]

The 1936 report of the Irvine Committee emphasized the search for harmony between academic and technical research and between pure and applied research, but this harmony was not achieved, according to some students at IISc. Ten years later, one student was told to look elsewhere to do "pure research" because it would not be done at IISc.[54] It appears from this evidence that no one yet had a conception or a program in which these two strains of scientific and industrial research could be harmonized. After so long abroad, Bhabha's introduction to science in India in 1939 was to the smoldering embarrassment of the conflict that Raman had with the institute.

What Homi Bhaha Found in Bangalore

The situation at the IISc in 1939 revealed to newcomer Homi Bhabha some of the fundamental problems in the growth of scientific institutions in India. With an early start in 1909, there had been a sound conception and vision behind a program of concentrated graduate teaching and research at the institute, intended to avoid the highly dispersed model of the existing universities in India. Funding was constant at the institute, and more generous than in any other scientific institution. Physical location and campus were ideal, if far from major cities. The expectation from the beginning had been that the IISc was to represent national, indeed, world excellence. This was a world Bhabha knew intimately, a successful Parsi world in India with cosmopolitan expectations and high standards. Yet when the 1931 Sewell Report warned that the IISc should not be viewed as a "South Indian" institution alone, the increasing strength of regional pressure had already appeared. That kind of pressure was hard to withstand even in cosmopolitan cities, whether that regional pressure was laced with the leadership politics of the Independence movement or with the later severe competition for good professional training and steady employment. How different was this warning about regionalism to the institute in 1931 from the concern of the Saha-led resistance to Raman at the IACS in 1931 in Calcutta? Was that resistance intended to make it an East Indian institution and not simply a South Indian one?

In the end, influences from Bengal in Bangalore were explained as part of the cosmopolitan "all-India" idea, though they were far more turbulent influences than anyone expected. Jnan Ghosh's role as director lasted from 1938 to 1948. Saha was a member of the IISc council for fourteen years, from 1930 until 1944. Appointments of British academic scientists at the institute continued, even after 1947. Resented as "colonial remnants" by some young well-trained Indians who wished to occupy those positions, admired by others for the good teaching and standards they (sometimes) embodied, British faculty constituted a significant number of "foreigners" who added to the complexity of the situation at the IISc. As chapter 5 demonstrates, Homi Bhabha worked six years in these conditions before moving to make his own institute in Bombay, and the evidence will show that he carefully planned and deliberately designed an organization to circumvent some of the problems that swirled around Raman, Saha, Bhatnagar, and the IISc in Bangalore.

So there they are in 1939, poised professionally—Saha to begin to build a new lab or institute within the prestigious University of Calcutta, Bhatnagar to negotiate his role in a new national council dedicated to scientific research for industry only a short distance from the viceroy's office in Delhi, and Bhabha, trapped by the war in Bangalore, to become a young member of the professoriate at the IISc, in a dynamic new physics department though one bruised by the conflict surrounding its head and senior professor.

At the same time, by 1939, Saha and Bhatnagar, riding partly upon Raman's energy, had helped to build a more effective ISC, spawning at least two academies of science, had turned to the media to explain and justify support for science and scientists, and had built relations with leading political figures. Raman and Saha had even created two regular science news "magazines" whose subscription lists were growing steadily and competitively. This whole apparatus was ready in 1939 to advocate the idea of scientific planning, with scientists directly involved in the process. It is into this particular context that Homi Bhabha stepped.

FOUR

Imagining a Scientific State: Nehru, Scientists, and Political Planning, 1938–42

Although the Bangalore affair around C. V. Raman and the IISc in the late 1930s was not a tempest in a teapot, it had little direct connection to the institutional face of the Indian demand for industrial and commercial independence. The methodology and interest underlying this demand was the new idea of "planning." For a variety of reasons, forces favoring scientific *and* industrial planning of national development pressed themselves into public life in the late 1930s. The cumulative experience of the Depression, in which much of Indian life and economy were treated with indifference at the official level, called out for a different approach, involving Indian organizations and capital in a new way. Notwithstanding this change in attitude, there was in fact little new room for Indian entrepreneurs, and the official attitude at the top really was of indifference and avoidance until the late 1930s. Tyabji, however, reminded me that "there was strenuous opposition to any efforts to increase the role of government, precisely on the grounds that this would disturb the principle of 'laissez-faire.' "[1] So the step toward planning and more effective deployment of public resources was not taken uncontested.

When the European war started in September 1939, everyone associated with the government quickly accepted national planning. The difference lay in a kind of strategic ambiguity at the heart of the doctrine of laissez-faire: one lay behind official indifference; the other gave first place only to the ambition of approved entrepreneurs and financiers. Those men played carefully, keeping as close to the Independence movement as possible without getting stuck in self-destructive "revolutionary" programs that would prompt authorities to shut down their business. Reflecting this complexity, both middle-level British and Indian officials advocated more planning for science well before the war started in 1939. Examples of planning in other countries—the United States, Russia, Japan, and Germany—were examined favorably in the newspapers. None failed to see that Germany was far ahead in "planning." The Congress Party was already divided on this issue, with a socialist faction that strongly advocated planning. Communists and socialists, though under police surveillance, advocated Soviet-style industrialization based on thorough planning. Gandhi's commitment to rural and agrarian development with a self-reliant village at its center frustrated and even appalled some "modernists" who advocated planning. But ultimately Gandhi's approach too had to be explained and justified in terms of its contribution to Independence. So there was a mixture of calls for sacrifice and renunciation (don't buy foreign-made goods), for inventing new machines and scientific techniques, and disciplined hard work—all of which were in short supply.

National development planning in India in the 1930s was generally a public discussion among intellectuals and a private one among colonial officials and business elites. There were very few politicians who could bridge the two and very little voice for scientists and technologists on this issue, because most of the scientific assemblies were too politicized to suit the elites. By the late 1930s farsighted people, even those not convinced there would be war in Europe, must have concluded that Britain would inevitably have to let go of India sometime soon. One effect of the economic depression of the 1930s in India was that people looked far beyond Britain for insight into how economies should be reorganized and redirected, and this search for insight included some (though not all) government officials. Besides the state-driven reconstruction of the American New Deal, there was a lot of attention paid to the USSR; for example, in both America and the Soviet Union, promoting rural electrification was a cornerstone of securing state power in an agrarian society and increasing rural productivity (and thus its contribution to state revenues). A fascination with Soviet progress spread among young people in the late 1930s, and middle-aged British officials (and, I suspect, many middle-class Indians of the same age) felt they had to understand this model: Zachariah suggests that the Soviet example was talked about by British officials even more legitimately after the USSR became an ally of the Allied powers in 1942, but it was certainly a common topic long before the war began. This was usually in terms of economic uplift, rural electrification, and reform, yet this planning enthusiasm coexisted, even before 1942, with a deep official distrust of communists and Communism in India.[2]

Practical Tendencies?

Advocating "industrialization" had two implicit meanings for people like Saha and Bhatnagar; one was a critique of the British colonial economic structures and recurring rules that inhibited or blocked the development of industry in India, and the other meaning a critique of the Gandhian groups within the nationalist movement, groups that took a "traditionalist" agrarian approach to development. In addition, advocating planning was a critique of the ad hoc administrative practices of a reactive, sometimes indifferent, always slow, colonial economic policy. Each of these emphases was a tactic in a long struggle and attracted support for industrialization across surprising class and political lines. The received British popular wisdom was that Indians were a bright but impractical people, so industrialization would prove the contrary, and this thesis brought about strange combinations. For example, Saha made working alliances with Sir Mokshagundam Visvesvaraya, the dewan of Mysore and member of the National Planning Committee (NPC), and he spoke positively about manufacturing automobiles in Bangalore with industrialists like Walchand Hirachand. When the war began, this automobile project proposal was pushed even harder as part of the war effort, though in the end it stalled over the problem of raw material. Bhatnagar was talking with the same individuals. An issue like this, cutting across such boundaries, was destined to become important to leaders of the Indian National Congress.

Other professionals preferred to push toward the one dimension that appeared open to them, which was technical education and practical training. The pressures building for this training in Bengal were strong, and Raina and Habib have laid out their origins using the debates in *The Dawn* in Calcutta in the context of the perceptions among the educated middle classes of Bengal concerning technology and science.[3] This was consistent with the gaekwad of Baroda's own efforts to build a technical institute focused on new skills, artisans, art, engineering, new machinery, their products, and the like—done in a mix of English, Gujerati, and Sanskrit—beginning in the 1890s.[4] Baroda trained skilled people and provided an opportunity for social mobility among artisan castes and classes, but this institute did not become a source of significant technical innovations. Tyabji described the origins of the Department of Chemical Technology at the University of Bombay, promoted in response to the problem that other institutions nearby such as Baroda were not providing the trained and capable people needed in the city's textile industry.[5] In 1928 the Bombay University's Act included for the first time reference to technology as an important branch of learning. But it was four years until a professorial position was created, partly on the basis of an endowment by a rich industrialist, and two years later the teaching of textile chemistry and chemical engineering began, and seven years later began the study of pharmaceuticals, chemistry of food and drugs, and dyes. This coincided with the foundation of the laboratories at the University of Lahore in the mid-1930s, where Bhatnagar trained his students in oils, emulsions, and alkaloids. In 1944, Bhatnagar's Council of Scientific and Industrial Research (CSIR) helped to organize the study of oils, paints, and industrial inks and dyes. All along the way cooperation was found among some British officials and experts (even some businessmen) who were frustrated with how the default mechanisms of macroeconomic colonial policy were always privileging British industry and technology and spare parts imports, thus implicitly steering Indians toward theoretical or more research-oriented kinds of work and away from practical applications. Theory and research, it was then argued, was just what Indians were best at. Very gradually the hesitation of major Indian capitalists to invest in this messy and dirty industrial work eroded, a process accelerated by the profitability of war production requiring new applications of technical knowledge.

Gradually the acceptance of the idea of industrial planning reached high office, but not until the entire war effort was shot through with planning, and planners like Cripps had intensive discussions in India about it in spring 1942. Soon after the Cripps visit, Leo Amery, the secretary of state for India wrote from London to the viceroy in Delhi to ask, "Might it not be our duty after the war to put ourselves in the position of a bold, far-sighted and benevolent despot, determined in a few years, in a series of five-year plans, to raise India's millions to a new level of physical well-being and efficiency?"[6] Zachariah has skillfully laid out the range of contradictions that such a plan faced, asking if there even could be "a reformed imperium?" There were long-range thinkers who imagined a restored empire under a new confederation, but at the end of the war they were in a small wishful minority.

But by 1942 it was rather late for the government of India to become a bold, farsighted, and benevolent despot: Japanese forces had captured British territories in Southeast Asia and had now overrun all but a tiny northern tip of Burma. Refugees from that country were straggling by the thousands into India, some very poor and some well off—for example, the entire Steel Brothers' establishment was transferred from Rangoon to Calcutta to continue business, though hardly "as usual." The viceroy was at this time "supervising" the Cripps Mission to India, whose purpose was to negotiate a satisfactory working relationship between Britain and India during the war, and he would already have known it could not succeed because the cabinet (in particular Churchill) was not prepared to grant relative autonomy now and independence to India at the war's end. Viceroy Linlithgow reacted to Amery's plan to pursue such a series of "restructuring" five-year plans by writing in the memo's margins, "Good Lord!"

Saha, Subhas Bose, and Nehru at Work in the Committee

Indian scientists now planned to build the institutions they had earlier imagined. Bhatnagar in Lahore was soon to direct a large program of research for national industrial development but was at the edge of this particular planning group until about 1939. When Saha left Allahabad for Calcutta in 1938, he entered the politics of national planning, thus initiating his science networking and organizing for his new Institute for Nuclear Physics. Much depended on Jawaharlal Nehru and the NPC and on the tangential effects of the activities of Subhas Bose in the Congress Party. Saha had met Bose, four years younger, while they were living in the Eden Hindu Hostel around 1916. They met again while students at the university and were in England at the same time in 1920. Together they raised Rs 2.3 million for flood relief in Bengal in 1922–23, under P. C. Ray's direction, before Bose was arrested without charges on suspicion of "collusion with revolutionaries" in 1923 and imprisoned near Rangoon in Burma for three years. Saha chaired a big meeting in Calcutta to celebrate Bose's release. Bose returned to public life and was elected mayor of Calcutta in 1930, holding other Congress Party positions when not otherwise in jail or abroad until 1938. With Gandhi's support, Bose had been elected president of the Indian National Congress in 1937–38, in which year the party won elections in seven provinces and began to hold a number of government offices, including portfolios for industry. Industry was a provincial subject in the new constitution, as distinct from, say, finance. Like others, Saha urged the formation of a national planning committee in science and culture, and Bose invited Saha to work on this committee when the Congress Party high command established it in 1938.

Saha was invited by Bose, as Congress Party president, to a meeting in October 1938 at Delhi, attended by ministers of industry in party-controlled provinces, "and certain other prominent men of India."[7] Among them was K. N. Katju. This meeting was intended to form the NPC, and Saha, arriving late, discovered that Sir M. Visvesvaraya had been asked to become the chairman of the committee. Visvesvaraya, trained as a civil engineer, was dewan of the state of Mysore and had supervised the building of hydroelectrical power dams, iron and steel mills, a university, and schools and libraries. Mysore was seen as a model planned princely state, but Visvesvaraya was better known in this committee as the author of a 1934 book, *A Planned Economy for India.*[8] Given his prestige, he was a logical choice, but Saha says he persuaded Visvesvaraya not to accept the position of chairman because unless an important member of the Congress Party was chairman, the planning committee would "be regarded merely as academic and would have no value in the eyes of Congress. The grand old man saw the force of the argument and readily agreed. It was my suggestion that Nehru, then in Europe, be invited to take up the Chairmanship of the proposed committee."[9] Saha then became the chairman of the NPC's Power and Fuel Subcommittee and a member of the River Transport and Irrigation Subcommittee, two subjects on which he wrote regularly, with authority, in *Science and Culture.*

Following this organizational meeting in Delhi, Saha went directly to speak with Rabindranath Tagore at Santiniketan. They had first met in Germany in 1921 through the physicist Arnold Sommerfeld. Nehru accepted the position of chairman of the planning committee when he returned to India, and in November 1938, two letters were written to him from Santiniketan—one from Tagore and the other from Tagore's secretary, Anil K. Chanda. Tagore wrote supporting the idea of thorough planning, supporting Nehru in his new role, and urging him to be strong-minded; Chanda wrote to say that Tagore was supporting Subhas Bose for reelection as president of the Congress Party because Nehru would be working hard on planning; in this context Chanda said Tagore "has been rather captivated by Dr. Saha's ideas of Rational Planning and he is hoping much for the Committee."[10] Subhas Bose campaigned for the presidency against Gandhi's wishes and was reelected president of the Congress in 1939. He resigned later in the year, because without support from Gandhi, he could not function effectively within the party high command. Tagore called for unity in Bengal behind Bose, after his resignation from the Congress presidency.

Thus it was that Saha came into frequent contact with politicians and industrialists who split off from the NPC and eventually formulated the 1944 Bombay Plan—a mixed public and private sector kind of economy. There was, however, another Saha involved, one who claimed to have persuaded Subhas Bose to set up the NPC in the first place and who also became a member of the NPC. This was A. K. Shaha, of the same caste group as Meghnad Saha and also from East Bengal, but with a different, Russianized spelling of the name. He was invited to Moscow as a foreign specialist, married a Russian woman, and achieved the Russian equivalent of the DSc degree in engineering. He advocated for India the same rapid industrial approach that Stalin was taking (e.g., the Stakanov movement), and like Meghnad he praised the role of science and scientists in industrialization.[11]

So influential did A. K. Shaha become that Subhas Bose wrote to Nehru to ask that Shaha be appointed as joint secretary of the NPC: Nehru replied that Shaha "might have been appointed an honourary General Secretary" but would have had to resign his unpaid position as member of the NPC if he accepted the paid position of joint secretary. Nehru wanted "to profit by his special experience in the sub-committees. . . . I have been trying hard to get a suitable place for him in some provincial government."[12]

Meghnad Saha had already discussed the issues of industrialization outside the NPC in 1938. For example, as president of the National Academy of Sciences in Allahabad, he chaired a session on power supply problems where he proposed the use of power from nuclear reactors in 1939. Nehru, whose home was in Allahabad, presided over that meeting and was thanked by Saha thus:

> It was in the fitness of things that Pundit Jawaharlal Nehru has agreed to preside over this annual gathering of scientists in India. His position in this country can be described by a phrase which Americans use with respect to Abraham Lincoln [read: George Washington], "first in war, first in peace." Next to Mahatma Gandhi, he occupies the first place in the hearts of his three hundred million countrymen. The time has come for him to give a lead in peace time work of reconstruction and consolidation of the country.[13]

Saha also invited Subhas Bose to preside over the third general meeting of the Indian Science News Association, which published his journal *Science and Culture*. In August 1938, Bose, now president of the Congress Party, came to its meeting in Calcutta and made his views on development clear:

> Though I do not rule out cottage industry and though I hold that every attempt should be made to pressure and also revive cottage industries whenever possible, I maintain that economic planning for India should mean largely planning for the industrialization of India. And industrialization, as you will all agree, does not mean the promotion of industries for manufacturing of umbrella handles and bell-metal plates, as Sir John Anderson would have us believe.[14]

Like Bose, Saha was equally critical of the so-called Congress high command and its support for cottage industries and for the mixed economy proposed by prominent industrialists in the discussions leading up to the

Bombay Plan (their first discussions occurred in 1938–39, and their final report, drafted by John Matthai, was published in December 1944).[15] Saha agreed with Bose's explicit view that India needed "a forced march to progress" like the USSR, supervised by a strong state. He was critical of Gandhi's policy of *khadi* (homespun cloth) strongly supported by one faction of the Congress—and, for example, by his chemistry teacher and ally, P. C. Ray. Saha went to lengths to explain he meant no disrespect to Ray but pointed out that students could and should disagree with their teachers on matters of principle.[16] Striking against Gandhians, he was angry when K. N. Katju, the Congress minister for industry in Uttar Pradesh, where Saha had been professor at Allahabad, opened a match factory in 1938, saying it was heralded as the start of large-scale industrialization. Saha quoted Bose extensively in his authorized biography because he said in 1954 that "in contrast to other political leaders, Bose's mind was absolutely clear on the post Independence problems."[17] Here he was alluding to Nehru as "other political leaders," establishing for himself a problematic relationship with the very person on whom he would have to depend. Saha and Bose disparaged John Anderson's views about umbrella handles and bell-metal plates for another year, and this irritated both Gandhi and Nehru.[18] Anderson had articulated an authoritative and safe position for Congress moderates like K. N. Katju, a position which neither Gandhi nor Nehru could criticize very strongly, given their dependency on the moderates.[19]

Disparaging Anderson's views meant simultaneously condemning the colonial government and dismissing a man who had, prior to leaving India in 1938, already been considered for the position of viceroy and would be again, by Churchill, during the war.[20] It also meant criticism of those in the party who agreed to serve under the viceroy in central or provincial governments before resigning when war was declared in September 1939 and those whose businesses profited from this situation. Subhas Bose was demanding thorough land reform, which most of the Congress Party did not dare to support. Saha's unabashed promotion of modernity and complete industrialization, and his backing Bose, exposed him to further criticism and suspicion, even though eventually in Bengal Bose was given hero status. Therefore Saha's support for Bose raised Saha's reputation among those Bengalis who had a romantic belief in Bose, a bright young man who went to Cambridge, a labor leader arisen from prison, a trained military officer coming to liberate the country, a visionary killed tragically, perhaps deliberately, just before his time.

Another Calcutta physicist trained in Cambridge and former secretary to Tagore now became involved in the NPC: Prasanta Mahalanobis was becoming a well-known statistician, having founded the small Indian Statistical Institute. Nehru invited him in 1940 to review the NPC's numerous reports "from a statistical point of view." This was what the committee meant in its first report by a role for "disinterested experts." This launched a new career, and within three years Mahalanobis was nominated in 1942 for an FRS in "mathematical statistics" by E. J. Russell and R. A. Fisher and elected a Fellow in 1945.[21]

Although Saha's friend Subhas Bose was reelected in 1939 president of the Indian National Congress for the second time against Gandhi's continued opposition, Saha must have begun to realize that his main political relationship would have to be with Nehru. That realization became clearer when Bose was forced to resign as president in the same year when the right-wing factions of the party refused to accept Bose: "Gandhi intervened and Nehru sided with the right."[22] Nehru was, after all, accessible to Saha in meetings of the NPC, and in 1940 Saha was appointed to the new Board of Scientific and Industrial Research (BSIR), under the first director, Saha's old friend Shanti Bhatnagar.

The Emergence of the CSIR

As we saw earlier, after the visit to Bhatnagar's laboratories in 1939, the initiative for the BSIR was taken by key viceroy advisors Sir Ramaswamy Mudaliar, Sir Azizul Huq, and Sir Ardeshir Dalal; Saha wrote to Mudaliar twice in early 1940 pressing him to establish the BSIR and to distinguish between scientific and industrial research, noting that setting up industries is different from protecting them.[23] Dalal was head of the Planning and Development Department in Delhi and a key Bombay leader close to the Tatas. The scientific members of the new Council of Scientific and Industrial Research were Nazir Ahmed, Cambridge-trained physicist turned to cotton and textile research, Saha, and Saha's old friend, chemist Jnan Ghosh. Each had been a member of the NPC too. It was first proposed that the Governing Body of the BSIR should not contain any other scientists, so all three scientists, plus Bhatnagar, threatened to resign until the decision was reversed. In spite of their previous disagreements, Saha's relation with Nehru in 1940 became crucial because Subhas Bose left the Congress. Seen another way, Nehru needed Saha's credibility in Bengal. Saha lost a most useful connection to the Congress Party when Bose resigned; he had been replaced at Gandhi's wish by Rajendra Prasad, one of Saha's acquaintances from his Eden Hindu Hostel days and now professor of law in the University of Calcutta. Subhas Bose meanwhile thought that Congress should confront the Raj during wartime and in 1940 he formed the Forward Block to do so. He was put under house arrest for civil disobedience around a demonstration against plans for a British memorial to the Black Hole of Calcutta. Though Gandhi and the Congress Party frequently used the same tactics, they had not authorized this particular demonstration, and it was not "legal" (i.e., it lacked a permit). This opportunity was used by the police to arrest Bose, who then made a daring escape in January 1941 from his house arrest across Central Asia to join the Axis powers, eventually taking command of the already-formed Indian National Army in Singapore and leading it in battle beside the Japanese against the British in Burma.[24] It is assumed that he died in a plane crash in Taiwan in 1945, though in Bengal this was doubted for many years later.

Saha, however, continued to promote Bose's point of view in the NPC, despite his departure from the Congress Party and Calcutta. Nehru, later instrumental in developing Saha's laboratories, had been exposed to Saha's attitudes on these topics from the beginning. For example, in September 1939 Nehru wrote to Krishna Kripalani, who was now Tagore's secretary at Santiniketan, about Saha's views on planning the economy. Bose was still president of the party. Nehru had read a letter from Saha to Kripalani and Tagore in which "he has referred to me repeatedly and made various statements regarding me which are bound to convey an entirely wrong impression of what I said in the Planning Committee."[25] Saha wrote that the planning committee had endorsed Gandhi's plan for cottage industries to the exclusion of large-scale industrialization, saying cottage industries would therefore become stuck on "ancient techniques." Nehru also said that Saha believed party leaders had already consented to foreign management and foreign investment controlling the industrial sector. This incident involving Saha's letter tested Nehru's complex relationship with the moderate and right wing of the party, and his reply to Tagore was that the planning committee had not encouraged Gandhi's ideas to the exclusion of others, that he personally believed in large-scale industries, that he didn't "represent Gandhiji's viewpoint to any large extent; in my mind there is no essential conflict between the two."[26] Finally, Nehru found Saha's view that party leaders were puppets in the hands of big industrialists, most of them foreigners,

> really extraordinary, and shows Professor Saha is not conversant with what has been happening in India. [It is] amazing and displays a lack of appreciation of the whole political, social and economic events in the recent history of

India. It is unfortunate that Professor Saha's letter has been written in a spirit which is far from scientific or dispassionate.[27]

Planning in India, War, and the Colonial State

The NPC fell gradually into inactivity when the war began in September 1939, and particularly when its chairman Nehru was put in jail in November 1940, not to be released until December 1941. Since this was the first time scientists had worked in a committee with national leaders, it left its mark on both sides. Nehru, who enjoyed the company of many scientists, probably found Saha's spirit too critical and manners too rough. As in his relationship with Bhabha, Saha was dealing in Nehru with a man from a diametrically different social background. Nevertheless, the first funds Saha received for his new lab were gained from private industrialists through Nehru while he was in prison and again in 1947 and 1948. It seems that they did not recognize, for ideological and temperamental reasons, the convergence of their personal preference for large-scale industrialization. Saha, responding to criticism about his preoccupation with industry, was anxious to point out that he was not suggesting a neglect of the agricultural sector in favor of industry, and said so frequently. In numerous speeches and editorials, he declared his interest in agriculture and favored industrial development that would make direct contributions to farmers, such as fertilizer factories and dams to provide power for rural electrification. But these were sideshows to his main preoccupation with the commanding heights, the steel, petroleum, chemical, and power/energy nexus, which he believed should be publicly controlled. Four years later, in 1944, he explained his position most clearly, in a talk in London:

> The successive famine commissions have rightly diagnosed the excessive pressure on land to be one of the causes of malnutrition and recurrent famines, and recommended that the burden should be taken off the land by providing a large section of the population engaged in agriculture with industrial occupation. But the small amount of industrialization which had taken place in India is totally inadequate for taking the burden off the land . . . There is therefore no inherent antagonism between industry and agriculture; and without development of agricultural industries, the rural population of India can never be pulled out of the dreadful medieval conditions in which they find themselves.[28]

Saha tended to explain all the causes of the notable gap between recommendations of many commissions and their actual implementation as arising from the "problem of political leadership":

> Is there any indication that the problem is being properly understood by the Central or Provincial Government circles? . . . The fact is that the planners lack direction from the center. . . . It is regrettable that the Indian leaders have so far paid attention only to the question of political freedom . . . in fact we believe that the only way to achieve unity of thought and purpose in the political field, which is now wanting, is to look at the problem of living for India's millions.[29]

In the planning process mixing scientists with private industrialists and state development ministers, there is a quality that is analogous to good musicians without instruments. Musicians can imagine music by reading a score, composers can have the music running soundlessly in their heads. They do not hear it like others, nor do they need to. It is analogous to good chess players without pieces on boards who can play the game in their heads, in their imaginations. Nehru appears to have been able to hear that silent music and to imagine that chess game. So did a number of others, like Saha and Bhatnagar. In that sense scientists engaged in planning imagined a state and imagined the development of scientific institutions in it, beginning in the late 1930s. They envisaged a laboratory state for which they did not yet have financing or the laboratories to realize it. They imagined the power and influence of science (and scientists) for "good," even though this was a sustained conceit at the time. Like music and chess, this was a cerebral activity that moved far ahead of the mundane practicalities that they faced. With Nehru, these people believed India had failed to develop a source of power and for this reason was a backward, not powerful, nation. Identifying and harnessing that power was their objective. They all said India needed electricity. Having done so, they believed they would rightfully inherit the other kind of social and political power they needed to keep the scientific enterprise growing. They were halfway now, after all, stocking committees and guiding commissions with their expertise. Deepak Kumar, historian of the period, says that they had almost established a "scientocracy" by 1940, or at least certainly the idea of one.[30] The need to justify was shifting to the need to explain and persuade.

This new confidence in science came from many quarters: there is an example from a physicist who was not engaged in Saha's planning process, yet who wrote his plea for "resolute optimism" in *Current Thought*, an influential Calcutta quarterly in 1941. Satyen Bose, like Saha, read German and followed news about the war; he knew Germany was poised to invade Britain as he wrote this piece. Linking science with civilization, again, Satyen Bose asked "what our civilization really stands for, and what immutable goal we should set before ourselves?" His carefully reasoned answer is that men of science are now being "called upon to devise means of fighting," referring to chemical gas warfare, while in times of peace they should "further the cause of mankind." And, referring to India, "the peculiar difficulties of any nation need not cause permanent despair."[31] At this stage there were no definition of Bose-Einstein statistics, nor was there a group of particles called "bosons" in his honor, and Satyen Bose was still sixteen years away from his election as a Fellow of the Royal Society, as a statistician! Not a "political scientist," Satyen Bose nevertheless gave us a contemporary example of the reasoned linking of science with the state on behalf of "civilization."

Bhatnagar Builds the CSIR

Shanti Bhatnagar arrived in Delhi at the beginning of the hot summer of 1940 and realized that there was nowhere to put his Steel Scholars and nowhere to build his promised research labs. Conscious that he had to have an institute as a base for research and training, he began negotiating for something better while he established a small office in the university. By custom, the whole government of India moved to Simla in April and stayed three months up in the cool hills. As the advisor and soon-to-be director of scientific and industrial research, Bhatnagar joined this annual escape from the heat. In this more intimate atmosphere of a small hill town, he made high-level contacts essential for his career and his new organization. It was decided that his labs and researchers would move to spacious buildings in Calcutta and that he would travel the country but return to supervise them in Calcutta once a month. Because "industry" was a provincial subject under the Constitution, Bhatnagar had to mobilize research and development at the level of provincial capitals like Calcutta, Bombay, and Madras. He flourished, living up to his nickname "Steamship Bhatnagar," rapidly starting projects that substituted Indian raw materials and skills for those that had previously been imported, such as oil, textiles, steel, glass, and the like. This was Bhatnagar's version of the ideal of *swadeshi* (made in the country) promoted by Gandhi and many others. Indian industrialists anticipated and got impressive profits through defense contracts and so supported Bhatnagar's program. The Department of Supply and Munitions was the client for most of his projects and so in a way the NPC's industrialization plan was now beginning, using war money. He publicized his activities in a newspaper article titled "War and Indian Industry," saying,

> There is a tremendous scope for India to make a really substantial contribution to the war effort by organizing a planned cohesion of labour and industry. In a restricted manner even now a great deal of work for the supply of goods to the army is done through approved contractors who get bits accomplished here and bits elsewhere. . . . The greatest scope for India lies in her ability to make good by indigenous production what now constitutes a shortage in Indian industry owing to restricted imports, and this presents a vast field of investigations for the technical men and the Universities.[32]

This was at a time when some Indians were asking whether India should help with the war effort and some British were asking, both in condescension and ignorance, whether India could actually *really* help, other than simply supplying fighting troops. Privately they asked themselves, is this simply a make-work project?

Bhatnagar now was in regular contact with the three most powerful men who were responsible for industry within the government and who had great influence in elite circles including the Congress Party: Sir Ramaswami Mudaliar, Sir Ardeshir Dalal, and Sir Azizul Huq. He reported directly to Mudaliar, the member for commerce of the Viceroy's Executive Council, and indirectly to Dalal. During these months, a decision was taken to nominate him for a knighthood, and in January 1941 he was made "Sir Shanti Bhatnagar." As important as his program of practical research was, equal effort was allocated to publicity; he lost no opportunity to speak on radio or write in newspapers, all the while keeping up his speeches to specialized groups around the country. He spoke about "the tide of industrialization," ringing the same bell Saha had been ringing since 1938. Unlike Saha, he publicly praised the efforts of the Tatas, Sri Ram, and Birlas, advocating that India should manufacture everything it could, should contribute massively to the war effort, and should reap the benefit when the war ended. He praised the Tatas for starting a major metallurgy laboratory at Jamshedpur, a lab he was later to "convert" to one of his national laboratories. This allowed him, in his skillful way, to support the grassroots small-business *swadeshi* movement, the big industrial houses, the government's war effort, and the call for large-scale industrialization made by Saha and somewhat more gradually by Nehru. Moreover, he felt an exhilaration in his marriage of "the pure and applied"; he said, "I remember the days when we used to consider greasy, fatty, oily, cement, leather, textile, coal-tar, bitumen, paint and varnish chemists as something removed from the realm of real chemistry, looked down upon by the Pundits of Chemistry."[33] Although coming from a bookish scholarly background, he had conquered the middle-class aversion to dirty work that was widely believed to be responsible for India's "backwardness." This added to his legend.

Japan's advance into Burma in January 1942 soon solved Bhatnagar's problem in having an office in Delhi and labs in Calcutta, because the government decided to move all strategically important functions away from Calcutta inland to Delhi. In February 1942 the CSIR, with a staff of fifty, moved into new buildings originally built for scientists at the University of Delhi.

Saha Starts an Institute

It was with a long history of political involvement that Saha began the reorganization of the Palit Laboratory at the University of Calcutta. At Allahabad the grant from the Royal Society in 1931 had encouraged him to continue the experimental work on thermal ionization. But of most of this earlier work, Saha's authorized biography has only this to say: "The work of the Allahabad Ionospheric School, attended with good luck at the very beginning (mostly due to the work of G. R. Toshniwal), fell to pieces after Saha's departure."[34] Concerned about the coming war, and keen to build physics in India, C. V. Raman proposed that his nephew S. Chandrasekhar leave the University of Chicago to take the chair in physics now vacated at Allahabad by Saha's departure. But Chandrasekhar would not think of leaving Chicago then.[35] Saha returned expectantly to Calcutta, the cultural center for Bengalis, the center even for those who, like Saha, originated from the isolated villages of East Bengal and were looked upon with a condescending urban elitism.

In 1938, Saha began a complete reorganization of the syllabus of the graduate physics course he had helped to write twenty years earlier when he was a lecturer. Raman's apparatus remained available to him in the nearby Palit Laboratory. D. M. Bose had occupied the Palit Chair for four years (1934–38) but had not reorganized or expanded the facilities, and now he had moved on. Saha made certain that someone would use Raman's spectrographic equipment and made plans for developing modern nuclear physics. Saha says he dropped his interest in ionospheric experiments at Calcutta because S. K. Mitra, with whom he had taught twenty years before, had already built a teaching laboratory in the Science College: "It would have been ridiculous to set up two rival laboratories for identical purposes in the same compound."[36] But there was to be rivalry, as we shall see. Saha shifted to a study of the new particles of the nucleus now at the center of physics, although according to a recent conversation, "he just does not seem to have understood quantum theory; it seems to have largely passed him by in the 1940s."[37] Saha set up an automatic Wilson cloud chamber in 1939 to measure the lifetime of the Mu-meson and another counting station in his house in Darjeeling in 1942, indicating his commitment to the popular field of cosmic ray studies. N. N. Dasgupta returned from London at the beginning of the war, with interests in biophysics and electron microscopy; Saha appointed him and decided to pursue biophysics, though it was not until 1948 that the first electron microscope in Calcutta was ready for Dasgupta to use.[38]

But Saha's keenest interest was to do research with particle accelerators, having seen Lawrence's Radiation Laboratory at Berkeley in 1937. His student from Allahabad, Basanti D. Nagchaudhuri, whom his American colleagues referred to as "Nag," was sent to Lawrence for a doctorate in 1938. "Alfred Loomis (at the lab for several months) was helpful in many ways; through Donald Cooksey he was able to help Basanti Nag get steel and copper for a Calcutta cyclotron. . . . The dedication of these lab men to their work and the responsibility which Ernest [Lawrence] felt for them impressed Nag."[39] The discovery of nuclear fission by Otto Hahn and Lise Meitner in 1939 moved Saha to include nuclear physics in the graduate syllabus. Saha had to raise money to pay for these projects, and the university had no funds for this at the time. He discovered the difficulties of raising money when he sought funds for a memorial on the eighty-first birthday of his teacher P. C. Ray in 1939.[40] Though he did get Rs 40,000, the money was mostly subscribed by Ray's former pupils: according to Saha, the businessmen who often utilized P. C. Ray's name for their personal profit showed little or no enthusiasm.

His own retrospective interpretation of the growth of nuclear physics was quite astute:

> Saha knew perfectly well that nuclear physics was very expensive, but at the same time he knew that asking the University and private bodies for very large grants would give them a shock and they would be scared away. He followed a middle path, but this later on led to great troubles, for the small amount of money with which he started on the venture was ridiculously inadequate, but people wanted quick results, and were disappointed when they were not forthcoming, and began to entertain serious doubts about his abilities. These were used against him in high quarters.[41]

The first funds Saha received between 1941 and 1944 for the cyclotron totaled Rs 210,000.[42] The university's senate was reluctant to accept the initial Tata grant of Rs 60,000, which Saha considered a seed grant in order to attract others. In 1941 Saha clearly had opponents to his plan to change physics in the university. Nehru himself had to intervene through Congress Party leader B. C. Roy and Asutosh Mookerjee's son Shyama Prasad Mookerjee, former university vice-chancellor in the 1930s, in order to stop the senate from delaying. Nehru continued his interest in the lab from prison, where Saha wrote to him, notwithstanding the criticism that had flowed between them,

> Dear Panditji, the cyclotron work is getting on rather slowly. We sent about 16,000 dollars to USA to my pupil B. D. Nag, who was working under Prof. Lawrence—the inventor of the machine. He placed all orders for materials and has returned and taken up the work in earnest, but owing to shipping difficulties we have received only 6000 dollars worth of goods. . . . I am so glad to learn of your interest in the work, which was rendered possible by your generous intervention.[43]

Without doubt Saha relied on the rivalry between the giant industrial houses of Birla and Tata, and their grants were structured to recur for five years.

In 1941 B. D. Nagchaudhuri, at age twenty-four, returned to India after three years' doctoral research at the University of California; the 50-ton magnet and other parts of the cyclotron followed him early in 1942 (at the beginning of the Quit-India movement). But the crucial high-vacuum pumps were lost when the freighter was sunk by a Japanese torpedo. This was the year that Japanese planes began to bomb Chittagong and Calcutta. Although the CSIR gave a special grant to Saha and although a system reaching a vacuum of 10^{-5} mms of mercury was constructed, Saha later admitted, "It was beyond the capacity of the university workshop to produce larger pumps. The work came to a standstill by 1944."[44] Nagchaudhuri later agreed that the capability of the lab was inadequate and failures in the vacuum system of the cyclotron were frustrating.[45] In fact the difficulties were caused by a wartime scarcity of expertise and by physicists trying to solve engineering problems in a context where no one really knew the answers.

The progress of the war did not really end the planning that had begun in the late 1930s. On the contrary, the war proved that planning was invaluable, though it also proved how often planning could go wrong in practice. By late 1944, when the war's end looked closer, planning had become an official presumption and state funds were flowing to the applications of science, to build the necessary infrastructure. By then Indian scientists were beginning their tour of important strategic-military laboratories abroad (see chap. 6). This tour clarified the potential of the nucleus for them, and by 1944 Homi Bhabha was proposing his own institute of nuclear physics and thinking of its role in atomic energy for national development. But between 1938 and the arrest of key leaders in August 1942 there was in the planning theatre a very public interplay between scientists, technologists, and political leaders of all stripes, even poets, confident that the tide was on their side, whichever side that was.

FIVE

Homi Bhabha Confronts Science in India, 1939–44

Homi Bhabha was born in Bombay in 1909, one of two sons in a very well-established Parsi family. Homi's grandfather, Dr. Hormusji Bhabha, had been the inspector general of education in the progressive feudal state of Mysore at the turn of the century. Homi's father, when a young lawyer, had married a granddaughter of Sir Dinshaw Petit, founder of the famous library in Bombay of the same name. The boy Homi, legally named Hormusji like his grandfather, grew up in an aristocratic world of books, music, and painting—all supported by successful international business and industry, well embedded in the thriving Parsi community.[1] In 1916 Homi began to attend Cathedral School in Bombay. It was then, as in the 1960s when he was expanding his nearby institute, attended by European children or by "westernized" Indian children and was operated on English lines with a number of foreign staff. Each day he went across the road from school to eat lunch at the home of his paternal aunt Meherbai, who lived in the ancestral house of the Tata dynasty. Here Homi saw and heard the nationalist politicians of the day, including Mohandas Gandhi, who were houseguests. He saw the relationship between politicians and merchant-princes of Bombay familiar with grand ideas and their implementation, investing very large sums of money in the risk of new industrial enterprise, and making profits.[2] Homi went to Cathedral School with children of a similar background; some of the faculty at the institute he founded in south Bombay were sending their children to the same school fifty years later.

Bhabha's maternal grandfather Dinshaw Petit had a fine library, to which his father added books on art during his days as a student at Oxford and London. He took painting lessons while a boy and continued to paint all his life. His mature style developed into a dark-colored melancholy around European symbolism, without much reference to Indian themes.[3] He frequented galleries and exhibits, taking advice on paintings quite freely, and took seriously the task of making beautiful the places he worked in, including the rather prosaic environment of a nuclear reactor. His father and maternal aunt both had classical record collections (Beethoven, Mozart, Verdi, Wagner), and his lifelong appreciation of symphonic and operatic music was firmly founded by age sixteen. He was a frequent attendant later at concerts, whether he was in Vienna, Boston, or Bombay. Walls of the TIFR building in Bombay became a gallery of very good paintings by living Indian artists, most of them chosen by Bhabha.

That institute, and its national role in building science and the nuclear program, began as an idea in Bhabha's head in 1942–44, when he was a young physics teacher in Bangalore, a place he found himself in by accident during the war. Though he started with clear advantages in social terms, it can hardly be said that the probabilities for success were on his side. He combined a reputation in physics with his social advantages, but the political economy was hardly ready for this plan, and the scope of Indian leaders to make its decisions were years away. Even in 1947, the numerous other priorities that awaited decisions, sometimes for a long time, made nuclear development a long shot. So how could this improbable thing happen at all? How did Bhabha's unconventional biography change the odds in favor of a national institute for fundamental research in physics and mathematics?

At age sixteen, Bhabha went to prepare for his Senior Cambridge examination at the Elphinstone College and the nearby Royal Institute of Science, whose science instructors did a little research along with teaching. William Penney said that his elders' wish for Bhabha's studies was not unconditional: "The intention of his father (who was a barrister-at-law) and of his uncle, Sir Dorab Tata, was that he should obtain an engineering degree with a view to joining the Tata Iron and Steel Company at Jamshedpur."[4] Homi's uncle Sir Dorab Tata was an alumnus of Gonville and Caius College at Cambridge in the 1890s and had given a munificent gift to establish Cambridge University's Department of Engineering. Bhabha therefore began his undergraduate studies in 1927 at this college in Cambridge, at age eighteen; his father had been a student at Oxford before doing a law degree in London. Bhabha disliked his mechanical sciences tripos subjects and wanted to study more theoretical fields, particularly mathematics, in which he was tutored by Paul Dirac, who later advised him on beginning a school of mathematics in the institute at Bombay in 1947. Homi stuck to his tripos on a promise from his father that if he achieved a first-class mark, he could study what he pleased.[5] He designed a cover for the college magazine and also continued painting, rowing, distance running, and tennis; he later had tennis courts made for his institute. Said one acquaintance, "Bhabha once told me in an unguarded moment that in Cambridge he had a choice of becoming a communist, or an artist, or a scientist."[6]

Having secured first-class marks in engineering in 1930, Bhabha began learning theoretical physics just at the time when Cockcroft, Walton, Blackett, Occhialini, and Chadwick were doing important work on the structure of the nucleus in the Cavendish Laboratory. Bhabha published his first physics paper in German in October 1933 in *Zeitschrift für Physik* at age twenty-four. The following year he was elected to the Isaac Newton studentship that enabled him to remain at Cambridge for the next three years, complete his PhD under the supervision of R. H. Fowler, and travel in Europe. During this time, he visited the groups of Pauli in Zurich, Kramers in Utrecht, and Fermi in Rome, then centers for both theorists and experimenters.[7] He also worked in the extremely active institute at Copenhagen that housed Niels Bohr's group; a photo taken there in 1936 records the first time that Bhabha met Saha.[8]

Most visible in the photo are Niels Bohr, James Franck, Wolfgang Pauli, Werner Heisenberg, Max Born, and Lise Meitner. Meghnad Saha sits in the second row talking to Marcus Oliphant, who was then assistant director of the Cavendish Laboratory, someone with whom Bhabha eventually became very friendly. Oliphant would meet Bhabha and Saha again, ominously or luckily for Saha, in 1955. Saha was forty-three when the photo was taken, with an established reputation in science and an influence in academic politics in India. Homi Bhabha, twenty-seven, a fresh postdoctoral fellow with nine years experience in Europe, was sitting back in the fifth row, still very much a junior figure. But Bhabha published very actively during this early period, including an influential collaboration with Werner Heitler, on the theory of electron cascade showers, work that established his international reputation. Werner Heitler said later that by 1934 Bhabha's work had secured him a permanent reputation in theoretical physics.[9]

Bhabha was a confident person, quite prepared to challenge more senior people who doubted his work, as in 1937 at a weekend conference in Manchester, where Patrick Blackett, then thirty-nine, insisted that the quantum theory of radiation must fail at higher energies because there could be no particles heavier than electrons in the penetrating component of cosmic rays at sea level. Bhabha, only twenty-eight, in the company of Heisenberg at the Manchester meeting, patiently persisted with Blackett in saying that there is a penetrating component that is a particle heavier than an electron. According to an observer, Blackett was challenged and was stubbornly reluctant to concede he might be wrong, but in a few months he agreed that an energetic electron could produce a cascade shower according to the Bhabha-Heitler theory and that the penetrating cosmic rays must therefore consist of a new type of particle with a mass intermediate between the electron and proton.[10] In 1938 Bhabha won the prestigious Adams Prize in physics for this work. The judgment of Heisenberg would have been influential in Bhabha's career, and, had the war not occurred, one can imagine a continuing relationship, perhaps focused through Copenhagen. But revelations of Heisenberg's role in the German nuclear project and his ambiguous meeting with Bohr in Copenhagen in 1941 meant a shadow fell over him among physicists outside Germany after the war. We know little about any continued relationship between Bhabha and Heisenberg when Bhabha became chairman of the Atomic Energy Commission in 1948.

Shortly afterward Werner Heitler, now a German refugee in Britain, and Bhabha were both considered for a position in physics at Liverpool University, where a 37-inch cyclotron had just been built at Liverpool. How ironic that these two rivals around the issue of scattering of the cosmic rays should be applying for the same position. They were interviewed serially in 1939 by James Chadwick, who had first sought them out for this position and then declined both candidates reluctantly. Of Heitler, he said, his being German was going to confuse the emotions of the students during the war. Bhabha, Chadwick thought, was "too good for Liverpool" students and much of the teaching "would be drudgery to a man like Bhabha, who was a most exceptional man. He was a painter and poet and had extremely wide interests—not merely interests but far more than that—and I didn't feel that however much I liked him it was fair."[11] In the end Maurice Pryce, a Cambridge student of Fowler and Dirac, was appointed at Liverpool. Chadwick, who became the most senior British physicist in the Manhattan Project, knew Bhabha in Cambridge in about 1931–32. In fact Bhabha did not get suitable employment in 1938–39 in Britain, though he was clearly seeking it.

When Bhabha was on holiday in India in 1939, war broke out in Europe. So he remained in India after spending twelve years abroad in the most active centers of the hottest field in physics. M. G. K. Menon, whom Bhabha had named as his successor at TIFR, stated that "the role of the environment in which he lived and worked cannot be minimized; in particular there was experimental work of the highest order around. This led to an extrovert type of thinking with theoretical work closely linked to experimental observations."[12] Stuck in Bangalore, Bhabha faced the slow building of the physics department, saw the troubles surrounding Raman, experienced the distance from his friends and colleagues abroad, and felt the censorship and delays in scientific communication out of India. Yet it was in India that he wrote the essay that won him the Adams Prize at Cambridge and where he did the work now known as the Bhabha equation.

Bhabha and Physics at Bangalore

Caught in India by surprise, Bhabha accepted a post created for him as reader in theoretical physics at the Indian Institute of Science, Bangalore; the institute's governors were strongly allied to the Tata Group of companies. Jnan Ghosh had become the director of the institute in 1938, but C. V. Raman was still in the Department of Physics. The American physicist Robert Millikan, who earlier played a role in negating Saha's chance of a Rockefeller grant, along with his colleagues Neher and Pickering, conducted a series of rubber balloon flights (3 balloons per flight) at Bangalore, Agra, and Peshawar to study the latitude variation of primary cosmic rays near the geomagnetic equator, which passes through South India.[13] This was reminiscent of Compton's experiment in the Kashmir's Dal Lake fifteen years earlier. Closer to the equator there are fewer background low-energy particles to confuse the photographic plates carried by the balloons. The Indian Meteorological Department was responsible for hydrogen supplies and laboratory facilities for the Millikan project. Flights were launched in 1940 from the observatory in the heart of Bangalore city. A photo hanging in the IISc Department of Physics shows Millikan of Caltech with Raman, Bhabha, Vikram Sarabhai, and K. R. Ramanathan. Raman felt much encouraged by the presence of these active younger physicists; Bhabha was thirty-one and Sarabhai was twenty-one.[14] Bangalore was now firmly in the middle of a disagreement among top physicists about the global distribution ("scattering") of cosmic rays. Here was Sarabhai, who would himself soon go up to Kashmir for his dissertation experiments, flying balloons with Millikan—Raman's old ally from Caltech—for cosmic ray studies. Bhatnagar and Nazir Ahmed had arranged the penetrating radiation experiments of Arthur Compton in Kashmir in 1926–27; subsequently the disagreement between Millikan and Compton about the latitude effect in penetrating radiation clouded the atmosphere when the 1936 Nobel Prize in physics was awarded for cosmic ray research.[15] Indian physicists at this time saw themselves to be contributing to the understanding of latitudes and the dispersion of background radiation. Another physicist working on high-altitude cosmic ray research was Piara Singh Gill, just arrived in Lahore and teaching at Bhatnagar's old college in 1940 after completing his doctorate with Arthur Compton at the

University of Chicago. Gill, like the better-known Jacob Clay of Bandung, Java, now Indonesia, had also conducted latitude studies on board ship in 1937–38, nine years after Clay's original work.[16]

Raman nominated Bhabha for an FRS in 1940, writing to Paul Dirac that "Bhabha is a physicist of exceptional ability whose work is of the highest quality."[17] Other signatories in India to the nomination of Homi Bhabha were paleontologist Birbal Sahni and physicist K. S. Krishnan, who had just been awarded a fellowship. Raman did not ask Saha, who might otherwise have been a logical choice. Bhabha began trying to secure funds from such philanthropic bodies as Tata Trusts to begin some serious research on nuclear physics. But he found that neither government in Mysore or Delhi wanted any large expenditures during wartime, and private funds were not forthcoming to him unless governments also contributed.[18] Not yet. He had already put aside the initial impulse that he should return to work in Britain during the war: "I have not made up my mind definitely," he wrote in November 1939, "and if I can be of any use will try to come over to England for the Lent Term," spring in Cambridge.[19]

But he was, as shown in correspondence in 1940 and 1941, seriously planning to return abroad when the war ended. He was not particularly involved in the political circles of Bhatnagar and Saha at this time; after all in 1941 he was only thirty-two years old. In a letter to Patrick Blackett he said:

> One becomes extremely isolated in India and this [is] one of the reasons why I want to build up a school. But I must confess that when such great issues are at stake as in this war, pure research seems of secondary importance, and one wishes one could take more share in helping on the cause in which one believes. I would certainly do whatever I could if I were in England. But in India this is unfortunately not possible as the attitude of the Government is as die-hard as ever. The mis-rule would astonish you. I look forward eagerly to being able to return to England.[20]

Despite his isolation he knew that Langevin and Bohr and Joliot-Curie had all decided not to leave their positions in Europe, and this troubled Bhabha. Like many others, he worried for their safety.[21]

After two years as reader at the IISc, Bhabha was promoted to professor of cosmic ray research in 1941, the year that he was elected Fellow of the Royal Society. The following year, possibly as a consequence of election to the Royal Society, Bhabha was offered a chair in physics at the University of Allahabad, "with especially favourable conditions." This offer was rejected, as was the offer, probably arranged by Saha, of a chair in physics at the IASC in Calcutta, also in 1942. Neither position offered him sufficient opportunity to build a school of physics as he now intended.[22] He had by this time accepted remaining in Bangalore until the end of the war. He published two papers in the *Proceedings of the Royal Society*, which, in his opinion, show that the only possible cause "of the anomalously high scattering of charged mesons can be removed if it is postulated that the heavy particles can exist in states of all integral charge, positive and negative, so that negative protons, double protons, etc. may exist. (This idea has also been adopted by Heitler but he first got it explicitly from me, and I feel he has not properly acknowledged the fact.) I have also worked out the probability for the creation of these new particles."[23] To Dirac he wrote in frustration about getting more acknowledgment, "I communicated the matter to Heitler as far back as the summer of 1939. . . . I would like the matter to be settled experimentally."[24] He scorned "the nonsense that is almost universally talked about re: the breakdown of quantum mechanics and Heisenberg explosions."[25] He told Blackett that he was completing new work, "the first time a complete solution of the problem with ionization loss has been given and shows that the previous attempts by Arley and others are not even quantitatively right."[26] Though his own work at the time was mainly theoretical, he began to form a small research group that eventually, in 1944, carried out cosmic ray measurements from US Air Force planes stationed at Bangalore for military purposes. In 1941, however, he obtained £1,200 to launch balloons to "investigate the formation of mesotrons and heavy particles in the high atmosphere à la Millikan." It is evident that he also enjoyed the practicality of this research and enjoyed its members too, like the young Ahmedabad-born student of physics Vikram Sarabhai, the very person who succeeded Bhabha as chair of the Atomic Energy Commission.

Bhabha Looks Far Ahead

Despite his recognition as an FRS and promotion to professor, there must have been frustrations in starting this experimental work and a longing for the company of other advanced theorists. In a letter in March 1944 to the chairman of the Tata Trusts, he refers to 1941–42 as a period when he had thought of returning to a place like Cambridge or Princeton when the war was over.[27] In early 1943, Bhabha was the president of the Physics Section of the Indian Science Congress at Calcutta, where he could see in a few days a cross section of all research conducted in India. "Bhabha stayed on [in Calcutta] to give, on the invitation of M. N. Saha, a course on collision processes in gaseous media."[28] Saha was instrumental in bringing about Bhabha's two offers of teaching positions at Allahabad and Calcutta, both of which Bhabha declined. Saha had not failed to see the potential in Bhabha's knowledge and connections and sought to bring him closer into his orbit.

Positive signals for the planning and support of research institutions were communicated from the viceroy's office even before Bhabha received his FRS in January 1944 in Delhi. These signals coincided with Bhabha's realization that he ought to try to do something creative in India and not leave when the war was over. Bhabha was now in the elite information network through which he would have known all about such high-level signals. He made his first approach for independent funds and used the example of the USSR during the war: in a letter to J. R. D. Tata, head of the Tata Group of firms and chairman of the Tata Trusts, Bhabha pointed out that, despite the orientation of science to rapid economic development in the Soviet Union, fundamental research was supported there. He expressed his view, which he also described with approval as the official Soviet view, that "there is no genuine knowledge of the universe that is not potentially useful for man, not merely in the sense that action may one day be taken on it but also in fact that every new knowledge affects the way in which we hold all the rest of our stock."[29]

Bhabha argued in the same letter that support for fundamental research should be "at a pace which the talent of the country would warrant." Bhabha had extensive discussions about physics in India with Archibald Hill at a ceremony in New Delhi (see chap. 6). On his return to London Hill advised him to look beyond physics: "All I want to urge is that India's fundamental needs are in the biological field and must be so regarded if a true picture is to be formed."[30] Hill wrote pointedly to Bhabha in June 1944, urging him to ensure biophysics be included in the new institute that Bhabha was planning.[31] But Bhabha was preoccupied with physics at this stage, and it was twenty years later that he started molecular biology at his institute. Saha, on the other hand, had already started biophysics in a small way in Calcutta.

Following these discussions, and encouraged by the favorable response from J. R. D. Tata that "the advancement of science is one of the fundamental objects with which most of the Tata Trusts were founded," Bhabha made a formal proposal to establish a research institute, reasoning:

> It is absolutely in the interests of India to have a vigorous school for research in fundamental physics, for such a school forms the spearhead of research, not only in the less advanced branches of physics but also in the problems of

> immediate practical application to industry. If much of the applied research done in India today is disappointing and of very inferior quality, it is due to the absence of a sufficient number of outstanding pure research workers who could set the standards of good research. Moreover, when nuclear energy has been successfully applied for power production, in say a couple of decades from now, India will not have to look abroad for its experts but will find them ready at home.[32]

This 1944 letter showed that he had now changed his mind about going abroad after the war: "In the last two years I have come more and more to the view that provided proper appreciation and financial support are forthcoming, it is one's duty to stay in one's own country and build up schools comparable with those that other countries are fortunate in possessing." He stressed two principles that he attempted to follow in his institution and built into the proposal; first, that research groups be built only around exceptional scientists, and, second, that government support need not entail government control.

Though Bhabha was carrying out cosmic ray measurements in US Air Force planes flying at 35,000 feet, he continued his own theoretical work, and a friend of Bhabha's said later: "It is very noticeable that the papers he published after his return to India were very much more mathematical and deeply introspective in character. There was a powerful mathematical streak in Bhabha's mental makeup, and he was deeply conscious of the aesthetic beauty inherent in exact mathematical solutions."[33] Bhabha had published joint papers in Bangalore with Harish Chandra (1944 and 1946), who later became a mathematician at Princeton. Bhabha may have anticipated then that the emphasis at the IISc would, for some time, be on the engineering departments, doing mostly applied research, with very little in pure research.[34] While Bhabha was still in Bangalore, the Tata Trusts agreed in June 1945 to establish an institute of fundamental research. The trusts approached another wealthy Bombay family to co-finance the institute with them (possibly the Godrej family), though the offer was considered and eventually declined. The credit thus fell entirely to the Tata family, their companies, and their trusts.[35] Bhabha started work on the institute right away and began to plan the meetings of the Atomic Energy Committee, setting the course for India's nuclear development beginning in 1946. But that was far ahead of the rest of India, still struggling with the war and the Independence movement. In fact the war provided the reason to show major Allied research facilities to a select team of Indian scientists, facilities from which they derived ideas for future organizations and research. Not being senior enough, and not much in Bhatnagar's orbit, Bhabha was not on that team, but he was surprisingly well tuned to the thinking that emerged in 1945–46. In a country where age counted for much, here he was, thirty-five years old, planning to guide the government's Atomic Energy Committee and its program of investments and development.

SIX

Indian Scientists Engage the Empire: The CSIR and the Idea of Atomic and Industrial Power

The Japanese military rolled swiftly through half of Burma by February 1942 and by April had cut off the Burma Road at Lashio, still moving northwest toward India. Hundreds of thousands of refugees walked to India and China for safety, and thousands died on the way. India was transformed from a sideshow of the Second World War to the front line of the American war with the Japanese. American entry to the war in December 1941 crystallized an abstract American interest in Indian politics. Indian political leaders were divided over the question of helping the British and their allies fight fascism in Europe and Asia, versus cooperating as little as possible in order to accelerate the demise of the Raj, or at least loosen its grip. The nearness of Japan's army in Burma focused this division right down into the population. In this context, the mission of Sir Stafford Cripps to India in March and April 1942 was intended to counteract these conditions and to offer Indian leaders the prospect of "self-government," though a very limited kind, until the war ended. In the optimistic view, India would become a dominion like Canada when the war ended. In the pessimistic view, reasons would be found to delay that status and continue a renamed colonial dependency. For complex reasons Cripp's mission did not meet with success, largely because of the question of what role the Congress leadership in government would play while the war continued and with whom Congress would have to share power.[1] The British expectation of full support for the war effort was unacceptable to Congress without a commitment to genuine independence, but Churchill, who wanted full Indian support for the war effort and opposed independence in a visceral way, appears to have sabotaged the Cripps mission too.

Cripps was made a member of the India Committee of the cabinet before this mission started, but Churchill preferred that Cripps not negotiate and just consult. During March and April 1942, Cripps was squeezed between two unready groups, Congress and the British cabinet. He made contact with leaders at the highest level, contacts that became essential in 1946 when, with greater cabinet support, Cripps negotiated the beginning of the end. But 1942 was different, and to signal that this phase of negotiation was over, Congress started the Quit India movement (not fully under Gandhi's direction) in August 1942, bringing noncooperation into every aspect of colonial administrative life. Following this announcement most of the leaders like Nehru were picked up suddenly and imprisoned, and others went underground. Advice given to scientists in late 1942, including students of science, was that they should continue to work and study in anticipation of Independence and keep their heads down; they would be needed as professional scientists later. While everyone had a wary eye on the possibility of a Japanese invasion of eastern India, propaganda flowed in every direction, particularly on the front line in Bengal.[2]

Bengal was faced with an increasingly serious famine in early 1943, resulting in the migration of hundreds of thousands of hungry people into Calcutta looking for relief and perhaps a place to die. Moreover, the US government had made its own assessment of the situation in India, now that they were involved in a war in Asia: the results of the US Technical Mission, called the Grady Mission, were widely discussed and usually with Indian approval, in places like Saha's journal *Science and Culture.* American talk of the right of self-determination was approved, because it was an implicit criticism of the British policy of "hanging on." Who knew what the relationship between India and Britain would be after the war? Should India not develop friends in other places, like America? In 1943 there was still no certainty about the outcome of the war, so the situation could really be described as a continuing crisis for the British in India. Indians were appalled by the inept management of the Bengal famine, and Indian and British publics were sobered by the movement of Japanese military forces to the very borders in Assam in May 1944. Thousands of British, Indian, Chinese, Canadian, and American troops were stationed in bases around Bengal, while the USAF and RAF launched strikes into Burma from nearby airfields, often flown by Canadian pilots.

The British carried out constant surveillance of all US activities in India because they were very suspicious of their intentions after the war. Americans were warned to be careful, particularly in their relations with British officials: in 1944 an American consul-general in Bombay described India as "a police state."[3] In this situation Indians perceived a unique opportunity, and industrialists had a boom in war production, although their Congress leaders were behind bars. Only the Communists were exempt from the ban on political parties since they did not belong to the Quit India movement and were defined as friends because of the USSR's new alliance with Britain and America. This strange hiatus turned into a planning season for scientists and their new institutions.

The CSIR Comes Out of the Shadows

In this new strategic context it became important in official circles to open and improve relations with unofficial and apolitical Indian groups, particularly among those believed to be disappointed, disenchanted, or disaffected, in order to inspire and mobilize popular opinion in favor of the government of India, to counteract the criticism by political leaders. Thus someone "unofficial" was to be sent to talk with scientists and industrialists, remind them of British reliance on their contribution, inspire them to do more, encourage them to build new institutions suitable to India, and assure them of British commitment to their interests when the war ended. The mood in London and Delhi was leading to bolder thinking in official circles in 1943; there was now talk of helping with industrial development, involving British firms that would work cooperatively in India with Indian partners, the very approach that Bhatnagar advocated.

It is unclear whether anyone else in Britain was considered for this delicate negotiating visit before Archibald Hill accepted it. Having had no previous contact with India, Hill was clear of previous alliances; the Cripps Mission was seen in London as having been complicated by Cripps's previous alliances with Nehru, but, ironically, previous alliances had made it possible. Hill had already spoken as a member of parliament in favor of more effective relations with Indian scientists in February 1942. Hill was deeply implicated in the war and widely trusted with official business, having been scientific attaché at the British embassy in Washington in 1940, chair of the Executive Committee of the National Physical Laboratory, and member of the war cabinet scientific advisory committee. And Hill consulted with other experts, like Sir Stanley Read, the still-influential retired editor of the *Times of India*. Hill's coming was signaled well in advance in India so that plans for meetings could be made; the ostensible reason for his coming was to confer Fellowships in the Royal Society upon Indian scientists who were unable to travel to London to receive them. The early signal of his arrival was accompanied by the Delhi government's move to increase the funds available to the new CSIR.

Archibald Hill wrote modestly in his autobiography that his inclination to do this job arose thirty years earlier, when he was a young tutor and lab demonstrator at Cambridge, where an elderly acquaintance, Mrs. Poole, who had lived most of her life in India, introduced a new student-boarder to him. The student was S. L. Bhatia, and the two men became friends: "I have always been fond of soldiers," wrote Hill, "and he looked like a soldier."[4] In 1914, Hill and Bhatia both went off to war but remained in touch until Hill came to India in 1943. Hill made a reputation during World War I for establishing a creative group called Hill's Brigands—mathematicians, physicists, and the like—to study new weapons and their effects. When he won the Nobel Prize in physiology in 1922, he gained very wide recognition indeed. When Hill and Bhatia met again in 1943, Bhatia was deputy director of the Indian Medical Service in Delhi, the most senior Indian physician in government service. "If I had not known Mrs. Poole and Bhatia in Cambridge in 1911, it seems unlikely that I would have been concerned in 1943 about the lack of liaison between British and Indian scientists. Nobody else seemed to be inclined to do anything about it."[5]

Normally viceroys paid not the slightest attention to science and technology, but news of this coming opportunity forced some of Viceroy Wavell's attention onto science and industry; thus the government of India decided in late 1942 to enhance funding and powers of the new Council of Scientific and Industrial Research. Though dominated by officials, four unofficial scientists were later appointed to the council at Bhatnagar's insistence. More important, there was official acceptance of the detailed proposal written by Ramaswamy Mudaliar, the viceroy's council member for Commerce, and Bhatnagar that the CSIR should not be a government department but rather registered under the Societies Act and operated from a separate fund established by the government. This autonomy was to protect it from excessive interference and allow it to frame its own administrative rules. But starting with a fund of Rs 1 million annually, it could actually do very little research, as most of these funds were tied up in salaries and maintenance of buildings and equipment. Mudaliar and Bhatnagar said the money was insufficient to create the new kind of industries in which people with money would invest. Council member Ardeshir Dalal, on the other hand, thought that industrialists would not invest now anyway because they did not believe they would be protected from foreign competition when the war ended. And so he argued first for a change in economic policy now, to establish the right climate in which the CSIR could operate. This disagreement brought the issues of scientific research, industrial development, finance and investment, and trade and economic policy—and the contradictions between them—sharply into focus at the time that Archibald Hill arrived in late 1943. A colonial economy in military crisis might not appear to be an environment in which to cultivate something new, but Hill appears to have seen otherwise. And so too did some Indian scientists and industrialists.

Not waiting for Hill's arrival, but in its anticipation, the CSIR made plans for the National Physical Laboratory and the National Chemical Laboratory in early 1943. Although securing the land and sufficient money to build these institutions took another three years, it is remarkable that in February and March 1943, at a time when the Allies were not succeeding in Burma against the Japanese, the CSIR met in Delhi, Hyderabad, and Bombay to discuss the plans for a National Physical Laboratory, including its size, function, location, and director. A planning committee was formed, which included K. S. Krishnan, the person who would be chosen its first director, and Homi Bhabha. There was a series of meetings in 1943–46, including one in which Meghnad Saha said he thought he should be chairman of the Planning Committee of this laboratory, but Shanti Bhatnagar decided the planning should be done by the president of the CSIR council Sir Ghulam Mohammed, later governor-general of Pakistan. Bhatnagar, supported by a majority of the CSIR council, thought the lab should be built in Delhi, but Saha said it should use the existing buildings in Calcutta left by Bhatnagar when his lab moved to Delhi with additional new buildings across the road.[6]

To try to keep control of the process, Bhatnagar appointed his old student K. N. Mathur as assistant director of planning for the new lab. Without securing his role as planner, Meghnad Saha soon resigned from the committee, and the other member from Calcutta, D. M. Bose, "did not attend a single meeting."[7] Finally, the job description of the desired director was written by Mathur, supposedly but in fact by Bhatnagar, to best fit K. S. Krishnan, a student and colleague of Raman widely believed to have been the major yet largely unacknowledged contributor to the 1928 discovery of Raman spectra. Nazir Ahmed opposed this preemption of the National Physical Laboratory director's post by protesting to CSIR president Sir Ghulam Mohammed. Perhaps Ahmed feared procedural wrangling or had someone else in mind; he himself, though trained at the Cavendish Lab in Cambridge, had not been a practicing physicist for more than ten years. Ghulam Mohammed finally directed that Mathur rewrite the director's job description under Nazir Ahmed's dictation, seeming to override Bhatnagar. In the end, Krishnan was appointed director in 1947 anyway.[8]

Hill's visit provided a deadline to mark out some stages of progress and recognize some achievements. For example, just before Hill's arrival in

1943, Tata Trusts announced a gift of Rs 830,000 to the CSIR's new National Chemical Laboratory in Poona. This was a very large sum of money for scientific research in India. Evidently at that stage the Tatas, who were making lots of money, could see their interests very clearly in chemicals, although they were also funding Saha's cyclotron in a physics laboratory. This grant to the chemical laboratory thrilled Bhatnagar, as this lab embodied his love of chemistry. Three other CSIR labs were planned near Calcutta, one for metallurgy at Jamshedpur, the second for fuel near the coalfields at Dhanbad, and the third for glass and ceramics in Calcutta. To some extent these three labs were to soften the sting of the location of the National Physical Laboratory and the National Chemical Laboratory away from Calcutta. No CSIR lab was planned for Bombay, though the National Chemical Laboratory at Poona was only a few hours away by train. Seeing all of this, that Indians were starting to organize research institutions for their industrial future, impressed Hill.

Hill's midwar journey was by flying boat around and then across Africa; it took nine days in November 1943 to reach Delhi. Carrying a letter from Winston Churchill to each of the Fellows of the Royal Society, Hill gathered them all together and awarded the honor to two new Fellows, Bhabha and Bhatnagar, together in Delhi in January 1944, where they signed the customary acceptance of the society's charter.[9] There were many speeches at the ceremony by and about the new Fellows. In 1943 Bhatnagar had been elected both vice president of the Society of Chemical Industries in London (the first Indian to hold senior office in it), and Fellow of the Royal Society. Hill had already studied Bhatnagar's life from the files for election to the Royal Society: in that file was a supporting letter saying Bhatnagar is "an absolutely first class scientist—he doesn't care a damn about the government."[10] Bhatnagar's nomination for FRS followed a powerful chemists' network; it was initiated by Donnan, his teacher in London, then seconded by Sir James Irving, on whose committee Bhatnagar sat in 1938 when Raman was being reviewed and removed as director of the IISc. Krishnan had been elected in 1940, on Raman's nomination, and Bhabha had been elected a year later in 1941, also on Raman's nomination. Bhabha and Bhatnagar were awarded Fellowships from Hill in the presence of Viceroy Wavell, with other Fellows present. The pomp of the Raj was now available for scientists.

Hill met all the leading scientists and industrialists of India, touring the major cities, laboratories, and factories. For example, Hill visited Calcutta's Indian Association for the Cultivation of Science, where he was awarded the richly endowed Joy Kissen Mookerjee Gold Medal for Scientific Progress in

India. This January 1944 event at the IACS would surely have been arranged by Saha, who was very active in its Committee of Management. But to indicate the loyalty of the institution to the Raj, its president declaimed the names of those who had received fellowships and awards recently. As if to remind Hill of the integration of science in India with British life, the IACS (nationalist in character but poorly funded by governments) had already appointed three honorary fellows who were already FRS, Sir Henry Dale, Sir Robert Robinson, J. L. Simonsen; awarded medals to astronomers Sir E. J. Russell and F. W. Aston (who had voted for Saha as FRS); and awarded medals or fellowships to people who were not British—namely, Niels Bohr, Arnold Sommerfeld, and Robert Millikan (in 1939).[11] No one could doubt the reputations of the men honored by the IACS, including physiologist Hill. But these men were also figures in the British scientific establishment, able to make significant appointments in India by recommendation letters, able to vote or choose to abstain in the election of Fellows to the most prestigious Royal Society in London, able to intervene for students coming to London. So the IACS was, whether during the war or before, poised on the edge—searching for funding, indeed the very kind of funding the Tata Trusts had made to the National Chemical Laboratory, funding well matched by the government of India.

"Poised on the edge" means the IACS was also a symbol of nationalist self-reliance, bottom-up science the way Raman had done it on a bench in the association's laboratory since the end of WWI. Hill was reminded by the IACS, as if he might forget, that in this institution "a research worker carried on his research which won him the high honour of a Nobel Laureate, and it has turned out two Fellows of the Royal Society" (meaning Raman and Krishnan). The laboratory sat in Bowbazar, a middle-class neighborhood of Calcutta where there was widespread sympathy both for the Quit India movement and for Subhas Bose, who was then in Burma leading the Indian National Army against the Allies, though under Japanese command. None of this was lost on Hill, as his correspondence shows.

Hill also met the chiefs of staff of the Indian military to discuss new ideas on operational research then being promoted by his Cambridge colleague physicist Patrick Blackett. The commander in chief of forces in India wrote a few weeks after Hill's arrival, asking Hill to provide advice *directly to him* when possible. Hill is remembered for his impact on scientific institutions and personalities rather than on military thinking in India, although he may have been the first to lecture there on "operational research," a hot wartime subject. Nevertheless he declared widely on his return that Britain had to change its relation with India or might "lose India" altogether. Hill wrote that "some of the Indian scientists were no doubt disaffected, but the great majority certainly were not."[12]

Hill and Bhatnagar established a friendship that lasted until Shanti's early death in 1954; their correspondence was regular (sometimes funny, always revealing), and their meetings frequent. Hill was seven years his senior, but lived much longer than Bhatnagar. Hill traveled very widely in India, speaking to everyone, and even counseled Bhabha to include biophysics research in his own planned institute, arguing presciently that India's fundamental problems lay as much in biology as in physics. More important to Hill than these other contacts, he found his old friend S. L. Bhatia, and together they planned a great center of medical research, which ten years later became the gigantic All-India Institute of Medical Sciences in Delhi. This major institute is thus partly the consequence of the introduction of two young men made by a Cambridge landlady in 1911.

Building a Grander CSIR

Notwithstanding their current military predicament, the British were thinking about reestablishing a key role in international science after the war. One of the practical outcomes of Hill's mission was a tour of Great Britain, Canada, and the United States by a group of Indian scientists. This tour was intended to provide models for organizing research and to encourage them to build effective relationships with scientists there. At the same time a plan was circulated in London to establish a commonwealth scientific office that would bring Indian, Canadian, and Australian scientific attachés together, in London. The year 1944 was a big season for planning: the United Nations, the International Monetary Fund, and the World Bank all had their gestation that year. In India there were three national plans published: the Gandhian Plan of S. N. Agarwal, with a foreword by Gandhi; the People's Plan of communist M. N. Roy; and the more influential Bombay Plan of the industrial network led by Purshotamdas Thakurdas (a network that included Visvesvaraya). At the same time, and with these plans in mind, the government of India established the Planning and Development Department, to be led by Sir Ardeshir Dalal.

Between Hill's departure from Delhi in March 1944 and Bhatnagar's arrival in London in October 1944, they had regular and detailed correspondence, in which Hill's expressed purpose was "doing something for the people" and Bhatnagar's regard for Hill was "as part and parcel of my house. . . . My wife and children look upon you as if you are a near relative."[13] Their joint objective was to get official action on Hill's recommendations and plan the tour of Indian scientists. Bhatnagar was optimistic, assuring Hill that "the war will be almost over by the time we arrive," but Hill cautioned him phlegmatically, writing, "Several of these flying bombs fell within earshot while I was dictating this letter. It's no good taking any notice. Work has to go on. We had one . . . about 150 yards from my house. It broke all the windows on that side and took off most of the roof . . . but fortunately no one was hurt. The great thing is to avoid the flying glass."[14] The question of India's loyalty to Britain subtly pervaded their correspondence. Bhatnagar assured Hill, "I have no doubt that the British government and people will help us in realizing our dreams of a prosperous and contented India. . . . Most of us, trained to think in an international way, will be glad to be members of the Commonwealth as equal and friendly partners." As for Hill's visit and recommendations, Bhatnagar was despondent that nothing was being done about it yet, and "men of science are treated like labourers"; still he was very glad that Sir Ardeshir Dalal was appointed member of the Viceroy's Executive Council for Planning and Development and was also made a member of Bhatnagar's Executive Council of CSIR. Dalal was concurrently a director of Tata and Sons; two years later he was made president of the CSIR council. Bhatnagar's relationship with Dalal was crucial for scientists more than once.

Partly as a result of the appointment of a new viceroy in 1944, partly because of the new importance of India, and partly as a result of Hill's support for the Indian mission abroad, Bhatnagar dined with the viceroy and learned informally of things to come. He heard about the possible "scientific" uses of the taxes unpaid by industrialists involved in war production and about support for the new national laboratories. As further evidence of Indo-British scientific ties, Bhatnagar invited physicist John Bernal, who would shortly become known as founder of the field of biophysics, to participate in the CSIR planning committee meetings in December 1944. Bernal had been appointed in October as Mountbatten's scientific advisor, visited the beaches of the Japanese-held Arakan coast of Burma in order to study them for amphibious landings, and tested bombs in the Sri Lankan jungles in November.[15] But he showed up for National Physical Laboratory planning meetings in Delhi, along with Meghnad Saha.

Bhatnagar wanted to be promoted from director to a position of greater influence in the bureaucracy, saying to Hill, "I hope you continue to be of the same opinion as before that if they don't give me the position of at least an Additional Secretary I should resign." He got the position and did not need to resign. Bhatnagar's approach in 1944 was not much different from his approach to the directorship of the Chemical Laboratory in Lahore years before. Bhatnagar asked Hill to look out for a good British director of the new CSIR laboratory on fuels, and, on behalf of J. C. Ghosh, asked for three British heads of department in metallurgy, aeronautical engineering, and applied mechanics at the IISc in Bangalore. He even succeeded in pushing Hill to persuade his old teacher F. G. Donnan (then seventy) to visit India as soon after the war as practicable. Bhatnagar also urged Hill's full support for the election of "my friend Kothari," who had been Saha's student at Allahabad in the late 1920s and later trained in the Cavendish Laboratory, Cambridge, as Fellow of the Royal Society.

Though the Kothari nomination was unsuccessful in 1944, the Royal Society continued to send strong positive signals to scientists in India, electing astrophysicist S. Chandrasekhar in 1944 and physicist-statistician P. C. Mahalanobis in 1945. The announcement in Chandrasekhar's case occurred while he was a "British subject" living at Yerkes Observatory in Wisconsin and was being urged by physicists Hans Bethe, Edward Teller, Victor Weisskopf, and Robert Oppenheimer to join the Manhattan Project at Los Alamos. Despite these prestigious entreaties, Chandrasekhar ultimately did not join the bomb project.[16]

Hill, for his part, received from Bhatnagar news clippings of "misrepresentations" of Hill's recent visit with equanimity: "I am quite used to misrepresentation. Thank you for standing up for me." The interpretation in question was that all British efforts, like Hill's, "are a trick to avert Indian self-government." Given his position and recent experience, Hill became involved in all the discussions in London about scientific staffing of South East Asia Command under Mountbatten and applications of science to the war in Asia, and he was also proposing Indian appointments to a new commonwealth office planned for Washington. Clearly Britain was trying to reposition an empire science network, soon to be called a commonwealth network at a new center of power in Washington. He urged Bhatnagar to decide on the Indian attaché to be sent, although this appointment was not actually made until after 1947. It is clear from this correspondence that it was Bhatnagar who decided on the final composition of the Indian delegation for the tour of Allied nuclear facilities, in consultation with Hill.[17]

Indian Scientists on Tour

Finally in October 1944, the official team of physical chemist and team leader Sir S. S. Bhatnagar, physicist M. N. Saha, physical chemist Sir J. C. Ghosh, radio researcher S. K. Mitra, agricultural development specialist J. N.

Mukherjee, medical education authority S. L. Bhatia, and agrotechnologist Nazir Ahmed went for a five-month tour of research facilities in the United Kingdom, the United States, and Canada as guests of those governments. Bhatia was Hill's medical friend from Cambridge days and, like Bhatnagar and Nazir Ahmed, came from Lahore.[18] Ahmed was a physicist trained at Cambridge, and Ghosh and Mukherjee were chemists trained in London, like Bhatnagar, in Donnan's labs. Mitra and Saha were colleagues in the same faculty in the University of Calcutta, though Mitra had completed his PhD and DSc in Paris, working first with Charles Fabry and then in the Institute of Radium with Joliot-Curie in 1923. C. V. Raman, K. S. Krishnan, and Homi Bhabha were not included, and there was no one from Madras or anywhere else in South India or Bombay. Four of the seven were from Bengal, three from Lahore and Delhi. Hill asked in August why no one from agriculture was included, and so Ahmed and Mukherjee, physicist and chemist by training, were quickly reclassified to this subject for the purpose, although they had been previously included for other reasons. Ahmed now worked in cotton research, proving the flexibility of elite training in physics at Cambridge. But most important, Hill identified Bhatnagar as team leader, and so his role and the CSIR's role were defined as preeminent.

In London as guests of the Royal Society, the team members were made temporary members in order to stay at the nearby Athenaeum Club, which rarely, if ever, had Indian guests. The team of scientists was received by the king and queen at Buckingham Palace. It was a curiosity of the contradictions of the situation that an academy (NISI) founded by Saha (hardly a fan of the king) was being classified as *the* national academy and considered a candidate for a royal charter, just at a time when knighthoods and royal connections were, in India, intensely suspect. Saha was a paragon of that suspicion. But public figures like Raman, Krishnan, and Bhatnagar all kept their knighthoods throughout this period. To be a Fellow of the Royal Society meant something quite particular in British society, rising with the society's illustrious origins in the seventeenth century. Indian Fellows were not elected in the foreigner category and were the only "foreigners" accorded this status as roughly equivalent British subjects.[19] In India the term "royal" was well known because Indians were surrounded with princes and princesses descended from kings and queens. This particular "royal" was different, however, and came from London, and regardless of what you thought about the British in India, the Royal Society had a social magic about it. Among scientists a fellowship was valued because it was rare and offered standing as a sign of genuine achievement, which they had begun to feel their own academies were not recognizing.

Winston Churchill invited the Indian scientists to lunch but at the last minute could not attend, so they lunched with the deputy prime minister Clement Attlee. They also met Sir John Anderson, former governor of Bengal and candidate for viceroy, in Saha's case by no means for the first time! Anderson was now an even more powerful figure in the British cabinet, in charge of almost all policy toward India.[20] Each team member gave speeches to their professional groups, and all attended the British Association for the Advancement of Science. Saha addressed the association, setting out his philosophy of the social purpose of science, the problem of poverty, and his plea for proper exploitation of natural resources like river valleys and power potentials, in the manner of John Bernal's *The Social Function of Science,* published in 1939. Even the secretary for India Leo Amery was in Saha's audience, as official companion of chemist-in-exile Marie Joliot-Curie. It is notable that a figure of her stature was present at Saha's talk. Her husband Frederic Joliot was in London to begin to renegotiate the French agreement with the British about patents on radioactive processes in order that France be able to build an independent atomic energy program. Saha and Joliot-Curie would soon cooperate more and more. Saha's talk was published in *Nature.*[21] He also spoke to the skeptical members of the Physical Society in London about his explanation that the high-temperature corona of the sun was heated by nuclear fission reactions in the photosphere.[22] Bhatnagar, on the other hand, gave a speech at the East India Society; he had been apprehensive about addressing such a "conservative" body (meaning pro-empire, in his words) and was relieved that Hill chaired the meeting for him. Bhatnagar said in his talk that he was among those Indians who wanted intimate friendship and intense cooperation with Britain but that he was critical of the management of Indian affairs at present and the limited role given to scientists and technologists. His talk was published, not in *Nature* but in the *Asiatic Review,* the official journal of the East India Society with large circulation in the UK and India.[23]

The mission visited a mixture of secret and open facilities—the Malvern radar laboratory, the National Physical Laboratory at Teddington, radio research at Slough, chemical industries at Huddersfield and Billingham, and the major universities like Cambridge, Oxford, and London. In North America they went to McGill University at Montreal, where part of the Manhattan Project already had British, French, and Canadian participation, the University of Toronto, and the National Research Council at Ottawa, where another part of the Manhattan Project was located. But most of their time was spent in the United States. These models of large military-oriented research institutions and universities presented to the Indian mission not only something their labs could resemble but also sources of new PhD training that members of the Indian team were seeking for the next generation of independent Indian scientists.

Hill's four months in India and the five-month tour of Indian scientists marked an important change in the relations of Indian scientists with the rest of the world. The Royal Society published Hill's report just before the Indian team arrived in Britain, substantiating things he had already said four months before, providing Indians with the first written feedback on their requests to him, and preparing British readers for their visit. Hill listed all the scientific initiatives that were under way in 1943 before he arrived and evaluated quite honestly the feeble and incoherent approaches that characterized official British responses expressed through the government of India up to that time. He finely balanced the problem of reform of archaic university environments with the fact that other industrial laboratories were only at the blueprint stage. Hill said later about his experience in India,

> At both ends politics tended to dominate the scene, to the exclusion of more sensible things; together, at our end, with the snobbishness of racial superiority. . . . My Indian scientific friends, particularly Shanti Bhatnagar, being quick witted, saw in my coming an opportunity of getting their needs attended to. I was given every possible assistance and encouragement, and constantly urged that discretion is not the better part of valour; but rather was invited to criticize openly whatever I thought wrong or stupid. Many things were wrong and stupid. . . . I admit I did not really expect much from the Report. I wrote to [viceroy] Wavell in 1945, and he replied sympathetically.[24]

The important change—not apparent to everyone they met—was that in London or in Berkeley was a group of Indian scientists who all knew their subjects very well, asking penetrating questions and understanding the answers. Whether on nuclear power, dam construction, oil exploration, astrophysics, chemical processes, the ionosphere, or anything else, members of this team understood the relevance of the subject to its applications in India. Moreover, they were the individuals who were destined to influence policy and power in the new India, regardless of what constitutional form India took and what its relations with Britain were to be. They were political scientists in waiting.

Moreover, the war was stimulating the CSIR and providing political justification for actually building its new laboratories. Hill's mission was intended to deepen commitment to the war effort by expanding commercial opportunities. Foremost in revenue generation was the Rs 50 million sale of vegetable oil blends as lubricants and fuels, a favorite project of Bhatnagar's, and Rs 10 million sale of cloth gas masks for soldiers. A valuable air foam solution produced from CSIR's formula successfully extinguished fires during Japanese air raids, and CSIR-designed jettison gas tanks for aircraft enhanced its reputation. The CSIR also earned Rs 198 million by leasing processes to industry, plus a share of royalties: this included payments from the US Air Force, Royal Indian Air Force, and the Defence Department.[25] To these were (notionally) added the large mounting sums of British pounds owed for India's costs in the war abroad, to be known as the sterling balances and paid at war's end.

War's End in 1945: The Tata Institute, the Atomic Energy Committee, and the Atomic Energy Research Board

Positive signals for the planning of research institutions were communicated from the viceroy's office even before the arrival of Archibald Hill on his official mission. These signals coincided with Bhabha's realization that he ought to try to do something creative in India and should not leave when the war was over. Bhabha was in precisely the right network to have known all about these positive government signals, and thus he reasoned that the time was ripe to propose a major project to the Tata Trusts. He knew all about the large Tata gift for the National Chemical Laboratory. Bhabha had extended discussions with Hill and Bhatnagar when he was awarded his FRS at the special ceremony in New Delhi in 1944.

In this context, well before India was an independent nation, scientists formed the Atomic Energy Committee in 1945. Clearly with official knowledge, it had Bhatnagar as its secretary and the CSIR as source of its funds. The committee supported nuclear research and allocated money through the closely allied Atomic Energy Research Board. Meeting in Bombay, the committee and board were chaired by Homi Bhabha; other committee members were Saha, of Calcutta; Krishnan, of Allahabad; and Bose, of Calcutta. This was the groundwork for the Atomic Energy Commission founded in 1948: it was the first intermediary for foreign interest in India's nuclear development. Negotiations with J. R. D. Tata representing the Tata Trusts were favorable and led to their agreement in June 1945 to support an institute focused on nuclear physics. So at this stage there was a plan for a major building in the suburb of Delhi for the National Physical Laboratory, wholly funded by the government; a plan for an institute of fundamental research in physics, one-third to be funded by the CSIR, one-third by the province of Bombay, and one-third by the Tata Trusts; and a plan for a functioning nuclear physics laboratory in Calcutta, where a US-built cyclotron was undergoing renovation and repair in order to carry out experiments. All this occurred prior to the testing of atomic bombs in New Mexico in July and prior to the bombs dropping on Japan in August 1945. Indians in India, forming a nucleus of expert advocates, thus had an advanced grasp of the potential power of the nucleus of the uranium atom.

SEVEN

Saha, Bhatnagar, and Bhabha in Contrast, 1944–45

> We believe that the only way to achieve unity of thought and purpose in the political field, which is now wanting, is first to look at the problem of living for India's millions.
>
> —Meghnad Saha, 1944

> I have no doubt that the British Government and people will help us in realizing our dreams of a prosperous and contented India. . . . Most of us, trained to think in an international way, will be glad to be members of the Commonwealth as equal and friendly partners.
>
> —Shanti Bhatnagar, 1944

> I have come more and more to the view that provided proper appreciation and financial support are forthcoming, it is one's duty to stay in one's own country and build up schools comparable with those that other countries are fortunate in possessing.
>
> —Homi Bhabha, 1944

These three men, near the end of the war, were poised at the beginning of their dual careers as scientists and politicians moving ineluctably toward and beyond an independent India: though aspects of their historic personal differences had already become evident, a comparison of the influence of their contexts is striking.[1]

Meghnad Saha was from a large, poor family of a powerless community in a village in East Bengal. Sahas are a caste in the *namasudra* group, very numerous in East Bengal, specializing more in business than in farming. This group of castes is among the lowest in the Hindu hierarchy in Bengal and was looked down upon by most others, though some households (and *gushtis,* "lineages") had valuable assets like land and high household incomes. At one time Sahas were associated exclusively with the brewing of liquor, a socially polluting activity for many castes throughout India. In some larger villages near permanent markets in Bengal, there was a *Saha para,* that is, a Saha neighborhood, consisting of some very substantial houses, and these families often owned considerable land, which they did not necessarily farm themselves. As a result of their success, and as a path to enhance it, some Saha families began to go into medicine, engineering, and law in the late nineteenth century. But this was not Meghnad Saha's origin. Picture the small house and shop Saha knew, the dilapidated school with coconut leaves for slates, the acres of mud in the monsoon, and the inevitable illness. After his first political experience in a school boycott at age twelve and subsequent expulsion, he was refused at age twenty-one the chance for government employment because of his political affiliations. In this way he recognized the state's ability to track his associations and movements, with consequences. He was well aware of his family's economic dependence upon him; for example, he brought his younger brother to study in Calcutta in 1915 and earned money by cycling south from North Calcutta to give tuition classes in big houses on Landsdowne Road and Elgin Road, preparing wealthy students for exams. He did not perceive research and an academic career as a means to fulfill his family's expectations until his access to a respectable career in government service had been barred to him. He had been taught by some inspiring teachers and then gained bright, active colleagues in a new, and malleable, institution, the Science College in the University of Calcutta. But he had to learn how to work through an obsolete syllabus, a poor library, inadequate equipment, and the personal clashes that interfered with day-to-day work; he also learned of the domination of scientific life in India by administrators. He also knew the pressure to publish abroad, where his work would be judged by scientists of international reputation, and he finally managed to arrive in London at age twenty-seven, without a prearranged location or supervisor for his work, and with few connections. But he landed successfully on his feet, having completed his doctorate and published from Calcutta, and that gave him a rapid rise to recognition.

Shanti Bhatnagar also came from an insignificant small town in the north and grew up in poverty brought about by the larger family's excommunication for his father's commitment to reformist principles in Hindu life, which others would have called the father's willfulness. But this nonconformism had a context, and the Brahmo community in North India to which the Bhatnagars belonged had established some mechanisms of mutual support. Although Bhatnagar never met his father, who died when he was born, he acknowledged the influence from his infancy of his mother's father, who was "a distinguished engineer and was employed on railway construction work." Bhatnagar, like Saha, left his mother as a teenager to pursue an education in the city, as his father and uncles before him had. Like Saha he had to work as a tutor to support his family. But he did so in Brahmo institutions that understood and accepted his father's principles. Like Saha he was married by arrangement in his early twenties to a teenage girl of minimal education. Bhatnagar's mother came from a distinguished, literate, but poor family, in the *kayastha* group of castes. His biography records (with Bhatnagar's acknowledgment) that his fellow caste members included his Bengali colleagues Jnan Chandra Ghosh and his "teacher" P. C. Ray, although there are differences in the social position of *kayastha* castes in Bengal and Uttar Pradesh. This caste group also includes Mathurs; his biographer writes, "among Sir Shanti's researchers are several Mathurs." Bhatnagar's son and daughter both married Bhatnagars: Shanti Bhatnagar is described in the authorized biography as "a heretic in matters relating to communalism, sects, ritual, customs, and orthodoxies of every kind, but it so happens in the matter of caste he has been a strict adherent."[2] As evidence of this, he was a lifelong vegetarian and never drank alcohol, though it is said that when he went to get Maulana Azad's signature on a file in the evening, he would take along a bottle of good whiskey and they talked more about poetry than administration. He went to London on a scholarship established by a Brahmo patron in Lahore and reached the center of his scientific universe, and the center of the empire, without a ready "contract" to study there. With diligence, he soon completed a DSc at the University of London and established a potential network running through London and Calcutta, linking academia and industry.

Compare Saha's and Bhatnagar's humbler origins with the large airy Parsi mansions of south Bombay, with full larders and servants effectively insulating Homi Bhabha from the nearby poverty. He grew up in a confident, successful community, and his family seems to have been concerned only about a career of appropriate respectability for him, as long as it was inclined toward technical matters. By the time Bhabha went abroad at age eighteen, he had personally seen little evidence that scientific research could be done in India and had little contact with scientists. But he understood a way of life that kept his Parsi community skillfully afloat in a tumultuous political world. In a colonial situation where collaboration with foreign firms was necessary for industry's survival, Bhabha met foreign business leaders along with famous nationalist politicians as houseguests and learned that nationalists too were adaptive and strategic in their relationships. Parsi life in India was flexible and exclusive, with a special language (a dialect of Gujerati), with its own cuisine, theatre, newspapers, and literature; Parsis mastered and managed many cultures and were not simply "stuck between two." They were supposed to live and die according to the dualist philosophy of Zoroaster. The Parsi Panchayat functioned as a kind of community court, where difficult issues and disputes were settled without involving outsiders. Rooted in Bombay, Parsis were found everywhere by 1947: Bhabha's voyage to England in the 1920s was repeated again and again, creating an international diaspora that sent Parsis all over the world. Early on Bhabha became associated with India's most successful Parsi family, the Tatas, to whom he was related, and their corporation provided massive support for his research institute.

In one of the world's most populous countries, there are millions of people with the name of Saha and Bhatnagar. One would be asked, which Saha and which Bhatnagar? But Parsis as a whole numbered a few hundred thousand, and Homi Bhabha was part of its own tiny cosmopolitan elite, comprising a small though influential fraction of the elite of India. Everyone knew which Bhabha he really was.

Whereas Saha could not afford to go to Cambridge in 1920, Bhabha's uncle had already been a patron of one of its colleges and helped to finance the engineering department in that university. Scholarships were essential to Saha and Bhatnagar; without them neither could have studied abroad. Bhabha did not need scholarships. Saha met old Calcutta friends abroad but made important new connections. He and Bhatnagar met some of the stars of scientific life for the first time in Germany and France, where they could also observe postwar reconstruction in the 1920s. Saha extended his network to include significant political figures and accepted a secret role with an underground political group that sought to overthrow the government. He also looked to Germany as a force against the British, particularly admiring the organization of science and industry. Before starting their careers as scientific organizers in India, Saha and Bhatnagar spent only two years abroad, whereas Bhabha's socialization into the scientific community occurred over twelve years in its great European centers. Saha and Bhatnagar returned to India to be responsible for their wives and families; Bhabha returned to India without such encumbrances. On his return Bhatnagar learned that the chemists in Calcutta were powerful and that there were other chemists ahead of him with claims upon the leadership of academic chemistry. Even with his good reputation, Bhatnagar did not become president of the Chemical Section of the Indian Science Congress until 1939, whereas Saha was voted president of the Physics Section fourteen years earlier, in 1925. Bhatnagar's long delay was because leadership in the Science Congress had greater importance to the chemists than to physicists, there were more chemistry members than physics members, and chemists were more visible as leaders among them; he had to wait in a queue. The chemical industry naturally had greater influence in India than physics just after WWI, certainly as great an influence as the engineering industry. Engineering and chemistry offered more local employment, particularly in Bengal, and more involvement in international activities, in part because of local investment in these activities. V. V. Krishna speaks of an "Indian school of chemistry" in this period, and engineering and chemistry were certainly two popular sites of Indian and British investment in India.[3]

By 1939, Saha had returned to Calcutta, the center of science politics in India and one of the two main centers of industry and capital. By 1939, when Bhatnagar was moving from Lahore to Delhi, the city was set to become the strategic capital of India and to be inhabited mainly by politicians and civil servants. Bhabha had arrived in India for a holiday just as the war started in 1939. Stranded involuntarily in Bangalore, he was taught much about the politics of science at the Indian Institute of Science, and his guide was none other than C. V. Raman, frustrated and bitter about his treatment in the previous year.

As for the biggest news in physics in 1939, the discovery of fission, both Saha and Bhabha learned about it from the same journals as everyone else in the world (e.g., *Nature*) and understood the theory and experimental requirements, if not all the strategic and military implications. Saha, Bhatnagar, and Bhabha were in regular communication, though we do not know how much they discussed fission. Nor do we know if they discussed the freeze on publications from 1940 about fission, organized in part by physicist Patrick Blackett in London. Bhatnagar too read *Nature* regularly and knew Saha's view that explosions of fissile material are possible under the right conditions, because like others he was a regular reader of *Science and Culture,* in which Saha expressed this view. In 1941 Saha wrote, "It is quite possible that a process may be discovered which renders the [chain] reactions to proceed with explosive violence . . . the idea that a tablet of U235 . . . may blow off a mighty Super Dreadnought [battleship] cannot but be an exciting one."[4] This was almost four years before Saha toured major North American research sites associated with the Manhattan Project and before he suspected that a major project involving U235 was well under way, as revealed in chapter 6 and examined in *Negotiating Nuclear Power*.

Saha tracked nuclear development regularly, including during his tour of Allied installations in 1944–45. *Science and Culture* carried an article about the bomb one month after the terrifying explosions in Japan in early August 1945, "The Story of the Atomic Bomb" by Saha and his student B. D. Nagchaudhuri.[5] Then Saha wrote the lead article "The Logic of the Atomic Bomb," and he also reproduced in full the detailed British document *Statements Relating to the Atomic Bomb* (September 1945), naming names and committees instrumental in British participation in the Manhattan Project (e.g., the MAUD Committee). Saha wrote that the project demonstrated "that if a team of well-chosen scientists be selected for studying a problem in an objective way, and be directed to find out the remedy, and if sufficient funds and power be placed in their hands to execute their plans, they can be trusted to solve problems of reconstruction which baffle the professional politician and centuries of neglect can be compressed into decades."[6] A month later Saha summarized Henry Smyth's lengthy study published in the *Review of Modern Physics* in October 1945, explaining in an official way the American participation in the Manhattan Project.[7] He published his own eight-page article "The Atom Bomb" six months later, complete with diagrams showing how one is made and four methods of uranium separation (U235 from U238).[8] A year later, in 1947, he wrote a long article entitled "Industrial Utilization of Atomic Power in India," arguing in public the sort of position he was taking with Bhabha and Nehru, pointing out that U233 is synthesized from thorium, of which India has rich deposits, urging the public to raise questions about India's choices in atomic energy.[9] Soon, however, he asked the obvious question about ends and means: "No greater vile or criminal application of a great and magnificent scientific discovery could have ever possibly been made. This has rudely shaken the conscience of the scientists today and they are gradually becoming alive to their responsibility."[10]

Not only as individual scientists, but also as organizers of research, these three men worked under quite different conditions. While Bhabha worked as an organizer almost entirely after Independence, Saha and Bhatnagar did most of their organizing in the colonial period, and their political energies and alliances focused on the achievement of Independence and strategic "positioning" in anticipation of it. Bhatnagar initiated and guided a network of research laboratories that became one of the largest in the world, with many thousands of scientists employed across the country. He had received his knighthood and election to the Royal Society before he was fifty, and the confirmation of his social status and scientific reputation helped him advance his prominence during the 1945–47 transition and justified the cost of this network. The budget of the Council of Scientific and Industrial Research was drawn directly from Parliament, and the prime minister of India chaired its council.

Saha established an institute, but it did not become self-sufficient in his lifetime. His early acclaim (age twenty-five for the ionization paper) and recognition (FRS at age thirty-four) did not provide him with skills in managing a large-scale organization. Saha's political views on the role of science in development were also decisive. Saha supported a strong anticolonial and socialist line in the Independence movement and assumed that there had to be complete reorganization of society, particularly around class and caste, and industrialization of the economy. This socioeconomic reorganization did not occur quite as he hoped because of opposition by entrenched interests, reinforced by international forces beyond the influence of the equalitarian movement Saha supported. But he did have considerable influence on nuclear, technological, and industrial planning. His pre-Independence difficulties were severe, but, in addition, he chose colleagues who, despite great effort, were largely unable to develop the institute after his death. Nevertheless, his protégés Kothari and Nagchaudhuri certainly did rise to positions of prominence, and the latter was director of Saha's institute for many years. Though Saha had wide contacts within the political system during the 1930s and early 1940s, he could not secure an independent or constant source of finance for his institute, which he built by transforming a laboratory at the University of Calcutta. Opening finally in 1950, Saha's institute became dependent, after his death in 1956, upon the Department of Atomic Energy and, ironically, on Bhabha, whom Saha had criticized. But what is most important is that it did not fade away.

Bhabha began his institute in 1945 on a small amount of money but with surety of an increasing supply. It grew rapidly and gave birth to the large Department of Atomic Energy and the Trombay atomic research establishment. With slower but steady growth, Bhatnagar's CSIR became responsible for subjects that did not attract major political interest—nuclear, agricultural, and medical issues were studied elsewhere. But Bhatnagar remained at the center of the CSIR because of his connections with Bhabha and with Nehru through his minister Maulana Azad. With Bhatnagar's death in 1955 the CSIR's relationship with the golden goose of atomic energy ended, and the two institutions drifted apart. The institute that Saha founded remained dependent on the department that Bhabha created but was seen as a scientific gesture by Delhi and Bombay to Bengal, as "Bengal's nuclear installation."

In aristocratic style, Bhabha combined his energy, his scientific reputation, and his family connections with extraordinary skill to obtain large sums of money first from private and then from government sources. The institute he founded became as financially autonomous as any project in India, established an independence that had a liberating effect on scientists, and enabled a freewheeling style of rapidly growing research teams. Such freedom was rare in science in India and was quite a contrast with conditions in the CSIR founded by Bhatnagar; despite the council's direct relationship with the prime minister, there was after Bhatnagar's death no such freewheeling atmosphere in most CSIR laboratories.

Their remarks quoted at the beginning of this chapter show a significant difference in Saha's, Bhatnagar's and Bhabha's intentions and directions in the watershed years just before Independence. Saha had become, at the moment when organizing science was about to change, concerned with "the problem of living for India's millions." He showed this large but diffuse concern eight years later by entering parliamentary politics, even while he continued his involvement in scientific associations and laboratories. Bhatnagar saw science not simply as the way to solve many of India's problems but more as the right way to relate to the rest of the world, with potentially competitive countries like Britain through reason, cooperation, and intelligence. Cooperation with powerful business and economic forces, rather than confrontation with them, appeared to him to be the wisest path for a country with little international power and little means to obtain industrial power. Meanwhile he envisaged a whole chain of related laboratories, all dedicated to building new industries, discovering new processes and products, protected by a vigorous patent regime. He believed that his laboratories could survive the tension between the fact of widespread private industrial ownership and the public expectations of nationalized ownership of production. Bhatnagar's way of achieving their survival was to promote public science in support of both public objectives *and* private industry, and he grasped the opportunity of mass media to achieve that, while carefully cultivating his own reputation.[11]

In 1944, Bhabha decided that more specialization and greater focus was absolutely necessary. He had worked in a Bangalore institute built on the grand design, one that had embraced "all of science." His concern became more instrumental and more focused, less concerned with Saha's "problem of living" than with the necessity to "build up schools" of pure and applied research that would be ultimately useful in dealing with the problem of living, but not immediately. There was a double risk here. There is little room for compromise in building good schools, but it is possible to create large groups of people in them who are more satisfied in their work if they are allowed to ignore others' problem of living. This studied and structured "ignoring" generated both distance and envy, if not resentment. On the other hand, a diffuse appreciation of the problem of living was also a prerequisite for any effective program of action. But Bhabha thought scientists were likely to be too diffuse, too unfocused, and would not grasp the strategic necessities unless they could address and overcome the obstacles in building a scientific institution with an international reputation. He wanted a "responsible elite." This tension remained strong in the work of Saha, Bhatnagar, and Bhabha, essential in the process of developing a scientific community and of changing a society. The tension in each case reflects their dual loyalties to both local conditions and world science.

At the end of the war in 1945, the British cabinet turned its attention to a process leading to Indian Independence. Eventually a three-member delegation came to Delhi in March 1946, led again by Stafford Cripps but this time with a larger team plus two other sitting cabinet ministers. While administrators prepared for the transfer of power, Indian professionals focused on establishing their influence in national institutions. Nehru, Gandhi, Viceroy Wavell, Stafford Cripps, the Congress Working Committee, Jinnah, and the Muslim League were locked into intense and lengthy negotiations that included moving the entire group to the hills in Simla in order to avoid the April heat in Delhi. Without any resistance from the British cabinet (where Churchill was absent, now being in the opposition), the delegation tried to find a formula by which Congress and the League would cooperate in a united India, acknowledging the need for creative forms of confederation. Some of the formulas of confederation and power-sharing were opposed by Gandhi and others were opposed by Jinnah. In a complex dance that involved the viceroy's Indian executive council becoming essentially a cabinet of a national shadow-government, the irreconcilable interests that led ultimately to Partition confronted each other in 1946, well before Mountbatten's appearance in Delhi in 1947.[12]

Realizing that the British cabinet ministers had not succeeded in finding a compromise and seeing the sudden change of viceroys from General Wavell to Admiral Mountbatten in January 1947, the builders of scientific institutions began to work quickly, out of the public limelight. It was in this context that Patrick Blackett was asked by Nehru in January 1947 to become his military and scientific advisor. The chaos of communal riots and uncertainty swirling around them was in fact their camouflage; scientists like Saha, Bhatnagar, and Bhabha were focused, knew where the money was, and heard the clock ticking as a countdown to August 1947. Two of them, Bhatnagar and Saha, were just about to lose homes and property in the Partition and made their decisions to remain in India and not return to "where they came from." At the same time they acted as a committee to build a nuclear program, agreeing just sufficiently among themselves to get things done. Saha was already taking an independent approach to Bhabha's and Bhatnagar's assumptions and plans. This independence soon led him to Parliament, where he could mount opposition to Bhabha and Bhatnagar. These institution-building strategies are the subject of the following chapters.

EIGHT

Restless in Calcutta: Meghnad Saha's Institution-Building

Given his active political correspondence, travel, and search for funds for his institution, you may rightly wonder whether Saha actually got any scientific work done. Remarkably he did, but his research experience confirmed the importance of separating the administration of his labs from the university. At the end of the war, Saha's vacuum problems with the 38-inch/5 MeV cyclotron had not been solved. Administrative obstacles in the Palit Laboratory were acute because accounting, purchasing of parts, and salaries were handled by the notoriously inefficient and distant university offices on College Street and their notoriously underpaid staff. His five-year grants, which started in 1942–43, from Tata and Birlas had not quite run out but would soon. His colleague on the tour of the Allied laboratories in the UK and North America, S. K. Mitra, had begun to form the separate Institute for Radio Physics and Electronics when the tour returned in 1945. Saha and Mitra knew all about Bhatnagar's plan to create a chain of research laboratories separate from universities because they both sat on the CSIR's planning committees. Though he was, after all, also a dominant voice in the planning of the glass and ceramics lab project, near the new nationalist university in Jadavpur, he realized that except for that Glass and Ceramics Institute the CSIR would not build another lab in Calcutta; Saha understood that he had already failed to have the National Physical Laboratory located there. The independent Tata Institute of Fundamental Research was already in existence in Bombay, and its director Bhabha chaired the Atomic Energy Research Board, from which the money for nuclear research, if any, was going to come. So Saha had to try to transform the institution where he was and consolidate his political relationships above Bhabha's head, with both Bhatnagar and Nehru, and by 1947–48 he had achieved considerable success.

Nevertheless, he was curiously restless. Saha began experimenting in 1946 with reorganization and a new building for the Indian Association for the Cultivation of Science (IACS); its council had passed the management of the new lab to him in 1946. The irony would not have been missed by his old opponent Raman, who had nurtured the association until his departure from Calcutta in 1932. Raman said Bengalis neglected this key institution since his departure. Perhaps they had, but it certainly had become part of Bengal's establishment. The new IACS building was to be near Jadavpur university in the booming suburbs of South Calcutta, across the street from another laboratory Saha was influencing, with Bhatnagar's approval, the new CSIR laboratory for glass and ceramics, where there was an optical physics group planned, and he was mentoring the scientist in charge of the project, Atma Ram. Saha was now a member of the Planning Committee and chairman of the Advisory Board of the CSIR's Central Glass and Ceramics Research Institute. Seen from Bhatnagar's perspective, was this a useful method of keeping Saha absorbed in Calcutta, away from other labs? Was Saha attracted away from his university institute because he saw the limitations on the money he could raise for it or because these other labs across town opened new avenues of political influence in the scientific community?

A constant investigator, he learned as a member of the council of the CSIR that the Ahmedabad textile mill owners had not paid tax on their profit between 1942 and 1945 because of policies that were applied to them during the war. This money, the state government had decided in Gujerat, would not be taxed if it was contributed "towards fundamental or industrial scientific research": about Rs 10 million was thus available for scientific research in Ahmedabad. Could this occur in Bengal? Saha approached Bengal-based industrialists during the previous year, he wrote later, but "they wanted a written assurance" about this money being untaxable before any commitment was given for its use in fundamental research. He thus asked his vice-chancellor to request the government to give written assurances of this tax exemption and also to invite "industrialists who may be willing to help us—N. R. Sarkar, N. N. Law, S. M. Bose—etc." to meet to discuss the policy.[1] (Nothing came of this effort, so far as I can determine.)

Saha also gained influence through those individuals who were once his students or colleagues such as R. C. Majumder, who became professor of physics at the University of Delhi, and D. S. Kothari, who was seconded from that university to become scientific advisor to the Ministry of Defence, with the intervention of Patrick Blackett and Sir Maurice Gwyer, then vice-chancellor of the university. Then there were his classmates from university student days who were, like Jnan Ghosh, well connected in Calcutta and influential in Delhi. His students held top positions in meteorology, defense science, military signal corps, railways, police, and all technical and administrative aspects of civil service. Saha's influence in physics was a function of four forces at work: first, students had been sent as investments by their families to study and gain secure employment in technical fields, preferably in government; second, most students came to physics because it was a "high-scoring subject," meaning that their performance could stand out in competition; third, Saha was an excellent teacher and his students performed well on competitive examinations; and fourth, universities at both Allahabad and Calcutta were prime sources for recruitment to cadres of government servants because these two cities had disproportionate political influence, and degrees from those universities had wide credibility and acceptance.

A 1967 editorial in *Science and Culture* explained how this network functioned as the Calcutta-Allahabad "research axis."[2] Besides explaining that classmate N. R. Dhar went to Allahabad and then encouraged Saha to follow him, the editorial chronicled the fortunes of four student-associates of Saha: Atma Ram, B. D. Nagchaudhuri, A. C. Bannerjee, and B. Srivastava. Ram completed his DSc at Allahabad on photochemical reactions with Dhar and Saha, and in 1952 he became director of the Central Glass and Ceramics Research Institute in Calcutta, which Saha had helped to plan and build. In 1966, Ram became director general of the CSIR. Nagchaudhuri was a student of Saha at Allahabad and at Calcutta before doing his PhD with Ernest Lawrence at Berkeley from 1938 to 1941. He went back to California in 1948 to work on cyclotrons for a year. On Saha's death, Nagchaudhuri became director of the Institute of Nuclear Physics in Calcutta, in 1968 he became scientific member of the Planning Committee, and then in 1970 scientific advisor to the Ministry of Defence, in essence a cabinet advisor (and an insider in the bomb project). Bannerjee moved from Calcutta to Allahabad to be a professor, then became the vice-chancellor of Benares Hindu University. Srivastava was Saha's student at Allahabad and became the director of the IACS in Calcutta in the late 1960s, an institution greatly influenced by Saha. Saha, Dhar, Atma Ram, and Bannerjee were all presidents of the Indian Science Congress. This shows a fairly tight interlocking network within the scientific community in India, particularly in its traditional university sectors most influenced by Saha. Add the presence of D. S. Kothari in Ministry of Defence and Jnan Ghosh in the Planning Commission, and one sees how well-placed Saha's network was.

Saha's influence was based on his work with and through universities, and he tended to oppose building other organizations that he felt might threaten the declining power of universities. This was the basis of his opposition to CSIR's practice of hiring good people away from universities, a subject he discussed frequently with Bhatnagar. At the same time, acknowledging the power of CSIR's budget, Saha argued in the University Grants Commission for more support for fundamental research in the universities. In 1948 his negative attitude to the creation of the Atomic Energy Commission must be understood in terms of this power base, though by this time plans for his Institute of Nuclear Physics were in their final stages and he had even managed to raise Rs 620,000 during the upheavals of 1947–48, from a number of sources.[3]

In Saha's experience the university was the only place where research could be supported, and he was raised in a culture where Calcutta University was at the very center of political competition for scarce resources.[4] He lived at the heart of an educational network where he could choose the very best students and build them into teams. A typical example is the following statement:

> When I stood first in the exams at Dacca University (1946), my *viva voce* was taken by Meghnad Saha himself. I naturally came to Calcutta to ask him what to do next. He gave me a CSIR research assistantship in the Palit Lab. I went abroad in 1953 only when I had worked for him for six years. And then I came back here after 18 months, as promised.[5]

These are the words of a young person who received the top marks in physics out of thousands of people writing a competitive examination, and Saha could meet him in person, give him advice, and provide him a job, quickly. Saha's objective was to recruit and retain people whom he thought were going to be his equals, or surpass him. History will judge whether he did this well, but there is no doubt that he seized every opportunity that presented itself and looked at new fields from every angle.

The foundation stone for the Institute of Nuclear Physics was laid by senior Bengal politician Shyama Prasad Mookerjee in April 1948 (Asutosh Mookerjee's son, prominent in the Hindu Mahasabha party), using the Rs 620,000 raised in 1947. In 1950 Saha managed to get another Rs 120,000 for furnishings for the building from Bhabha and the Atomic Energy Commission (AEC).[6] The building was officially opened in January 1950, not by Prime Minister Nehru as in openings of most of the ventures of Shanti

Bhatnagar and Homi Bhabha, but by Frederic Joliot-Curie, head of the French Atomic Energy Commission, with whom Saha felt both scientific and political kinship because Joliot was a member of the Communist Party of France. Joliot-Curie was invited to visit India to attend the Indian Science Congress and came, with his wife, Irene Joliot-Curie, to Calcutta as a side trip and opened Saha's institute. This occurred a few weeks after the Communist Party came to power in Beijing and the USSR exploded its first atomic bomb (in October 1949). But Saha could not secure regular funding for the institute because of his poor relations with the AEC and the slowness of the Senate and Syndicate of the university. Near the end of 1950, correspondence between Nehru and the vice-chancellor, S. N. Bannerji (also a justice of the Calcutta High Court), showed that the central government would finance the institute regularly only if it were to take on an "all-India character," meaning the employment and admission for training of non-Bengalis, people from outside Bengal.[7]

The compromise solution struck with the central government was that the chairman of the institute's Governing Body be the vice-chancellor of the university and that 40 percent of the seats in the post-MSc nuclear course be reserved for non-Bengali students (or students originating outside Bengal as there were/are large non-Bengali populations within the state of West Bengal). There was no similar provision for institute faculty, and in 1970 well over 75 percent of the scientists were Bengalis, with a proportion of 90 percent in nonresearch staff. The draft constitution was debated early in 1951 in the university Senate, where Saha was supported again by Shyama Prasad Mookerjee and by S. N. Bannerjee, the vice-chancellor. Mookerjee had, in 1941, persuaded the Senate to accept the Tata Trust seed grant for Saha's cyclotron and laid the foundation for his institute in 1948.[8] The Senate opposed the construction of a semi-autonomous institution within the grounds of the Science College of the university, land which it had fought hard for, but Mookerjee helped Saha overcome this opposition although they each supported opposing political parties. The university wanted full control of the land, governance, and admissions/employment in the institute, asserting the right of the city and West Bengal against the central government. Mookerjee's constitutional compromise between local and national forces allowed the institute to begin functioning as an all-India institution in July 1951. The complete dependence upon AEC/DAE finances coming from Bombay and Delhi did not occur until 1955, but it had begun with the construction of this fine new building. Saha had been drawn into a national system whether he liked it or not.

Cyclotron Research Group Struggling in Saha's Institute

An Institute of Nuclear Physics report in 1955 explained Saha's earlier difficulties in the following terms: "In those days (1944–46) research work developed around individuals and for want of personnel strong groups could not be formed." In most cases groups consisted of one or two research workers. That was the tradition of research in India and in the universities, constrained by a colonial government's impecunious way of thinking about science and by the need to cover many subjects in the syllabus, usually with one instructor per subject; it was the time of the heroic scientist, alone in a lab against a hostile world. Saha had not been surrounded by many competent collaborating colleagues, because he worked mainly with his students, avoiding colleagues of his own age. (His friends Mitra and Mahalanobis both created their own institutes nearby, so the avoidance was mutual.) Saha's institute's research work was dominated by the cyclotron, around which the largest group formed. The cyclotron continued to give trouble in 1948, and Nagchaudhuri went again to Berkeley to get more experience to operate it. Ernest Lawrence again assisted them in the acquisition of a new working vacuum system, though an oil leak persisted. Lawrence may have heard about these later difficulties, but on his 1953 tour to India he did not even visit Saha's laboratory nor his former student Nagchaudhuri.[9] But experimenter Emilio Segré from Berkeley did come to the institute a year later, offered advice to Saha and Nagchaudhuri on the cyclotron, and continued to do so in writing. Blackett also came to the institute in Calcutta, but, given his close relation to Bhabha, Saha may not have been able to form much of a relationship there; however, Saha well knew that his student Kothari in Delhi owed much to Blackett's favorable appraisal of him.

Nagchaudhuri was widely seen as responsible for the Calcutta cyclotron and was the only one who had been trained in Berkeley. Like Saha he was born in a village outside Dhaka, but in a rich family. His early education in physics was at Allahabad, and he said he "tagged along" when Saha moved back to Calcutta in 1938. With Lawrence's support to Saha's recommendation, Nagchaudhuri arrived in Berkeley, California, in late 1938 to do his PhD at age twenty-three and said, "I did not have a clear idea of what I wanted to achieve, even then."[10] He began working with Emilio Segré's cyclotron group and remained until just before Pearl Harbor in late 1941. By then the Radiation Laboratory had become one of the focal points of American physics for the cyclotron and bomb projects. Segré was the central figure in the building of cyclotrons, from which plutonium was first produced. He and Lawrence designed and then financed larger and larger cyclotrons with higher and higher energies, so that the Calcutta cyclotron was eventually left behind in the cyclotron world. But the Calcutta cyclotron was certainly built by Donald Cooksey and Ernest Lawrence in 1938 and disassembled for its long sea journey to India.

Oppenheimer and Lawrence were both very active in an uneasy partnership at Berkeley when Nagchaudhuri was a young doctoral student.[11] The arrangements for the cyclotron to be purchased and moved to Calcutta had been made by Saha and executed by Nagchaudhuri, so Nagchaudhuri returned to Calcutta to await its arrival. Saha wrote to Nehru in 1941 explaining that $16,000 raised by Nehru from various sources had already been sent to Berkeley as part payment (some items were gifts) and that due to shipping difficulties less than half of that value had been received.[12] The cyclotron produced little but trouble for the next ten years: the demountable oscillators turned out to be very difficult to build, and evacuating the "Dees" and creating a vacuum took a long time. The vacuum system, crucial to particle beam production, was lost in the torpedo attack off Japan, and the new one resisted many attempts to improve it. Finally, after Segré's 1954 visit it produced an internal beam, and four years later an external beam—for which it had been designed. By that time, engineers had finally been appointed to look after the Berkeley machine, and the institute was building a new cyclotron. In 1966 a continuous uninterrupted beam was achieved. In 1968 Director Nagchaudhuri left the institute to work as the first scientist in the Planning Commission in Delhi.

At the famous meeting of Saha and the AEC in Delhi in 1955, a proposal was approved from the Calcutta institute for a 50 MeV variable energy accelerator (VEC), which was ten times the energy of the existing one. But after Saha's death the project was sidelined for the next ten years. Nagchaudhuri later said that if there had been a representative from Bengal at the Delhi meeting, "the VEC project would have moved much faster."[13] In fact actual construction of a new accelerator began in collaboration with TIFR, DAE, and SINP only in 1970. But notwithstanding these protracted difficulties the cyclotron group predominated in the institute; the successor to Nagchaudhuri as director was D. N. Kundu, an early cyclotron worker and head of that division. His role in the cyclotron group was probably the only reason he became the director. Kundu's directorship was then followed by Ajit Saha, son of Meghnad, also from the cyclotron group. The institute's move to a new building coincided with the construction of the new accelerator at Salt Lake in northeast Calcutta, and SINP began to operate it as a facility for the Department of Atomic Energy in 1977. This group thus formed the largest division of the institute, with the largest number of staff. One of its main difficulties in the 1950s seems to have been that Saha delegated the vacuum engineering job to physicists, not to engineers. Only in 1966 were engineers appointed, twenty-five years after its arrival in Calcutta.[14] An effective research beam was first produced from the cyclotron within a year of appointing these engineers.

Competition among Colleagues

There were many cross-cutting initiatives starting up at the same time—rivals for attention and funding. Even before opening his own institute in 1948, Saha promoted the development of a biophysics group involving N. N. Dasgupta. In 1945 he used grants from his businessman-relative R. P. Saha (Rs 45,000), and businessmen B. C. Law (Rs 7,500) and industrialist G. D. Birla (amount uncertain but probably Rs 200,000) for this purpose.[15] Dasgupta went from Saha's group in Science College to Stanford to work on electron microscopy with Marston, returning in 1948 to construct the first electron microscope in India.[16] Though Dasgupta's group was small (in part a function of cramped quarters), the demand for use of the microscope by scientists in other medical and biological institutions in Calcutta was great. When the institute was founded in 1950, on top of funds from industrialists Birla, Law, and R. P. Saha, Meghnad managed to get Rs 60,000 from the Congress chief minister of West Bengal B. C. Roy and Rs 60,000 from Rajkumari Amrit Kaur, the minister of health in the government of India, for the purchase of a second electron microscope. This money came to Saha even after his refusal to run on the Congress ticket in the 1948 elections; he had been warned that his refusal might backfire on his institution-building, but the government appeared ready to support new technology and new research.

When Saha was preparing to retire and go to Parliament in 1952, he made it clear that Nagchaudhuri was to succeed him; while Nagchaudhuri was nominally directing the institute in Saha's absence, Dasgupta was preparing to split off and establish a separate laboratory at Belgachia, five kilometers away. Nagchaudhuri and Dasgupta had equal seniority in the Palit Laboratory, both having arrived there in 1938. The tension of cramped quarters and leadership competition was resolved by Dasgupta's move to the new lab where the biophysics group increased its dependence upon the CSIR and University Grants Commission for funds and moved away from the DAE. By 1954, Saha thought that "the work of the Biophysics division had been important enough now to think of raising it to the status of a separate institute."[17] Until Dasgupta left the institute to become Nagchaudhuri's successor as the Palit Professor of Physics in 1969 in the university, his laboratory was ambiguously related to the older part of the Institute for Nuclear Physics in Science College—isolated from Science College and the institute, yet ultimately still dependent on it for governance and financial decisions.

Cosmic ray research, the low-cost method of studying high-energy particles, begun in 1938 with cloud chamber studies at the Palit Laboratory in Calcutta, was moved to a higher elevation in 1942 at Saha's own house in Darjeeling. At the 1955 meeting with DAE in Delhi, Saha proposed a five-year plan with cosmic ray research continuing at 8,000 feet in Darjeeling, with two new stations at 7,200 feet at Lalapahar and on the Darjeeling-Lhasa road at 16,000 feet. But Bhabha's cosmic ray research group was now in full bloom at TIFR, with Oppenheimer's student Bernard Peters leading it. The Saha Institute's proposal was subject to the approval of the Cosmic Ray Committee of DAE, and Bhabha gave no immediate reaction to the plan.[18] Former SINP cosmic ray workers in 1968–69 said that they believed Bhabha turned down the plan after Saha's death because he intended to consolidate all cosmic ray studies in India under the control of TIFR. Small research studies in cosmic rays were also conducted at the nearby Bose Institute under D. M. Bose and at the Indian Statistical Institute under P. C. Mahalanobis, without any local effort to coordinate or consolidate these other projects, though Saha must have known of these small efforts. Neither of the other institutions was formally constituted to do research in physics: the Bose Institute continued Jagdish Chandra Bose's traditions in biology, and Mahalanobis, the founder of the Indian Statistical Institute in 1931, became the chief advisor to Nehru on economic planning. What this shows is that cosmic ray research was popular and not expensive. This deep uncooperative fragmentation of scarce resources typifies the evolution of scientific institutions in Calcutta at the time and signals the sensitivity of three physicists in the same city to international trends.

Confident of his sense of direction, Saha continued to publish on primary cosmic radiation and began writing on the theory of the solar corona; papers appeared in 1942, 1945, and 1947. In 1946 he read an account of a lecture given by Canadian astronomer Henry Plaskett in London, and thus he sat down and wrote his famous letter to Plaskett, which in turn Plaskett circulated widely. Plaskett had explained in his lecture that Indian physicists could perform very well when they moved to supportive environments abroad, so Saha reminded Plaskett that his first important work had not been done in London as was often thought. This confusing assumption paralleled Satyen Bose's experience that most people thought that the Bose-Einstein statistics were developed by a German called Bose, working with the famous Einstein, a German-speaking Swiss. Saha's letter reminded everyone of the fact that while he had made good progress in his work when he came to London in 1920 and then to Berlin, he had indeed already written the first key papers in solar ionization in Calcutta long before he had personal contact with European scientific centers, and these papers were very widely cited though none remembered their origin in Calcutta.[19] DeVorkin calculated that there were one hundred citations to Saha's work by 1930, the year when he was first nominated for the Nobel Prize.[20]

Locally, Saha's work was closest to the research of his physicist neighbor, Sisir Kumar Mitra. Three years older than Saha, Mitra won a gold medal for his MSc work in 1912, worked as a lecturer with Raman, and was granted a prestigious DSc from the University of Calcutta in 1919 before going to France as a postdoctoral fellow in Madame Curie's lab in 1920–23. When Saha left Calcutta University in 1923 for Allahabad, Mitra was appointed to the Khaira Chair, which Saha vacated. When Saha came to Calcutta in 1938, some students were attracted away from Mitra's wireless radio physics lab to work with Saha during the war, when the Berkeley cyclotron arrived. One example is B. M. Bannerjee, later professor of physics in the Saha Institute.

> Bhatnagar had given Mitra a grant to develop a loudspeaker/microphone system on which I worked. Since one had to make an appointment to see Mitra, there was little interaction and I got little encouragement from him. Since Bhatnagar himself was not personally interested, I knew the system had no real future. I was becoming very interested in Saha's work. He knew I was working with Mitra so wouldn't give me a job at first, but he didn't stop me from coming to cyclotron design meetings. Mitra eventually became angry with Saha because I got a job with Saha trying to fix their vacuum system.[21]

But farsighted Mitra had earlier found in Saha an ally in proposing a radio research committee in the 1930s and, through the influence of Saha and Bhatnagar, becoming chairman of that committee of CSIR in 1943, prior to going on the scientists' tour in 1944, along with Saha. Readers should not underestimate the importance of radio in India at the time; though India eventually missed the transistor revolution, radios and radio wave physics were seen to be tied to a coming revolution in communication, made clear by the nationalist movement's use of radio despite its complete control by the state. Its military application was easily observed in India, and radio research received a huge boost during the war. From 1939 to 1944 Mitra was also the general secretary of the Indian Science Congress and thus had a big network of contacts throughout India. Mitra wrote a book that completely reviewed work from all over the world on the physics of the upper atmosphere. Begun in 1935 as a review paper in the *Proceedings* of the National Institute of Science (NISI), the book became part of the research of Mitra's group, and students contributed to it as part of their responsibilities. Though there were many delays not of their making, the manuscript was finally completed in 1947, resulting in a book that was very widely quoted internationally. Eventually a belt in the ionosphere was subsequently named "the Mitra belt."

Despite their uneasy relation, Saha was still useful to that book, according to Mitra: "It was fortunate that, at this time [1948], Professor MN Saha was the President of the Asiatic Society of Bengal. He realized the value of the book, and, in spite of the discouraging reactions from publishers with world-wide experiences, he persuaded his Society to publish it as one of its Memoirs, regardless of cost."[22] Having gained an independent institute in 1949 and a new building in 1952, Mitra was elected FRS in 1958 largely as a result of his early research and the international impact of this book. The buildings established by Saha, who died in 1956, and Mitra, who died in 1963, faced each other boldly in Science College. Both institutes had autonomy within the university: it is characteristic of the competitive and uncoordinated climate in the Science College of the university that these two entrepreneurs and exceptional scientists did not cooperate though they worked in closely allied subjects. It is also characteristic that no local forces influenced them sufficiently to cooperate. Saha even made later attempts to begin radio astronomy while he was a member of Parliament, and, even though the design of a radio telescope was discussed and studied, nothing was built, as we shall see.

Because Saha's own reputation was in astrophysics, he could not remain indifferent to changes in his own field. When Mitra's new radio physics institute was finished in 1952, it opened its ionospheric research field station at Haringata, then out in the country but soon a suburb of Calcutta. Saha had something similar in mind and had already begun work on plans for radio astronomy, visiting the famous radio telescope at Jodrell Bank near Manchester in early 1954. Saha was very much impressed: M. K. Dasgupta remembered Saha saying in Manchester, "I wonder when we'll be able to see such a laboratory flourish in India."[23] Saha became so committed to the idea of radio astronomy that when he returned to Calcutta, he changed the research of two junior members of the institute to begin to work on these plans. One student, J. K. D. Verma, stated that he was more interested in nuclear physics, but Saha obliged him to begin developing the radio astronomy project, using plans from the Dutch research group under

Hendrik Van der Hulst.[24] Unless it was to be located in the same site at the Ionospheric Research Laboratory at Haringata, Saha's radio telescope would have been constructed in direct competition with the plans of Mitra. Saha suggested this move in August 1955, when he visited Mitra's facilities at Haringata.[25] (This was after the DAE's approval of the Five-Year Plan for Saha's institute, but it is not known how Mitra received it.)

The Dutch plans for the telescope were delivered to Saha, but he was so busy that he looked at them only once before his sudden death in early 1956. He did, however, oblige this same researcher Verma to accept a scholarship to work with the Dutch group, though Verma himself did not want to go. The plans were eventually dropped and the scholarship was changed. Just after Saha's death, the editor of the Calcutta University Physics Alumni 1957 *Annual Report* stated: "At a meeting of friends, admirers and students . . . a proposal to set up a Radio Astronomical Institute in the memory of Professor Meghnad Saha was unanimously accepted. A sum of Rs 500,000 is to be raised from the public for the purpose." Apparently it was never felt that anyone except Saha was energetic or politically powerful enough to complete such a project: the memorial gesture revealed the almost complete dependence upon the honoree, for neither the memorial institute nor the telescope was ever built.

Theory and Experiment

This brush with radio astronomy directs us to consider not only the projects that Saha began but also those he neglected. The clearest example of this tendency is the delay in development of a group of theoretical physicists. The founder of this group in the institute said Saha's main interest in theory was as a phenomenologist: he utilized available theories to explain new experimental data and to bring new data to bear on theory. In the view of three physicists who knew his work, Saha was not a theoretical physicist per se.[26] Saha's own son Ajit went abroad in 1948 to do postdoctoral theoretical work with theorist Max Born in Edinburgh. The first PhD thesis produced in Saha's institute (1950) was done by S. Biswas working between nuclear theory and cyclotron experiments, although the cyclotron was hardly operating efficiently at that time. Sharing his father's diversity of interests, Ajit Saha noted this impact of his father's outlook on everything around them in the institute:

> My father could enthuse a person about anything. If I had not gone into physics I would have gone into archaeology. He was so knowledgeable about so

> many things, and he was such an exception in his generation of the family. Among his siblings none of the others shifted away from business or became much educated; but every one of my own brothers and sisters are doctors, professors, and engineers.[27]

Nevertheless his son's experience abroad did not result in creation of a theoretical physics group even when Ajit returned from studying with Max Born in Edinburgh. S. Biswas soon left for postdoctoral research in the cosmic ray group at the University of Melbourne, Australia. When about to return in 1952, Biswas wrote Saha to see if he could return to the institute (he wanted to be near his family from North Bengal). Saha advised him that with his interest in cosmic rays and nuclear theory it would be better to go to TIFR in Bombay, and he wrote to Bhabha recommending Biswas. Typically, after reading the published work, Bhabha cabled an offer of appointment without an interview, so Biswas went to TIFR, where he did important work on the theory of solar radiation, which was Saha's original field.[28] Manoj Bannerji's dissertation, guided by Ajit Saha in 1955, also had, he said, about 75 percent theoretical content. Expected to construct and use a beta ray spectrometer for his thesis, Bannerji began to read theoretical journals in the library because of difficulties in getting the cyclotron lab "to produce an experimental beam." When Bannerji obtained a scholarship for postdoctoral work in Britain in 1955, Ajit Saha was head of the Nuclear Physics Division, oriented to experimental physics. Still there was no theoretical physics group in which to include bright theorists. While he allowed his son Ajit to work in Edinburgh with Max Born, Meghnad Saha insisted that Bannerji renegotiate his scholarship to enable him to go to Princeton. When Bannerji returned to the institute from Princeton as reader "for nuclear structure theory," he was still responsible to the Nuclear Physics Division; its experimenter-leader Ajit Saha had to approve all expenditures, such as the purchase of sophisticated electronic calculators. Only when Bannerji was made professor in 1959 was the Theoretical Physics Division officially established, reportedly in the spirit of satisfying the expressed wishes of the now deceased Meghnad Saha, who had curiously put little effort into it himself.[29]

The Wider Context

The early days of Saha's institute can only be understood in the context of Science College and its Department of Physics. It is significant that the University of Calcutta's Science College had changed little since its establishment in 1915 and rapid expansion about 1920. Fifty years later there were, in 1970, no facilities oriented to the majority of its students and staff, no bookshop, few toilets, nor any adequate place to eat. The decline of Science College was a function of some of the forces already noted, such as reluctance to expand into new fields, a decreasing amount of support, accompanied by a uncooperative spirit among some its faculty. The college had lost its monopoly of the production of India's best young scientists and technologists during the 1920s and 1930s, though it continued its influence through its students who were highly placed elsewhere, at least throughout northern India. Science College and the University of Calcutta had benefited from Bengal's domination of the national political struggle through to the 1940s, but by 1950, with the capital moved irrevocably to Delhi, those days were over. Could they have seen that, as Bengalis sought to confirm their achievement of a deeper knowledge economy? They were conceding that their manufacturing industries (with one or two exceptions) were organized and financed by non-Bengalis and that those industries were based on the prevalence of coal in the region. The jute supplied to the big mills of Bengal, coming now from East Pakistan, might be cut off at any time. The oil of Assam was limited and far away. Hydroelectricity in the foothills of the Himalayas was very limited too. So Bengal was poised to take advantage of the region's coming coal and steel economy. There was a proliferation of research and development institutions such as the Indian Statistical Institute, the Institute of Nuclear Physics, the Central Glass and Ceramics Research Institute, all planned to modernize Bengal's economy once more. When Independence came in 1947, the plans were already in place for the first Indian Institute of Technology, built as a British investment and opened to fanfare in 1950 in the premises of a prison at Kharagpur previously used for the British government's Indian political prisoners. This completed a long push for advanced technical training begun at the turn of the century, intended to contest the British assertion that Bengalis were *babus*, very good as clerks and poets but not as technical thinkers or workers. This was one of the outcomes of pushing from the Bengal-based members of the National Planning Committee under Subhas Bose, Nehru, and Saha.

But all was not well in the implementation of these ideas. Calcutta University's Department of Physics, in the Science College, was weakened by a process of institutional schizmogenesis, beginning with the establishment of the Department of Applied Physics in 1920. There was only one applied physics professor, Phanindranath Mahanti, from 1946 to 1956. Courses in this department excluded study in the Pure Physics Department. The size of Pure Physics was further decreased when Mitra's radio physics institute split away in 1949, and when Saha took away the only active research laboratory in 1950, the Palit Laboratory, and installed it in his autonomous institute. Despite their proximity, the drifting apart of these three units was driven by financial and personal tensions, which the university administration did little to counteract. The existence of other "distinguished colleagues" in these nearby institutions seems to have made little difference. In the Science College anyone not a full professor was virtually powerless, and professors became like lonely rajahs of small kingdoms. Posts given long ago on merit became sinecure rewards for tenacity. Of course there were exceptions: in 1946, Satyen Bose returned to Calcutta from Dacca just ahead of Partition, to become Khaira Professor of Physics, the post Saha held in 1921. Given his international prestige, Bose soon became dean of the Science College, but any efforts he might have exerted to raise the quality of ideas and research, he told me whimsically, had little effect: he explained the difficulties of Science College as only a part of the larger malaise of the university and of Bengal.[30] By common agreement Bose did not do any further physics comparable to his papers in the 1920s, certainly not after returning to Science College in Calcutta in 1946. Nevertheless, it is impressive that at the end of the twentieth century there was a consensus in India that more than half the best physics students applying for positions in India and abroad still continued to come from Science College in the University of Calcutta, and numerous people across India have asked me "How is it still possible, under those conditions?"

Saha attempted to change this adverse situation from within: it seems his allies and his energy just outweighed the forces pressing down on the creation of his new institute. Before beginning reorganization at Calcutta, Saha's experience was set in a world of science as heroic individual effort. It was easier to rely on the cooperation of one or two students than create a colleague; most theory and experiments were kept quite separate. The new Institute of Nuclear Physics, said one senior member, "was run like the old days of the early 1940s, like a small family." Despite Saha's concern for planning, he allowed the city to take haphazard care of his staff's transportation and housing needs. He himself traveled to work by bus in the early 1950s. Since it was run like a small family, when Saha was in Calcutta, he probably felt he could handle any problems that arose among his staff. Becoming busy with politics in the 1950s, Saha delegated more and more of his responsibilities and apparently did not check up on the outcome of such delegation, though when he discovered some new discrepancy or problem months later, his famous temper exploded. D. M. Bose has recalled Saha's style at this time:

> At any one time he could work on more than one project. When he got tired of being tied up with a project after it had passed the foundation stage, he took up another. Saha was getting tired of the close attention he had to give to the organization of the Institute of Nuclear Physics. With the formal opening of the Institute, he handed responsibility over to his former pupil B. D. Nagchaudhuri, and then accepted the offer from the Council of the IACS in 1951 to become Director of that Association. This gave him sufficient freedom to divert a part of his time to other problems that more and more attracted his attention.[31]

Saha was very conscious of the difficulties of poor people, and said so frequently, but in building the institute's research program he seems to have been unable to provide for transportation, housing, or a cafeteria, the concrete needs of his workers. This was in a congested metropolis that provided little of this kind of support itself. Some of his institute's workers lived in poor households, though his institute gave better working conditions than they could usually find elsewhere. It was these outward conditions that Saha hoped to change through his political work. Well ahead of the publication of the results, he knew the depressed living conditions of students of the university being studied by its own Department of Anthropology in 1954.[32] These physical, economic, and social problems were quite similar to the university's problems he described to the Sadler Commission of the University of Calcutta in 1919, when Saha began teaching physics. Just as the curriculum had changed little, so the problems resisted feeble attempts to solve them. Saha hoped to tackle them in a different manner. He wanted to change the entire system.

NINE

Bhatnagar Builds a Chain of National Laboratories and Steps Upward

Shanti Bhatnagar began planning a chain of laboratories prior to leading the overseas tour of scientists in 1944–45. Using CSIR committees for advice and his mandate and mobilizing his considerable international network, he sought and found money outside his small annual government budget allocation in 1944 of Rs 1 million. For example, he helped to arrange for the grant from Tata Trusts for the National Chemical Laboratory long before it was built. Pointing out that in 1943 alone the processes released to Indian industry accounted for goods valued at Rs 50 million, he knew very well which industrialists had profited from this war-oriented production. He espoused a vision of harmonious collaboration between big industry, state needs, and an autonomous scientific community and agitated among these industrialists for more money. By 1945 his budget had been increased ten times. In 1948 he wrote an account of his wartime and postwar activities:

> Since 1940 my work is mainly in connection with war work particularly with respect to industrial development by organizing scientific and industrial research throughout the country. I am also a member of the Supply Development Committee of the General Headquarters and in my laboratory we have liaison officers from the Master General of Ordnance, the Royal Air Force and the American Air Force and we are conducting war research-work with all sorts of organizations including those in China.[1]

He thought a transition to Independence might not require the immediate departure of British capital and know-how. As he said in 1945, "I am convinced that if the Indian and British businessmen had cooperated in enterprises brought into being by common capital, even the political factors which now sometimes impede our industrial progress would have long ago disappeared."[2] He observed in 1944 how this cooperation was organized in wartime Britain and America and, encouraged by Archibald Hill, was given the confidence of Ardeshir Dalal and Nehru that enabled him to establish the CSIR chain of laboratories. This confidence was reinforced when Dalal hired Bhatnagar's eldest son, Anand, to be his assistant industrial advisor in his government office. Bhatnagar was soon to become fully aware of the political issues that face scientific managers—the protection of inventions by patent, the need for secrecy, the cost of training young scientists abroad, the difficulty of ensuring their return and risk of losing them, the appointment of top-quality first-time directors for new institutes, the problem of their succession, and so on.

Bhatnagar's relations with the Attock Oil Company, a subsidiary company of Steel Brothers of Rangoon, became more important during the war. The company was a refugee from Burma's oil fields because its refineries and wells there had been destroyed in March 1942 in order to deny their use by the advancing Japanese troops. The Steel Brothers conglomerate took up residence in India, and Attock Oil continued to prospect for oil about 90 km south of Rawalpindi, finding an important new well in 1942–43. The Attock district's importance rose sharply when Japanese troops were poised briefly in 1944 to seize the oil fields of Assam. Sir Louis Dane, former lieutenant governor of the Punjab and a person well known to Bhatnagar, was now ex-governor and chairman of Attock Oil's board. The business of making kerosene, petrol, paraffin, and asphalt/bitumen was very profitable during the war, selling both to a growing Indian market and to Allied military forces in the Middle East.[3]

Bhatnagar held patents in his name for work done on emulsions and lubricating oils at university laboratories in Lahore, although in at least two cases his name was attached to the patents after the application had been filed by others in the lab.[4] He was not particularly oriented to patents or secrecy and was comfortably settled at the end of the war, and so in 1945 he decided to place his own personal royalties at the disposal of the new CSIR Governing Body. Perhaps this was done as an example to others. The Governing Body then turned these royalties worth Rs 250 over to a CSIR staff member in difficulty, as an honorarium.[5] Bhatnagar, reviewing the patents held by Indians in 1945, concluded that "weaker countries like India suffer particularly as their scientific work is sometimes exploited even without acknowledgement."[6] A year later he instituted a Patent Expert Office in the CSIR with responsibility for patent legislation and trusts and placed it under the supervision of his former student, friend, and assistant director K. N.

Mathur.[7] Mathur was already acting for him with respect to planning the NPL and was one of the Bhatnagar-Mathur team that developed the magnetic balance, by then enjoying profitable sales to laboratories. But patents were still a problem for the CSIR when Blackett began to meet Bhatnagar in 1948, as we shall see.

And in addition to work on patents, Bhatnagar faced the famous problem of the drain of educated Indians abroad, a drain that showed up immediately when the war ended in 1945. He had to answer Hill's direct questions from London about unsupported and unauthorized Indian students arriving in Britain in a context of rationing of all food and supplies and an economic crisis for the British: women stood in long lines in the street waiting for the same commodities that had been rationed in the war. Evidently the requirements and requests of these students had become a problem, because Hill was asked by authorities to inquire about them through the government of India. Bhatnagar found out about it and wrote immediately to Hill saying, "You have complained to [the high commission] of certain students whom we had permitted to go to the UK without securing admission. . . . Gov of India took precautions, but it is extremely likely that some people have managed to go abroad as business travelers and eventually offered themselves for university education."[8] One must imagine the expectations of young Indian men arriving in an exhausted country, having traveled below decks in troop ships for two weeks on poor food, subjected alternately to kindness and contempt. This sea experience was, by all accounts, formative. People thrown together from all parts of India met others destined for greatness (among them Indira Gandhi, future presidents, and the like), and many young scientists were among them. These relationships started on the docks and on board ships. Seeing Britain in a decimated state awed those who had grown up impressed by its great power. They began building networks in British universities, finding satisfactory accommodation, tolerating the dull and scarce food, enjoying the thrill of new freedom from family and community oversight, discovering how to study and work on their own in a postwar environment, learning to lower their expectations.[9] Some of these students were on scholarships that were given by the UK in lieu of cash reparations or instead of the sterling balances. As head of the CSIR, Bhatnagar became the channel for Indians abroad seeking further support and changes in policy; for example, metallurgist G. S. Tendolkar at Sheffield University told him in 1947 that most Indian scientists found it impossible to get into factories in the UK for practical training after completing their theses—whether in textiles, glass, or metallurgy.

> Why should the Indian students not be sent to Germany so that they would learn the necessary techniques there? Why should we depend upon Britain? If India sends something like 500 students to Germany just for the sake of practical training that would be more profitable in the long run than accepting reparations in kind . . . a University degree is of no use if it is not backed up by practical experience: since this is denied, most of us feel frustrated in Britain.[10]

This was the start of a long political preoccupation with training Indians abroad, coaxing them back to work in the right place, promoting or not promoting them, and managing their disagreements with their institutions. This letter also stood in a legendary Indian tradition in which bright students wrote directly to influential leaders, even Nehru, asking for their intervention. In many cases, including Bhatnagar's, they got results if they said the right things, wrote to the right people, or were lucky with timing. In the cases of Bhatnagar and Bhabha, these students were approaching men with the capacity to employ literally thousands of young scientists, and knew that these top scientists were prepared to hire students who were abroad.

Equally important, Bhatnagar began in 1945 to make national science policy and to represent India in international circles. Before the war was over, Sir Edward Appleton wrote to him to solicit his cooperation in holding the Empire Scientific Conference after the war; Bhatnagar received the same letter as did heads of the departments of scientific and industrial research in Australia, Canada, New Zealand, and South Africa and was thus considered one of the scientific leaders of India. The purpose was "to consider the best means of ensuring the fullest possible collaboration between the Civil Government Scientific Organizations of the Commonwealth and to make formal recommendations for the approval of the governments represented."[11] Presumably in light of British efforts toward an atomic bomb project, it was agreed that the conference was specifically not to consider "matters of liaison on secret defence science." Such matters were indeed discussed at a separate defense science conference that ran concurrently. Although the conference planners called it an "Empire" conference in 1945, when it occurred in 1946 it was called a "Commonwealth" conference: times had changed.

The eleven-member Indian delegation to the conference finally traveled to London in June 1946, as guests of the Royal Society. India had been asked, "which one of the Indian delegation would be its leader," noting that Sir C. V. Raman was the most senior Fellow of the Royal Society in the list. On hearing of the invitation to come to London, Raman stated bluntly that he objected to the empire and did not care about the Royal Society and would not attend; this was reported in the press, as were most of Raman's public utterances.[12] In writing, however, Raman made a polite excuse for avoiding the conference. Bhatnagar and others spoke in favor of persuading Raman to attend, but clearly did not want him as leader of mission. Not surprisingly, Raman finally decided he could not be left out and traveled to London as a member of the delegation too. "In view of certain delicate personal problems of which Professor Hill will be aware," the government of India wrote to the Royal Society, "we have decided not to appoint any one as a leader of the Delegation. Bhatnagar will be leader only of the official group," meaning only those who were civil servants like him.[13]

The CSIR had to get the per diem rates for Indian government officials increased because of the high cost of staying "in such swanky places as the Savoy or the Mayfair Hotel," as Bhatnagar put it; less money was available now than during the scientists' tour in 1944. Bhatnagar traveled in deluxe diplomatic style by ship to London, stopping in South Africa on the way to meet chemical industry contacts, as he had just become vice president of the Society of Chemical Industries, an important international body. It is possible he also inquired about South Africa's capacity to sell uranium, as he was already secretary of the Atomic Energy Committee of India. The Indian mission prepared position papers on all subjects relevant to development—energy and fuel, food and agriculture, chemicals, minerals, and the like. A week each in Oxford, Cambridge, and London kept the conference moving upward to a level of official intergovernmental debate and approval. With four members of the Atomic Energy Committee present (Bhabha, Bhatnagar, Saha, Krishnan), there were many informal contacts with such physicists as John Cockcroft, Edward Appleton, W. B. Lewis, Marcus Oliphant, Patrick Blackett, William Penney, and James Chadwick who were deeply involved in British strategic nuclear planning and research. The mission was composed of people who had achieved international reputations (such as FRS), and national prominence, the very people who formed the nucleus of a scientific elite on the eve of Independence.

When Bhatnagar returned to Delhi from England, the sky was darkening. He was a man whose wife had died just as he was preparing to leave for London, but, having attended to her funeral, had decided to travel to London anyway. The prospect of Partition was on the horizon. Bhatnagar even worked as a volunteer after office hours for the Red Cross in Lahore when he visited that city in late 1946. He was about to see his beloved city cut off from him and, like many thousands, to be sent "into exile" by Partition. He would lose his house and property.

But he had children well-married, he knew Muslim and Hindu culture well, and his students continued to write faithfully to him from Lahore while he worked in Delhi.[14] Professionally things were going well. Two years before Independence, Bhatnagar had been finding the land and architects for and starting to build and staff large laboratories—separate from universities—in the same way that this was being done in Canada, Australia, France, the United States, and Great Britain. His work was part of a grand design. There had been, however, occasional crises, as in 1945, when he and Ardeshir Dalal felt the government was about to abandon the CSIR or change the department that managed it. Dalal talked of resignation, and Bhatnagar proposed to follow suit, but when these rumors were heard, Bhatnagar was told to stay in his post, because things would soon change. The CSIR grant was then raised in 1946 from the 1945 level of Rs 10 million to Rs 16 million. Their ultimatum had paid off for Bhatnagar and Dalal, and the budget was increasing dramatically from the Rs 1 million in 1943. Dalal and Bhatnagar could soon show that the CSIR had earned Rs 10.68 million in royalties from industries utilizing CSIR patents between 1944 and 1948, attached to processes producing hundreds of millions of rupees profit for those industrialists.[15]

Bhatnagar was not shy about using his credentials in contests with authority, and he earned a good reputation for his boldness. For example, in 1946 he raised his reputation dramatically in the civil service when a Delhi traffic accident killed Dr. R. K. Pillai, a young biochemist on his way to work at the CSIR laboratories. Finding that the city's chief medical officer had not conducted a timely postmortem at the accident site, he intervened with Delhi's most senior (British) police officer, going above the head of the chief medical officer, an Indian surgeon: Bhatnagar announced he would take the body away for cremation before sundown as Hindu custom for cremation then required. If this postmortem were not done, Bhatnagar accepted legal responsibility himself for removing the body prior to the postmortem. The British police officer came to the mortuary to see the body and observe. Finally the chief medical officer arrived, very late and near sunset to confront Bhatnagar in the course of removing the body for the promised cremation, despite the absence of an autopsy. The surgeon "threatened him with dire consequences for interference in the performance of his duties." Dr. Bhatnagar then revealed his identity as secretary to the government of India and a knight and told him he was leaving for London that very night but would see to it that the surgeon was suspended "before the sun rose the next day." He started the process of removing the body for cremation and said that after cremating it he would meet the viceroy, who had invited him for dinner. Shocked, "the surgeon then apologized for his discourteous behavior, introduced himself as a former student of Dr. Bhatnagar in Lahore, and requested a few minutes time to complete the formalities."[16] Bhatnagar thus mastered the infinitely subtle distinctions in hierarchy upon which the Indian administration thrived. The drama with the police officer, the absent medical officer, and the reference to dinner with the viceroy were not forgotten. As the accident victim was a CSIR employee from South India, this event raised his reputation inside the rapidly growing organization, confirming its cosmopolitan character.

Bhatnagar left just after this accident for the Commonwealth Science Conference in Britain as head of the official delegation. While at the conference in Oxford, Bhatnagar "saw flames coming from the organic chemistry laboratory, when he passed it on a walk. Finding the door locked . . . Sir Shanti climbed to a window, forced his way in, and put out the blaze." Herbert Morrison, senior member of the Labour Party announced Bhatnagar's timely act to the whole conference, saying it saved Britain many thousands of pounds.[17] This occurred precisely at a moment when there was public criticism in Britain of the large amount of sterling balances the UK owed India, which Britain, while acknowledging the debt, was currently unable (and unwilling) to pay to India. The second Cripps Mission had just returned from India following inconclusive constitutional talks, and the relationship needed this kind of boost. This vignette about putting out the fire in the lab added to British elite approval of Bhatnagar and to his legend in India: he was just the kind of man to climb through the window of a burning lab and force his way in! Just what India needed, said senior officials taking their early morning walks among the tombs in Delhi's Lodi Gardens.

Appointing Laboratory Leaders

As CSIR director Bhatnagar could now make significant appointments in his chain of laboratories by using his influence beyond it. For example, he appointed to the CSIR in 1948 S. Bhagavantam, a physicist who soon became the scientific liaison officer in the Indian High Commission in London.[18] Raman's student in crystallography, Bhagavantam eventually left the CSIR to become scientific advisor to the minister of Defence in 1961, succeeding Saha's student D. S. Kothari. More important, beginning in 1945, Bhatnagar selected a number of foreign scientists to be CSIR directors and used the same method by which he was himself selected as professor of chemistry at Benares in 1921 (the "English committee"). Following Sir Alfred Egerton's advice, he appointed Professor Whitaker at the salary of Rs 3,000 per month for a five-year contract as director of the Fuel Research Institute outside Calcutta. The intermediary between Bhatnagar and Egerton in this case was Archibald Hill,[19] who was simultaneously acting in a similar manner for Jnan Ghosh, director of the IISc in Bangalore, also keen to find experienced new heads for some of his departments. Bhatnagar and Ghosh offered similar high salaries. The IISc was looking for prestigious professors and researchers outside India. In 1948 Bhatnagar appointed George Sachs, an American metallurgist, as director of the National Metallurgical Laboratory and appointed British chemist J. W. McBain, FRS, as director of the National Chemical Laboratory, and Sir Edward Mellanby (briefly, for five months) as director of the Drug Research Laboratory.[20] A year later the Austrian-American glass scientist N. J. Kreidl was appointed director of the Central Glass and Ceramics Research Institute in Calcutta, the lab in which Saha was so deeply involved. Clearly these Indian institutions were authorized by government to pay competitive international-level salaries, and few appointments lasted longer than two years. This would have been approved by Nehru himself, president of the Governing Body of the CSIR; because the council was legally registered as a society and was not strictly a government department, it enjoyed some autonomy from the public service regulations of that time.

These short international appointments were, in the end, the result of complex and unpredictable negotiations: the National Chemical Laboratory (NCL) and the National Physical Laboratory (NPL) are good examples. After long preparation of the NCL's site at Poona, the construction slowly began using the large Tata gift. Chemist Salimuzzaman Siddiqui had been circulating in jobs within Bhatnagar's influence for some time, in Calcutta, London, and Delhi. He was head of the CSIR's chemical laboratories in Calcutta starting in 1944; he urged Bhatnagar to move quickly into atomic energy in 1945, writing to Bhatnagar from the office of the High Commission in London, where Siddiqui was scientific attaché. His work in the laboratories of the Unnani Tibbi and Ayurvedic College of Delhi identifying indigenous plants and processes impressed Bhatnagar so much that he appointed him director of the NCL in early 1947. In fact the CSIR chemical laboratories, soon to become the NCL, were temporarily located at the Ayurvedic and Unnani College in Delhi, where Siddiqui was actually in charge. There was a strong interaction between researchers and students in this lab, just as Bhatnagar had achieved at Lahore. Siddiqui's was precisely the kind of appointment long called for by Indian industrialists and chemists because it would enable the lab to interpret and apply to modern problems the rich potential of Indian pharmacopoeia, natural products, and indigenous knowledge.

There was also another implicit policy at work, to demonstrate that India was hospitable to Indian scientists who were also Muslims. This policy anticipated the mutual accusations and retaliations that occurred when Hindus or Muslims left senior positions in both countries, after Partition. Siddiqui was a long-serving insider and appears to have built a network of research colleagues and students in natural products chemistry; there is evidence of widespread satisfaction at the announcement of his appointment as director when the NCL foundation stone was laid in April 1947.

But by August 1947 the political situation had changed and Siddiqui had to flee from his house near Delhi University and take refuge in the home of a minister of the government, Rafi Ahmed Kidwai, whose house had regular police protection, accorded to all cabinet ministers. Siddiqui's elder brother was a high-profile leader of the Muslim League, and when this brother "suddenly went away to Pakistan," Siddiqui's future went with him. According to a witness, as a result of his brother's sudden departure Salimuzzaman Siddiqui's "position in a key post in CSIR became untenable. . . . Perhaps because of this, the appointment of Dr. Siddiqui was cancelled and after some time, J. W. McBain FRS was appointed instead."[21] After Partition and Independence, Siddiqui remained in Delhi with the CSIR, sustained by Bhatnagar as an assistant director, working with two or three researchers at the Tibbi College, where he had always been. But he was marginalized. In April 1951 he moved to Pakistan and became the first director general of the Pakistan CSIR. The Muslim and Hindu communities of North India were being torn apart. Three years later, in October 1954, Siddiqi and the Pakistan CSIR began to administer a training program for nuclear scientists and explorations for radioactive minerals.[22]

The story underneath the McBain appointment (or part of it, anyway) suggests that not all the best choices, or at least those available to Bhatnagar and the CSIR Governing Body, were viable under the political circumstances. McBain insisted on making his own appointments for the new chemical laboratory in Poona. And of course, his appointment, like others, was not uncontested. According to minutes of the council of the CSIR, "M. N. Saha objected to the appointment of Dr. McBain for several reasons."[23] This is quite consistent with Saha's disagreement with the involvement or appointment of non-Indians in Indian scientific institutions. Nevertheless Saha gave his grudging agreement to the appointment of N. J. Kreidl of Bausch & Lomb USA as director of the CGCRI in Calcutta.

The appointment of the director of the NPL was also complicated. It appears that Saha thought he had reason to expect the job, which may be one of the reasons he advocated it be located in Calcutta. Nazir Ahmed challenged Mathur and Bhatnagar's definition of the post, thinking Saha was senior enough to merit it, and he certainly was. We do not know if Bhabha or even Raman (age fifty-eight in 1946) were considered for the job, but its sheer prestige suggests that they were. Yet Raman's student Krishnan, ten years younger, got it. K. S. Krishnan had left Raman's lab in Calcutta in 1928 to join Satyen Bose at Dacca University and returned five years later to Calcutta to the physics chair that Saha himself coveted, in 1933. He stayed in Calcutta almost ten years, before becoming professor of physics and department head at the University of Allahabad, where Meghnad Saha had previously been. (The Allahabad chair taken by Krishnan had also been offered to Bhabha in 1943.) It was from Allahabad that Krishnan moved to become NPL's director in 1947, having already established an international reputation for work on scattering of light in liquids, magnetic properties of crystals in complex structures, and electrical conductivity of metals and alloys at low temperatures. He was particularly an expert in the properties of graphite, and by 1946, when graphite was clearly a strategic commodity as the moderator in the fission process, Krishnan became deeply involved in the atomic energy program with Bhabha, Bhatnagar, and Saha.[24] Unsuited to the kind of administration a director simply had to do, Krishnan was content to let each group "do its own thing"—a practice that ran counter to the growing culture of administered government science, but one consistent with a philosophy of unrestricted play of curiosity and inquiry in science. When Bhatnagar died in 1955, Krishnan no longer had an immediate superior at the head of CSIR who tolerated his style. Inquiries were subsequently aimed at the management and organization of the NPL until Krishnan retired in 1961.

The labs for chemistry, physics, metallurgy, pharmaceuticals, and fuel were considered the more prestigious of the CSIR projects. Later Bhatnagar turned his attention to the more prosaic subjects like building and road research. After his death the National Aeronautical Laboratory was opened in Bangalore, acknowledging the concentration of the aircraft industry there and presaging the development of space research. At the same time, Bhatnagar was able to fund institutions outside the CSIR chain, such as in 1946 when he approved continuing support for the new institute of his old friend and ally Birbal Sahni, distinguished paleobotanist, elected FRS in 1936. In 1942 and 1943 Sahni signed Bhatnagar's nomination for the Royal Society fellowships: Sahni himself had a supporting signature from C. V. Raman in

1936. Sahni complemented Bhatnagar's knowledge of the chemistry of petroleum with a prodigious knowledge of the geological circumstances and location of possible oil resources, and, like Raman, Sahni invested his own personal savings in his research. The Institute of Paleobotany in Lucknow became India's premier international research center in the subject before Sahni's sudden death in 1949. Geologist D. N. Wadia and paleontologist Birbal Sahni constituted a powerful team in minerals and petrochemicals in 1947 and 1948, and we shall see that Bhatnagar, as well as Bhabha, needed just such a team.

Bhatnagar also turned his hand to medical research, helping S. L. Bhatia to persuade Archibald Hill to speak to Lady Mountbatten about the creation of the All-India Institute of Medical Sciences in Delhi. Edwina Mountbatten's strong ties to nursing and very close relationship with Nehru were well known to Hill and to Blackett, who was also consulted on the medical sciences institute. As Hill said later, "I asked his [Mountbatten's] wife who was keenly interested in medical and particularly nursing services to keep an eye on it. That may have helped in the implementation of the Bhore Committee's Report which had been published in 1946."[25]

But not only did Bhatnagar have to staff them; he also had to equip the laboratories, so every trip abroad was a voyage of procurement. He alerted Hill to the visit of a senior official being sent to London to buy equipment in 1948. Hill explained to the CSIR official how to do his shopping through the office that disposed of surplus government equipment: "We have profited very greatly in my laboratory by tapping this source of supply, and indeed the whole University has profited too; I hope that some profit may result to your laboratories."[26] Bhatnagar had become, like Bhabha, a major spender—building large buildings, equipping and staffing them, providing vehicles, gardens, and the like; his empire was growing as quickly as any other government program in India.

Bhatnagar was now in a position to invite influential foreign experts to visit and work in India. At the Empire Scientific Conference in 1946, P. C. Mahalanobis and Bhatnagar had proposed a scheme called Short Visits of Scientists from Abroad, and Mahalanobis followed up with a list generated by all Indian scientists attending the conference. This list included Robert Oppenheimer, Norbert Wiener, and Niels Bohr as well as the British names Patrick Blackett, J. B. S. Haldane, Henry Dale, Henry Tizard, and Joseph Needham, more familiar to Indian audiences. Nehru, who had by 1947 met all these people, agreed to sign the invitation letters personally to give these visits prominence.[27] Eventually each of the individuals on this list, with the exception of Oppenheimer, came to India and got involved—some more, some less—in the development of their field and Indian research institutions. Weiner was Mahalanobis's choice, Oppenheimer and Bohr were Saha's and Bhabha's choices, and so on.

This worldliness in Bhatnagar and the "nobility" he cultivated in dress and speech (after all, was he not a knight?) were intriguingly balanced with his taste for the dirt and smell of industry, a familiarity with leather, paint, oil, mud, and dyes—all the things that elite Indians in "the right circles" were supposed to abhor, according to their own self-stereotype. The millions of workers whose trade and livelihood depended on that dirt and smell could therefore relate to Bhatnagar, and he to them, in ways that were difficult for K. S. Krishnan, Homi Bhabha, or even Meghnad Saha to contemplate. Bhatnagar had established himself as a good friend of industry, and he was now receiving the benefits.[28]

Bhatnagar Becomes Secretary of Natural Resources and Scientific Research

Six months after Independence, the administration of scientific and industrial research changed. Bhatnagar was made secretary of the new Department of Natural Resources and Scientific Research, and the minister to whom he began to report was the prime minister, with an intermediary minister Maulana Abdul Kalam Azad. There was also, by early 1948, a minister of State (of lesser power than Azad) for Natural Resources, K. D. Malaviya, Saha's student at Allahabad. It is unlikely that these ministers would have been able to push Bhatnagar around because the CSIR remained "autonomous," with the prime minister as president of its Governing Body and Bhatnagar as director general. This autonomy mirrored the autonomy sought by Homi Bhabha, who became chairman of the new Atomic Energy Commission and the Atomic Energy Board; the minister to whom he reported was also the prime minister. Besides being the minister for Foreign Affairs, Nehru was now the minister for Science and Atomic Energy. Bhatnagar and Bhabha were essentially like deputy ministers, top civil servants with direct access to the prime minister or cabinet. Nevertheless, Bhatnagar hastened to explain to Hill, "the new arrangements do not centralize science in one department. They give a chance of some coordination at a high level."[29] As a Conservative opposed to nationalization under way in Britain, Hill remained cautious about the risk of centralization: "we are in danger of too much centralization here, indeed one cannot help feeling, perhaps ironically, that the Socialist Party [i.e., Labour] is so opposed to the 'profit' motive that they are anxious to replace it by a 'loss' motive, which seems to be likely to result from nationalization. This comment, however, does not apply to India where conditions are so different, and where rapid development is so necessary which can scarcely take place by individual enterprise and initiatives."[30]

As the new central government took shape, Bhatnagar now was in regular contact with ministers. He had more experience in Delhi than any other scientist and more experience with the civil service than most new ministers. He had greater knowledge of industrial and scientific policy than any Indian Civil Service officer. Though he was not one of them, he had managed to achieve a secretary's position without qualifying as an ICS officer. Surrounded by higher-status administrative ICS officers, some of whom, like him, had been educated in London or at Oxford, Cambridge, and Edinburgh, he often stated that he believed "in scientists on top not just on tap." That was one reason why he gave full support to Bhabha. Most other influential figures, of course, preferred quite the reverse, "scientists on tap."

Skilled as an intermediary, Bhatnagar arranged in 1948 for Bhabha to meet with the new minister of Defence Baldev Singh, in order to negotiate transfer of the navy's land in Bombay on which TIFR now stands; in this meeting Bhatnagar also acted as secretary of the new AEC. Bhatnagar now introduced the new minister of Finance to Archibald Hill in London, who then guided the minister to meet the right people in the new UK government. He also promoted the cases of physicist S. Bhagavantam, Raman's student, for the position of scientific attaché in the High Commission in London, and chemist Atma Ram, Saha's student, to a visiting scientific position through the embassy in Washington. Both Bhagavantam and Ram were to play important roles in the scientific elite in the 1960s, the former becoming defense advisor and the latter the director general of CSIR. On the personal side, Bhatnagar got help from Hill in 1948–49 in securing admission for his son Devendra to University College, London. He also asked Indian High Commissioner Krishna Menon to lend his son Devendra a hundred pounds if he was ever in an emergency.[31] Nevertheless, Devendra is found two years later, back in medical school in India.

Until his death in 1955, there was no one with the knowledge and daily influence on the range of industrial and scientific issues that compared with Bhatnagar. Probably because he was secretary to the Ministry of Scientific Research, Bhatnagar became secretary of the newly formed cabinet Committee for Coordinating Scientific Work. Chaired by Nehru, this committee had by 1952–54 two cabinet members, Minister Maulana Azad, for Natural Resources and Scientific Research, and Minister K. D. Malaviya, of State for Natural Resources, and unelected scientific members Homi Bhabha,

D. S. Kothari, and P. C. Mahalanobis, statistical advisor to the cabinet, plus the director of the Indian Agricultural Research Institute B. P. Pal and secretary of the Indian Council of Medical Research C. G. Pandit. This committee eventually became the Scientific Advisory Committee to the Cabinet (SACC), whose work will be analyzed in detail below. This was the political side of Bhatnagar's role as director general of CSIR, and for these reasons he had direct access to the prime minister. His chain of institutes and laboratories was flourishing and all this reflected well on Nehru.

Though they were both poets, Bhatnagar's relationship with Maulana Azad was quite instrumental, Azad being the intermediary to the cabinet if necessary. Bhatnagar could also see Nehru alone, if on AEC business. In effect the two files converged. Azad was born of an Arab mother in Saudi Arabia and spoke his father's Punjabi and Urdu and his mother's Arabic. He came to Calcutta in the 1930s and learned Bangla there. This is where Bhatnagar would have learned about him. A newspaper publisher and supporter of the revolutionary Jugantar party, Azad was also known to Saha, particularly after the British banned his newspaper and he was arrested and imprisoned in August 1942 while in the middle of writing a letter to President Roosevelt.[32] As a government official Bhatnagar would probably have been very cautious in communicating with Azad until about 1946. Azad had no background in his portfolio of natural resources and therefore relied greatly on Bhatnagar, who did.

At the same time as he ran the CSIR, Bhatnagar was secretary of the AEC under Bhabha's chairmanship. Although its offices were in Bombay, effective separation from Delhi was neither practical nor wise and thus Bhatnagar had, as AEC secretary, much to do within (and against) the bureaucracy on Bhabha's behalf. Between them Bhatnagar and Bhabha commanded most of the funds for scientific research in the country. Although this satisfied Bhatnagar's need for effective power, it alarmed Saha, who wrote in 1948: "Power hunters have gathered around Panditji [Nehru] and are trying to mislead him. Recently a proposal was brought before cabinet of concentrating all research under the premier, with Bhatnagar as Secretary . . . so there would be a partition of research, just like the partition of India?" And later, "I think you know that the Council [of CSIR] has been deprived of almost all power which has been given to Dr. Bhatnagar."[33]

In Nehru's opening address to the cabinet's Committee for Coordinating Scientific Work in 1948, he emphasized that its job was to recommend action to ministers, but it soon appeared that there were very few matters on which ministers knew enough to contradict members of the committee. In answer to a crisis in scientific supplies, Nehru said the bureaucratic procedures must be simplified, including new foreign exchange coupons for the importation of scientific supplies and equipment. Later, the committee decided that researchers sent abroad for training at government expense were to be given meaningful positions quickly, even if of a temporary "supernumerary" nature. Bhatnagar immediately tabled his "Note on Physical and Psychological Effects of Working Conditions" to explain why regulations for employment of scientists had to be unlike those for civil servants and why scientists needed different rules governing their appointment and promotion. Finally, the committee pointed to the CSIR procedures on patents, which had been crafted by Bhatnagar, as ideally suited for India and said they were to be followed by all other government departments. With Nehru thinking of uranium and oil and leading the committee, members agreed that the Geological Survey of India should be greatly strengthened and that cooperation with both the Canadian and Australian Bureaus of Mines should be extended for joint exploration and training. Other fields to be strengthened were road and building research, for which Bhatnagar wanted and got special laboratories, food preservation and aviation medicine which Kothari wanted for defense, and marine biology.[34]

The reference to mines and minerals had a very great importance for Bhatnagar and Bhabha. Bhabha was preoccupied with minerals like beryl ore, thorium, and uranium. Bhatnagar was preoccupied with oil. International resource corporations were dramatically extending their reach after the war, and an independent India offered new opportunities, particularly to American firms, hitherto blocked by colonial trade protection policies. Since there was a policy vacuum and Indian legal regimes were not yet in place, the corporations reasoned that they might secure a good starting position during this ambiguous period before new trade and legal regimes were in place. Bhatnagar personally reviewed all proposals for exploration and development that required government permission and had an unofficial network involving geologist D. N. Wadia and paleobotanist Birbal Sahni to evaluate most of these proposals. Although Bhatnagar and Wadia separated their secret work on uranium and thorium for the AEC from their work on other resources like oil, they both required exploration in the field.

India was about to enter the petrochemical era in a big way, and Bhatnagar explained in 1952 why and how it had to happen.[35] After reviewing the long history of prospecting for oil in India over the previous ninety years, Bhatnagar's lecture concluded:

> Compared to world production, production of petroleum in India is very meagre. . . . In 1948–49, world production was of the order of 3.4 billion

> barrels of which India's share was of the order of 1.9 million barrels. Important fields were taken away from us when Burma, which is now producing over 300,000 barrels, was separated, and now the creation of Pakistan has deprived us of the Khaur and Dhulian fields in the Attock district of Punjab where production is . . . 490,000 to 746,000 barrels. The only important oil field that we have at present is in Assam at Digboi . . . and production there is of the order of 1.9 million barrels.

Bhatnagar went on to explain that in 1951–52 India had to import petroleum products valued at Rs 707 million, and imports, mostly from Iraq, had risen steadily each year since 1947. The search was therefore intensified for new oil fields, using the very aerial magnetometer that he and Saha had seen in the United States during their wartime tour of 1944–45; it was flown suspended over the shoals of the Bay of Bengal behind a Standard Vacuum Oil Company aircraft in 1948. "No other survey would have been so cheap and yet so expeditious," said Bhatnagar, anticipating criticism from people like Saha. Burmah-Shell Oil was also conducting ground surveys in Assam. Moreover, he said, the major refineries under construction by Burmah-Shell, Standard Oil, and Caltex were "potential harbingers of new industries and besides stimulating Indian science and technology, they will, by providing raw materials, assist in the development of many chemical and other industries." Moreover, he said, "Indian taxation on the profits earned will add to the national coffers." These were the very companies and practices that Saha criticized so publicly, yet everyone knew that Indian imports, mostly from Iraq, had a large negative effect on the economy: it was not just the rich who were interested in petrol. The poor depended totally on kerosene for light in their lamps at night.

Bhatnagar and Bhabha thus knew intimately and empirically just how constrained the Indian economy would be by the absence of fuel and electrical power. Bhatnagar was not particularly gleeful about having to make contracts with powerful oil companies but saw no Indian capacity, either private or public, to go ahead alone, as Mexico had done in the 1940s by nationalizing both production and refineries. Refining and marketing were left in private hands, but exploration and production were taken over by the state; the state then built its own refineries to compete with the oil companies in marketing. This oil crisis convinced Bhabha, and Bhatnagar too, that nuclear power was now even more important: Bhatnagar worked with Bhabha until the end to secure the finances, equipment, and raw materials for the nuclear program, in addition to his responsibilities for natural resources and scientific research. Far from being the somewhat marginal player that he is often thought to be, Bhatnagar was in fact central to the government's energy and nuclear policy and practice.

The issue facing the advisory committee to the cabinet, the council of CSIR, and the AEC was how to identify the right people to lead all these new facilities and how to ensure they would do the right thing. Until about 1950 the opportunities had been sufficiently limited that the evolving "old-boy networks" were the means by which most appointments were being made. But these networks had to be enlarged and new and confident methods brought into place to find competent people for new subjects for which the network was unprepared. Put the other way, potentially competent people had to be identified, sent away for advanced training, and brought safely back (tiptoeing past the door of the brain drain). The expansion of laboratories that had no explicit training or teaching function proceeded much faster than the enhancement of existing universities and their laboratories where training was supposed to occur. This became an issue among scientific elites long before it emerged as an open public concern, and Bhatnagar was at its center. He spoke about this issue when the NPL foundation was laid in 1947, reminding listeners that it was essential to guard against a doctrinaire distinction between applied and fundamental research. He pointed to CSIR support of the Palit Research Laboratory (under Saha's direction) at Calcutta University; this was, in Bhatnagar's opinion, a good example of government funding for fundamental research in universities.[36] When the NPL opened before a crowd of five thousand in 1950, Nehru and Bhatnagar choreographed and animated the whole show, "a piece of theatre," according to participants. Bhatnagar announced that the NPL would create an "industrial physics" division to stimulate industry as the NCL had stimulated the chemical industry.

That the CSIR labs would do fundamental research added to the anxiety among university-based scientists about the consequent dilution of research in universities. Clearly the CSIR labs had to do some fundamental research, and that was why Sir K. S. Krishnan had been brought from Allahabad University and made NPL's director, said Bhatnagar.[37] One of the first cautions to reach him came from his old friend Archibald Hill, who wrote,

> I am always a bit alarmed when I see the great developments in government research laboratories in India, lest by getting all the best people away from the Universities you may dry up the source of scientific talent, or at least training, for the next crop of scientists. There is even some danger here where the universities are relatively very much stronger than in India. I am not

criticising the development of the government research institutes which indeed, in a minor way, I suppose I have helped to promote. But I do urge most strongly that at least a similar effort ought to be put into scientific education and research in the universities, or else the set-up may become sterile and self-perpetuating.[38]

Bhatnagar was evidently sensitive about this and wrote back the day after he received Hill's letter (letters between London and Delhi took five days at this time):

> As much as Rs 12 lakhs [Rs 1.2 million] a year is being spent on subsidizing research schemes in the universities in this country, and for giving scholarships and technical aid by the CSIR. The universities in this country have not suffered for want of government help but the public interest in the universities has declined largely because the universities are having vice-chancellors not on the consideration of their attainments but of their political affinity. The net result is that the public hesitates to come forward and the Government has to finance all the universities, which it cannot do alone with its meager resources. Unfortunately, in the educational institutions in the country there are very few at the helm of affairs who have got any dynamic qualifications and strength of persuasion which we have in the Ministry of Natural Resources and Scientific Research. I am not trying to take this compliment, but many of my colleagues have shown great stamina and the joint pulling together of us all has helped a great deal in creating these magnificent laboratories which have been acclaimed in every part of the world.[39]

Nevertheless, while writing to each other about the universities, both Bhatnagar and Hill were searching outside India for potential leadership in India's CSIR laboratories. In this effort he enlisted the help of Patrick Blackett in London. Hill by this time was sixty-five and still working in his laboratories in University College, London. He now learned about the current situation in India from Patrick Blackett; people in India like Daulat Kothari knew that the CSIR was being criticized for taking people from the universities and "drying up the source." They knew that a message about this problem from Hill would reach Bhatnagar's heart and take effect sooner than anything else and that Hill could be reached through Blackett. Nevertheless, Bhatnagar's response to Hill's criticism reveals that he was not prepared to take responsibility to greatly change the situation. He was apparently so happy that the magnificent laboratories were receiving their deserved acclaim that the consequences in universities, if any, were not really on his agenda, having himself left universities in 1940. This was the sort of disagreement that Saha had with Bhatnagar, one that continued fruitlessly until they died. But an effect of this pressure on Bhatnagar was that CSIR continued to support graduate student training and research in the universities, and this included finances for Saha's new Institute of Nuclear Physics. Saha knew this contradiction well but resented it: power and money were beginning to amplify the tensions inside the nucleus. But there was an overriding centrifugal force keeping it all together, and that was their realization that they *were now* where they had imagined they should be many years before, embodying the glow of the future of the new nation and making decisions about the new state.

TEN

Bhabha Builds His Institute in Bombay

After the war was over, Bhabha moved to Bombay in December 1945, to set up the Tata Institute of Fundamental Research in the small bungalow in which he had been born on Peddar Road. Until now Bhabha had commuted by train between his job at the IISc in Bangalore and his home. In 1944–45 the government of Bombay had invited Bhabha to accept a chair of physics in the Royal Institute of Science, but in early 1945 Bhabha proposed to the governor of Bombay that that money be spent on his new institute instead, since Tata Trusts had already made a commitment to it. Soon the government of Bombay agreed to support the Tata Institute, instead of creating a teaching chair for Bhabha at the Royal Institute.[1] The first TIFR council meeting in May 1945 recorded the start-up income to TIFR as Rs 45,000 from Tata Trusts and Rs 25,000 from the government of Bombay stretched over the next three years. In addition the Tata Trusts paid Rs 50,000 for the cost of equipment, including what had been taken or purchased from the Cosmic Ray Research Unit at Bangalore, set up originally by Bhabha. Then in mid-1945 the CSIR Governing Body received a request from TIFR for Rs 75,000 annually for the institute, though apparently the decision on the request was postponed.[2] In 1945–46 the first budget proposal from TIFR was made to both Tata Trusts and the government of Bombay for Rs 80,000, and in 1946 the CSIR contributed both a block grant of Rs 75,000 for training and Rs 10,000 for "cosmic ray and meson experiments." This pattern of CSIR support for training for physicists to run Bhabha's betatron purchase, though the purchase never occurred, was repeated in 1947 but reduced to Rs 25,000 in 1948.

In 1946 the Atomic Energy Committee began to meet in Bombay, with Bhatnagar as its secretary and Bhabha as chair. The exchange between CSIR and TIFR was constantly strengthened. In 1946 CSIR considered a request from TIFR for a continuing block grant of Rs 75,000, and their decision was predicated on a representative of CSIR on the council of TIFR (as this required government of India approval, it took time). In 1946 the Atomic Energy Board, a subset of the committee, accepted proposals for funding decisions to be made in 1947–48, and Bhatnagar began to represent the CSIR on the TIFR council. The Atomic Energy Board made some grants to TIFR (e.g., the betatron grant) in 1948, and CSIR gave a grant to TIFR for training of nuclear and high-energy accelerator scientists. But in 1948–49 CSIR increased its support and gave a grant of Rs 10,000 for meson experiments to TIFR and Rs 86,000 for development. And in the same year the Ministry of Natural Resources gave TIFR a block grant of Rs 100,000. This support from the central government gradually increased until 1954, the year in which the Department of Atomic Energy was created; the Tata Trust grant was Rs 100,000 and the government of Bombay contributed Rs 40,000. In 1955–56 the famous Tripartite Agreement came into force, with continuing but not equal funding by the Tata Trusts, the government of Maharashtra, and the government of India through the DAE; all these donors had continuing representation on the Tata Institute's council.

Bhabha began with a small research budget and decided that the first program of research was to be a continuation of the cosmic ray studies begun at Bangalore using higher-altitude balloons to measure the penetrating component of primary cosmic radiation (recalling his debate on the penetrating component with Blackett in 1938). This work was not fully organized until 1947, and the first TIFR balloon flight was done in Delhi and then at Madras Christian College in 1948 under the direction of A. P. Thatte, an expert in balloons and electronics from the Meteorological Department, whom Bhabha had known since the Millikan experiments in Bangalore.[3] At the same time there were a few TIFR balloon flights from Bombay airport, organized by P. S. Gill. Bhabha was already interested in elementary particle accelerators, knowing they were in operation in many other laboratories in Japan, Europe, England, and the United States, as well as in Saha's laboratory in Calcutta. The CSIR accepted his proposal late in 1946 and granted him Rs 40,000 for a "high-energy accelerator," though this sum was probably insufficient for the purpose.

Bhatnagar knew of the difficulties plaguing the Calcutta cyclotron and, like other committee members, hoped for an Indian breakthrough in this important nuclear field. Bhabha knew that Bhatnagar was giving Saha a lot of money through the CSIR for this purpose, and, at the very least, Bhabha could not disappoint Bhatnagar, his new ally. Sir Ardeshir Dalal, one of the Tata directors, appointed by the viceroy as the member for planning and development to the viceroy's executive council, was then appointed to the Governing Body of CSIR, and by 1946 he was its president. Dalal's presence in Bombay and in Delhi made the CSIR's approval of Bhabha's accelerator proposal much easier. In the middle of 1947, ten young scientists, most still working on their PhD's, were appointed for work on the high-energy accelerator that Bhabha tried to buy, a betatron (beta particle accelerator) from General Electric, New York.[4] This was exactly what Enrico Fermi was trying to do at Chicago and Marcus Oliphant at Australia National University, precisely at the same time. All three scientists failed in their attempts: the US government prohibition on the export of accelerators eliminated Bhabha and Oliphant, and GE sharply increased its price, which prompted Fermi to build his own accelerator. Had Bhabha obtained the betatron, it could have been tuned up for a few years of experimentation until other machines would supercede it in terms of flexibility and energy level. Just as other small accelerators were being left behind in the early 1950s, Bhabha's would probably have been left behind too; he seems to have been persuaded that India could not afford to compete with accelerators, yet with the very flexible balloon technique, TIFR could adapt and still survive. There was a role for India in the fast-moving search for new particles, especially because higher-energy particles were, at low latitudes close to the equator, more accessible for study since the background low-energy particles were attracted away to the earth's poles.

Mathematical Recruitment Drive

Bhabha also turned his attention to a subject closer to his aesthetic tastes—mathematics. He wrote his last theoretical physics paper on multiple meson production in 1953 at age forty-four but continued to work on a book on spinor algebra during his holidays in Bangalore. M. G. K. Menon, who joined the institute in 1954, said, "Homi often told me that though the work which he did in India was less recognized internationally, it gave him much greater intellectual pleasure than the earlier and more celebrated work done in Europe."[5] His old tutor, Paul Dirac, advised him that a school of physics could not grow without a school of mathematics, and Bhabha agreed. In 1945, Bhabha appointed Damodar D. Kosambi to begin a school of mathematics at the Tata Institute and added Fernand Levi shortly later. Levi had been teaching at the University of Calcutta since the 1930s and Kosambi, also a committed archaeologist and Marxist historian, was already teaching mathematics in a college in Bombay. Bhabha apparently became quite dissatisfied with the slow growth under these two senior mathematicians and began to try to attract younger people. Bhabha discussed his dissatisfaction with John von Neumann at Princeton in 1947, who introduced Bhabha to a young Indian mathematician, K. Chandrasekharan. Before responding to Bhabha's offer, "K.C." asked both von Neumann and Herman Weyl whether or not he should return to India at this time. Apparently they cautioned Chandrasekharan that it would take ten years to build a strong group of mathematicians and that unless he obtained autonomy to develop as he saw fit, it might not be possible at all. Chandrasekharan requested and got this autonomy from Bhabha and resisted later efforts to compromise it. He had been cautioned by Weyl that he was working with a good physicist and a powerful personality, and so he would have to be equally strong and fight for his autonomy.

Chandrasekharan arrived at Bombay in 1949 as reader in mathematics, and it is said that high-quality mathematics papers began to appear from TIFR within about a year.[6] But Chandrasekharan's efforts to reorganize the group were resisted by Kosambi, according to TIFR mathematicians and scientists in the late 1960s. Bhabha tried to assuage his old friend Kosambi but backed Chandrasekharan's ideas. Interviews for new appointments in mathematics were begun immediately at TIFR. Until Bhabha became too busy, he attended interviews for mathematical appointments with Chandrasekharan in 1951 and 1952. Even after he was no longer able to attend, Chandrasekharan kept Bhabha's name on the interview committee list, to keep him informed. Levi eventually returned to Germany to accept a chair in mathematics, and a dissatisfied and suspicious Kosambi, now on a shorter annual appointment, returned to Poona and famously commuted to Bombay by train, reading mathematics and mystery novels; a recent biographer concluded that "his whole mathematical career appears to be one long clash of values."[7] In 1953, mathematician K. G. Ramanathan was also appointed to the institute. Selected foreign mathematicians were invited to lecture at TIFR for between two and four months. Lecture notes were recorded and published (mostly unedited). The series was widely distributed in India and abroad; some of the best-known lecture notes from the institute, it is said, were pirated and reprinted in the United States. This method fueled the training of younger mathematicians to a high standard without leaving India.

Constant Recruitment in an Expanding Pool

Bhabha's first choice seems to have been to work with young colleagues, as he himself was only thirty-eight at this time, and he continued his search for talent at home and abroad. He directed the physicist Bernard Peters and mathematician K. Chandrasekharan to do the same. Bhabha recruited young scientists from Kashmir to Kerala. In the 1960s I met dozens of scientists who in the late 1940s and throughout the 1950s had been inspired to speak to Bhabha, at colleges and universities, on railway station platforms, or anywhere it seemed possible. Always hesitant and intimidated before their interview, they recalled their surprise that Bhabha listened: if they were invited for an interview, many of them ended up working in Bhabha's expanding network and eventually became colleagues. None of them were considered by Bhabha to have been "his student" quite like Saha would think of someone as his student. Bhabha specifically tried to overcome the class and rank distance between him and these young men although it was not easy, neither for them nor for him. And he went to great efforts to retrain most of them for his purposes and so created a special "school" that met alternately in Bombay and Bangalore.

Bhabha also spread a wide net abroad. For example, he identified potential in PhD student Raja Ramanna, whom he met at Kings College in London, and arranged for him to stay on in London for another year to study nuclear fission after completing his doctorate in 1948, at age twenty-three. Ramanna had entered Madras Christian College at sixteen and had met Bhabha during a concert at Mysore during the war, probably in 1944. Ramanna entered his doctoral studies in London at twenty, in 1945. Bhabha met Ramanna again, in London, and so Ramanna began working at TIFR in 1949. He worked on the first research reactor named Apsara, and then, at age thirty, he traveled in 1955 with Bhabha to Moscow and then to Chalk River in Canada to explore new reactors and agreements.[8] At the same time Bhabha interviewed B. V. Sreekantan, later TIFR's director, a student from the IISc in Bangalore. Unsure whether he should declare himself as an experimenter or theorist, the young man hesitantly asked Bhabha to make the decision. Bhabha reasoned that he knew electronics and could make a real contribution as an experimenter, and that decided it.[9] In the case of physicist-historian G. Venkataraman, who worked variously on the physics of reactors and advanced computers for missiles, he was a student at Madras Christian College and met Bernard Peters by volunteering to assist with early morning balloon flights at the college outside Madras in 1950. This balloon launch led to an interview, and so another young physicist was brought for training at TIFR, resulting in Venkataraman's helping to build the Trombay atomic energy establishment.[10]

Bhabha recruited relentlessly, meeting and interviewing on the spot, or asking new people to come to Bombay for an interview; but not all of them were persuaded to stay, even at the golden beginning. For example, P. S.

Gill, who worked in Lahore, was offered a position in TIFR in 1945, which, curiously, he did not accept. After spending 1946 on secondment working in the United States with M. S. Vallarta at MIT on solar flares and studying penetrating component intensity on board airplanes at much higher altitudes (40,000 feet), he returned to India. Gill, only two years younger than Bhabha, decided finally to accept a position at TIFR, where he arrived in July 1947, only weeks before Partition. He flew some high-altitude hydrogen balloons with TIFR from the airport at Santa Cruz in 1947–48, continuing the work he had begun from Lahore in 1945, in the mountains and on airplanes. But relations with Bhabha were uncomfortable and he concluded that Bhabha would prefer that he not stay, so Gill was looking for an exit from Bombay; "I was unable to operate freely," he later recalled. He was surprisingly mobile; after a brief stay at TIFR, Gill left India again on a full salary, authorized by TIFR's council, saying, "I have an offer from the United States Government which I wish to accept."[11] TIFR finally (and perhaps gratefully?) accepted his departure, and reminded Gill to vacate his institute housing promptly. A recruitment story is also told of R. R. Roy, who had been offered a TIFR position on award of his PhD from King's College, London, in 1949. He came to TIFR in 1950 and was interviewed by Bhabha, though only after being kept waiting a number of hours, in the company of Raja Ramanna, also awaiting an interview. During the interview something went wrong and Roy "stormed out of the room in a great fury saying he wanted to return to Europe at once."[12] TIFR attempted to reach him, even on the ship, to persuade him to return, but did not succeed. Roy was en route to America, where he apparently remained. This was a golden era for young well-trained nuclear physicists in terms of expansive employment in the United States, comparable to the employment gold rush experienced by trained Indian computer programmers in the mid- to late 1990s.

But not all the recruits were physicists, nor could they be.[13] And not all stayed in Bombay; for example, Bhabha appointed engineer Homi Sethna to work for the AEC in 1951, but he soon went off to manage the rare earths factory in Kerala and then to study fuel reprocessing in France. In September 1955 engineer M. R. Srinivasan was appointed to DAE while he was still in London, and without going back to India he was sent immediately to the UK's atomic energy research center at Harwell to study reactors. Then when he arrived in India he went immediately to Nangal, where he worked beside the new Bhakra dam in the foothills of the Himalayas to build India's first heavy water plant alongside German engineers (see *Negotiating Nuclear Power*). Both Sethna and Srinivasan would eventually become chairmen, along with Raja Ramanna, of the Atomic Energy Commission.

Cosmic Ray Studies, Secrecy, and Bernard Peters

Bhabha found more than enough money for research but still carefully selected the few people suitable for work. Although the libraries, adjacent laboratories, and number of other scientists available in Bombay were simply much fewer than he would have found in Calcutta, it was his own city, where the Parsi community and his network had great influence, and his institute had no precedent and no equal, neither in Bombay nor anywhere else.

When importing an American betatron appeared impossible, Bhabha concentrated all the institute's experimental efforts on cosmic ray studies with balloons. Even people recruited for other purposes often first began active work on the balloons and experimental apparatus. The first launches, sixty-five small rubber balloons per flight carrying the payload, occurred in Delhi in 1948, but their altitude was quite limited. Bhabha heard of high-altitude Skyhook balloons rising to 90,000 feet when he was visiting Professor Salant in Paris in November 1947.[14] H. J. Taylor, then teaching physics at Wilson College, Bombay, was made professor of experimental physics at TIFR to help build the cosmic ray group. Taylor noticed an article in *Popular Mechanics* (May 1948) on very large balloons and suggested that Bhabha get in direct contact with the Visking Corporation in the United States. As in the case of the betatron, it was soon discovered that these plastic balloons too were banned for export from the USA. The reason for the ban was secret, but by 1950 the Americans were organizing a vast high-altitude balloon surveillance project to track Soviet nuclear tests using high-resolution cameras. The consequence of this ban was that the TIFR group developed its own mixture of imported and Indian-made plastic balloons. This was the first impetus among many to push Indians to produce their own technologies following an embargo imposed by other countries. All over the world physicists were experimenting with balloons, sometimes supported by military funds, sometimes not: cosmic ray physics was now a hot subject.

Using high-altitude balloon reconnaissance in July 1950, an experiment from a balloon launched for the US Air Force at the University of Minnesota airport took pictures from 28,000 meters. Enlarged photos showed a fisherman with a fishing rod in his hands on the Minnesota River.[15] In September 1950 the Air Force Scientific Advisory Board concluded that balloons were best for photographing USSR industrial heartland, and in October 1950 Air Force Intelligence concurred. Thus Project Gopher began in 1951, leading to some successful flights, followed by a rash of failures resulting from the low quality of polyethylene. The best poly was then being used for signal cables in the Korean War. So a new Project Moby Dick started in 1952 to improve balloons, and the decision was taken in 1953 to catch the balloon and gondola in midair before they hit the ground or sea and thus save the expensive camera. In 1953–54 there were 630 balloons launched! So TIFR was experimenting with a strategic technique guarded jealously by the US military, and a leader of its cosmic ray group was a Danish physicist Bernard Peters, who had just been described as a communist sympathizer to a committee in Washington. The air force and its contractors continued to experiment with new balloons and cameras, which was what TIFR was doing, only they were taking pictures of incoming cosmic rays (mostly neutrinos), whereas the United States Air Force and the Central Intelligence Agency were taking pictures of the earth's surface in Russia and China. In January 1956 the United States launched nine balloons from Turkey, and three were recovered successfully in the air near Japan. Next they released 448 balloons(!), 44 of which were recovered; 40 of these had photos of the Russian and Chinese earth surface. This is how the United States discovered the vast nuclear refinery at Dononovo in Siberia.[16] A forty-fifth gondola recovered one year later in Alaska, still floating in water, provided the best images of the Soviet Union in 1957. That was the year that the USSR launched *Sputnik*.

By 1950 the Tata Institute was bursting at the seams; most of it had moved in 1949 to the Old Yacht Club in downtown Bombay, to quarters six times the size of the bungalow on Peddar Road. Bhabha worked hard on forming a relationship between applied research and pure research. Two groups concerned with nuclear chemistry and metallurgy moved into the old Peddar Road building. By 1952, besides cosmic rays and mathematics, there was theoretical physics (in which Bhabha himself participated), electronics, and nuclear physics. Bhabha had also been named chairman of the Atomic Energy Commission, and he formed research groups in TIFR to take care of all aspects of production, research, and development for atomic energy. It was not long, however, until these groups started to move out to Trombay, and the scale of everything was doubled.

The most active group in the Tata Institute in terms of early published results was the cosmic ray group. Bhabha had published two joint papers with TIFR colleagues R. R. Daniel and S. K. Chakravarty before 1950, and the following year the team published a paper by Bhabha, Taylor, Daniel, Heermaneck, Swami, and Srikantia on experimental work. TIFR hosted the important International Conference on Cosmic Rays and Elementary Particles in 1950, and in 1953 the group found the K-meson on their nuclear emulsion plates, giving the world's first accurate experimental confirmation of this particle, in hot competition with other groups elsewhere.

One delegate to the 1950 cosmic ray and elementary particle conference in Bombay was Bernard Peters, then assistant professor at the University of Rochester, New York. Bhabha and Peters first met late in 1949 in the United States at a conference.[17] According to Peters, they discussed the possibility of conducting an experiment that required stratospheric balloon flights near the equator. Bhabha agreed to have TIFR collaborate in this project, and Peters spent five months with the cosmic ray group in 1950. Peters wrote: "I enjoyed the experience so much that I accepted Bhabha's invitation to spend a longer period (we talked about a two-year period) at the Institute."[18] Peters formulated research projects with the group in 1950 and then returned to Rochester to complete his work and arrange for his family's move to Bombay. The atmosphere created by the House Un-American Activities Committee and McCarthyism "compared unfavorably" with the atmosphere reigning at the time in TIFR, Peters told me. The freer atmosphere in Bombay encouraged Peters to accept Bhabha's already attractive offer. He did not, therefore, accept a late-arriving offer of tenure (an associate professorship) at the University of Rochester in 1951, he told me. Bhabha encouraged Peters to remain at TIFR until 1958. Peters relieved Bhabha from some of the work of organizing the cosmic ray group because at this time he had numerous other pressing tasks. The "Peters solution" appeared just when missionary physicist Professor H. J. Taylor had been transferred by the Scottish Church Mission away from Bombay to Calcutta and from there to teach in Assam.

The Bernard Peters story and his productive relationship with India was evidently more complex than Peters explained to me in 1970. When he first came to Bombay in 1950, he had already been named along with two others by Robert Oppenheimer as a "communist sympathizer" when Oppenheimer testified before the House Un-American Activities Committee in June 1949. A Danish citizen, Peters appears to have been squeezed out of the United States along with Oppenheimer's other student David Bohm, anticipating a negative decision from the government. Peters's name was blurted out to the HUAC committee by Oppenheimer in 1949, and a newspaper in Rochester picked it up and named Peters as a communist.[19] Oppenheimer had years before named Peters's wife as a communist in a conversation with a senior FBI agent monitoring the Los Alamos part of the Manhattan Project. Now Oppenheimer was casting doubt on Peters all over again. Shocked, Peters wrote to Oppenheimer twice, asking why he had named him in this way in Washington and asking him to intercede for him to clear his name, as he had "done nothing wrong." Whether Oppenheimer ever replied to Peters is not known. But "this betrayal of his own student deeply disturbed a number of Oppenheimer's scientific colleagues," according to Thorpe's authoritative biography, and they wrote to him in 1949 to advise a statement which served to protect Peters.[20] Then, losing hope, Peters began to write pleading letters to Dean Acheson at the State Department and to his former supervisor at Berkeley, E. O. Lawrence, first in June 1949, referring to a series of attacks on him for eight previous months, bringing, in Peter's words, "my name into connection with the so-called 'Berkeley spy ring.' . . . I went with great hopes to Washington when I was asked to appear before the House Committee on Un-American Activities. But again no suspicions were voiced or charges made, so nothing could be cleared up."[21] No effective intervention appears to have come from Oppenheimer or Lawrence either, though Peters continued to write to Lawrence through 1950.[22] The reference to the "Berkeley spy ring" suggests a reason why the Indian mission was so closely questioned after its contacts among physicists in Berkeley in early 1945.[23]

At the time of the Peters affair in 1949 and early 1950, the debate over the hydrogen bomb was taking place, with Fermi and Rabi taking positions against, and Teller vigorously for, the superbomb. Oppenheimer tried to work it both ways, it seems, so that when President Truman overruled the AEC's caution and ordered production of the H (thermonuclear) bomb, Oppenheimer was eventually able to support the idea after Teller and Stanislaw Ullam made an innovative new design for that bomb a year later in 1951. Peters, not a US citizen, also was approaching Dean Acheson for assistance in the very season when noncitizens Klaus Fuchs and Alan Nunn May were being watched and revealed as spies for the USSR in the United States.[24]

Oppenheimer and Bhabha were old friends, dining together in New York in the company of Oppenheimer's very beautiful wife, Kitty, in 1947 and 1949. Oppenheimer was five years older than Bhabha, had passed through Cambridge just before Bhabha arrived, and was teaching in Berkeley by the time Bhabha received his PhD at Cambridge. But by the mid-thirties they were moving in the same orbits in Europe. Abraham points out that Nehru eventually invited Oppenheimer more than once to visit and stay in India following his loss of security clearance in 1954, and presumably it was Bhabha who arranged this invitation from Nehru, as Bhabha and Oppenheimer were in regular communication, both at Princeton and elsewhere.[25] The extent to which Bhabha understood Peter's predicament at Rochester prior to his first visit to TIFR in 1950 or to which Oppenheimer and Bhabha discussed Peters is unfortunately not known. The offer of tenure from Rochester reported by Peters came after Peters had been given a US passport or travel document in order to leave for Bombay. It appears then that a path was being cleared for him away from the United States, though the University of Rochester did an honorable thing by offering Peters a tenure-track position, albeit at the last minute. Peters's cosmic ray research at TIFR was supported by Bhatnagar's ministry in Delhi, with grants valued at Rs 7,700 in 1952–53 and Rs 4,800 in 1953–54, as well as workshop development grants.[26] Moreover, he had the resources and staff of the entire institute at his disposal, with support from Bhabha, and he made very good use of them.

New Opportunities and Money for New Groups

During this period Bhabha had become increasingly busy, but he continued to spend time with his young TIFR colleagues. Many of them were bachelors living in a hostel and eating in a nearby Gujerati vegetarian restaurant on monthly passes. Bhabha himself was a bachelor and gave them lifts in his car and invited them to his home on Little Gibbs Road on Cumballah Hill. Early obstacles to group cooperation, some due to the sharp differences in their class, cultural, and linguistic backgrounds, were reduced by Bhabha's energy and enthusiasm for work and also by the special qualities of Bernard Peters. As an outsider, Peters could diminish or go around these differences, and he soon played an instrumental role in changing the group's work patterns, to a more industrial mode so that publications came out quickly. The youthfulness of Bhabha and Bernard Peters was frequently remarked upon, and so was the cohesion that came from working in their successful result-producing groups, full of very bright young people. It was a difficult time for large-scale experimental work, and most Indian scientists labored under the widespread conviction that good experiments could not be conducted in the tropics. Most nonscientists held the same belief that experiments were too difficult to do in India and therefore one should focus on theoretical research.

While Bhabha was abroad negotiating the first steps in building a nuclear reactor program, the institute was building new research groups like nuclear chemistry and metallurgy until they could be transferred from Colaba to Trombay, providing the basis for a 1958 plan to integrate atomic power plants into the Indian electric grid system. The attitude toward reactors, however, was to use them for both pure research and plutonium production and for development applications. So while Bhabha was thinking about the growth of theoretical physics and mathematics, he also paid attention to new workshops for reactor development and experimental physics, interviewing the candidates himself. A typical example shows how interrelated the reactor program and TIFR were:

> In July 1955 I was interviewed by a committee chaired by Dr. Peters. Along with me about half a dozen people were selected. Some of us were assigned by Dr. Bhabha to TIFR while the others were assigned to Trombay; I was in the latter category. I was asked to report to Dr. Ramanna and assist him in research activities related to atomic energy. My place of work was the barracks in Colaba. Though I worked for the Atomic Energy Establishment at Trombay, for all practical purposes I was a member of TIFR and that is how H. L. N. Murthy of TIFR built some of my equipment.

Murthy was originally a glass blower in the labs at TIFR but rose to become one of India's top builders of research apparatus. For physicist Venkataraman, Murthy built a neutron spectrometer that still functioned in the early 1990s, thirty-five years later.[27]

Bhabha was by 1952 having tea or dinner with Nehru almost every two weeks.[28] These men shared the same patrician backgrounds with proximity to wealth and political influence; both had been to Cambridge, lived like bachelors, considered themselves connoisseurs of art, music, food, and the like. Through their mutual attraction they developed the ability to speak the same language. The meaning of Bhabha's remark about government support not necessarily entailing government control was becoming clear. If the atomic energy program and institutions were made an integral part of the government, and controlled by people committed to them, but who were not "government officials," then, despite given the routine conditions of political interference in India, the necessary autonomy could be maintained. These encounters became crucial on the eve of the first step in the institute's construction, symbolically marked by Nehru's coming to lay the building's foundation stone: only fifty days before the opening date, Nehru wrote bluntly in November 1953 to the minister for Defence to say that "many international scientists of repute" were coming to the event and the ground had to be cleared of the military huts that occupied the institute's land in Colaba.[29] "Defence have not paid too much attention to this and have not made any real effort," said Nehru to Minister Mahavir, "and we cannot have important national work held up because some arrangement cannot be made." The site was cleared but only just in time.

In 1954 the atomic energy program expanded its status from commission to department of the government, and Bhabha became a secretary to the government of India; the DAE was created to execute AEC policy. He now had the constitutional power to maintain the autonomy of the program he started. But Bhabha did not lose sight of the development of his own institute, for which he cared most deeply. By the time the DAE was created in 1954, there was an electronic production unit within TIFR and metallurgy and chemistry laboratories at Peddar Road. There was already a program of prospecting for raw materials, and the DAE continued extraction of pure thorium salt and uranium salt at the Indian rare earths plant at Alwaye. Bhabha had already selected Trombay in north Bombay as the site of the reactors and development laboratories when the DAE was created. In his capacity as secretary of the AEC and trustee on the Council of Management of TIFR, Bhatnagar worked closely with Bhabha on all these projects.[30] Bhabha knew he needed Bhatnagar's influence and support, which is why he kept him on the TIFR council; moreover, the government of India's contribution to TIFR flowed through Bhatnagar's ministry. From someone who wanted to leave India in 1942, within ten years he had become the architect and builder of a new type of scientific organization and had the status with both government and the private sector to finance it on a continuing basis.

Besides his attention to such lofty objectives, Bhabha also addressed the mundane world of institutional life. He established the code that loose or colorful Indian style clothing was not to be worn by men at TIFR or Trombay. White shirts and dark trousers became the norm. This was followed until his death, but after it this practice was not followed: the new chairman of the AEC was Vikram Sarabhai and he wore Indian-style clothes. But styles changed slowly, and gradually more of the men wore Indian clothes. Bhabha's edict could not apply to women, as they would simply not have worn what was then called "western dress," although there were few women at TIFR at the time. When women did begin to work at TIFR, they chose their own mode of dress, so that Goan Catholics wore long skirts and dresses, South Indian Hindu women wore *saris,* women from the north wore *salwar-kameez,* and so on. Bhabha was also insistent on the building's cleanliness, and so TIFR stood apart from most other public buildings in terms of its appearance. He was so fastidious about the hygiene of its bathrooms and other facilities that memos were circulated regarding the standards to be set and kept. The spitting of betel nut juice, common in public corridors in big Bombay buildings at the time, was forbidden. Senior staff of the institute were required to supervise this hygienic regime. The quality of health services of TIFR for staff and researchers was legendary.

Although not his explicit intention, Bhabha developed a style and code that drew his institute apart from "India" and created a new cultural "island." This is how a majority of Indians perceived TIFR even during his lifetime, whether scientists or not. Gradually Bhabha spent less time there and was involved less in its research, but his eye for detail at TIFR remained. The institute became known internationally as *the* place to go and work, whether for a week or a month, and as a place from which good research came. Few people grasped how it became that way or what deep role it played in incubating the Indian nuclear program of reactors and then a bomb.

ELEVEN

The Politics of the Early Indian Atomic Energy Committee and Commission

At the European war's end in 1945 there was a rush of activity among scientists everywhere, and no less in India. Although the battle around Japan was not yet concluded, the pent-up desire for reestablishing contacts, exchanging ideas, acquiring new equipment, restarting publications, going abroad for doctorates, all flourished in a brand-new environment—a post–atom bomb culture. Though control of all matters atomic remained in military hands, the idea of civilian control was soon in circulation. By 1947 governments were forming atomic energy commissions, and India, not yet a sovereign state, already had an official committee disbursing funds for research and development from its own budget. This chapter describes the work of that committee and its transition to a full Atomic Energy Commission in 1948, setting the stage for an understanding of the early years of the Indian nuclear program.

Saha was the delegate of the Indian government, at Soviet expense, to the 220th anniversary celebrations of the Soviet Academy of Sciences in July 1945 in Moscow, just before the atomic bombs were dropped. This was the same conference, organized by Peter Kapitsa, to which Patrick Blackett and other British scientists were invited, namely, J. D. Bernal, Charles Darwin, Paul Dirac, E. A. Milne, and Neville Mott, but which their government (in this case Sir John Anderson himself) prohibited them from attending. The explanation given was that these British scientists could not be spared until the end of the war with Japan; this prestigious group tried to get the order overruled by the prime minister but did not succeed. It was feared they might be held in Moscow if the war's conclusion in east Asia did not suit Stalin. The Cold War was coming, but the idea had not yet sunk in. The fact is that these scientists, collectively, carried a great deal of the strategic and secret knowledge of their country in their minds; some, like Bernal, had been committed communists, though others, like Blackett, were not. "By the beginning of 1945 Soviet intelligence had a clear general picture of the Manhattan Project," according to David Holloway.[1] The Russians knew about the Hanford plutonium production unit, the Metallurgical Laboratory at Chicago, and the Oak Ridge isotope separation projects, as well as the Anglo-Canadian heavy water reactor in Montreal, information passed from Chadwick's student Alan Nunn May in Montreal through the Soviet embassy in Ottawa. They also knew about Los Alamos through Klaus Fuchs and others, and Holloway confirmed that they expected a bomb test within two to three months. The Soviet Academy conference occurred in this context, and Saha, who had just visited US nuclear installations, was an honored part of it.

The academy's invitation to physicists around the world to come to Moscow was intended, in part, to restore communication for Soviet scientists with the world and confirm the academy's reputation. Most of all it was a signal that the long war in Europe was over. It was just before this time also that the British Tube Alloys Consultative Council, originally the Maud Committee, gave birth to the Atom Bomb Committee, also known as Gen 75. British scientists and policymakers knew much more than the mere outlines of the Manhattan Project and knew also that they were being steadily denied more and more information. Convinced of the scarcity of fissile material, they were looking for all avenues of future nuclear development, and the 1943 UK-US Declaration of Trust had identified Indian thorium as essential to the Allies' futures. The British were thus rapidly assessing their options, including even the mobilization of empire (or commonwealth) scientific and natural resources. Though the Indian team presumably had known nothing of the secret trust, it appears from their interviews with the FBI, the team of Indian scientists touring the United States and Great Britain in early 1945 realized that a large special research program was under way on nuclear physics, and so they were probably not completely surprised by the bomb.

There are some unusual contradictions in all this in 1945. Saha supported Subhas Bose and criticized Nehru and the rest of the Congress leadership in 1941. Bose disappeared to raise an army to fight with the advancing Japanese forces against the British in Asia. Saha criticized moderates who would cooperate with the Raj and support the "retrograde" thinking of Gandhi. He also criticized Bhatnagar's CSIR because most of its policies were intended to bring India more productively into the war effort and to encourage private industry and capital. Yet Saha met and planned with industrialists, sat on the CSIR council, and sought to bring its laboratories to

Calcutta. He obtained Nehru's help in getting financing for a new cyclotron from the two strongest private industrial companies cooperating with the Raj, the Tatas and Birlas. The most important part of this cyclotron was sunk by a Japanese torpedo, the navy favored by his hero Subhas Bose. Saha went on an official British government tour to the laboratories of Allied war production, learned a great deal, and was put under FBI surveillance. Three months later he attended a conference in Moscow, literally while Truman, Stalin, and Churchill were setting the stage for the Cold War at the Potsdam Conference. At that moment the Soviets were planning their surprise attack against the Japanese in east Asia and their occupation of one of the northern islands of Japan. Yet the next year Saha was involved in the Atomic Energy Committee of India, sitting in membership with people to whose political positions he was opposed and whose decisions he publicly criticized.

On his return from North America in the spring of 1945, Saha began to explore ways of transforming part of a university department into a separate institute of nuclear physics. This move was in parallel with similar efforts by other physicists at the same time, for example, Enrico Fermi, who was transforming the university's Metallurgical Laboratory at Chicago into an institute. This paralleled Bhabha's initiative to build a separate institute in Bombay. Saha was inspired by his exposure in 1945 in Great Britain, Canada, the United States, and the Soviet Union to the latest developments in physics. In August he followed the consequence of the explosions of atomic bombs in Japan and an impending Cold War arms race. After returning from his tour, Saha published his enthusiastic report on Soviet support for science and industry in *Science and Culture* and also wrote to Nehru: "I am very anxious to meet you and relate to you all my experiences in the UK, USA, and Soviet Union." He also reiterated: "I do believe the time has come when the Congress should formally announce their programme of work in case they get power, its present programme is too much tied up with old world ideologies like spinning wheels and homespun, division of power on medieval basis, etc. etc."[2]

Memos and Money: Starting the Committee

It is not clear what discussions occurred prior to establishing the Atomic Energy Committee as a subcommittee of the CSIR. But the CSIR already had more than two dozen money-dispersing committees in 1946, and Bhatnagar had been named chairman of most of them. Though he was not, in the case of atomic energy, an expert in the subject, he was appointed the committee's chairman anyway. Yet by 1946 he was clearly deferring to Bhabha's wisdom, judging from the minutes written by him as "secretary of the committee." Bhabha had been just one of the three proponents of the formation of an Atomic Energy Committee in 1945 and still had no secretariat. It is likely that Bhabha pressed for the committee through the voice of Sir Ardeshir Dalal, president of the CSIR Governing Body and member of the viceroy's executive council. Dalal, a senior figure in Tata circles, would have easily understood the value of this initiative both for a soon-to-be independent India, for Bombay, for the Tata connection to new strategic developments, and for the Parsi community's transition to a new role in India.

Within six years of returning to India at age thirty-seven, Bhabha had joined forces with Bhatnagar, the most powerful person in the organization of science at that time; Saha was the other member of the committee. Membership does not appear to have been exclusive; the records also show occasional participation in the committee by D. M. Bose (nuclear chemistry), D. N. Wadia (geology), Nazir Ahmed (nuclear physics), and K. S. Krishnan (nuclear physics). The Atomic Energy Research Board was, however, just the smaller group of Bhabha, Bhatnagar, and Saha; it met with the committee in Bombay and disbursed funds on behalf of the committee, funds that came to it from the CSIR through Bhatnagar. Some of the funds went to board members like Saha and Bhabha, other funds went to committee members, like D. M. Bose. It was important that almost all the meetings were in Bombay, where Bhabha lived and worked: this is the first sign that the Atomic Energy Commission was not going to be confined to the labyrinths of Delhi.

Saha had already spoken and written publicly about nuclear energy and nuclear explosions in early 1941, about twelve months after papers by Hahn and Strassman and Meitner and Hahn announced the achievement of uranium fission in *Nature* in 1939. After *Nature* arrived in Calcutta, Saha specifically mentioned "the possibility of a chain reaction" and said experiments were currently being done in Germany on this, although nothing was known about progress, if any. In 1941 Saha had already pointed to the possibility of chain reactions and explosions through fission of U235. The idea of an atomic bomb was in circulation soon after the fission papers in *Nature*, at least by February 1939, according to Joseph Rotblatt. He said that by the summer of 1939 he had articulated and communicated "a theory of nuclear deterrence" to explain how the bomb could be used to stop Hitler and Nazism.[3] Scientists talked and wrote to each other about this possibility through 1939, and it was not until early 1940 that they began the movement to censor themselves, a plan put into action by Blackett and others. Saha would have known some of this, even though after August–September

1939 travel to Europe became difficult for him and other Indians. After that Saha pursued the cyclotron more vigorously, announcing his interest in experimental nuclear physics more strongly than in any other experimental possibility. Scientists in the Soviet Union, Great Britain, and the United States had written individually to Stalin, Churchill, and Roosevelt, warning each of these leaders that it was likely that a very powerful bomb could be made from a fissile reaction and that—to varying degrees—they thought it likely that such a bomb project was under way in Germany. A number of physicists had also asked Patrick Blackett in 1940 to act as intermediary in producing an Anglo-American consensus for a moratorium among scientists on publishing in the physics of nuclear fission, particularly in journals like *Nature*. Blackett achieved this consensus and moratorium, after considerable negotiation. Thorough readers of scientific journals in India would thus have noted an abrupt absence of the debate that usually accompanied concentrated activity in physics. This was like Sherlock Holmes's observation about "the dog that did not bark," leading experts in Moscow to think that something big was going on among physicists in the West.[4] It is fair to conclude from available evidence that physicists in India were excluded from communication about an atomic bomb project until early 1945.

Bhatnagar and Saha learned a great deal about the atom and secrecy in their 1944–45 tour. Always discreet, Bhatnagar nevertheless pronounced himself against secrecy in 1945: "that the USA is not prepared to share the scientific knowledge regarding the atomic bomb with the rest of the world is indicative of how statesmen and politicians, even in advanced countries, can hamper science in solving international problems."[5] And more prophetically, in a radio speech broadcast throughout India, three months after the Hiroshima-Nagasaki bombs, he said, "If scientists were ever forced to submit to secrecy in such vital research [atomic power and weapons] a Galileo would arise from among them and smash political interference in matters of intellectual freedom."[6] Moreover, he knew that India would have to move quickly on atomic energy. Responding to a question about nuclear research from the Indian high commission in London, Bhatnagar said, "We raised the issue in the last meeting of the Governing Body [CSIR], and I am raising it again. I fully appreciate that India cannot be allowed to completely ignore this research and that we shall have to take very active steps to do something as it is going to be a very potent factor in industrial development of the world, and India cannot be a cipher in this direction."[7] This was precisely when CSIR physicist S. Parthasarathy was sent by the government of India to accompany the British Mission to Japan, to investigate the effects of the two atomic bombs dropped there. On his return, Parthasarathy met with Viceroy Wavell and Field Marshall Auchinleck, as well as with his scientific colleagues in the CSIR, to brief them in January 1946 on his disturbing findings of the devastation at Hiroshima.[8]

Thorium at Stake

The concern of the great powers now was to establish and maintain a monopoly on fissile resources like uranium and rare earths like thorium and beryl. The 1943 Declaration of Trust formalized this strategy among the United States, Great Britain, and Canada. Just after the war ended there was a secret mission to India undertaken by D. E. H. Pierson, later secretary of the British Atomic Energy Commission; his purpose was to evaluate India's thorium reserves independently.[9] Persistent well-grounded rumors suggest that French prospectors disguised as archaeologists also visited India at the same time to appraise the potential. The minerals attaché at the US embassy in Delhi was given permission to tour the area of Travancore State, in part because the state's shrewd dewan, C. P. Ramaswamy Aiyer, invited him, hoping American interest would result in proposals to build a processing plant in Kerala. An American interest was sure to stir British fire, he reasoned correctly; the state of Travancore was staking its claim to the thorium. British firms had a strong grip on the trade and processing of Indian minerals. According to Abraham and Helmreich, sales of Indian monazite (thorium) had totaled about 3,000 tons annually before the war (sold to the United States, Great Britain, France, and Germany), mainly for refining thorium into thorium nitrate, used in the manufacture of incandescent gas lamp mantles.[10] These mantles were, in turn, sold back and widely used in India. C. V. Raman himself owned and operated a mantle factory at the time, making a very good income from it, and said he looked forward to a cheap and reliable supply of the nitrate within India. Raman exerted his reputation to try to achieve this end—which was his kind of homage to Gandhi's insistence on *swadeshi* industry. The gas mantle was essential for those who could afford gas lamps, because there was little electricity available in India, and thus few alternatives. Millions, however, could even not afford gas lamps.

In late 1946, however, the dewan first blocked and then yielded to a CSIR survey of Travancore's fissile minerals, which led Bhatnagar to mobilize the Atomic Energy Committee to persuade Nehru to place an embargo on the export of monazite, beryl, and all radioactive sources. This was to be done immediately, in the event of Independence, the moment when Travancore State acceded to Indian jurisdiction. The American government even intervened with an American company, Lindsay Light and Chemical, which had threatened to stop export of thorium nitrate mantles to India in retaliation for India's own embargo on monazite; the government explained to Lindsay that the sales should continue because it was good for American business but could not say why. What the US was really doing, secretly, was positioning itself with the new government in India to get access to Indian thorium and beryl ore and to obtain Indian guarantees not to sell radioactive material to "unfriendly nations." Nevertheless, in 1946 British Titan Products and DuPont were negotiating to build a processing plant for monazite in Kerala.

Since the three-nation trust was secret, firms like Lindsay Light and individuals like the dewan did not know that the Americans had preassigned a responsibility for Indian rare earths to the British and that this division of spheres of influence was part of the overall struggle between Great Britain and the United States concerning American attempts to dominate atomic energy and atomic weapons development across the globe. Because they were intervening vigorously elsewhere, the Americans were not going to appear to intervene in India. But they would permit American firms to make proposals for export or build processing plants, as we shall see, around the time of Indian Independence. In the end, despite British prominence in the thorium field in India and despite the fact that negotiations between India and Britain about the processing plant continued through 1948, it was La Société des Terres Rares of France that built the processing plant in Travancore, starting in 1950. This was followed by the 1951 bilateral agreement on atomic energy between France and India, the first for either country, though the French official involved suggested that the French commercial and strategic approaches to nuclear development were, in this case, unconnected.[11] I think this "unconnected" thesis is slightly disingenuous, given that the American and British interests were so strong in the same resources and the French were fully aware of their opponents' obsession with those same Indian fissile resources.

The Committee's Early Decisions

This was the context of the first meetings of the Atomic Energy Committee; impending Independence, though no one knew exactly when, was accompanied by jockeying for new positions on a new playing field. Indians realized there was a new opportunity that could not be missed, and Britons felt squeezed out of American global nuclear activities. There was already sensitivity among Indian scientists to the new secrecy surrounding nuclear physics, because after a few meetings among the Calcutta-based members of the Allied tour group (Ghosh, Mukherjee, Mitra, Saha), a confidential report on the 1944–45 tour was finally written by Saha and circulated in early 1946. It is striking that no reference is made to nuclear physics in this report, although it discusses all the other relevant subjects: geophysics, radio physics, fuels, power supply, electrochemical industries, meteorology, and land reclamation.[12] The missing reference to nuclear matters may be a clue to their realization that some things were better left unwritten.

Thus it was that the AEC of India was quite well informed about conditions in other countries when it carried out it first two years' work, up to India's Independence in August 1947. More than a handful of scientists in India understood the physics of the fission process and grasped the implications of successful projects in the United States and Canada. They probably guessed that Great Britain and the Soviet Union had decided to launch their own separate bomb projects—and knew for certain that atomic power reactor projects were under way in each country. They knew each country was forming its own atomic energy commission. Before "the stroke of midnight" in August 1947, to use Nehru's poetic phrase, the initial plans and connections for nuclear development in India were established. There was virtually no British interference in these developments, and I speculate that the British attitude rested on three assumptions: (a) that the politics of withdrawal from India mattered more; (b) that Britain could reconnect more smoothly to an independent Indian scientific community by not interfering now; and (c) probably widespread among segments of the Raj, that Indians would not get very far with their nuclear ambitions anyway.[13] The British were not completely indifferent as the end drew near, however; after his intensive meetings in India with members of the Atomic Energy Committee and Nehru in early 1947, physicist Patrick Blackett immediately briefed Viceroy Wavell and Field Marshall Auchinleck in Delhi and met Prime Minister Attlee, Sir Stafford Cripps, and Lord Mountbatten (about to be the last viceroy) in London. The subject of all these high-level meetings was "the atomic energy setup in India," which Blackett had learned through Bhabha, Nehru, and Bhatnagar.[14] Cripps and Mountbatten knew personally all the people Blackett had met, so British leaders were well informed of Indian nuclear plans prior to Independence, and British officials approved of Canada's shipment of uranium oxide to India in July 1947, as we shall soon see.

In May 1946, however, members of the Atomic Energy Committee had decided on four issues: that the Tata Institute in Bombay should be "the centre of all large scale programs of atomic research in future"; the Tata Institute should be funded to purchase a 200 MeV betatron accelerator and establish a team of ten scientists to operate it; capital and recurring grants would be given to Saha "towards the expenses for the operation of a cyclotron"; and a capital and recurring grant would be given to D. M. Bose "for research on the trans-uranic elements" at the Bose Institute. A subcommittee was created to study uranium-bearing minerals in India, including the "vast deposits of thorium and associated minerals in Travancore."[15] There is no mention of the National Physical Laboratory, which was in the design stage and whose director Sir K. S. Krishnan would soon be appointed as one of three members of the new Atomic Energy Commission in 1948. Krishnan was little involved in the pre-Independence work of the committee.

The decisions of the committee were not unanimous: Saha objected to the assignment of large-scale programs to Bombay and refused to sign the minutes. The first indication of Saha's position came in June 1946 and continued for the next seven months. In his typical style, while obstructing the whole process on principle, Saha asked Bhabha for special funding from the committee for his son's postdoctoral research abroad. Bhabha had to meet specially with him and negotiate a suitable wording in the minutes that allowed Calcutta a more significant role.[16] Annual conferences on atomic research were to be held either in Bombay or Calcutta. Bhabha worked hard to persuade Saha to sign the minutes, particularly before the next meeting of the committee in February 1947, when he reported on his recent meetings in Britain and North America and D. N. Wadia reported on thorium reserves and prospecting.[17] Bhabha and Bhatnagar even approved funding for Saha's son's research in August 1946, but not Saha's request for an expensive mass spectrometer. Saha meanwhile had to send a long letter to Bhabha explaining why his cyclotron was not functioning properly, in order to justify certain aspects of a grant he was seeking from the committee (he made the same approach to Nehru). Saha was also seeking and got separate and additional funding from the Tata Trusts at the same time, on which Bhabha would doubtless have been consulted by the trusts.

When direct communication was difficult between Saha and Bhabha, each wrote about the other to Bhatnagar, who then mediated. In any case, Bhatnagar was the source of the money for the work of both men. At this stage (1947) the CSIR funded 33 percent of the Tata Institute's budget and more than that percentage of the budget of the Institute of Nuclear Physics in Calcutta. After the second meeting of the committee, Saha wrote to Bhatnagar to complain, again, about Bhabha and the committee's plans. He said Bhabha's ego stopped him from letting Calcutta (read Saha) fill its true national role in atomic research and that the committee "has no well defined ground to cover," unlike commissions in the United States or Great

Britain that focused on "large-scale release of fission energy and problems associated with it."[18] Saha made a detailed analysis, criticizing the attention paid to (and money spent on) "general problems of nuclear physics and cosmic radiation . . . and the production of high-energy particles (100, even 200 MeV). There seems to be no close link of either cosmic rays or very high energy particles with fission phenomena." He quoted Eugene Wigner's estimate that the cost of a small experimental pile was about $200,000, "which is the same order as the price of a high-energy accelerator," thus attacking Bhabha's plan to purchase an accelerator from US General Electric. He pressed Bhatnagar, and through him Bhabha, to concentrate on prospecting for uranium rather than relying on thorium and to study large-scale separation of isotopes.[19] This was Saha's consistent critique of Bhabha's plans until he was elected to Parliament in 1952, where he expressed them on the floor of the Lok Sabha and in his journal *Science and Culture.*

Foreign Nuclear Relations Require Legislation

Bhabha was, however, thinking about uranium and isotope separation and had just traveled to meet John Cockcroft, director of the British Atomic Energy Commission at Harwell and W. Bennett Lewis, director of the Canadian Atomic Energy Establishment at Chalk River. In a sense he was prospecting for uranium; both these men knew Bhabha when they had all been together in Cambridge in the early 1930s. Given the revelations of the defecting cipher clerk Igor Gouzenko in 1946, intelligence gathering for the Soviet Union in Ottawa had been very active in the circles Bhabha visited, and it is probable that Cockcroft and Lewis were more guarded than they might have been with Bhabha in the past. While in Ottawa, Bhabha made one of his first direct requests for assistance from any government, though this was well before India had its own government and before Bhabha was formally appointed to it.

In June 1947 Homi Bhabha asked the National Research Council of Canada for a ton of crude uranium oxide, to enable Indian scientists to start experiments, just until Indian minerals could be used in the construction of an atomic reactor. With the blessing of the United States and Britain, Canada made the shipment in the hope that it would ensure future Western access to India's thorium supplies. Curiously, Canadian records show that the request came not from Bhabha in Bombay but from the Bose Institute in Calcutta; dated 7 June 1947, the request was to purchase uranium, and the shipment was made secretly at an "undetermined cost" (according to the file) in October and November 1947.[20] The shipment may have been requested by or for the Bose Institute because, as decided by the Atomic Energy Committee, D. M. Bose received research funds to study the separation of U235 from uranium oxide (coming to India in yellow cake form U_3O_8), after the ore is crushed, ground, and leached in sulphuric acid, then concentrated and purified by centrifuge and heat. Unlike Saha's laboratory, the Bose Institute was not a public building but a privately owned institute, perhaps removing a problem posed by the fact that India was not yet a sovereign country, and therefore its public institutions were not really under the jurisdiction of the committee or the CSIR. D. M. Bose was no ordinary scientist—he was the nephew of renowned Jagdish Chandra Bose, who already had a slightly deified status in science and culture, not just in Bengal but across India. In any case, the Bose Institute had more space than the Tata Institute in Bombay, which frankly had no appropriate buildings. The Bose Institute had been one of the best-funded labs in the country. Again, the shipment, which actually occurred just after Independence, may not really have been delivered to the Bose Institute anyway.

The only reason offered by Great Britain, the United States, and Canada for agreeing to the secret shipment of the ton of uranium oxide before India's Independence was that all three were interested in the very large deposits of thorium in India.[21] Though commercial interest in India's radioactive resources was also strong in all three countries, by 1950 Indian government controls restricted both trade and use. Nevertheless, the trust partners were playing the same game as the private sector in other fields, assiduously and legally building a cartel, judiciously applying geological and mining intelligence supplied by people who already bridged the private and public sectors. In this tight context, Indian scientists were gradually opening a strategic partnership with France, the country frozen out of the closed though uneasy alliance of Anglo-American-Canadian atomic energy commissions. This exclusion occurred despite the fact that French research in Paris, Cambridge, Chalk River, and Montreal had been a major and acknowledged contributor to the progress of the other three.[22] French resentment was not forgotten and subtly informed their involvement in India and in other countries such as Israel and Iraq for years to come.

Seven months after Partition and Independence, Nehru led a debate on atomic energy in April 1948 in the Constituent Assembly and introduced legislation drafted by Nehru, Bhabha, and Bhatnagar that eventually created an Atomic Energy Commission. Perkovich and Abraham describe in rich detail this debate among elected members, only one of whom, Dr. B. Pattabhi Sitaramayya, had technical training. Clearly some members were familiar with the new British bill on atomic energy, on which the Indian final draft legislation was modeled, and they expressed concern about the secrecy required in the bill, and its haste. When Nehru and some elected members stressed the need for speed, for the state's total commitment, and for secrecy, the link between peaceful uses of atomic energy and possible military applications was very clear and open to parliamentarians in 1948.

In the debate, Sitaramayya, elected from Madras, pointed out that the demand for complete state monopoly over nuclear research and the practice of secrecy might actually promote insecurity and signal preparations for war rather than peace. He said that in the future, when atomic energy is a more domesticated and routine daily matter, the assertion of a state monopoly would appear laughable. He questioned the need for so much secrecy if it were not the intention to prepare for war? Chiming in about secrecy, S. V. Krishnamurthy Rao asked why India should take on powers greater than the British or American acts allow? (In fact, the Indian draft was based very closely on the British legislation, and the effective powers were equivalent, as the British were to discover to their chagrin.) Perkovich described Rao as the act's "only forceful critic." Professor S. L. Saksena rose and asked why India should hesitate to pursue all aspects of atomic power, having just failed to obtain a seat on the Security Council of the United Nations, calling it a "humiliating spectacle." It is necessary, he said, to actually *have the capacity* to use atomic energy for destructive purposes in order to say meaningfully that India will not use it for destructive purposes.[23]

As Abraham describes this parliamentary debate, Nehru, when challenged, could not and did not "deny, in effect, that the Indian atomic energy programme has a military component from this moment of inception."[24] Nehru explained in the debate that he did not know in a practical sense how to distinguish between peaceful and military applications of atomic energy. In my opinion, this ambiguous blurring was essential for Nehru's psyche from the beginning, for Bhabha's politics, and for the eventual bomb-making program years later. There was no coyness here: neither of them embraced nuclear weapons at this stage, though Bhabha moved more quickly toward them than Nehru did as the years rolled by.

Forming the Atomic Energy Commission of India is interesting for each of the principals. Even with his severe criticism of Bhabha's plans, Saha was thought necessary to the commission. This may have been due to his bosslike role in science in Calcutta, a city where Congress and Nehru needed to retain an influence amongst their leftist-socialist-communist critics; Saha appeared to Nehru as a bridge over to this opposition. Saha had already been asked in

1947 by a faction of the Congress Party in Bengal to run for election in 1948. Though Bhabha might have preferred to have Saha outside the commission, Nehru preferred him to be inside it. When approached by the government in 1948, however, Saha "unequivocally expressed himself against" the formation of an Atomic Energy Commission.[25] His own reasoning for this refusal, written expressly for his biography in 1953, was that India first needed to grow an independent industrial strength and also to train personnel in nuclear physics in universities. He knew the proponents of the AEC, neither of whom had jobs in universities, and to him that was a deficit. Bhatnagar was already a secretary in the central government and Bhabha did not work in a university. He thought Bhatnagar's and Bhabha's approach was wrong, but his reasoning is hardly persuasive. Nehru then asked Saha to be a member, and Saha's refusal was first countered by the prime minister phoning Saha in Calcutta to try again to persuade him to change his mind.[26] When Saha again declined the request that he join the commission, speaking directly to Nehru, finally the third member appointed was Sir K. S. Krishnan, who had left Saha's old department at Allahabad to become the first director of NPL in Delhi in 1948; Krishnan appears to have played a more passive role in the AEC than Bhatnagar or Bhabha, and certainly a more passive role than Saha would ever have done. What is interesting is that for generations afterward people thought Saha had been snubbed and excluded, but in fact he had been courted by none other than the prime minister.

In replying on the AEC question, Saha "advised the government to follow the French model as the conditions were similar in the two countries": Saha's reasoning about France was that "on termination of the war, they found themselves in an unenviable position. The government of General de Gaulle appointed an AEC at the head of which they proposed Professor Joliot-Curie, a professed communist, and Roger D'Autry, a civil servant. They set to work with great energy but found themselves handicapped on account of dearth of raw materials (uranium, thorium, graphite, beryllium), instruments and trained personnel."[27]

Saha spoke about France from firsthand experience; Frederic Joliot-Curie had invited him, along with Bhabha, to the tenth anniversary memorial conference honoring Joliot's hero Lord Rutherford. Top British scientists such as Patrick Blackett, J. D. Bernal, R. H. Fowler, and others such as Leo Szilard were invited to this deluxe conference in Paris in early 1948; Saha stayed as guest at the prestigious official Palais d'Orsay and Bhabha at the luxurious Hotel Georges V. This is where Saha, in conversations with Joliot-Curie, began his study of the French approach to atomic energy, continuing when Joliot-Curie visited the Saha Institute in Calcutta to lay its foundation stone.[28]

Saha was well aware of the conflict in the United States for civilian control of the AEC, of the movement among scientists against secrecy and monopoly (for example, Bhabha's unsuccessful attempts to obtain two items, a betatron and plastic balloons). Saha's view was that the 1947 McMahon Act had "given force to monopolistic American tendencies." He decided against the AEC and, instead of supporting it, remained outside. The other members perhaps breathed a sigh of relief, given his habit of criticizing and contesting almost everything. Saha eventually felt the chagrin of becoming financially dependent upon the CSIR (and later the AEC), meaning that he was personally dependent on Bhatnagar and Bhabha. Though his sources of private sector funding were not drying up, even in 1947, they were limited and he may have already realized that this limitation would keep him out of the race for a top nuclear laboratory. Perhaps he thought refusing membership in the AEC would send the most critical signal possible, or he felt better remaining formally unconnected, a position from which he could more easily be critical. Although his rejection of membership in the AEC cut him out of some important information, his curiosity about nuclear physics and atomic energy was constant; even while declining Nehru's invitation to join the AEC, he was searching in early 1948 for a map of French territories in Africa that might show their uranium deposits and naïvely asked his young research associate B. D. Nagchaudhuri, then traveling in the United States, to purchase heavy water for research, either there or in Norway.[29] He had some funds convertible to foreign currency, and despite his exclusion from official channels, by using a network of former students and friends, such as D. S. Kothari and Jnan Ghosh, Saha kept remarkably well informed about the AEC and the CSIR. When the Atomic Energy bill was passed in 1948, Bhabha and Bhatnagar were already rolling ahead in the commission, pulling K. S. Krishnan along with them. Saha's chances of buying heavy water anywhere, however, were zero.

In January 1948 the Atomic Energy Commission was formally established out of the Atomic Energy Committee, now with Homi Bhabha as its chairman, Shanti Bhatnagar as secretary, and K. S. Krishnan as third member. This was really a continuation of the 1946 committee's work, and the new money continued to flow through Bhatnagar, as before, from the budget of the Ministry of Scientific Research. Despite its legitimacy created by the hurried passage of the 1948 Atomic Energy Act, it was still just a commission, with a limited staff and budget, and not a formal department. It was to take six more years for Bhabha to establish a new government department. Bhatnagar was now a secretary to the government of India, polished, with superb networks in Delhi, and a reputation for getting things done. He was also a senior officer of the powerful international Society of Chemical Industry, based in London, and already had years of informal access to the large corporations (Imperial Chemical Industries, DuPont, etc.) that dominated the growing nuclear contracting business. Bhabha was established in Bombay, wealthy and well connected to both business and government. There was no apparent tension between the two in reputation and prestige—they were a balance of powers. Both were FRS. Both were building a different chain of laboratories, at big expense. Though some people have spoken of a tension, based in part on Bhabha's assumption of his intellectual superiority with respect to Bhatnagar, we have as yet no evidence for such tension. Though Bhatnagar was a strict vegetarian and nondrinker, quite unlike party-loving bachelor Bhabha, they got along quite well. It was a relationship of accommodation for mutual gain.

The Triangle in the AEC and Its Missing Member

Bhabha and Bhatnagar had each had only one visit to North America before starting the commission in 1948, and their strongest relations were really with Great Britain and Europe. But both were building American networks and worked together, in tandem, with Nehru. In secret operations for the AEC Nehru signed the letters and checks as minister of Atomic Energy or of External Affairs. This was the active triangle. The third AEC scientific member, K. S. Krishnan, does not appear to have been a part of this activity, and his occasional influence, though not zero, is difficult to weigh. The real missing member was Meghnad Saha.

In the relationship with British, and then Canadian, and perhaps even with some American and French scientists and officials, the fact that two of the AEC members, Bhatnagar and Krishnan, were knights was also of subtle social advantage. Krishnan was present in meetings but not active in operations unless they called for his considerable scientific reputation as a physicist. Bhabha had great respect for Krishnan's achievement as a physicist, as can be seen from the sensitive memoir-obituary he and physical chemist Kathleen Lonsdale wrote about him for the Royal Society (see Biographical Notes).[30] Now that he was director of the NPL and a commissioner, he was a sign that the NPL could be used to remind national political leaders and their allies that physics really matters and to justify the money being spent on the nucleus; in short, it could be used to manage leaders' expectations and keep the public quiet.

A number of these people had joint appointments and interlocking relationships. Bhabha now had a secure official position in the government of India as chairman of the commission, besides the strong informal ties he had to Nehru and others. He was not, however, a secretary to the government in the way that Bhatnagar was (a secretary was more like a Canadian deputy minister or an American assistant secretary). Nor was Bhabha a knight who could address Sir John Cockcroft or Sir Edwin Plowden, of the British Atomic Energy Authority, or Sir William Penney, head of Britain's nuclear weapons program, as equals in that particular sense of rank, as Bhatnagar could. Thus Bhatnagar's social status in Britain was essential to successful bureaucratic politics there and in India, notwithstanding the limits of being "an Indian knight" and the prejudices directed toward Indians, not only in Britain but also in Europe. What Bhabha had, however, was a Cambridge background, an FRS in physics, and a very good current reputation as a physicist.[31] This too was indispensable. Finally, he had throughout an indefinable advantage through his friendship with B. K. Nehru, Jawaharlal's cousin, after they met in Bombay in 1947–48. Brij Kumar Nehru enjoyed a string of ambassadorial appointments putting him in the flow of the best intelligence, and this must have been very useful to Bhabha. Moreover, B. K. Nehru had an intellectual wife with her own important network, particularly among scientists.[32]

It is likely that there would have been four members of the AEC had Saha agreed to Nehru's invitation in 1948, and Nehru would then have found symmetrical representation for the commission—north with Bhatnagar, east with Saha, south with Krishnan, and west with Bhabha. After all, Saha had been a full member of the Atomic Energy Committee in 1946 and 1947, and Saha's inclusion would have satisfied both Bengal and the left politicians with whom Nehru had to deal. However, though Saha rejected Nehru's personal invitation to join the AEC, he maintained an extensive nuclear information network despite his official self-exile and used this network quite strategically before and after coming to Parliament in 1952. From time to time he received additional assistance from D. N. Wadia in geological explorations or D. S. Kothari in nuclear physics, who was very well connected to Blackett and was the scientific advisor in the Ministry of Defence in Delhi. He was also well informed by Jnan Ghosh, chemist and Delhi political insider with a strong Bangalore and Calcutta base. It is unclear how much Saha's interventions actually changed the course of the AEC's work, if at all, but fewer questions would have been raised publicly had Saha not persisted, even before he went to Parliament in 1952. Most important he posed questions for the elite public's mind, and his editorials in *Science and Culture* were particularly widely read and influential at this time, especially after his election to Parliament. Journalists are known to have begun to pursue certain stories after first reading about them in *Science and Culture.* Some of the early information about the atomic energy program would have remained confidential but for the heroic efforts of Meghnad Saha.

Secrecy in India, Information, and the Nuclear Powers

Recalling Bhatnagar's pronouncement against secrecy in science in 1945, it is important to observe secrecy creeping into the commission in 1948. Of course the Atomic Energy Committee itself already had pre-Independence secret undertakings, one of them being Bhabha's visits to Ottawa and Washington in 1947 to secure a shipment of Canadian uranium. As evidence of the efficacy of this secrecy, the written histories of the nuclear program and international cooperation are silent on this Canadian uranium shipment to India; silent too is Bothwell's exhaustive writing on Canadian nuclear history.[33]

One must also remember that India was excluded from information shared by all the nuclear powers, just as the three allies, Canada, Great Britain, and the United States, were keeping things hidden from each other though they were compelled by their agreement to share. (This is why informal networks are so important in nuclear history.) So India's AEC imposed the same embargo on its own information with respect to Indians *and* foreigners that India was subject to by other "friendly countries." Nehru had also worked as a journalist and so knew the art of releasing a little information to the press but not releasing too much, and never all at once. Having learned to circumvent British surveillance during their struggle for Independence, having developed secret codes in their imprisonment, could members of the Congress Party like Nehru have abandoned the relative safety that secrecy gave them through the 1940s? But the creep of nuclear secrecy continued unabated in India, as in other countries. Bhabha had already been told that agencies of other governments could not communicate with him or his commission unless legally enforceable secrecy governed his own commission. Secrecy is a condition requiring at least two willing partners. The desire for secrecy was not simply a result of Bhabha's and Nehru's appraisal of its necessity inside India but was more the result of the conditions imposed by foreign agencies on which India would necessarily become dependent for the negotiation of nuclear power.

This nuclear secrecy applied even among politicians in the cabinet, among those who took an oath of cabinet secrecy. In a 1952 letter Nehru explained to his minister of Finance, C. D. Deshmukh, that the AEC had to function in secrecy because it was inevitable. The minister of Finance was obviously troubled by the high cost of the AEC and frustrated that he was provided with so little information about it, so he asked Nehru, as minister for Atomic Energy, for more. "The work of the AEC is shrouded in secrecy," wrote Nehru defensively and disingenuously. "I try to keep in touch with it and get reports from time to time. I do not know how else we can proceed in this matter." On the contrary, Nehru was in regular and intimate communication with Bhabha and Bhatnagar on this very subject. Nehru provided his Finance minister with a summary of the AEC's report and asked him not to show even this summary to others. If he wished to read the whole report, said Nehru, "I do not want copies made of it." He sent the letter to Deshmukh and a copy to Gulzarilal Nanda, Home minister: "I do not propose to send these papers to anyone else."

In this letter Nehru also provided the Finance minister with the AEC budget estimates for the next four years (Rs 920,000 in 1953–54 and almost double to Rs 1.74 million in 1956–57), saying, "This is a substantial sum from our point of view although it is about a thousandth part of what the USA is spending on atomic energy development." He mentioned twice that the Planning Commission would have to form its own opinion about the AEC's program but left the clear impression that such an opinion would have limited weight.[34] About a year later, when laying the foundation stone for the Tata Institute in Bombay in 1954, Nehru addressed the nagging question of secrecy again: "Science in fact does not flourish in secrecy," but India's association with other countries demanded it. "Those other countries are more advanced than we are, and if we have any association with them in regard to this work, they want us to keep it secret, even if we do not."[35]

This is the basic agreement imposed on the cabinet about the whole AEC operation: Nehru is saying, read our summary version of the atomic program plan, return the document, don't ask further questions in cabinet, and approve the money. This situation does not differ much from the British, Canadian, or American government's management of information about atomic weapons and power at this time. Less than a year before Nehru wrote this letter, Churchill had returned to power in Britain as prime minister and was surprised (Churchill, given to hyperbole, said "shocked") to discover that a very few members of the previous (Labour) cabinet had already approved spending £100 million on atomic energy and weapons without consulting their cabinet colleagues, not to speak of Parliament.

Although nuclear secrecy was imposed and directed from the top down and although it mirrored and repeated familiar practices received from

British administration in India, scientists were also quite familiar with it. This applies not simply to the generation of Bhatnagar and Saha, or even Bhabha, but to the young generation of the late 1940s. For example, in 1946 Raja Ramanna was being guided in his doctoral research by the head of the Department of Physics at Kings College in the University of London, Alan Nunn May, until one day his supervisor suddenly disappeared from the college. Nunn May had been tutored in physics at Cambridge by Patrick Blackett, worked in Chalk River in Canada on the Manhattan Project, and was supplying detailed information to the Soviet embassy in Ottawa during and just after the war, when Gouzenko revealed his knowledge. Nunn May was put on trial in London in April 1946 for treason and espionage, did not return to work, and was sent to prison for ten years. Shortly thereafter, Ramanna was visited while working in Nunn May's lab by none other than the key British scientist in the Manhattan Project and current chair of the British bomb project, neutron theorist Sir James Chadwick; they discussed Ramanna's ongoing work and Chadwick recommended that he use U235 (an unobtainable "secret" isotope) in his dissertation research.[36] Ramanna was twenty-two years old at this time, and this was an abrupt introduction for a young Indian scientist to secrecy and its transgression, though he could hardly have known the background to this drama. It also conveyed the contradictory nature of nuclear secrecy—that it does not permit or enable the intellectual development of the very people it is designed to protect—scientists—and so must necessarily be circumvented for the discipline to evolve, as Chadwick's advice to young Ramanna showed. Other young scientists were going to experience it by exclusion from secret facilities abroad or by trying to obtain equipment for their work that was on someone's embargo list. Though secrecy could be avoided, circumventing secrecy required scientists to engage in reciprocal exchange of information; this was an essential task of India's nuclear networks, just as it was for the scientists of other countries.

India's scarce hard currency reserves and the strong demand for their alternate uses (e.g., in armaments or food purchases) compelled Bhatnagar and Bhabha to employ the AEC's monopoly on fissile materials for profit. Instead of simply purchasing foreign equipment and material using the sterling balances, they shrewdly used thorium and beryl ore in negotiations for uranium, reactors, and heavy water. They correctly surmised that cash would not be enough to engage in this game; there had to be a strategic interest developed in the minds of these foreign powers. They had no bargaining chips in this game other than thorium, beryl, and the inherent rivalry among the allies. The American, British, and Canadian rivalry and their group fear of the French and the Russians were, however, just tangible enough to provide Bhabha, Nehru, and Bhatnagar a hand to bargain with. India had neither oil nor a strategic location, which appear to have been the only other bargaining chips that would have qualified India to participate in this game. (Though Indian airbases had been considered in 1948–49, forward anti-Soviet bomber bases had already been chosen by the United States and Great Britain in northern Pakistan in 1949–50.) Though its gold, diamonds, and opium had greatly interested others in the eighteenth and nineteenth centuries, India did not have gold, narcotics, or diamonds in any appreciable volumes now. Nor did it extract or manufacture anything else that appealed to foreign strategists at that moment.

The result was that even with limited nuclear resources Bhatnagar and Bhabha were given a relatively free hand in negotiations, access to sufficient hard currency, and cooperation with India's diplomatic machinery abroad, such as it was. They learned not to trade India's thorium and beryl for cash; they learned that equipment, designs, training, and other materials were more useful than cash. There was no market for these materials, no formal brokers, no prices that were not secret (it was, after all, a cartel). Moreover, these matters were precisely what the men negotiating with them had control over in their countries. And Nehru was quite prepared to continue the secrecy and to justify it and got angry when there were leaks that implicated him. But his vigilance was quite insufficient, starting in his own office. For example, in 1955 he wrote to his whole cabinet about the appearance in the *New York Times* and in an Indian newspaper of a secret Soviet document containing suggestions to India about solutions to "the international problems of the Far East." After inquiries Nehru couldn't decide which office was responsible for this leak of a secret Soviet document but surmised it was the External Affairs ministry, of which he was the minister. "I would request members of the Cabinet also to be particularly careful not only about Cabinet papers but also about any talks with pressmen. Some of these journalists openly brag about their contacts with members of the Cabinet, their frequent visits to them and the opportunity given to them sometimes to see papers. Probably this is exaggerated, but it creates a bad impression among other journalists as well as the public."[37] Nehru needed timely outlets for his atomic energy publicity and asked for a minimum of questions that might lead to interference. Not everything was secret, however; Bhabha wrote and sent part of the draft text of the Canada-India reactor agreement on his department's letterhead through the open postal system and had people hand carry documents. Not only was this sometimes expedient; it also avoided Indian government channels and embassies when he wished to. Though he may have believed that his own channels would keep his decisions and communication more secure, not everyone accepted his claim on power.

All of this nuclear planning and negotiation clearly needed an effective administrative structure. Just after Independence (1948–50) top Indian scientists had arrived at levels comparable to diplomatic and banking elites in terms of their access to foreign exchange and travel. Bhabha, Bhatnagar, and a few others were traveling, usually expensively, by air, to London and European capitals about three times a year during this period, and to North America at least once or twice a year. As the minister responsible for both Atomic Energy and Foreign Affairs, Nehru was involved in negotiation for nuclear permits and equipment, traveling to Ottawa in 1948 to discuss nuclear cooperation, and to Geneva and Brussels in 1949 in order to obtain centrifuges for uranium separation. Contrary to popular conception, Nehru was deeply involved in all atomic energy issues, dissembling when he thought necessary, engrossed in the thrill of this direct contact through which he was, to use his own phrase, "making friends with science." But, unlike top "political scientists" like Bhabha and Bhatnagar, bankers had banks and banking acts, diplomats had embassies and foreign policy—so scientists would have to get a similar apparatus, one that resembled the authority of nuclear scientists in other countries to do their jobs. Was this "resemblance" born out of a desire to look like others, to look modern? Yes, the look of modernity was important, and still is. But more important, resembling other nuclear powers administratively was a prerequisite to interacting with them, to coming to the table with them, to being able to bargain. Such resemblances were spreading, the international system alternately inviting them or compelling them, and nuclear communities, commissions, legislation, and ways of thinking were on the leading edge of this process.

TWELVE

Scientists' Networks, Nehru, and India's Defense Research and Development

Thinking about a nuclear future ran parallel with plans for an Indian military establishment and in that context development of defense research and its integration with India's small industrial capacity. Although this involved the creation of new institutions and attitudes crossing many boundaries, like parallel lines the nuclear future and military development remained separate. Scientific questions and scientists' networks played a key role in spanning the boundaries between nuclear and military development, though their formal separation remained a characteristic of India's development for another twenty years. It was probably inevitable that an outsider with insider connections should have been a catalyst for such changes during the immediate postcolonial period. Such a person was Patrick Blackett, new to India in 1947 but very experienced in British research and military affairs.[1] Key Indian figures found him an excellent conduit for getting things done and made effective use of Blackett as a consultant and intervener.

This conduit was important because senior Indian scientists and military officers had no common experience and little established means of communication. Indian leaders had opposed the regime in which Indian senior officers were trained and embedded and which they were now going to command. Indian scientists had no sustained interaction with military people unless they were part of the same extended family. Professionally and intellectually speaking, their worlds were miles apart. The military had no significant capacity of its own to conduct research in 1947, and nothing in its culture suggests that it would do so unless there was a catalyst from the outside. If the small nuclear community was going to resemble the larger, more powerful American and British scientific communities far away, so too the Indian military establishment was going to resemble military setups elsewhere just as it was trying to negotiate deals with them for the supply of conventional armaments.

Invited to lunch at the Nehru home in January 1947, Patrick Blackett and the acting prime minister were seated beside each other. Jawaharlal Nehru knew of Blackett's experience in war and military affairs and asked him how long it would take "to Indianize the military," meaning both its command structure and weapons production and supply. He was not yet the prime minister and India was not yet an independent nation. Blackett's reply was a challenging one, obliging Nehru to explore two different kinds of strategy and thus two different military setups. For the "realistic" strategy Blackett preferred, he told Nehru that Indianization could be completed in eighteen months; this would prepare India for conflict with other similar powers in the region. It would also require an indigenous research and development capacity. For the "unrealistic" strategy, in which India would prepare for conflict with major world powers, Blackett predicted it would take many, many years to "Indianize." Nehru liked this approach and wrote to Blackett soon afterward to ask him to advise him on military and scientific affairs. From this invitation much followed.[2]

Paradoxically, Nehru also did not personally wish to build a military state. He did, however, want to build a scientific state. He knew he stood in the shadow of Gandhi's impact on the public, pointing India in a different nonmilitarized direction, but he was ultimately commander of its forces. According to a section of its people and their international supporters, India's forces should be unlike those in other countries. According to D. K. Palit, Sir Robert Lockhart, first commander in chief of the Indian Army, presented a paper on the growth of the army and defense to Nehru in 1947. Nehru is said to have retorted, "We don't need a defence plan. Our policy is nonviolence. We foresee no military threats. Scrap the Army. The police are good enough to meet our security needs."[3] As he was keen to forget them, Nehru was avoiding thinking about the numerous deployments and confrontations in 1945–47 between troops of the Indian Army and civilians in Burma and Malaya.[4] He did not want India to gain the reputation as a newly independent militarized state, certainly not one that would carry out occupations and confrontations with others.

But Nehru was right at the center of a decision to separate the political and professional aspects of military development when he agreed that "based on the recommendations of Lord Ismay, and upon the advice of the first Gov Gen Lord Mountbatten, service headquarters opted to function apart and separate from the Defence Ministry. The result was no horizontal integration between the service HQs and the Defence Ministry."[5] The Defence Ministry would therefore be kept closer to the cabinet and the prime minister than to senior military staff. Thus was established a tradition in India in which senior generals of the joint chiefs of staff did not engage directly with politicians, mindful that until 1947 there were two kinds of politicians—those who could be pulled in by the viceroy's office (and were British) and those whom the viceroy might have trouble pulling in (and were not British); professional military officers had to be wary of both types but for different reasons. Although the theory of "martial races" was officially repudiated in 1947, the Indian Army became a ground of plate tectonic tension (and release) among major social groupings—Muslim, Sikh, Hindu, Christian—and regional-linguistic populations.[6] The objective was to make the army cosmopolitan and to have the new air force and navy follow suit.

Blackett's value to India lay partly in his skill in the comparative analysis of military systems and research systems. Despite the chaotic changes going on around them, his friends in India correctly appraised his value to their objectives. After his intensive meetings in India with Nehru and Bhabha, Bhatnagar, and Saha of the Atomic Energy Committee in early 1947, Blackett briefed the viceroy and British commander in chief Field Marshall Auchinleck in Delhi, and Prime Minister Attlee, Sir Stafford Cripps, and Lord Mountbatten (just appointed to be the last viceroy) in London upon his return. The subject of all these meetings in early 1947 was "the atomic energy setup in India" well before there was any official atomic energy setup in India.[7] Blackett established his position on atomic weapons quickly after their first use in August 1945, arguing in a memo to the chiefs of staff three months later that an atomic bomb might decrease rather than increase Britain's security. The readers of this memo were not amused and made sure Blackett understood that. He visited the United States to receive high honors in September 1946, a month after passage of the McMahon Act: there he saw, in the final weeks leading to the election of Harry Truman as president, the seeds of the hysteria about communists and atomic secrecy that led to his being labeled persona non grata, Blackett's American honors notwithstanding. Coming home to London, and just prior to his first journey to India, he arranged two meetings with Prime Minister Attlee to lobby him on the virtue of the international control of atomic energy, including a prescient call for a "world system of inspection and control." When Attlee decided, along with one or two others, to prepare for construction and testing of an atomic bomb in January 1947, Blackett was in India and was not informed officially, though one can hardly believe he did not learn about it via his friend Henry Tizard, who was chair of the Committee on Future

Weapons.[8] These are all reasons why Nehru would find Blackett's presence and opinions particularly timely and useful. That he had a huge scientific reputation helped Nehru deflect any skepticism from colleagues.

On Nehru's simple invitation, Blackett became the prime minister's advisor on military and scientific development. In the interstices of that relationship they also debated foreign policy because scientific relations and foreign relations were connected in almost all strategic questions, and the connection between these relations were Blackett's passion. In 1967 he remarked, "I had no official status in defence matters except as an advisor to Nehru." Since his reports were marked secret and never available in India and his military consultations were not widely known, it is understandable that his public reputation was largely in the field of scientific research institution building, and not in strategic development.[9] This is in marked contrast to Indian perceptions of him, which focused mainly on his influence with large scientific research organizations. There is truth in both views. Blackett rightly pointed to his relationship with Prime Minister Nehru as crucial to his influence in both spheres.

Nehru identified with and trusted Blackett because both had been to Cambridge, held favorable attitudes to political socialism, were cautious about economic socialism, and cautious about the same kinds of people, including Americans. In turn, Blackett found Nehru to be unusually receptive to his ideas and approaches, more receptive than his own British prime ministers of the period (such as Attlee or Churchill). What Blackett does not say is that Nehru acted on his ideas, not just because of Nehru's personal receptivity, but also because Blackett's ideas were acceptable to a handful of other influential people in India, namely, Homi Bhabha and Shanti Bhatnagar, senior military officers like General J. N. Chaudhuri, and to a lesser extent physicist D. S. Kothari. Blackett also had direct access to high offices in Whitehall and to people like Lord Mountbatten, through naval networks. Few of these people would have been untouched by Blackett's multitude of interests and torrent of energy, and all would therefore have paid some attention to his presence. More important, Blackett articulated ideas the Indian leaders had, supported them in their efforts, and made connections to people outside India for them. And finally, his influence coincided with and extended the work of his old Cambridge colleague Archibald Hill, who had been deeply involved in India since 1943. To this nexus of trusting and increasingly familiar relationships should be added the force of a number of material factors and ideas explained below, making what Blackett said and did even more attractive in elite Indian political and scientific circles.

Blackett, this British expert, was sought precisely as the sun was setting on the British Empire, as physics was taking a strategic turn, and as military development was becoming both a national and commercial enterprise in which British university-based scientists gradually played a less important role. In 1947–50, Indian adjustment to the United States, the new world power, was complicated by the sudden emergence in 1947 of Pakistan and its increasingly close relations with the United States. British adjustment to American power was also difficult and uncertain, and there was considerable tension before the formation of NATO in 1949. The United States was a major power in Asia too, with bases in Japan, Taiwan, and the Philippines. When the first Soviet atomic bomb was exploded in August 1949, three or four years before the Allies predicted it, military airbases in northern Pakistan quickly acquired a new significance to Britain and America. When the victory of the Communist forces in China was complete in October 1949, vast and diverse populations on a great arc of land from the Baltic to the Pacific lying just to the north of India were separated and segmented from the world of the West that India understood so well. The Anglo-American approach to India would have been complicated enough after 1947 without Pakistan, a Soviet bomb, and a communist China. But with these additional factors it was very complicated indeed. India was usually a secondary consideration to the great powers as the Cold War evolved, but it was occasionally a sharp and pressing consideration. Yet as a stage on which to demonstrate *another kind* of economic and political development, both from the rivalrous view of the great powers and from the view of the Indian government that Blackett served, India became more central.

The Prime Minister and the Physicist

When Blackett received an invitation to come to India in 1946, signed by Nehru, he accepted. He was asked to address the 1947 Indian Science Congress, of which Nehru was president, and the Association of Scientific Workers of India. Blackett had already established a friendship with Homi Bhabha, had Bhabha to stay as a houseguest at Manchester in 1946, had examined Vikram Sarabhai's dissertation, and knew people like S. Chandrasekhar and D. S. Kothari from Cambridge. Blackett had been a committed member of the Association of Scientific Workers in Britain since the 1930s, and the objectives of the Indian association were similar to his British association, to increase the application of scientific rationalism in politics and planning and to improve the working conditions of scientists. His old

Cambridge colleague, and head of the association in Britain, the physicist J. D. Bernal, was influential in getting international recognition for the Indian association, but the impetus for its development came from India.[10] The British association experienced a great political shift from 1947–48 onward, when disenchanted British Communists like Bernal began to leave it, and thousands of lab technicians joined to make it a kind of trade union from the early 1950s onward. In India it retained its original intellectual, planning, and agitational character for the next thirty years and was an irritant of directors of the CSIR laboratories, where its branches were permitted.[11]

During his 1947 visit Blackett received an honorary degree from the University of Delhi, an event requiring months of planning and Blackett's foreknowledge. So evidently he had decided in the autumn of 1946 to go to India. Traveling to the meetings of the Indian Science Congress in early January 1947, Blackett and Nehru found themselves on the same plane and managed to talk.[12] Shortly afterward, Blackett went for lunch and a talk at Nehru's home in Delhi. Much of Blackett's engagement in India resulted from those two personal conversations.

Blackett's paternal uncle had been a missionary in India, his mother was the daughter of Sir Charles Maynard, an officer in the Indian Army around the rebellion of 1857, and his mother's uncle a tea planter in Assam. Perhaps because of these associations, or in spite of them, Blackett had not really wanted to go to India before it achieved Independence, and though he was proud to be there during that great change, he soon discovered how old lines of dependence were maintained and new lines of interdependence were established. As a broker in negotiating a new relationship between India and Britain, Blackett was an early harbinger of the whole discourse on the proper role of science and technology in newly independent developing countries.

Few prime ministers anywhere at that time had the appreciation of science and scientists that Nehru did, precisely at the time when his prime ministerial influence in the Indian scientific community was profound. He was the direct minister with portfolio for Atomic Energy and for Natural Resources and Scientific Research. Nehru was intimately involved in two further ministries—Defence and Economic Planning. He was also minister for External Affairs, until he appointed Krishna Menon to the post. He kept these five files close to him and met directly with the people responsible for them. The indirect evidence is that Nehru discussed many issues in his responsibilities with Blackett, judging from comments in letters written by Bhabha, Bhatnagar, Mahalanobis, Saha, and others. Scientists and officers soon learned that one way to Nehru, and simultaneously to Bhabha or

Bhatnagar, was through Blackett's ear. From his second visit in 1948 onward, Blackett usually stayed in the prime minister's residence, often for weeks at a time, seeing the ebb and flow of high politics in India. In 1948 his wife, Constanza, stayed there with him, on and off, for many weeks. He received his correspondence and phone calls there. If Blackett got inspired about your project or problem, his energy knew no bounds and he believed in the possibility of long-term projects. This must have appealed to Nehru. Having Blackett around was part of what Nehru meant when he said India "had to make friends with science." Of course he meant something grander too, but Blackett's presence was a tangible expression of this friendship, one which suited Nehru himself.

During the years when Nehru was formulating the movement of non-aligned nations, he tried his ideas out on Blackett, who was very receptive to them and probably contributed to them, even by disagreeing with them. Blackett later spoke approvingly of this part in India's foreign policy and explained that India could not possibly win an arms race, not even with nuclear weapons. He said in 1967, "I think you [India] would have split up if you had to fall into the western orbit in the first five years of Independence. I think it was [Nehru's] great contribution of showing that non-alignment is a thing that is feasible. . . . I did not think it was feasible at the time."[13] Nehru was already preparing for nonalignment before India was created and before he became prime minister and had organized the 1947 Asian Relations Conference in Delhi on 23 March 1947, the day after Louis Mountbatten's arrival as viceroy of India. At the conference, Nehru extolled the idea of an Asian confederation of states, just as India was about to be torn apart.[14] Ten weeks later, on 3 June 1947 the British decided to partition India. This partition occurred only a few months after the vivisection of Europe into "blocks," in recognition of which the United States declared the Marshall Plan on 5 June 1947. Even then, with Partition, the new India was to include not just old British India, but the large territories and populations of about five hundred kingdoms and principalities, some poor, some wealthy, that the British had governed "indirectly." In the three cases of Hyderabad, Kashmir, and Junagadh these territories were incorporated by use of the force of the Indian Army, under the ultimate command of Nehru, and in the case of Kashmir the situation has been unstable ever since.

Blackett was candid about Nehru in 1967, saying,

> He was a superb leader. But he did not know how to get things done very well. He believed in science in a rather naive way. We all did at the time. He was not more naive than other people. It was enormously valuable that he should

> put science first in making Indians scientifically minded. But science is only part of a game and the real effect of science comes from producing wealth. . . . Now India is finding out that the problem of turning science into wealth . . . is very much more difficult than just doing science. It is not his fault that he did not fully understand this. . . . We were all scientifically naive. We thought science was the solution to everything. I do not think I was very conscious of it explicitly earlier in this period.

Blackett's deepest critique was about the gap between idea, implementation, and action. "Nehru did an enormous amount to get nonscientists to understand what was scientific. But his regime did not do nearly as well in implementation. What he lacked were hard-headed industrial-minded Ministers who could push on the agricultural program, the industrial program."[15] The idea that "science was the solution to everything" was not held only by senior leaders but was in circulation among new recruits too, such as the young Canadian-trained engineer M. R. Srinivasan, who would start to build nuclear reactors and heavy water plants for Homi Bhabha. Years later he recalled that in the 1950s everyone in India believed that "the great and powerful countries of the world had moved ahead and that we should catch up as rapidly as possible. . . . Strangely we did not hear a well-reasoned argument for the development strategy we should adopt . . . many of us had no clear idea of how science and technology would transform society."[16]

Blackett went on to say about Nehru that

> he had a bit too much intellectualism to solve the problem. He spent, from one point of view, too much time talking. . . . He liked intellectual company. (And he did not get it except in Homi Bhabha and people like that.) He had extreme informality and charm; his physical presence was extremely attractive; he was very engaging, with a shy sort of smile. He was sort of light-hearted. I liked this about him. But he spent too much time, I think, on science anyway. Considering the amount he had to do, running a country of that size, the amount of time he did spend with us was indeed surprising. On the whole, he liked me and others more as companions than as consultants.[17]

The atmosphere surrounding Blackett's work for Nehru in 1948–50 was turbulent following the violence of Partition. Indian scientists like Bhabha, Bhatnagar, and Kothari were thrown into work that had immediate strategic implications. The conclusion of military operations directed by the Indian Army against the princely states in 1948 were also quite violent, particularly against Hyderabad. Chiefs of staff were still British, but no longer did foreigners direct these activities: Home Minister Sardar Patel and Indian generals handled these operations on the ground. The political scene in India also changed dramatically. Less than a year after Independence and Partition, a young Hindu zealot assassinated Mahatma Gandhi in his garden in Delhi, creating a martyr and settling a leadership question. The people who surrounded Nehru would now be fully in charge, and with the death of Patel in 1950 it was Nehru alone who guided the Congress Party. Gandhi's rootedness, like many of his followers, became a progressively more nostalgic force; Nehru's cosmopolitan, patrician, and elitist leadership was without serious challenge, even from the Hindu extremists who privately approved of Partition and Gandhi's death. The Partition and Gandhi's assassination were a dramatic closing and opening of possibilities.

Indian Scientists and Patrick Blackett

Nehru did, of course, have two scientists who were like hardheaded industrial-minded secretaries of the portfolios for which he himself was minister. They were not elected ministers but had, as secretaries to the government, direct access to him, sometimes more than any elected minister in the cabinet. They had deputy minister status but were as powerful as ministers, in my opinion. One was Homi Bhabha, whom at the time of Bhabha's death in 1966 Blackett called "my best personal friend." Blackett went to parties at Bhabha's house, had his portrait sketched by Bhabha, had Bhabha to stay at his own house in Manchester. The other powerful secretary who also reported directly to Nehru was Shanti Bhatnagar, head of the CSIR, which arranged the invitation and paid for Blackett's first trip to India. With Sir Shanti, Blackett also formed an ongoing personal relationship, until Bhatnagar's early death in January 1955. They all had frequent meetings in London as well as Delhi.

In addition, Blackett was friendly with Prasanta Mahalanobis, Cambridge-trained physicist turned statistician, building the influential Indian Statistical Institute in Calcutta. Mahalanobis, elected an FRS in 1945, had quasi-ministerial status, saw Nehru regularly, and shaped Nehru's entire approach to economic planning through creation of an elaborate system of empirical data collection and analysis at the district level. He was soon titled cabinet advisor on statistics and built a vast system of statistical intelligence across India (prices of essential commodities like rice and lentils, etc.), thus enabling the central government to know (if it cared) what was happening in rural and urban areas. More junior than Mahalanobis was D. S. Kothari, another Cambridge-trained physicist, and Blackett's advice shaped his career when he became the scientific advisor to the minister of Defence in 1948 and headed the Defence Science Organisation when it was established in June 1949, modeled on the one Blackett had just prescribed for the United Kingdom. Kothari eventually gained the most important position in the politics of education, research, and universities in India, becoming chairman of the University Grants Commission. Together these men controlled larger budgets for construction and for employment than anyone else in the scientific community. It will be noted that Nehru's nemesis, Meghnad Saha, was not among Blackett's notable physicist-confidants: ironic to think that it was precisely their contacts with leftist and communist scientists that gave both of them as much professional difficulty as it did political appeal.

These Indian scientists were traveling regularly by 1950: Bhatnagar to Norway to negotiate a heavy water deal, Bhabha to Ottawa to negotiate uranium, Kothari to Moscow to purchase troop transport aircraft, Mahalanobis to Washington to look at computers. As they passed through London they all met with Blackett, who in turn visited their institutes in India, gave lectures there, examined their doctoral students, helped select candidates for appointments, appraised new research programs, and then promoted them if he liked them. The Blackett friendships were an extension of Archibald Hill's friendships in the professional sense of advocacy within the scientific community. Hill knew Bhatnagar very well and advised Bhabha on the establishment of his own institute. As personal friends of Blackett's, these men and their wives also asked him to watch out for their children and other relatives when they studied or worked in London, which he did, as Hill had done. The relationship of Bhabha and Blackett was, however, quite different from the others. They already knew each other well at Cambridge, though Bhabha was eleven years younger than Blackett. (Their friendly 1938 confrontation in Manchester over particle scattering is described in chap. 5.)

Bhabha did not depend on a lab for his reputation in physics, but he liked the practical experimenter's gleam in Blackett's eye and was delighted when Blackett got the Nobel Prize in 1948 for experimental work with the cloud chamber in the early 1930s in Cambridge. Bhabha had been there at the time and knew that the chamber demonstrated the existence of particle showers by photographing the splitting of cosmic rays and demonstrating pair-creation of negative and positive electrons.[18] This is the subject on which Bhabha built his reputation. Bhabha had even asked Blackett to lend him the big magnet Blackett built at Cambridge for cosmic ray research in Bangalore in 1941; it was never sent because of wartime shipping and controls. When Bhabha built the Tata Institute of Fundamental Research in Bombay, Blackett was in on every step. Every time Blackett came to India, often at someone else's expense, Bhabha would command part of Blackett's schedule and arrange meetings for him, including introductions to the captains of industry with whom he was very well connected. Blackett could not have had a more powerful and effective set of intermediaries in India, nor a more secure base from which to criticize and challenge establishment thinking. That some of his friends constituted part of the establishment and wanted similar changes within it only enhanced his influence.

During this entire early period as military consultant, Blackett was actively promoting his own scientific projects in India. He was already in transition away from the 1930s subject for which he won the 1948 Nobel Prize. Now he turned to the question of the reversal of the earth's magnetism. He lectured in India on both "the origins of cosmic rays" and "reversely magnetized rocks" and proposed projects on white dwarf stars, the subject of D. S. Kothari's work at Cambridge. Blackett supervised collection of lava and rocks in India in the 1950s and 1960s for his project on geomagnetism. He presented—for his friends and for casual observers—a unique combination of the theoretical and the practical. He was what could be called an engineer's physicist because of his mechanical ingenuity in designing and building equipment. The combination in 1948 of winning the Nobel Prize, launching a book critical of Cold War logic, and being put on the US government's non grata list won many hearts and minds in India.[19] In Britain the book gained notoriety because Blackett won the Nobel Prize just as the book appeared in print. Briefed by the Indian high commission (particularly Krishna Menon), Nehru, and other Congress leaders as visitors to Britain every year, Bhatnagar and Bhabha were well informed about British politics, and they would have known that some British leaders did not entirely approve of Blackett, whether for his opposition to a British bomb project, already started but still secret, or his support for some policies of the Soviet Union.[20] Why this was so and how they tried to marginalize Blackett in 1946–49 is discussed in *Negotiating Nuclear Power*.

India's New Defense Establishment and Britain's Reactivated Influence

India was trying to build a scientific and technological community, to apply existing skills to socioeconomic problems and to build up a scientific military infrastructure. Blackett realized that much of the effort was drained in the importation of costly weapons systems and replacement parts. Thus he argued that India should define very carefully what armed conflicts it would face and should choose its weapons for those conflicts with equal care. He correctly realized that the military forces would be used often in conflicts inside India, particularly after the 1947–48 experience in Hyderabad and Kashmir. He faced the need to address the imperative of the era, which was then, as now, to reduce poverty and inequality. Conflict between the competing interests of industrial and rural socioeconomic development and between defense and science was unresolved throughout this entire period and lies at the heart of the rise of the influential community of scientists in this story.

That there would be a continuing British influence in Indian military policy is perhaps counterintuitive for those who expected a clean break between the two countries. Hadn't they each had enough of the other? But the Indian forces were perhaps the most truly Anglo-Indian creatures and their separation overnight would be improbable, to say the least. Senior Indian officers ready for command were still being trained in British institutions by British instructors. Royal commissions or king's commissions carried great prestige among officers. The chiefs of staff in India were British in 1948 and 1949. The chief commander of the Indian Navy was British, right through to 1956, appointed jointly by India and Britain and paid by the British. The air force was built out of remnants of the Royal Air Force. This is how and why Blackett's anomalous or dual standing—scientist and naval officer—made sense, and made him influential.

His advice ranged from the most concrete, like creating jobs and arranging for specific appointments of individuals, to a general concern for the proper administrative development of new bodies like the Defence Science Organisation. Then there was his wider commitment to cultivating certain ways of thinking, ways we would now call "systemic," to do with how systems operate and how they can be understood dynamically in the field. Blackett was, after all, among the earliest practitioners of operations research. This wider commitment was explained in terms of strengthening Indian strategies for economic development, industry, and defense. Blackett later reflected that he had not been conscious of the oversimplification in his approach to science and development and began to rethink what he and his closest allies in India believed about science in the 1940s and 1950s. Though a leader, Blackett was simply resonating a widespread overoptimism about the scientific state (and the scientific socialism on which it should be founded). In this Nehru, Bhabha, Saha, Mahalonobis, and to some extent Bhatnagar concurred with him.

Blackett was not just a physicist who happened to understand nuclear strategy but also a naval officer with practical experience in war. In this he had something in common with the new viceroy of India, Lord Mountbatten, with whom he would gradually become friendly. Blackett's meeting with Mountbatten in London on return from his first trip to India in 1947 was not a great success because he could not isolate Mountbatten from the surroundings; "I wish I had been able to see Mountbatten alone," he wrote to Stafford Cripps; "he was obviously much more sensible than Ismay but still disappointed me a bit. He seemed very keen on finding all the arguments on why India should remain in the Commonwealth. It may be that this will in the end happen, but to over-stress our desire for it to do so would seem to me a mistake."[21] Blackett had been convinced, in January and February, by people in Delhi who said Britain should leave India within three months, not six or nine. This relationship between Mountbatten, Blackett, and Nehru became important; if Mountbatten had not generally respected Blackett's work, he could have undermined his effectiveness in India. From the friendly letters they exchanged in 1970–72 ("Dear Dickie," "Dear Pat") it appears that Mountbatten maintained throughout a respect for Blackett's activities in India, each treating the other respectfully like the two aging naval officers they were.

In 1967 Blackett reminisced, with characteristic confidence, about the experience of coming to India to try to influence the defense establishment and why he was selected: "Nehru spoke to all the scientists, but I was the only scientist there with professional military experience; five years at sea in the first war, and four years in the second amidst the application of scientific methods to modern warfare. So it was not very accidental that Nehru chose me to advise him."[22] Five years later he said, with continuing satisfaction, "On the whole I think that my views about the Indian armed forces expressed in 1948 have not proved too incorrect."[23] Blackett's reflections in 1967 were consistent with his first report in 1948. After the Indian conflict with China in 1962 and Pakistan in 1965, Blackett said to an interviewer that his effort from the beginning had been to prevent the unnecessary and costly introduction of weapons and strategies that would not have practical value and to focus attention on the military risks India did face. The most pervasive problem he faced was, Blackett said, that "Indian officials and advisors were thinking purely from a Whitehall angle. There was an appalling psychological dependence on every word that Whitehall speaks. I understand in the beginning, in 1948, there was very little time and experience to think for yourselves. But a great many of your problems are due to imitative adoption of Whitehall habits. Actually a lot of that thinking should not be exported anywhere. Some of it is not even good here [Britain]."[24] It was in this context, as military consultant, that he began to think about building a scientific research and development capability within the military, where there was then almost nothing.

Blackett met with the chiefs of the armed forces every time he went to India and with the minister of Defence and the minister's scientific advisor. He gave a talk to the chiefs of staff in the war room each time, toured armaments and aircraft factories, appraised candidates for strategic analysis positions, and interpreted strategic implications of the Cold War in the Indian context. He also got into the details of building specific equipment like tanks and rockets in India. He said, with great satisfaction in the 1967 interview, "I like to think that . . . I saved India a lot of money by discouraging her from buying too much big and expensive western equipment." He distinguished carefully between the Indian and British military risks to which new weapons were to be the solution. For example, in 1948 he advised the British forces to follow a rapid program to develop supersonic fighter planes, whereas for India he took a more cautious approach and did not support integrated production of jet fighters under license in India until seven years later when he proposed a light versatile transonic fighter, based on an Anglo-Dutch model designed by people whose reputation he knew very well. Yet he advocated immediately the acquisition of large military transport aircraft.

Underlying Blackett's 1948 report, said Abraham, is "the understanding that it is the intention to make India as nearly as possible a self-supporting defence entity as may be at the earliest possible date," noting the remarkable similarities to the objectives of Archibald V. Hill's work during the war. "Yet this understanding was neither invented by Blackett nor did it come from Hill," according to Abraham; Blackett took the quotation on self-reliance verbatim from the "Report on Defence Science" (1946), written by Dr. O. H. Wandsborough-Jones, a British defense scientist advising the colonial Indian government.[25] The ideal of self-reliance was immediately attractive to Blackett, in part perhaps because he was sympathetic to Fabian socialism, which stressed self-reliance, and also because experience during WWII required it of Britain.[26] Blackett's report, submitted just weeks before his Nobel Prize award, gained considerable reputation through judicious circulation, even though it was marked "secret." Wandsborough-Jones's phrase became powerful in Blackett's hands because of the physicist's timing and authority. A stress on self-reliance resonated perfectly with a core value of the Independence movement articulated over the previous sixty years and with the fiscal imperative of the present budget situation.

But more important for Nehru than the issue of self-reliance, Blackett forced chiefs of forces and Defence ministers in India to define what kinds of war they expected to fight, who the probable enemies were, and what the risks were. He gave shrewd appraisals of his first adversaries, the admiral with whom he disagreed over the navy's future, the air marshal, and the chief of staff of the army. These British officers still commanded Indian forces. Admiral Parry "was a very nice man," said Blackett in 1967, "but he tried to sell India, or make her buy four fleet aircraft carriers which would have required sixteen new destroyers to protect the carriers." (These were light carriers of about 15,000 tons, then in great surplus and being sold cheaply by the UK to friendly navies to raise money.) Blackett might also have known that it was the Royal Navy's policy not to sell a carrier to India that year, because it felt that the Indian Navy was not ready to maintain one. His advice and the Royal Navy's position coincided. This kind of carrier was eventually obtained by India and renamed INS *Vikrant.* The chapter in Blackett's report on the Indian Navy is the only subject on which there was cabinet disagreement in India, and Blackett had to strengthen his argument by obtaining, from friends in Whitehall, secret estimates of costs of ships currently under construction by the Royal Navy in 1948. The military situation in India was quite volatile; Blackett had heard about the mutiny in the Indian Navy in the previous year. Admiral Parry returned fire a few days after seeing Blackett's report; "I personally think you are being unrealistic in your fundamental assumptions—particularly that India should only prepare for a local war against an imaginary opponent of comparable overall strength to herself."[27]

The Indian Chief Air Marshall Elmhirst "tried to make India buy long range bombers," said Blackett, which would ruin India while being useless in local wars. Worse, he said, long-range bombers would have been dangerous to India, inducing massive and uncontrollable retaliation. During 1948 and particularly after August 1949, the British were looking for forward staging airbases from which to conduct preemptive first strikes into the industrial heartland of the Soviet Union. Northern Indian and Pakistani bases were their choice, to which Elmhirst was responding. They knew the Americans were searching for the same bases, but the uneasy and uncooperative relationship between Great Britain and the United States meant that Britain pursued its own strategy.[28]

For the chief of the army Blackett had more respect. General Bucher "was less intellectual but wiser than the others. He was born and brought up in India, and understood what it was about, he knew the terrain. He immediately spotted that the arms salesmen were trying to sell us things like these tanks without telling us they were too big to cross our bridges."[29] Blackett learned about these salesmen through Whitehall rather than through Delhi. It cannot have escaped Nehru's attention that it was very useful to have an expert like Blackett with a network in Whitehall to appraise the plans of India's British defense chiefs with their own networks in Whitehall. And containing or cutting military costs was a top priority. Nehru knew about increases in expenditure from Rs 1.57 billion in 1947–48 up to Rs 1.71 billion in 1948–49. "In effect there was great and growing pressure on Nehru to effect economy in military outlays from the beginning of his tenure in office as Prime Minister."[30]

Blackett also discovered at firsthand the postwar "Colonel Blimp culture" surrounding India, which enraged him, and he reported to Sir Stafford Cripps that he had just met General H. L. Ismay in 1947 with Lord Mountbatten in London. Also born in India, Ismay "produced more 'Blimpisms,'" said Blackett, "than I have heard from anyone for ages. He did not seem to me to have a clue as to the real situation in India. He just doesn't know the facts."[31] Indicating how much the old Raj suited Churchill, Ismay became secretary of state for the Commonwealth in the Conservative government led by Winston Churchill in 1951. Blackett could not have had a more distant relationship.

A New Indian Defense Research and Development Organization

Indian leaders knew that a new institutional arrangement was needed in order to link science and the military effectively. Blackett and scientists he knew began to propose a research and development organization with civilian leadership. In his 1948 report to the minister of Defence, Blackett listed those weapons that should *not* be on India's list for development, as follows: atomic weapons, chemical warfare, supersonic jets, high-performance jets, and guided missiles. All of these were unsuitable for India's strategic situation, he said. At the same time Blackett grasped from the beginning what few others fully realized—that, despite two hundred years of a deeply intertwined military development, British forces would not likely play any further role in conflicts involving India and that foreign troops and equipment would not likely be based in India. Blackett soon learned how Indian troops were deployed inside India, in Hyderabad and Kashmir. Nehru wrote a long personal letter in 1948 to Blackett at Manchester University to thank him for his work and advice, praise the Indian military success in Hyderabad, and say, in a spectacular understatement, that the war risk over

Kashmir had subsided: "I think definitely that there is hardly any chance of war between India and Pakistan. Of course the Kashmir issue remains and it is a difficult one."[32]

Concerned about the costs of licenses for defense production in India using foreign companies and anxious to capture benefits in India and elsewhere from Indian innovations, Blackett asked Shanti Bhatnagar in 1948 to provide him a list of all patents held by Indians that might be applicable to defense production in India. In 1950 Bhatnagar sent a list of all "the various projects which have been patented, exploited, or under consideration for exploitation."[33] Bhatnagar and Blackett then began to examine production facilities, one by one. Their later tours of armaments factories were an observational displeasure for Blackett, who recalled a factory set up north of Bombay by the Swiss arms manufacturer Oerlikon: "Absolutely four million pounds went down the drain. India did not want new prototype weapons like Oerlikon did, Indians wanted to manufacture existing weapons. The factory had some refugee-Germans trying to invent recoil-less guns, under the charge of a charming [Indian Civil Service] man who had been to Oxford and who did not know anything about machine tools in the first place." In 1967 he went on, "You ran your Bangalore electronics factory down. It is running all right now. But it took ages to get it going, because the people in charge had no knowledge of it. One of the Defence Minister's followers was a poet. He was so embarrassed. He did not know one machine tool from another." "Don't misunderstand me," Blackett said elsewhere and getting a Cambridge dig in at the same time, "I am deriding my civil servants just as much. They thought they could run anything, being at Oxford."[34]

As part of "Indianization," Nehru accepted Blackett's advice in late 1947 to establish a new research capability within the Ministry of Defence. Until this time most research had been carried out in Britain. There were a few technical development establishments under the Indian Army with the purpose of providing inspection and quality control for ammunition in ordnance factories. Although the officer corps was well trained and educated for this work, scientists and engineers were not involved in these establishments, and their capabilities were very limited. The individual being considered to direct this research effort was D. S. Kothari. A few weeks after giving his first major report to Nehru on defense in 1948, Blackett wrote to the minister of Defence, "I am delighted with your choice of Dr. Kothari to be scientific advisor to the Defence Ministry. I am in complete agreement with his views on these matters."[35]

Daulat Singh Kothari, who became the scientific advisor to the minister of Defence with Blackett's help and who headed the new Defence Science

Organisation (later the Defence Research and Development Organisation), had first met Blackett in the Cavendish Laboratories in Cambridge in the early 1930s.[36] On Kothari's return to Allahabad, Saha heard of an opening at the university in Delhi and encouraged his appointment there in 1934. The vice-chancellor of the University of Delhi, Sir Maurice Dwyer, supported the choice of Kothari to the Defence Ministry in 1947, as Dwyer had known about him for the ten previous years as a faculty member at the university. With Blackett's advice, in 1948 Nehru appointed Bhabha, Bhatnagar, and Krishnan as members of the new Scientific Advisory Committee to the Defence Ministry, where they remained almost permanently, accompanying Kothari.[37]

All three members of the Atomic Energy Commission were now advisors to the government on defense and also Fellows of the Royal Society, but Kothari was not among them; he had first been proposed for election to the Royal Society in 1944 by Sir Arthur Eddington, seconded by Sir James Jeans and countersigned by E. A. Milne, C. G. Darwin, R. H. Fowler, and from India S. S. Bhatnagar, B. Sahni, and M. N. Saha. Saha was probably the prime mover in this nomination, which focused on "astrophysics and statistical mechanics" and referred to Kothari's work in the 1930s on formation of neutron stars in white dwarfs. The unsuccessful Kothari nomination certificate, though supported by a galaxy of famous Fellows, was suspended a total of fifteen times between 1944 and 1959.[38] In the first year Kothari was competing for a physics fellowship against astronomers like Fred Hoyle and Hermann Bondi; S. Chandrasekhar was elected in 1944 for his work on astronomy too. Though senior military officers would have cared little for this recognition from the Royal Society, it mattered among those scientists who were commencing their professional lives in the Defence Science Organisation.[39]

Blackett was also encouraged to intervene in defense training and planning. Having overseen the creation of the Defence Science Organisation, in 1950 he urged creation of functional groups, such as the Weapons Assessment Team and the Operational Research Group. By 1951 he was clearing the way for Indian defense scientists to spend a year at Cambridge and in the UK Operational Research Group. He also acted as intermediary for the appointment of a British RAF officer as the first director of the Indian Institute of Armament Studies. At that time Kothari discussed a naval research laboratory for Bombay with Blackett. Although Blackett wanted a focus on real problems, he also said to the minister of Defence, "It is most important to realise . . . that a research and development establishment must often keep a considerable number of its personnel employed on work which promises no immediate or tangible results."[40]

Amidst Indian excitement about new weapons, Blackett never lost sight of the importance of conventional weapons in India, like tanks. To Nehru in 1951 he wrote, gently promoting Kothari's influence, "I have heard from Kothari that he is carrying on energetically the investigations we started on tank and anti-tank gun performance."[41] At the same time, however, Blackett was talking to people in London about new weapons like rockets, as he did with H. A. Sargeaunt, deputy scientific advisor for the Army Council in the War Office, Whitehall. "Kothari's group," reported Sargeaunt on a visit to Delhi, "is certainly making great strides, and I think there is no doubt that when you next come you will be impressed. My feeling is that they must now think in terms of specific Indian problems rather than copying the problems of other nations."[42] This "copying" was the very issue that Blackett was striving to address—he perceived a predisposition to adopt the solutions developed elsewhere to problems which were not India's. Nevertheless, Blackett was supportive of new initiatives like rockets, which did not even have to be British: "when I was there at Christmas [1950] the Minister was particularly keen on a French rocket weapon which sounded very good."[43] In fact he was rethinking his position on atomic bombs as tactical weapons, so that he revealed that he had previously underestimated the rapidity with which missile systems would become more accurate than they had been in WWII and could be equipped with smaller hydrogen bombs.[44]

What emerges is a picture of competitive access in India for European technologies and experts, particularly in the military and engineering fields. Gustav Tank and Willy Messerschmitt, successful designers of German fighter aircraft, both came to India as consultants after 1949 to discuss jet fighters. Aware of British efforts to sell surplus equipment and design or build new equipment, Blackett spoke of French rockets, just at the moment when Nehru and Bhabha were considering a bilateral nuclear agreement with France. European experts were reported to be charging lower fees than the British, Americans, or even Canadians. It was in this competitive environment that Indian leaders began to think about the cost of conventional weapons and new unconventional weapons.

The new Defence Science Organisation in Delhi was modeled on the one Blackett had just prescribed for the United Kingdom. Since it was first housed in the new National Physical Laboratory of the CSIR and borrowed scientists and equipment from it, there was during 1950 a deeper integration of personnel in defense research and industrial research. The close relationship between Bhabha, Bhatnagar, Kothari, and Blackett—and all of them with Nehru—reinforced such structural integration. Kothari now joined the group of scientists who had institutes to build and positions to fill; within months of starting work, Kothari received a letter from his teacher, Meghnad Saha, inquiring about a job in defense research for one of Saha's sons.[45] By 1951 Blackett was channeling requests for employment on defense matters in India directly to Kothari. Kothari often paid the cost of Blackett's visits and scheduled his time. In 1953, for example, Bhabha heard of Blackett's visit to Delhi and phoned Kothari from Bombay to ensure adequate time was set aside for a lengthy visit to TIFR in Bombay. Kothari then informed Saha and Mahalanobis in Calcutta of the Blackett visit to TIFR, and they immediately wrote to Blackett to have their institutes put on the itinerary. Curiously, Blackett does not seem to have developed a similar relationship with K. S. Krishnan, the distinguished physicist who was director of the National Physical Laboratory, where Kothari worked, and also member of the Atomic Energy Commission.

Defence, the CSIR, and Patrick Blackett

Another connection between defense research and industrial research was also made through Blackett. The CSIR started in 1942 by developing prototypes for Indian industry to manufacture for the Allied forces, but it had no ongoing connection with the new Indian forces. Though Bhatnagar was on the defense advisory committee and despite the fact that the military labs were housed in the CSIR's new National Physical Laboratory, there was a structural disconnection between CSIR and the military. Blackett was the person who could alter that, and, as he did with Bhabha and atomic energy, he did as much as he could to alter it—whether in India or in Britain. But Blackett did not have any elaborate network in the United States, and that is what the younger generation of scientists in India knew was needed.

In the eight years they knew each other (1947–55), Bhatnagar and Blackett helped each other considerably, carrying on the productive relationship of Hill and Bhatnagar. Until Bhatnagar's sudden death in January 1955, they were in regular communication and sometimes toured CSIR laboratories together, often far outside Delhi. They were the same age, Blackett liked Bhatnagar, and he approved of the general direction of the CSIR's evolution as the state's applied research system, even when he saw research in labs that was not actually being applied in the way he thought it should be. Bhatnagar valued Blackett's influence with Nehru and supplied Blackett with the list of patents he needed for a study of defense production contracts.

Nehru's influence in cabinet had been necessary to initiate that study and policy review. Bhatnagar worked hard to find good appointments for his laboratories both outside and inside India. There is no clue in the Blackett papers whether Blackett advised for or against the appointment of non-Indians, but he certainly assisted Bhatnagar, Kothari, and Bhabha to identify non-Indian candidates abroad and appoint them. Therefore in 1951 Blackett interviewed a Canadian metallurgist working in London, who wanted to "get a job in India and settle there permanently." Bhatnagar said the new National Metallurgical Laboratory was "in the charge of a brilliant young Frenchman," and he welcomed the arrival of the Canadian metallurgist W. K. A. Congreve from London, accompanied by Blackett's favorable opinion based on a colleague's expert recommendation.[46] Hill played a similar role at this time, looking internationally for good candidates for Bhatnagar's labs.

It was not all about rockets and the nucleus: Blackett was excited by a research project on solar power at the National Physical Laboratory in Delhi and went to considerable length to obtain the French reports on solar power for the Indian scientists in 1951. It is curious that Bhatnagar could not obtain these directly, for an agreement was signed that year between France and India on scientific cooperation, including rare earths and nuclear power.[47] The NPL solar-powered cooker project did not deliver on its promise, much to Bhatnagar's embarrassment. Nehru was always asking Blackett and others about progress on work at the NPL in northwest Delhi and had begun to talk about the solar cookers there. Bhatnagar praised the project, and Nehru began promoting it. A paper was published on it, though in the CSIR's own journal.[48] But the cooker turned out not to be viable. Then Nehru's displeasure rose at having promoted something unsuccessful, and apparently he scolded Bhatnagar, who thought of resigning but didn't. Both leaders had been caught up in its premature promotion; Bhatnagar's known reputation for promotion (and a bit of self-promotion too) had landed him in embarrassment and his ally the prime minister was embarrassed as well.

I had the idea long ago that Nehru must have been so busy with Kashmir and Korea, Kerala and cabinet, that he might have turned to scientists only in the mid- to late 1950s and then only to receive and coolly appraise proposals from stars like Bhabha; I thought that perhaps his fascination with scientists was reached in his old age. I also thought the separation of the military and scientists in India was so strong that it dissolved gradually only at the time of the first nuclear bomb test in 1974. I recall my naïveté to remind us that this was the received popular perception widespread after Nehru's death in 1964, largely because no realistic version of history was publicly available. But Nehru's active involvement with scientists and strategy occurred at the very beginning and was always international in character. And there was some military and civil integration from the beginning, through scientists. Although one should not exaggerate the amount of influence these scientists actually had in defense spheres, there was some flow of information between these separate spheres from the beginning.[49] But the flow was through individuals and their networks. There was not an articulate interaction of structure and organization.

Since it was a profound influence on his choice of the people he listened to and the practical projects like atomic energy that he promoted, we should reflect on Nehru's more personal view of science given in ad lib fashion on his feet in Parliament in December 1954. Challenged to define his science, Nehru responded, and we should note the critical reference to Meghnad Saha:

NEHRU: It becomes necessary to think out these problems, not academically but scientifically—not like Professor Saha—but scientifically, I say.

SHRI S. S. MORE: What *is* your science?

NEHRU: My science, if I may say so, is essentially based on social statistics; not wishful thinking—except wishful thinking in the sense of the objective—but essentially based on social statistics; how we can gain something and how we can have a balanced economy, heavy industry, medium industry, light industry, cottage industry; how we can provide employment within a short space of time, and how we can generally raise the level of human happiness in the country and national strength. . . . It is quite possible . . . that there has been lopsided development. There has been. And, if I may say so, there has been lopsided development in most other countries too, even in trying to plan.[50]

This was not the place for Nehru to try to explain his romantic and aesthetic attraction to science, to describe how genuinely and emotionally moving he found the discussions among scientists whom he could follow but only partially understand, whether about biology or astronomy. Nor was it the place to describe with fervor his support for the role of scientists in the defense of India. Many of his listeners in Parliament had guessed at those attractions, as had the scientists who managed to catch his attention. This was the occasion to talk of social statistics and admit to lopsided development. Ten years later, one big consequence of that lopsidedness would present itself to India's leaders and scientists in its fullest form.

1*a*

Courtesy of the Tata Institute for Fundamental Research, Mumbai

1*b*

1*a* and *b*. Homi Bhabha launching hydrogen-filled rubber balloons for high-altitude cosmic ray studies at Bangalore, about 1948. The physics group established at the IISc in 1934 by C. V. Raman included Bhabha and young student Vikram Sarabhai and had done some modest experiments under wartime limitations. The group had its particle detectors carried aloft during 1943–44 training flights of US Air Force B-24 Liberator bombers. Balloons capable of climbing to 30,000 meters became a favored technique of Bhabha's cosmic ray group in Bombay, enabling them to study high-energy particle interactions at lower cost than in cyclotrons. Despite the move to Bombay, the group continued to launch at Madras, Bangalore, and Hyderabad; by 1950 high-altitude and balloon cosmic ray research was growing in India.
Courtesy of the Tata Institute for Fundamental Research, Mumbai

2. In the cyclotron room at the new Institute of Nuclear Physics, Science College, University of Calcutta, 1948: *(from left)* A. P. Patro, with two unnamed students behind him, B. D. Nagchaudhuri, B. M. Banerjee, and a beaming Meghnad Saha. This 38-inch cyclotron was partly a gift and partly purchased from the University of California, Berkeley, when Nagchaudhuri completed his PhD there in 1941; its vacuum system was lost to a Japanese torpedo in 1942, and it still had not produced an experimental beam in 1948. Seven years after this picture was taken, Saha accepted the support of his rival Bhabha (at Nehru's insistence) and agreed to long-term funding for his institute by the Department of Atomic Energy. After his death in 1956 this institute was named after Saha, and eighteen years after this picture was taken the cyclotron finally produced a low-energy experimental beam and research results. A new cyclotron was planned and partially built during the 1960s. Nagchaudhuri left the institute in 1968 for senior positions in Delhi, where in 1970 he became scientific advisor to the Defence minister, playing a key role in the May 1974 nuclear bomb test.

Courtesy of the Saha Institute for Nuclear Physics, Meghnad Saha photo archive, Kolkata

3. Meghnad Saha speaking at an election rally in 1951; before and during his time in Parliament, Saha took part in public demonstrations and shutdowns (*hartals*). Though he had lived in and out of Calcutta since he was twenty, he could still speak to crowds in the rural style, and this made him more familiar to the millions of in-migrants in the city. This was a reason why they and others elected him to Parliament for the Revolutionary Socialist Party in North-West Calcutta in 1951. Though deeply involved in the politics of science of the era, Saha was known better by the public as a "famous atomic scientist" who advocated economic planning, flood control, hydroelectricity, and the elimination of unnecessary imports; in short, he called for revolutionary changes in Indian conditions. Having lost much in the Partition himself, he empathized with the refugees who flowed into West Bengal after 1947 and frequently spoke for them in public. He died at age sixty-three in January 1956 while playing his role as an elected parliamentarian in Delhi.

Courtesy of the Saha Institute for Nuclear Physics, Meghnad Saha photo archive, Kolkata

4. A confident young Vikram Sarabhai welcomes scientists to the start of his new laboratory in Ahmedabad in 1952, among them *(from left)* the secretary of the Atomic Energy Commission and most powerful official in Indian science Shanti Bhatnagar, the chairman of the Atomic Energy Commission Homi Bhabha, and the dean of Indian physicists Nobel Prize winner C. V. Raman (both seated), who considered Sarabhai to be his student. Bhatnagar was a member of the 1937–38 committee that recommended Raman's dismissal from his post of director of the Indian Institute of Science. Now he was director general of a system of twelve laboratories and constructing ten more. Bhabha had decided nine years earlier not to stay in Raman's sphere of influence in Bangalore but to build his own laboratories in Bombay. Sarabhai, founder of India's space program, rose quickly by combining business and science, until he became at age forty-seven the chairman of the Atomic Energy Commission, following Bhabha's death in a plane crash in the Swiss Alps in 1966.
Courtesy of the Physical Research Laboratory, Ahmedabad

5. *From left*: Nehru, Bhatnagar, unknown, Bhabha, Bombay Chief Justice M. C. Chagla, Gujarat's Congress leader Moraji Desai, and J. N. Choksi of Tata & Sons viewing the new Tata Institute model in January 1954. The Tata Trusts contributed significantly to the costs of the institute, which was now the "nursery" for the atomic energy program managed by the new Department of Atomic Energy. J. R. D. Tata, chairman of the company, became a member of the Atomic Energy Commission. Moraji Desai opposed some of the expansive activities of the DAE after Bhabha's death, when Desai became minister of Finance in the late 1960s; when he became prime minister in 1977, Desai criticized the DAE atomic bomb makers and initiated a thorough and unfriendly review of the structure and operations of the DAE.
Courtesy of the Tata Institute for Fundamental Research, Mumbai

6. Homi Bhabha (*right*) giving directions about construction of the uranium metal plant and work on the gardens at the atomic energy facility in Trombay in 1955; he became passionately involved in the layout and composition of gardens, consulting gardeners far and wide. He discussed the gardens of the Tata Institute regularly and in detail with his close friend Pipsi Wadia; he even sent one gardener to see gardens in Europe, to inspire him with new ideas. Here he is seen with engineer Homi Sethna (*center*), who succeeded him as director at Trombay, and Colonel G. R. Menon (*left*). Bhabha often intervened when he saw trees being cut in other parts of Bombay and saved them to be transplanted in the fine gardens surrounding these utilitarian structures. His relationship with Nehru protected him from objections by financial officials to these "unnecessary" expenditures, but earned him a reputation for "lavishness," to which others replied that he was just "doing things the way they should be done."
Courtesy of the Tata Institute for Fundamental Research, Mumbai

7. Discussing modifications of a British design for India's first swimming pool reactor, Bhabha is shown with his team of young physicists and engineers in 1955; they met frequently with visiting experts in Bombay, building confidence for the construction of the larger more complex CIRUS reactor three years later. When this photo was taken, Bhabha had been appointed as chairman of the first Geneva conference on peaceful uses of atomic energy and was involved in secret negotiations with American, Canadian, French, and British atomic energy authorities for reactors, uranium processing, and heavy water. In the following seven years, he articulated opinions with Nehru and wrote policies that became the official Indian position on safeguards and inspections. Although wealthy and at times distant, Bhabha inspired young colleagues by his example of constant hard work and his ability to communicate with them.
Courtesy of the Tata Institute for Fundamental Research, Mumbai

8. Standing near the construction of the cooling tower about 1958, Nehru, Bhabha, and a foreign visitor (probably Patrick Blackett) discuss the future of the Trombay atomic energy establishment. In a suit, with glasses, and standing at the rear, is Bhabha's comptroller and project manager E. C. Allardice. Bhabha and Allardice were preparing for intense international negotiations with the British, Americans, and Canadians about new reactors and control of the nuclear fuel cycle, and Nehru, contrary to popular perception (a perception he cultivated), knew what was going on in the nuclear program from direct observation and was consulted on every major step. Throughout the 1950s Bhabha met Nehru about twice each month.
Courtesy of the Tata Institute for Fundamental Research, Mumbai

9. Homi Bhabha showing Prime Minister Nehru the work of glass blowers at TIFR after the inauguration of the new TIFR building in 1962. Though Bhabha's own work had little to do with glass for research, he appreciated the art inherent in this skill and went out of his way to find good glass blowers for his labs and nurture them, including sending them abroad for advanced training. He enabled one particular glass expert, H. L. N. Murthy, to work in other groups such as the Trombay atomic energy reactor group and travel to see and learn glass techniques elsewhere. Though these conversations with Nehru wandered off the official script, this wandering evidently pleased both Nehru and Bhabha, and official handlers had to wait patiently until the two of them were finished talking about art or music or writing.
Courtesy of the Tata Institute for Fundamental Research, Mumbai

10. Homi Bhabha, M. G. K. Menon, and E. C. Allardice showing M. C. Chagla a new mural by painter M. F. Hussein at the Tata Institute in 1963. Justice Chagla was now India's minister of Education, soon to be minister of External Affairs. Hussein's international reputation was taking off when Bhabha gave him this mural commission in the late 1950s, and it was one of the admired legacies in the building after Bhabha's sudden death in 1966. Bhabha himself was painting and drawing regularly at this time and deeply enjoyed this kind of contact with Indian painters. He also insisted on the integration of art and music in scientific projects and scientists' work. Allardice was an influential British former ICS officer with years of experience in India who helped Bhabha manage his many projects as a kind of comptroller and executive director, including completion of this famous building designed by Chicago architect Helmut Bartsch.
Courtesy of the Tata Institute for Fundamental Research, Mumbai

11. John Kenneth Galbraith looking at the special plastic for a high-altitude balloon at TIFR, in the company of Homi Bhabha and M. G. K. Menon of the cosmic ray group, about 1963. The capacity to extrude heavy-duty heat-resistant and cold-resistant plastic (Mylar) was brought to and developed in India for the first time by the cosmic ray group in the late 1950s. Without these balloons the cosmic ray group, to which Bhabha too belonged, would not have been able to do the quality of research on which its reputation was built. Following India's conflict with China in September 1962, US Ambassador Galbraith, a renowned economist, helped negotiate for TIFR the first big powerful computer in India, at the same time as US officials were negotiating a valuable contract for India's first (and only) light water reactor, to be built by a US company (General Electric) at Tarapur, north of Bombay.
Courtesy of the Tata Institute for Fundamental Research, Mumbai

12. Indira Gandhi in conversation with Vikram Sarabhai in the fuel rod control room of the CIRUS reactor at Trombay, outside Bombay, in July 1967. Prime Minister Gandhi was the minister responsible for atomic energy, and Sarabhai had, as chairman of the Atomic Energy Commission, just returned from a frustrating tour of foreign capitals seeking a nuclear defense guarantee in case of attack. The conclusion was that great power talk of support was not a guarantee, and the tour was declared unsuccessful. Six weeks before this photo was taken, Indira Gandhi said, "We may find ourselves having to take a nuclear decision at any moment." Although Sarabhai had already suspended a research program to design and plan a small nuclear bomb at this facility, the weapons-grade plutonium for the 1974 atomic bomb test was being produced near where he and Gandhi were standing, as a by-product of this reactor started with Canadian fuel and designs and with American heavy water. Sarabhai was not an advocate of building and testing a bomb. Had he not died suddenly in December 1971, Sarabhai would have been persuaded by Gandhi to leave his atomic energy position in 1972 and concentrate on launching a satellite through a new space ministry.
Courtesy of the Bhabha Atomic Energy Centre, Trombay, India

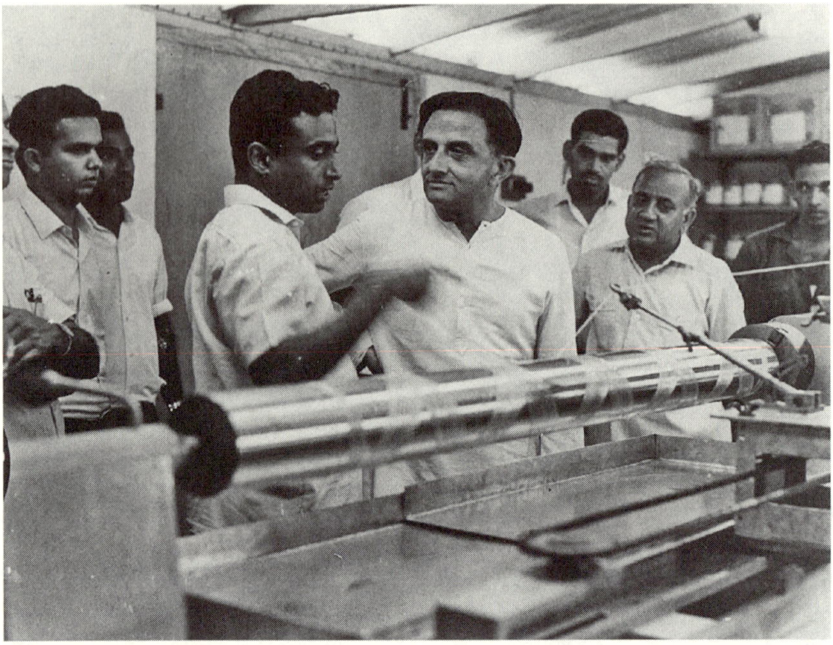

13. Sarabhai discussing a filament winding machine with Abdul Kalam, program leader for the satellite rocket and later president of India, Madhavan Nair, later chairman of ISRO, rocket engineer S. C. Sathya, and H. G. S. Murthy, later director of the Thumba equatorial rocket launching station in January 1968. Though a hand-turned filament machine had just been rigged up, Sarabhai asked the group to develop a mechanized one to produce nonmetallic fiberglass nose cones needed to encase and fly magnetometer payloads. Sarabhai promised to have the new winding machine, incorporating an old car gearbox, switched on by the prime minister in February 1968 at Thumba, but prior to that formal occasion, he wanted to make sure that the machine worked properly. Sarabhai involved these colleagues in reengineering a Soviet rocket engine and reengineering French and American rockets; he was briefed regularly by Kalam concerning a Defence Ministry missile project. A site for eastward long-range rocket and missile launches was being chosen and developed at his time, just north of Madras. Sarabhai died in 1971, before he could see an Indian satellite go up—something he worked on since 1962.
Courtesy of the Indian Space Research Organization, Bangalore

14. Having ordered and applauded the bomb test of May 1974, Prime Minister Gandhi finally visited the blast site at Pokhran in the Rajasthan desert seven months later, accompanied by DAE chairman and engineer Homi Sethna (*middle, gesturing*), cabinet minister K. C. Pant (*left*), and member of the Atomic Energy Commission industrialist J. R. D. Tata (*far right*). The bomb had exploded 100 meters below their feet with a yield of 8–12 kilotons (roughly up to the yield of the 1945 Hiroshima bomb). Almost sixty scientists and engineers had built the bomb for testing, backed by a team of a hundred soldiers who prepared the site, undetected by foreign intelligence organizations. India's delegates at the IAEA in Vienna began to justify the test in terms of peaceful nuclear explosions. Missing from the picture was chief BARC physicist Raja Ramanna, unable to attend on the short notice given him by Sethna; Ramanna felt strongly that he had lost "face" and had been "upstaged" by being absent, adding to an uncooperative relationship between them. Also absent was physicist B. D. Nagchaudhuri, student of Meghnad Saha and advisor to the minister of Defence. Now in an economic crisis, Gandhi would soon confront her opponents and declare a State of Emergency six months later. The end of nuclear cooperation with Canada and the US in joint reactor programs followed this first test and delayed India's nuclear reactor development.
Courtesy of the Photo Division, Ministry of Information and Broadcasting

15. While engineers and physicists were planning a bomb test, others were completing a new high-precision radio telescope at Ooty in the early 1970s, at an altitude of 2,150 meters in the Nilgiri Hills in southern India, far from TIFR. This radio astronomy research group, established in Bombay in 1963, developed all the capabilities and techniques essential to this precision-engineering project. Its location was because of the ingenious idea that the long axis of the telescope could lie parallel to the low latitude of ~11 degrees N on an Ooty hill slope, enabling the astronomers to track the moon from its rise to its set, and gave them high angular resolution <1 arcsecond for extragalactic astronomy research; TIFR built twenty-four parabolic frames on twenty-four towers, resulting in a half-kilometer-long parabolic dish antenna. Research on occultation, pulsars, and the solar wind established a world reputation for the group, and experience gained in building and operating the Ooty telescope helped TIFR to design the larger and more complex great meter radio telescope in the countryside outside Pune, operational in 1999–2000.
Courtesy of the Radio Astronomy Centre, Ootacamund

16. The nose cone of a rocket being wheeled carefully to the launch site. When Prime Minister Gandhi formally opened the Thumba equatorial rocket launch station in February 1969, there was inevitably a large entourage of prestigious visitors, all expecting special treatment. Consequently all the space agency's jeeps and light vans had been commandeered two hours before the 6:00 p.m. launch. "I had been totally forgotten by our Administration in the VIP rush for vehicles," recalled C. R. Sathya. The nose cone for this French Centaure rocket had to be walked with its sodium vapor payload, carefully for two kilometers. The famous French photographer Henri Cartier-Bresson also found a bicycle as his "official" vehicle, rode around the site before the launch, and stopped to click this photograph of rocket engineer C. R. Sathya and instrument maker Velappan Nair (holding the nose cone). This rocket flew to a height of 145 km. Ten years later the space agency ISRO was ready to launch its satellite on an Indian rocket, but the second stage of the rocket failed on the first attempt in August 1979. A year later in July 1980 an Indian satellite was successfully lifted by an Indian rocket to an orbit beyond 300 km.
Courtesy of the Indian Space Research Organization

17. Technicians assembling the coil on the 88-inch variable energy cyclotron next door to the new building of the Saha Institute in 1976. When energized, this coil and accompanying 262-ton magnet produced the required magnetic field within the cyclotron's "Dee," and particle behavior in that field could be controlled and studied. Even after the 1955 meeting where the VEC was discussed with Saha, Bhabha did not authorize the project until after lobbying in 1964 by a number of physicists in various cities in India. Construction began in November 1969; an internal alpha particle beam of 8 MeV was achieved in June 1977 and an external one in 1980. Components were fabricated at a number of engineering facilities across India, and the final cost of construction was Rs 270 million. Like its smaller predecessor, this 88-inch cyclotron was based on cyclotrons being built in California; it is a medium-energy, room-temperature machine in which by changing the magnetic field and the frequency of the electric field, researchers can vary the particle energy at will. Thus physicists can conduct experiments at much higher-energy regimes than available with the old 38-inch cyclotron and so perform atomic and condensed-matter physics as well as heavy ion experiments by varying the particle energy.
Courtesy of Vaishali Nayek, with permission of Variable Energy Cyclotron Centre, Kolkata

THIRTEEN

A Scientist in the Political System: Professor Saha Goes to Parliament, 1952–56

In 1947, the first Constituent Legislative Assembly election was held in West Bengal. India was not yet independent. "Netaji" Subhas Chandra Bose's brother Sarat Chandra Bose was a member of the first interim cabinet, trying to unite internal opponents of the Gandhians who then dominated the state-level Congress Party. Bose persuaded Saha to stand for election along with him, having known him for many years. But soon after that, Saha was visited by "a very prominent Congressman" who flatly stated that since Saha had criticized Gandhi's symbols of "chakra and khadi" (looms and handwoven cloth), Saha's views were unacceptable to the party.[1] Saha was invited to alter his position, and then run. Saha refused the invitation to recant his position and thus to run for the Congress Party, in his words, "because I believe and have proved that this insistence on primitive technology shows a very retrograde and anti-scientific mentality, and persons who are wedded to this mentality would bring disaster to the country when they are in power."[2] Saha kept on building his institute and engaging in nuclear policy politics. But three years later, largely because of his disappointment with the government record between 1947 and 1951 on education, industrialization, health, river valley development, and planning, Saha "decided to offer himself" as an independent candidate in the constituency of North-West Calcutta in the 1952 elections, he said mainly because of his interest in planning.[3]

This chapter examines Saha's work as an elected politician and the last two years of his life as a nuclear planner and negotiator with Nehru, Bhatnagar, and Bhabha. Information concerning his relations with the Russians and Americans will be found in *Negotiating Nuclear Power.* What the preceding two chapters have shown is Saha's gradual and subtle exclusion from official nuclear and military opportunities that were opening up. His students and colleagues were more involved than he was, and he relied on them to keep his journal and political networks up to date. This chapter allows us to see how and why questions about atomic energy planning and its leadership were raised in parliamentary debate.

Saha already knew that planning would be dominated by Mahalanobis, now appointed honorary statistical advisor to the cabinet in 1949. When the Congress Working Committee proposed a planning commission in 1950, Mahalanobis and his Indian Statistical Institute were chosen as the focus of the new developments and given major responsibilities. Mahalanobis was named chief advisor to the prime minister. While very critical of the design and actual execution of specific plans, Saha was in general accord with planning based on statistical analysis and with Mahalanobis's proposals, which he quoted with approval in *Science and Culture* (1950). He intended to reinforce those plans in Parliament while criticizing the government's implementation of them. This was, in Saha's mind, the purpose of contesting the election in 1952: dissatisfaction with the government's performance and a conviction that he could make a difference on implementation through Parliament.

It is ironic, to put it mildly, that in the copious writing about Saha's political views and extensive reprinting of them, no scholar has examined his working methods as a politician or even the strategy or mechanics of his election. Though such an examination is beyond the scope of this book, it would be rewarding. Notwithstanding this gap in evidence, we know he did not seek the Congress or Communist nomination as a candidate in 1947–48 or 1952, but instead chose the Revolutionary Socialist Party founded by M. N. Roy, whom he had met in Berlin in 1921.[4] Saha's contact with Roy had been more than casual. It was in Berlin that he was probably made the keeper of the secret code for the Jugantar Party in Calcutta, whose leader Jatin Das he knew when a student around 1915. Codes were essential because of the increasing government surveillance of such political parties and because of continuing conflicts between the revolutionary parties. Saha's role was described in interviews of individuals engaged in politics at that time, particularly the interview of B. K. Dutta by Gautum Chatopadhyay. Dutta, and other members of Jugantar, organized the reception in Bengal for Nalini Gupta, Roy's emissary from Berlin. "Towards the end of 1921 or early 1922 . . . M. N. Roy's courier [Nalini Gupta] reached Dr. Meghnad Saha, who was keeper of our secret code."[5] Chatopadhyay describes the complex relations among the revolutionary groups in both Calcutta and Dacca and their young leaders; Nalini Gupta had learned how to make bombs while working in a munitions factory in Great Britain during the war and so was in demand among Bengali revolutionaries. Saha and Gupta met with others at a sympathetic dentist's chambers in Wellington Square. Presumably when Saha left Calcutta for Allahabad in 1923, his secret role in this Bengali political party ended.

Although there were a number of other candidates in the 1952 election, Saha's chief opponent was a well-funded Congress candidate, P. Himmatsingha, who was connected with large business interests. North-West Calcutta was growing rapidly, with many rural in-migrants, small business merchants, middle-class professionals like government clerical staff, and traditional artisans in the constituency. Very large old palaces near the river were being surrounded by densely packed lower-middle-class houses and growing slums. Saha was not well funded for his campaign and wrote to his publisher in November 1951 to ask for a Rs 5,000 advance against the sale of his popular textbook *Treatise on Heat* "because I am standing for election in the house of the people from NW Calcutta."[6] Just before the election Saha gave his clear public support to a *hartal* ("lockdown") against the Congress state government and wrote, "The political parties outside Congress have combined in a manner which has not been witnessed since election of the present Parliament. . . . The *hartal* was completely successful . . . the Government has been shaken. . . . We feel the agitation must go on."[7] Saha won a large majority in the election and took his "independent" seat, soon retiring from the University of Calcutta. His twenty-one-year-old daughter Chitra acted for his campaign at a polling station on election day. After his success, his Congress opponent came to Saha's house "to congratulate him and pay his respects."[8]

That was the year in which he had just persuaded the university's Senate to accept the definition of the Institute of Nuclear Physics as an all-India institution and thus qualify it for central government money. Saha was not the only Calcutta scientist in politics at this time. Physicist Satyen Bose was appointed to the Rajya Sabha in 1952 until 1958, but there is no evidence that the two old friends had more than casual interaction in Delhi, no evidence that they acted as a team. At the outset of Saha's political career in Parliament, K. N. Katju, whom Saha had criticized in 1938, warned him to confine his attention to the laboratory, saying that he did not really belong in Parliament. Katju was now Home minister in control of police, intelligence, passports, and the like and was speaking to him in Parliament, under customary privilege.[9] Previously governor of Bengal, Katju was now a force to be reckoned with.

Katju's warning was unheeded, and Saha continued the criticism that *Science and Culture* was known for. Saha was one of the few who stood up to Nehru, particularly in questions relating to atomic energy. Nehru was himself not always clear or well prepared and showed at times an uncertain or confused grasp of the technicalities for which he was responsible. The following is an example from Nehru's letter about secrecy to Finance Minister C. D. Deshmukh: a reactor "is the basis for atomic energy work. It is not, I might mention, a pile, which is used ultimately to make atomic bombs and the like." Nehru's confusion about reactors and piles may have been purposely sown in his mind by Bhabha, who would have found the distinction useful, or this may have been his own confused grasp of details (or both), or he may have been dissembling to Minister Deshmukh, who would not have known that a pile and a reactor are the same thing. Saha knew Nehru's intellectual limits on this subject—and the limits of most other cabinet ministers too—and Nehru could not have Bhabha or Bhatnagar always present to help him in the Parliament. It is likely that Nehru did not feel his customary self-confidence when entering the Lok Sabha to discuss nuclear or technical matters. The parliamentary record shows that Nehru took pains to answer most of Saha's questions, though often with sarcasm or irony. Saha needled Nehru regularly and successfully and was, it seems, meticulously prepared. Having to edit technical reports for nontechnical readers of the popular *Science and Culture* probably helped greatly with that preparation.

Before long Saha experienced the limitations of being an opposition member in a small and poor political party; late in 1952 he wrote to Archibald Hill asking about his experience as an independent MP in London, saying, "I find it difficult, with my training as a scientist, and accustomed to build my views on a hard core of facts, to identify myself with any of our political parties, and therefore I cannot do anything very effective. But I find that the Public likes my views. I feel that if I were given some ministerial job, I could have done it far better than the professional politician. But I see that I can only be an 'Irish' member."[10] Hill told Saha in his polite reply, as he had Bhatnagar, that he was much happier back in his laboratory and away from politics and administration.

Despite his belonging to the Revolutionary Socialist Party, Saha was always described as an "independent" member of Parliament and was the RSP's only representative. But another new member elected from Baharampur constituency in Murshidabad district West Bengal looked to Saha for guidance and helped to make "a caucus of left groups other than the Communist Party of India and the Praja Socialist Party." A letterhead was printed for this small group, with an office in Paharganj in north Delhi. This MP was Tridib Chaudhuri, whom Saha invited to share with him the MP's allotment of half a house in Delhi from January 1953. Both spoke Bengali, though wrote to one another in English. Chaudhuri wrote, "I have come to look upon you as my guru in scientific social thought."[11] Later Chaudhuri said that they were brought together by the East Bengal Refugee Committee and the Food and Famine Relief Committee, of which Saha was president and Chaudhuri a very active member in 1950–52, and said that "both organizations functioned under his leadership for about two years in perfect harmony and mutual cooperation and amity, and they were wound up smoothly when their work was over." Curiously, Chaudhuri remembers that Saha was "not a member of any particular political party as an MP" though the RSP claimed and supported him, but they worked together until Chaudhuri was arrested and detained in Goa during the anticolonial movement in 1955.[12] When Chaudhuri was released from prison, Saha had already died. Although the letterhead for a caucus of left groups in 1953 was not to include the Communists or Praja Socialists, they claimed him too. According to Hirendranath Mukerjee, who was elected for the Communist Party of India at the same time as Saha in 1952, "it was not that he was in complete agreement with the Communist Party or the RSP, but like communists he was in favour of an exploitationless society, i.e., a classless society."[13]

River Valleys, Fertilizer, and Oil

Although Saha carried on his criticism of the atomic energy program through meetings and correspondence, his approach in Parliament to river valley development, fertilizer production, and oil exploration provided a clear sense of how he thought the country should develop. Saha began with a criticism of the implementation of the Damodar River Valley scheme that used the Tennessee Valley Authority (TVA) as one of its models. Saha's village of Seortali near Dacca was on the edge of a large plain that flooded annually; he had direct experience of the benefits of water management and the costs of floods, something most East Bengalis experienced every year. He had worked in flood relief in 1923 under P. C. Ray, was consistently interested in this subject, and presented a scheme for flood control to the Damodar Flood Enquiry Committee during 1943–44. Saha then became a member of that committee, founded after serious floods in 1943, which inspired the study by the US Army Corps of Engineers that led to the Damodar scheme. Saha published his own scheme in *Science and Culture,* which involved dams, irrigation canals, and afforestation to check erosion. Saha was also responsible for the temporary appointment of W. L. Voorduin, planning officer for the TVA, to the Central Technical Power Board of India.

He was essentially the most prominent protagonist for the Damodar project but became the best-known critic of its execution.[14]

Saha basically wanted a TVA-type project in Bengal, and the Damodar River project would both control floods and supply rural electricity. So he already knew the history of the Damodar project when he entered the Lok Sabha. He decried the waste and the recruitment of non-Bengali workers for the huge project, which included the Sindri Fertilizer plant on the banks of the Damodar River in Bihar. This plant was recommended as urgent during the famine in 1943 but due to interminable delays was not completed until 1950. It was the first big venture of a government committed to socialism, said Saha, but the plant was built by an American firm contracted by a British company acting as agents for the government of India. That government far away in Delhi had simply not paid attention to the project, having other concerns. Saha's complaint about wastage was compounded by his feeling that the public sector nature of the plant was threatened by Nehru's invitation to Sri Ram and Sir J. J. Ghandy to assist in its operation and sit on its board. Sri Ram was a Delhi-based industrialist with plants in both Uttar Pradesh and Bengal. Ghandy, a Bombay lawyer, was manager of the Tata Steel Plant at Jamshedpur in West Bengal and knew about Saha through his own association with the IISc in Bangalore in the 1930s. Though for reasons different from Saha's, Sri Ram was also insistent on economy measures "as the higher costs will not enable the management to sell its products at a rate which could tempt the poor farmers."[15]

Following Nehru's appointment of Sri Ram and Ghandy, Saha criticized the working of the Damodar Valley project in the Lok Sabha: "It is wrong to hand over the administration of the fertilizer factory to a body of private industrialists, who had no share in its conception or erection, and who will try to use it for their own benefit."[16] Unlike Ghandy, Sri Ram was not interested in advising; he fought with those who tried to run Sindri like a government department: he wanted it run as a business by someone like him. Cost estimates for the factory had been Rs 10 million but actually rose to Rs 23 million. In imposing economy measures, Sri Ram had the backing of both Nehru and Bhatnagar, the same people who praised Ram's industrial venture in South Calcutta where sewing machines were produced largely for export at Jay Engineering Works.[17] In Saha's eyes, Sri Ram was part of the small group of non-Bengali industrialists like G. D. Birla (both were Marwaris) who had long dominated the economy of Bengal and who had become rich during the war. The decline of Bengal's political importance in the Independence movement further revealed this "outsider" domination, in Saha's opinion. Saha had, however, been able to begin his cyclotron research only on money raised from the trusts of non-Bengali capitalists in Calcutta like the Birlas. Nevertheless, Saha felt sensitive about the Damodar scheme because it was so personal an idea. He had conceived of the Damodar scheme as bringing power, of both kinds, to Bengal and had loudly advocated it because he believed deeply in the success of the TVA, having toured it in early 1945, with the 1943 Bengal floods in his mind.

Another instance of Saha's opposition to the established powers in government and their allies in business was oil exploration in the delta of the Ganges. During his 1945 tour of laboratories in America, he visited and received information on exploration geology from his friends, P. D. Foote and A. Fleming. Foote, whose early research was stimulated by Saha's work on thermal ionization, had been the chief editor of the *Review of Scientific Instruments* and was then director of Gulf Oil Company's research laboratory outside Pittsburgh (later Gulf and Western). Fleming was director of the Carnegie Institute in Washington. Foote and Fleming told Saha as well as Bhatnagar in 1945 of exploration of subsoil geological conditions by magnetometers flown in airplanes, flights that were kept secret during the war.[18] Saha had long been interested in energy sources for India's industrialization and in 1946 spoke as president of the Royal Asiatic Society of Bengal on the possibility of oil deposits in Bengal's delta of the Ganges. Saha had been told by Gulf Oil that the rich deposits of Kuwait and Saudi Arabia had been found at a depth of 1,800 to 2,700 meters by using the same airplanes. These were not impossible depths: oil wells in nearby central Burma were drilled to 3,000 meters, something Bhatnagar knew through his contacts in Attock Oil, a company owned by Indo-Burmah Petroleum.

Both Saha and Bhatnagar learned about these geological exploration technologies while together in the United States in 1945: Saha fought to have them carried out by a national enterprise, developed in India expressly for the purpose, while Bhatnagar facilitated international oil companies in completing these surveys. In 1952, Saha found that the government had agreed to exploration of that same offshore region by American oil companies. He pointed out in the Lok Sabha that the United States was using up its natural resources very rapidly and thus needed more; he demanded a refusal to their application for prospecting licenses. In 1953 he wrote about foreign oil companies:

> They want to envelop India in their octopus tentacles. The mentality of the American capitalists towards India will be clear from the fact that when they carried out airborne magnetometric survey of the Bengal delta, no Indian scientist was allowed to go near their apparatus nor allowed to see how they

> carry on the work. They have taken away all the data of their survey to their own country. This country does not know what results they have got, except by hearsay or what they please to disclose. When they discover oil they will bring all the technicians from their own country for the running of the wells and refineries, and this country will become helplessly dependent on them for the running of this industry as the Iranians have become.[19]

In this criticism, however, Saha was directly opposing the interests of the CSIR. Bhatnagar, whose reputation had been founded through a spectacular relationship with a foreign oil company in Punjab in 1936, had himself appointed as the one-man commission in 1951 that negotiated the establishment of oil refineries in India and induced both Burmah Shell and Assam Oil to create scholarships worth Rs 200,000 annually.[20] Saha had written to the minister of Finance only six months earlier, enclosing his article "Oil and Invisible Imperialism"; now he argued for more funding for the Central Geophysical Institute, and his outlook was more negotiable:

> If the government had listened to my advice we would be in a much better position to negotiate with the Standard Vacuum Oil Company. . . . I may add that I do not object to the Americans prospecting for oil in India or setting up a refinery plant here. In fact we have neither the technical efficiency nor the funds for such highly specialized and expensive work. But . . . Indians should be in commanding positions in administration and the technical side—in both exploration and refinery.[21]

Saha was focusing on a big issue. In 1949 Bhatnagar accepted US and UK offers of technical missions to India to plan the location and construction of refineries. In 1950 companies like Standard Oil offered to build two refineries on condition that they be permitted to market oil at 10 percent above the world price. The Indian government would not accept this condition, and the question was stayed until Iran nationalized its oil reserves in 1951, after which the three companies (Caltex, Burmah-Shell, Standard Vacuum) became more conciliatory. Saha's former student K. D. Malaviya became minister of Mines and Fuel in 1954, when import of oil cost $200 million, then about 15 percent of all India's import costs. A commission was established to investigate oil price mechanisms. Finally India announced it would build three refineries of its own, with foreign participation in refining and marketing, but dependence on imports increased until the discovery and development of offshore oil near Bombay in 1974.[22]

Saha's critical approach to government policy continued undiminished; in the few months before his sudden heart attack, he asked many questions in the Lok Sabha. The following are some of the topics on which he debated between August and December 1955, in effect, during the final months just before his sudden death.[23]

25 August: Questioned whether importation of heavy trucks was necessary, or whether these could be built in India.

15 September: Inquired about petroleum prospecting in Assam.

16 September: Challenged Nehru on a foreign firm's (Krebs) involvement in oil refinery at Bhavnagar in light of 1948 industrial policy. Nehru responded with evasive nonanswer.

16 September: Questioned necessity of importation of steel when steel was manufactured in India.

21 November: Questioned why tubing for manufacture of bicycles was imported when steel tubing was produced in India.

21 November: Questioned the value of continuing resettlement of East Pakistani refugees outside Bengali-speaking areas.

21 November: Questioned electricity output estimates for Bhakra Nangal dam project as exaggerated: this exaggeration may have been due to the fact that the dam was planned to produce heavy water, which requires large amounts of electricity. [It is not known if Saha knew this, since he was not a member of the AEC and this was an AEC project, but Saha was usually very well informed and prepared. This question was not answered.]

23 November: Drew attention to inadequacy of archaeology team sizes in relation to work planned by Archaeological Survey of India. (Saha was a passionate amateur archaeologist, having visited sites in Iraq on his journey to Europe in 1936.)

23 November: Pointed out conflict of interests in having new University Grants Commission composed of vice-chancellors of universities, who are themselves seeking funds from the UGC for their universities.

20 December: Requested ceiling on electricity rates that could be charged by private electric companies to the public (like those electricity companies that served Calcutta and Bombay).

It will be noticed that there is no direct question concerning atomic energy on this list, and it is therefore essential to know how much discussion was going on offstage, most of it raised by Saha, as the following evidence shows.

Saha Challenges the Prime Minister at the Informal 1954 Atomic Energy Meetings and Gets Replies

In a parliamentary style, Saha and Nehru sparred back and forth, provoking the prime minister to drop in Saha's name in the middle of a speech, saying, "as Professor Saha would have us believe" or similar words. Saha publicly criticized Nehru, sometimes obliquely, but everyone knew what he intended. Behind these public faces the two men corresponded directly and talked about each other in third-party letters and conversations. Saha was after bigger game, namely, the structure of the economy and government priorities, consistently challenging a licensing system that offered exclusive access to Indian markets for foreign capital and firms; this was later called the "license-permit-raj." He scrutinized government project budgets to determine the ratio between estimates and actual expenditure. He specifically criticized Indian firms seeking foreign collaboration that Saha said was either unnecessary or humiliating. He criticized the government for acting as broker in these collaborations and asked whether these licensing, trading, or producing patterns furthered the self-reliance of India or decreased the system of inequalities within the country. As he usually found these patterns were the very basis of India's continued dependency and thwarted growth, his criticisms were really of government policies, and atomic energy was a part of that, though the part on which he rightly claimed greater expertise. He envisioned something entirely different for "nuclear India."

Saha uniquely combined the roles of parliamentarian and professor dependent on government financing for his most important projects. His relationship with Nehru presented that combined face, and he criticized the administration and policies of the AEC and DAE even while waiting for a decision from them on sustained funding for his institute. Nehru disliked the criticism intensely, one can see, but persevered with Saha for mostly political reasons and partly for personal reasons. Saha kept the pressure on from the beginning; writing to Bhatnagar for advice about Saha's criticism, Nehru said of Saha, "According to him, this [secrecy] only prevents atomic energy workers in India as well as the public here from knowing what we are doing. This criticism must be met."[24] But his response to Saha was, "I did not like the tone and approach of your note. . . . I decided to send it to all three of our Atomic Energy Commissioners. There is nothing very secret in their replies, but at the same time, I can hardly broadcast them."[25]

As for the atomic energy program, Saha understood most of it to be outside the public discourse and discussed it in the Lok Sabha only occasionally. Counseled by his friend Jnan Ghosh in the Planning Commission, he decided to exercise his influence privately or through his editorials in the monthly *Science and Culture*. What he may not have understood is why countries presumed to be cooperating with India did so at such a slow pace. Saha rose in Parliament in May 1954, following creation of the new Department of Atomic Energy, to criticize the AEC's lack of progress over the previous five years. He made an indictment of Bhabha, Bhatnagar, and Krishnan, all, like him, Fellows of the Royal Society. He pointed out that there was really nothing unique about atomic scientists, that they were simply ordinary physicists and chemists who had studied the nuclear subject thoroughly, that's all. "So," he said with sarcasm, "the atomic scientist is not a new race that has come into existence."[26]

Later that year Saha faced the contradiction of public opposition to Bhabha's nuclear program and his own private negotiation. He crystallized his negative judgment of the atomic energy program at two special meetings and yet a few weeks later had to accept the scrutiny of the AEC on his own institute, in order to secure continuing financial support from the very people he had criticized. This suggests that either he did not always have a good sense of political timing, or he considered the fate of his institute less important than the country as a whole, or that he dared the AEC to take retribution upon him for his critique. The first of the two special meetings, held in September 1954 at Saha's request, was at Nehru's house, where Saha and others had been invited to a seminar. Saha spent a social evening debating the virtue and consequences of the policies being followed. The prime minister and Bhatnagar were present as well as some of Saha's allies, including chemist Jnan Ghosh.[27] Bhabha was absent. Saha extolled the French atomic energy model for India, emphasizing its self-reliance. Notwithstanding the changes made after Joliot-Curie's sudden removal, self-reliance remained the French objective. He said that India should now be where the French were but that India was now ten years behind them. He wanted big administrative changes and asked for a scientific board that would advise the prime minister directly, with full-time officers who would, unlike Bhabha, have no other job. And he wanted less secrecy. As Abraham comments, "The political leadership was hearing, probably for the first time, a sustained attack on one of Nehru's vanguard projects from a source whose nationalist and scientific credentials were impeccable."[28]

A few weeks later Saha led the attack again, with the prime minister, Bhabha, and the entire AEC present, before a hand-picked audience at a conference in the NPL. This November 1954 meeting, called the Development of Atomic Energy for Peaceful Purposes in India, had as its intent to present the new face of the Department of Atomic Energy, to prepare a national policy statement for the forthcoming Geneva conference on peaceful uses of atomic energy and to help sort out political differences in time for the AEC to make choices in pending reactor negotiations. Nehru had decided to stage it on hearing the criticisms at his house two months earlier. At the November conference Saha wanted to influence the composition of the Indian delegation going to Geneva the following year and to have himself included. By the time of this meeting in late 1954, Bhabha's secret talks with the British, French, and Canadians about reactors and fuel were proceeding quite well. Nehru opened the NPL conference with a classic statement of his reasoning why atomic energy was strictly a national central government responsibility and why a department with full powers, including secrecy, was necessary.[29] Bhabha, along with others from the Bombay nuclear project, like the young physicist Raja Ramanna and engineer Homi Sethna, spoke about the plans they were making but without giving any details. Bhabha insured that his young assistants from Bombay were present and active; this in itself was unusual in a policymaking environment like Delhi. Bhabha announced a three-stage plan, starting with natural uranium reactors to produce power and plutonium, then moving to a second stage with a new type of reactor using Indian thorium mixed with Indian plutonium (generated from the first stage), and this would produce a by-product, U233. The third-stage reactors would burn that earlier by-product U233, mixed with thorium to *breed* more U233 than it would consume; this would create a large (unlimited?) supply of mixed thorium-U233 fuel, all of Indian origin. At that point so-called breeder reactors were only in conceptual-design form. Though the phrase was not in use then, this would later be called "mastering the fuel cycle" (fully treated in *Negotiating Nuclear Power*).

Bhatnagar and Krishnan had to sit there and weather the storm of Saha's criticism that the DAE plans were unrealistic and premature for Indian conditions and could not possibly be achieved in five years as Bhabha claimed even for the first stage.[30] According to Ramanna, Krishnan managed to wrap up the conference by softening the thrust of Saha's criticism by saying that India, like any man contemplating marriage, had to start somewhere and should thus start here, and start now.[31] This meant that the present arrangements were confirmed, but Ramanna remembered a "mood of discontent" because the rate of progress was thought to be poor, as Saha said. Homi Sethna gave a paper on rare earth processing, and the famed physicist Satyen Bose made an unrecorded cynical comment casting doubt on the realism of the DAE's goals, but national decisions were being made. Bhabha seems to have held his tongue, presumably on advice from Nehru; he certainly said nothing about his secret negotiations abroad.

A week after the conference Nehru announced in Parliament that India would build a research reactor at Trombay and said that discussions were under way with the United States, Great Britain, France, and Norway to do so. He expressed this with careful vagueness and did not mention Canada. Of course Nehru was quite ambivalent about how much he should reveal and how much conceal. Eleven months earlier he had reminded his audience when opening the TIFR buildings in Bombay that "so far as this Institute is concerned there is going to be no secrecy about it at all, unless some very grave emergency arises, that is a different matter. It is like any institute. Science in fact does not flourish in secrecy. It is a bad thing to have too much secrecy about scientific work."[32]

On the last day of 1954 Saha wrote to Nehru about a statement Nehru made ten days earlier criticizing Saha's grasp of national income statistics, per capita income, and particularly of whether or not industrial production had risen 18 percent or 13 percent. This arose in the context of a debate in which Nehru described "my kind of science" (see chap. 12). Saha much resented the suggestion that he had not done his homework and wrote to Nehru about their entire working relationship going back to 1936, beginning with "I am sincerely sorry at the deterioration of good relations between us," and closed with "I have not asked nor shall I ask for any personal favour from you. But as you are the Prime Minister, I have come to you for the sake of institutions with [which] I am concerned."[33] Saha's timing was crucial, for he knew Nehru would meet Bhabha soon, just before a most important meeting on the future of Saha's institute. In spite of what he wrote about not asking for anything, Saha well knew he *was asking* a personal favor. He reminded Nehru that in 1942 after Nehru was sent to prison Saha named him as the chairman of the Science Congress meeting to be held in January 1943 and said, "most of the prominent men of science in India today, except the Calcutta men, thought that discretion was the better part of valour. Raman, Bhatnagar, Bhabha, Krishnan etc all found one excuse or another to keep away," but, Saha reminded Nehru, he had put Nehru's picture on the presidential table at Science Congress in early 1943 to "demonstrate to the world that we scientists were solidly behind our national leaders in their great struggle." Finally he closed this letter by evaluating the two people on whom Nehru most relied in atomic energy matters: "I have been put to one humiliation after another. I have been asked to take orders from Bhatnagar, whom I consider a very poor scientist, and from Bhabha, who though a good scientist is 18 years my junior and the conferment of enormous power on him has made him extremely bumptious."

Saha's letter was written on New Year's evening, the very evening that Shanti Bhatnagar died. Showing their relationship was not casual, Nehru wrote promptly two days later to Saha about their disagreement over industrial production statistics. In his reply Nehru explained the path that the recent negative appraisal of Saha had taken to reach Nehru—it came, said Nehru, from Bhatnagar to Maulana Azad and then to Nehru: "You ask me to name the person who spoke to me. There is no point in my doing so now because the person is dead." Then Nehru gave Saha his own appraisal:

> After making a strong attack on everything that Government has done and running it down, you were good enough to compare us to Chiang Kai-shek and his failure. It seemed to me that your criticism was not only unjustified but completely lacking in objectivity and therefore most unscientific. You were evidently angry and lost sense of perspective. If you attack the Government, surely you do not expect them to remain silent. I can hardly judge myself. It may be that you are a better judge of me than myself.[34]

Within weeks of these announcements of the grand plans of the DAE and his critique of them, Saha had to prepare for a meeting with the AEC/DAE about his own institute in Calcutta, crafting its necessary dependency upon Bhabha, literally "grafting" it on to the new DAE. Saha had retired from his university professorship in 1953 when he turned sixty and had accepted a full-time appointment as director of the IACS, meanwhile remaining honorary director of the Institute of Nuclear Physics. Saha now knew that this unavoidable financial dependence on the DAE was inevitable, and so early in 1955 he met with Bhabha and Krishnan of the AEC, J. C. Ghosh, now vice-chancellor of the University of Calcutta and therefore chairman of Saha's institute's Governing Body, and B. D. Nagchaudhuri, whom Saha had made director in 1953 and whom the university had appointed Palit Professor at age thirty-four. Bhatnagar had died a few weeks before this meeting, or he too would have been present. Included at the meeting were two joint secretaries in the Ministry of Finance, C. S. Menon and N. K. Dravid.[35] Sir Marcus Oliphant was also present as Bhabha's guest; Oliphant was from 1950 the director of research, School of Physical Sciences of Australia National University, Canberra (Oliphant is seen talking to Saha in a 1936 photo taken in Copenhagen). He and Bhabha knew each other from their Cambridge lives in the early 1930s, along with Blackett.[36] He had been deeply involved in the Manhattan Project and in British nuclear developments after 1946 until he went to Australia in 1950. He was involved, with Bhabha, in the planning of the Geneva conference on atomic energy.

The meeting reviewed the progress of the institute since 1948, causing Saha to explain some of the slowness in getting started, and discussed teaching and library facilities. Debate occurred over whether it was more appropriate for the DAE or the University Grants Commission (UGC) to support fundamental research in universities, or if it was appropriate for the DAE to support general basic physics education at all. Bhabha stated that balanced long-term funding was certainly needed for all basic education in the universities, as Saha said, but that DAE should concern itself more with specific "special projects." Jnan Ghosh, Saha's longtime ally with experience in the IISc in Bangalore, the Planning Commission, and the University of Calcutta stated that the UGC and the ministries of education of central and state governments were too poor to help expensive programs in nuclear physics. Ghosh urged complete DAE support for the institute. C. S. Menon, of Finance, argued that since ultimately the money came from his ministry he did not mind if DAE decided on special projects and the UGC decided on general education in the same institute. This offered Saha a platform to make specific requests.

Saha requested three more posts for professors, to which Bhabha agreed at the meeting. Bhabha said the number of academic posts would be fixed but could be circulated or allotted to different departments as needed. B. D. Nagchaudhuri, the new director of the institute, was delegated to negotiate details of the annual pay scales and salary increases. Annual budgets were to be submitted by heads of divisions in December 1955 and discussed and finalized with the director. Bhabha did not contest the idea of a fixed number of posts though it was not the way he felt his own institute should develop. He requested, moreover, that there be a "cushion grant" to help the institute's budget flexibility. C. S. Menon, revealing a difficulty that most Indian officials had in understanding research, demanded to know why a cushion grant was necessary. He agreed to it only if there was a representative of the Ministry of Finance on the institute's Governing Body. Saha played a deferential role here because he knew where the money had to come from. Oliphant's role seems to have been to give pertinent advice from his experience gained in Cambridge, Birmingham, and Canberra. Bhabha must also have reasoned that the presence of a distinguished and influential foreign physicist, whose strong reputation Saha knew well, would forestall any impression Saha might circulate that the DAE had "ganged up on him." Oliphant was like a witness.

With an annual budget at stake of Rs 5 million for the next five years, more money than Saha had ever seen in his institute, he was very cooperative, according to his own papers. But this meeting resulted *only after* the intervention of Nehru himself, many months before. Nehru had received a number of letters from Saha, asking for help with the institute. Finally he instructed the Minister for Scientific Research K. D. Malaviya to find the appropriate funding for Saha's institute from the AEC budget in spite of the fact that Bhabha preferred the money be found elsewhere. Nehru detailed the common criticisms of Saha in an April 1954 letter to Malaviya, long before Saha spoke in the two special meetings against nuclear policy:

> The real fact of the matter is that Saha has been very casual in his demands. They come at odd times, there is no planning and no one knows exactly how much he will require and when. This upsets our budget and our calculations. Also I might inform you for your private ear that some of our colleagues here do not think too highly of his Nuclear Institute. The private report of the Sir Alfred Egerton Committee was not very much in favour of it. Nevertheless I want this Institute to continue. It can undoubtedly do good work.[37]

Malaviya soon heard Saha's conclusion about Bhabha and the AEC: "had the Govt taken my advice in 1947 and allowed me to organize the work, we could have built 2 reactors by this time, just as France has done."[38] Still Malaviya had Nehru's instructions to persevere and arrange Saha's institute's continuing finances from the very institution he so firmly criticized.

The 1955 meeting achieved Nehru's objective of helping the Calcutta institute, a curious act of solidarity with Saha on Nehru's part. Malaviya, at the same time, could relieve himself of the burden of having to criticize his teacher across the floor in the Lok Sabha: "I have nothing more to say except that Dr. Saha is my old professor. I have been an old student of his. I know he was very obstinate. He is very good and useful when he is not obstinate, but not as good when he is obstinate. His sticking to politics, Sir, I think, is one of those occasions." There are (or were) few prescriptions as difficult to transgress in India as criticism of one's teacher, and K. D. Malaviya was sometimes in that uncomfortable position, something he always tried to do with courtesy.[39] That is the background of the interdependence of the newly created DAE and Saha's institute in Calcutta; Nehru remembered his own shrewd reasons for offering Saha a seat in the AEC in 1948, and the calculations had not changed.

Why could Saha achieve this unlikely influence? Was there so much respect for an accomplished older scientist who now worked in Parliament? There indeed was respect for his scientific reputation, and his remarkable biography had just been released, but his influence really had more to do with Parliament and his publication *Science and Culture* and the warming of the Cold War after Stalin's death in 1953. As editor Saha turned the magazine's perspective outward to comment on international, Soviet, and American nuclear politics (see precisely how in *Negotiating Nuclear Power*), but more pointedly he treated national and regional issues too, from a scientific point of view. This publication and this kind of message were unique, in that context at that time; Saha's rival Raman had started a similar magazine in southern India, but it did not handle political issues in this way.[40] For example, scanning the 1955 and 1956 issues only, some of which he edited before his death, I found that Saha celebrated the birthdays of Bengal's science gurus like Jagdish Chandra Bose; provided obituaries for national figures like S. S. Bhatnagar and international figures like Albert Einstein; provided the speeches of Nobel Prize winners and short biographies, for example, for the 1955 physics and chemistry prizes; described the work of scientific conferences, for example, the high-altitude research symposium in 1955 up at 3,000 meters in Kashmir; reviewed the Second Five-Year Plan using a panel of economists; reported speeches in Calcutta by Bhabha's close allies, like geologist D. N. Wadia addressing geologists and metallurgists or lectures by Bhabha's teacher Paul Dirac or friend Sir Marcus Oliphant; published a popular piece by TIFR cosmic ray physicist Bernard Peters; and published his own article "A Research Reactor for the Eastern Zone," somewhere "not too far from Calcutta." The range of subjects published made Saha's media reach and effect quite disproportionate to the number of his subscribers; everyone knew that each copy was being read by five to ten people, particularly in the elite and left circles in eastern and northern India. The typography was good and almost error-free, the illustrations and photos were precise and clear, and the news, often lifted from *Nature* or *Science* but not always, was seldom more than four to six weeks old. The reach was considerable and increased after Saha was elected in 1952. And by late spring 1955, after the Bandung Conference, Nehru knew that Khrushchev and Bulganin would be visiting India and seeing Saha in Calcutta, where he was chairman of the visit's organizing committee. It just would not do to have Calcutta's senior communists or leftists criticizing Nehru for failure to support a major institution in Bengal.

Final Calculations

Saha was a powerful force in a state and city where his many supporters had wide influence *outside* elite circles, and his presence was a stimulant for others on the left, others with whom Nehru needed to maintain a working relationship. It is probable that Nehru's frustration with Saha concealed a modicum of residual respect: in some ways, Nehru met his match in Saha. Not long after the 1955 meeting at which Saha finally received core funding for five years, Saha wrote to Nehru in July 1955 and criticized the way in which Bhabha excluded him from the delegation to the Geneva conference on peaceful uses of atomic energy, reminding Nehru that he (Saha) could have been there officially, if he had accepted Nehru's invitation to become a member of the AEC:

> Now let me close with some final remarks. You may remember that before our AEC was formed in 1948, you called me by phone to Delhi and were kind enough to offer me a seat on it. I declined the offer, with the remark that the arrangement would not work. I wished that my forebodings proved to be incorrect. It was my firm conviction then and even now that we should have done as well as France. We are far from it, at least by ten years. Why has it been so? . . . Our failure in atomic energy work is due to the fact that our Government has committed all the blunders which the German Government committed during the war [in its nuclear reactor and bomb project] and something more . . . the analogy is too transparent to be missed by anybody. But I cannot help remarking that the latest administrative measures taken apparently to improve administration have made Dr. H. J. Bhabha a Fuhrer in atomic energy development in this country as Heisenberg was made in 1944 in Germany. The analogy is almost complete and I have no doubt that it will lead to the same disastrous results.[41]

Saha suffered from chronic high blood pressure and died suddenly in Delhi in January 1956, at age sixty-three, walking up the steps of the national planning center.

In reflecting upon Saha's special mixture of science and politics, it is important to consider the evidence of the reaction to his death in New Delhi. Bhatnagar himself wrote, not long before Saha's criticisms of him at the September and November 1954 meetings, that perhaps their relationship was misunderstood:

> Very few people in India know that Prof. Meghnad Saha and I are very old friends. I first met him at the University College, London in 1919 where he came to see Prof. F. G. Donnan FRS, under whom I carried out my early research work. . . . When in London, Dr. Saha and I used to meet together nearly every day. When I was a professor at Lahore, he always stayed with me when he visited the place. . . . As a member of the Governing Body of CSIR

and the Board of Scientific and Industrial Research, Dr. Saha has shown great enthusiasm and has always supported good causes. . . . Prof. Saha is a veteran nationalist. Some people consider him to possess a very critical faculty but his criticisms are usually actuated by a desire to get more help for scientists and scientific institutions in every field, not against personalities.[42]

Despite their opposed political orientations, it is clear that each man recognized in the other a comparable enthusiasm, determination, and energy for organizing the scientific and technical community. But the critic in Saha caused the government to curb him, and he sought an alternate path. He had been a consistent sympathizer with the causes and people that had been most repressed by the colonial government. He believed they were being repressed by the new independent government and opposed that publicly. Bhatnagar, who never seems to have had his vision of the social order challenged (except during Partition), offered to the government a "safe" scientist, someone at ease with industrialists and international business, someone very constructive in the war situation. Saha exerted his intelligence and influence where he could, including in the opposition in Parliament. Bhatnagar was empowered to establish a whole network of permanently funded government laboratories. He had quasi-ministerial rank for natural resources and scientific research and was very active in the secret development of the AEC. Saha was kept, warily, at arms length. Bhatnagar watched how Nehru handled Saha carefully and reflected that relationship. Nevertheless, a witness at Saha's deathbed in February 1956, arriving from a defense laboratory a few minutes away from the hospital, recalls that Nehru himself jumped in his car on hearing the news of his heart attack and came immediately to see Saha.[43]

Ten years after Saha's death came a judgment from an established figure in the politics of science in India, Jagjit Singh. Chairman of the Railway Board and popular writer about science, Singh was an old acquaintance of Saha. He stressed that Saha's attitudes were formed by the poverty of his youth and that his "contempt for jaded democracies and admiration for the strict measures of Soviet transformation did not leave him, even in the moderate affluence of his later years." Jagjit Singh's opinion typifies a majority of the public's attitude to Meghnad Saha:

No wonder all these diatribes against capitalism and democracy made Saha unpopular in many circles. Some even accused him of fellow travelling. . . . He was in too great a hurry . . . he had an inadequate appreciation of the

> handicaps under which a democratic regime has to work to maintain its ideals . . . [but it was] his sea green incorruptibility and versatile intellectual powers that endeared Saha to the ordinary men and women of Calcutta.[44]

Incorruptibility remained a topic of constant discussion for the rest of the century among young scientists all over India, and Saha was often used as an exemplary figure in debates about corruption. Did people know he had to borrow against his textbook royalties in order to contest the 1952 election? Saha's protégé D. S. Kothari noted the contrasts in Saha's impact on other people: "He was extremely simple, almost austere, in his habits and personal needs. Outwardly he sometimes gave the impression of being remote, matter of fact, and even harsh. . . . It was not in his nature to placate others."[45]

Yet Kothari also describes the great warmth and compassion that Saha showed to those around him, and most particularly to his students—for whom he appears to have spared no pains. He wrote to them regularly, getting them to do things for his projects, and doing things for them. They all replied promptly "my dear Sir." Even Minister K. D. Malaviya, the former student whom Saha had criticized in Parliament, arranged for Saha to give a seminar in Nehru's house in September 1954, thus enabling Saha to speak informally yet directly to Nehru in the presence of others who supported him, like Jnan Ghosh of the Planning Commission.

And his relative simplicity of lifestyle was well known. Like others, Saha had supervised construction of large buildings and other public undertakings in Calcutta. These were commonly a source of informal and unrecorded income, flowing back from grateful contractors. But in Saha's case there was little extra money, so his "incorruptibility" was widely seen. Indeed, when Saha finally built his own house in Calcutta in 1954, he accepted a personal loan from his friend Vice-Chancellor Ghosh (paid back a year later when Saha finally received his pension money from the university). Saha also asked for another advance from the publishers of his *Treatise on Heat* to build his house.[46] It was a time of connections, not entitlement, for this member of Parliament: Saha had already asked his publisher's brother, H. K. Das, to use his medical connections to get him a bed in the famed Calcutta School of Tropical Medicine, where he spent more than a week in December 1952 for observation and tests.[47]

Notwithstanding the remarks about fellow-traveling and the FBI reports on his political views, Saha does not seem to have had a significant base in any of the factions of the Communist parties of India in this later period, even after they were legalized in 1950. But he knew all the Communist leaders and was known to thousands of their supporters. An MP who was elected at the same time as Saha in 1952 recalled that "Saha was of the same view as [Subhas Chandra] Bose in regard to 'forced march' (for the advancement of our poverty-stricken backward country), and this coincided with the view-point of the communists. This is why Saha maintained good relations with us, and we could work together till his death in 1956."[48] The autobiography of the retired leader of the party in Bengal, Jyoti Basu, dealt with the whole sweep of politics in Bengal since the 1940s, but Basu mentions Saha only once, as when "on the death of the eminent scientist Dr. Meghnad Saha we contested his vacant seat."[49] But at a time when Bengalis of his generation were learning to define themselves in terms of their loss of prominence ("Calcutta was the light of Asia") and to submit to the capital, objectives, and ideas of others, Meghnad Saha pointed to an alternative future for them.

FOURTEEN

The Indian Cabinet and Scientific Advice in the 1950s and 1960s: Bhabha, Atomic Energy, and Reforming Scientific and Industrial Research

The interaction between Nehru and scientists began in the 1930s and with other political leaders soon afterward, but not until these men were ministers in charge of the Indian state were they required to measure and appraise the value of scientists and their ideas. We have already seen numerous contacts between scientists and ministers of finance, defense, natural resources, and home affairs. The minister in charge of almost-everything-else—foreign affairs, atomic energy, planning, scientific and industrial research—Jawaharlal Nehru himself, supervised the other relationships with scientists if necessary and chaired the cabinet committee on scientific affairs. But scientists looked for another way to bring their influence to the table of power, while ministers looked for another way to reduce the risk around complex technology and science decisions. That way was through the Scientific Advisory Committee to the Cabinet, and it was dominated by Nehru and Homi Bhabha. This chapter examines its early workings and sets the stage for the contest over the role for technical expertise following the deaths of Nehru and Bhabha, leading up to the "war over self-reliance" in the 1970s (see chaps. 20–23). The contrast between the two large systems of research and development—the Council of Scientific and Industrial Research and the Department of Atomic Energy—and internal reforms conclude this chapter. There is a striking role of persons and personalities in administering, changing, and resisting change throughout these very large and complex structures. But this role was recapitulated on the laboratory floor, proving the importance of the interplay of policy and personal relationship (see chaps. 16 and 17).

Shanti Bhatnagar's Role in Science and Technology Policy

The hardworking secretary of the Ministry of Natural Resources and Scientific Research, Shanti Bhatnagar, was very popular with national elites in politics, industry, and science; was very knowledgeable about India's position in world trade in such special commodities as oil, heavy water, uranium, and thorium; and by 1950 was starting to negotiate these commodities in trade. Despite his not being from a trading family, Bhatnagar seems to have had the business in his bones and learned its craft on the job. His experience before and during the war was the basis for that training. But although he was losing control of atomic energy with the creation of a separate DAE in 1954, he continued to take part as a member of the AEC in the special conference on atomic energy at Nehru's house and at NPL in 1954, and was high on the list of the Indian delegation to the Geneva conference on peaceful uses of atomic energy in 1955. There is no doubt that by now he understood the issues confronting India and some of the physical chemistry of uranium and its modification to plutonium. Bhabha, with Bhatnagar and backed by Nehru, moved the AEC in the direction of cooperation with a number of countries, themselves scarcely able to cooperate with each other, such as France, Britain, Canada, the United States, and the Soviet Union. All of these connections began to bear fruit in 1955.

By now there was some international recognition of India's modest importance, partly as a reflection of its neutrality during the Korean War and Nehru's championship of the nonaligned nations movement, leading to the famous 1955 Bandung Conference in the highlands of Indonesia. This recognition ran through Nehru but was dispersed upon a few other scientists too. Bhabha was already in the UN planning committee that led to the formation of the IAEA in 1955. If Bhatnagar was embarrassed in the 1951 solar cooker affair or was criticized by others, as he indeed was, for example, by Saha, that embarrassment did not put much of a shadow over his legend, which he was always careful to shape. He was, instead, promoted to secretary of the Ministry of Education in addition to his other official duties, so his influence was very broad though he would have fewer hours for details. When Bhatnagar died on the last night of 1954, there were twelve functioning CSIR laboratories and twelve more in the planning or construction stages. This had been his personal dream, to which Nehru and the cabinet subscribed, and his death did not diminish it. Apart from the research supported by the AEC, Bhatnagar's CSIR laboratories had the largest research budget in the country, although they did not include medicine or agriculture. Ten years before, in 1944, Bhatnagar had supervised the small beginnings of state-oriented industrial research around the invention or creation of substitutes for products and materials that were unavailable because of the war. This substitution policy was to become India's general industrial strategy for the next thirty years.

From 1947 Bhatnagar pursued a similar policy with respect to the research and development agenda for CSIR labs, looking for products or processes that could be substituted for imports. He did all this although he seemed unready to confront the international oil companies and follow a truly self-reliant petroleum path. (I think he believed that path was impossible for India.) He had created what scientists everywhere wanted, a separate ministry for scientific research, and he reported, on behalf of hundreds of scientists, directly to the prime minister, and in this case a *sympathetic* one. But just as important, he had a direct hand in atomic energy policy from 1945 and negotiated some of its important early agreements.

He anticipated his mandatory retirement at sixty from the high position of secretary to the government of India and announced it by letter in mid-1954. But he seemed to be relieved when the CSIR's Governing Body at the specific insistence of its members Meghnad Saha, Jnan Ghosh, and Satyen Bose asked him not to retire but to stay on as director of the CSIR, and he was probably not surprised. (The CSIR was not subject to all the rules of the Union Public Service Commission.) No reference was made to retiring from the AEC, and one doubts that he would have.

In the end he did not have to face the question about what he would do after retirement. To the shock of everyone who had become accustomed to his indefatigable drive, "Steamship Bhatnagar" died suddenly and peacefully on the last night of 1954. Nehru praised him and so did most other leaders who knew him.[1] Bhatnagar's sudden absence on the first day of 1955 left a major hole in Indian science and technology policy, particularly in energy, nuclear materials, manpower management, and relations with industry. Much information died with him, for little of it was written down. There was no civil servant who was his equal in knowledge, network, or influence. Nehru acknowledged this in a letter to all the chief ministers of India's states shortly after Bhatnagar's death (but note that he does not mention his role in atomic energy):

> Dr Bhatnagar can truly be said to have built up our great structure of scientific laboratories in India. He has left as a memorial thirteen magnificent national laboratories and institutes. I am quite sure that but for him these laboratories

> would still have been in some early stage of building. I have seldom come across a more dynamic person who had the capacity to get things done. His loss to us is great.[2]

A few lines later, in the letter about Bhatnagar, Nehru wrote, "We in India . . . are on the eve of an Industrial Revolution."

When Meghnad Saha died a year after Bhatnagar, there was then *no scientist other than Bhabha* with the status, knowledge of the issues, and political base from which to assess, defend, or criticize India's atomic energy program. Sixteen years younger than Bhatnagar and Saha, Bhabha now became the sole power in Indian science, and Nehru's already great reliance on Bhabha increased proportionally. At age forty-six he was considered rather young to have such influence, but that was what separated science from the political party system where older men ruled. In another sense the cabinet's reliance on Bhabha increased, as he was now spending larger and larger amounts of money, on Nehru's authority, without giving more than rather general, and at times evasive, explanations to the cabinet. At the same time, the growing CSIR system of laboratories no longer had privileged access to the cabinet through its advisory committee, as none of its directors general in the next ten years lasted long enough to build networks in Delhi equivalent to Bhatnagar's.

Nehru soon had to acknowledge the roles Bhatnagar played, including cushioning Nehru from the work expected of him as president of the CSIR. Five months after Bhatnagar's death, Nehru revealed how dependent he was on Bhatnagar; in a case concerning salt research, an example that reveals that Nehru had already seen the "salt disharmony" file before, had interviewed two of its principals in Bhatnagar's presence, and had just heard through his personal grapevine that the two senior officers of the Salt Research Institute at Bhavnagar in Gujerat continued to quarrel and escalate their feud. Nehru now warned that peculiar things could happen to the files and records of the institute and instructed that these now should be obtained and removed away from Bhatnagar. He pointed out on 15 May 1955 that no meeting of the Salt Research Committee had been held for two and a half years. Salt had always been a politically sensitive commodity in India, symbolized in the famous "salt march" of 1930, and the government had to take a keen interest in its production and trade, as it was a monopoly, and traders paid a "salt development cess," a tax that was supposed to go for research and development. But the old subject of salt research was hardly high on the CSIR list of priorities, and here was the CSIR's president, the prime minister himself, dealing with the salt file, saying, "This matter should not have been brought to my notice at all at any stage."[3] He wrote about "the incessant and unseemly conflicts" between the director and assistant director of the CSIR's Salt Research Institute. Here is Nehru's dilemma. He knew he should not have been involved in this issue in the past "at all," but he had been involved, had read the file and interviewed the two scientists. Either Bhatnagar needed Nehru to solve it or someone unseen in a minister's office had got the conflict onto the prime minister's agenda because Bhatnagar could not deal with it, or Bhatnagar's death provided a new opportunity for the conflict makers.

The appointment of Bhatnagar's successor in 1955 was particularly difficult for the CSIR. Before Maneklal Thacker was finally appointed, both Vikram Sarabhai and K. S. Krishnan were considered for this position. Less than a week after Bhatnagar's death, the industrialist Kasturbhai Lalbhai, a member of the CSIR Governing Body, wrote to Bhabha to suggest that Thacker, and not Krishnan, replace Bhatnagar. Lalbhai asked Bhabha to write to Nehru with this recommendation. Clearly Krishnan, director of the prestigious NPL, was seen as the front-runner from within CSIR; Lalbhai knew that, but Krishnan was not perceived by outsiders as an effective administrator or conflict manager.[4] Lalbhai's approach shows that Bhabha was seen to be the effective intermediary with Nehru in this appointment. Nine weeks later the matter was still not settled, and the minister for Scientific Research, K. D. Malaviya, wrote to Lalbhai proposing Vikram Sarabhai as both director general of the CSIR and head of the National Research and Development Corporation, a body little known at that time. As I will show below, Lalbhai was a kind of patron for Sarabhai in Gujerat and was an active member of the CSIR council. It is a sign of the unusual nature of this appointment that a minister was writing to a private member of the council. Malaviya wrote that Sarabhai "will have to come and stay in Delhi. . . . Incidentally the work of the NRDC will also improve, and the undesirable rivalry between CSIR and NRDC will to a great extent vanish."[5] Sarabhai, only thirty-six, was unusually young for this appointment and was not chosen. So there was still no new director general in the CSIR office until eight months after Bhatnagar's death! As it turned out, when Thacker was appointed, he stabilized the CSIR and its institutes but did not, over the next six years, play a significant personal role in national science policy or politics. This is hardly surprising because Thacker had his hands full inside the CSIR.

When Bhatnagar died, the marriage of CSIR and the Ministry of Natural Resources and Scientific Research in atomic energy decisions ended. While it is true that Krishnan was a major CSIR lab director *and* a member the AEC, and thus gave a senior member of the CSIR an insight into DAE business, Krishnan's connection was much less important than the role Bhatnagar had played.[6] Bhatnagar's successor at the CSIR, Thacker, had until 1955 been the director of the Indian Institute of Science in Bangalore and thus quite uninvolved in CSIR or DAE affairs. As it turned out, new Director General Thacker stabilized the CSIR over the next six years, during which time the CSIR was being eclipsed in the national politics of science.[7] It became more and more self-absorbed, and because of its sheer size and organization it had much to be absorbed with. Thacker eventually became, as Bhatnagar had been, secretary of the Ministry of Education and Scientific Research. By then Natural Resources (Bhatnagar's old portfolio) were disconnected from Scientific Research.

Homi Bhabha and the Formation of Science Policy

In 1954 the new Department of Atomic Energy was allowed to move out from underneath the umbrella of the Ministry of Natural Resources and Scientific Research. Atomic Energy, now a separate ministry, had a secretary in Bhabha, who was a science policymaker without equal in the country. The minister to whom he reported directly was none other than Jawaharlal Nehru. In 1957 Nehru asked him to draft the Science Policy Resolution, or more precisely Bhabha persuaded Nehru to adopt the idea and accept Bhabha's own draft of that influential resolution. It was rare to see him challenged by anyone in the ensuing nine years until his death. By that time Bhabha was also making his version of energy policy through Nehru, arguing that reactors of a power output of 150–200 megawatts would soon operate at 80 percent efficiency and deliver electricity to the grid at lower costs than other fuels. There were people who doubted these calculations, but they did not get much recognition in India: one exception was the visiting British economist Ian Little, who wrote in the *Economic and Political Weekly* that the DAE had exaggerated the cost advantages of building nuclear reactors in India.[8] This was a direct criticism of Bhabha and an indirect criticism of Nehru; in some subtle form within India, doubts like Little's were crystallizing, not only about the nuclear plans but about the growing department that was executing them. That these doubts were published in a remarkable biweekly journal whose reach and influence among elites was growing steadily made them more important. Even someone inside TIFR was expressing doubt about the nuclear program. A wide-ranging intellect and astute statistician, Kosambi made his own appraisal of Bhabha's plans and by 1960 was expressing skepticism about the cost and feasibility of nuclear power; he favored small reactors for rural electricity (say, 40 MW) and opposed large reactors of the kind Bhabha was planning to build (say, 200 MW). Well known for his opposition to nuclear weapons, in concert with figures like Bertrand Russell, Kosambi spoke publicly about reactors and weapons, for instance to the Rotary Club of Poona in July 1960.[9] Nevertheless, Bhabha's speech to the 1955 Geneva UN Conference on Peaceful Uses of Atomic Energy conveys a subtle influence from Kosambi's Marxist way of seeing things; others have noted that Bhabha's powerful speech was structuralist and materialist, confirming his attraction to and grasp of these ideas going back to the early 1930s at Cambridge.

Nehru countered in the Lok Sabha in August 1960 by reading from Bhabha's script that India would be able to produce nuclear-powered electricity more cheaply than the US or UK and confirmed Nehru's own perception that "no other department" of the government "has had this unanimous appreciation."[10] Doubts continued, however, and so Bhabha took the trouble to reinterpret and repromote his strategy in 1961 through newspaper interviews that were soon answered quietly but skeptically in editorials in the *Times of India* and *Hindustan Times,* reminding readers about the reliability of conventional fuels and the higher costs of nuclear power—echoing debates that were occurring then in Britain.[11] It was characteristic of India's openness that these criticisms found a serious home in newspapers, which were privately owned. But I agree entirely with Perkovich that we should not overestimate the impact of these doubts on the political life of the middle class or reputation of the nuclear strategists.[12] To the poor people this was all irrelevant or unintelligible; their energy needs lay in more cow dung and firewood for cooking and more kerosene for light at night.

One year after Bhatnagar's death, the former members of the Committee for Coordination of Scientific Work were reconstituted into the new Scientific Advisory Committee to the Cabinet. This is the way the cabinet operated for the next twelve years, until 1967. SACC began life small, but grew to be large. From 1956 Nehru chaired it, sometimes delegating the role to Bhabha. Just as he told the CCSW in 1948, when he chaired its first meeting, Nehru told SACC it could not pass orders but would make recommendations to ministries through the cabinet. The other SACC members were statistician P. C. Mahalanobis, physicist K. S. Krishnan, electrical engineer M. S. Thacker (CSIR), physicist D. S. Kothari (Defence), agronomist B. P. Pal (Indian Council of Agricultural Research), and physician C. G. Pandit

(Indian Council of Medical Research).[13] It will be noticed that Mahalanobis, Krishnan, and Kothari were, like Bhabha, in Nehru's inner circle, and all physicists.

One of the first activities of the SACC was to discuss and recommend the Science Policy Resolution drafted by Bhabha in 1957 and approved by the cabinet and Parliament in March 1958. This was followed four months later by the First Conference of Scientists, University Vice-Chancellors, and Educationists. The resolution gave official commitment to the orderly development of science and scientific institutions, all in the *new* spirit of national self-reliance. The conference was intended to build a constituency for the objectives of the resolution, and later conferences were intended to test and confirm the achievements of the resolution, if any, providing a special opportunity for political, educational, and scientific elites to interact and offering a kind of accountability in various sectors of research and development. It was an important opportunity for members of SACC and their chosen colleagues to lobby with cabinet ministers in the open.

Nehru sustained an astonishing interest in improving the configuration of the scientific and planning structures and achieving a more powerful articulation of their talents. This has often been said, but one has to see his letters and memos to various officials and politicians to grasp his engagement with the details, nominating and promoting or demoting specific individuals whom he knew or of whom he had been advised. For example, in a series of memoranda starting in early 1955, Nehru prodded the Planning Commission and the heads of laboratories and institutes to work together and to use more effectively foreign experts such as those at Mahalanobis's institute in Calcutta, among them J. B. S. Haldane. He handpicked committees, reviewed their terms of reference, and advocated "a scientific civil service" with enhanced salaries that would encompass most of CSIR. He stuck Bhabha into committees of the CSIR (whereas Bhabha preferred to remain apart) and cross-fertilized ongoing panels and reviews. When he did this, he explicitly referred to the coming formal resolution on science policy.[14]

With Nehru and many members of the cabinet present, the 1958 conference gave precision to the new Science Policy Resolution, recommended new facilities and opportunities for research work, pushed recruitment and retention of science professors and teachers through improved salaries, instituted a search for talent in high schools, and approved the provision of newly written textbooks. The most challenging recommendation for the government was the creation of a scientific pool for people with talent and training yet without permanent jobs.[15] This recommendation of an employment pool echoed back to discussion in the first meeting of CCSW with the cabinet in August 1948, namely that scientists sent abroad at government expense for training should be given meaningful work quickly on their return through a system of "supernumerary posts" until they could be placed in regular employment. This problem had not been resolved during the ensuing ten years and was now to be called the Scientific Pool. Local scientists were uneasy about this—better trained, younger, and ambitious returnees were a new kind of force—and insistence on seniority and rules protected them against that new force. Many of them had not had the advantage of class, location, or family that might have boosted them into this pool.

In 1958 Nehru presented to SACC a report he had received, "Problems of Indian Scholars and Technicians in USA," written and sent directly to him by K. M. Mukerjee in Chicago. The problems listed in the report were all quite familiar to Nehru: the inability of trained people abroad to know what employment was possible at home, clear disadvantages for newcomers in salary and benefits, and poor fit between training and eventual work. Frustration with scientists' underemployment or misemployment spilled over at the First Conference in 1958, as it did at every conference thereafter. D. S. Kothari worked hard, on behalf of SACC, to deflect this criticism, pointing out the progress in hiring at Departments of Defence and Atomic Energy. But it was immediately noted that the 1958 resolution was for "science," not "science and technology," and this dissatisfaction worked like an undercurrent through the First Conference in 1958, leading to arguments for more emphasis on technology. Within a few months of the conference SACC was dealing with a request from the cabinet to define and develop a system of eight regional engineering colleges and twenty-seven "polytechnics."[16] These colleges and polytechnics sprang up like mushrooms in 1960. They were to be allied with a growing network of well-funded, though not yet prestigious, Indian institutes of technology, each with its bilateral international partner and patron.[17] The scale of the plan for technology and technical education was now huge, but its realization was quite far off.

Frustrating Working Conditions in the Mid-Levels

There were other employment frustrations that compelled the intervention of SACC, in unexpected ways. The suicide death of scientist Dr. D. Parthasarathy in August 1960, having been widely reported in the newspapers, was discussed in SACC, focusing on the dead scientist's frustration with the obstacles to professional advancement. That suicide was followed five months later by the suicide of M. T. Joseph of the Indian Agricultural Research Institute in Delhi, unable to annul his bond with his institute to accept a better-paying and more suitable position at the Haffkine Institute in Bombay.[18] Nehru himself wrote on this occasion a memorandum asking scientific institutions to "improve conditions of work," and instructed the Home Affairs Ministry to prepare a new policy called "Measures to Be Taken for Ensuring Proper Utilization of Scientific Personnel." Notorious delays in promotions plagued most research institutes, though this occurred less in DAE-funded institutions than in CSIR-funded laboratories. The subject of timely appointments and promotions had already been discussed at the first meeting of CCSW in 1948, when Bhatnagar wrote his "Note on Physical and Psychological Effects of Working Conditions." This problem of improving conditions at work was not going away. Nehru's memo was written in 1960 while he and Bhabha were commissioning the new CIRUS reactor and opening the bids for the nuclear power reactors for Tarapur and Rajasthan. It was easier to borrow, buy, or copy glamorous new technology than to make effective use of all the skills that the country (and individual families) had already invested in over the previous one or two generations. It was found easier sometimes to fit whole technologies into India than to find the fit for three or four kinds of skill that had already been built up but not specifically for those technologies. And when working conditions were intolerable, there were serious consequences, some even pathological, like suicide.

The Parthasarathy case in 1960 is the first reference I have found to suicide among Indian scientists.[19] Not long after Parthasarathy's death the Home Ministry presented a note to SACC, "Conditions of Employment for Junior Scientists" (January 1961) and offered to "relax the rule that no more than 5 percent of the scientific staff may receive merit promotion," stating there should be "effective arrangements for independent and impartial evaluation," and "scientists may apply twice a year for promotion instead of once." When approved and implemented, said SACC, these regulations "should not require detailed scrutiny in the Ministries."[20] Obtained with Nehru's influence following the suicide, these changes were nevertheless applied piecemeal and slowly to existing institutions and taught a lesson over and over again to younger scientists—join the DAE labs and things will be different. This was the prevailing attitude from the 1950s through the 1970s. The CSIR was seen as a backwater in this light.

There were a number of later suicides that caused public outcry and demand for improvement of working conditions for scientists and laboratory workers. Although a certain number of suicides would be probable, given the large size and heterogeneity of the scientific community in India, some of these cases have had major political and public effects. Five years after the Parthasarathy affair a research worker in the Technical Physics Division in the Atomic Energy Establishment at Trombay (with a joint appointment in TIFR) committed suicide in January 1966, by drinking potassium cyanide in his laboratory.[21] Bhabha came to see the body on 22 January 1966, just prior to his departure for Vienna.

Improved working conditions for scientists in laboratories had been the battle cry of the Association for Scientific Workers of India since the late forties, an implicit objective encouraged by its patrons like Nehru, John Bernal, and Patrick Blackett. SACC discussed these working conditions regularly, including the process of actually getting passports, visas, and foreign currency when going abroad for study (most of this had to be done in Delhi), and providing employment quickly to qualified returning scientists, in case they left India again. Cabinet members discussed ways of searching for unusual talent among them and asked for regular assessments for promotion, with necessary lateral movement (through another institution or between central and state government institutions). Even failure to give visas quickly to visiting foreign scientists was discussed in SACC. Small problems sometimes loomed large, such as a government regulation that prohibited more than one person (scientist or not) to attend any particular international meeting; the result of this was that two Indians in the same field could not attend the same meeting; this policy in a country with 500 million people was, scientists pointed out in 1960, absurd, but it was one method of restricting the use of scarce foreign exchange. Larger problems followed, such as a research institution's need to make its own decision regarding which foreign collaboration and collaborators to accept or reject, without interference.

Though few were permanent, a large number of foreign scientists worked in India, and SACC—that is, Bhabha, Mahalanobis, and Kothari—fought to secure their exemption from certain Home Ministry rules when they worked for Indian institutions. Employment security and mobility (or insecurity and immobility) were constant themes, for example, when scientists needed for agricultural research were blocked when moving from state agricultural facilities to central government institutions or, when doing so, were made to give up their seniority and pensions. SACC set up a study group that insisted on the "mobility of scientists without loss or interruption of service benefits."

But the Home Ministry and Union Public Service Commission objected to this autonomy and flexibility, not simply that they (officials in Home or the Public Service Commission) did not have it themselves, but more that it did not conform to their conception of administrative hygiene; after all, why would one move unless one's performance was not satisfactory? The unorthodox approach to administration taken by scientists was, by the late 1960s, a common topic of conversation among many administrators whom

I met. Scientists' freedom to move was notoriously contested. These rules were not all post-1947 inventions; many had been well established in colonial administrative culture. By 1960, however, Indian officials had ample time to elaborate upon that earlier culture and invent new restrictions on the room to maneuver among scientific laboratory managers; they were not simply puppets of the past.

We should not think Nehru or the cabinet came upon these questions suddenly. There was a lineage behind each one: Nehru had expressed himself as early as 1952 about how scientists had to be treated separately from others, "because they are different." In his 1952 letter to Finance Minister Deshmukh, speaking particularly of atomic energy scientists, he wrote:

> The Atomic Energy Commission is an ad hoc body carrying on from year to year. The result is that appointments by the Commission have to be from year to year and it is difficult to get competent scientists on this yearly basis. At least a five-year term ought to be given to them. It is desirable therefore that we consider the Commission to be a more or less permanent body and permit them to give a five year contract to some (not all) principal scientific workers. . . . The work of scientific workers is rather peculiar. It is not routine work. A bright individual with initiative and vision might go far and do extraordinarily useful work. Another person might simply plod on. Hence it is necessary not to follow the normal rules of promotion, etc., but to judge by sheer merit and work done. Presumably the only persons who can do this are those who are competent to do so, that is, the members of the Commission.[22]

One can hear the voices of Bhabha and Blackett in Nehru's letter, and this is the kind of subtle influence on Nehru that I think Bhabha and Blackett had, encouraging him to think of science and scientists as something quite separate and special, yet essential. They all realized that scientists had worked independently during the Second World War, and that a few top scientists had spoken directly to prime ministers and presidents and had been listened to. After all, a "laboratory state" should have real autonomy, shouldn't it?

In this 1952 letter Nehru is laying out the plan for Bhabha's permanent and autonomous scientific administration everywhere, based on Bhabha's model for TIFR and approved by Bhatnagar, who was on the TIFR board. Bhatnagar had been trying to institute some of these criteria in some CSIR labs, though not the five-year contracts. The Public Service Commission had not agreed. Almost all of these scientists' jobs were known in common speech as "government service" and were in principle held for life. That was what families paying for the education of the scientist had intended;

in India merit was commonly attributed to seniority, even in the military, and variation from this principle was extraordinary (and potentially conflictual), so how could they justify the special treatment of younger experts?

Equally important, the frustration so eloquently described by individual scientists concerned their lack of autonomy and freedom *to do their work,* as much as about their salary and promotion. This lack of autonomy was troublesome for whole institutions too, and SACC members regularly spoke about the need to achieve that autonomy (known as freedom from interference) as a precondition for innovation and discovery. Composed of scientific leaders and heads of institutions and agencies, SACC was divided about this frustration: members knew this was a recurring problem and they would have to be more systematic in their approach, but they were also deeply implicated in the system as it operated now. They themselves had come through scientific institutions and themselves administered these rules. They had themselves set many administrative precedents. Some thought a little frustration was productive of better work. From about 1960, Bhabha seems to have decided he could use his influence, through SACC, to try to establish this freedom from interference from government rules and regulations on behalf of a large number of other (non-DAE) research institutions. He had tried unsuccessfully to assist Bhatnagar to do this with CSIR labs years before and prided himself in the fact that Nehru had at least told Bhatnagar to organize constitutions for CSIR labs on the TIFR model.

In general, the "bureaucrats," as they were called with suspicion, all senior members of the Indian Administrative Service (IAS), were opponents of this approach who wanted orderly change, if any, and accountability to themselves alone. The IAS viewed scientific autonomy as antithetical to their interests; they were, after all, inheritors of the supreme power of generalist administrators achieved by the ICS under British administration. In 1960 Bhabha prepared his "Note on the Autonomy of Scientific Organizations Working under Non-Scientific Departments." Choosing his example carefully, he spoke about the Tata Research Hospital for Cancer in Bombay, describing how greater autonomy would benefit such organizations and the nation.

Bhabha's Attempt to Strengthen Kothari's Profile

Throughout this account of sixteen years of SACC's work, it is evident that Bhabha turned to Kothari for help, presumably with Nehru's promotion. When Kothari would leave Defence in 1961, as he must after ten years, he would be going on to even more influential jobs and would need something to enhance his prestige. As if to recognize Kothari's enhanced role, Bhabha decided in 1959 to renominate him for election to the Royal Society, because the 1944 certificate expired in 1959 after fifteen tries.[23] Kothari was then in the middle of India's rocket research project, spurred on by Defence Minister Krishna Menon (see chap. 20). Bhabha and Kothari were not close friends, and there had been an earlier sense of rivalry around credit for the official 1956 study on the effects of nuclear explosions (see *Negotiating Nuclear Power*). Bhabha asked Kothari for a complete list of his scientific papers and telephoned round an elite Indian FRS network—K. S. Krishnan, D. N. Wadia, P. C. Mahalanobis, and S. N. Bose—to sign the nomination with him. This year the bid failed on a procedural flaw, when Mahalanobis failed to get his signature to London on time. The following year (1960), working through Blackett, Bhabha asked Sir Harold Jeffrys to second the nomination with him, but Jeffrys declined, saying he would sign but not be the seconder, not knowing Kothari's work as well as a seconder was expected to. Moreover, Jeffrys reported that Fred Hoyle said "he does not think there is much case for re-opening the question." Bhabha also asked S. K. Mitra, astronomer Henry Plaskett, and mathematician T. G. Cowling to sign for Kothari: Plaskett wrote back saying that "his exciting work is of an earlier period" and Cowling that "the recent work has appeared to me like a lot of Indian work, of largely formal interest. It does not appear to have a great deal of contact with nature." The 1960 bid did not succeed, and after 1961 Bhabha appears to have let the matter rest, advised of the difficulty of this case by Blackett. By 1961 the deaths of Bhatnagar and Krishnan had collapsed the scientific capacity of the AEC; Kothari might have played a role there for a while had he been elected a Fellow. Kothari's role was now as chair of the huge national funding agency, the University Grants Commission but his office became the Delhi secretariat of SACC.

Central Government Rules, Audits, and Control

Through the early 1960s, as chair of SACC, Bhabha led a discussion culminating in a confrontation with the government's comptroller and auditor general over audits of research grants. That official insisted on *his* audit of all grants *after* the completion of the annual commercial audits required by law in the institutions themselves. SACC declared this plan "a waste of time and energy, and vexatious," but the comptroller and auditor general insisted on it, seeing clearly that SACC's declaration was part of the quest for autonomy, and to the auditor general that quest was itself vexatious. Frustrated administrators pointed to the troubles of the NPL in Delhi, a lab that already had autonomy but had grown unwieldy, a lab whose budget lacked controls, they said; evidence for this was that Bhabha's friend Patrick Blackett was asked in 1963 to investigate the NPL and recommend reorganization and accountability. Bhabha's choice of the cancer hospital in Bombay directed SACC's attention to a facility that administrators themselves could understand better than nuclear research and development; perhaps they would use it if they contacted cancer (though not common in India then). That hospital stood for "science with social benefits" and reminded bureaucrats of the munificence of the private sector at the same time. The example was quite effective in SACC, but not with the auditor general or the ICS.

Bhabha therefore asked TIFR and other DAE-funded institutes, and the whole CSIR system, to prepare "a list of audit queries and objectives which could be considered unnecessary or wasteful of time" in 1963.[24] The long list was impressive in its detail, but there was soon a kind of power stalemate over this. It was understood that it would be up to the prime minister and the cabinet to confront the bureaucracy on this large question, but Nehru's age and poor health did not augur well for a winning move for either side. Nehru needed more allies than he then had to force such a change through the entire administrative system. The result was that changes were done piecemeal, and new constitutions were installed only if new institutions were created, but not in existing ones. The profound inertia of rules, regulations, and double-audits in CSIR and other government research institutions (except DAE labs) continued for many more years, well into the 1970s.

Not to be defeated by regulations, SACC kept up the fight. In 1963 a "Model Constitution for Research Institutions" was presented to SACC; Bhabha asked the noted mathematician K. Chandrasekharan, deputy director of TIFR, to prepare and present this model, based on TIFR's experience. The prime minister was present, and SACC persuaded him to speak directly to the auditor general and thus end the fight over the double audits. A model constitution evolved through discussion, and when approved by SACC in 1965 it provided that there would be a governing council with about seven members, of which the head of the institute would be a voting member, that this council have the power to scrutinize budgets, make appointments, and delegate limited but clear powers to the head of the institute, and that every institution should have a qualified accountant who can anticipate the standards and methods applied by auditors. In particular the model constitution said that the primary objective of the institution is scientific research and "the purpose of purely administrative officers . . . is to relieve the Head of the Institute of routine work."[25]

The other kind of freedom attractive to scientists was mobility between institutions and between states in India. In the 1961 census there was a "special enumeration" of scientists and technologists and this allowed an assessment of the flow out of states to other states, and confirmed the public knowledge that the southern states were major contributors to the technical classes in the area around Bombay, or that West Bengal was a major contributor to the economy lying north and west of Calcutta, and finally that the southern states received little scientific inflow, and that Gujerat and Maharashtra essentially exchanged the same proportion of scientists and technologists.[26] But the Public Service Commission fought hard against mobility for scientists, seeing it as an additional burden to the already large amount of time taken up with appraising and negotiating individual requests for transfers within the same department or service. On the other hand, some junior staff sought protection from arbitrary and unplanned movement and transfer from one place to another, to strengthen a new project or institute.

Another kind of autonomy from government regulations was approved by the cabinet in 1963, namely that R&D institutions be exempt from "Reservation Orders for Scheduled Castes and Tribes." These orders normally required government departments to reserve positions for qualified individuals originating from these "backward classes," as they were then known and defined in the Constitution. SACC was unanimous that no criterion other than merit should be used to make an appointment to or promotion in research. Because of systemic discrimination in educational institutions, there were few qualified candidates for scientific posts from these social groups (later called Dalits). It was many years until the number of scientists from "backward classes" was sufficient to test this exemption rule, and years after that until the resistance to such appointments gradually broke down. In 1965 this regulation was again amended by the cabinet, on SACC's recommendation, to allow research positions, but *not all* positions in laboratories, to be exempt from the regulation, and that is where the eventual "scientific" employment of previously excluded groups began.

New R&D Policy after the 1962 China-India Conflict

An entirely different kind of crisis entered SACC's work in 1962, when immediately following the conflict with China an emergency subcommittee was formed to report directly to the cabinet on how to reorganize science and technology institutions and mobilize scientists for India's defense.

Within a few months of the end of the war with China, physicist S. Bhagavantam, Kothari's successor at the Defence Research and Development Organisation of the Defence Ministry (DRDO), stated in SACC that "defence is mentioned only once in the Science Policy Resolution . . . and such an imbalance should be corrected. We cannot separate scientific output into water-tight compartments, separating defence labs from universities and civilian labs." Soon the director general of CSIR Husain Zaheer replied, "we have found recently that there has been a great rush of scientists from the CSIR to appointments in the Defence Science Organization."[27] So mobility between organizations *was occurring* after all, when strategic requirements were strong enough! SACC discussed and approved the rush to develop radar at the DRDO and recommended its immediate application to the Himalayas following the China conflict. Contributing to this mobility was the rumor about big increases in the defense budget; this turned out to be strikingly true, with Defence expenditures in 1962–63 of Rs 4.774 billion and the next year 1963–64 of Rs 8.117 billion.

SACC pointed to the need for radar along the Himalayas, and so the small microwave engineering project at TIFR was expanded using Defence Ministry funds, reflecting Bhabha's commitment to involve his institute like others in defense research, in keeping with strong public sentiment that all resources should be mobilized for this task. Microwave engineering had been started to complement TIFR's project far away from Bombay, the Ooty radio telescope in the Nilgiri hills in southern India. Now, after 1962, microwaves had new strategic importance, needed to control and calibrate radar systems. Immediately following the China conflict, Bhabha was appointed by Nehru to the National Defence Council and soon after was made chair of the Electronics Committee in 1963. Bhabha was now also starting to build two brand-new research groups, on molecular biology and radio astronomy in TIFR, but was also personally involved in intense negotiations with the United States and Canada on control of the nuclear fuel cycle in four new reactors, ready for construction.

The effects of the 1962 conflict with China explain why the atmosphere surrounding the Second Conference of Scientists, University Vice-Chancellors, and Educationists in 1963 was more charged with strategic issues than the first. With Nehru and cabinet ministers present, the scientific leaders held the floor to explain and justify expenditure on science. Husain Zaheer asked, "Why are we doing scientific research anyway?" and answered that it is for national development. "But," he agreed, "many Indian scientists are tied to concepts and ideas which were current 25 or 30 years ago, and are still attached to them for various intellectual, psychological and material reasons." A new energy filled the room, pumped by the surprising weakness of Indian military approaches to the 1962 conflict. Mathematician K. Chandrasekharan of TIFR complained, "there is insufficient recognition of the value of creative work here," and J. B. S. Haldane, renowned biologist working for the CSIR, also present at the conference, echoed that "scientific talent dries up at age 35 because the scientist is overwhelmed with administrative work." Satyen Bose (still appointed to the Rajya Sabha in Parliament) said that it is "quality not quantity" that counts, so that the significant fact "is not the number of papers published by a scientist but the number of times a paper is cited." (This is truer in *his* case than for any Indian scientist, *ever*.) And P. C. Mahalanobis said that "there is a great need for public debate among scientists, because it is this debate which will, in the long run, contribute to better choices in research." The conclusions of the conference were by now familiar: rebuild the universities (the constituency of many of those present), free heads of departments from routine administration, avoid making any policy or building any institution that could weaken universities, select 6–12 universities, give them lots of money, and let them flourish in research, and take the University Grants Commission on a tour of major American universities to see how things should be done. Kothari responded wisely and proposed first doing "case studies to find out where the trouble actually lies, and where the difficulty is only superficial."[28] But it was pointed out that there was no capacity in India for such studies; in this context the importance of the founding of the CSIR's new policy and planning unit under A. Rahman can be understood, and 1963 is where it received its mandate. From this beginning rose the eventual National Institute for Science, Technology, and Development.

Nehru's Role in the Politics of Science

It is difficult to see how, when faced with this litany of complaints from scientists, Nehru sustained his famous "charm for science." Since the charm for science was part of his persona, and essential to his internationalism, he persevered. What more could Nehru do? Had he not given as much as possible to science? And did he not believe in most of what he said in his speeches? There is little doubt *he did believe* what he said. He wrote these speeches himself, and we have his verbatim extemporaneous statements about the value of scientific temper (see chap. 12 for Nehru's 1954 answer to the question "What *is* your science?").

After his death in 1964, a myth emerged about Nehru and his charm for scientists.[29] This myth had important uses and served the purposes of a number of people, not least his daughter and the Congress Party; like most myths, it is certainly based on an essential truth.[30] It is why Nehru encouraged SACC to organize a strong role for India in the first UN Conference on Science, Technology, and Development in 1963, where the whole discourse about scientific institutions for development came onto the international stage and was made "official" and was eventually translated, on a modest scale, into UN agency activities. Four months after Prime Minister Nehru's death in May 1964, Bhabha became chair of SACC, no longer just an acting chair. But the direct link with the prime minister's office had been broken with Nehru's death, and it was difficult for Bhabha to reestablish it, as the character of the new prime minister was so unlike Nehru's or Bhabha's.

The new prime minister, Lal Bahadur Shastri, continued the practice that the agenda and minutes were to remain secret but the discussion papers circulated within SACC were not to be classified. Clearly there were some criticisms of the closed exclusive character of SACC, and some acceptance of this critique within the committee, so Kothari was asked by Bhabha to review the list of policies accepted by the cabinet after SACC's recommendation, to strengthen an appearance of relevance and better defend the committee. Although it is not clear that the *full* list of cabinet-adopted measures was compiled before Bhabha's death, there was evidently some charge made that SACC had no real influence in the cabinet, particularly after Nehru died; it was asked, what are the new policy outcomes of its advice?

At the same time, Bhabha and Kothari began to reorganize SACC in 1964–65 by creating disciplinary national committees—in astronomy, statistics, geography, and the like. This brought in dozens of experts across the country, gave the government access to their intelligence, and provided them with a clearer sense of involvement—at relatively low cost. This practice continued in 1966 and 1967, with new committees on water and hydrology, computer-aided decision making, instrumentation, and transfer of science to agriculture. The latter committee was chaired by M. S. Swaminathan, who was beginning to make his reputation through the green revolution; chairing these committees also gave honor and recognition to those scientists previously excluded from SACC membership and raised the importance of emerging specializations.

The Cabinet, Bhabha, and the Nuclear Questions

There *was* a hot policy question that SACC discussed in 1964, though it is not recorded in the SACC Review, namely, the question of planning for an atomic bomb. Five months after Nehru's death the anticipated first Chinese atomic bomb test occurred on 16 October 1964. Bhabha quickly broadcast on national radio his estimates of the low cost of building a bomb, while pointing to its benefits of deterring attacks on India. He now let the newspapers do much of the work of agitating for the bomb, all well described in Perkovich's remark that "six weeks of debate shifted India's nuclear policy."[31] Though led by right-wing Hindu-nationalist parties like the Jan Sangh, it was gradually followed by others who shared few of the Sangh's narrow Hindus-first conception of India. I speculate that had Nehru lived another six months, he too might have reluctantly agreed to the kind of conditional plan for the design of a bomb that emerged late in 1964. Faced with the crisis of 1965–66, however, he also would have put it on hold, as such a bomb design would not have played any role in dealing with the elements of that crisis.

But the conflict with China in 1962, the death of Nehru, and the Chinese bomb in October 1964 had altered the framework within which this decision was made, and the debate took shape in a six-hour cabinet meeting on 29 October where Bhabha was an observer throughout. This led up to the All-India Congress Committee meeting where Shastri marshaled the support of other parties as well as his own to hold firm on a no-bomb policy. Criticism of Bhabha's radio speech was in circulation at that meeting and reached the newspapers, so Bhabha skillfully put the responsibility back on the cabinet and on the prime minister—saying that India should not rush into this for political reasons, even if it was technically possible now and not expensive. Then the question came to Parliament in the last week of November, where Bhabha's position and cost estimate were the evidence for much of the debate in the Lok Sabha. Some members proposed the bomb, watching Prime Minister Shastri's response. Shastri stepped beyond criticisms already made of Bhabha, having shown that he and the cabinet would make policy, not scientists. The "inconclusive conclusion" was, however, that DAE could only study the use of nuclear explosions for peaceful purposes. "By bringing Bhabha onto his side, Shastri could ward off criticism on the nuclear issue, which in any case was understood superficially and debated hypothetically and/or symbolically compared to more pressing and familiar domestic issues."[32] In a context of increasing food and agriculture problems, an approval of a study for a bomb's design was *definitely not* one of the outcomes in the policy list attached to the SACC.

In retrospect the list of cabinet approvals of SACC recommendations does not seem momentous, but SACC had numerous other useful, if unintended, and informal functions. In fact some of its work was to bring influence from one scientific leader or agency to another, not to the cabinet. The list of cabinet approvals developed by Kothari is as follows:

1962 Establishment of a government-run "science press"
1963 Liberalization of the rules for working scientists, finding jobs for promising young scientists, facilitating exchanges between universities and research institutions, appointing distinguished scientists as heads of institutes
1964 Adoption of a model constitution for research institutions, better use of foreign experts, teams or expeditions in India
1965 Exemption of positions in R&D from rules on employment of Scheduled Castes and Tribes, establishing a Himalayan Institute to include geology and hydrology.

One should remember that many other changes were achieved through negotiation with specific ministries and ministers, changes that did not need formal approval of the whole cabinet; for example, Prime Minister Shastri and Bhabha referred very few questions about atomic energy to the cabinet. Bhabha clearly kept his colleagues favorably informed about the DAE and AEC, except when they wanted to talk about the scale of his budget increases. And information about major issues like the 1962 war with China, the agricultural crisis of 1965 and 1966, and devaluation of the rupee in 1966 flowed the other way, from the cabinet to the Scientific Advisory Committee, so scientists were beneficiaries. The finance member of the AEC during Bhabha's last four years (1962–66) informed me that he could only check the cost and feasibility of DAE projects, and nothing more—Bhabha set the priorities, and for most of that time there was no other scientific member.[33] Meanwhile Bhabha was so extremely busy that he did not have time to check everything as he used to do.[34] For example, an accident occurred in mid-1965 when an underground storage tank used to store irradiated fuel rods from CIRUS developed a leak and flooded, so the water had to be pumped into the sea, irradiating the shoreline. Waiters who were using the beach for their toilet were contaminated by touching that irradiated water. "Unaware of this they served in the canteen and passed on the contamination to the technicians who carried it into the labs." Bhabha did not know until the last few months of 1965 that the plutonium plant was not working well, probably as a result of this accident. Clearly out of touch, Bhabha asked why Homi Sethna had not explained the whole situation to him.[35]

The CSIR Tangles with SACC

This chapter began with the effects on Nehru and Bhabha of the death of Shanti Bhatnagar, and his loss on CSIR itself. It ends with a consideration of the entanglement of the CSIR and SACC. We should ask whether being a member of SACC was invariably an advantage for an ambitious scientific leader, particularly in terms of the efforts to reform the CSIR from within? Beginning with Bhatnagar's death in 1955, the CSIR was constantly reviewed and then admonished by SACC. These public reviews usually discovered the same ailments that were discovered by the previous review. A site of intense internal political maneuvering itself, the CSIR Governing Council had no protection from this "outside" interference, often from the political guardians of regional centers and institutes. Its hapless director general had to sit there in SACC while his huge empire was criticized by other scientists who heard about its weakness through the newspapers. Directors of CSIR labs, such as the NPL in Delhi, were even more vulnerable, with so many big political fish swimming around them and their laboratories. In contrast there is no record that Bhabha's institutes or projects were reviewed in the same manner, and one doubts that they were until the 1970s (see chaps. 20–23).

A special opportunity to increase the influence of the CSIR arose with Thacker's retirement in 1962 and the appointment of Husain Zaheer, just when the war with China demonstrated the need to mobilize science and technology for defense. There was wide popular support for this defense objective: women gave their gold bangles to the defense fund and knitted warm socks for the soldiers, who famously did not have cold-weather clothing for fighting in the mountains. Zaheer established "defense cells" in all CSIR laboratories, and there was a Defence Coordination Unit in Delhi to make CSIR's research amenable and available to defense applications. Zaheer came from within the CSIR system and seemed determined to shake things up. He started the first program about scientific research planning, including the history and sociology of science in India, called the Research Survey and Planning Organization in 1963. That organization soon embarked on a study of the costs and expenditures in all the thirty-one national laboratories—apparatus and equipment, chemicals, libraries, furniture, staff housing, salaries, foreign visits, publications in India and abroad, and the like.[36]

Zaheer gave some important autonomy back to lab directors and encouraged or pushed their executive committees to give more support and broad direction to their labs. In so doing, he addressed the chronic problem of structural reorganization of the very large laboratories; for example, he approved Bhabha's suggestion that Patrick Blackett be invited to investigate the situation in NPL and recommend fundamental changes, including severing parts of the lab and sending them off to other places. Clearly he knew that Blackett's recommendations would be carefully read by Nehru himself, because he was the president of CSIR. Now the SACC and CSIR would be face-to-face.

A series of leadership problems arose in the NPL after the death of K. S. Krishnan in 1961, and Patrick Blackett was appointed by the minister of Scientific Research and Cultural Affairs, Humayun Kabir, on the request of Zaheer, to conduct what the newspapers called a "Full Enquiry at NPL." The publicity surrounding Blackett's 1963 inquiry at NPL prompted his old friend biologist J. B. S. Haldane, now living and doing research in Orissa, to invite Blackett to visit for a discussion about CSIR. With seven years' experience working as a CSIR researcher in India, Haldane drew Blackett's attention to what he must learn before he formed an opinion about the CSIR and provided Blackett with the following advice:

> I don't envy you your job. Considerable efforts will be made at the top to prevent you from finding out the truth, and when you do get through to unofficial sources you will hear some outrageous lies from people who cannot imagine an uncorrupt reason for doing anything. I advise you to get hold of the contract offered to junior scientific workers (such as myself) agreeing to go anywhere at a month's notice.[37]

Blackett worked fast at the NPL and submitted a report after a month's inquiry. He recommended reorganization of the whole laboratory, moving some or most of those who worked largely in basic research to a new center for advanced physics at the University of Delhi, moving those in rain and cloud physics out to the Meteorological Department, moving potential manufacturing units in radio, glass, and ceramics out to join their industries, and defining the work of remaining divisions more clearly while enhancing communication among them. Blackett found a culture of administrative rigidity in the NPL where most people fought to define and protect the boundaries of their work. In effect, he said, the NPL lacked unity of purpose.[38] Blackett had earlier criticized the fact that the CSIR pressured K. S. Krishnan to abandon his basic research in a university, offered him double his professor's salary, brought him to Delhi, and told him to try to build a research tradition applied to industrial problems. Blackett said he had seen this same phenomenon at Britain's National Physical Laboratory at Teddington.

Both Blackett and Hill criticized this practice in India and Britain. D. S. Kothari criticized Bhatnagar for taking scientists away from the universities, but his own Defence Science Organisation did the same thing in the early 1950s and then took them away from the CSIR when it needed them after the conflict with China in 1962–63. Bhatnagar and Kothari had argued that there was no source of competent people other than the universities, and of course—unless their labs trained such people themselves—they were both right. The source of competent young people was the universities, and government laboratories were not going to train them from scratch. Although Bhabha established an advanced summer school for training his physicists and engineers in 1956, he too knew that the universities were the first line of contact with science. The debate had long been loud about universities being deprived of good researchers. Universities argued that this outflow meant they could not demonstrate that good research was itself a form of training and that this deprivation turned them into mere teachers. On the other hand, there was criticism that universities were inhospitable to good researchers and that government laboratories actually were the source of most good research in India.[39]

The CSIR struck a committee to review Blackett's report, tour the labs, and recommend how the Executive Council of NPL should respond to it. In April 1963 this committee reported to the Executive Council that "the Committee finds itself in general agreement with the spirit of the Blackett report." Its own assessment was that "there should be enough flexibility so that when an individual project is taken up which requires that scientists from different disciplines be brought together, then it should be possible to do this without difficulty. This aspect has been sorely neglected hitherto." There was one exception to their approval, however, in that "the Committee was not at all in favour of one possibility indicated by Professor Blackett, that of completely amalgamating the proposed Centre for Advanced Physics with the Delhi University." The report suggested instead an institute like the Saha Institute in Calcutta that was autonomous but very close to the university, with university participation in its Governing Body. The members of this committee included M. G. K. Menon of TIFR, who authored the report; R. C. Majumdar, of Delhi University; A. K. Saha and B. D. Nagchaudhuri, of Calcutta's Saha Institute of Nuclear Physics; all three students of Meghnad Saha; and scientists J. C. Kapur, V. M. Vaidya, and L. C. Verman.[40] This committee proposed changes that did not always satisfy Blackett, and on his copy, beside the issue of the center for advanced physics moving to the university, he noted in pencil, "All this does not agree with my recommendations." In the end an advanced center was created, an amalgam of the Blackett and the Menon committee plans, at arm's length from Delhi University. Blackett's report on the NPL had a long life, with review committees dutifully referring to it over the next four years; it was dusted off when Krishnan's replacement physicist P. K. Kitchlu, Saha's student in Allahabad, was not renewed as director.[41]

Director General Zaheer wrote to Blackett again about the saga of the CSIR and NPL late in December 1964, saying, "I am afraid matters have not improved very much during the past year. The Director, whom we appointed last October, had eventually to be removed from service. The main defect was that he could not get on well with his younger colleagues. Now we are on the lookout for appointing a Director." This new director of the NPL would have to implement the changes recommended by Blackett and Menon's committee.[42]

Now that SACC had swung to focus on electronics and Bhabha chaired an electronics committee, Blackett, having just been to China, wrote to Zaheer that the Chinese were far advanced in electronic instruments. This was not the first time someone influential had said this, but in 1964 people paid more attention to it because Blackett was now more famous in India, because China embarrassed India militarily in 1962, and because it had just now conducted its first atomic bomb tests.[43] Zaheer agreed with Blackett's warning about China and electronics and quickly outlined the steps two CSIR labs were taking—one in electronic engineering in Palani and the other in scientific instruments at Chandigarh. He did not admit, however, in his 1964 letter that the underdevelopment of these labs was retarding economic development in India. The bigger issue at stake was not simply electronic instruments for researchers or electronic engineering for military systems, or even consumer electronics; the bigger issue was the supply and delivery of electricity itself.[44]

How foreign experts should best be used was indeed a complex problem for SACC, but not all were a waste of money as the candid observations of some of them showed. Blackett was an illustrious example, no longer classed as a "foreign expert," but there were many others. For example, Blackett and some members of SACC received a copy of the confidential report written by C. G. Wynne, of Imperial College, for the director general of CSIR about optical designing.[45] At the same time, Wynne also wrote Blackett a long personal account of his visits to National Aeronautics, Ltd., the Glass and Ceramics Institute, and commercial firms lamenting the amount of foreign money wasted (in his opinion) in technical assistance to India and the frustration of good foreign-trained talent working in CSIR research centers like the Glass and Ceramics Institute in Calcutta. Wynne said everyone in India recognized the strategic necessity of developing a high-precision optical glass industry, everyone thought India was proceeding too slowly toward implementing that goal, but everyone explained the delay by blaming someone else. Wynne's report matched the negative views of a wide range of Indian and foreign opinion-makers and, when communicated through Blackett to the SACC, encouraged a wider review of the success and failure of CSIR in the scientific and industrial community.[46] This eventually was transformed into the Sarkar Commission (see chap. 20).

There is a loaded phrase in Zaheer's 1964 letter to Blackett about the NPL director Kitchlu being unable to get along with his junior colleagues. Zaheer was setting about to abolish some hierarchical terminology in the scientific ranks in CSIR in 1964 and is thought to have received Nehru's approval of the plan before his death. The terms "junior" and "senior" scientist were abolished and replaced by an alphabetical system. Then Zaheer persuaded his Governing Body to accept another change, one that obliged directors to review the progress of individuals every five years and to promote them on merit, whether or not there was a vacant position available. "This marked, thanks to Zaheer, the beginning of the end of the 'no-vacancy' syndrome which could have blocked recognition in terms of upward mobility of many a scientist in CSIR."[47] Nevertheless, these changes disturbed many people in the CSIR, and there were reactions and perturbations for years to come. Hierarchical distinctions were carefully conserved in most spheres of social life in India in this era, and although they were often resented by those classified as inferior, the distinctions were conserved for years; scientific institutions only mirrored a widespread social and administrative practice. Zaheer's "modernist" and "merit-based" plan disturbed some senior scientists and directors sufficiently that they organized a political movement, involving some executives on laboratories' governing councils to question Zaheer's decisions, protect the status quo, and have Zaheer prematurely removed from the post of director general in late 1965. This took influence right into the prime minister's office (for more on the outcome of this process, see chaps. 18 and 19). In the end Zaheer did not leave until his normal age of retirement in 1966, but the movement set a precedent. Yet another new review committee was established and became the new political context in which laboratory managers and senior scientists worked, with very thin lines of communication up to the committee that advised the cabinet.

How much the cabinet really understood what was going on underneath them must await further studies, but oral evidence from directors of CSIR laboratories suggested that it was "not very much." The new prime minister in 1966 was soon to be in a position to make science policy more or less by herself.

FIFTEEN

A New Scientific Elite: Sarabhai Builds Another Atomic Energy Network, 1966–71

Homi Bhabha's sudden death in late January 1966 created a large and unprecedented opening in the political leadership of the scientific community in India. Indira Gandhi had only just become the new prime minister, Lal Bahadur Shastri having died only two weeks before. Her relations with scientists, derived largely from her father's influence, were familial and cordial but untested. She had liked and trusted Bhabha, having first met him in the company of C. V. Raman, thirty years earlier, on board a ship traveling to Britain in 1936.[1] There was, however, a lot of uncertainty in January 1966, and the new opportunities were very large. The list of Bhabha's responsibilities included the secretary of the DAE, the chairman of the AEC, the director of the Atomic Research Centre at Trombay, director of TIFR, chairman of the SACC, and of the Electronics Committee. Bhabha had designated the deputy directorship of TIFR to M. G. K. Menon, and there was an expectation that the post of director would be Menon's next step at TIFR, the best-funded and most prestigious scientific institution in the country. Menon had entered an elite world largely uncontested. Bhabha's other responsibilities were open for negotiation. When the new Atomic Energy Act was passed in 1962, Nehru and Bhabha enlarged the commission from three members to five, a change under which industrialist J. R. D. Tata, chairman of the large Tata Group of industrial companies, was appointed to the AEC. Bhabha did not fill the other vacancy for another three years, until December 1965, when he appointed Vikram Sarabhai, the first scientist to be appointed to the commission after Bhatnagar died in 1955 and Krishnan died in 1961. The commission had the resources and freedom to create new scientific programs like space and industrial programs such as satellites.

The cabinet secretary Dharma Vira was first put in charge of both the AEC and the search committee for Bhabha's replacement as chairman, but

AEC members wanted the new prime minister more directly and personally involved: they asked her to begin seeking advice very widely.[2] This opened her role onstage as conductor of the orchestra of scientific leadership, playing to its foibles and projects, limiting its demand for money, continuing to cultivate new and younger players. Involving the prime minister in the search led to an invitation to S. Chandrasekhar to return to India and become chairman of the AEC. It was not the first attempt to bring him back to India. In 1945 the minutes of the CSIR Governing Body recorded that "the services of S. Chandrasekhar of the University of Chicago have been secured," noting that "he has recently wired his acceptance of the post."[3] Bhabha also invited Chandrasekhar to join TIFR in 1946 and 1951, without success, and Nehru and Thacker had tried to make him a national professor in 1961. They had no more success than his uncle C. V. Raman had, when he began trying to bring him back to India in 1938.[4]

After much consideration in 1966, he declined the offer of the AEC, saying that he really preferred to concentrate on his work, that it would be very difficult to invite an American citizen (which he had become in 1953) to do the job, and that the job's various demands would be difficult for him personally to deal with. He was now fifty-six; his refusal of the offer played on his memory, so he sought an interview with Prime Minister Gandhi later, in order to explain his refusal of her offer. Chandrasekhar's view was that "it was clear that to be in a position like that required familiarity with Indian politics and required administrative capacities. I wanted to decline it, so I wanted to explain to her the reason." When he did so, Indira Gandhi said, "Well, all of this was not made clear to me," suggesting that she had been both disappointed and unsure why he declined it.[5]

In retrospect Chandrasekhar was a curious choice but an important symbolic one. It was widely thought that someone of his great international stature would inspire scientists (and others), immunize research centers like TIFR from the risk of mediocrity, and keep the AEC genuinely international. Making the offer to Chandrasekhar was itself important to Indian scientists, as it was a message to be read by other scientists abroad that India would welcome their return. The term "brain drain" was now in common speech. Indeed, Vikram Sarabhai soon supported an AEC study of the brain drain, and an article about it was very popular and widely discussed in the late 1960s.[6] Chandrasekhar managed a famous journal in astrophysics but was unlikely to accept the load of administrative and diplomatic work entailed in being chairman of the AEC. He said he even had an aversion to administrative work around him at the University of Chicago.[7] Even if protected from some of the routine administration by others, by arrangement with the prime minister, he was unlikely to have accepted the obligations of the atomic energy appointment with equanimity.

Meanwhile Indira Gandhi and Dharma Vira appraised physicist Vikram Sarabhai, physicist Raja Ramanna, and engineer Homi Sethna for some of Bhabha's positions. Beyond TIFR, his other positions commanded huge budgets and shaped India's nuclear and scientific policy. Each position entailed interaction with economic and strategic developments and with the country's military and industrial leaders; they offered opportunities for signing agreements and contracts with large Indian and foreign corporations and for being the intermediary in other international agreements. Bhabha's positions all presented an opportunity to appoint experts, all of whom then could appoint other people to governing boards and research positions.

Each of these positions entailed a lot of international contact and travel and provided opportunities for post-appointment international employment as well. In the end, Sarabhai was chosen for the senior roles of chairman of the AEC and secretary of DAE. Homi Sethna was chosen as director of the Trombay atomic energy research establishment, soon to be called the Bhabha Atomic Research Centre (BARC). Menon was confirmed as director of TIFR. Bhabha's four main jobs were divided among three people, not four. All this coincided with Gandhi's role, as the new prime minister, of President of CSIR where the post of director general was being filled by Atma Ram, offering her an opportunity to satisfy another constituency, namely, the Congress old guard led by Moraji Desai, who might not be happy with her appointment of Sarabhai to so important a role.[8]

Indira Gandhi was a sort of loner in her situation. Her mother had died when Indira was eighteen, and her father was in and out of prison up until 1945. Admitted to Oxford in 1937, she did not complete her studies there (or anywhere else). She herself was jailed for eight months in the Quit India movement of 1942. When she was forty-six in 1964, her father died, her husband was already dead, and she had long been the single mother to her two sons Sanjay and Rajiv. She had economic security and evident political determination and hidden inner strength.[9] She arrived in the prime minister's office with intimate knowledge of its functions but no real political support.

Indeed she was "placed" there by people calculating that she would respond well to their signals, despite the fact she had already been president of the Congress Party for a few years. She was, after all, a "young" forty-eight-year-old woman surrounded by much older men. From this unlikely beginning in 1966 she emerged ready for an election in one year, and two years after that she confronted her most powerful opponent in the

Congress, Moraji Desai. The absence of a formal university education gave her a kind of "negative identity" with respect to scientists and science because she was literally surrounded by people who had been to Oxford or whose children were going to Cambridge. She could not be like her father, who prided himself on his natural sciences degree from Cambridge. She knew many of the scientists and their reputations but could not share much of their intellectual ethos nor their engagement in the networks in India as graduates of elite universities. She may have already developed some impatience with those networks. She exhibited an interest in modern technologies but not an obsession with developing all of them in India. She looked for and chose strong leadership in the scientific community and had a number of top appointments to make. Mrs. Gandhi eventually picked Vikram Sarabhai to be the DAE's leader after Bhabha's death and named Atma Ram as the head of CSIR a few weeks after that. Sarabhai had only recently been made a member of the AEC and thus was an outsider to the DAE, but Ram had been a CSIR insider for twenty years. Sarabhai was perceived to be a strong leader, but Ram was not; the difference of DAE and CSIR was to deepen in the next few years.

Vikram Sarabhai's Background and Atomic Energy

Between 1966 and his sudden and unexpected death in 1971, Sarabhai became the most influential person in the Indian scientific community. What qualities did he bring to his work, and what explains his unusual prominence? Upon placement at the AEC and DAE, Sarabhai had to sever his relationships with all the family's businesses in Ahmedabad. He worked, as Bhabha had, in Bombay, not Delhi. He asked for, and received, a salary of one rupee per year, reflecting his belief that his family was wealthy enough to enable him to work for his country without pay. He told me he did not have any sense of dependence on the government while growing up and did not wish to start now. "I should be able to walk away from this job without financial considerations," he told me. Already strongly identified with space research and director of the Physical Research Laboratory in Ahmedabad, Sarabhai had been considered by some to be capable of running the CSIR in 1955, at age thirty-six. Being part of a very political Ahmedabad family, he had a different and deeper understanding of Indian politics from Bhabha.

Unlike Bhabha, Sarabhai had personally built an industrial company in textiles and chemicals and was a businessman as well as a working physicist. In fact, his father's opposition to Vikram's acceptance of the position of chairman of the DAE in 1966 might, in part, have been due to a reluctance to lose Vikram from the growing Sarabhai Group of industries. This reluctance was prescient: the Sarabhai Group declined in size and effectiveness when Vikram left it. Economist L. K. Jha recalled, "He had to make a difficult choice: whether or not to cut off his connections with his family business in order to become a full-time public servant."[10] The Sarabhai family was, says Amrita Shah, "thrown into a tumult by the event." Vikram had to sever all his business ties, but older brother Gautam, who had also completed his degree at Cambridge, had to carry out new commercial activities he had not personally planned; no one really seemed to approve of this appointment, not his wife nor his children.[11] Some insiders, including members of the Tata Trusts, knew that Bhabha had already picked Sarabhai as his successor and had spoken about him favorably to others. That was why Bhabha had Sarabhai appointed to the AEC. While the search for Bhabha's successor was genuine and the invitation to Chandrasekhar was genuine, Bhabha's wish concerning Sarabhai was satisfied in the end.[12] So he began his official relationship with some insiders on the premise that many knew of Bhabha's preference for him, and they were bound to be disappointed. But that did not remove tension from the situation, as we shall soon see. For example, engineer Homi Sethna thought he had been tipped for the top AEC position and is said to have produced a letter from Bhabha saying he would be in charge in Bhabha's absence, but when he failed to get it, he maintained his contest with the new chairman, "a bitterly adversarial position."[13]

The Sarabhais had eight children and educated all of them privately, using a staff of thirteen teachers, three of whom had PhD's from European universities.[14] Following the Montessori methodology, Vikram, born in 1919, was well trained in mathematics and physics and other sciences: he had access to a small private workshop and chemistry and physics laboratory. Through the Ahmedabad political network, he knew (or knew about) most of the people with political power in India. Mahatma Gandhi stayed in the spacious family compound when convalescing, and Motilal Nehru and his son Jawaharlal, Rabindranath Tagore, and Maulana Azad all stayed with the Sarabhais in one of their houses on the river. Tagore wrote a letter to Cambridge in 1935 recommending Vikram's admission to the university. "I know him personally and his people. He comes from a wealthy and cultured family in the Bombay Presidency, and he has a brother and sister studying at Oxford at the moment."[15] Sarabhai was admitted to St. John's College and completed an undergraduate degree in natural sciences at the University of Cambridge in 1939; he was to have continued his graduate studies in physics but was stopped by the war. At his father's insistence he returned to India in 1940 but waited four months into the war so he could write the exams for his bachelor's degree; he even negotiated an agreement that he could work toward his PhD if he studied with C. V. Raman. So he was sent to Bangalore to the IISc to study cosmic rays with Raman, where he met Homi Bhabha in 1940; both were exiled from Cambridge by the war, and this relationship proved to be very important for the development of scientific institutions in India.

As soon as the war was over, Sarabhai returned to Cambridge for two years with his wife and completed a doctoral thesis, in which he states that the experiments involving the dimensions of latitude and diurnal variation were undertaken "to supplement Millikan's results from high altitude cosmic ray experiments in US military planes based at Bangalore during the war."[16] Bhabha and Sarabhai had both participated in that work and expanded it to use balloons. Robert Millikan became head of the California Institute of Technology in 1922 and was awarded the Nobel Prize a year later for determining the charge of the electron, before specializing in cosmic ray studies. Sarabhai's dissertation research was planned with the assistance of meteorologists at Poona and conducted not from bombers on training runs but in the mountains at an altitude of 4,200 meters in Kashmir in 1943: "I evolved the method of shower anti-coincidences . . . to measure for the first time the intensity of the slow mesons with Geiger counter apparatus. . . . Before extensive observations were accumulated I had to leave Poona to return to Cambridge on the termination of the war."[17] There was great interest in the backward flux and disintegration time of slow mesons in cosmic radiation, particularly near the geomagnetic equator, thus privileging research in India. Sarabhai took his new wife Mrinalini to the mountains, mobilized a large group of researchers and assistants, and even called on the governor's help with logistics.

But these measurements were plagued by coincidence in which data were counted more than once, which is what Sarabhai worked on for his thesis at Cambridge, arriving in 1945. Because the war had changed the composition of the physics department so greatly, there was no one present at Cambridge to examine Sarabhai's thesis in cosmic ray physics. So Patrick Blackett was asked to be the examiner, and Sarabhai, at twenty-eight, went to Manchester, in early 1947, for the examination, the year that the pi-meson was discovered. According to Blackett's children, they got on well, because Sarabhai became known in the Blackett household as "my father's student."[18] Little could Sarabhai know that Blackett was then being considered for the 1948 Nobel Prize, based on measurement of cosmic radiation in cloud chambers he built, the experimental identification of a new particle, the "positive electron," and the confirmation of electron pair creation.

Before he was awarded his doctorate, Sarabhai had taken part in a flurry of experiments on slow mesons concentrated in Bangalore and published in the *Physical Review* from 1944 onward, of which Sarabhai's was the earliest. Researchers in the USSR Academy of Sciences immediately leaped on the anomaly in rates of disintegration detected by Sarabhai and his group (referring to Bhabha's formula), and a theoretical physicist in Argentina referred soon to that same research at the IISc in Bangalore in discussion of "new representations of types of nuclear forces."[19] Just as Sarabhai was defending his dissertation before Blackett, John Wheeler at Princeton was writing a paper, also for the 1947 *Physical Review,* on the mechanism of capture of the slow mesons. The 1949 Nobel Prize was awarded to Hideki Yukawa for his work fourteen years earlier on mesons. Sarabhai was on a wave in physics research, given his good timing.

The teacher and student met again in India when Blackett went to Ahmedabad in 1958; he dined at the Sarabhai house and toured the Physical Research Laboratory that Sarabhai was building. Blackett recognized that Vikram Sarabhai, who was then thirty-nine, would be an important force in science in India. After the dinner, Sarabhai wrote to Blackett to discuss their interest in anomalies in the divergence of the actual terrestrial magnetic fields from an ideal single dipole field, something that Blackett was very excited about that year. Sarabhai was beginning to direct his gaze to India's role in space and geophysics, the latter being Blackett's current fascination. A further dinner in London was proposed, to talk about how to obtain state support for scientific research.[20] Blackett helped Sarabhai with the organization of the 1963 Pugwash Conference in nearby Jaipur.

On his return to India from Cambridge in 1948, Sarabhai established two rather different institutions. One, the Physical Research Laboratory (PRL), was to be the home of his professional life as a physicist. To help him build the PRL he appointed K. R. Ramanathan as director. Ramanathan was the physicist-meteorologist at Poona named in his Cambridge dissertation as having been essential to the cosmic ray experiments in Kashmir. In the 1950s, his PRL received funding from the government of Bombay State, the CSIR, the Ahmedabad Educational Society, and the Sarabhai Group. This was rather similar to the way Bhabha first financed TIFR.

The other institution, the Ahmedabad Textile Industry Research Association (ATIRA), articulated his relationship to his father's industrial community and provided the start for his career in business. To help build ATIRA he found an alliance with Kasturbhai Lalbhai, a powerful mill owner who spotted in Vikram a young person who could and would use his high social status and new educational credentials to make significant changes. The mill owners were looking for someone effective but probably had no idea about how to build an innovative textile research institution. "During my long career," said Lalbhai in 1974, "I have not come across another person of such great vision."[21] Lalbhai was well connected through the Congress Party leadership to national committees for CSIR, and he controlled the large Ahmedabad Educational Trust, which funded both PRL and ATIRA.

ATIRA was operated as a cooperative among mill owners, raising money for research from them. Sarabhai worked as ATIRA's honorary director for the first few years. Sarabhai persuaded S. S. Bhatnagar to accept an appointment on the Board of Directors of the PRL first and then persuaded Bhatnagar, K. S. Krishnan, and Kasturbhai Lalbhai to accept appointments to the board of ATIRA. Lalbhai and Sarabhai made sure Nehru saw ATIRA early and frequently as it grew, and Nehru advised others to see ATIRA. An industrial psychology division was planned, and Kamla Chowdhry, a young Harvard-trained woman from the Punjab, was interviewed and appointed, thus starting a famous and fateful relationship with Sarabhai and his family but one which provided Sarabhai with astute advice about institutional development.

Starting with these two public ventures, Sarabhai then bought or developed a series of joint ventures from 1950, when he took over the management of Sarabhai Chemicals; these were Suhrid Geigy, Sarabhai Merck, Synbiotics, Sarabhai Engineering, and Sarabhai Glass in Baroda, where he traveled each week by train, Swastik Oil Mills in Bombay, and Standard Pharmaceuticals in Calcutta. Nearby in Baroda's new industrial area was Kasturbhai Lalbhai and his company. To these new Sarabhai units were later added the Operations Research Group (thinking of Blackett's influence) and the Sarabhai Research Centre in Baroda. The long delay in confirming his appointment to the AEC and DAE in June 1966 was due to the time required by the government to perform a background check on Sarabhai's private holdings and involvements. He had already had a lot of interaction with government departments in his effort to get permission for and build joint ventures with international companies, including an unsuccessful antibiotic project.

While building the Sarabhai Group, he became involved in the politics, design, location, and curriculum of the new Indian Institute of Management (IIM) at Ahmedabad, in the mid-1950s. It was located there because he objected to the Ford Foundation's plan for only two such institutes in India, one in Calcutta and the other in Bombay. In this maneuver he skillfully played all the Gujerat cards available in Indian politics, changed the minds of planners at the Ford Foundation, and twinned the new planned

Ahmedabad institute with the Harvard Business School. In this effort he collaborated with his close friend Kamla Chowdhry, now at ATIRA but soon to move to the Institute of Management. She showed him how to articulate his ideas about institutions in the current terminology of management studies and to link his projects together in a more coherent whole. After the failed attempt to become the vice-chancellor of Gujerat University, he was, briefly, the director of the IIM. Going beyond his Cambridge connections, he made the essential step, along with Indian politicians and planners generally, to build an American network of advisors during the 1960s, mixing business and science. A glance at his address book and diary for this period shows the extensive connections he had with such American universities as MIT, Harvard, and Chicago, with US and European multinational pharmaceutical and chemical industries, and with major US institutions like NASA, the National Science Foundation, and the Ford Foundation.[22]

So great was his support for Kamla Chowdhry that he proposed her to be his successor as director of IIM in 1963. Family friend Prakash Tandon was chair of the board, but the board resisted this move because the Harvard partners would not support Chowdhry as director. Prakash Tandon said Sarabhai told him he thought he should resign in order to demonstrate his recognition of Kamla's abilities but also felt unwanted by his own board (something he felt he had built). His frustration grew into "a total and utter lack of ability to confide," according to Tandon, and Sarabhai chose to "withdraw."[23] Chowdhry never became the director of the IIM but remained involved for many years.

Sarabhai's own intellectual path as a physicist led him, in the early 1960s, to focus on work in two fields—geophysics and space research. Around 1960 Sarabhai, and members of his board presumably, succeeded in persuading Bhabha to fund PRL using DAE funds. This move enlarged the pattern that began with Saha's institute in 1955, a pattern by which the DAE gradually became the main source of funds for physics research and development in the country. In 1962 Sarabhai was made chairman of the Indian National Committee for Space Research, and a year later he established the rocket launching station at Thumba, just outside Trivandrum in Kerala—all funded by Homi Bhabha at the DAE. He published his research and supported others in studies of cosmic rays, was a frequent participant in cosmic ray conferences at TIFR, and pushed for Indian development of rocket building, launching, and guiding capabilities. In 1963, Bhabha gave him extra recognition by an appointment to the Electronics Commission and then in late 1965 by an appointment to the AEC itself. He was now fully "in the big picture" at the age of forty-four and was spoken of as Bhabha's successor. When he wrote in 1966 to Prime Minister Gandhi about the top atomic energy position being offered to him, he explained that he already had "a regular association with the Laboratory of Nuclear Science of MIT (Senior Visiting Scientist). This involves collaboration in scientific research pertaining to cosmic rays, solar activity and physics of the interplanetary space, particularly with Professor Rossi."[24] At MIT he talked regularly with presidential science advisor Jerome Weisner and frequent State Department advisor on India Myron Weiner. His working connection with two of the most prestigious institutions in American science, MIT and Harvard, neutralized any doubts about him outside the Indian scientific community.

It was in 1965, just before Bhabha died, that the Space Science and Technology Centre was opened in Trivandrum. Being Sarabhai's favorite project, the stage was now set for the continuous expansion of "space" within the "atomic energy" family of organizations. Sarabhai's influence with Indira Gandhi, his readiness to promote space just when it was nationally and internationally desirable to do so, his skill at explaining and justifying domestic applications of space research long before they were actually realized (forecasting monsoons for agriculture, satellite broadcasting for educational television in villages, and the like), his approval from Homi Bhabha, his wide American network of scientists and industrialists, including those with political influence in Washington, and his accomplishments in organization-building inside and outside Ahmedabad—all this contributed to his successful appropriation of the positions left vacant by Bhabha. His calmness and modesty, regular use of North Indian clothes (unlike Bhabha, who almost never wore them), vegetarianism, and national "reach" (starting with his marriage to a successful South Indian dancer and artist, and his building of a research organization in South India)—all made him a very different kind of person to take over from Bhabha. Moreover, Sarabhai worked about eighteen hours a day, normally with five hours sleep; his employees and colleagues joined him on trains and planes, or in airport waiting rooms, traveling the country and the world—relentlessly pushing his objectives, day and night. Abdul Kalam records being asked to an early morning meeting in 1968 with Sarabhai on rocket-assisted takeoff for military aircraft at 3:00 a.m., and a site visit at 4:30 a.m.[25] On another occasion, his last night it turned out, he signed papers even after having started to go to bed, approving for a late visitor a ten-ton liquid oxygen plant for a planned semicryogenic rocket engine.[26]

In Bhabha, Nehru found a set of familiar assumptions about governing and living. As for Indira Gandhi and Sarabhai, it is usually said that this was a case where Nehru's daughter felt a charm for a scientist, and he certainly knew how to charm her, occasionally handing her roses on arrival to see his projects. She recognized in Sarabhai someone who, like herself, grew up in a "fishbowl" atmosphere of economic and political success and social obligation. Indira Gandhi said soon after his death, "I am a person who is not often down-hearted, but I can tell you that sometimes I was a little discouraged. Vikram would come and lift up my spirit and say there was no reason to be a pessimist."[27] Gandhi sometimes had to confront Sarabhai over specific practices (e.g., his launch of the Ten Year Profile in 1970) or the troubled relationship between space and atomic research in his huge portfolios. The prime minister was not alone in Sarabhai's beneficial effect on her spirit, according to conversations with his colleagues—he did communicate a sense of optimism. But for a contrary view, see the observations of Bharat Karnad derived from an interview with an anonymous theoretical physicist on Sarabhai's staff at PRL; Karnad says Sarabhai was abstemious and excessively frugal, in part to compensate for his great wealth, and "all of these negatives ended up alienating and demoralizing the Trombay community."[28]

Sarabhai as Leader among Menon, Ramanna, Sethna, and Dhawan

Sarabhai had now to deal with a group of scientists, slightly younger than he was, in a country where minor age differences invoked strong protocol and political distinctions. Minor differences in seniority, both in chronological age and time of appointment to a job, had very strong effects on scientific institutions, in policy and in practice. The youth of this generation of scientific leaders in 1966, like the youth of the prime minister, was consistent with the youth of Indian star scientists in the 1920s and 1930s like Saha, Bhatnagar, Bose, and Krishnan—though none of this new group were Fellows of the Royal Society. There was the new thirty-eight-year-old director of Bhabha's institute, M. G. K. Menon, a cosmic ray physicist trained in Bristol. Menon's relations with American scientists and institutions were not as developed as Sarabhai's, but they were developing. It was not long before inquiries were made with Blackett about an FRS for Menon, as we shall see. There was Raja Ramanna, nuclear physicist now aged forty-one, trained in London in the late 1940s and made director of the large Physics Group at BARC. Though not a visitor to the United States until 1969, Ramanna had often been to Canada and Europe since 1955. And there was Homi Sethna, not a scientist but an engineer trained at the University of Michigan, who worked on major DAE projects from the early 1950s like the Indian Rare Earths' monazite processing plant in Kerala and the plutonium reprocessing plant at Trombay, which had just opened. Sethna, age forty-one, had frequent international exposure from the early 1950s. Satish Dhawan, Caltech-trained aeronautical engineer, was the new director of the IISc in Bangalore in 1966 and very interested in space. Now age forty-six, the same age as Sarabhai, Dhawan began to work closely with Sarabhai and "inherited" a new position as leader of the space program, created just at the time of Sarabhai's death. In the end Sethna got one of Bhabha's key jobs and became director of the newly named Bhabha Atomic Research Centre at Trombay, reporting to the chair of the AEC Vikram Sarabhai. The choice of an engineer at BARC was symbolically wounding for most physicists, including, most poignantly, Ramanna. This was considered by them to be quintessentially a physicist's job because it always had been, but in fact much of BARC's work had become a complex blend of engineering, physics, metallurgy, and chemistry.[29] Seen from the outside, in media and politics, these were all scientists following a scientific method, but seen from the inside these were two epistemic communities, preferring different thinking about problems, different kinds of understanding. Not for the first time in India, two groups were in a struggle over "who or what kind of person would prevail?" This was not simply about Ramanna's Brahmin origins in Mysore and Sethna's Parsi origins in Bombay. Though such regional and social status considerations continued to influence perceptions and attitudes throughout the scientific community, the professional training and the epistemic presumptions that go with it were perceived to be just as important.

Sarabhai thus began a new life in June 1966 as chairman of the AEC, secretary of the DAE, and chairman of the Electronics Committee. He had to make appointments to boards and committees among these people, keeping them in his sight, tapping their capabilities, and making use of their networks. This situation was entirely new for all of them and included a new prime minister whose control of her own party was unsteady in 1966–67. So Sarabhai found himself spending a lot of time and effort to find effective working alignments between these other members of the scientific elite and the SACC to which he now belonged. What he also discovered, to his regret he told me, was the amount of time that had to be spent intervening in the boards and committees that governed research institutes and laboratories funded by DAE: he disliked doing this kind of intervention, he said, and wanted to do it, if at all, indirectly. After all, he said, "I am the director of my own lab too; I know how it feels."[30] But he realized that the loose ways in which most institutes governed themselves was problematic and said that it was certainly necessary to improve working conditions in them, including simple matters of their daily life. Ministers at the state and national level liked to associate themselves with smart modern scientific laboratories and liked to make contributions to their work quite publicly, so there was a continuing politics around scientific life.

During this intense and final five years of his life, Sarabhai had three other major professional questions to face: Would he tolerate, if he did not support, development at Trombay of a test bomb? Would the power reactors so long in the making finally produce electrical power? And would India be ready to launch a satellite ("his" satellite) by the mid-1970s? Big though these questions were, they were set in a context of larger questions concerning India's food self-sufficiency, reducing poverty and internal instability, earning enough foreign exchange to establish a balance of payments, avoiding debilitating debt, and preventing armed conflict. The power reactors, a satellite, or the bomb would have little influence on those larger questions. But they were most important to the scientific, technical, and industrial communities, and these upwardly mobile communities had developed an intense politics of their own. This new elite were the guiding stars, not only on the laboratory floor but in the cabinet, as we shall see in the following chapters.

SIXTEEN

A Day in the Life of Two Research Institutes in Bombay and Calcutta

The scientific leaders at the top in the DAE were all working in institutions that were no longer brand-new. Their founders were dead, routines were established, and budgeting was a protracted and difficult affair. During the late sixties these leaders saw the arrival of younger foreign-trained colleagues, and in many cases the buildings were simply full. There were also well-trained colleagues who had not been abroad. Few were working "alone in the wilderness" anymore. There was little room to expand, unless there was dramatic new funding for buildings, salaries, and equipment. The top research leaders were often out of town, moving from committee to conference or project to project. Though they worked hard to maintain a working relationship with their students or groups, this was becoming increasingly difficult. These are the working conditions that had exercised the scientists on SACC through the 1950s and 1960s.

Senior scientists loved to talk with me (1967–70) about conditions in their laboratories and institutes and were generally very concerned about how to make changes and improvements in a completely uncertain environment, without friction. What follows is a thick description of daily life in two quite different institutes funded by the DAE, neither of which had much *directly* to do with nuclear power or weapons. Seen as useful pools of talent, both institutes supplied important leadership to most of the DAE-funded programs and projects. The chapter begins with the Saha Institute in Calcutta and ends with the Tata Institute in Bombay. Their surrounding cities evoked very strong loyalties among scientists, even those who migrated there from communities far away—all arriving people became "indigenous," eventually, in these cities of such mass and such strong gravitational pull. This happened no less to scientists than to their neighbors; the search for good schooling for children, a useful role for educated spouses, and affordable shopping and entertainment united them.

A Day in the Life of the Saha Institute in 1968

The Saha Institute buildings could scarcely be distinguished from the other departments of Science College that surrounded them. Like the applied physics and radio physics institutes and departments founded by S. K. Mitra, the Saha Institute was discolored by the monsoons. One approached it on a dirt road, dodging puddles if it rained, past a small flower garden that bloomed very gaily in the winter. People arrived at the institute any time up until noon. In Bombay I was frequently asked, "What are your timings?" reflecting both Bombay's preoccupation with time and the lack of familiarity among ordinary people in that city with the flexibility and irregularity of academic habits. But in Calcutta it was assumed someone who did research would not keep regular hours, and the question about time seldom arose. Although nonscientific staff members were expected to sign the Saha Institute's attendance book before 10:30 a.m., or face questioning (the book was removed after 10:30 to the director's office), scientists arrived when they needed to. Most people arrived after an ordeal in a lurching bus or packed jerking tram. They required a few minutes to shed the numbness that comes from traveling in such a dense crowd. Some of them, including men, had already done the household shopping in the early morning, had a walk, read the paper, and completed some professional work. Attendance at the institute was not required on a hundred days per year, those being the days when the institute was closed officially. This included approximately twenty-three national or Bengali holidays, Sundays, and half the Saturdays of the year. Sometimes, however, a general strike (*bandh* or *hartal*) had effectively closed it because there was no public transportation, and most of the huge city was shut down. So in practice, in the late 1960s, the institute would be officially closed more than a hundred days a year. On those days, one could do research by special arrangement with the guards, but it was not always easy. That clearly limited the possibilities, particularly for experimenters.

Meghnad Saha's statue looked down upon the staircase from the upper floor where the director's office was. At the wide entrance to the building, two or three people were usually chatting around the telephone switchboard, under a sign that forbids chatting. Some lounged at the door talking with the mustached guard (*durwan*) dressed in khaki. (Although the building was protected, the compound around it was not.) A few bearers in shabby off-white clothes carried open cups of tea up to professors. Some of the institute's corridors were dimly lit and lined with dusty bookcases largely unopened, but there for storage. Floors were sometimes dirty, because of the ubiquitous monsoon mud or Science College dust, raising for experimenters the problem of dust-free labs. Nevertheless, there was a valiant effort to keep it clean. As in any giant industrialized city, the air of Calcutta was dense with industrial waste, and it settled down everywhere. Doors and walls were marked by a whole generation of hands. For an institute of this size there were few toilets; they were small (unlike the Tata Institute), and all in the Indian style. A few rooms were air-conditioned and thus had large flush doors that closed tight; otherwise junior scientists glided in through half a door into their hot crowded rooms. A few senior people sat in their large offices behind vast wooden desks (like those that characterize tycoons in the cartoons of the *New Yorker*). In those rooms, the professors' typists worked in a little cubicle near the window, sectioned off behind a wooden screen. Where only desks were needed for work, there may have been more than four occupants in a room sitting as close under the fan as physically possible. In labs, however, there were mountains of equipment, and stools were scattered around.

Distractions were inevitable. One large room had eight students preparing their theses, without any partitions. Most of the faculty had a separate room or shared one with a colleague; some had desks in air-conditioned areas intended for the preservation of apparatus. Writing and reading were generally done in these rooms, not the library. Only very senior members had a telephone extension, so messages were relayed through a bearer coming with a piece of paper from the switchboard or through someone else's nearby phone. This was taken lightly; unlike TIFR, there was no telephone directory in the Saha Institute. Owning a home phone in Calcutta was rare, for there were 110,000 phones in a population of 8 million people in 1968, and an enormous waiting list of about 90,000. The building was not designed to let in sunlight. If the weather was nice in the winter the shutters were flung open, but in the monsoon, with the shutters closed to keep the wind and rain out, glaring and humming florescent lights had to be turned on, their frequencies establishing a new sort of electro-acoustic music. Sounds and loud conversation reverberated from the hallways, and the fan whined or groaned.

The library was cramped and had only a dozen desks for reading and writing. Unlike other overguarded libraries in Calcutta, access to the institute's stacks was free and easy, and the important journals were well displayed in a fan-blown area. The books, however, were kept in an ill-lit, dusty, and disorderly manner. Expensive sets of volumes, like a new four-volume history of organic chemistry, were breaking apart from improper shelving. New books were packed into small display cabinets. Only younger faculty or students could be seen browsing through the journals or general interest items like *Minerva, Soviet Life, Time, Bulletin of the Atomic Scientists, Scientific American,* or *Discovery.*

Younger students sometimes read the Bengali periodicals on a shelf past the industrial magazines. Though journals and books were often "late," it was agreed that this was the best physics library in Calcutta, even though its members used it infrequently. To envious outsiders, the collection of current journals and crisp new books was worth a fortune. Through the library windows one looked out into the upper-story bedroom in the equivalent of a Satyajit Ray film set, in a house that leaned more and more crazily against the institute. One's eye could swing, in a glance, from the bright cover picture of the latest *Nature* to the old gentleman snoring in his ancient four-poster bed next door, only a few meters away.

In this way the Saha Institute was completely open to the world of city sights and sound, unlike TIFR, where everything was secluded and hushed except for the quarrels of birds in the garden or the crash of the sea rolling against the concrete wall. On two sides, SINP was bounded by the walls of old houses in an old neighborhood crossed by narrow lanes. Out of these houses came the sounds of the whoops of family life or mourning of death. The metallic clunk of the bell of the thin *rickshawallah* marked his disappearance down the road that joined SINP through Science College to the city. Sometimes one could hear the hammers of the cobbler, squatting with his back against the throbbing machine-shop wall, mending a scientist's pretty slippers while she attended a seminar.

Research groups did not have their own seminar rooms, nor was there any single room for this purpose. Consequently, there were few of these kinds of scientific discussion. This was curious in a place built for scientists, whose work proceeds by conversation. There was simply one large lecture hall where classes were held and where guests gave lectures. This theater was socially divided, like a Calcutta tram, with one soft bench in the front row, and hard wooden seats stretching up and back for twenty-five rows. Senior professors sat on the front row, and others sat behind. This was the scene for all meetings and ceremonies. It was a very official space, formal and unyielding. The very structure of this one lecture room created and maintained distinctions between speakers, leaders, and passive lower-status listeners. There was almost nowhere, except in a person's own room, where a scientist could be "off stage," on neutral ground without every move being determined by the status differentials between the actors involved.

Bengali was used for a great deal of the interaction in SINP. Even non-Bengalis used the words of courtesy and adopted the naming system that has, like the verb endings of that language, a hierarchical deference built into it. Older or senior persons would be referred to either as elder brother (e.g., Meghnad-da) or gentleman (e.g., Meghnad-babu). Even in speaking English the Bengali naming system (Meghnad-da) was retained. Subtle distinctions in age, like a difference of three years, or in rank could be expressed through Bengali, and it is not a leveling language, as Bengali speakers of English say that the English language is. But there is a leveling term to describe a person who exceeds the conventions of hierarchy, or thinks herself or himself more important than she or he really is, and that is *bahaduri.* Institute members were alert to little illustrations of *bahaduri* and found ways to mention them when the time was right.

Typically, members of the institute ate a large brunch before arriving and subsisted on tea and snacks during the day. Without a canteen or cafeteria there was nothing else to do. Work was expected to proceed throughout the day, and classes to be attended by students. Some people had teaching responsibilities, in SINP or in Science College, and others augmented their income outside the institute by taking private tutorials, as Meghnad Saha did before 1920. There were no regular seminars or colloquia, though an occasional guest lecture at 4:00 p.m. always attracted a crowd. Some scientists remained to work in the evening, "until the traffic rush is over," as one optimist put it; frankly, there was no real pause in crowding on buses or trams or electric trains until 9:00 p.m. But the evening was cooler and quieter to work in, and most people gave that reason for remaining after all the support personnel had left. Unless there was a good beam up on the cyclotron or an important meeting, senior people tended to leave about the same time as administrative personnel, well before those who were struggling to finish theses or publish papers. In the eyes of younger people, these senior people had largely become administrators. Only one or two of these senior scientists were ever seen in conversation among younger scientists or were described by them as models of exemplary leadership. In their defense senior scientists stated that as heads of groups, they had to give up almost all their own research work so others would have a well-run group to work in. The point of this is to show that the rank system of SINP was based on a great number of distinctions among a relatively small group of persons.

The only occasion when I saw social interaction between people at the high and low ends of the institute's social ranking system was the celebration of Visvakarma Puja in September 1968. As in most industries and workshops across Bengal that day, this annual ceremony was held in the institute workshop amongst the lathes and benches of machinists. Leaning against a drill press that towered over his head, with his head occasionally bumping a "Danger—440 Volts" sign dangling above him, the priest (*purohit*) squatted and poured purified butter (*ghee*) on the small fire of wood and leaves in front of him. He had been coming to do this puja here for the past seven years. Said one of the lathe-turners with a sweep of his hand, "We are worshipping the engineer of the world today, the world architect. The one that makes all this happen." The workers who operated the machines every other day rubbed their eyes from the stinging smoke of the leaves and oil. The usual grease had been swept up and Sanskrit verses (*slokas*) were whitewashed onto the dark oily floor, and on top of the verses were the little mounds of coconut, sweets, and green plantains lying on wide leaves neatly placed around the floor. The *purohit* muttered and chanted something inaudible even to the lathe and drill machinists who squatted nearby in a ring opposite the fire. A workshop apprentice, himself a Brahmin, assisted the priest by fanning the pot of burning incense and by throwing flowers and petals onto the machines standing above him and onto the pipe-wrench, hammer, calipers, and rulers that lay arranged in a row beside the fire. These tools of work were being blessed, renewed for another year. The atmosphere in the high-ceiling workshop was light and easy.

The life of the laboratory continued; it was not one of the institute's many closed holidays. The director had sanctioned institute money for the priest's costs and tasty sweets that were distributed around the offices and laboratories by the workers. Other low-paid staff came and took their places, like me, on the benches by the workshop door. Despite a general invitation for everyone to come at 11:00 a.m., the action in the workshop started earlier, at 10:00 a.m. Some workers brought their children and sat on their work benches in the background, trying to restrain the playful ones. Meanwhile the clerk of the workshop enabled another clerk to come in and obtain some data from one of his files, then returned to his place beside me on the bench. The whole gathering moved in and out easily, talking happily.

Suddenly the director, registrar, and a few senior professors entered as a procession, without warning. Benches were cleared for them to sit down, amidst the machines. The director exchanged greetings with familiar workers, while the priest and his assistant continued the puja, unaffected. Climax soon came—those workers personally engaged nearest the fire knelt their foreheads, eyes closed in devotion, right down to the oily floor. Gaunt-faced and burning-eyed from the smoke, the priest had now stood up to sprinkle "water from the Ganges River" onto all of us. There was giggling among the gathering about the coming of the rains. Everyone relaxed. I turned to see that the director and his party had just as suddenly gone, having stayed about six minutes. The mounds of fruit and sweets were now distributed among workers present, including me. Flower petals were scattered everywhere. Stacks of little string-tied boxes from a caterer were now presented, and amidst considerable confusion, people were assigned to distribute them to the three hundred members of the institute who were not present. Said the foreman: "Yes, I suppose we do trust the priest not to make a mistake. None of us knows the correct way anyway. If this puja were not done, I don't think anything bad would happen—we just like to do it. If our shop was not so crowded, maybe more visitors would come. Look how small it is, even to work in, it's dangerous. The director is almost expected to come for a little while, but the others—we don't meet them any other time."

Only a few scientists made an appearance at the puja, their indifference to it outweighing their wish to demonstrate solidarity with the shop workers. This was months before the confrontations with the institute administration (described below), confrontations that suddenly gave some scientists a sense of solidarity with institute workers. People remained in their place, probably as a result of two forces: one was that so many rank and salary distinctions between people encourage them to stay separated, and the other was that in the world outside the institute's wall it would not be expected that a puja put on by and for a small group of machinists would have been fully attended by professionals, scientists, office workers, or students. The symbolic presence of the director and his party, and the extreme brevity of the visit, suggested to me something about the quality of relations between those who governed and all the others in SINP, especially those who were not scientists. Of course this was a little different among experimenters who depended on the machine shop for their work and reputations.

I think that confrontations that occurred later in SINP were exacerbated by the very restricted communication and limited social relations between various groups and ranks in that institute. This restriction and limitation was not the only factor but made the quality of the confrontations more intense. When it moved years later to the expansive compound at Salt Lake, the institute came under the direction of people who tried to find solutions to social problems that plagued it. But the marked contrast of conditions at SINP with conditions at TIFR in Bombay played a significant role in the appraisal by scientists of where they would prefer to work if they had a choice. Everyone observed and weighed these differences in their own way, but they were part of the individual calculations nevertheless.

A Day in the Life of the Tata Institute in 1967

Most scientists had to rise before 7:00 a.m. in order to complete the journey through Bombay to arrive at TIFR on time for work, about 8:30 a.m. At least in the early morning, the city's meager water supply was running, and they could bathe while wives prepared a small breakfast. A few couples went out to work together. Because most scientists with a family could not afford to live near the institute in Colaba in South Bombay, they stayed where housing was available, often in DAE-owned buildings, in distant neighborhoods to the north where rent and ordinary food prices were less expensive. But from those neighborhoods in the north of the city, it took an hour and a half hours to reach the institute. The 140 respondents to the TIFR questionnaire spent between two and three hours traveling on the world's most crowded buses and electric trains each day. Even in first-class railway compartments there was rarely a seat; reading or conversation was impossible, and everyone hung on in grim determination simply to reach the Churchgate or Victoria stations. The institute had wisely provided buses from there. Some senior faculty who lived closer (perhaps on Peddar Road, the original site of the institute) were part of a carpool. Young bachelors living in the institute hostel in the city center caught a special bus at 8:30 a.m. that brought them to breakfast in the West Canteen.

The institute is a very distinct modernist glass and steel structure, designed by Helmut Bartsch, of Chicago. It stood in a fine spacious garden, cared for by thirty-five gardeners in 1968, each earning Rs 85 monthly. The approach roads were paved, and the grounds beautifully landscaped. The institute had a slightly paramilitary look because there were uniforms for gardeners; workshop, canteen, and transport people also wore their own uniforms of brown, gray, or blue. The building and grounds were supervised by an ex-commander of the Indian Navy, and a large naval base surrounded TIFR. The institute was kept, with rare lapses, hospital-clean by constant polishing and washing of the vast glass areas and marble floors. Many of the employees were ex-military personnel, and the institute was being guarded under the Defense of India Act, invoked first at TIFR during the conflict with Pakistan in 1965. No one was allowed to lounge or loiter around the front gate on the approach road where important visitors and senior staff were likely to come. The whole atmosphere was antithetical to lounging and loitering. Buses delivered institute staff at the rear of the buildings. There were few reminders of the outside world—only a glimpse could be had of the skyscrapers of Bombay, seen over the roofs of the shantytown, huts without water or latrines thrown up by poor fishermen and construction workers on the scarce high shore next door to the institute. The law and politics of land expressed themselves harshly at this tranquil spot.

By 10:00 a.m. everyone was expected to be at their desks or in their labs working, though many had been there, working, earlier. The library and workshop shifts came at 7:45 a.m., and some scientists, seniors driving and juniors by bus, followed soon after. Unlike most of the staff, scientists did not have to sign an attendance book or pick up their keys from the security officer in the cool lobby. Some scientists went immediately upstairs to their offices, or they were caught by a group of friends to go and drink an early coffee, gaze out over the Arabian Sea, and draft a reply to a new paper for the *Physical Review Letters*. There were certainly those senior scientists who were wringing their hands about the time spent by others in the canteen, but it was often an essential part of the work. Nobody could deny, however, that the canteen was also used as a place to evade work. From 10:30 there was steady work in the laboratories, in the library, or in offices for two hours. Chance conversations in the corridor between colleagues often ended up at a blackboard in an office. Equipment was built, output was retrieved from the computer, programs were debugged, ideas were tested, and committees met.

By noon there was a line for habitual early diners, many of them mathematicians, waiting before the glass doors to the West Canteen. Others went to the cheaper, more boisterous East Canteen, where only Indian food was served (chappati made of wheat, no rice served because of the 1967–68 shortages). Generally only scientists and administrators went to the West Canteen; staff with lower incomes did not eat there. In both canteens, the lathe-turners and biologists, bus drivers and gardeners, all waited in line for the same cafeteria-style service. Some work groups got time off from work at specific times, and that regulated their movement to the canteen together. Gardeners ate at the same table in the East Canteen; mathematicians ate at the same table in the West Canteen. But strict vegetarians ate happily at the same table as consistent nonvegetarians, though this was not the case everywhere in India at the time. At snack time, friendships were revealed across many boundaries.

Talk at the table ranged widely: politics in the newspapers, physics or mathematics, new projects, prices, experience in Canada or Calcutta. Scientists expected others to take these discussions seriously. Occasionally I managed to get a senior faculty member to eat with junior scientists who did not know each other. Some of them later told me it had been a "new" or "unexpected" experience; generally younger people interacted more closely with faculty members of their own group; there was little regular incentive or mechanism to do otherwise. The question of the role of the institute in the development of science in India was an almost daily topic of debate, thus raising the touchy question of the justification of this new institution, built to house and promote a new class of people.

Not everyone was comfortable about becoming a member of that class. Because it was a neutral space where status differences could be overlooked temporarily, the canteen played an important role. People paid for each other's coffee, establishing a little reciprocal indebtedness within their professional group. One could also meet someone by "chance" when it was too difficult to meet him in his office; on the spot one could ask for a favor and get a reply to an old request. The director was not seen much in the West Canteen: he was often busy and often out of town—but it was possible to see every other senior person in the institute there at least once in a week, and usually more.

After lunch there were strolls in the gardens (except in the soggy monsoon), walking at the edges of the waves, playing cards in the casuarina grove, glancing toward the beautiful women in the biology group, or talking and more talking in the breezy colonnade of pillars that hold up the building. Sometimes there was a classical music concert in the auditorium, produced by a music-loving sound-system expert in the computer group. Records were borrowed from a library of long-playing records operated voluntarily by scientists and technical personnel. Homi Bhabha would have liked that, given his love of records and concerts. After lunch there were more walks to the little post office nearby, often to send part of a salary home to parents, by postal money order. Some of the people went for strong tea brewed in the café within the naval yard, needing that to get through the warm afternoon.

Trainees, technicians, and accountants strolled with their friends, perhaps speaking Konkani or Tamil, providing the relief of slipping into one's own language after working only in English all morning. Among scientists, even if two Bengalis were speaking together, they might not have used more than one or two words in Bangla, particularly not within hearing distance of other people. I noticed sometimes that conversations that began in one language in the elevator usually switched to English as more people got on. Occasionally a young woman belonging to a research group would join the stroll, but mostly the men ate and walked alone. There were occasional alliances formed, but to the North American eye these alliances were very discrete and very subdued. Marriages between people within the institute occurred about once each year.

By two o'clock in the afternoon the garden was quite deserted. A few people remained at lunch, including those who became so involved in their work that they forgot to come earlier. The canteens were closed until tea time at 4:00 p.m. No one sat in the lounge at the library door reading magazines, and the corridors were empty. The atmosphere was like a boys' private school, where students were doing their homework. Many people worked hard in the library; some even left their offices to work there because it presented fewer distractions. It was a spacious room, cool and quiet, with hushed voices and huge carpets. People could browse in the dozens of well-displayed new books. A brand-new edition of G. H. Hardy's *A Mathematician's Apology,* about Ramanujan, had a waiting list of twenty-one readers before its release, while the Bhabha Report on electronics, published a year earlier, had only eight readers in the past six months. It was easy to monitor what was being written in many other fields—from brain structure to cosmology. The journals and books compared favorably with libraries in rich countries. The library encouraged this browsing, and some people spent almost the whole day there. Disenchanted young scientists from BARC in Trombay (many of whom trained at TIFR) came to read at TIFR instead of working in the vast Modular Laboratory building at Trombay, welcoming the excuse that they had "to check up on some literature." Access to the TIFR stacks was easy, and both faculty and students could be found searching for things there or browsing in the racks of journals along the wall.

The afternoon was the time of the weekly seminars, and the BARC people from Trombay liked coming to those too: colleagues reported on new work or on recent papers by a rival group abroad. There was also a weekly Institute Colloquium where invited lectures of general interest were held. Sometimes the room was full, even with nonscientists present. Questions came from both scientific and nonscientific staff. Most seminars ended by about 4:00 p.m., though the Institute Colloquium attracted people until 5:00 p.m. Debates begun in seminars continued all through the evening, over tea in the canteen or back in offices. Many good papers had been written at night at TIFR as a direct result of late afternoon seminar discussions; according to interviews, the number of such papers was a significant fraction of the institute's total output. Other researchers, of course, utilized the seminars as a relief from the boredom or stalemate of other work.

At 5:00 p.m. the institute buses were crowded with people going home, though in principle standing passengers were not allowed: thirty-six drivers, earning between Rs 100 and Rs 200 a month plus benefits, operated in relays between the institute and a terminus equidistant from the two railway stations. Some people got off earlier in Colaba, but most went to the trains, clutching books or flowers from the institute gardens for their families or wives, ready to do battle in the crushing sea of tired hot bodies. If the following day was a holiday (sixteen closed holidays were observed by the institute), families made traditional sweets and complained about the price of milk and rationed sugar. Scientists planned to finish reading a new book on such days or to take their families for a walk on Juhu beach.

By 6:00 p.m. anyone left in the institute was likely to stay through the evening to work, pausing just before the 7:30 dinner in the East Canteen in order to go out and experience the glorious sunset on the sea's edge. From then as many as thirty to forty people worked through the evening, some of them beyond the time of the last institute bus at 11:30 p.m., and even the last city bus at midnight. If they lived in the hostel at the Gateway of India, they were provided special institute transportation over the three miles. Everything was done to encourage hours of productive work.

The idea of one big family at TIFR was stressed in the celebration of Ayuda Puja in October 1967, comparable to the Visvakarma Puja at SINP the following year, just described. It occurred on a Friday afternoon before a weekend holiday, and the institute was in a festive, nonworking mood. The director toured the institute's various sections, accompanied by an entourage of two senior administrators and his personal secretary. With the exception of a few individuals, behavior in the institute was relatively informal. But today it became really informal. Large groups strolled around the buildings and gardens. Signs were up requesting visits to various group areas and labs. The library staff served sweets to all its visitors. Although the director's tour here was "touristic," its informality revealed a conscious effort to convey the idea of the lab as a family having fun. In this, perhaps, lay a basis for potential solutions to gathering problems.

Ayuda Puja consecrates tools and rededicates skill for the coming year. Whether they were students of mathematics or bus drivers, the day was in honor and praise of technique and tools. Flowers and symbols decorated the shiny air-conditioned computer in its new lab. Though TIFR was an institution with high Christian and Muslim involvement, this puja was recognized and enjoyed by everybody. A one-day generous atmosphere was created, producing some light humorous interactions between people normally kept apart in their own spaces. I strolled with an experimenter through the large workshop. Lathes were decked with flowers, and new constructions were displayed with their builders standing proudly beside them. Workshop mechanics had built an impressive shrine (*pandal*) for their deity. A striking likeness of Bhabha's face had been drawn in chalk by a draftsman, on the drafting workshop floor (this was a year after his death). My physicist companion said, "This is very nice—a nice surprise—I haven't been over to this part of the institute for years. Really this is nice."

The opening of the Homi Bhabha Auditorium at TIFR in 1967 was also intended to be an expression of the community of feeling within the institute as well as a means of providing hospitality to a select and influential public. The concerts by India's best classical musicians, like M. S. Subbalaksmi or Bhismillah Khan, were all full. Care was taken to balance the program with South Indian and North Indian music, catering to the large northern and southern constituencies. In general, the families of lower-paid members of the institute seated themselves voluntarily at the rear of the hall, while senior people sat nearer the front. Important guests sat in the two front rows. The Bhabha Auditorium was not designed to seat 1,300 people, so the TIFR family would not be able to sit down together in it. But these gatherings certainly promoted a sense of familiarity between various ranks and a stronger sense of participation in institute life. This was in strong contrast to the absence of that sense in SINP.

But beyond the effort of the director or faculty to sustain a one-family image, what encouraged these interactions? There were numerous regular classes and seminars and frequent visitors bringing outside influences. Lots of neutral space encouraged casual meetings away from offices and labs, so that people of a number of separate ranks were involved each time. Younger scientists did not seem to feel at a disadvantage in the canteens, and the senior scientists did not seem always to need the protection of their desks and secretaries. People of different ranks were accessible to one another, although from other evidence I knew that status and rank were becoming contentious among some scientists in the institute. There were lounges and gardens for a stroll, and elevators, bus queues, and corridors where meetings and conversation promoted familiarity. All scientists interacted in the library, and there were planned occasions such as picnics, concerts, and evening lectures that became the settings for social familiarization. The relative neutrality of English as a means of scientific and social communication encouraged the interaction of widely separated people, and English was usually defined as neutral by these persons of essentially middle-class orientation. This included not only those for whom English was a language since they were children, but also those for whom it was a vehicle to greater aspirations. My experimenter companion on puja day, who depended for his professional career on the workshop, acknowledged that it was remote from him socially and geographically. It was not class alone that was being transcended at the puja: it was also the difference of regional, linguistic, and religious-community origin. Time and again in the history of Bombay those differences have been used to disrupt the fabric of social life, but then those differences had to be integrated again, because the strength of the city lay in its cosmopolitan origins. Somewhere in people's minds the institute's success—and the idea of what a successful institute could do for them—was connected to keeping those differences in perspective and in check.

Launched and Burst: The Night Life of the Tata Institute and High-Altitude Balloons in Hyderabad, 1968

Picture the predicament of an American stunt pilot held by authorities at an abandoned airport somewhere on the west coast of India because the police think he is part of the foreign smuggling gang caught the preceding week. Picture the pilot trying to convince the police that he is really chasing a balloon—at two o'clock in the morning—and the Tata Institute people holding a shouting match with the police over a poor telephone line from Hyderabad 320 km away, trying to explain that there really is a lost balloon! "Balloon" is not a word rural police patrols understand easily as part of their work, and not over the telephone.

That is one of the dozens of stories produced by twenty years of cosmic ray work in India (1948–68). Some stories are hardly printable; others may be products of bright but bored men waiting out a day of unfavorable winds while the temperature on the verandah is 41 degrees. These flights (there were about ten each year) in February and March were the dramatic peak of a year's preparation in the lab, a cycle that had been going on uninterruptedly since 1948. The pattern I describe here was fairly general, though I personally observed only this flight in March 1968. Though these impressions might vary with the memory of other participants, I have the support of my field notes written that day to guide my memory. The world of scientific research is usually a cross-cultural world, and, being young too, I tended to see it through the eyes of the younger participants in Indian laboratories. Had there been a group performance at the Saha Institute comparable to balloon launching, I was told I was as welcome to participate in it as I was at the Tata Institute.[1] A young anthropologist got some acceptance because so many people in the labs were sympathetic to the effort needed to complete a dissertation.

After returning from dinner in Hyderabad city, we set to the evening task of getting the payload ready for tomorrow morning's flight. Some people were carrying on an argument that began in a lusty beer-drinking Punjabi joint we went to, not the alternate neon-green South Indian vegetarian hotel across the street; some of the flight group knew their parents expected them to go to the neon-green place, but, having lived abroad, they had different and more flexible tastes—they liked the beer too. The workshop was littered with wires and packing cases on the floor and crowded with shining clean electronics on the benches. As people worked they glanced at the electric clock on the wall, which now showed just before midnight. The leader of the guest group from Physical Research Laboratory Ahmedabad enviously caressed TIFR's new imported multichannel analyzer, but told one of the Tata Institute professors that his lab discouraged importing such a thing if it can possibly be made in India. His director in Ahmedabad, explained the guest, "is wearing two hats," as he put it, because Sarabhai was the chairman both of the whole of the atomic energy program *and* chairman of the Electronics Committee. So he had to push limitations on imports like these and encourage substitution with Indian instruments.

The bare lightbulb glared on the shiny aluminum gondola waiting for its payload. A student searched for his personal crescent wrench lost amid the equipment. He began repeating echoes of a showdown earlier in the day between the TIFR flight manager and the leader of the guest group about overcrowding in the workshop and the mess on the floor. The flight boss, who had no PhD but was soon promoted to associate professor, asked that the guest group move to an empty warehouse almost two miles away, to help reduce the crowding. The guests, who had come previous years, countered that they had no transport, there was no water there, and the workshop was primitive and without a telephone so they could not coordinate their work with the hosts.

The guests brought their payload and a few tools but seemed to borrow everything else—from oscilloscopes to wiring tape—from TIFR. The student finally found his wrench beneath someone's foot and stopped grumbling. The guests' experiment was now ready to fly piggyback or to hitchhike on the host balloon. Though the source of funds for both labs was the same (DAE), this host-guest ambiguity had generally been their relationship for some years. What was new was that Sarabhai, secretary of the whole DAE, was the (absent) leader of the guest cosmic ray research group, so TIFR people had to be a little bit more receptive, and had to watch their language.

A few sleepy people disappeared at midnight, but the other thirty men still had too much work to do under the glaring lights. The temperature dropped so the old barrack rooms were cool enough to sleep four hours until countdown. The rest of us, lightly clad in pajamas, stayed on. I was invited to paint the gondola with a sign: DO NOT BE AFRAID—THIS IS HARMLESS APPARATUS—REWARD Rs 100. TELEGRAM "BALLOONS HYDERABAD." Everyone had visions of the previous year when the balloon landed in a quarry where playful or frightened workers axed the expensive gondola open to see what it really contained. I had already speculated out loud on whether anyone would know what "harmless apparatus" means. Then someone rewrote the message in Telegu, hoping that the balloon would come down in this linguistic area around Hyderabad. Other balloons have flown as far as Nagpur (almost 500 km) and the balloons at Fort Churchill Manitoba drift as much as 880 km. Both students and technicians chatted easily among themselves, and rarely when they addressed a professor was the word "sir" used, unlike other places in India; at most there was twenty years difference in age between them (the youngest about twenty-four, the eldest in mid-forties). Tea was prepared in the dead of night by the faithful cook brought from Bombay (with his faithful kerosene fridge).

At 4:30 a.m. a final forecast came in from the Meteorological Office at the airport, and senior people made the decision to fly that morning. Even after all this preparation, flights have to be postponed, sometimes two days. Some people loaded tools and emergency equipment onto the truck, and a few chilly ragged coolies, who seem to hang about knowing there was a chance of making a few rupees, helped lift the 100 kg gondola with instruments locked inside onto the truck. All four men with their PhD experiments riding on this balloon looked at one another, now acutely anxious: for one, it was the first time his type of experiment on X-ray sources had ever been flown. Senior researchers had done this many times before (ten times each year approximately; an average of a hundred flights since 1960). But in the end, near 6:00 a.m., as the morning light became just visible, everyone—young and old—looked anxiously at the details of the balloon and its payload.

Rumbling slowly and carefully on the way to the university's football field, the headlights of the truck caught the eyes of stray cats. The flight manager, shouting above the motor's roar, announced his calculations of the precise time for the final decision to fly or not fly (7:30 a.m.), time needed to fill the 90-meter-long balloon with hydrogen, time needed for tracking, and time needed for the actual experiment at a required altitude. Though the stars and moon were fading and dawn approached, we still needed the bright headlights to see leaders of the cosmic ray groups wrapped in scarves against the chill, to see their sure hands darting to carry out the barrage of orders from the flight manager. Slowly the expensive plastic balloon came crackling off the roller on the truck to lie in a gray sheen on a long cloth in the dewy grass.

Now came the morning glow of the sun, the pounding feet of men running, the balloon filling by a tube passed through a tub of hot water so that the hydrogen warmed up slightly. The gray sheen within the plastic was the carbon extruded with the plastic film, carbon that absorbed the solar radiation and retained heat in order to pass through extremely cold (–50 degrees C) layers of the atmosphere. Hot water was constantly replenished by a man riding a bicycle back and forth from the launch spot with a milk can to a little fire under a big pot at the edge of the field. I too was anxious now, but kept busy following the flight boss's instructions to me (my task) while trying to circulate to see as many performances as possible—bodies running, sun rays dancing warmer now, drowsy eyes focusing on the balloon, truck headlights turned off, bodies running again, flight supervisor marching around efficiently, spotting little errors in unseen details, neighborhood urchins appearing first at the field's edge, now creeping in from the sidelines, curiosity bringing them ever closer, before being shooed away.

Why hadn't the payload truck arrived? Decision for lift-off time was 7:30! Our aircraft-clearance window lasted only fifteen minutes—the flight boss telephoned the airport to get it extended half an hour from 7:45 to 8:15; someone muttered, "It's their job anyway." But all else was very deft and efficient, and people seemed good-natured about the hard work. There was joking; things were not just coldly mechanical, not just military style. Now at 7:20 a.m. the payload truck came bouncing across the grassy field. Hands clapping and shouts to clear the field of urchins, but to no effect—a boy raced off on a bicycle somewhere; everybody moved back, of course. I was nervous to think about all that warm hydrogen, the flight manager shouted to clear the field, the urchins scurried away—one minute to launch now—I marveled that such a complicated action was done by unassuming people I had never seen work in a paramilitary fashion.

At 8:18 a.m. clapping was drowned by a cheer; the balloon rose three minutes late and swayed away from the empty truck, on which one man squatted with his knife ready to slash free the gondola should it be caught on anything. The gondola was worth a fortune as was the balloon. I followed it rising high, out of the corner of my eye, as I now rode on the running board of the truck, careening off to the library, the highest point on the Osmania University campus. At 8:25 a.m. we all stormed up the library steps past puzzled students at a summer institute, up to a room on the roof where homemade radar was tracking the disappearing balloon. Soon clapping was heard from the next room: Ahmedabad was beginning to transmit its first sputtering data. A veterinary student climbed gingerly up the ladder to the roof to ask what kind of guinea pig was sent up, quite sincere—bringing tired chuckles and a brief explanation—and, quite disappointed, he disappeared down the ladder again. The balloon vanished into the atmosphere—only a bright sun remained. It should have risen to 30,000 meters where less than 2 percent of the atmosphere is left above it. Now the data were streaming in.

Then, only 72 minutes after launching, the radar lost track of it and a new signal came that the balloon burst at 27,000 meters. Faces were frozen in grim silence, hardly caring to follow the safe descent of the gondola on the auxiliary balloon, down to earth. People went inside to the shade, a bearer appeared with coffee for the professors while students slunk away in disgust. Everyone had come down to earth with a bang.

Hardly anyone talked now until the balloon's landing quarter could be determined—someone suggested the recovery team should leave soon but somehow no one moved. There was some weary talk of the next flights, but both teams were lying on their beds angry and absorbed in the heavy heat.

An hour later the recovery team was chosen, and people were voicing opinions on what went wrong—"no one can say until the data-recorder is seen" was the flight boss's repeated verdict. No disagreement was heard now, and speculation ended abruptly. There was no outright statement that is so common elsewhere in India: "Oh, it's an Indian balloon, what can you expect," but there were hints of that. Again references to the flight recorder. The history of balloon successes and failures had been revealed the night before by some senior people in the eating and drinking in the fancy nightclub downtown. Meanwhile the flight boss, with one postdoctoral fellow and one student, drove off in the truck, 85 miles, to recover the undamaged gondola in a Telegu-speaking village. My sign had been effective; people were guarding it carefully.

Launched and Out of Sight: An Evening at the Thumba Equatorial Rocket Launching Station, 1969

In February 1969 I observed a rocket launch at Thumba; with the planned launch for 5:30 p.m., one hour before sunset, it was still hot at 5:00 when onlookers like me were admitted to the site. I sought the biggest coconut tree to stand behind, looking up carefully to avoid the spot on the sand where a big coconut could drop on my head. Because I am not a "somebody" in the rocket community, I found some shade and obscurity behind the tree. Most of the people watching the launch with me had spent the day working feverishly in hot concrete sheds and workshops. The entourage that arrived at 5:15 around the leader Vasant Gowarikar swelled the crowd. He had been backstage making sure everything was going to work satisfactorily, backed by others to ensure that he would be closely identified with the launch; was this to make it a team success or team failure? Right on time at 5:30 a man on a bicycle arrived with the nose cone, and it was carefully lifted off and fitted delicately onto the Rohini rocket, with occasional shouting to overcome the noise of the wind from the sea.[2] Then the flight boss Gowarikar looked around and whispered something to the man next to him, and the same guy who brought the nose cone on the bicycle was dispatched to chase the small herd of goats away. This was done quickly, in sharp rude Malayali, and everybody laughed nervously. This was, after all, the first fully Indian test rocket. More checks and double checks, Gowarikar and team looked at each other; it was 5:45, and the bright light of ignition and roar of the fuel consumed us. In a few seconds it was gone past the trees and gaining speed, every second. People held their breath, literally, "hoping it would not crash in the sea," as one engineer told me later. Ten seconds more and it was almost out of sight. Clapping and shouts ended the launch, just like the end of a cricket match. Gowarikar wiped his brow, the goats returned to graze as the crowd faded away in the sunset, and, ignoring the goats, the bicycle man wheeled his bike away, laughing with friends.[3]

Conclusions

Though it is easy to pick out a few hours from hundreds of days of observation and conversation, it is not easy to reconcile these hours with the whole. We all talk about the "social relations of production," but shown here are concrete processes and sites of research production—workshops, libraries, seminars, rocket and balloon launches—which are all maintained through larger structures of social relations. These structures were not bounded and limited by the institutes in which the production occurred, but resonated with much wider patterns in society. Having said that, it is equally obvious that it was individuals and the quality of their social relations that made these performances possible, what they put into them and what they took out of them. In the coming chapter these institutes are compared again in the late 1960s in terms of their approach to contentious and potentially divisive questions. So we return again and again to issues that attracted the attention of India's most senior and influential "political" scientists sitting on SACC, and at the same time seized the imaginations of India's dynamic, younger, and newly recruited scientists in these same institutes.

SEVENTEEN

Governance, Management, and Working Conditions in Research Institutes Founded by Saha and Bhabha

When Saha and Bhabha established research institutions in the late 1940s, they could not foresee how these places would be governed after their deaths. Since neither reached retirement, they did not have time to predict what might happen to their institutes when they left them, during their old age. At the time of Saha's death in 1956, the image of SINP was "small and struggling," whereas the image of TIFR was already "big and growing" when Bhabha died in 1966. It is important, therefore, to examine the way their institutes operated without them. In this chapter the governance of a scientific institution means the system adopted for the continuous exercise of authority and direction, regulating the conduct of research, recruitment, and retirement from the institute, and all working conditions relevant to a scientific institution: a focus here is on strains and tensions arising from the research process, pointing to the different ways and means through which the system of governance addressed those tensions.[1]

This chapter compares (i) the structure of the governance in these two scientific institutions, (ii) the administrative influence of the institutional ranking system and the resulting nature of the hierarchy, and (iii) the influence of governance upon the morale and motivation of scientists. To illustrate the complexity of the situation in each institute, specific cases are examined. It may be thought odd to say that scientific institutions are *governed;* is it not more accurate to say that scientists are more self-governing than other kinds of people? Well, management is the performance of tasks in accordance with the authority that is granted through the governance of scientific institutions, so the focus should be on governance more than just management.[2] This analysis reveals underlying questions about the evolving political culture of scientific institutions. How did scientists view different kinds of authority that impinge upon their work? Were there institutional restraints (or "checks and balances") on the exercise of rank and authority? What was the degree of influence on institutional policymaking among members of research groups? What were the prevailing concepts of viable governance and models of its proper functions, and did those concepts enter the creative process? These questions are examined with reference to the perceptions and opinions of the scientists in Bombay and Calcutta in the late 1960s. Preceding chapters showed how working conditions in laboratories were discussed in the cabinet, and the following chapter will show how the very public debate about the award of the 1968 Nobel Prize to Har Gobind Khorana, and the brain drain affecting India, resonated perfectly with debates about conditions inside these two elite institutes. The intention among scientists to go abroad was very strong, as evidence in the final tables show, but the reasoning leading there was complex.

Governance affected almost all the conditions under which scientists were socialized into the culture of science and were trained in research, including the critical conditions under which research groups perform and evaluate their work. The sum of these conditions was defined by most scientists to be the morale of the institution, its atmosphere, and climate for work. Morale level was used by most scientists to explain the quality of the work done by individuals and groups and to explain, even justify, scientists' intentions to work there in the future. More precisely, most scientists thought low morale would lead to poor-quality work and/or a desire to leave (and conversely too). On the one hand, some maintained that researchers might try to maintain a *higher* quality of work in order to be able to leave an institution if it had low morale, but if that person worked in a group, it would be very hard. In that view, good research work was a pass to mobility out of an undesirable situation. Scientists thus agitated for participation in governance because of its effect on the very crucial conditions that they perceived to enable them to create or maintain their scientific reputations. They also spoke about the importance of a strong productive institution both in terms of their own futures and of the national development of science and technology. Even among those with no plan to move anywhere else, there was an expectation that the institution should provide the rewards of rank and status at least commensurate with their productivity and seniority. The interplay of seniority, age, status, and professional rank was very complex and influenced both their attitudes to governance, and their actual administrative practice.

On the other hand, those who governed these institutes understood that they exercised different kinds of authority, conditioned by the historical origins of the institution, their relationship with the founders (if any), its financial fortunes, and by the degree of autonomy it enjoyed from its source

Table 1. Comparing institutional size in 1968

Staff	TIFR personnel		SINP personnel	
	N	Percent	N	Percent
Faculty scientists (reader and above)	39	3	38	10
Nonfaculty scientists	244	18	104	28
Scientific and technical staff	564	42	93	26
Administrative staff	193	15	39	11
Maintenance staff	290	22	90	25
Total	1,330	100	364	100

Source: Annual reports of TIFR and SINP, 1968–69.

of funds, in both cases the DAE. Each institute's governance also reflected the political culture of which scientists were members in the late 1960s, the political culture of the city that surrounded it. In Bombay and at TIFR there had always been a pattern of negotiation and accommodation, while in Calcutta and SINP there was more a pattern of pressure, reaction, and confrontation. Though these institutes had similar formal systems of academic ranking, the role of scientists in administration and the hierarchies of scientific esteem were quite different in each place.

A common reaction among scientists to my plan to compare TIFR and SINP was that there was insufficient difference between them to warrant the comparison or that the study would miss the real problems of science in India because it did not include any other, less prestigious institutions. Other scientific leaders suggested that these comparisons would be invidious and would have a negative effect on the Indian research system as a whole. These common reactions warned me to look carefully at these two "unrepresentative" institutions where the potential for superior research was thought to be so high. My approach was confirmed when I visited about a dozen other scientific institutions in India in the late 1960s.[3] Most of the problems reported in conversations with scientists elsewhere in India were found to some degree in TIFR or SINP. Comparisons between TIFR and SINP were made regularly by scientists when researchers from both institutions visited each other and also when they met on neutral ground. SINP members were not alone in envying conditions at TIFR, for those conditions in Bombay were the envy of most researchers throughout India, and some from outside India too. These ongoing professional comparisons were an essential part of scientists' self-definition and were the basis of their judgments about productivity and scientific morale. My motive for this comparison was not to suggest or draw invidious distinctions but to see the consequences (if any) of the history of different governance in scientific institutions.

The first comparison between TIFR and SINP made by scientists was usually the difference in size, meaning budget. The main difference between the size and structure of the two institutes lay in the much greater number of administrative and support staff in TIFR, and in the higher ratio of non-faculty scientists to faculty in TIFR when compared with SINP. The total staff (all categories) of TIFR was four times as large as SINP and its budget was almost six times bigger in these years.

Two Patterns of Governance

The system of governance of each institute appears to be similar, though their practice was rather different. TIFR was presided over by a Council of Management, working through the offices of the director, the deans of the School of Physics, and the smaller School of Mathematics. In SINP the chief policymaking board was the Governing Body, working through the office of the director and the faculty. The formal structure of each institute remained fairly constant over the second half of the 1960s. In each institute the main governing council was composed of members whose allegiances or attachments were to other institutions, supportive government agencies, or industries; few "directors" were personally engaged in scientific research in physics or mathematics. Moreover, council members were very busy with numerous other responsibilities on boards and committees and were thus seldom able to allocate much time to this work, as evident in the membership of the two governing councils (1968–69).[4]

The Tata Institute's ten-member Council of Management consisted of three members appointed by the government of India: Vikram Sarabhai, PhD, secretary to the government of India, Department of Atomic Energy, and chairman, Atomic Energy Commission, Bombay; I. G. Patel, economist, secretary to the government of India, Ministry of Finance, Department of Economic Affairs, New Delhi; and Satish Dhawan, aeronautical engineer, director, Indian Institute of Science, Bangalore. Of the ten, one member was appointed by the government of Maharashtra: S. E. Sukthankar, administrator, secretary to the government of Maharashtra, Agriculture and Cooperation Department. Two members were appointed by the trustees of the Sir Dorabji Tata Trust: J. R. D. Tata, industrialist, chairman of the Council of Management, and Professor Rustom Choksi, managing trustee, Sir Dorabji Tata Trust. A director of the Tata Institute was an ex-officio member: M. G. K. Menon, physicist. And finally there were three holding trustees: B. Venkatappiah, economist, Planning Commission, government of India;

S. E. Sukthankar, administrator, government of Maharashtra; and Naval H. Tata, lawyer, Sir Dorabji Tata Trust.

The nine-member Governing Body for the Saha Institute consisted of two members representing the University of Calcutta: S. N. Sen, economist, vice-chancellor, University of Calcutta, ex-officio chairman of the Governing Body, and Satyen Bose, National Professor of Physics. Three members represented the Department of Atomic Energy: Raja Ramanna, physicist, director of the Physics Group, BARC; M. A. Hadi, administrator, BARC; and Y. S. Das, administrator, DAE. Two members were appointed by the Ministry of Education: N. C. Chakravarti, administrator, Secretariat for Education, and N. R. Tawde, physicist, vice-chancellor of Marathwada University, representing the interuniversity board. Two members were appointed by the government of West Bengal: H. K. Maitra, accountant-general of West Bengal, and D. N. Kundu, physicist, director of SINP, ex-officio.

The chief influence of these two governing councils was through their scientific members working through institute directors. The director of TIFR at this time, M. G. K. Menon, had been identified by Bhabha to be his successor when he was appointed TIFR's deputy director about 1963, and there was no open contest over Menon's succession to Bhabha's position as director of TIFR. The director of SINP, D. N. Kundu, was nominated by his predecessor B. D. Nagchaudhuri, who had been appointed to SINP in 1951 by Saha, taking over major responsibilities when Saha moved to Parliament in 1952. The founders of these institutes had established and exercised charismatic leadership over most scientists working there, particularly those in the early stages of their careers. The founders began work by taking risks, in the heroic mode, and their successors and their institutes ended by making rules. In terms of Weber's concept that was elaborated by Shils, the founders' charisma had been dispersed, and what charisma remained attached to the position of director had become increasingly routinized in the perception of most working scientists, and their authority had been diminished. Neither succeeding director had yet established major international reputations; while both Saha and Bhabha were heroes of science in India and entered the history of physics through Saha's equation and Bhabha-scattering, their successors Menon and Kundu had not attained these distinctions, though their work was certainly respected. In 1971, Menon was elected an FRS for his work on the K-meson.

In both TIFR and SINP the important power was ultimately concentrated in the office of the director, having been largely delegated these powers by the governing councils.[5] The network and alliances of the directors, therefore, were important for the future of their institutes. Menon enjoyed prestige in national circles, but Kundu was thought by his colleagues to have reached his top position, and he did not move up in the administrative or political hierarchies during the 1970s. Both directors were present in their offices only irregularly; combined with the concentration of powers in their positions, this sometimes made administration and quick decisions difficult. As Menon's other national responsibilities increased, his own attendance to work at TIFR necessarily became irregular; this became most pronounced after January 1972 (see chaps. 20 and 21).

Regardless of their own career ambitions, these directors were expected by their scientists to encourage creative research, secure and maintain a relatively stable and increasing flow of funds for their institutes, and resolve disagreements before they became disputes. The governing councils above them were expected to support this effort and the directors to commit their networks and alliances to the future of these institutes. Equally important the role of these councils was also inherently regulatory. They were expected to guide these institutes to conform to national policies, and their directors to conform to financiers' objectives. Appointments were made to the councils on this assumption. In fact, some members of the councils were on deputation from governments explicitly to see that financial conduct conformed to official practices, and their roles were limited to this subject. There was some old tension between financial members of these councils and others over the unpredictable nature of research spending (what financial people also termed "the irresponsibility" of scientists). The role of scientific members of these councils was therefore particularly important in the minds of scientists in managing or even reducing any tension in the institute created by trying to conform to or to avoid government personnel rules or financial regulations (for examples, see chap. 14).

Comparing Institutional Financing

As the history and size of TIFR and SINP differed, so did the financing of research. SINP's Third Five-Year Plan for 1969–74 was being scrutinized by the DAE in Bombay during 1968–69, recalling the complete review in Bombay of Saha's plans for his institute in 1955. The Bombay member of SINP's Governing Body, Raja Ramanna of BARC, still had the greatest influence over this budget process: four senior faculty from TIFR and BARC visited Calcutta during this period of review, ostensibly to look at progress of groups working in their own special research fields but actually to prepare independent appraisals of the institute as a whole. Results of this DAE review on the Governing Body's decision for the future were not openly communicated in SINP, and Director Kundu simply instructed each group leader to reduce their budget projection in order to prepare for the new Five-Year Plan, in some cases by almost 50 percent. As there was no faculty-wide meeting in SINP, there was no place for an explanation of this instruction or discussion of it. One research group's leader in SINP eventually telephoned Bombay early in 1970 in order to have the complete financial situation explained and received—contrary to expectation—considerable useful information that was unavailable to them in Calcutta. Doubts about the director's commitment to communication were raised, and there was a contest in SINP for any new money. The director retained his power by restricting information about the budget, thereby keeping himself in a position to bargain with various research groups, but it appears he had little bargaining room left in 1969. This contest was sharpened by the enlargement of each research group and by the fact that many improvements and repairs were being postponed until SINP could be moved in 1971 to new land and buildings at Salt Lake in the suburbs of Calcutta, where it would also manage a new and larger cyclotron. This long postponement produced inevitable physical and social stress that was supposed to be relieved during SINP's Third Five-Year Plan, stress sharpened by further postponement of the move when refugees from Bangladesh took over Salt Lake in 1971 and did not leave.

In contrast, TIFR in Bombay was not obliged to prepare a five-year plan for the DAE. It had representatives on most national committees of the DAE and had regular and convenient contacts with senior DAE and AEC officials, thus obviating this need. TIFR was named the national center of the government of India for nuclear science and mathematics. The size of its budget and its relative freedom on spending reflected its high status, its convenience in securing scarce foreign exchange, and its unimpeded access to state and central government officials who otherwise had sufficient power to regulate a research institute's affairs. Financial difficulties had not been serious at TIFR because Bhabha was *both* the director of TIFR and secretary of the Department of Atomic Energy and tried to secure autonomy for most of his projects by having the same officials and scientists fulfill overlapping responsibilities, seen clearly in the composition of the councils that governed TIFR and SINP, providing a network of people responsible (and loyal) to the DAE.

The ultimate financial control over research in these two institutes lay in the DAE, as table 2 shows. In Bombay it was common to hear discussion of the question how can TIFR get more out of the DAE, whereas in

Table 2. TIFR and SINP financial revenues, 1968–69

TIFR	SINP
	Rs 17,000 Miscellaneous and Donations
Rs 100,000 Sir Dorabji Tata Trust	Rs 36,000 University of Calcutta
Rs 50,000 Government of Maharashtra	Rs 42,000 Government of West Bengal
Rs 20,650,000 Department of Atomic Energy	Rs 4,130,000 Department of Atomic Energy
Total Rs 20,800,000	Total Rs 4,225,000

Sources: TIFR; personal communication from J. V. Kotwal, secretary, School of Physics, 1 March 1971, EFR/PF1594 (26/67 26615). TIFR sum excludes costs of TIFR's Radio Astronomy Centre in Ootacamund; also SINP: Annual report, 1968.

SINP it was more common to hear about the interference of the DAE in the Governing Body and the restrictions on their development. The deep interdependence between the DAE and the Atomic Energy Establishment, Trombay, formally BARC after Bhabha's death, has already been established in preceding chapters of this book.[6] A system of overlapping responsibilities and informal contacts at all levels was established to preserve BARC's autonomy from the head office of the DAE and AEC. But because the DAE and BARC both evolved through long assistance from TIFR, TIFR had a special privileged relationship to them. The informal contacts of TIFR with DAE and BARC, as well as interlocking membership of people on TIFR's Council of Management and on the AEC in the late 1960s (e.g., J. R. D. Tata and I. G. Patel), inevitably strengthened this special relationship. The perception of this integration and interlocking was that TIFR could withstand almost any conceivable budget or personnel crisis.

Because of SINP's distance from Bombay and its complete dependence upon the DAE for financial planning, it held a comparatively subordinate position. It had no significant financial support other than from DAE, notwithstanding the ongoing participation of the University of Calcutta and government of West Bengal. I discussed the role of the DAE in SINP with four members of its Governing Body and with faculty members, and not surprisingly the views in Calcutta and Bombay were quite different. Essentially the SINP Governing Body members said that there was a balance of interests from Calcutta and from Bombay. One member from Bombay said (and this was typical): "Our Governing Body simply has to work within the Calcutta limitations. Calcutta can outvote outsiders if it really wishes to, though we try to avoid such confrontations. If DAE interferes, Calcutta members will get angry about Bengal being pushed around by the Centre."[7]

While SINP's director pointed to the slowness with which the Governing Body works, he and the three most powerful faculty members remained evasive or obscure about the question of DAE's role in the Governing Body. But other SINP faculty members were clear in their minds that the Calcutta-based members of the Governing Body were powerless to confront the DAE in Bombay about funding cuts. They pointed to "a weakness of our director in relation to Bombay" and to "our lack of any internal decision-making strength." They said that because the chairmanship of the Governing Body belonged to the University of Calcutta's vice-chancellor, it was "symbolic of a severed relationship," a separation intended by Saha himself to enhance autonomy in the late 1940s. But by the late 1960s the enfeeblement of the University of Calcutta left the SINP council chair without national influence. Further, it was believed in Calcutta that SINP's dependency upon the DAE could not be altered because *all* institutions in West Bengal were suffering from central government disapproval as a result of the rise, and then fall, of the anti-Congress United Front government and the imposition of President's Rule in 1969. Local scientists explained that the political culture of Calcutta infused the SINP-DAE process with wider, uncontrolled forces of center-state relations. Scientists in Bombay were equally conversant with center-state power relations but did not explain disagreements in TIFR by using this idea. Bombay was seen to be on the rise, above such things; Calcutta was not.

Rank, Esteem, and Authority Compared

One of the more striking differences between these institutes was the strength and coherence of TIFR's internal organization and the weakness of internal organization in SINP. In addition to a prestigious Council of Management and director, TIFR had an active and well-organized faculty meeting. For instance, the School of Physics had a weekly meeting of its twenty faculty members, including everyone above the position of reader; it could also include readers who were appointed to it. There were close working relations between the office of the director and the faculty in TIFR, in part because the director was also an active cosmic ray physicist with his own research group, and in part because the director wished to retain professional contact with his field. The SINP director maintained working relationships with his "parent" cyclotron research group, but there was no regular faculty meeting for discussion of general issues.

Notwithstanding these arrangements, TIFR experienced tensions (described below) and so there was evidence that scientists placed considerable value in the procedures already established. Disagreements were discussed, often after members of the faculty were pushed to channel communication upward from below. At SINP, on the other hand, the director's office was the locus of all decisions. There was no faculty meeting, and the scope of membership and terms of reference of a faculty meeting were in constant informal debate. The relative complexity and strength of the system of governance in TIFR, and its simplicity and weakness in SINP, had important consequences for scientists' perceptions of the proper functions of the governance of scientific institutions and their need for access to decision-making.

While the discussion about governance so far could have referred to any large complex government-controlled institution (an airline or a military force), what made scientific research institutes distinct was a mixture of a routine system of ranking by salary and seniority with a unique scientific hierarchy related to individual merit and the shifting importance of certain fields of research. This combination of the routine and the unique presented a continuing problem for the governance of scientific institutions, emerging most clearly in inherently scientific funding choices, which scientists thought they should have access to through the decision-making process to resolve these very issues. This process was not simply about competition for scarce material resources but also about maneuvering for enhanced influence by categories of scientists previously excluded from governance. This maneuvering was underpinned by a conviction that scientists would make better choices that relate to their research and should be involved in the process.

Salaries in TIFR ranged from Rs 3,000 monthly for the director to Rs 85 for the gardeners. In 1968 there were eleven different academic ranks in TIFR, from professors (Rs 1,400–1,950) through research fellows (Rs 500–750) down to research assistants (Rs 325–450). Salaries in SINP ranged from the director (Rs 2,500 monthly) to the maintenance staff (Rs 100). There were also eleven academic ranks, from senior professors (Rs 1,300–1,800) through lecturers (Rs 350–900) to research assistants (Rs 270–425). The number of opportunities for fine distinctions of rank in both institutes was much greater for technical and administrative staff than for academic staff. But the consequence of these numerous distinctions was everywhere to be seen among scientists, particularly those committed to the collective and collegial ethos, which had come to pervade the conduct of science in the 1960s. Scientists in SINP continuously referred to this ethos, and so did TIFR scientists, occasionally. This was a moment in European and American history that influenced young Indian intellectuals; some were very knowledgeable about the spring events in Paris and Prague in 1968; some had lived in the United States in 1964–68 in times of the free speech movement and the police riots in Detroit, Chicago, and Los Angeles. Their collectivist outlook was not merely derived from influence abroad but was also supported by their youth and a long Indian tradition.

Apart from the status and power associated with seniority and salary, there was also a unique scientific hierarchy based on perceptions of and communication about an individual's scientific merit. This merit accumulated and was calculated in esteem, which was granted or withheld with reference to international recognition by the leaders of research fields, publishing important results in major foreign journals, and being known for contributions to their field of physics. Where people worked in research groups the merit of the group attached to them, but they could also potentially transcend the group's reputation. Unlike the ranking system (lecturer becomes reader), this hierarchy of esteem was not fixed, and though there were disagreements over the relative merits of individual contributions (potential or actual), everyone acknowledged that *any* reputation could rise or fall. Such disagreement about esteem was not limited to scientific communities in India, as the history of science everywhere has demonstrated.

There were constant references to people whose rank in the institute system did not correspond to their place in the hierarchy of esteem. This potential asymmetry between rank and esteem did not negate the role of esteem in the governance of these institutes. In fact, the hierarchy of esteem was shored up by the example of a few important individuals in whom high rank and high esteem coincided. When an esteemed individual was assigned an appropriate place in governance, lower-status people said that he or she was "finally being taken seriously" and the qualities of esteem prevailed again.

Governance and the Importance of Research Groups

Many scientists in these institutes, however, were really better known as members of research groups. Appraisal and appreciation of individual merit in groups was inevitably combined with perception of the relative importance of the group's work and of the importance of its field of research to the natural sciences as a whole (mathematics, biology, physics, etc.). This referred directly to relationships among research groups: each group in TIFR and SINP had developed for different reasons, under different leaders, and with different relations to their founders. Particularly in experimental work where results are produced by and attributed to the group as a whole, individuals were identified with their groups—their personal "politics of reputation" lay primarily in the future of the research group. This was largely a function of an individual's investment of time in the group. While these institutions were established to further the national interest and science policies, the changing importance of research fields was for the most part appraised in terms of international fashions and prestige. The effect of foreign training and foreign advisors was strongest in influencing the evaluation of the importance of different fields and different groups. All these influences led to a slow shift and redefinition of the importance of fields, including the abandonment or adoption of certain paradigms, as Thomas Kuhn[8] showed; some fields were seen as "hot," others as "cooled off." Some fields were subjects of "big science" funding abroad, whereas others were just struggling along hoping they might "become important again, in a new form." Earlier chapters are full of the stories of the attempt to establish viable and continuing research groups, almost all of them revolving around an expensive piece of equipment.

Some idea of the disjunction between the perception of international fashion and national priorities can be seen in the example of the microwave technology group at TIFR. Created to enable the progress of a large radio telescope project in the Nilgiri Hills of South India and the linear accelerator project in Calcutta, this microwave expertise was ready when TIFR responded to requests for microwave technology useful to the Indian military and expanded using funds from the Ministry of Defence in 1963 after the conflict with China in 1962.[9] The group was applying physics to interesting new engineering problems which, if solved, could have had breakthrough effects in completely new fields in India like radio astronomy. The application of microwave technology to defense radar and communication systems (see below) was seen as of secondary importance, for it lay outside the scope of the institute.

But the microwave work was rated as important enough to have TIFR respond to military pressures and accept its first formal defense contract. Members of TIFR were advisors on microwave installations across India, and some members of both TIFR and SINP had worked on defense contracts abroad, mostly in the United States; a few continued to do so while in India, and there was prestige in the institute attached to this work. But the microwave technology group had lower prestige in the institute than its national importance might suggest because the work of this group was rated as "not interesting" and "not fundamental" by most physicists at TIFR in terms of its contributions to physics.

It is inevitable that a few groups in each institute enjoyed an esteem and prestige greater than others. But what were the consequences of the ranking system and of the hierarchy of individuals and groups for the actual governance of the institution? In TIFR the two largest groups in 1967 were cosmic rays (111 members) and nuclear studies (108 members), with computer sciences third in size (76 members). The cosmic rays group was, in fact, a cluster of groups in nuclear emulsions, high-altitude studies and plastic balloons, extensive air-showers and cloud chambers, and geophysics. Cosmic rays was represented by six members on the faculty of the institute plus the director of TIFR and, after 1970, by the new dean of the School of Physics (a total of eight). Cosmic rays had been the first major research group in the institute, a founding group, had generous funds for emergencies and special purposes (including foreign exchange) beyond its budget, and had a strong local reputation that matched its international reputation. Its six faculty members were also on national committees like the Physics Committee of the DAE, which advised the AEC on future policy and research directions. Other cosmic ray group members were involved in TIFR as head of the library committee, head of the food services committee, head of the employees' credit union, and faculty overseer of workshops. Commenting on the role of the cosmic ray group and group hierarchy, the leader of another large group said: "It is commonly felt that cosmic rays get almost everything around here—money, privileges, and the pick of the best recruits. This is partly for historical reasons. It was the first group. It gained an international reputation the other groups did not have. It has become the largest group. I'm convinced that a hierarchy of groups is inevitable."[10]

Given the strong position of cosmic rays and nuclear studies in TIFR, other groups worked toward autonomy *from* them. Mathematics was constituted as a small separate school from its beginning, but the newer computer science and molecular biology groups were located within the School of Physics. The computing facilities were actually most heavily used when the cosmic ray and nuclear studies groups did analysis of large amounts of data; so the computing group functioned both as a service center for others and a research team on programming. These groups were all considering moving "out" on their own, but there was no "other" category in the institute that could receive and administer them; the molecular biology group eventually moved, with its founder, out of the Bombay institute to Bangalore, where it became a national center for the study of biology, though still part of TIFR.

In Calcutta at SINP the situation was quite different. In 1969 there were seven groups and divisions, the largest being the nuclear physics group with 45 members, and the accelerator group with 35 members: four groups had more than 20 members. The two most powerful men in SINP, one of them the director and the other Meghnad Saha's son Ajit Saha, were each heads of one of these two groups. SINP grew up around the cyclotron accelerator in the 1940s, though an effective external beam for research was achieved only in 1966. Whereas Bhabha was not considered by colleagues to have had many students and none in TIFR described themselves as "Bhabha's student," seven out of nine heads of groups in SINP identified themselves as "a student of Meghnad Saha."[11] The rewards of rank in SINP had, in some cases, been given for the greatest tenacity over time rather than for virtuosity in a particular field or for the production of successful colleagues and students. In 1968 all but one group leader had been in SINP at least since 1958, some for many years before that. They had, in the absence of administrative staff, absorbed more and more routine administrative responsibilities themselves, so that being group leader made them, as one put it, more "administrator than scientist." The larger number of administrative staff at TIFR tended to inhibit this process of "scientists becoming clerks," protecting their scientific productivity better than SINP did. It was common to hear the motive of empire building ascribed to some group leaders in Calcutta, even from members of their own groups.

Because there was no regular faculty meeting at SINP, each group had to articulate and struggle for its needs separately, thus expending considerable energy on the politics of reputation. This emphasis on time and tenacity appears to have increased group leaders' reluctance to take on risks in research because there was absolutely no funding or administrative cushion that would absorb "failure" in risky work. The desire for a quick return operated as strongly at that time as it did in much of the business community in India. SINP's group leaders often made quite cogent arguments against taking risks in research by pointing to the need to conform to the Five-Year Plan, to their own group's poverty, or to the bad state of their equipment.

The caution of group leaders in Calcutta stood in stark contrast to the exhortations in 1968 by the Saha Institute's previous director B. D. Nagchaudhuri: "It does not mean that scientists in India should not take risks; on the contrary, a certain amount of risk has to be encouraged and risk-taking is all to the good."[12] This contradicting view is cited because Nagchaudhuri, writing then as a member of the Planning Commission, had been, until that year, the director of SINP and has thus helped to establish the climate of thinking in which group leaders addressed themselves to my questions about risk. In addition to the system of rank and the hierarchy of groups in SINP, there was the regular use of the Bengali language with its underlying grammatical patterns of deference to traditional social hierarchy and age (already referred to in chap. 16.)[13]

Internal Decisions and Finances

There was little controversy in TIFR over financing ongoing research of established groups or the purchase of a general facility like a photocopy machine. If a group wanted a small capital purchase like a Rs 10,000 laser, the request would be reviewed for its utility and probably approved. There was no vote taken in the faculty meeting in these matters; decisions were reached by compromise and consensus. A senior member of the cosmic ray group said that that big group took its proposals to the faculty meeting like all other groups. "Often facilities which benefit the whole institute take precedence over our group's requests, and we accept that," he said. The leader of a smaller research group said that there were some financial decisions which needed a little faculty discussion with the director beforehand:

> When I need something which is essential to our work, I ask the director about the chances of getting it. He eliminates a lot of doubt. He tells me immediately if it is likely or not. He never says, "I will speak to X and he will discuss it with Y." That's what happens in the rest of India. Our budget is clear to us, once it is set there can be no other major interference from other groups. If there was, a little group like ours would be in trouble all the time. Once we get the money, we are better able to produce research here than in any other institute in India.[14]

Bigger issues coming before the faculty meeting in the School of Physics had already been reviewed by one of the five special group committees. A few members of faculty plus a few readers (by rule, not members of the faculty) comprised these committees and dealt with topics ranging from theoretical physics (Group Committee 1) to "courses, students, University of Bombay, etc." (Group Committee 5). Created as a "forum for detailed and realistic evaluation of research proposals," these committees were intended to "put up new consolidated recommendations," and not simply review research projects already approved by the faculty. According to TIFR's director, "It is important to assign priorities to the various programs and to decide on what programs need to be stopped, curtailed or have their emphasis changed."[15] This acknowledgment of the need to change direction, and the methods developed to cope with changes (imperfect though their critics found them), showed a significant contrast with the governance of SINP in Calcutta.

In Calcutta funds in SINP were disbursed from the office of the director on the basis of annual requests from group leaders, with reference to the

Five-Year Plan established by the DAE for the Governing Body to administer. There was no wide discussion of the priorities of the Five-Year Plan, except between the director and the Governing Body. There was no faculty meeting in the late 1960s for discussion of group requests between the cycle of annual requests: any variation was privately and individually negotiated by group leaders with the director. Some group leaders consulted their colleagues; others did not. Without a wider understanding of Five-Year Plan priorities, let alone their debate, only a few senior scientists could guess (using "leaks") the criteria on which funding decisions would be made in a five-year budget of Rs 4.25 million. In the late 1960s most repairs and facilities for SINP were being postponed until SINP moved to new buildings at Salt Lake, next door to the new Variable Energy Cyclotron. Library, cafeteria, washrooms, and other physical needs were "on hold." However, renovation and air-conditioning of the director's office (the room in which the Governing Body usually met) was completed in 1969 in the face of strict budget controls. Director Kundu's explanation of this slightly secretive state of affairs was:

> This lab used to be run, when we all started, like a small family. When I worked in America, all department business was conducted through committees. It was boring and slow, so I used to sleep through it all. But here it moves faster because we can take decisions among ourselves, quickly. [August 1968]
>
> We have no formal faculty meetings here. Anyone can walk in anytime, and I'll talk with them. Do you think that after so long in the States I would not be informal? [November 1968]

This arrangement in SINP allowed for governance by rumor, which, in the opinion of many observers, left the institute open to disagreements and confrontations: these had become the preferred forms of communication by 1968–69, in this very place that continued to be run "like a small family."

The Saha Institute had just gone through a frustrating experience in trying to build a variable energy cyclotron on its own, a long promised project that was closed down in a disappointing way in 1968, but after the expenditure of a considerable amount of its own funds. That disappointment coincided with the move to Delhi of its longtime director Nagchaudhuri and probably contributed to the poor relations between SINP and the DAE (though in the end the DAE and BARC "rescued" this project). Nagchaudhuri and others in the cyclotron and nuclear physics groups moved ahead on a new cyclotron in spite of the indifference shown by Bhabha to his earlier commitment to a new cyclotron. In 1963 Aiyagiri P. Patro and Binayak Basu of SINP proposed to fabricate a sector-focused cyclotron using indigenous methods and equipment. This project was included in the institute's Five-Year Plan (shown routinely to the DAE, from which funding came). Having assembled the necessary funds, they built the required space in a newly constructed laboratory building on land donated to the University of Calcutta by Messrs Koley Brothers in the suburb of Pagladanga in Calcutta. A vehicle was also purchased to transport workers between this new laboratory and the old institute buildings at Science College. By 1965, the magnet with its coils and pole-pieces with spiral ridges was erected, a four-inch-thick aluminum block was purchased to scoop out the chamber, magnet power supplies were assembled, and the ion source and other accessories were in the process of fabrication.

Bhabha learned of this effort and, after years of indifference or preoccupation with other projects, was pressed and persuaded by about thirty nuclear scientists to assist it. So a meeting was convened at the newly opened building of the TIFR in August 1964. His observation was that nuclear scientists needed two machines in India: one to be fabricated (AVF) and another to be brought from abroad on a turnkey basis (Tandem). However, he commented that "the Tandem and the AVF accelerator are complementary. It may be possible for an AVF machine with the help of elaborate equipment to do what a Tandem does. If we decide to go in for a Tandem, it seems clear that we should buy one. The technology of an AVF is entirely different and this is the field, except for the cyclotron at Calcutta, which has been neglected in India. This is the field which I think should now be developed. We are in a position now to enter in a new field of building of accelerators. Therefore, I, myself, favour the idea of going in for an AVF machine."[16]

Nothing had been done by the DAE and BARC, however, when Bhabha died in January 1966, and so Vikram Sarabhai began to work through Ramanna from BARC (on SINP's Governing Body) and B. D. Nagchaudhuri and Satimoy Chaterjee from SINP to have plans developed in 1967–68. Though Nagchaudhuri was leaving Calcutta at the time, the plan for the VEC was signed and financed in August 1969. Meanwhile, trouble had arisen for the SINP cyclotron project at Pagladanga with the rise of urban terrorist activities; the Naxalites had been largely a rural movement until 1967, when they became an urban movement, and the isolated place where the laboratory was situated became a hiding place of fugitives from the law in 1968, including but not limited to young men committed to armed struggle in both the city and the countryside. The security guard and caretaker at the new lab was attacked and further violent attempts were made on him, so that the laboratory was left unattended for months. Finally late in 1968, SINP decided to abandon the project because no suitable alternative sites could be found, the equipment was removed, and the property went back to the University of Calcutta. The experimental physicists at SINP were, in 1969, thus in a trap not of their own making, forced to wait for their new Salt Lake home with no firm date in sight.[17]

Restraints on Authority

While rank, seniority, salary, and the hierarchy of groups were the main sources of influence and authority in both institutes, there were certain restraints at work upon the exercise of this authority. These restraints were stronger in TIFR than in SINP at this time.

In TIFR both the cosmic ray and nuclear studies groups could have dominated the institute. But there were no allegations among other groups that they did; for example, it was not said that either of these two groups had unfair priority in the allocation of official institute housing in Bombay. This was in spite of widespread knowledge of unfair practices in the fierce competition for Bombay housing, which normally included very large unrecorded payments above the fixed rent *(bakshish)*. Though these two groups had the largest research budgets, they could not appropriate all powers to themselves. Research groups in TIFR were interdependent because their leaders, who made most decisions, constituted the faculty meeting. Some groups like cosmic rays depended upon others like computer sciences for the successful completion of their work. The faculty showed considerable solidarity in spite of the disparate social and linguistic backgrounds of its members (or perhaps because of it). These twenty men, and all were men, had come to know each other well over the preceding twenty years. They were held together by common objectives—the maintenance of their positions and privileges, along with the stable growth of their groups and enhancement of the reputation of the institute. Whether because of or in spite of their different origins (from eight linguistic groups and from classes ranging from son of a wealthy dewan to son of a poor village priest), the faculty were now all members of the urban, literate, upper-middle-class minority. They were responsible for large sums of public money and for a significant part of the nation's scientific development. They shared a common technical and scientific culture.

Another restraint upon hierarchy and authority was the assertion of a new attitude about work: many people in TIFR drew on the egalitarian ideology of science, particularly as it was expressed in North America and

Europe in the 1960s. Conversations were sprinkled with talk of "equal opportunity" and of "we are all workers here, there should be no bosses." There was a stress on collegiality that observers marked as new in India, even if collegiality was not always easy to sustain. These restraints on hierarchy and limits on abuse of authority appeared to some scientists as a major blessing, while to others, restraints were seen as a source of frustration or delay. This frustration was stronger in those whose previous experience was in the hierarchical, often authoritarian decision-making patterns of the surrounding culture. The implications of these questions for governance will be analyzed in the final section of this chapter.

In SINP the restraints on hierarchy and authority were comparatively weaker. Though they drew support from the same budget, used the same building and library, and so on, SINP's research groups had minimal interdependence. Decisions were taken by the director along with one or two unofficial deputy directors who were heads of major groups, joined with the registrar, whose office acted as the comptroller of funds. There was no faculty meeting in which groups had formal representation, and this allowed for almost complete control by the office of the director. Privileges could be extended or withheld at his discretion, though he could not, by himself, hire or fire anyone with a monthly salary of over Rs 200. All such personnel decisions had to be approved, he said, by the Governing Body.

Each research group, indeed each individual, had become a client of SINP's director, thus encouraging divisions between groups that were overcome only by great personal effort. To some extent all tenured members of SINP were interdependent through their joint need for the functioning of SINP as whole, even under these limiting conditions. For example, the privilege of a lift home in an institute car, an important benefit in Calcutta where public transport was very crowded, could be offered or withheld. With no institute bus, no adequate canteen, and no institute housing, the efforts of most scientists were as much directed toward physical shelter, food, and transport as to research. There was a widespread feeling, and some specific allegations, that authority in SINP was abused by those with rank, seniority, and salary; whether true or not, the persistence of these allegations revealed the reasons for the poor morale among active scientists in SINP, in the judgment of the scientists themselves.

Strains and Accommodation in TIFR

The council and faculty of TIFR grasped some of the implications of the structural strains beneath them, shown by the remarks of its director M. G. K.

Menon at the second memorial service on the anniversary of Bhabha's death in January 1968 and repeated at the funeral service of TIFR's senior gardener in July 1968. This was an assertion that "we are all one big family." It was somewhat unusual that a gardener's death would be celebrated, and though this was no ordinary gardener, it is unlikely that a gardener's death would have produced a similar symbolic event many other scientific institutions in India. At these events, the excitement of Bhabha's era was recalled and senior members of TIFR were asked to reaffirm the values for which the institute stood. Published shortly after these ceremonies, the director's remarks acknowledged the strength of hierarchy and authority and so emphasized the need for dialogue throughout the scientific community.

> How is this dialogue to be carried out? The scientific community exists in universities and in various research institutions and national laboratories; it is a widespread community. It consists not only of older scientists who occupy high positions, but of a large number of younger scientists, full of enthusiasm and energy; and into this latter group there is a constant stream of newcomers, freshly emerging from their university training. I must strongly emphasize that I have no intention of drawing artificial distinctions between older and younger scientists; such distinctions would be totally alien to the spirit of science, for in scientific life there are, and should be, no such hierarchies. My pleading would be to bring the younger and junior scientists into the mainstream of national endeavour; and they will, I am sure, provide powerful support to the few senior scientists who are already in the picture.[18]

The need for this dialogue in this big family of over 1,300 people came up during conversations and in meetings or statements: there were about twenty-five "young" scientists who had been in TIFR for at least ten years, and who had also held responsible faculty positions in foreign universities. Some were fellows, and along with others they were pressuring for more access to decision-making, suggesting, for example, that that minimal access could be given by circulating a copy of the minutes of faculty meetings. Others said that the confidentiality that surrounded most decisions helped to "cloak the mediocrity" of the research of some groups and their leaders. There was, in 1967–69, a series of informal evening meetings organized by one senior fellow; generally the discussion was about the growth of science in India and revolved around the institute's contributions to it. At one meeting in 1968, a senior fellow spoke on the structure of TIFR and its long-term goals. He said that "the faculty is sitting on a volcano." They hired younger people who had held responsible positions abroad, he said, but did not understand that these people were determined to influence policies at TIFR, asking them, "Are you going to give these people a voice, or will you continue to sit on top of a volcano?" This meeting generated discomfort and even hostility among a few senior scientists. The convener of these meetings felt discouraged by this reaction and eventually cancelled this series in 1969.

At this stage, in mid-1969, another senior fellow characterized the problem by saying, "The faculty are not antagonistic to us. They just don't know how to respond." One physicist joked after the director's 1968 speech at the memorial service by saying, "Yes, no distinctions between older and younger scientists, at least not above certain ranks and not below certain others": everyone standing nearby laughed. One senior fellow who had been in TIFR fifteen years explained how many people of his age and rank perceived the problems in the governance of TIFR:

> The director emphasized Bhabha's tradition to the senior people because one feels now that they are in a struggle against one another for power. When Bhabha was alive, he was such a powerful personality, there could not be any opposition or struggle. Competition was simply inconceivable. It is difficult to tell but I think that now one rises and the other feels jealous. We juniors are helpless. That is why the director urged the seniors to listen. A lot of growth took place here in the '50s, and very young men were involved. That was one of Bhabha's values. That is why the director said, "Bhabha was like a tree, and we others were like children playing underneath." But those men are still young, and are now in positions of power. There is less scope for the rest of us. Other institutions are out there, but you know about those. It is hard to say what our fate will be.[19]

Some of the pressures for change were apparently supported by members of the faculty. One senior fellow said the director's remark about not drawing artificial distinctions was "just pointless rhetoric." One year later, however, in April 1969 he told me that he had been wrong about the rhetoric:

> There is a power struggle. But I never realized it, and I've been here for years. I think it is new, but I could have missed it even now. It is not quite in the open, not apparent to casual observers. You see we never learn about some decisions and policies, how they are made, I mean. Of course there are some rumors, but that is not much to go on. I was surprised, and I have been here for years. It seems the only way to have any effect on a faculty decision, I mean to force them to make a decision at all, is to use some kind of moral blackmail. At least it is for those of us on the outside.

One form of "blackmail" was the ultimatum in 1969 by a senior fellow that he would accept an offer of a permanent position in the United States unless he was offered a promotion to reader. The faculty postponed its decision for months, apparently due to the realization that acceding to his demand for promotion would "open the flood gates" to about one hundred other fellows with PhD's and publication records (the number varied with the speaker, this is an average). Finally the decision was taken, his ultimatum was refused, and this physicist left in 1970. At this time he wrote, "What seems to be holding back the bosses is that they are not sure of the value of these young men, and they take too long to decide about them."[20]

TIFR Example One: Page Costs for *Physical Review*

But what policies were these young men determined to influence? By examining three of the controversies that arose in TIFR in the late 1960s, it is possible to see that these policies affected the conditions under which a strong scientific reputation could be created or maintained.

When the prestigious *Physical Review* journal began to insist on payment of $65 per published page from authors, the faculty had to decide whether these foreign-exchange payments were justified. There was no question among TIFR physicists whether it was desirable to publish articles in this American journal, but there were long debates about the length of articles that could be submitted and approved for payment. Two members of the theoretical physics group who had had their articles accepted for publication and were then awaiting a decision argued that even though theoretical articles tended to be longer than experimental articles, the payments were justified because "theorists did not request foreign exchange for experimental apparatus." While this was largely true, the faculty struggled with the nature of exceptions to any rule they made and were (rightly) nervous about the introduction of invidious distinctions between theorists and experimenters. After months, it was decided that page costs for *Physical Review* (or any other journal) would not be paid by the institute. It was at this time that the director reiterated his request that TIFR groups try to publish in Indian journals as well as in inexpensive foreign journals like the well-reputed Italian journal *Nuovo Cimento*. People immediately began to look for alternative ways to publish in *Physical Review*, particularly when they were on leave abroad where they could find the $65/page more easily from their host institutions. This decision, made on grounds of budget and in the face of the difficulty of administering the delicate relationship between groups with different-sized budgets and different lengths of articles, contributed to the anxiety among younger fellows who sought prestigious publication to enhance their mobility; they now thought that mobility inside TIFR would be more and more difficult. While senior people were publishing in Indian journals, younger people did not feel they could afford to.

TIFR Example Two: The Bubble Chamber Group

As a result of the widespread use of the bubble chamber in experimental physics in the early 1960s, many physicists returned to TIFR with experience in this kind of research. A proposal was sent up from a group committee to create a bubble chamber research group. The cosmic ray group was already analyzing emulsions exposed in balloon flights and had experience in scanning for particles, and the like. Scanning pictures taken of emulsions was a standard technique in bubble chamber research. It was proposed that the new group would first analyze film exposed at accelerators in America, Europe, and the USSR. Then it would collaborate with those institutions on designing new bubble chamber experiments. Because scanning was done by relatively untrained young women at advanced laboratories like the University of Chicago, the high cost of exposing and scanning bubble chamber film in foreign institutions would encourage them to collaborate with TIFR where labor costs were lower. Proponents of the bubble chamber group already had promises of such collaboration with foreign institutions and planned also to use the new variable energy accelerator in Calcutta (this is one of the early types of outsourcing so common at the end of the twentieth century).

But this proposal demanded a great deal of foreign exchange and required a number of new permanent positions. One group leader said in 1968, "We try to avoid acrimonious arguments in the faculty meetings but this bubble chamber issue has been a long hard fight." The fight involved the knowledge that bubble chamber work was declining, even at the laboratories where the proponents had worked earlier. It was already being called, by some, "a dying field," contradicting those who had built their careers on the technique. Two people suggested that scanning might not be done by human eyes for long, nodding to computer-readers. Its value in India, with only one cyclotron still under construction, was questioned. Questions were asked, in response, about the future of cosmic ray experiments. Introducing a large new bubble chamber group might mean expanding the size of the institute and particularly the size of the cosmic ray group to which it would logically be attached. Even though special tours were taken by TIFR to visit the big accelerators at Dubna in the USSR and CERN in Switzerland to see what forms of collaboration could be negotiated, the bubble chamber decision was finally postponed, because agreement could not be reached. Success in this field obviously involved a large number of contingencies and uncontrollable factors, some of which were beyond the scope of the faculty decision-makers at TIFR.

TIFR Example Three: Graduate MSc Studies in Physics

A proposal for a graduate MSc program at the university expressed the concern that TIFR scientists felt for national scientific development; at this time the Department of Physics at the University of Bombay was not organized to teach at the graduate level, and most training of young physicists in Bombay was done in the BARC Training School at Trombay, started in 1956. TIFR members were involved in planning the curricula and selection of candidates in both the university and the training school but wanted more involvement in the university, whose impressive buildings they passed every day on the way to work. But active experimenters and theorists disagreed about the capacity of TIFR to manage an MSc program on its own; theorists required little more than a good library, colleagues, calculators, blackboards, and chalk—all readily available at TIFR, but for experimenters an MSc program meant an intrusion of students into their laboratories or the creation of separate teaching laboratories. The institute already had a space problem and crowding was a common topic of discussion in some research groups.

The debate not only concerned those who would be teaching but also those who worked in library, transport, and canteen facilities, for this meant more work for them too. Though the decision was postponed because agreement could not be reached, at a deeper level this example touched fears that TIFR would expand, become "too large," and exhaust its capacities and itself, thereby losing its reputation and financial support and weakening individual reputations along with it. Having seen the poor training of people who already had an MSc and who applied for positions in TIFR, there was a strong desire to make the institute's program more relevant to the needs of the whole of India. Through negotiation with the university, members of TIFR were learning that it would not be as easy as they had previously thought to transfer whole research groups to other institutions or to have well-respected TIFR members appointed elsewhere. The whole Indian system resisted this kind of mobility to which some TIFR members were accustomed abroad. As a result of their experience, TIFR physicists brought more pressure on the University of Bombay to establish a full teaching Department of Physics, to upgrade the standards. After resistance, the university finally agreed, then TIFR members agreed to teach part-time in that program, and eventually a graduate physics department was established at the university in which some TIFR scientists became adjunct faculty members.

Integrating scientists with such different educational experiences was difficult everywhere, as the experience of a special summer institute for physics at North Bengal University in June–July 1966 had shown. By that year summer institutes had been organized by Indian physicists since at least 1951, but a special US-India initiative produced this North Bengal institute and others, each funded by "counterpart funds" accumulated by the sale in India of US food aid. An American participant in the North Bengal University summer institute wrote a detailed and very critical account of its planning and operation, pointing out problems that were the common subject of conversation among physicists in SINP and TIFR when discussing the predicament of college-based teachers of physics—for whom the institutes were intended.[21] The institutes, however, were an experiment in crossing another cultural divide, one not quickly understood by foreign scientists—the divide between the bleak and unsupported life of the college classroom where teachers had to supply their own chalk, the more secure and better resourced universities where students were expected at least to have a textbook, and the luxury of institutes such as TIFR and SINP. Exams set by the summer institute for the college teachers disturbed them, and their inability to comprehend much of the difficult technical English unnerved them, and lectures without apparatus or assigned problems as homework left them apprehensive. Yet scientists at SINP and TIFR knew very well that each year a thousand unrecognized and unvalued college lecturers would quietly identify a thousand potential young scientists in their teenage years and guide or push them toward better training somewhere else, in all fields. That guiding and pushing had already resulted in the population of India's best and brightest arriving at the door of its best laboratories, including those whose families could hardly afford this step and did not consider "scientist" to be a serious occupation.

Pressures for these educational objectives revealed that there was some limited dialogue at TIFR, but frustration arose when some scientists thought they could initiate the changes and continue their regular "paid" work. Their concern for productivity and institutional decline were part of the feeling that there were already competing forces at work in the institute and that they still needed to shape and affect the conditions that influenced their own personal scientific careers and reputation.[22]

Communication and Confrontation at SINP

The governance of SINP was by a form of selective secrecy, and this created rumors about decision-making, which in turn created the opportunity (and perceived necessity) for confrontations involving arguments for structural change. As a reflection of these conditions, the director said that there were no "slow democratic formalities" and that SINP was still "run like a small family." Writing on the need for dialogue among scientists, Nagchaudhuri, the previous director of SINP (1951–68), stressed the obstacle of hierarchical authorities:

> There has not been enough time to build up a strong and independent tradition of scientific criticism and evaluation. The administrative machinery of Government continues to be highly centralized with a great deal of secretariat control on details, partly because the system was taken over from an alien government and partly because of the lack of other forces of development of science. In this situation there are serious dangers of scientific progress being hampered by the pervasive pattern of hierarchical authorities.[23]

The lack of an independent tradition of scientific criticism and evaluation in SINP were the reasons given by scientists for the problems of its governance. This lack was not due, however, to the shortness of the life of the institute but to the policy of denying independence and responsibility to scientists. Scientists pointed to the absence of a legitimate structure for responsible action. There were published rules and bylaws in SINP as there were in TIFR, but given the powers delegated to the director by the Governing Body, the autocratic style he adopted was predictable. Even when speaking of something as minor as the early closing hours of the library, which he decided unilaterally, the director spoke of it as an "irreversible executive decision." Aware of the pressures for expanded governance that were building up on him and his small staff, the director responded in August 1968 by appointing several committees:

> To decrease the work of the Director, I have created some committees to look after things like the library and equipment purchases. These are voluntary. People can resist or refuse if they wish to, but few do. Well, the truth is I have to assign work around here. I tell you, this is India, and if it were left really voluntary, nothing would get done.

Because the director termed his compliance with their recommendations "discretionary," these committees were not powerful or effective. His philosophy at that time was summed up in his statement to me in September 1968: "There is simply no limit to the problems we'll have if we extend ourselves beyond our research. I know there is nowhere around here to eat, I know this city has terrible transportation and a shortage of housing. But I cannot do anything about these things. If we go in for transport, housing, and feeding here, we'll never be done."

Basically four factors accounted for these conditions: first, the social and political distance of the Governing Body from scientists, and its disconnection from (and perceived indifference to) the daily affairs of SINP; second, the extent of the powers delegated by the Governing Body to the director; third, the absence of any faculty participation structure in SINP; and, fourth, the long-term acquiescence of most senior people in SINP to these conditions. Following the 1968–69 confrontations cited below, the working relations in SINP deteriorated still further until the director stopped coming regularly to the institute in 1971: "he has been very irregular in attending the institute and at the same time wants to ensure that every trivial decision in the institute be taken by him."[24] As in the rest of Calcutta there was violence in Science College in 1969–71, including an attempt on the life of a senior professor of pure physics in the Science College, all caught up in the struggle between the young supporters of the (sometimes suspended) United Front, the Congress Party in West Bengal, and their extreme left opponents (some of whom referred to themselves as "Maoists"). The three leaders of SINP, the director plus two group heads, were named *zamindars* (landlords) in the poster style reminiscent of the big character posters of the Cultural Revolution period in China, pictures of which were seen in India at this time. But these later outcomes were but enlarged reflections of the kind of governance in SINP in the preceding few years, as the following three examples show.

SINP Example One: Creation of a Faculty Council

There was no full faculty or broad institute meeting in SINP in the late 1960s, so there was no clear understanding of the institute's goals or budget to meet them. Scientists said the five-year plans were just frameworks because it was not possible to plan research five years in advance. Though the previous plans had not been strictly adhered to, the new plans were used in order to "repeat what was done five years ago" or "to justify remaining in a rut." One criticism was that new research was supported with budget "if it bolstered the prestige of our leaders" (hardly unique to Calcutta). Characteristically, in a situation without open communication, ordinary working scientists were unaware of the director's claim that the Governing Body limited his discretionary powers. In February 1969, five months before the confrontation with students, a meeting was organized by scientists to discuss formation of a faculty council. A broadly representative committee, not restricted to heads of research groups, was established to prepare an agenda for the next meeting. Many questions were posed for this agenda, such as, why were the buildings painted last year at a cost of Rs 45,000 instead of fixing the library? And what kind of computer do we *really* need, and why isn't there any discussion of the computer? When this agenda committee again met the whole group, there was a tension between keeping the agenda focused on common facilities like the library and computer, and new interests such as postdoctoral fellows who wished to be recognized as members of a proposed council. (This latter proposal to recognize postdocs was voted down, 31 to 12.)

Having seen how many votes were in support of recognition, the director spoke individually to scientists who had voted *in favor* of including the senior fellows (postdoctoral fellows) in the council, in order to discourage them from further action in this direction. Rumors then circulated that the idea of the council itself was in disrepute among some of the middle-rank scientists, the colleagues and peers of the very people who had organized the meeting and the agenda committee. Realizing that a faculty council might become a source of difficulty for him, the director came to the next meeting and said that his obligation to transmit decisions from the council to the Governing Body was discretionary and that it was not clear that the Governing Body would listen to the council's views.

In April 1969, however, a third meeting was held, allowing a vote on whether or not the new council's deliberations were to be binding on either the director or the Governing Body. This vote decided narrowly that the council should have no binding powers. The director and senior colleagues felt they had successfully inhibited the enthusiasm for change. In trying to secure the legitimacy of the council itself, the questions of housing, food, transportation, the library, computer, and budget were not pressed much further. Finally when the confrontation over the diplomas occurred in June 1969, the director did not turn to the broad faculty council for support but to an older and more restrictive faculty committee that had not met for many years.

When asked in June 1969 about the faculty's role, a senior physicist from BARC on the Governing Body, Raja Ramanna, said, "The director himself has the power to air-condition the library. He has the power to address all day-to-day working conditions. But just remember the special situation—we are not spending any more money on that old building in Science College; things will have to wait until they move to Salt Lake."[25]

Pressure continued, however, for a properly (meaning "fully") constituted faculty group and meeting. In 1970 the director promised to have a staff committee recognized by the Governing Body, giving it "as far as possible the constitution and rules of a faculty council." By now the quest for recognition of the workers' union and recognition for the postdoc fellows had enlarged the scope of the "council," now to be called the Staff Committee. While the Governing Body ratified this Staff Committee, the rules were changed to empower the director to appoint the secretary of the Staff Committee who would in turn decide on the acceptability of any agenda subject. Following the appointment of the secretary with this agenda-setting power, this new committee failed to obtain a quorum on its first meeting. The director reportedly took this as a show of no confidence and began to come less and less to the institute and to do more work at home.

SINP Example Two: The Installation of a Computer in Science College

Because theoretical calculations were becoming more complicated and because experiments involved more data storage, physicists at SINP wanted a computer of their own. During 1968–69 the total monthly computer time used by SINP scientists was over 200 hours per month, with the Theoretical Physics Division alone using 150 hours. Computation was being done on computers at five locations: the closest two machines were in Calcutta at Jadavpur University and the Indian Statistical Institute, but the rest involved long journeys. The computer at Jadavpur was plagued with operating problems and did not use the common FORTRAN computer language, and the Statistical Institute computer not only cost twice as much per hour as other machines (Rs 500 compared with Rs 250–300) but also had "such a small memory that we need three programs to run through one integral," as one theoretical physicist said. The third computer used by SINP was at IIT Kharagpur, an hour away by frequent train, but too small for very complicated work and notoriously slow in processing work. Most SINP computation was thus done at IIT Kanpur, 1,000 km away, and TIFR Bombay, 1,800 km away. Theoretical physicists from Calcutta were thus often locked into long journeys on the train carrying their precious boxes of computer cards. Of the 69 computers then in India, TIFR had the one with the largest memory (a CDC 3600 with a 13k memory). This was the big machine Bhabha bargained for with Ambassador John Kenneth Galbraith just after the 1962 conflict with China, when American military assistance to India was accepted. It took up an entire air-conditioned clean room in Bombay. In spite of their location in research establishments, these machines were already used for administrative purposes of government and private business, including the advanced calculations of banks, which could afford to pay the higher fees. The access to computers needed by scientists, therefore, was becoming more competitive and the electricity and personnel costs of maintaining these big machines was being distributed among those who could pay, both state and private institutions.

The purchase of its own computer had been discussed in SINP in 1968, in cooperation with other nearby departments of the university's Science College. At first the new computer was expected to be a gift from the USSR, but the Science College computer committee had not appeared to know that the Russians then had no software for that computer. Although a few Soviet loyalists in the college were prepared to try the Russian machine (even to write software for it), this idea was set aside when the software situation was made fully known. Meanwhile the SINP director was recommending an IBM computer, familiar to everyone who worked in North American and European labs in the mid- to late 1960s. In 1969, the chairmen of IBM, Honeywell, and International Computers Ltd. (ICL) were all visiting India to compete in urging the government of India to adopt their new technology, and eventually ICL secured an agreement for the assembly of their small computers in India as a joint venture with Bharat Electronics (see chap. 20).

Finally IBM was selected for the contract in the University of Calcutta. When I asked SINP scientists in 1968–69 about the new computer, they were particularly vague about how the IBM machine was chosen. One other senior professor said, however:

> This institute has already chosen an IBM 1130. The director called me in to ask me about one technical point. I explained it. He handed me a booklet on the 1130. Later they took advantage of my absence to proclaim they had consulted me. Now nothing can be found out about it. They have taken a vow of silence. That is the way they consult and involve you around here. I don't think they realize that they have chosen a computer more or less useless for research.[26]

The public announcement of this purchase came while the Calcutta University Employees Union demonstrated in Science College against the installation of the computer, saying that it would be used for clerical work and reduce their employment. There were a number of anti-automation protests going on against the installation of computers in Calcutta and other cities, though by a combination of stealth and force a few computers were finally installed at banks, government offices, and major firms during President's Rule in 1969. The press release about the installation of the Science College computer therefore had to stress its use for research:

> It was being imported from the USA at the cost of about Rs 900,000. The Vice-Chancellor, during Monday's discussions, repeatedly pointed out that apart from the Syndicate's resolution, the terms under which the machine was being procured also prescribed against the use of the machine for purposes other than research work. Departments sponsoring the project, apart from the Saha Institute of Nuclear Physics, were Radio Physics, Applied Mathematics and Statistics.[27]

The director of SINP denied to me that any of his employees took part in the anti-computer protest. However, the protest was one source of inspiration to workers in SINP when forming their own union in the following months. Most scientists in SINP predicted that the computer would be installed in spite of protests, at gunpoint if necessary; so it was installed, 33 months later, under police guard. The first public criticism in SINP of the chosen computer came from a physicist who used computers in his work.

> Is it because of the increased complications in managerial problems that the employers and Government are compelled to install computers in their offices? Is it impossible to work out those problems without computers? One gets the answer by looking at the type of computers being installed for commercial and administrative purposes. These are, without exception, small-memory computers with fast input-output, such as the IBM 1401. Thus, unless equipped with auxiliary devices, they are capable of doing only those jobs in which the processing of very small input datum involves only a small amount of information storage.[28]

Dasgupta's statement was presented to the alumni physicists of the Science College, coinciding with the movement in SINP for a faculty council and with the post-MSc diploma controversy in June 1969. Leaders of SINP anticipated that opposition to the computer would be coordinated in SINP and organized in all other departments and would increase the strength of the clerical workers' protests (and by and large they were right). In spite of demands in late 1969 for full disclosure of information about the computer and its location, SINP's director remained silent. Subsequently the vice-chancellor of the university again gave his personal assurance that this computer would be used only for research and not administrative purposes. In 1971 the Calcutta journal *Frontier* published a letter signed by thirty-four scientists of SINP pointing out that in spite of such assurances the other major computers in India were used for nonresearch purposes and therefore research work often got second-class treatment compared with banks at these computer centers.

The director of SINP then issued instructions that anyone wishing to use the computer must sign an undertaking not to interfere in its installation or operation and that members of SINP should not write letters in newspapers which were detrimental to the interests of SINP, TIFR, or DAE. A meeting of the revived (but still "unrecognized") faculty council was then called to demand that the director withdraw these instructions regarding use of the (still not installed) computer. The director, working at this time from the safety of his residence, eventually withdrew his instructions. The IBM 1130 computer was finally installed during the holidays of October 1971, when the university was closed and staff and students of SINP and Science College were away—though as a precaution, a police guard was called in. This month was also the height of the Bangladesh crisis, when half of the *10 million* refugees from the east in West Bengal were camped on land at Salt Lake, where SINP was to be rebuilt. The new computer was installed next door to SINP at the Institute of Radio Physics and Electronics at a total cost of Rs 700,000 (about $100,000) of which SINP's contribution was Rs 200,000. The use of the machine began early in 1972, running about three hours per day, at a concessional rate for educational institutions of Rs 150 per machine-hour.[29] Use grew steadily after that. The process from discussion in 1968 to installation in 1971 took almost three years. As later chapters will explain about the "new" computer business in India, IBM was asked to leave the country two years later.

SINP Example Three: The Diploma and DPhil Confrontation

In 1969 there were twenty students at SINP studying in an Associate Course for Higher Studies in Nuclear Science. This ten-month course for students who already had an MSc was SINP's (sole) claim to be a teaching institution and to qualify for certain grants. Though there were no assurances given that these students would be absorbed into SINP on completion of the course, the students who performed well also had inside knowledge of the few internal opportunities for future employment. The night before their diplomas were scheduled to be conferred, these students met to decide that they would not accept their "Associate" diplomas unless senior students in SINP had their own applications to complete their doctorates (DPhil) processed "satisfactorily." In fact, the DPhil committee of the University of Calcutta *had been delaying approval* of theses being done at SINP; the head of the Department of Pure Physics, whose life was later threatened, was chairman of this university DPhil Admissions Committee. SINP was dependent upon the University of Calcutta for granting degrees to its best students, just as TIFR was dependent on the University of Bombay to grant degrees to its students. Neither institute was yet a "deemed university" in the parlance introduced in the 1980s.

The late-night decision by students to refuse the master's diplomas was inspired by the desire to bring about an exposure of conditions in SINP, to demonstrate solidarity with their "older brothers and sisters" in SINP who applied to work on their DPhil, backed by a youthful expectation that conditions could be quickly changed through confrontation. In his capacity as the chairman of the Governing Body of SINP, the vice-chancellor of the university came to confer these diplomas on students who identified him as the top official with the right authority to move the university in the right direction. When the gathering occurred in SINP's big lecture room, the students dramatically stopped the ceremony to read out their demands, and the director of SINP still offered the diplomas to them in surprised disbelief. When the students still refused, the vice-chancellor personally offered to investigate the matter, which is what the students wanted, and so the diploma presentation was over, and the story was carried in the major daily newspapers.[30]

Students were subsequently called individually into the office of the SINP registrar to be told that they should never do anything that detracted from the dignity of their teachers. Organizers of the protest were threatened with withdrawal of support in the search for employment in India or in applications to study abroad. Most of the students reported being wary of being tricked into agreement and so met to compare their offers and their threats carefully. They expected that the SINP director would drive a wedge between the post-MSc diploma students and the more senior DPhil students in order to avoid further embarrassment. On one occasion Kundu is reported to have said, "I don't eat grass [*ami ghash kachi na*]. I know which teachers are behind all this. I'll abolish the whole diploma course." To which the students then replied: "If you allow the formation of a legitimate faculty-staff association, we will accept the diplomas." The surprised director then revived a dormant Faculty Committee that had been appointed a few years before in order to give him support in "dealing with the students." Finally the vice-chancellor gave assurances that he would investigate the DPhil Committee's performance. The students accepted the diplomas at the beginning of June 1969, and an official guarantee of "no victimization" was given to them when they signed in acceptance of their diplomas. Though this confrontation gained attention in the newspapers, there was no mention of it in the 1969 *Annual Report* of SINP.

A number of important patterns were evident in this confrontation. The least secure group of young student scientists had united to accomplish something on behalf of more senior students, raising their status in everyone's eyes. They skillfully used an issue which involved the University of Calcutta and its relation to SINP, thereby trapping the vice-chancellor in his two roles. But in this context they could also indirectly embarrass the leadership of SINP and call attention to their own inadequate stipends and difficult living conditions. The directors of SINP attempted to weaken relations between these two groups of students and the faculty who were guiding (or supposed to be guiding) their theses. The "dignity of the elders" was invoked, as was customary, but the weakness of SINP as a teaching institution and its vulnerability were exposed. Few of "the elders" were actually contributing to teaching in a classroom because it was perceived as an either unnecessary or undignified task. Others therefore contributed disproportionately to the teaching program, producing a quiet resentment. Finally, because of economic class divisions between the students and lower-paid maintenance and technical staff, others in the institute were largely unaware of this confrontation until it was over. But in the following week these very workers (including those in the workshop described in chap. 16) began demanding that the institute and its Governing Body recognize their union (a pattern followed by workers throughout Calcutta). This new coincidence of interests—the goal of representation and legitimacy—increased both communication and cooperation across this deep status and class barrier. In the end, the union gained recognition long before the faculty council was actually constituted and began meeting.

Conclusions

In the perspectives of working scientists in the late 1960s, conditions in their institutes had to be changed. Convinced of the value of change in the intellectual life of science and sure of the power of scientific thinking, younger or less senior scientists hoped they could accelerate change in the political life of scientific institutions. They wanted to remove conditions that they thought inhibited their personal and professional growth as scientists. This, they expected, would build institutions where they could enhance their scientific reputations (thus enhancing their mobility). Many also wanted these institutes to serve their country in a different way, to make them more self-sufficient, and the country more self-reliant. All this seemed to them to require their participation in the governance of their own institutes. These were sophisticated young people, in close touch with events in Prague and Paris in the spring of 1968 and with American, British, and Canadian student movements of the 1960s—using shortwave radio, word of mouth, and the very open press, long before the Internet. They were, at least in the Calcutta case, surrounded by declarations of revolutionary intent and quiet, sometimes desperate, administrative reaction, before, during, and after the United Front government led by Jyoti Basu. This was a time of high political turbulence in West Bengal. The use of force was commonplace, students were targets of and actors in violence, and a threat on the life of a professor was not considered improbable.

In this desire to change their governance, many scientists said they encountered leaders who sometimes preferred to inhibit active scientists than inspire and coordinate research. Mid-level scientists often said, and only half-jokingly, that they looked toward the pseudo-relief of becoming bosses themselves, if they were patient: "that is how our system works," said one, with irony. Their immediate alternative was either to demand active participation in the governance of the institution in order to determine the local conditions under which research was done or to go abroad, if possible. Going elsewhere in India seemed—to almost everyone in these institutes—the wrong step. Some of these scientists already had an international reputation sufficient for mobility but lacked necessary local influence with authorities who controlled departure currency-control letters, who signed for passports, wrote letters of recommendation, and the like.

How did the scientists explain their intentions and plans, given these conditions in their institutions? The information presented here is the result of a questionnaire I distributed in TIFR, in May 1968, and SINP, in April 1969. In TIFR, 120 of 300 scientists replied; their average age was 30 years, and 42 percent of them had worked abroad after receipt of their doctorates for an average of 4.5 years. In SINP, 72 of 145 scientists replied; their average age was 31, and 34 percent of them had worked abroad for an average of 4 years. These response rates were (and are) considered acceptable and fall within the range of significance for populations of this size—192 out of 445 possible respondents. The majority of nonresponding scientists were either away from the institute or were too busy on those days; a handful politely said that they wished not to participate.

In SINP the opportunity to go abroad was more limited than in TIFR, though the desire to do so was slightly stronger. Just as important to scientists was the question, What influence do conditions of work have on the morale of the institutions and the willingness to do creative work in them?[31] Had all scientists been planning to remain in their institute, the character of their involvement and commitment would presumably be somewhat different from the answers shown here, where *almost half* planned to leave the country, if possible. The intention to move elsewhere was connected to a realization that it might be necessary to shift one's field of research, as shown in table 4.

From the perspective of those involved in the governance of these institutions, there was a choice between the philosophies of "less government is best government," and of "law and order through the full exercise of authority." By asking scientists questions about each institutional system, the answers to these questions were varied. On the one hand, leaders of TIFR denied that the formation of representative group committees was in response to pressures from senior fellows for more participation; the fellows believed the opposite, but the committees were formed, nevertheless. On the other hand, the leaders of SINP were convinced that it was an advantage to have SINP run like a small family, where decisions could be taken swiftly; other faculty and fellows believed the opposite, and little evidence of swift decision-making can be seen in these three cases.

In the life of scientific institutions the issues of size, growth, mobility, reputation, and decline were of greatest concern. Scientists thought that if the institution grew too large, it might become saturated and be unable to grow when a hot subject came along. Missing a hot subject was an abiding anxiety. A saturated institute would not absorb new recruits, and this

Table 3. Scientists' plans for five years, 1968–73

Plan	TIFR (%) (N = 120)	SINP (%) (N = 72)
Go abroad permanently	7	3
Work abroad for 4 or 5 years	37	43
Work elsewhere in India	7	6
Remain in this institute	43	39
Don't know	6	9
Total	100	100

Table 4. Scientists' research intentions for five years, 1968–73

Intent	TIFR (%) (N = 120)	SINP (%) (N = 72)
No change from present work	41	67
Switch within present field	30	22
Leave research altogether	3	3
Switch to entirely new field	4	0
Don't know	14	8
Total	100	100

would indicate exhaustion, because—the reasoning went—new ideas and work come from younger people. Though some did not see such decline as inevitable, it was accepted that saturation would in turn cause a loss in reputation, limit the individual mobility of scientists within India and abroad, and *probably* lead to decline.

All TIFR and SINP scientists had the example of stagnation in some nearby universities in India in mind, and feared decline; for this they had sufficient experience, they said, pointing to specific examples, naming names and places. Some members of SINP apprehended the stagnation caused by unproductive scientists around them. This fear in the 1950s of saturation and stagnation in TIFR had been strong enough to prompt Bhabha to put a special clause in the Rules and Bylaws of TIFR. Clause 4.2.1 stated that no research appointment would be made for more than five years, after which time a review would be held of an individual's progress. If the review was satisfactory, "a continuing appointment" was offered. Though this clause had been used in TIFR's School of Mathematics to limit its size and "eliminate non-working passengers," as one mathematician said to me, there were demands in the School of Physics that the clause should be used in a number of cases, "more directly." Bhabha's preference for efficiency and the occasional departure of a colleague or two had still not inspired the use of this clause. After his death this formal review was reported to have been used "only with mutual understanding and in an indirect manner," as one senior professor explained.

While the leadership of TIFR made a response, however awkward, to this "independent-minded evaluation" of their peers, the directors of SINP resisted it. The relationship of a scientific institution to its city surroundings—what physicists called in their work the "neighborhood-effect"—was ambiguous; though an institute's work was mysterious and alien to most people outside it, its workers were embedded in urban social life and were influenced by the political cultures of their surroundings. The proximity of TIFR to the DAE, its larger budget, and premier status in the country were enhanced and amplified by the optimistic and energetic atmosphere of Bombay. The distance of SINP from the DAE, its acute internal problems, and the prolonged uncertainty of moving to new space and buildings at Salt Lake were amplified by the high unemployment, poverty, and political unrest that contributed to the general crisis in Bengal. Continuity of employment ("a job for life") was a common expectation in Bengal's industrial sector.

The United Front government came to power in Bengal in November 1967 but was removed by President's Rule four months later, and there was no election for one year, until in February 1969 the Congress Party (in coalition) was elected and that remained quite steady until President's Rule was imposed by Indira Gandhi in June 1971. After ten months an election was held and won by the Congress Party, which held power until 1977. This oscillation between local power and Delhi kept West Bengal's relation with the center in continuous tension. That tension reach right inside the university and the Saha institute.

In SINP there was not just saturation but actual overcrowding. Nevertheless, according to observers of the culture of the city, there was a greater tolerance of overcrowding because it was so common in Calcutta. No TIFR-style five-year review clause for senior SINP people existed, and the extraordinary levels of unemployment in Bengal (where single offers of employment received thousands of applications) made people cling to any position available under almost any conditions. Even with the promise of moving the institute to Salt Lake, it was acknowledged that SINP was already saturated and would immediately fill even those new buildings. Resistance to necessary structural changes confirmed scientists' perceptions that the governance of SINP was reaching a state of stagnation and the institute might be near exhaustion. Plans to work abroad partly reflected this, particularly among those who had already done so. Obviously the nurture of a creative climate in scientific institutions required inordinate patience, considerable luck and organizational skill and versatility; the evidence in both TIFR and SINP for 1967–69 supports this conclusion.[32]

Some of the people active in these issues rose to power and influence: physicist B. D. Nagchaudhuri, second director of SINP, moved in 1970 from the Planning Commission to be advisor to the minister of Defence, where he played a key role in India's missile program and in the design of the 1974 bomb test (see chap. 22); physicist Raja Ramanna, member of SINP's Governing Body, was, in 1969, designing the test bomb and, after 1972, played the organizing role in completing that project, and moved to the Ministry of Defence. Both enjoyed direct access to the prime minister. Knowledge of the tension over these conditions in India's labs filtered upward: Prime Minister Gandhi spoke in November 1970 at the Third National Conference of Scientists, Technologists, and Educationists concerning the need for "the *democratization* of decision-making in laboratories, universities, scientific agencies, and indeed in the scientific system as a whole."[33]

EIGHTEEN

Governance and Influence in the Research Institutes Bhatnagar Built

The conditions in DAE laboratories described in the preceding chapter foreshadow the difficulties reported by scientists in CSIR research institutes. CSIR scientists from many laboratories assured me that problems in labs like TIFR and SINP were minor at that time compared with most CSIR labs and that their problems were to be found "in miniature" in DAE labs; only in CSIR labs, they said, were these "Indian problems" experienced in full form.[1] These two separate systems, "scientific and industrial research" and "atomic energy," were like evolving galaxies, drifting further and further away from each other, absorbing more and more of their own energy and information and having less and less to do with each other, though oriented to the same national questions. Nevertheless, India's elites knew what was going on in CSIR, or could know if they wished to; for example, the list of those attending the Governing Body meeting, chaired by Nehru himself, in March 1964 is a "who's who" of the scientific and industrial community.[2] The concentration of power, knowledge, and wealth in the room would have been the envy of those in the 1940s who had said, "If only we could concentrate our resources and focus them on our problems, we would surely succeed."

CSIR's labs continued to be excluded from atomic energy work through the 1960s, and this added distance between these galaxies and is another reason that the CSIR does not figure more prominently in this book on the nucleus and nation. DAE's large contracts brought it closer and closer to big industry, not where CSIR had built or could build its alliances. This does not ignore the effort being made by CSIR labs to be as close as possible to the industries and markets of their subject areas, for example, food technologies, leather, aeronautics, and the like, but only the National Chemical and National Physical Laboratories could have been good partners with the DAE at this time, and they were not. The ties of the CSIR with atomic energy ended with the 1961 death of the only CSIR member of the Atomic Energy Commission, Sir K. S. Krishnan. So scientists working at DAE labs in the late 1960s routinely described anyone who thought of leaving DAE to work in the CSIR as committing "professional suicide," giving an idea of the distance between these two large organizations. Though both were in effect "state conglomerates" in the business sense, there was a marked difference between them from the scientists' point of view. In 1967 a survey was conducted of 9,000 CSIR scientists, and on the basis of 2,300 questionnaires received the conclusion was drawn that 52 percent of them preferred to move abroad, 30 percent to another research lab, and 22 percent to a university lab. Foreign-trained scientists (438 of them in the study, almost 300 with PhD's) were particularly sensitive to the two variables of economic reward and good working conditions.[3]

The CSIR had become a conglomerating organization, expected to inherit and support research institutes of many diverse types, some of them new and others established well before 1947. (An example of this in Hyderabad is given below.) By the late 1960s there were dozens of CSIR institutes with thousands of staff, all managed through headquarters in Delhi. CSIR changed its leadership in 1966, at the same time that the DAE changed after Bhabha's death. Atma Ram (Saha's student and collaborator at the glass and ceramics lab in Calcutta) now became CSIR's director general, and he announced he would sweep away many of the changes inspired by Husain Zaheer, implying that there had been irregularities in the appointment of directors of some laboratories, though he did not offer proof.

The prime minister was still the president of the Governing Body of CSIR. After patiently listening to public criticism of the CSIR and after great delay, a Committee of Enquiry into CSIR was appointed in May 1968 to be chaired by Justice A. K. Sarkar.[4] There had been three review committees before this one, in 1949, 1954, and 1964, all chosen and/or approved by the prime minister.[5] The Sarkar Committee, whose members were physicist M. G. K. Menon (TIFR), geneticist M. S. Swaminathan (Indian Agricultural Research Institute), statistician C. R. Rao (Indian Statistical Institute, Calcutta), plus elected members from the Congress Party, the Communist Party (India), and the Jana Sangh. They interviewed fifty scientists, industrialists, and senior administrators, including foreign experts like Patrick Blackett. By the time it presented its two reports, Part I in 1970 and Part II in 1971, CSIR laboratories had 8,000 scientists (and many more support staff and daily workers) and a total budget of Rs 200 million. The Sarkar Committee's prolonged three-year discussions apparently had a chilling effect within the

CSIR, because its work was used to delay many crucial decisions, pending the receipt of the committee's report.[6] When Part II appeared, Atma Ram was already scheduled for retirement in 1971. So the regime for the reforms that the Sarkar Committee had been established to define, was now passing. The committee even recommended winding up the Governing Body, which was composed of some people who seldom came to meetings. During Atma Ram's tenure (1966–71), the Governing Body had not even met for the minimal two meetings per year required by the constitution (it was not wound up, however). So serious (and so complicated) had the situation become in 1971 that the Sarkar Committee asked that the Ministry of Finance's representative in CSIR's Governing Body should cease to have regulatory powers on expenditure and keep a focus simply on the budget. In effect the Sarkar Enquiry Committee became involved in trying to direct the reforms they were calling for. Meanwhile the CSIR had already instituted a new post, the director general for vigilance, located within its own headquarters. This move anticipated correctly where the criticisms would point. Moreover, said the committee, any future expansion of the CSIR should depend on contracts and agreements with industry or government. This was a demand made by most previous CSIR review committees and often repeated by Indian industrialists, but without effect.

Enter Atma Ram, with Patrick Blackett in the Shadows

Having earned his doctorate in physical chemistry at Allahabad University in 1937, where he also met Saha, Atma Ram worked in Bhatnagar's CSIR research team in Calcutta during WWII and then was handed responsibility for developing the Central Glass and Ceramics Research Institute (CGCRI) laboratory in South Calcutta in 1945. Here he came again into regular contact with Saha. Much later, when he became the new chairman of the National Committee on Science and Technology (NCST) in 1977, he explained that he had long before concluded that India had "not created in our people what I would call *machine sense*" (his emphasis). He recalled, "While I was unemployed after completing my doctoral work at Allahabad, I joined a sugar factory as an unpaid apprentice." At that factory and later at the CGCRI, Ram met and was influenced by "a sugar technician without formal education" and "a very ordinarily educated potter." The technician taught him how to grow sugar crystals in vacuum pans and the potter how to make very large ceramic pots. He worked, he says, with the technician and applied some chemistry to improve the quality and quantity of sugar crystals, and the potter made valuable suggestions about the lab's ceramics kiln. "We must enthuse our scientists with the spirit of mastering *skills*, and not merely considering skills to be the job of *mistris* [mechanics] and operators."[7] He remained underexposed to foreign institutions and laboratories and retained his earthy, somewhat Gandhian persona, which must have attracted him to "traditionalist" elements of the scientific community, and even to the deputy prime minister Moraji Desai at this time (1966–69). Even suave and complex Patrick Blackett, who agreed deeply with the importance of "machine sense," found something valuable in Atma Ram, despite the cultural gap between them.

Shiv Visvanathan's astute analysis of NPL builds upon the 1963 Blackett report and other official review committees, examines the change in lab directors and their relations with directors of the CSIR, and studies intergroup relations through case studies of ceramics, telephones, ferrites, and mica, all involving Indian (Bharat Electronics) and global manufacturers (Philips), and even the US military. In previous chapters it has been shown that the NPL was the lab most exposed to political scrutiny because it was so readily available, just down the road from the centers of government and power. Visvanathan's work concludes that

> the controversies in India surrounding the reorganization of the CSIR reveal the need for the sociological approach to science. The debate demonstrated that little attempt had been made by policy-makers to understand industrial research within the wider structures of the innovation chain. The disillusionment with industrial research is an inversion of the euphoria of the early Nehru era. But it also reflects the deeper dissatisfactions with current paradigms of science and technology in India, and shows the import-substitution model may be at the end of its tether. The cracks of doubt may one day inaugurate a regime of new possibilities.[8]

The situation facing CSIR was given some exposure by Patrick Blackett after Atma Ram's appointment in 1966. Blackett chose "innovation" as his subject for the new Nehru Memorial Lecture in 1967. His approach was provocative and may have helped Ram a bit. Attended by Prime Minister Gandhi and reviewed in all newspapers, Blackett's lecture called for new thinking about "the innovation chain" from research to production, revealing a rather linear way of thinking about innovation. Blackett wanted industry in India involved in the process of innovation from its inception. He criticized the spurious "sanctity" of basic research, knowing full well that there were CSIR labs and research groups in them that were largely oriented to basic research. Blackett's Nehru Lecture was received gleefully in some government and industry quarters, because it shook things up a bit. Some researchers, however, including those who respected him at NPL, whose lives depended on the "sanctity" of basic research, felt he had attacked them, and some younger "basic" researchers felt betrayed because Blackett's lecture seemed to undermine their vulnerable position, even in the elite laboratories founded by Homi Bhabha and visited by Blackett himself. With Bhabha and Nehru gone, they felt they had no protection.

There was a negative reaction to Blackett's lecture among researchers in labs in Bombay and Calcutta, but in CSIR labs it was more acute. Blackett probably never heard about this adverse reaction and would not understand that researchers "in the ranks" thought the support for their basic research was fragile, nor how vulnerable they felt. The lecture would, younger scientists thought, embolden an anti-intellectual tendency, showing resentment toward research that did not have evident applicability. Blackett's most recent direct contact was in 1963 when he chaired the committee of inquiry into the National Physical Laboratory, and he was now communicating at a level very far removed from researchers "in the ranks." He thought that India's support for basic research was now sufficiently strong, and so he could press harder for practical applications. This was also what Prime Minister Gandhi was thinking, as she would soon explain to him. Atma Ram was moving in the same direction, so Blackett pushed him further. A few days after the 1967 lecture, Blackett wrote to Ram, "It has been quite clear in recent months that our ways of thinking are very similar. . . . I do realise the difficulty you will have in guiding Indian policy in a new direction."[9] Blackett had not before so publicly attacked what he viewed as the irrelevance of much research in India, including basic research. Given his prestige and access to the new prime minister, it is no surprise that scientists in some labs felt uncomfortable about Blackett's lecture in 1967. Ram continued this same line of reasoning in his January 1968 opening address to the Indian Science Congress; protests from scientists there also greeted his call for more application of research to development, and more technical innovation. They said that fundamental research was too fragile in India to brush off attacks, even indirectly, from authority figures who should be supporting it too. But Ram's view was, after all, the prime minister's view of the role of science and technology too, and Ram took some of his cues from meetings with her since his appointment to the CSIR eighteen months before.[10]

The influence of lab directors and senior scientists extended right from their laboratories to the Royal Society in London at this time. Some were sounding out prospects for nominations as Fellows, as we shall see. Others sought to involve Blackett in their struggles in India, and it appears Ram fought them back, sometimes through Blackett. For example, writing to Ram in 1967, Blackett criticized "the otherwise excellent man Varma, talking lightly about starting to design an electron microscope at NPL. It seems that all you and I have been saying has not been taken in at all."[11] Ram and Blackett were talking about the relevance of CSIR investments in research which were supposed to lead to development.

The University of Delhi asked Blackett about a British offer to transfer a small (obsolete?) accelerator from the Cavendish Laboratories at Cambridge University to Delhi, in 1966. It was to be accessible also to researchers at the nearby NPL. At first Blackett thought it would be a good machine for training purposes, but he examined the project carefully when he visited India in 1967, only to realize "that the machine was asked for not by physicists at all but by engineers who wanted some way of playing around with high voltages. I do not think they should be allowed to do this, as the techniques have almost no relevance in any other branch of engineering. The fact that they should want to do this shows, I think, the underlying drive towards prestige subjects, which is quite a danger in India as well as here."[12] Was this remark about disdain for engineers? About preference for serious research versus "playing around"? About conquering the impulse for prestige and high technology? I think this letter is about Blackett trying to help Indian directors like Atma Ram and A. J. Kidwai fight entrepreneurial "chaos" and bring control to the laboratories, as much as it is about avoiding unnecessary importation of equipment. In fact, CSIR laboratories like NPL wanted to make their own deals and import their own equipment without the constant need to seek approval of the director general's office in Delhi, but that office is where "customs clearance without duty" certificates were issued. Directors were constantly referring to the limitations placed on them by headquarters and the special powers of its growing administration.

Conditions in Two CSIR Laboratories

Although a comprehensive review of CSIR laboratories before or during Ram's tenure is an undertaking beyond the scope of this book, two small illustrations provide some understanding of the difficulties in the organizational culture of the CSIR. The two illustrations are the Indian Institute of Chemical Technology and the National Geophysical Research Institute, both in Hyderabad.

The original home of the Institute of Chemical Technology was established by the nizam's government, surprisingly, during the war in 1944, and Husain Zaheer became its director in 1948. It was taken over by the CSIR in 1956, to be called the Regional Research Laboratory. Its deputy director Bharat Bhushan recalled the introduction around 1956 of a system of headquarters-style management at each CSIR laboratory. This top-down approach coincided with M. S. Thacker's arrival as director general. However, said Bhushan, under Director General Zaheer, the laboratory began to retrieve its original autonomy about 1962. Zaheer obviously knew Hyderabad better than many other labs in his control as director general, having spent much of his working life there. From then on a tug-of-war evolved, between CSIR headquarters in Delhi, the executive committee in Hyderabad supposed to govern the lab, and the laboratory director. Opportunities for projects, including international collaboration, were sometimes blocked by CSIR headquarters in Delhi (local people usually added the word "arbitrarily"). In the case of this lab, a condition of transfer of the building and institution to the CSIR in 1956 had been that the chief minister of the surrounding state of Andhra Pradesh would be the chairman of the institute's executive council. This stipulation opened the lab's governance to a central vs. state government tension that came to a head during Atma Ram's administration.[13]

The other illustration in Hyderabad is the National Geophysical Research Institute. Founded in 1963 in the basement of an Osmania University building, it remained there for six years before inheriting a new building. Said its first director Hari Narain,

> Indian science has been notorious for its "crab" mentality, and NGRI could not be an exception. . . . A number of committees were set up by the CSIR, the Ministries and the Planning Commission to confine NGRI in a cocoon, to carry out research in geophysical exploration and interpretation. I was carefully excluded from these committees and wasn't even called for a hearing. Things grew so bad in 1969 that I was forced to resign. Soon Directors of a few other CSIR laboratories, who were also under similar stress, resigned or threatened to resign. The President of CSIR Indira Gandhi had to intervene. My resignation was not accepted. I was provided full support, and the Executive Council of the Institute was held in abeyance and later reconstituted.[14]

Among the contentious issues were foreign geophysical surveys in India and use of bilateral aid in paying for them, as well as protests about undercutting Indian expertise in this field; in fact, some bilateral aid for mineral or water development was tied to the use of foreign consultants who participated in these surveys. This linkage was common in other fields. Narain's resignation clearly drew the attention of the prime minister's office: Hari Narain was appointed a member of the NCST in 1971.

Hyderabad was, by the late 1960s, a city with regional economic ambitions and political power. Some new labs and industries were coming, like the Defence Research and Development Laboratory in 1972, to rise from the university or official buildings built originally by the nizam. The industrial and research importance of Hyderabad by 2000 attests to this strategy, and it would compete with Bangalore for prestige. The appearance of political figures in both these stories, and the inability of CSIR leaders to use their influence in order to resolve "internal" CSIR issues, and the unstable result of their powerful intervention were features of life throughout the CSIR system. This shows how complex the job of the director general was and how little in the way scientific training or experience would prepare anyone for it.

Changing or Tinkering with the System?

This systemic instability was just what the Sarkar Committee was supposed to address. The Blackett papers in the Royal Society archive show that a loose CSIR network was still running through Blackett's office in the late 1960s. Just before the appearance of the final report of the Sarkar Committee, Blackett made his final visit in 1971, at age seventy-three, a nine-week tour in the hot months of March–May. While there, in March, he observed the reelection of his old friend Indira Gandhi as prime minister. The end of Atma Ram's term as director general of the CSIR had been signaled to Blackett, and the talk was of appointing someone much younger. Blackett toured all his familiar scientific sites like TIFR and the National Physical Laboratory, plus new ones like the Space Research Centre in Kerala and Bharat Heavy Electronics. He went to a dozen CSIR institutes. He met again with the most powerful person in Indian science, Vikram Sarabhai, and with M. G. K. Menon at the TIFR, Y. Nayudamma at the CSIR's leather research institute in Madras, and Pitambar Pant of the Planning Commission. He toured the Trombay Atomic Research Centre (now renamed after Bhabha) in the company of astronomer S. Chandrasekhar from the University of Chicago, someone he had known since the early 1930s at Cambridge. Although the CSIR paid the expenses of Blackett and his wife Constanza on this last journey, he was also invited to advise the Indian Statistical Institute by P. C. Mahalanobis and the University Grants Commission by D. S. Kothari. Blackett remained ready to try new things in India. As an example, while president of the Royal Society, he promoted a new School of Genetics on the campus of the ISI in Calcutta during 1969–70. This institute was where Blackett's friend J. B. S. Haldane had first gone to work when he left Britain to live in India. His conversations with Indira Gandhi showed Blackett that although she retained her father's respect for science and scientists, she had also learned a lot after his death and now linked "science" more strongly to "technology."

Indira Gandhi was determined to see more practical results from the state's massive investment in industries relying on science and technology. Whereas her father had aligned science with "education" and "culture," Indira Gandhi reasserted the link of "science" to technology and industry. She was also prepared to redesign institutions to achieve more synergy between science and technology and, like her father, took a direct interest in key scientific appointments. She was preparing to move the CSIR out of the education portfolio to a new ministry, the Ministry of Planning, and to give the portfolio to C. Subramaniam, one of her most effective and powerful ministers. She not only had the Sarkar Committee studying the CSIR's problems, but also a "scientific departments committee" within the Administrative Reforms Commission. In writing the report on Scientific Departments for the Administrative Reforms Committee (1968–69), M. G. K. Menon, of TIFR, and an Indian Administrative Service secretary wrote the first draft on reform of the CSIR, but Ram did not like it and would not approve the draft, so they had to turn the CSIR report writing entirely over to him! Ram was unfriendly to the DAE, though it can hardly be said to have been friendly to him. (This would be replayed, in the inverse, when Ram investigated the DAE; see chap. 24.) Other agency and department heads wrote their own reports too, with little cross-cutting and cross-checking. According to Parthasarathi, the only members to come to all meetings of the committee were Sarabhai, Menon, Thacker (now at the Planning Committee), industrialist Arvind Mafatlal, and its chair V. K. Mathrani.[15] Most department heads left each other alone. It was the Sarkar Committee that provided a basis for dealing with some legal difficulties in which the CSIR director general had got himself, as we shall see. Taken together the effect of these two inquiries was powerful.

Various lab directors wrote to Blackett in 1970–71 to ask him to persuade Atma Ram to promote their objectives, improve liaison with CSIR headquarters in Delhi, and provide more freedom for their scientists to innovate. They did this in spite of the fact they knew Atma Ram was supposed to retire and knew Blackett was very critical of the whole CSIR system and its labs. Some directors had just drawn back from the brink of a mass resignation against Ram's administration. Without being opposed to their pleas for more autonomy, Blackett said in his confidential 1971 report that the central problem facing some CSIR laboratories is that they have

> shown a tendency toward rather basic research without always having any very definite practical goal. Some of the work seems to me to be more suitable for university departments than for government research stations, which were undoubtedly set up with the general goal of producing practical results of use to industry and agriculture. This tendency towards purity is by no means only to be found in India [he then mentioned the tendency in the UK]. I think there are two main reasons for this drift to purity. One is that it is often much easier to do good basic research than good applied research of interest to industry. The second reason is the widespread misapplication (in my view) of the principle of the importance of the "freedom of science." . . . In the field of basic curiosity-driven research, it is fully established that able research workers should be given the greatest possible freedom to plan and execute their own research. . . . What has gone wrong, I think, in some government research stations is that the principle of scientific freedom has sometimes been applied to mission-oriented R and D, where it is largely inapplicable, instead of only to curiosity R and D where it does apply.[16]

In particular, he said that some labs "do not always seem to understand the importance of directing work toward a market requirement and close collaboration with industry, and so are surprised when their work is wasted. Then there are some individuals in certain stations who don't seem to be trying very hard to do anything useful but think it is more important to advance basic research." Blackett again recommended increasing the powers of the Governing Body of the CSIR *over* the strategic direction of its labs (leaving the tactics "firmly in the hands of the Directors"). He said India should study how cooperation of the Harwell atomic research center and industry actually worked and how government funds were available to industry scientists (not just to government scientists) for this cooperation. Y. Nayudamma was praised at the Central Leather Research Institute for appointing four economists to its staff, thus providing it with a better understanding of its markets and clients, unlike other CSIR laboratories. Nayudamma was soon to have begun work as the director general of CSIR, succeeding Atma Ram, and was also appointed to the Board of Governors of the International Development Research Centre in Ottawa (en route to one of its meetings in Canada he was unfortunately killed in an airplane crash).[17] But Blackett did not exempt industry from criticism, saying that "the backwardness of much of Indian industry itself" was partly to blame. He pointed to an internal CSIR study of ten years' experience with £10 million expenditure on twenty-three industrial pilot plants, a study that concluded that the plants "brought in little financial return."[18]

Blackett felt the scientific community in India had reached a certain maturity and should be able to tackle the practical problems that confronted the country, which were in his opinion poverty and population. He was not alone in his criticism of the scientific community. Writing from New York in 1971, P. C. Mahalanobis warned Blackett that he also did not like what was going on in India: "Indian science is in a state of confusion. . . . The immediate future does not look too bright . . . because we still remain a structured hierarchical system."[19] Sensing this concern among many reputable scientists, and some science leaders, Indira Gandhi spoke a number of times in 1970 and 1971 about undemocratic practices in laboratories.

The Blackett-CSIR relationship in the late 1960s was also seen as an opportunity by friends of Atma Ram to approach Blackett about an election to fellowship in the Royal Society, before Ram retired; Blackett then sounded out people in London about Ram's eligibility. One replied, "I understand Atma Ram is a friend of Moraji Desai, so with the present state of Indian politics his election might be regarded as an affront to Indira Gandhi. . . . No great harm would surely be done by letting AR wait a year or so until further enquiries are made."[20] Blackett replied in a week, conceding that there are differing views on Ram and said that his name had been "withdrawn from this year's list" and he would collect various opinions about the merits of the case when he next visited India.[21]

This was Blackett's last journey through India, at age seventy-four. In 1972 he began to experience the fatigue of a debilitating illness from which he died in 1974. Few foreigners played a role in independent India such as the role Patrick Blackett played in military and scientific affairs. Blackett understood early how the decline in India's foreign exchange reserves would influence major strategic programs, determine the level of imports, influence relations with rich countries, and lower the volume of subsidies sought by the Indian government. This reinforced an attitude of "self-reliance" in circulation before the 1940s, articulated by Indian scientists (among others) long before any real prospect of independence for India. This attitude of self-reliance in science and technology traded on the fading memory of Gandhi's idea of a self-reliant industry, made-at-home goods (*swadeshi*), and independent-minded politics. The argument for "self-reliance" became stronger, although paradoxically it was increasingly difficult to put into practice because powerful new forms of economic and technological dependence were steadily building up. While Gandhi would not have imagined this idea applied to military and industrial development, Nehru and his political advisers learned that it would be a necessity because they could see few alternatives. At the same time Blackett's independent views on nuclear weapons and "the arms race" were a stimulant to discussions with Nehru, Bhabha, Kothari, and other scientists. He was a regular guest of the chiefs of staff, discussing both disarmament and armaments, and he promoted the idea of science applied to military affairs. In this sense there was an official acceptance of his presence and advice. During this period Indians studied and learned how to optimize the potential of great-power rivalry to obtain favorable terms for commercial and military contracts and agreements.

The Scientific Workers Movement and the CSIR

If half the CSIR scientists wanted to move abroad in 1967, was there something about working conditions that made them so dissatisfied; after all, this was a "dream job" when they were younger, a job for which they had studied hard and felt themselves prepared.[22] The story of the scientific workers movement in India starts in 1946, though undoubtedly people in India had thought of this long before, and there is an important British connection.[23] The World Federation of Scientific Workers (WFSW) was formed in London during the Royal Society's Newton Tercentenary Celebrations in July 1946, at which Meghnad Saha represented India, accompanied by his student Atma Ram, who was studying in London. On returning to Calcutta, Meghnad Saha started the Indian Scientific Workers Association (ISWA) as a part of the movement, and propagated the idea in his journal *Science and Culture*. The membership was limited to the circles close to him in Bengal. Other scientists in India thought that it was better to launch the movement on a much larger scale at the forthcoming Indian Science Congress in Delhi. Accordingly the Association of Scientific Workers of India (ASWI) was founded at this Science Congress by a resolution moved by Homi Bhabha and seconded by biochemist Bires Chandra Guha, who had been president of the congress the preceding year. Patrick Blackett and John Bernal had both been invited for this purpose. Then Nehru inaugurated the association because he was the general president of the Indian Science Congress and was elected as the president of ASWI. The association was to be supported by two general secretaries, B. C. Guha and physicist P. K. Kichlu. Guha was a professor in Calcutta and Kichlu in Lahore. The task of this body was to draft a constitution to be ratified at the next Indian Science Congress in Patna. It was decided on the spot in 1947 to start a publication called *Vijnan Karmee*

(Knowledge Worker), and the responsibility of editing this publication was given to Kichlu while Guha worked on the constitution. Thus Guha was building a scientific workers association (ASWI) that contradicted Saha's strategy in forming the ISWA. Though Saha suspended further expansion of the membership of his association, he did not dissolve it because he waited to see if the new association would be sustained. Because India was becoming independent, he wondered if Nehru would find time to look after ASWI as its president. But surprisingly Nehru did find time to look after the ASWI, and the main task, finalizing the constitution for ratification at the 1948 Patna Science Congress, was achieved; thus the ASWI became functional, and Saha's association continued to run parallel with it. The first general assembly of ASWI was held in 1948 at Patna. Because some of the original executive members now lived in Pakistan and were no longer members of ASWI, the General Assembly chose a new president in Major General S. S. Sokhey and reelected general secretaries Guha and Kichlu; along with new members of the executive committee, the organization began to function. In Delhi director K. S. Krishnan was given responsibility to assist the new ASWI at NPL, making that laboratory central in the growth of ASWI for the coming years. But during the 1950s this association remained small in its membership. According to Visvanathan, the 1961 suicide of agricultural scientist M. T. Joseph "was to have a remarkable impact on unionism in science and emphasized also the particular problems of science operating within a culture of scarcity."[24]

Also in 1961, K. S. Krishnan, the renowned director of the NPL, died; he was hard to replace and it was not until three years later that P. K. Kichlu was appointed to the job. A former student of Saha, Kichlu had a position teaching physics at Lahore and Chandigarh, before moving to Delhi University in 1950, where he worked to produce optical instruments by creating an optical workshop in the physics department. That department did not consider this workshop as a legitimate academic activity, putting him in a constant problem with the university and his colleagues. He was therefore interested in a move, perhaps to the already established Optics Division in NPL; it had already done some internationally recognized work on interferometry. Within this big optics group were colleagues vying to rise to senior positions and struggling for promotions, a matter over which K. S. Krishnan had been quite tough. When Krishnan died, V. M. Vaidya of this group took over as the acting director of the whole lab. As Vaidya had little professional reputation, it was evidently necessary to find a "permanent" director, and S. C. Jain, who had wanted to replace Krishnan as acting director back in 1961, hoped for this opportunity when the position appeared open. Kichlu was waiting in the wings, not far away at the University of Delhi. But this was the beginning of the era of a reformer director general of CSIR, Husain Zaheer.

It is important to understand how Zaheer became the director general of the CSIR; he was one of the sons of a judge with whom Nehru's father, Motilal, had friendly relations, and so Husain had known Jawaharlal right from childhood. Like Nehru, Zaheer considered himself a leftist, and his brother Sajjad Zaheer was a communist. So his credentials were good, though little is ever said about Zaheer's research and development abilities. Atma Ram was among the senior directors within the CSIR and was hoping to become director general when Thacker retired in 1961. But Nehru frustrated Ram's hopes and asked Zaheer, then director of a CSIR regional laboratory in Hyderabad, to take the responsibility of making CSIR work better and fulfilling the purpose for which it was created. This is why he tried to build a new team of directors of CSIR laboratories, and Kichlu was one of them.

Scientists like S. C. Jain in NPL may have hoped to become director of the lab, but instead Nehru accepted Blackett's recommendation to appoint P. K. Kichlu. and Jain lost out. That created tension because Kichlu as director from 1964 did not give Jain an "out of turn" promotion to soften his disappointment, despite being advised to do so by Zaheer. At about this time, Zaheer took a decision, in keeping with regulations elsewhere in India, that a person could not be a director of an institution in the CSIR after he or she reached the age of sixty-five. Having almost reached that age in 1966, Kichlu himself resigned under pressure, circulating a document explaining why he had done so; on retirement from NPL he organized his own company known as the Optical Instrument Company. The same age rule was then applied to Zaheer by CSIR's Governing Body, and he was required to retire as soon as he reached the age of sixty-five. Thus in 1966 CSIR had to find a new director general and the NPL a new director. Enter Atma Ram, Saha's protégé and director of the Central Glass and Ceramics Research Institute in Calcutta, chosen by the Governing Body in March as the director general, to take charge six months later. Such moves and maneuvers were not unique to the scientific community, nor were they peculiar to India, but they took on a particular intensity because of the long time required by scientists to reach a level and age where they could enter the stage of struggle for leadership, offering a very short window in time before the requirement to retire.

Atma Ram had to build a power base where he had none, and he well understood the culture of rank and status envy that ran through CSIR. He started working to identify Zaheer supporters and to discipline or punish them. Moreover, being a senior director in the system, Atma Ram knew some financial irregularities had been committed in the name of Zaheer. The budget, after all, was large, almost Rs 200 million, and when Ram came to office in 1966 he ordered inquiries against these "Zaheer" appointments; this information he passed to the Central Bureau of Intelligence (CBI) for further investigation and he asked them to file specific cases with charges. He also decided to bring pressure on the ASWI, saying that it could not register as an association because it was essentially a trade union, and moreover its members were not all within the CSIR system of employment, so it could not be "representative." Based on his years of observation of both men, Prasad states that "the differences between Husain Zaheer and Atma Ram were personal and each tried to weaken the hold of the other."[25]

The following account of the scientific workers' situation in the NPL is based on communication with Ram Prasad.

> The ASWI General Secretary was Y. R. Chadha, who had been given the position of Director of the Publications Branch of CSIR, which was working as an independent organization within CSIR. He was put on probation and Dr. Atma Ram was taking action to demote him. When Nehru was alive in 1964 Zaheer gave a grant to the ASWI for organizing an international seminar, but some of the money had been used in sending some people abroad to take part in WFSW activities. After Nehru's death Zaheer had lost his prime ministerial support. Atma Ram asked the CSIR to take legal action against ASWI, declaring that this was an unauthorized use of the grant. This is why Y. R. Chadha took a long leave and why ASWI officers appointed me as their acting General Secretary. At the time I was not aware what ASWI was. Its membership was confined to CSIR and their counterparts in agricultural research, and the members seemed more interested in day-to-day activities than the broader aspect for which the WFSW was created in 1946. I called two general assemblies side by side, one for the ASWI and the other to create a Scientific Workers Association of CSIR (CSIR-SWA). So K. R. Bhattacharya was chosen as the General Secretary of CSIR-SWA, and Narendra Singh (who had done remarkable work as a makeshift Secretary during the early time around 1948–50) was chosen as its President. Both were located in Central Food Technology Research Institute Mysore, also a CSIR lab. I was asked to become regular General Secretary of ASWI. A noted (and uncontroversial) scientist was chosen as President of ASWI. As the status of ASWI began to decline, the status of CSIR-SWA began to rise. Thus I became acting General Secretary in 1969 of the CSIR Scientific Workers Association, and then President of the CSIR-SWA in 1970.[26]

In all this activity, the Delhi location of the NPL made it the center for these political changes: Visvanathan described the rising relevant and increasing membership of CSIR-SWA from 1973 onward, including staff, technicians, engineers, and scientists. This too was a part of the war over self-reliance. The inclusive nature of the association was new; the Staff Association of NPL had been restricted to those classified as scientists and mostly organized seminars. "The leaders of the association realized that unionism in the laboratory was still new and that the scientists had to be convinced that unionism was not infra-dig."[27]

Scientific Temper in a Litigious Climate

During this time the legal cases against laboratory directors appointed by Husain Zaheer were being developed by the CBI, and these cases came to the prime minister's desk in 1970, about a year before Ram's retirement. They are discussed here, not because the CSIR would have been *alone* in facing a litigious confrontation (litigation over working conditions was then common in India), but because these cases involving leading scientists in their field were particularly notorious.[28] While he was director general (1961–66), Zaheer had promoted three of his former colleagues at the Regional Research Laboratory Hyderabad to the position of director in three separate CSIR laboratories. Ram leaked some of this information about appointments of directors in 1967, attacking "Zaheer's boys" at headquarters and in the directors' offices where Zaheer had made the appointments. According to Parthasarathi, "The Zaheer group hit back in kind. This was soon extended to Parliament [in 1967–68]. As a result the whole atmosphere in the CSIR got vitiated."[29]

Though the Sarkar Committee "basically dismissed most of the allegations" against these directors, according to Parthasarathi, Atma Ram had built up enough pressure inside and outside government to make Prime Minister Gandhi agree to these CBI inquiries against five CSIR directors in early 1970 (a year before his retirement in June 1971), and the cases were to be decided before Sarkar completed his work and before Nayudamma took office. The cases were all eventually dropped, with the prime minister's approval, but the CSIR was harmed by damaging publicity around a few of the allegations which were not dismissed. One exception to dropping these cases may have been the case of physicist Piara Singh Gill, director of the Central Scientific Instruments Organization in Chandigarh, who was required to resign in 1971 after a CBI investigation.[30]

This was the period, in 1969–71, of the conception and creation of the new Ministry of Science and Technology, and its active agent, the new National Committee on Science and Technology. This ministry was a creature of neither CSIR nor DAE, but a kind of holding company for some of the institutes and programs which did not fit well into either of those two other huge organizations. When Electronics and Space both became separate ministries, the scope of the Ministry of Science and Technology was even more circumscribed. As one observer noted dryly, the other agencies "derive their budgets directly from Parliament, and it is the Ministry of Finance and the Planning Commission which largely decide on the resource allocation among the various ministries. The Department of Science and Technology has little share in decisions on resource allocation even among the scientific departments."[31] And because the Sarkar Committee reports in 1970 and 1971 were addressed to CSIR's internal financial situation and governance, the commission was unable to alter that vast organization's relationship with cabinet or the DAE. So when Ram left the CSIR, it was not much stronger with respect to the cabinet than when he started. Moreover, the procedural and personnel problems that plagued it were largely unresolved, including the notorious requirement that in order to promote a scientist performing well, a new post had to be created for them into which they were appointed, leaving their old one vacant for someone else.[32]

But the prime minister had heard enough and understood that a major problem was that CSIR's financial autonomy was limited to issues below a certain cost, and above that all files had to be approved by the Ministry of Finance. Unlike DAE or Space or Electronics Commission, there was no member for Finance in the Governing Body. However, in 1971, through negotiation involving the prime minister, the new Governing Body was now smaller, chaired by recently appointed Director General Nayudamma. The minister of Finance approved a position for a finance member on the Governing Body, thus giving the CSIR important powers of government—administrative and financial.[33] The minister in charge (under the prime minister) was now the influential Planning Minister C. Subramaniam. So the Sarkar Committee and the Administrative Reforms Commission had provided the cover and stage to make changes in the governance of the CSIR. When Nehru had hoped to be less embroiled in the affairs of the CSIR around 1955, his daughter fifteen years later was engineering a major change in its operations and structure.

NINETEEN

Articulating Science and Technology Policy for Indira Gandhi's Cabinet

A new cabinet, with its new formal advisors and informal networks, presented a challenge for the new scientific elite of which Vikram Sarabhai was a sort of a young crown prince. Nehru and then Bhabha had so dominated this process until 1966 that their absence opened up a completely new situation. There had been an agreed method to influence policy, but there were so many unseen and complex routes to ministers that this Scientific Advisory Committee to the Cabinet (SACC) was frequently outguessed or outmaneuvered by others. Its effectiveness was called into question so often that it was gradually, and then abruptly, replaced by another body in 1970, operated out of the prime minister's office. This chapter describes the structure of power around Indira Gandhi while Sarabhai was alive and explains the situation that leading scientists found themselves in when they tried to establish control of the science and technology agenda.

During the short limited war between India and Pakistan in 1965, a more serious crisis was emerging, demanding complete attention in 1966 when the new prime minister took power. The brief and expensive thrill of the war began in April during routine border skirmishes in the desert of the Rann of Kutch, during which commanders on both sides agreed to use only infantry and to avoid use of the aircraft against ground troops.[1] Emboldened by their performance, Pakistan's commander in the region persuaded President Ayub to permit the infiltration of troops into the region on the Kashmir border and then to begin after the rains in August a tank-based war on the border in Punjab near Kashmir. The Pakistani intention was to "liberate" Kashmir. Both armies were using American equipment, and, together, with literally hundreds of tanks in combat, they created the largest tank battles in the world since World War II. Pakistan was then surprised by American disapproval of its adventurous approach and by the imposition of an American arms embargo on both countries. Nevertheless, by agreement of the chiefs of the two air forces, no bombing of civilian populations occurred.[2] Indian strategic elites could now point (largely for their own benefit) to two new risks: the Chinese atomic bomb in 1964 and the Pakistani military conflict with India in 1965. This "war" might have been forgotten had not the Russians intervened and brought the two countries together for mediation four months later in Tashkent. But it all was *really* forgotten in India because of a wider crisis of foreign exchange reserves in the Reserve Bank of India, the drought and famine in eastern India, and the unexpected death of Prime Minister Shastri while negotiating a conclusion of this war, under Soviet supervision in Tashkent.

The official approach to the first problem was the secret rupee devaluation, planned during the spring of 1966 just after Shastri died. Foreign exchange reserves had fallen to their lowest level since Independence. International advisors had long been trying to persuade India to devalue the rupee. Basically the postwar sterling balances had been exhausted, and India's dollar reserves had never been large enough for India's ambitious import-led industrialization program. Users of foreign exchange were put on a short list and had to submit to Cabinet priorities, and ministers competed to get their important projects funded. Following discussions with the British, the Americans, the World Bank, and the International Monetary Fund, Indira Gandhi's inner group secretly planned and managed a 57 percent devaluation of the rupee in June 1966 from Rs 4.76 to Rs 7.50 to the US dollar (Rs 13.33 to Rs 21 to the British pound), and this caused substantial curtailment of importation of industrial and scientific supplies and equipment among other essential items. As an offset to this curtailment, a large IMF loan was also arranged and this brought even louder calls in 1966, against the IMF, for India's self-reliance and import substitution.[3] As an offset the World Bank established a requirement of $750 million in nonproject aid for 1967–68, but in May 1968 the bank's Aid India Consortium offered only $642 million. The Indian expectation was for generous aid to cushion the inevitable effect of devaluation, and the shortfall between their official requirement and the bank consortium's offer was among the causes of Indira Gandhi's loss of confidence in "pro-market policy voices."[4] In December 1965 the Minister of Finance T. T. Krishnamachari resigned, having delayed official response for more than two months on a report by the World Bank advocating devaluation: he considered the bank's advocacy of devaluation an unacceptable intrusion in national policy, but by the time he resigned the bank's loan to India had been held up for three months.

An entirely new version of the Fourth Five-Year Plan was called for by Prime Minister Shastri but was given committed financing only for its first year—indicating the extent of the financial crisis. Bhabha knew all about this complex situation because it profoundly influenced his projects: not only was it vigorously discussed in the national press, but also Bhabha's close friend B. K. Nehru, the ambassador in Washington, was a strong advocate of devaluation, along with senior officials of the World Bank and IMF. With a good early warning system in place in Washington, Bhabha was doubtless able to protect some of his projects from the adverse effects of devaluation. But a clear indication of the importance of this move, and of the strength of the debate about it in various sectors (despite its secrecy), is that devaluation did not actually take place for another six months after Krishnamachari's resignation.[5] At the beginning of 1966 this currency crisis would have been SACC's biggest preoccupation, bigger than the cabinet's search for a mediated solution of the Pakistan-India conflict.

We have no account of the effect of the June 1966 devaluation on the nuclear program, nor of the effect of the July 1967 devaluation of the pound. Years of British balance of payments deficits and an "overvalued" pound were thought, even inside the Labour government, to be inhibiting British exports.[6] This is relevant because the rupee was still pegged to the pound, the rupee having just been devalued in order to export more easily to India's traditional markets like Britain. Rumors and speculation against the pound preceded the British devaluation, and the Reserve Bank of India had to make its own adjustment as best it could, given that its currency did not really trade in exchange markets (the rupee was not then defined as a "hard currency"). Given the uncertainty of Indian politics between January 1966 and August 1967, the currency question loomed larger than normal, inside the cabinet and the Congress high command: some significant figures opposed the deepening engagement with the World Bank.

The second big problem was recurring drought and famine in Bihar and eastern Uttar Pradesh, and the consequent rising food prices led to confrontation with American president Lyndon B. Johnson in 1965–66. Johnson had decided that India was not serious enough about its low food production increases or high population increases and wanted to stimulate another approach. In the context of suspending US aid during the India-Pakistan war in mid-1965, Johnson began to manage the "India file" personally: the only aid he did not stop was food aid, but he placed that on a month-to-month basis. He began to look at weekly Indian rainfall maps, weather forecasts, and district out-migration statistics. Loans over $5 million had to be submitted to him for approval. Johnson understood from the embassy that C. Subramaniam, minister of Agriculture, welcomed this external pressure (and may have partially orchestrated it), thus enabling him to convince the cabinet to direct more resources to agriculture.[7] On the other hand, Subramaniam privately disliked Johnson's interventionist approach on food aid and loans, saying in 1991, "Johnson thought he was driving Indian agriculture. We had already changed our policy, but implementation was still important. Perhaps he thought that unless this pressure was there the policy wouldn't be properly implemented."[8] India was criticizing the US for its conduct of the Vietnam War, yet wanted loans and food aid. A month before the minister of Finance's resignation in December 1965, all buffer food stocks had been exhausted. Within a month of becoming prime minister, Gandhi faced a food crisis in Kerala (not normally a food crisis area) because the government was trying to block speculators transporting grain across district boundaries, and in Kerala this led to confrontations. Subramaniam was getting an unwanted reputation as a pro-American, as minister. Food production in India actually fell 17 percent between 1964 and 1965, and prices rose 14 percent: Johnson forced India to bring on its buffer stocks by approving smaller food aid shipments than normal and asked for a serious commitment on agricultural reform and family planning. Famine was declared in districts of eastern Uttar Pradesh and Bihar, in part because of severe drought and because planned food production increases had not occurred during the monsoon months while the buffer stocks were being eaten.

Good relations between Gandhi and Johnson had actually blossomed during her first Washington visit as prime minister in March 1966, but gradually deteriorated. "The United States delayed and reduced food shipments—even during the tight food supply situation that prevailed in 1967. During this period Johnson was, in effect, his own food-for-peace officer."[9] In this context, the Ford Foundation (backed by the Rockefeller Foundation) had already agreed to assist Subramaniam to identify the potentially most productive districts all over the country, install new irrigation facilities in these districts, and cultivate the new fertilizer-responsive rice and wheat varieties—in short, to promote what came to be known in 1968 as the Green Revolution. This strategy became known as the Intensive Agricultural Development Program and culminated in other kinds of American involvement in rural and agricultural development planning in India. The emphasis in this program was speed and capital, intended to get the highest possible increases in production in the shortest possible time. The famine exposed a weak supply and transportation system, and a breakdown in what economist Amartya Sen later called "entitlement" for those without food. All of this occurred in very poor food-producing rain-fed farming districts, in areas where the Congress Party had traditionally done well in elections. Prime Minister Gandhi was conscious of the election approaching in 1967. None of these poor districts, however, were defined as "potentially most productive" in either rice or wheat, and so none were included in the intensive agricultural development program, largely because they lacked irrigation potential.

A currency crisis and new concentration on rural development through applications of science and technology was, therefore, squeezed onto the agenda of the scientific elite that informed the cabinet and prime minister's office. As agriculture increased its stature in the national hierarchy of issues, so did agricultural scientists. It was at this stage that M. S. Swaminathan became director of the Indian Agricultural Research Institute in Delhi and began a rise in status and career previously unthinkable for an Indian plant geneticist.[10]

Who Was Indira Gandhi in 1966?

An impression has long circulated that when she became prime minister in early 1966 Indira Gandhi had little political experience or judgment. This impression may have been something she herself liked, concealing her experience and toughness, useful to her in the struggle about to begin. But she actually had much more experience than her critics thought, beginning in the late 1930s. Every politician had walked through the door of her father's house, where she had seen them all. In January 1959 Indira Gandhi was elected the president of the Congress Party (via the patronage of Home Minister G. B. Pant), so a third Nehru was now president of the Congress! (Nor was she the first woman president—Sarojini Naidu had been president in 1925.) She was also personally inclined to tough positions and decisions.[11] As president of Congress she pushed for President's Rule in Kerala until it was imposed by her father in July 1959, ending the first Communist government to have come to power by election in any country. This made her enemies in the Communist Party of India, something that took her ten years to overcome.

She was also seen to be quite capable of running things even in the final days of her father's rule. For example, in January 1964, when Nehru collapsed at a meeting, rumors had Indira running things from behind the scenes, though in fact Home Minister Gulzarilal Nanda and Finance Minister T. T. Krishnamachari were running the country, and a week later Lal

Bahadur Shastri was brought back to the cabinet. In a way Indira needed politics: a widow after 1960, with young children, she had no formal credentials. When Nehru died in May 1964, he had few assets and had made no provision for his daughter's future in his will. On his death she was soon asked to leave their government house Teen Murti, which was to be turned into a museum. She was appointed minister of Information and Broadcasting but so often made initiatives outside her ministry that she was seen as meddling in the business of others. In October 1965 Shastri was so angry with Indira's meddling that he decided she should become Indian high commissioner in London (because he soon died this did not happen). Gandhi had gathered around herself a circle of advisors—Inder Malhotra, journalist; Romesh Thapar, journalist; Dinesh Singh, rajah of Kalakankar and minister of State for Foreign Affairs; Inder Mehta, deputy chairman of the Planning Commission; Inder Gujral, brother of painter Satish Gujral. In December 1965 when Indira dined at Romesh and Raj Thapar's house, she was openly critical of both Congress and Shastri.[12]

When Shastri died suddenly in Tashkent in January 1966, Indira smelled power, but six other hopeful candidates for this high office appeared within a few days. Her key backer turned out to be renowned philosopher and revered President of India Sarvepalli Radhakrishnan, who consulted and advised behind the scenes. So many people wanted to block Moraji Desai that Indira became a top candidate because others thought they could control her, unlike Desai, whom they could not control. Her Nehru and Gandhi names were also highly valued in terms of public awareness, despite the fact that she was not related to the mahatma. The final vote was Indira Gandhi 355 votes, Moraji Desai, 169—she was the Syndicate's choice, and now Dinesh Singh was her main ally in the cabinet through 1967, backed from outside it by her old friend P. N. Haksar.

During the February 1966 food crisis in Kerala and eastern Uttar Pradesh and Bihar, Indira tried to modify official food zones (between which zones one could not move food) in the face of many congressmen's calls for their abolition and open food trade. This began her long struggle against the power of the private grain traders, who tended to be stronger supporters of the Congress Party. Observers recall Indira as not being good on her feet in Parliament and not an effective Parliamentary speaker, so "as time passed she attended Parliament less and less and increasingly overruled or circumvented it."[13] In March 1966 Indira went to the United States (aid had been suspended during the fighting with Pakistan in 1965): the US, the IMF, and the World Bank were demanding devaluation of the rupee, and a new expert committee made a secret report for her supporting devaluation. In

Washington she was looking for wheat, and apparently President Johnson was charmed by Indira Gandhi that time and decided to support her. She had to agree to creation of a US-India educational foundation for the use of "counterpart funds," but what was really needed, according to Ambassador Nehru, was wheat, and so that was why some of the cabinet agreed to devaluation.[14]

Heavily criticized on her return from the States (including by Krishna Menon), Gandhi went ahead with devaluation, despite cabinet resistance, on 6 June, but her own "kitchen cabinet" now urged moving away from the US and making closer ties with the USSR. Thus she could be seen to move further away from the pro-US Moraji Desai group and the old-guard Syndicate. She criticized the US bombing of Hanoi in July 1966 and let her US ties weaken, though the wheat continued to come, linked to monthly performance targets. President Johnson no longer found her quite so charming.

In December 1966 Indira Gandhi identified her opponents as Desai and Kamaraj. Preparing for the 1967 election, she began speaking of herself as the fifty-year-old mother of two sons, playing left and right politics simultaneously. She favored nationalizing the banks; Desai opposed it. The 1967 election was, in part, a contest between factions inside the Congress Party, and the young Gandhi was uncertain of her control. She knew she was in power because she was Nehru's daughter and knew that the other leaders (all men, all older) could agree only on their aversion to Moraji Desai. Women had been important in the Congress Party in the past, but India had never before had a female prime minister (nor had any Indian state yet had a female chief minister).

Changing the Official Approach to Technology and Science

One of the few avenues of new action open to Indira Gandhi was to change the government's approach to applications of science and to talk more about technology and technical solutions in the economy. This would allow her to appropriate one dimension of the struggle for self-reliance as hers. But she had to do this in an environment without foreign exchange, the one factor that deeply affected and united the technical and business classes. If she had thoughts of carrying through on a nuclear test, following Bhabha's agreement with Shastri in 1964–65, these early years in her regime were hardly propitious for it.

In this particular climate the SACC tried to get the cabinet's attention. This incarnation of SACC was under a new prime minister and had no one of Bhabha's status in its membership. Before Bhabha's death Kothari had been asked to form a committee to review SACC's performance; he conducted this review at a slow pace between 1964 and 1966, introducing his report with the remark that "difficult and sometimes unpleasant choices have to be made among many competing claims on the use of available talent and professional manpower." Then he stressed that SACC could not be objective because it was composed of members who are directly involved and should be reformed with the inclusion of "detached members."[15] Anticipating a critical review from above, SACC sought immediately to establish itself with the new prime minister. In 1966 it announced a Round Table conference for 1967 and decided that CSIR's three-year-old Research Survey and Planning Organization would become SACC's secretariat for the purpose of the Round Table. This was a brave CSIR attempt to limit the adverse exposure it was regularly receiving in the cabinet. Immediately, and referring to the rupee crisis, the new organization conducted "a study on the conservation of foreign exchange by the national laboratories."[16] At the same time Sarabhai decided to establish his own program analysis and planning group to serve the DAE and AEC. CSIR had such a group, and the planners now were calling for empirical evidence to back scientists' calls for change and reform.

It might have been logical to ask the more experienced Kothari to be chair of SACC, but Sarabhai inherited Bhabha's role here too. Perhaps sensing change, Sarabhai took the lead in 1966, as chair, in agreeing to investigate SACC's charge that the Planning Commission had not allocated adequate funds for science and technology in the coming Fourth Five-Year Plan, a charge that the whole of science was being devalued. This is practically the first mention of the important Planning Commission in the SACC summary. The foreign exchange crisis had hit research and development in India too, and the cabinet secretary was delegated to discuss solutions with the Department of Economic Affairs. An official study of foreign assistance to scientific research in national labs was completed in 1966, but the study did not include any DAE institutions.[17] On the ground, there was more and more concern about the brain drain, and Sarabhai presented a preliminary paper to SACC showing that about a thousand highly trained people were "staying away" from India because of unfavorable conditions. There also appeared to be an internal brain drain, through which discontented scientists were moving from research to administration, perhaps attracted by the higher pay; this was particularly evident in agricultural research, resulting from donor pressure (e.g., World Bank) to expand and improve the quality of administration of agricultural institutions. This wish among scientists to migrate became even more of an issue in 1969, as we shall see.

SACC's swan song occurred in 1967, ending with the planned Round Table and the special meeting with the cabinet in December. The Round Table was a summit of top scientist-administrators; attendees heard Vikram Sarabhai reiterate his call for massive investment in science and technology, Homi Sethna say that "an indigenous capability has been forced on us, for example, in atomic energy," and S. Bhagavantam claim that "investment in science and technology is not paying off, and India may have to import technology which is ready outside instead of spending years to build it inside the country." Ashok Metha, minister for Petroleum and Chemicals, advised that Japan's successes in fertilizers and atomic energy should be lessons to India, and India should, like Japan, be focusing on a few key subjects and leapfrog past the obstacle of "doing everything at once." There was a special session on Nehru's favorite phrase, "the scientific temper." People at the Round Table admitted that there was too much talk in India about the "wonders of science" and not enough about its methods and empirical outlook. It was now essential, experts said, to infuse a *scientific temper* into the cultures of India, through films, radio, science journals, better textbooks, and the like. Some in the Round Table even said that strengthening the use of the English language would better connect young Indians to international science and to its temper. They pointed out that education, being a state responsibility rather than a central one, lurched from the vernacular of various states to the English medium, and back again, regularly. Sarabhai must have recalled his unsuccessful shadow-contest with Moraji Desai in 1961–62 for the vice-chancellor's job at Gujarat University, his purpose being to push for English medium education, and thus he kept quiet about it in SACC.

Despite the fact that SACC's days were numbered, it actually established policies and committees that survived its demise. Immediately after the Round Table, the cabinet delivered its ranked list of national problems to be dealt with by SACC; the problems were, in this order, (1) use more science to achieve maximum agricultural production, (2) find "quickest and most economic ways of making the country self-sufficient in aircraft and allied equipment," (3) develop nuclear capability, (4) develop ability to deal with biological warfare, (5) improve hydrology, (6) develop armament and equipment to meet operational defense needs, (7) arrest the brain drain, (8) create incentives to promote research in the private sector. Nuclear capability meant both nuclear power and preparation for nuclear bomb tests; "ability to deal with biological warfare" meant developing defensive strategies and protective technologies.

SACC had an immediate response to the ranked list of national problems early in 1968, creating a Committee on Aeronautics in response to priority 2, like the Committee on Electronics previously chaired by Bhabha from 1963. This new aeronautics committee was chaired by former minister C. Subramaniam and would report directly to the prime minister. It focused on the vexed "indigenous jet fighter" question; just why, people asked, has the famous Indian jet fighter not flown yet? But there were other less glamorous and expensive changes that SACC promoted too, so that the cabinet approved a pool of uncommitted "supernumerary posts" to be used by lab directors to recruit promising young Indian scientists now working abroad: these posts would be bid for directly from a central pool and not influence the labs' budgets until they wanted to make an individual scientist permanent. This was the latest method to address the brain drain. The Round Table conference on science and technology was considering the reform of SACC itself, in light of Kothari's review. But just before this conference, the Study Team on the Machinery of the Government of India concluded its work and recommended the abolition ("reconstitution") of SACC. The Round Table, composed largely of SACC members, proposed that SACC stay the way it was and that a Ministry of Science and Technology be established to coordinate everything. After all, SACC was an effective bridge to the cabinet for powerful scientists who were heads of agencies. With Indira Gandhi's agreement, however, SACC, founded and chaired by her father, was replaced in early 1968 by the Committee on Science and Technology (COST), although with the inclusion of many of the former SACC members. Meeting with the cabinet for the last time in January 1968, SACC heard physicist B. D. Nagchaudhuri of the Planning Committee say that although SACC was supposed to coordinate, "it was necessary to first decide on the functions envisaged for SACC and then reconstitute the Committee accordingly."[18] Prime Minister Gandhi said on the same day to SACC, "It is useful to remember that establishing new bodies by itself solves no problems." Then she asked how much SACC *should* really be concerned with planning. This was a loaded question coming from the prime minister, but SACC members did not know it yet.

No matter how much official study teams might find the method for advice inefficient, there would inevitably be a body through which the scientific and technical elite could interact with members of the cabinet to secure their interests and at the same time regulate the access of other experts to the same cabinet members. In the new Committee on Science and Technology, the inclusion of the term "technology" was taken as a signal of Mrs. Gandhi's intention to shake up the scientific community, and perhaps move beyond her father's fascination with scientists. Still, she was careful too, saying to the CSIR Council in August 1968 that science had not yet made its full impact on the masses and could not do so unless "our culture is cross-fertilized by what my father used to call 'the scientific temper.'"[19] Actually COST soon began to discuss and debate the same issues that SACC had traditionally discussed. Like SACC, the new COST had no power to issue orders, only to recommend. It was chaired *not* by the chairman of the Atomic Energy Commission, although he was a member, but by the new scientific member of the Planning Commission, physicist Nagchaudhuri. This was a triumph for the Planning Commission, which had previously played only a minor role in SACC and did not direct science and technology funding in any practical way. It could not do so during Bhabha's life because he evaded any stricture that the Planning Commission would like to have applied.

Though he brought in fresh air, Nagchaudhuri was neither a Planning Commission insider nor a Delhi insider, and it is hard to see the relevance of this renaming: its issues were very similar and the participants little changed from SACC. COST's members were, besides the new chair, Atma Ram (physical chemist, director general of CSIR), S. Bhagavantam (physicist, Defence), D. S. Kothari (physicist, University Grants Commission), B. P. Pal (agronomist, Indian Council of Agricultural Research), P. N. Wahi (physician, Indian Council of Medical Research), Vikram Sarabhai (physicist, Atomic Energy Commission), Homi Sethna (engineer, BARC), and the cabinet secretary. Two and a half years later COST's membership, though little changed, was augmented and broadened by the addition of C. R. Rao (statistician, Indian Statistical Institute), A. R. Kidwai (Union Public Service Commission), S. K. Mukherjee (vice-chancellor of the University of Calcutta, West Bengal), and Dasarathi Banerjee (Rubber Manufacturers representative). These were all the scientists at the commanding heights, with an industrialist and vice-chancellor thrown in.[20] To open the proceedings in 1968, Prime Minister Gandhi gave her government's instructions to COST: "In view of the financial constraints and precarious balance of payments position, it is necessary that scientists and technologists demonstrate the social and economic benefit of their work."[21] Moreover, she stated, "the Planning Commission will determine the national priority for projects."[22]

So here on COST's first day was the deeper meaning of Mrs. Gandhi's remarks on the last day of SACC's work: "the Planning Commission will determine the priorities." Agency leaders had sought skillfully to minimize the commission's influence over their projects. Now the pendulum was swinging back toward planning and priority setting. COST also signaled the rise of B. D. Nagchaudhuri, a physicist trained by Meghnad Saha at Allahabad and Emilio Segre at Berkeley, the first scientist with an American degree to hold an important role like this. He was appointed in 1968 to the commission as member for science, and to COST as its chairman.

International Influence in the Indian Scientific Community

The first major debate in COST concerned the role of foreign scientists and foreign grants in Indian institutions. Ending foreign dependence had not lost its attraction among scientists and technologists in the twenty years since Independence in 1947 but had gained importance. Stricter foreign exchange controls forced choices even in minor issues, resulting in personal disappointment and dissatisfaction; this was the ground for a debate about the way science and scientists contributed (or didn't contribute) to self-reliance. Very large networks had been built up between foreign researchers and Indian labs, mostly through the hundreds of individual scientists going back and forth, but also through institutional efforts; through these collaborations there were exchanges of visits from foreigners, materials, equipment, and data. This was a mode of reentry for some Indians abroad, bringing with them the prospect of a secure link back overseas if their decision to return to work in India proved too difficult for them or their families. It was also a mode of escape for other foreign-trained scientists already in India and, finally, was exactly the right option for those well-established scientists who wished to pursue their work elsewhere from time to time. Through these networks and through government projects—whether in agriculture or oil exploration or space—researchers came from abroad to India and stayed for weeks or months (or years, like Bernard Peters in the 1950s at TIFR). COST discussed how to make life better for scientists visiting and working in India (including exempting them from taxation) and how to enable research institutes to pursue foreign grants independent of government control. Because of strict controls on foreign exchange, trips abroad were being classified as "official," "professional," and "personal," and COST discussed ways that professional trips might also benefit the official agency or institution to which the traveler belonged. Few "personal" trips were being allowed, but "professional" travel was more common. No longer was such travel restricted to Sarabhai or other elite members of SACC and COST, and the culture of scientists and technologists in India was changing to reflect that.

But attitudes contrary to this trend promoted the cutting edge of self-reliance, and some people described these networks, collaborations, and grants as forms of dependence, based on privilege. Particularly when they were American, expertise and grants were being questioned just as they were becoming more accessible and routine. Attitude and practice were quite contradictory: there were allegations that these foreigners had "imperialist" designs to keep India dependent or that "capitalism" needed to have India dependent, otherwise the whole system would ultimately collapse. These allegations were made, however, by people usually ready to embrace those aspects of the "whole system" that furthered their interests too. In addition, the use in India of the huge pools of rupees called "counterpart funds" was being debated. Since about 1951 such "counterpart funds," generated from the government's sale of American, Canadian, Australian, and other foods and commodities like cooking oil through its food ration systems, had accumulated into large sums. These were being used for research and education—as agreed by the original donor agencies—though few ordinary people knew how long this had been going on. So large was the American fund generated through Public Law 480 food and commodity sales that this became the normal reference to all such funds (including similar counterpart funds from other countries), "a PL 480 grant."

How Indian scientific institutions and international donors interacted was now a matter of public discussion: COST concluded that each ministry or department would process PL 480 projects through a coordinating committee with cross-representation to COST, that there would be no restriction on foreigners' involvement in projects using these funds, "including those who were visiting establishments concerned with nuclear physics and related fields," referring directly to TIFR and BARC. Further, states the report, "Dr. Sarabhai said that as a matter of policy, it was wrong to say that no foreign scientist should be invited to participate in PL 480 projects."[23] Indeed, the scientific community was fighting back: for example, this right to use PL 480 "opportunities" was what CSIR laboratory directors wished to have on their own terms and to involve foreigners in their labs and projects without constant reference to officials at headquarters in Delhi. Finally, COST reviewed the employment of foreign consultants and experts in aerial surveys, mineral and oil prospecting, searching for water, acknowledging that large fees were being paid in hard currency for this expertise. But COST justified this practice as necessary and unavoidable. Sarabhai argued that his remote sensing program using satellites would reduce the need for these services to identify resources, which it eventually did. This debate, though muted in the summary document, echoed vigorous discussion in the media and elsewhere in the late 1960s about dependence on foreign experts. There were Indians who said they were equally qualified to do this contract work, were often trained in the same universities abroad as the consultants, and were ready to be paid in rupees. In addition, they also knew the scarcity of the currency in which foreigners were being paid in India. All of this suggests that in a muted form, India was beginning to experience a cultural revolution which took its form as intergenerational struggle: well aware of events in Prague, Chicago, and Paris, young scientists (some recently returned from abroad) articulated a new view of an Indian competence and self-assurance that bumped into the standard ways of carefully following foreign trends through foreign experts. Scarcity of foreign exchange was bumping into the careers and institutions of the richest and best-trained generation of scientists India had yet seen. As I shall discuss below, this cultural revolution began to take a clearer form in the early 1970s, which I call "the war over self-reliance," but as early as 1967–68 some of its issues were being defined.

The Khorana Question and the Brain Drain

Questions about the brain drain and international influence crystallized in the public debate following the award of the 1968 Nobel Prize in Medicine to Har Gobind Khorana, for nucleotide synthesis. Coming from Raipur of pre-Partition Punjab, Khorana went to school in Lahore, like other physics prize winners Abdus Salam and S. Chandrasekhar. Immediately after the war ended he was sent abroad by the government of India to study at Liverpool on the Overseas Research Training Scheme. In 1948 he did postdoctoral work in Zurich but as he was "bonded" to the scheme, he came "home" to a partitioned India (his home in the part of Punjab now lying in Pakistan). While looking for a job in Delhi in 1949–50, he lived for the first year in the courtyard of servants quarters of his uncle's house attached to the Red Fort. Essentially he was an academic refugee, among thousands who were born in Pakistan and now camping out in Delhi. In the end no employment was forthcoming, and the bond he signed with the scheme was annulled, as specified by the rules. Therefore, Khorana, now twenty-seven, was permitted to accept a fellowship at Cambridge, and his brother collected money from relatives and friends to lend him enough for the ticket on a ship. He returned to Britain to work there, married a Swiss woman a year later, and moved to Vancouver in 1952.[24]

A storm of delight and chagrin accompanied the 1968 announcement of the Nobel Prize Committee. It was the subject of regular discussion in the laboratories. One typical assessment of Khorana's prize was published in the top mass-circulation weekly (25 million readers):

> it is shameful that his potential genius remained undetected and that he had to face humiliation during his short stay in India. Group rivalry prevailing among members of the teaching staff, in the executive councils of the universities, and in the managing councils of the research institutes was the main reason for Dr. Khorana's bitter experience. The CSIR, UGC, and the Central Government should jointly appoint a body of experts to hear the genuine complaints of the younger scientists about their seniors or would-be employers. If their seniors are found to be in the wrong, they should be punished. This will create self-confidence among the younger scientific workers who can then serve their country well.[25]

This was the same question raised in Canadian newspapers about Khorana that month, where people also asked why he arrived from India and Cambridge in 1952 but left the University of British Columbia in Vancouver for Wisconsin in 1960. Though Khorana's original work on nucleotide synthesis (which laid groundwork for understanding the genetic code) began in Canada, the finishing touches at the University of Wisconsin clinched the prize for him. He was the first to isolate DNA ligase and to synthesize oligonnucleotides. "Why was his potential genius not detected?" asked both Canadians and Indians in late 1968, and "why did he succeed in Wisconsin?" Husain Zaheer, former head of CSIR, immediately called for a revival of his proposal to establish a national biological laboratory and to invite Khorana to direct it: Zaheer had spoken to Khorana about this in 1965 and 1966 and was told that when he completed this sequence of research, Khorana (an American citizen by 1966) would consider returning to work in India. This too did not occur.[26] The tea break at COST meetings would have been buzzing with this question. Rao's recommendation in the *Illustrated Weekly* to appoint a body of experts to hear complaints and mete out punishments reflected a growing awareness of the injustice of conditions of work in some institutions and a rising sense of the need for a legal method of addressing those conditions (in a way an appreciation of the good effect of the rules). This legalistic response was grounded in disappointment and even embarrassment that this official indifference to his talent and potential had pushed Khorana back only fifteen years before. But this call for legalism was viewed inside most laboratories as likely to lead to an atmosphere of "work-to-rule" quite antithetical to productive research. These specific hearings did not occur, although in later chapters I will focus on challenges to authority in SINP and TIFR, providing evidence of that sense of injustice in DAE-funded laboratories.[27] A legalistic approach to the working conditions in CSIR laboratories, chaired by Justice Sarkar, has been described in the preceding chapter.

Sarabhai clearly wanted to keep a focus on the sociology of science and technology and how it influenced policymaking, and so he authorized and read drafts of studies done by a Programme Analysis Group in his AEC office, studies authored by radio astronomer and policy analyst Ashok Parthasarathi. As SACC had in 1967, COST took the brain-drain question very seriously in 1969 and even sent researchers to the United States to study it.[28] In the end the brain-drain discussion was linked in the public's imagination to the Khorana question. The group looked at the way in which the imagination of bright young Indians abroad was captured during their dissertation: doctoral work had to be made more attractive and meaningful *in* India. COST therefore focused on the perceived source of some of the push toward the drain, the CSIR research institutions, and began studying the effect of rules in the CSIR like "automatic promotion," a term that Sarabhai wished to have "struck out" of CSIR contracts, according to the SACC Review. It also focused on the nonportability of pensions that limited mobility between institutions and the fact that many supernumerary posts available to the CSIR labs and other technical ministries were unfilled even though these posts were established as a solution to the problem of dissatisfied junior returning scientists. All COST members took turns in criticizing the CSIR in 1969, and the DAE-funded labs appear not to have come in for the same kind of appraisal by COST. Had the DAE labs been included, dissatisfaction there would have been known too (as chap. 17 showed). But some COST members were currently appearing before Justice Sarkar to contribute to his official review of the work of CSIR and others were members of his review committee. Opinion within the scientific community was tilted against the leadership at headquarters of CSIR, and this tilt appealed to many of the CSIR scientists themselves. Indifference to Khorana's potential was recalled publicly, over and over again, but the sharpest minds focused on why his potential was hard to detect. Sarabhai, Menon, Ramanna, and Sethna managed to keep dissatisfaction in DAE institutions out of the light, in part because this kind of dissatisfaction was clearly not equal to the problems in CSIR and also because they worked hard to keep DAE's problems apart from the official business of COST. Although they agreed on little else, they could all agree on the DAE's "untouchability."

COST debated and established national priorities for the cabinet in 1969, using 1968 Planning Commission guidelines: in the rural sector, COST's priorities were water resources, irrigated farming, agrochemicals, and dry farming. In industrial research the priorities were power generation and distribution and monomers/polymers and ferrous alloys. COST recommended to cabinet the establishment of a tissue culture laboratory and supported a greater emphasis on biological research. This represented a long journey for biology and agricultural research in India, finally contesting the supremacy of the fields of chemistry, physics and engineering while also contesting the prevalence of influential individual physicists and chemists. Agricultural research was perceived to have finally arrived at the policy table. COST's electricity power generation and distribution statement may be understood as an indirect criticism of the fact that DAE had still not delivered one megawatt of electricity to an electrical grid. Taking a more strategic and focused approach, in 1969 COST made an inventory of the materials of national importance, and these included magnesium, semiconductors, nonferrous metals, silicon, silicones, high polymers, ceramics, zinc, copper, and gold. India's weak position in the international trade or manufacture of all these materials was reviewed. Their scarcity and importance should have been a matter of concern to the cabinet. COST again pointed out, if it had escaped anyone's notice, India's precarious situation with respect to its supply of oil.

In a series of prominently published articles, the Programme Analysis Group addressed "the sources of technological growth" just after Patrick Blackett's 1967 Nehru Lecture in Delhi on the innovation chain and the crucial role of savings and capital formation in technological growth. Then Parthasarathi concluded that "one cannot help but feeling that for those scientists, engineers and managers, born in this country or abroad, who believe that the mere availability of [capital-intensive] technologies is what sets apart the development prospect for the poor countries of our time from that which faced those of earlier epochs, such realism from a leading scientist [Blackett] is a good corrective."[29] This was followed in 1969 by a high-profile launch in London of the distinction between appearance and reality in Indian science policy. Noting the disconnection of the thinking of the Planning Commission and SACC and carefully comparing Japan and India in their policies, particularly their utilization of investments and adaptation of industrial technology research, the group proposed (through Parthasarathi and co-researchers, authorized by Sarabhai) to rely on "individuals and groups outside government science," even the creation of an autonomous institute of science analysis.[30] From two studies released in 1969, Parthasarathi published an analysis of "the sociology of science in developing countries," citing the difficulty among young scientists to find inspirational leaders among their seniors, who were themselves in conflict with others: in examples of unattractive "wrangles within and between scientific institutions," he named Bhabha's public attack on the operations of the CSIR at the January 1966 International Congress of Scientific Unions (ICSU) conference in Bombay, and the unceremonious (and possibly unjust) dismissal by CSIR's Husain Zaheer of P. K. Kitchlu as director of the NPL in Delhi in 1965.[31] Parthasarathi also presented a paper at the Pugwash conference on the "brain drain from developing countries," highlighting one of Sarabhai's main concerns every time he met with Indian scientists abroad; the paper, published later when the author was in the prime minister's office, showed that in a survey of three hundred alumni of ten American universities graduating between 1956 and 1966, among those who obtained a doctorate degree, only 18 percent were likely to return immediately to India, the rest having fallen into "the PhD trap," that is, wanting to continue with the research they were now doing, thinking it was impossible in India.[32] A very different kind of responsibility for and effective communication with young Indian scientists abroad was now proposed, to be led by embassies and other official bodies and backed up by visiting scientists from India.

SACC and the Bomb?

There was an unspoken zone of silence around the bomb test planning in 1967 and 1968; the chairman of SACC said little about his awareness that scientists at BARC were now studying the ways and means to make a bomb for testing (called Study of Nuclear Explosions for Peaceful Purposes inside BARC, or SNEPP), even in spite of his known disapproval of the project. Sarabhai had instructed Ramanna to stop the SNEPP project in June 1966, when he took over the AEC and DAE and became responsible for BARC as a part of the DAE. But BARC's day-to-day director Homi Sethna had been appointed as director of BARC soon after Bhabha's death, a full five months before Sarabhai came to the office; a climate in BARC for the SNEPP project had already been quietly established. Sarabhai's instruction to Sethna and Ramanna that it be stopped was administrative and political but not decisive. Indira Gandhi was unlikely to have challenged him at this early stage, engaged as she was in a dramatic struggle within her own party. With this instruction to stop, Ramanna and others realized now that they could not go to Sarabhai to ask for more funds for the project, and since Ramanna did not like asking Sethna for anything, the bomb design project had to be low-key. Yet, because the equation of state for a small test bomb was not known, except secretly among experts in other countries, and because other technical questions had to be clarified before building and assembling the components, there was a lot of work that could be done on paper in 1967–68, describing in mathematical terms the temperature, density and pressure of the plutonium and other components in the bomb, what composition would make it critical, and therefore what intensity of high explosives had to be packed around the bomb's nuclear core to compress it to the desired density.

Moreover, it is unlikely that the bomb-test preparations could have been completed and the technical obstacles overcome by 1967, *even if* there had been adequate plutonium. There was thus little chance at this stage of actualizing Bhabha's prophecy of readiness "within eighteen months." But sensing a mood, anticipating official intelligence, and signaling an issue that she would make her own, Indira Gandhi said in early May 1967, "we may find ourselves having to take a nuclear decision at any moment."[33] When China's first hydrogen bomb was tested in June 1967, it inspired new calls for an Indian bomb, even within the Congress Party. A Chinese long-range missile carrying a 1,200 kg payload had already traveled almost 900 km only eight months before in late 1966, though it is not clear how many Indians knew that, as it was not publicized by the Chinese; that distance was far beyond the Indian capability in missiles or rockets in 1967.

People were gradually drawn in along with other work they were supposed to do for reactor development.[34] But the number of researchers remained small and the objectives limited at this stage; as George Perkovich reasoned, "once the preparations got to the point where the scientists were ready to build, and, they hoped, test the device, they then sought prime ministerial authorization."[35] SACC was not asked to give specific advice on the bomb and said nothing beyond "build nuclear capability," by which they meant to include reactors. In retrospect, readers interested in nuclear proliferation and how it occurred in India might be surprised that something so important could evolve with so little oversight, but in 1968–70 Indian leaders had so many larger questions to answer that their inattention to this one should not be surprising, given that almost everyone thought it was a long way off.

But on the subject of nuclear weapons tests and international treaties, SACC/COST was not considered the most competent group: when Bhabha was alive he had many opportunities to influence the debate, both as chairman of SACC and in his other capacities (including as member of the Defence Science Committee). The official decision not to sign the nonproliferation treaty was announced in Parliament in April 1968, though insiders knew negotiations occurred between January and March 1968 at

Geneva and Vienna to find language that would satisfy India's objections to the treaty, after the failure of the official attempt to establish for India a protective "nuclear umbrella" to be provided by some or even one of the nuclear powers. Given India's historic position, the decision not to sign in 1968 was hardly surprising and provided full continuity with Bhabha's era and the Nehru-Bhabha pronouncements on its discriminatory effect.[36] The prime minister did consult scientists about the bomb, but not through the mechanism of a cabinet committee.

All Change? Exit COST, Enter the National Committee on Science and Technology

Despite this diligent and energetic start, there were bumps on the road for COST. In November 1969 a preliminary report to the cabinet, written by the cabinet secretariat alone, set out COST's research and development priorities. The only mention of a specific technical project (to illustrate and explain COST's work to the cabinet) was about "design-making machinery." The cabinet secretariat sent the preliminary version of the report to COST, and immediate Notes of Dissent, written by Vikram Sarabhai, Homi Sethna, and K. N. Raj, were attached to this preliminary report as they were too late to stop its transmission to the cabinet. That COST's members were unprepared for this report and/or could not stop it is curious: it is almost as if someone wanted to sabotage COST, to make it look inept or ineffectual in the eyes of the cabinet. In that climate of crisis, reference to design-making equipment was seen as almost frivolous. But not quite a year later, at the Third National Conference of Scientists, Technologists, and Educationists (note inclusion of the new category "technologists") in November 1970, COST felt the wind of change. COST planned this conference around the theme of the implementation of the Scientific Policy Resolution passed ten years before in 1958. Among the 130 people attending the third conference, "there was a strong feeling that on several important counts the implementation of the Scientific Policy Resolution had been highly ineffective." The chief reason given was that India had really not managed to "associate scientists with the formulation of policies," as called for by the resolution. The prime minister spoke boldly at the conference in 1970 about the goal of "the democratization of decision-making in laboratories, universities, scientific agencies, and indeed in the scientific system as a whole."[37] Moreover, COST itself was criticized, and there was talk of forming a new national committee on science and technology to replace it. According to an insider at the third conference, TIFR physicist B. M. Udgaonkar said:

> Diametrically opposed views were expressed as to whether Agency heads should be members of the national apex body [like COST]. The value of their advice on various aspects of policy making and plan formulation was recognized, but a question was raised as to whether heads of large organizations (AEC, CSIR, ICMR, ICAR, BARC, DRDO, Cabinet Secretariat, etc.) who would be already heavily burdened with their executive responsibilities, would find sufficient time to devote to the tasks and responsibilities of the national apex body, especially where innovation is called for, and whether the national apex body should continue to consist of such very busy people.[38]

One insider's conclusion is that COST "did not use its chairman's Planning Commission member's status to shape the pattern of financial allocation of major scientific agencies as it could have and was meant to."[39] Now the replacement for the chairman of COST would be replaced by the minister of Planning himself.

The wind of change now turned direction again, pushing against the two-year-old COST. Within a few months of the third conference, the prime minister approved the creation of the National Committee on Science and Technology and the dissolution of COST in 1971. As hinted at the conference, the new National Committee was composed of laboratory directors rather than agency heads. The original plan, supported by most scientists, was to have full-time members of the NCST with a competent secretariat, capable of doing its own research and analysis. Justice Sarkar's report on CSIR (completed by Menon, Swaminathan, Rao) recommended creation of NCST in January 1971. The prime minister soon created a Ministry of Planning and put C. Subramaniam in charge of it. In April the prime minister created a Standing Group on Science and Technology (later called the Cabinet Committee on Science and Technology), which she herself would chair. It immediately empowered DRDO and the Department of Science and Technology to "evolve their own personnel practices" to resemble the DAE and avoid the Public Service Commission (for which Gandhi was the responsible minister).

But two problems soon arose, one among bureaucrats and the other among scientists. The plan to enable these new scientific departments to evolve their own personnel practices ran into "strong views of the traditional civil service on the recruitment of all government personnel through the UPSC," and the state minister was dragged in front of the prime minister, vainly trying to protect the administrative rules. But Indira Gandhi rode right over the objections. (She had already been talking of "a committed civil service," though, like her father, she knew that scientific practice, performance, and promotion could not follow the stereotype rules of advancement in the public service.) The Ministry of Personnel, however, delayed implementation of this instruction for three more months until an official threatened to take it back to the prime minister.[40] The second problem was that in the search for full-time members NCST bumped into reluctance among active scientists to leave the work they were doing and give most of their time to NCST, no matter how competent and authoritative the secretariat, no matter how prestigious the appointment. Clearly no one could predict how long this part of the game between scientists dispersed throughout this large country and planners at its center would last. So nine NCST members were announced in October 1971, soon followed by the chair, cabinet minister C. Subramaniam.

Membership of NCST, 1971–74

- C. Subramaniam, minister of Planning, and Science and Technology, Delhi
- Hari Narain, director, National Geophysical Research Institute, Hyderabad
- B. D. Tilak, director, National Chemical Laboratory, Pune
- M. M. Suri, chairman, M. M. Suri and Associates, Delhi
- A. R. Kidwai, member, Public Service Commission, Delhi
- V. Ramalingaswamy, director, India Institute of Medical Sciences, Delhi
- R. V. Talmankar, director, Defence Metallurgical Research Laboratory, Hyderabad
- A. K. Malhotra, general manager, Offshore Engineering, Engineers India, Delhi
- C. N. Rao, dean of research, Indian Institute of Technology, Kanpur
- A. B. Joshi, deputy director general, Indian Council of Agricultural Research, Delhi

Essentially this meant a delay in starting NCST and winding up COST. So the government continued to use COST as a test basin for new initiatives for cabinet and in April 1971 announced that COST would study present incentives to industries that should be involved in R&D. The government told COST that if private industry would increase spending on R&D in India and adopt technologies developed in India, new tax incentives were possible. A new legal environment was also established, limiting patent protection at fourteen years in India, except for food and drugs, which would have only five years' protection, thus establishing an environment in which India became one of the world's largest producers and exporters of generic pharmaceuticals. Gone too was the protection to enjoy exclusive importing rights established under British colonial policy. This ruffled many feathers in the Commonwealth and America.[41]

That Mrs. Gandhi was immersed with her advisors, like P. N. Haksar, in all this is evident. She sought advice from outside the country too: for example, the prime minister asked for discussion of his view of innovation with Patrick Blackett, a person whom she had known for twenty-four years. She must have asked him to act for her, because before leaving India in May 1971, he wrote to Atma Ram to propose a new group for the prime minister, to advise her about

> the areas in which efforts should be most concentrated—in a way, a "task laying down body." In order that R&D would have a relation to economic growth, this body should have an economist on it, and to ensure implementation of the assignments should include a high official. . . . This work should not be hampered by the plea of freedom for scientists. It must be appreciated that scientists are not free to do whatever they wish, certainly not in the field of applied science. . . . In order that this evaluation of present activities and assignment of future tasks is done unhampered and expeditiously, no lobbying should be allowed and the issues should not be bogged down in those superficial discussions and to and for arguments which can always be produced. The stakes are high, time is short, and there is no room for irrelevant arguments. Now when conditions have been created in the country, thanks to the thinking generated recently, this is the opportunity to give concrete shape to the R&D related to economic growth. Even in allocation of resources, at the most 10–20% may be earmarked for basic research, the rest should be devoted to applied work bearing on economic growth.[42]

Of course the government was already a major investor in new industry, using, for example, the funds of the Life Insurance Corporation and the Industrial Credit and Investment Corporation. The government now also had available the capital of all nationalized banks; the private sector was understandably cautious about paying for R&D when it had no experience with it, having relied for decades almost exclusively on government research. In December 1970, a month after the third conference, these big and wealthy state corporations held their own major conference on R&D expenditure and benefits in Bombay. Criticism of COST was aired there too. So COST soon responded by persuading cabinet to approve a COST study of the real effect of Indian incentives on Indian industry. (The allegation was common that incentives had no effect on Indian industry.) This proposal ran counter to the behavior of some firms owned by the government itself, state-owned firms that were maneuvering to obtain foreign collaborators even while COST was studying new incentives to make use of Indian skills and innovations. In this sense the comparative cost of indigenous technologies was going up: COST's study showed that R&D expenditure in India had risen from Rs 27 million in 1958–59 to Rs 136 million in 1969–70 but that private industry still accounted for less than 10 percent of that total increase in spending. Most of the increase was from government sources, but it was still not being translated into increased productive capacity or comparative industrial strength for exports. The new NCST, when it was finally inaugurated, worked on this question in its first "Approach" document.

But how did the cabinet really see and act on this, over the long run? The list of issues below is drawn from twenty-three years of summaries of the meetings of SACC/COST (1948–71), and separates technically specific objectives (e.g., modernizing telecommunication) from process and structural outcomes (e.g., improving the use of foreign exchange for research). This distinction between substantive and structural or process issues, though not exclusive, is useful, and anticipates the next three chapters:

Substantive issues

- Building electrical power systems based on nuclear, thermal, and hydro generation
- Strengthening defense with new weapons, including rockets, fighter jets, and submarines
- Building a petrochemical industrial base, with exploration, refining, and distribution
- Creating a capacity for space research and exploration, including satellites
- Building a heavy electrical production system, with components for grid-building
- Ending famine and malnutrition by transforming agricultural production
- Modernizing telecommunication throughout India, using satellites
- Introducing television in rural society for education, using satellites
- Establishing an internationally competitive microelectronics industry
- Establishing a computer-production industry, with programming capability
- Red ucing communicable diseases through research

Process and structural issues

- Raising the international prestige of India's scientists and research institutions
- Improving the working conditions and systems of promotion for researchers
- Stopping the brain drain of trained people leaving India

- Attracting talented researchers to come to or return to India
- Raising the quality of new Indian scientific institutions
- Making Indian scientific publications significant in world science
- Establishing and cultivating a "scientific temper" in Indian society and politics
- Raising Indian recognition of talented researchers, "popularizing" science
- Deploying researchers in more productive ways
- Making the best use in India of foreign scientific advisors
- Deciding how to use foreign funds for research in India

This account of the rise and fall of issues in the debate just outside the cabinet door, and changes in the composition and mandate of the committee that advised the cabinet from 1948 through 1971, is important because it reveals the interplay of international, national, and regional forces in the politics of the scientific and technical community. These changes, however, should not obscure the continuity in underlying factors, factors like the brain drain, which appeared as early as 1948 when SACC was founded. Nevertheless, the cabinet acknowledged most of these issues, did study some of them, and did adopt a few policies based on them. The ten-year review of the Scientific Policy Resolution showed insiders that grand goals were hard to achieve, so some of SACC's more important influences were not new policies but new specific projects and new justifications for the state to finance them. But as always there was an important personal dimension; at this stage the prime minister showed she was as seasoned and tough a player among the scientists as she was among politicians, facing up to offstage muttering that she had not succeeded at Oxford, could not play the Oxbridge card among that big and influential network, and was not quite capable of fully understanding the technical issues and questioning the experts.[43] How the personal and the political intertwined through the nucleus of the scientific community is at the heart of this history.

TWENTY

Building a High-Technology Economy through Atomic Energy, Space, and Electronics

During the second half of the 1960s the idea of a national industrial "high-technology" infrastructure came to the front of the stage in India, fueled by the ambition of its protagonists and backed by the desire to leave the stolid sectors of cement, steel, and railway carriages behind. New products and skills with new applications in technical assemblies received more political attention, although the phrase "high tech" was still in limited circulation. It must be remembered that the building of railway carriages and steel mills had been seen around 1950 as the way to surpass the cultivation of oil seeds and cotton, indeed to surpass the entire agrarian political economy. By 1960 there were railway carriages and steel mills being built in India, but the agrarian economy still predominated. New industrialists who had no base in money-lending, land, or the grain trade, who cared little about exports of coffee, leather, or spices, these industrialists were now pushing for their interests and economic growth in 1970. As the state too had invested in these new technologies, its agents pressed for a satisfactory return.

Derived from "high technology," and referring primarily to new electronic technology and new design, this term "high tech" appears to have first been mentioned in the journal *Social Forces* in 1955. This is the same year that the verb "automate" appeared in English. Derek de Solla Price used the distinction between high and low technology in *Science Since Babylon* in 1961, and quickly the *New York Times* (1961) and then a Kodak ad in the journal *Science* (1962) began to popularize "high technology." Irving Horowitz used the term in *New Sociology* in 1964, and the *Physics Bulletin* described high-technology industries in 1971 as requiring huge capital and R&D investments. The journal *Nature* first used "high technology" in 1972. Indians abroad and in India delighted in new English terms and were quick to adopt popular ideas, as well as naming new terms themselves. An early

Indian reference in writing is by economist Ashok Mehta in 1968. I remember Indian scientists and engineers who had lived in North America in the 1960s using the term in conversation from around 1968–69 in India.[1]

Not only was the Indian economy becoming more "commoditized" in terms of world trade, but levels of expert competence were, in some fields, thought to be on a par with other countries. Self-awareness and confidence had grown along with the solid and stolid sectors (steel, cement, etc.) but there were articulate interest groups who hoped for a bigger share of the Indian market and opportunity to earn foreign exchange. Among these new technologies and new fields were those over which the DAE/AEC had a monopoly: reactor technologies, space rockets, electronics, and miniaturized electronics and computers: the state precluded others from entering these spheres. Though each one had strategic and military applications, the government first emphasized the civil developmental benefits of these atomic energy, space, and electronic investments being made by the DAE; this was a key platform in the justification of state support for science and scientists. In addition the DAE was constructing equipment and facilities that had no parallel in India: it was building specialized structures, such as radio telescopes and cyclotrons, which were intended to put India on the world research map, and the specialists involved in that building process too were expected to contribute to this new high-tech competence and confidence.

But India was not going to export giant radio telescopes or cyclotrons! The reasoning was that this kind of project would establish the capacity that underlies other high-tech exports, in particular high-precision engineering using advanced materials. The DAE's relationship with the top industrial firms was established through these projects, with the firms asking for more and more of the government's contracts. The purpose of this chapter is to describe the initial stages of this high-tech economy, all of them sensitive to the politics of disagreement and opposition in the planning process. These stages were dependent—to some extent—on the DAE. Though its own budget was not inviolable, the greater importance of "high technology" showed up in discussions in the cabinet advisory committees SACC and COST in the late 1960s, and it would have been noted that most of the cabinet priorities listed at the end of chapter 19 were "high tech" in the definition of that time.[2] The prime minister too stepped in sometimes to defend large and ambitious programs even when they looked uneconomic. But ambition and desire could not possibly swing a change as profound as this one, even if backed by powerful and rich people like Vikram Sarabhai. India's political economy was still predominantly agrarian, and its capital investments (including those of the nationalized banks) were channeled toward "capitalizing" agriculture. But other interests were also working their way through and beyond the "old economy," and these interests were behind the push for a high-tech economy.

High-technology advantage lay behind the discourse in the late 1960s about another "great leap forward," since the first leap ten years earlier was widely thought to have been poorly timed and underfinanced. But talk of India's high-tech advantage in 1969–70 was confronted by chronic food shortages, poor health, low average life expectancy, and widespread poverty—in fact, an international reputation as a country in crisis and in need. Moreover, massive investment in new agricultural techniques in the preceding five years still showed none of the promised quick increases in food production, though the political status of agricultural research and development had been raised significantly, in tune with a new "modernist" kind of medium-sized farmer prepared to experiment with new technology, including new fertilizer-responsive varieties (known as high-yielding varieties or HYV). The high-technology idea was still quite new: the idea of "nucleus and nation" could not be nourished without new symbols, as Nehru often said. India already had an international airline and steel mills, and was building nuclear reactors and assembling jet fighters. So why shouldn't electronics, atomic energy, and space now carry the hopes of scientists and technologists to a new level of international competitiveness, and the whole nation with them? Why shouldn't the state invest in these things, not just to keep up but to leapfrog ahead, surpassing some difficult intermediate steps? In the process shouldn't the state continue to pay for the enhancement of the status and well-being of scientists and technologists associated with these technologies?

If this was not done, it was reasoned, some essential scientists might leave the country like the 1968 Nobel Prize winner Khorana had done in 1951. Surely they could not be compelled to stay? Were they not the children of upward-moving classes, interested in international mobility? So a high-tech formula gained currency in the highest places because it was understood by and appealed instantly to elites in every sector. There were good political reasons that the prime minister's office would take this formula seriously, beyond the fact that it was the next logical step of a science and technology building process begun by Indira Gandhi's father. Because she and her advisors had adopted a risky venture to break with the ruling syndicate of her party, she had to find new and young allies and adopt new strategies to appeal to a broader yet different range of interests—some on the left and some in the center-right. But it was not clear that this program could succeed, despite its appeal to business, civil service, and financial elites.

After all the prime minister was not invulnerable in her sphere. Mrs. Gandhi had become prime minister in January 1966, pushed there suddenly by K. Kamaraj, a Congress deal-maker politician from South India; the syndicate of the party expected her to do largely as she was told. But in the March 1967 election Congress lost 83 seats in the Lok Sabha, leaving it with a smaller 25-seat majority. This poor result, and Gandhi's apparent indifference to the views of the Congress old guard, irritated the older leaders in the party. Now they remembered having seen her headstrong ways before, despite their 1966 assumption that she would obey them. She had been elected president of the Congress Party in 1959 and had shown excessive enthusiasm for other people's projects when she had been a cabinet minister from 1963. After the 1967 election Gandhi had to appoint her rival Moraji Desai both deputy prime minister and minister of Finance, just to accommodate that powerful faction of the party. These people were not interested, in her opinion, in a high-tech strategy in atomic energy, space, or electronics; this disinterest offered a distinction between them. She began quietly talking to the communists and other left-socialist groups. The Congress Party finally split in 1969 between Indira Gandhi and Moraji Desai. However, Desai had to wait eight more years to come to power as prime minister, in coalition with his opponents, in 1977.

Just after this split Indira Gandhi courted the left socialist and communist parties more deliberately, removing Desai from the Finance portfolio in July 1969 just before nationalizing fourteen commercial banks (with assets of Rs 5 billion). Bank nationalization supplied capital to the state for the major investments deemed necessary for the Green Revolution, compelling banks to lend to the agricultural sector, which they had largely avoided before, having left that sector to moneylenders, one of the power bases of the Congress Party in rural society. By November 1969 there were two big Congress factions; one was the syndicate leaders who accused Indira of selling India to the Soviets, openly criticized her new chief advisor P. N. Haksar, and tried Indira in absentia for indiscipline and defiance of party leadership. The other was Indira Gandhi's faction. She called the others "fascists"; they called her a "dictator"—the split was characterized as a party vs. Parliament conflict. The Nehru family too was now split; with disapproval, her cousin Nayantara Saghal described it as a "new unprincipled era . . . of personal rule," and her aunt Vijaya Lakshmi Pandit spoke privately and critically to Indira. Annoyed, Indira had the Allahabad home of the Nehrus "given to the nation," and her aunt was unceremoniously asked to move out of it on the day before the handover ceremony. In December 1969 Indira took over

the Home Ministry herself, reorganized the Intelligence Bureau, created the Research and Analysis Wing for external intelligence, and took over both these branches herself.[3] Indira Gandhi also spoke about "a committed civil service" (Haksar's phrase), meaning that administrators with hopes for advancement should serve her goals.

In this national context Gandhi began to strengthen a rather passive relationship with the USSR, urged to do so by Haksar. But we have no evidence yet about Russian involvement in the nuclear program at this time; a shift begins in 1971 during the Bangladesh war for a clear break with the US, leading to a new (secret) treaty between India and the USSR. The consequences of this shift for the nuclear program appeared in 1973 with the early discussion of quasi-legal heavy water shipments for new reactors, probably through West German brokers.[4] We have no evidence about the Russian appraisal of India's high-technology strategy, but their approach was consistent with official Soviet policy for their economy. The Russians conceived of themselves as the first socialist high-tech economy; they were meanwhile urging India to adopt their computer systems and other technologies (as will be shown below).

At this same time the prime minister courted the industrial and technical elites and gave them a greater political role if they could be drawn away from their connections to the Congress old guard (essentially away from the people who had tolerated her father in his later phase). Vikram Sarabhai embodied this new thinking, one of the many reasons why the prime minister placed such confidence in him. Under Sarabhai there was a convergence in the DAE of the new spirit of private corporate enterprise (taking risks) and the older spirit of public service (making sacrifices and holding on to security). Contracting more work and responsibility out to the private sector, the DAE asked large companies like Tatas and Larsen & Tubro to execute more DAE work in its projects, balancing the role of state corporations. This was precisely what Indian industries had long been demanding, and who was more likely to be listened to by the private sector than Sarabhai? So there was gradually more DAE technology transfer within the Indian economy, including in high-precision engineering and new materials, and this was before the US and Canadian technology-transfer obstacles were set in place by 1974–75. Expected to *export* their products abroad, these firms also clamored to be the intermediaries in the potentially profitable transfer of technology from outside India into the country, but the government held tighter control of that option.[5] Sarabhai himself embodied this convergence between state science and technology, state industry, and private capitalist industry.

In a 1968 speech Sarabhai said, "There are those who preach that developing nations must proceed step-by-step following the same process by which the advanced nations themselves progressed. One is often told that such and such a thing is *too sophisticated to be applied* [emphasis in original]. This approach disregards what should perhaps be obvious—that when a problem is great, one requires the most effective means available to deal with it."[6] But Sarabhai was not ready to import high-tech facilities where he thought Indians could do the work; for example, in 1968 India's cabinet was in the process of deciding on an earth satellite receiving station and awarded the contract to a Canadian firm (Canada was at the that time the leader in nonmilitary satellite technology). I. K. Gujral, future prime minister, was then minister of state for Communication and recalled that Sarabhai "came to see me in great agitation. He was very upset about the decision to bring in outsiders to build the earth station. 'Where will Indians experiment if not in India?' he asked me." He also went to P. N. Haksar, principal secretary to the prime minister; "Vikram was diffident, nervous, angry, red in the face, and trembling almost." Sarabhai and Haksar had struck a working relationship about 1968, with respect to the Electronics Committee. Sarabhai specially objected to importing soon-to-be-obsolete technologies from abroad—he wanted to leapfrog. "The government was forced to relent and let him build the earth station after he had rushed around piling up the forceful intervention of Haksar, Gujral, and (to leave no stone unturned), the prime minister herself." At the same time Sarabhai was in negotiation with Hughes, GE, MIT, Ford, Robert McNamara, trying to persuade the US to lend India the new communication satellite ATS-6 for one year in September 1969 for a new television experiment for rural Indian audiences, just when the DAE was saying India would not sign the nonproliferation treaty.[7]

There was an additional reason for this convergence of the state and private industry: the attraction of DAE contracts in 1969–71 was great not only because the money was paid (if not always promptly), but also because the techniques and equipment were valuable to own and master. They could lead to repeat business, perhaps outside India, something these companies did during the oil boom in the Middle East after 1973–74. Firms requiring new equipment or parts involving foreign exchange probably got them more easily for these state-sanctioned high-tech projects. Firms were also delighted to be working on projects for which the prime minister was the minister and Sarabhai or one of his directors the principal negotiator of contracts. They expressed this delight in their advertising, announcing their relationship with the DAE on joint projects, putting them in a good position when bidding on other projects for other ministries. Being in the

prime minister's eye was, for an industrial corporation with international ambitions, a good thing in itself, particularly in a controlled and regulated economy.[8] But despite a continuing discourse about innovation, building the elements of the high-technology economy turned out to be quite difficult for the DAE. Despite its support at the very top, its plans encountered opposition at the top too.

Reactors, Sarabhai's Big Plan, and the Grasping Atom

The *Economic and Political Weekly* gave the DAE's problems top billing in a May 1969 editorial. Though well circulated among elite audiences, this was an unfriendly review, titled "The Grasping Atom."[9] The editorial strongly criticized Sarabhai's explanation of DAE's current difficulties and its budget secrecy. Sarabhai had said that reactors under construction were incomplete because of inconsistent "stop-go policies" of the Planning Commission, combined with the slow scrutiny of the Public Accounts Committee, which delayed expenditures. But the explanation for these difficulties lay elsewhere, the *EPW* insisted, particularly in the choice of an enriched uranium reactor for Tarapur and the consequent delays in its construction; the editorial called the Tarapur reactor "a technological dead-end." The *EPW* editorial also noted that the reactor projects in Rajasthan "required" full Canadian participation and countered that this "does not inspire confidence on the score of self-reliance"; the editorial ignored the fact that the Canadians asked for full participation. Finally the editorial criticized two proposed agro-industrial complexes in UP and Gujarat by saying "atomic power cheap enough to make nuclear-powered agro-industrial complexes a serious proposition will, it seems, remain a pie in the sky." With its brilliant title, the editorial then laid out the obstacles in Sarabhai's political landscape:

> The DAE's estimates of costs of nuclear power—as well as the costs of relevant conventional energy possibilities—will have to be examined by an independent, outside agency. If this is what the Planning Commission has in mind when it says in the Draft Plan that "the operation of these nuclear stations (those already taken up), their economics and other factors will be studied before further programmes are taken up" the Commission is quite obviously right. If to the DAE this smacks of a "stop-go" approach, then it has to be in the national interest.[10]

The DAE's situation was paradoxically both rosy and threatening. The DAE budget in 1969–70 was Rs 630 million (revised estimate) and was increased dramatically to Rs 920 million in 1970–71, a 49 percent increase. But eight months after this critique in the *EPW*, the Planning Commission late in December 1969 recommended a cut in the DAE's allocation for the 1970–75 period, from a requested Rs 3,981 million ($531 million) back to Rs 2,689 million ($359 million).[11] This was still a phenomenal amount of money for a department that had not yet delivered one kilowatt of electricity to the national grid after twenty years' commitment, but the cut was very big. In the end the Planning Commission, where B. D. Nagchaudhuri now worked, only got some of the cut that it wanted. Nuclear politics had come of age.

To fuel a high-tech revolution and an agricultural revolution, Sarabhai and the DAE planners decided that larger nuclear power plants were required to produce more electricity. This was the heart of the sweeping Ten-Year Plan announced in June 1970.[12] These plants could be used to incubate socioeconomic development in poor, neglected areas of the country lacking electricity. Electricity would give electronics social meaning. There could be agricultural and industrial development side by side ("as one sees in Japan," Sarabhai told me in 1969). His vision for an agrochemical project to surround a nuclear power plant in hot, marshy, and remote Kutch best indicates what he meant. The Rann of Kutch and its marshy wasteland ecology was the subject of a skirmish in the 1965 war between Pakistan and India, and it was thought (militarily at least) to be a poor place to site a nuclear reactor. But Sarabhai nevertheless wanted to build the reactor in this virtually uninhabited place, providing the nucleus for a new population to move in to a deserted region, use the electricity, operate fertilizer plants, desalinizing plants, and other chemical plants, and rehabilitate land that was considered barren using subsurface water.[13] The reactor would be used to desalinate salt water if necessary. This new larger reactor would thus attract and hold a new population around the fact that nuclear power, unlike other sources of power, can be generated anywhere (so long as water is available). If people moved to these remote areas with a potential for growth, such a plant would relieve population pressure in dense areas. Changing spatial relations was characteristic of Sarabhai's thinking.

The movement of people in order to draw workers for economic development (now called "displacement") was already a well-established principle in the Indian economy.[14] DAE engineer M. R. Srinivasan says Sarabhai discussed the agro-industrial complex with Alvin Weinberg at Oak Ridge, Tennessee, within the TVA: "Looking back the whole idea appears so contrived as to be laughable. Yet the American scientists got Sarabhai excited about this project" until practical considerations led to its abandonment, he said.[15] Again the TVA, the US Army Corps of Engineers, and rural electrification had stimulated thinking and planning in India, as it had Meghnad Saha in the 1940s, and this idea was also presented in 1967 in the film made by the US Atomic Energy Commission *No Greater Challenge*.[16] So many other contrived ideas were in the air (like a ring of rockets around Delhi) and taking shape on the ground, that I would not single this one out as laughable. Though their sites were not chosen for being wasteland, the area around other planned reactors too was supposed to be transformed. The plan for an agrochemical and nuclear complex embodied a necessary boldness; as Itty Abraham wrote about the Ten-Year Plan, "Sarabhai suggested that the means to save the atomic energy program was to integrate it into an all-encompassing ensemble of technologically-sophisticated artifacts. . . . Instead of producing energy for social consumption, as originally planned, the social world was being brought into the domain of atomic energy in order to be serviced by it."[17]

While Sarabhai talked about remote areas, he quickly learned that populated areas have their imperatives too. Just as Bhabha discovered about Meghnad Saha and Calcutta, regional pressure for a share of the DAE budget was strong, and Sarabhai received steady pressure from governors or chief ministers of states who wanted a nuclear power plant in their state. Even research projects had to be well distributed. Sarabhai continued to support the Variable Electron Cyclotron for Calcutta (despite its slow progress), and planned an interuniversity center for physics at Delhi, something proposed by Patrick Blackett six years earlier. The demand that a DAE project be located in South India, particularly Madras, was equally strong. True, a rocket launch site was operating at Thumba in Kerala, a new launch site for missiles was selected just north of Madras at Sriharikota, and a radio telescope was being built at Ooty in the Nilgiri Hills—all DAE projects in the south. But what southern interests really wanted was a nuclear power plant supplying electricity to Madras, a huge growing city with increasing electrical failures. Eventually a site was chosen just south of Madras at Kalpakkam in Tamil Nadu, and a major BARC-directed research center began to evolve there, next to a reactor. And, if Madras was going to have a nuclear power plant, then Uttar Pradesh would have to have one too, so a plan was produced to announce two sites, one at Narora and one at Kalpakkam.

Sarabhai proposed this plan by identifying India's greatest need as willingness to take risk and innovate, saying his job was to promote lateral thinking and internal technology transfer. Here he was speaking to the prime minister. He had long been thinking spatially, having begun in the 1960s to move some of Sarabhai Group companies to Baroda, having moved space research and development out of Delhi and Bombay to Ahmedabad and Bangalore. He was deeply interested in the history of Ahmedabad, a very old city that had grown up organically and somewhat chaotically. He told me he tried to think more organically and less linearly about his planning. A clue to his philosophy was his reproduction and personal distribution in India of Christopher Alexander's classic paper "A City Is Not a Tree; It Is a Semi-Lattice" because, Sarabhai argued, it showed that the ideal city (and its economy) would have a more complex structure than a simple "trunk and branch" model would ever permit.[18] He was also interested in the planning ideas ("ekistics") of C. A. Doxiadis. One had to rethink the uses of space and economies, he said, and the agro-industrial nuclear complex was an embodiment of that rethinking, in his mind. Though this troubled its critics, I think these ideas infused his approach to the organization of his favorite project, the rocket launching center in Kerala and its neighborhood effects.

The budgets for Space and Electronics, Radio Astronomy, and numerous other high-tech projects were embedded in the atomic energy budget, because there was the least political and administrative interference and the greatest abundance of cash; most important, the DAE almost guaranteed foreign exchange, on which most R&D relied heavily. Moreover, the DAE had a reputation throughout India (whether still deserved or not) for "getting things done." According to a young colleague, in the eyes of Vikram Sarabhai, "the Department of Atomic Energy was 'the Ministry of Advanced Technologies.'"[19] As in any large well-funded and well-insulated bureaucracy, however, there developed, as he observed earlier, "an in-breeding and self-confidence which went along with the monopoly of knowledge in nuclear matters."[20] This confidence was about to be challenged.

Pressure Behind the Scenes Becomes Public

If the DAE's ten-year budget plan for 1970–80 was "large," it was because so many varied interests wanted an increased share of it, reflecting the politics of state-to-state and regional competition. The budget also reflected private industry's increased demand for participation in major projects, coupled with the DAE's realization that it could not do everything itself. So it was not DAE leaders' ambitions alone that inflated it. But having had rich budgets and few controls for many years, DAE was not producing the results expected of it.

In fact, there was an internal disagreement in 1969 about the cost of electricity from nuclear reactors, and this upset the political process surrounding the preparation of the ten-year profile and reflected on the cost and regularity of electricity from the Tarapur reactor. In May 1969 Sarabhai and Sethna had been asked about the price of a unit of electricity from Tarapur: Sarabhai said it would be 5.6 paise, which was the generating cost, and Sethna said the generation *cost* would be 5.6 paise but the sale *price* would be 6.5 paise; "as a result there was considerable confusion in Parliament, in the scientific community, and in the media."[21] Parthasarathi questioned Sethna again about the cost of power and learned that the AEC's calculation was made without considering the cost of enriched fuel to be imported from the US and with the Tarapur reactor capital cost held at Rs 770 million (what the Parliamentary Committee was told, but lower than was actually spent). Sethna then said that if the International Atomic Energy Agency cost methodology, which covered all costs including fuel, safety, waste handling, and disposal, as well as decommissioning, were used, the price would go up: "he implied that it would raise the cost of power from our reactors to 'unacceptable' levels," at about 8 paise per unit, higher than the cost of thermal power in the grid, which was then close to 5.6 paise and the reason Sarabhai quoted that price.[22]

Seeing these higher costs while preparing his ten-year profile, Sarabhai thought they could be driven down by larger output reactors, of a 550 MW size, so "without consulting Sethna or anyone else at BARC," according to Parthasarathi, Sarabhai proposed to the AEC members bigger reactors. The profile was then written in early 1970 "almost entirely by civil servants" like T. N. Seshan in the DAE secretariat, without BARC input. The only other scientist on the AEC, Sethna, opposed this plan, saying that India should master the 230 MW reactors it was now building, make many of them, and drive the cost down that way. India was not ready, Sethna said, to build subsystems of such large reactors, let alone the reactors themselves. But Sarabhai briefed the prime minister, saying that what he wanted was agreement on the broad objectives, and she did not have a basis to question it.[23] Sarabhai assured her the other members, including J. R. D. Tata, P. N. Haksar, and I. G. Patel, were ready to approve this; she knew of the split between Sethna and Sarabhai.[24]

The three other members of the AEC, calculating for the Sethna-Sarabhai division, "took the easy way out" and endorsed the proposal, cleared it with Haksar rather than the prime minister, and released it, calling for bids for construction of the 550 MW reactors, which were immediately forthcoming in 1970! "Both Mrs. Gandhi and Haksar were keen, for both domestic and external reasons, to project a large programme on atomic energy, the feasibilities and practicalities notwithstanding." So the prime minister called a meeting of the political affairs committee of the cabinet, attended by

Sarabhai but not by Sethna, and the committee approved the profile. "The DAE was special and always a law unto itself, as it remains even today," concludes Parthasarathi.[25] So, the question was, would the minutes of this cabinet decision show "conditional" or "unconditional" support for the profile? If this looks like a minor semantic matter to outsiders, it took six months for insiders to settle and to finalize and circulate those minutes among insiders.

In March 1971 the DAE estimated a 3 percent rate of return on capital invested in heavy water plants like the one proposed at Madras, and this again attracted the attention of the prime minister's secretariat. "The DAE methodology of treating heavy water costs and construction costs as zero constituted a huge subsidy to the nuclear power stations." According to Parthasarathi, when properly calculated "the real cost of [nuclear] power came to much more than the cost of thermal power from the nearest coal-based station at Neyveli using its own nearby coal or even at Ennore in Tamil Nadu," using coal from Bihar-Bengal.[26] As a result P. N. Dhar, advisor to the prime minister, and P. N. Haksar persuaded Prime Minister Gandhi that this was so serious that the Madras proposal should be withdrawn from the next day's cabinet meeting, to be more carefully studied first. Sarabhai and the DAE/AEC "were furious," as this analysis challenged the economic viability of atomic energy. Dhar and Haksar wrote to Sarabhai to say "these things need clarification" before they come to the cabinet again.

An engineer with long expertise in India concluded that "heavy water was certainly our Achilles heel."[27] The initial work by Harold Urey separated heavy water from light water using the hydrogen sulphide–water exchange method, which was patented in the US. But hydrogen sulphide is corrosive and poisonous. Planners of Bhakra dam in Punjab in the late 1940s decided in the 1950s to build a nitrogen fertilizer (urea) plant at nearby Nangal so that it electrolyzed water to produce electrolytic hydrogen, as was done in the Norsk Hydro plant in Norway, from which heavy water could be recovered. The Nangal fertilizer plant, planned and built by a German firm (see *Negotiating Nuclear Power*), bought 150 MW electricity for a few paise, and Nangal had purchased a very long bank of electrolytic cells from the Italian firm Di Nora. DAE engineers like Srinivasan were brought on site to supervise, and Nangal then produced 10–15 tons of D_2O annually. Eventually this power-expensive method was phased out, but Nangal was India's only working heavy water plant until the 1970s.

Realizing that the output of heavy water at the Nangal plant was fading, the AEC planned a new heavy water plant. Sarabhai had just seen a new heavy water technology process built by the Gelpra firm at Mazingarb in

France, and the AEC decided it should be installed on a large scale in the ammonia production stream of a new state fertilizer plant at Baroda, the same city as Sarabhai's earlier industrial projects. In 1971 steel equipment for the plant was fabricated by Babcock-Atlantique at Saint Nazaire France and in 1972 transported to Baroda with a successful unloading via pontoon up an estuary.[28] It began to operate only in 1977. Using the same Gelpra-Baroda process, in 1971 a heavy water plant opened at Tutticorin, Kerala, within a Japanese fertilizer plant in Kerala built by Toyo Engineering. "The production of heavy water at Tutticorin suffered from frequent interruptions of power and the low-load operation of the fertilizer plant."[29] All this occurred under Sarabhai's AEC chairmanship, while he was proposing bigger heavy water reactors with higher heavy water requirements than those under construction in Rajasthan. This was in addition to the D_2O requirement of the plutonium breeding research reactors CIRUS and PURNIMA.

While the Tutticorin heavy water plant was being prepared for operation in February 1971, the DAE asked for approval of the heavy water plant for the Madras atomic power plant. The Indo-Canadian design called for one ton of heavy water for each megawatt of power, so the 235 tons required up front, at a cost of Rs 500 per kilogram, would total about Rs 120 million (assuming 5 percent loss annually). DAE had just created a special noncommercial entity called the Heavy Water Board, in order to identify and isolate heavy water problems, but did not publish a balance sheet; in the plan that came up to the AEC the cost of the initial 235 tons of heavy water (Rs 120 million) was not even included in the projected cost of the reactor! This was brought to the attention of Gandhi herself, and the calculation showed that the annual financial charge for the use of heavy water was not 3 percent, as DAE said, but close to 7 percent. Even with this projection, Haksar and Dhar decided with the prime minister in April 1971 that the Madras project should proceed because "there are larger objectives to our atomic programme than nuclear power and those objectives cannot be compromised at any cost."[30]

Throughout this heavy water period, a very different type of reactor not requiring heavy water, chosen by Bhabha in 1961, was nearing completion. In 1963 the Tarapur project began by planning a whole small town for Indian and American workers on the Gujarat coast; a separate part designed and built by Bechtel at one time housed 80 expats with swimming pool, club, guesthouses, shopping, and the like, all turned over to Tarapur when the expats left. American families also lived in Bombay as well as in Tarapur. But why did the reactor produce no electricity until 1970, and after that only fitfully? In 1963 the contractors for GE invited Indian engineers of the

DAE to view the start-up of a similar boiling water reactor in Italy south of Rome. Bechtel, the architect-engineer for the reactor, was responsible for the concrete on the bottom of the foundation called "the leveling course." It did not develop prescribed concrete strength due to errors in mix design, so Bechtel had to reconstruct the foundation, and their concrete specialist "was promptly sent back to the US." The DAE set up a well-equipped concrete-testing lab with well-trained technologists, and it was used widely in other projects in India.

In 1964–65 workers at Tarapur were divided between two union groups—Hind Mazdur Sabha (under the young and ambitious George Fernandez) and Indian National Trade Union Congress (led by an older S. K. Patil), which the US contractors had chosen to work with. These two rival unions were fighting for control of technical trade unions, and there was confrontation over credit and food services in the canteen as well as threats to some by other workers. The police were called, a confrontation resulted, and the police fired their guns, killing seven workers. During 1966 some tensions between project engineers led to transfers out of the project. DAE started to send engineers to San Jose, California, for several months to review designs and ensure they met specifications.

The Tarapur reactor pressure vessels developed hairline stress-corrosion cracks in the carbon steel walls, probably "enhanced by saline atmosphere," according to Srinivasan. All the stainless steel tubes of the steam generators had to be replaced in 1968, causing a further ten-month delay.[31] The reactor went critical in October 1969, and at that time refueling first took 4–5 months, then it was reduced to 2½ months, then 50 days. But these speedups were achieved only with expensive concentrations of extra manpower (even army personnel), so that eventually refueling was reduced to 41 days. Each time failed tubes had to be plugged, the human cost of the speedup was too great: "the exposure to radiation of the personnel carrying out the work was unacceptable."[32] Then the steam generators were eliminated, and the reactor was run on a single cycle, reducing the power output capacity from 210 MW to 160 MW. All the boiling water reactors built by GE in the 1960s and 1970s were shut down in other countries, said Srinivasan ironically, except Tarapur, which still functioned in 2000, "following our tradition of conservation."[33]

In January 1970, when the Tarapur reactor was opened by the prime minister, she said India "should work progressively toward self-reliance."[34] But it started to break down in late 1970: instability and surges in the Maharashtra electrical grid caused tripping of the reactor and shutdowns. In April 1971 grid malfunction tripped the Tarapur reactor, and there was loss of power, but during the shutdown a mistake caused seizure of turbine bearings. Moreover, leaky valves caused a shutdown resulting in a high concentration of radioactivity in solid and liquid wastes: fuel rods were damaged, and the reactor could not operate even at half burn early in June 1971. By then one unit had been shutdown 41 times, and the other shut down 33 times. Haksar and the prime minister were briefed about Tarapur in July 1971, so Indira Gandhi said, "shut it down until it is safe." At the cabinet committee where the ten-year profile was discussed again in July 1971, the issue of Tarapur's shutdowns was aired and "several MPs raised the question of the serious situation at Tarapur and what government was doing about it. Sarabhai merely asserted that everything was under control."[35] However, in August 1971 the DAE offered to create an Atomic Energy Regulatory Authority (AERA) out of a previous Directorate of Radiation Protection, and then the prime minister's office objected to the DAE regulating itself, a captive regulator. Sarabhai opposed letting it leave the DAE family, whereas Sethna and Haksar accepted that it should go outside. Gandhi sided with Sarabhai, and the AERA remained inside DAE until it was made an independent statutory body in November 1983 (as the Atomic Energy Regulatory Board), finally resembling the atomic energy regulatory instruments in other countries.

The uncertainty about the minutes concerning the cabinet's approval of the profile continued until May 1971, providing us a useful insight. Sarabhai had already written an unconditional approval version of the cabinet committee minutes, and the cabinet secretary drafted a conditional approval version. Parthasarathi was sent to meet Sarabhai in Bombay to assess the profile and the minutes critically, and felt the tension with Sarabhai, who was irritated by his arrival from the prime minister's secretariat, notwithstanding that the two had been working together at the AEC in the Old Yacht Club in Bombay. The minutes of the meeting five months before had not been circulated even in May 1971, so Gandhi, who had to sign the minutes, was consulted again, and she and Haksar decided to support the Sarabhai unconditional version, perhaps because they knew that Sarabhai would soon leave the DAE anyway! Meanwhile BARC's Sethna, Ramanna, and P. K. Iyengar had developed their own critique of the profile, and in 1971 arrived at a fifteen-year profile using smaller reactors costing Rs 8,000 billion for 4,500 MW installed capacity (not Sarabhai's Rs 12,500 billion for 2,700 MW in the profile).[36] This new plan was brought by Sethna into an AEC meeting, after it had approved the Sarabhai profile! Haksar tried but was unable to hold a joint meeting with Sethna and Sarabhai to reconcile these differences.[37]

In the end, the AEC approved the DAE's annual report for 1970–71, saying Sarabhai's ten-year profile has been accepted and "involves a commitment to a firm programme." The prime minister showed she was keen to project a large program but this made Sethna very unhappy; the prime minister was squeezed because "she liked Sarabhai and had a long family history" with him.[38] She had something else to show Sethna, quite soon, concerning the tenders offered by American, Canadian, German, and British firms for the 550 MW reactors, reactors that Sethna believed were too big and unwise. The 500 MW enriched uranium reactors that Sarabhai planned were to be modeled on Tarapur and were more than twice the size and power of the Canadian reactors then under construction in Rajasthan. They were to run on plutonium-to-thorium breeder cycles, as originally envisaged by Bhabha around 1959, and would produce sufficient plutonium to "breed the thorium" into a fissile fuel. The cost of this Ten-Year Plan was to be Rs 12,500 billion or about $1.67 billion, equivalent to the *annual* defense budget at that time.[39] Homi Sethna said that enriched uranium was costly to develop in India because uranium separation and enrichment also require a lot of energy. But the alternative, he said, is an undesirable dependency on imports.[40] Raja Ramanna likewise said that the cost of developing such large plants in remote areas would be too great, and India's industrial logistics and electrical grid system were unfavorable to large reactors. With irony, and the 1971 war between Pakistan and India fresh in his mind, Ramanna said in 1972, "Such expenditure is perhaps more justified in a peaceful world than, say, expenditure on border roads." Another physicist with long reactor experience said, "The plan for two 500 MW reactors was probably too ambitious; we were not ready for a heavy water reactor program of that size. It showed he was thinking about the big picture, but in 1970 it was probably a step too far."[41] Hearing some of this kind of criticism, Haksar and Parthasarathi decided in July 1971 to seek the prime minister's intervention to cancel the tender, "and she conveyed her decision to Sarabhai in no uncertain terms" (date unclear, probably July 1971). The cancellations were immediate; according to Parthasarathi, "Sarabhai was mortified; he had no option but to face the consequences of his extremely unwise action. The Sethna group in DAE and BARC was ecstatic."[42]

Sarabhai's plan was questioned also because there was only one functioning electrical power reactor in 1970, and it did not yet deliver electricity steadily to the grid system; experts were beginning to count the costs of reactor construction and maintenance quite accurately, including the cost of depending on other countries for heavy water. India's problem in 1970 was that it needed heavy water for the natural uranium reactors (for example, Rajasthan), but produced little heavy water, and it needed enriched fuel for the light water reactors (for example, Tarapur), but produced little enriched fuel. As for heavy water, the Parliamentary Estimates Committee published in July 1970 actual and estimated DAE expenditures on heavy water: in 1966–67 and 1967–68 the planned expenditure on heavy water itself was Rs 7.5 million (roughly $1 million both years), and the total budget for heavy water plants and deuterium was estimated for 1970–71 to be Rs 50,994,000 (roughly $7.2 million). Although it is difficult to compare these costs with steel mills, fertilizer plants, or jet fighters—the heavy water problem alone put atomic energy among India's highest-cost projects.[43] This will be examined in detail in chapter 22.

Sarabhai became a conduit for pressures against the development of bigger reactors, against the continuing combination of atomic energy and space, and against the elitism of the DAE establishment. He also was a conduit for demands for more investment in projects with little relationship to atomic energy, having to defend to outsiders these investments in terms of the eventual long-term payoff in the capacity for basic science to spin off new small or light industries (space was the paradigm for this critique). This was the context in early 1970 when Sarabhai's Ten-Year Plan for atomic energy received such a mixed reception.

Electronics and Computers as High-Tech Boosters

Though Sarabhai took over Bhabha's role as the chairman of the Electronics Committee in 1966, he brought the committee's work to a conclusion by negotiating a new Electronics Commission in 1970, soon followed by a separate Department of Electronics in 1971. Like others before him, including Bhabha, Sarabhai argued that the new electronics program had to serve every sector of society (from farming to defense) and that India was woefully backward in this direction. But the fact remained that electricity was scarce, and still in 1970 no electricity from any nuclear power plant had entered the grid system. How would electronics production fit into an economy in which there was a scarcity of electricity? Nuclear power was still not looking like a savior, and in fact the DAE was in financial trouble inside the government by 1969–70, with critiques from both the Parliamentary Estimates Committee and the Planning Commission. The Planning Commission was gaining enough confidence to review reactor projects, something that would not have been possible in Bhabha's and Nehru's period. Measuring the public doubt about the DAE, Sarabhai thought if there could be both electricity and electronics in the same place, it would demonstrate the logic of the investment and be revolutionary. But how would they overcome the limits on the supply and distribution of electricity, limitations that appeared as severe as ever?

The Electronics Committee, chaired by Sarabhai from 1966, was already composed of members from Defence, DAE, and some civil servants.[44] Soon the committee turned from electronics in general to India's need for computers. Although defense needs for computers were not then considered to be very great, at least in the opinion of some military planners, nuclear reactors and space projects clearly needed computers. Because every agency wanted to hire foreign-trained researchers who had used computers in other countries, electronics policy inevitably became computer policy. Large American mainframe computers were installed in 1965 at TIFR and at the two IITs at Kharagpur and Kanpur, in companies like Esso Oil, a subsidiary of Standard Oil, and in a number of banks. In a country where there was a widespread perception that clerks, thousands of them, were the proper keepers of records and accounts, people asked who would need computers to do that? Or, as a corollary, if computers did that work, what would these thousands of clerks and accountants do? The CDC 3600 (13K memory, Control Data Corporation) in Bombay, installed in 1965, filled a large air-conditioned room at TIFR. Installation of these computers was opposed by a well-organized and well-publicized anti-automation movement, rooted in labor unions (usually clerical but not exclusively). That the computers were American also gave the anti-automation movement added attraction and energy.[45] The Planning Commission established a special committee to study manufacturing computers in India; appropriate uses would be routine tabulation of exam results, railway bookings and tickets, tax checks, and the like. But the major manufacturers too scented a very large market: in 1968 International Computer Limited (ICL) of London formed a joint venture with India's Bharat Electronics (owned by the Defence Ministry), and the heads of American firms Honeywell and IBM visited Delhi to lobby politicians in 1969.

In 1967, with Sarabhai's full support, the leader of the TIFR computing center R. Narasimhan created a Working Group on Computers: a year later the group proposed a policy of indigenous development through prototype building and import copying. Along with large mainframe computers, Narasimhan prophetically emphasized the need for small computers. Such an opinion was not widely held, neither inside nor outside India, because there was a strong conviction that a few very large machines would be sufficient.[46] Though the Working Group's policy on self-reliance and technology transfer was the standard DAE approach, the Electronics Committee had already transferred responsibility in 1968 for actual implementation of its policies to the Department of Defence Supplies, which "had no belief in self-reliant development. . . . The Department did not take the views of the Electronics Committee seriously and considered them academic."[47] Defence Supplies was given the responsibility for implementation of electronic policy through Bharat Electronics: this military company sought to corner the market, and there was a marked "strategic" scent around electronics in India, opposed by the DAE. Because of this impasse between Defence and DAE, in December 1969 Sarabhai and the Electronics Committee planned a national conference to "mobilize support for their own position and to bring about changes in the administrative setup at the government level."[48]

Behind this lay some scheming: in November 1968 Sarabhai advised the prime minister to move the Electronics Committee away from Defence, saying electronics is "a most powerful handle to leapfrog our economy." Then some committee members succeeded in getting Haksar to promote this objective with her, to stop being a catalyst and become a driver; "there were serious gaps in the identification of demand and creation of supply." Japan after 1952 was used as an illustration; Japanese legislation in 1957 should be followed by India, said some committee members. As a result of this lobbying, the Electronics Corporation of India was created, and DAE insider A. S. Rao became its first managing director while still at BARC. Defence replied quickly to protect its turf, and the prime minister and Defence minister convened a meeting to listen to all sides. In October 1968 Gandhi said she was ready to create an Electronics Commission and Department of Electronics placed directly under her, but by mid-1969 this still had not been done and was causing uncertainty. Was the prime minister preoccupied with struggles inside Congress Party? Was Defence resisting effectively?

In March 1970 the National Electronics Conference was held at TIFR, chaired by Vikram Sarabhai, with five hundred participants from business, investment, industry, research, and government. According to an inside observer, commenting on a leaked report, "it became very clear that the Electronics Committee, which was supposed to direct and coordinate the development of electronics, had failed miserably." "The problems are both structural and inter-personal," the report went on. The Committee had no competent secretariat to produce analysis, no permanent members, and represented

> little warring kingdoms with the agency heads, who are represented on the Committee, refusing to take an overall view and instead trying to protect and expand the interests of their own agencies. . . . The Department of Atomic Energy is fighting the Department of Defence Supplies, small-scale sector is

> fighting big business, private sector is fighting public sector, and younger scientists are fighting old die-hards. All this fighting is mutually destructive as all these groups are highly interdependent. . . . But, alas, this is a behavioural problem and hence difficult to resolve.[49]

These criticisms of DAE were also implicit criticisms of Sarabhai's leadership and probably reflected overcommitment to his numerous projects. After all, they were not all really "his" projects, and problems like those described in chapter 17 are difficult to resolve everywhere. The entire scientific elite was overcommitted; some evidence lies in the fact that each member of the DAE's Electronics Committee was estimated to be serving on twenty other committees in 1970. When ministers and science advisors pushed for AEC-style autonomy for the electronics initiative, the cabinet approved the creation of the Electronics Commission and dissolved Sarabhai's committee in mid-1970. The prime minister took responsibility for electronics away from Defence and gave it to a separate new Department of Electronics, and at the same time she approved a committee composed of the secretaries of three departments (Atomic Energy, Communications, and Information and Broadcasting) to consider all aspects of satellites and educational television (the public objectives of the electronics strategy). The cabinet also created its own Joint Communications-Electronics Committee attached to the cabinet secretariat, chaired by a lieutenant general of the army. The Electronics Commission and new Department of Electronics, with responsibility for computers, was created in June 1970, and cosmic ray physicist M. G. K. Menon, still TIFR director, became secretary of the department, reporting directly to Gandhi. When a decision was taken in December 1970 to locate the commission in Bombay and the ministry in Delhi, Gandhi was trying to satisfy both the DAE and Defence. Members of the commission were Menon as chairman, A. S. Rao, M. S. Phatak from the Planning Commission, T. Swaminathan from the cabinet, B. D. Pande from Finance, plus two officials from Defence Supplies, no industry people, but staff secondments from TIFR and BARC, plus Defence Research and Development Organization's computer specialist Colonel N. Balasubramanian.

Though a large computer group was established within the commission, led by Balasubramanian, Atomic Energy still fought to remain the lead agency in the electronics field. "With these appointments the domination of the regulation, licensing, control and development of the electronics industry and its technology *by scientists* commenced. The computer industry and its technology became subject to 'scientific' treatment. . . . The group of scientists and engineers who manned the Department of Electronics had no industrial background or marketing outlook."[50] Others explained that the difficulties in electronics were due to its close accessibility to Delhi's bureaucracy: "The Government no doubt formed a Department of Electronics (DoE) but this did not blaze a trail like [atomic energy and space] had done earlier. Thanks to its Delhi base, DoE functioned like any other bureaucratic department."[51]

During this process of "integration," conflict continued between the Electronics Corporation of India, affiliated with the DAE, and the older Bharat Electronics, affiliated with the Defence Ministry, both in Hyderabad. After Sarabhai's death the DAE's role did not greatly change, and the new secretary of the AEC Homi Sethna became chair of the board of the Electronics Corporation of India. Secretary Menon, of the Electronics Commission, was a longtime DAE insider, as was Homi Sethna. The link between atomic energy and electronics, with both secretaries reporting directly to the prime minister, remained intact. What was relatively new was the close presence of the military at the working level.

So the scientific and technical elite addressed the electronics question, but the unmet demand for electronic appliances and equipment among the large middle class produced even stronger market and political pressures. In February 1971 a new government regulation was issued that citizens returning to India could import personal items like television, tape recorders, refrigerators, amplifiers, hair dryers, movie cameras, projectors, computer parts, and the like as duty-free baggage. This rule benefited, among others, members of the scientific community who had become accustomed to these "luxuries" while living abroad where they were commonplace. Resale value of this equipment was high in India and might even have to be paid for in US dollars; "almost new" electronics were prestigious and very valuable, as classified advertisements in major daily newspapers of Bombay, Delhi, and Calcutta showed. No other commodities except foreign cars held such fascination. The insatiable demand for imported electronics of this sort would not have been lost on the chair of the Electronics Committee, Vikram Sarabhai, who grew up in a business community.[52] On the other hand, there was an understandable attitude in India that consumer electronics were an unnecessary luxury given India's poverty, although the desire for them was without class boundaries.[53] Entertainment electronics had a huge potential market in India, though in the view of planners these were "completely nonessential." In 1972 the Electronics Corporation was licensed to produce up to ten thousand television sets, half of which would be for villages selected for the satellite-television experiment, and the consumption expanded exponentially from then.

Meanwhile the Electronics Commission appointed a study team in 1971 to negotiate the role of IBM and ICL in India, as both were assembling components locally, renting out used computers from abroad on an "as is" basis, and servicing them. Basically these computers were being sold twice: as the popular IBM 360 machines got older, the commission found that the cost of repairs increased, and price of parts increased too. IBM profits alone were "a massive monopoly in our computer market and one of the primary objectives . . . was to break that."[54]

In the context of a more permissive import policy in 1971, and as an apparent balance to it, Secretary of Electronics Menon announced in March 1972 that IBM and other foreign computer manufacturers operating in India were expected to find Indian sources of supply and expertise and stop dumping unsold previous-generation computers from their other markets into India. This was the start of India's confrontation with IBM, one of the world's largest and most successful corporations, running in parallel with India's ongoing confrontation with the Coca-Cola Corporation. Both were equally well-known American multinational corporations at the time, with head offices in the US. It was part of a mounting commercial cold war with the United States, crystallized by US cancellation of aid following India's 1971 war with Pakistan and liberation of Bangladesh. Menon's IBM announcement at the convention of the Computer Society of India caused tension between the private-sector computer users (including those, like banks and other companies, who now depended on IBM machines in operation at facilities they did not own) and government policymakers. Relying on instructions from his minister Indira Gandhi and the cabinet, Menon said at the conference that the use of computers that displace labor could not and would not be encouraged. Government agencies, however, such as the Life Insurance Corporation and the nationalized banks already relied heavily on computers for routine operations, and their managers doubted that India was really ready to ask IBM to leave, knowing India's software and equipment manufacture were completely inadequate for their needs.

As the new profession of computer scientist emerged around 1970, it was expected that others, particularly electronics engineers, should yield some power and prestige to them. There was therefore resistance and struggle over this change in India, pitting two professional communities against each other. But there was a continuing perceived domination of scientists in India's computer development. According to C. R. Subramaniam, former chairman of Bharat Electronics:

> The AEC/Bhabha school of scientists who had assumed power wanted a self-reliant development approach where all the computer systems and their components were to be indigenously developed and manufactured. Development was to proceed in parallel. The approach was neither pragmatic nor realistic. Technology was to be bought only for peripherals and some components. Foreign companies were not to be allowed to operate in the country either. . . . Menon had another brief—to reorient the administrative setup so that the aspirations and activities of the scientists would not be curbed by the bureaucracy.[55]

In this last objective Menon was following his mentor, Homi Bhabha, to give technically trained people autonomy from general administrators, just as they always had in TIFR, where Menon was still the director. Adding to his influence and to the autonomy of scientific institutions, Menon was elected at age forty-three to the Royal Society in the spring of 1970, after a first unsuccessful nomination in 1960. When Blackett proposed the election of a British physicist in 1968 in a letter to Clifford Butler at Blackett's own Imperial College, he commented that C. F. Powell of Bristol (Nobel Prize in Physics, 1950) may be supporting Menon's candidacy. Soon Blackett heard from Powell asking for support for Menon, whom Blackett knew very well, but Blackett replied, "Menon is a strong candidate but I feel he can wait a year or so. He will then have had time to show what he has done with the Tata Institute." Powell was persistent and quickly wrote back to remind him of the "master's wish": "Bhabha had written to me several times stressing the importance of an Indian member of the Royal [Society] and emphasizing Menon's claims." In the end Menon was elected for his work in cosmic ray physics, not on the first vote but on the second, in 1970.[56] He had no time for cosmic rays now, for he had become the "Raja of Electronics." Here was the direct link back to Bhabha, the master who had declared that electronics are the nervous system of a modern industrial economy and society. Menon was now simultaneously responsible for India's most important research laboratory and the nation's electronics "movement": little did he know that in four years he would be responsible for India's missiles too.

Rockets and Missiles: Space as a New High-Tech and Administrative Frontier

Indian intentions in space were clear when Indira Gandhi reconfirmed the government's commitment in 1966 while appointing Vikram Sarabhai to the AEC and DAE. Even while Bhabha was alive, the DAE had announced there would eventually be an Indian satellite system for broadcasting, remote sensing, weather forecasting, and telecommunications. Bhabha and Sarabhai stood together to announce these goals in 1963, though few believed it would happen in the next ten years. Given its huge size and growing influence, the DAE provided an ideal context for the expansion of space research and development. Although Bhabha had favored this expansion, it was Vikram Sarabhai who carried it out, and when he moved into the DAE in 1966, he emphasized the horizontal movement of ideas, resources, and personnel across administrative boundaries. His motive in this horizontal movement, he said, was to achieve technology transfer and new structures.

India had long experience with space, in terms of high-altitude cosmic ray and ionospheric research, beginning in the 1920s. Scientists like Mitra, Saha, and Bhabha supported laboratories focused on this work, were strongly interested in high-energy particles that could be studied this way, and moved on to high-altitude balloons even before the technique was perfected. In 1957 a tracking observatory was established at Nainital in the foothills of the Himalaya mountains in collaboration with Harvard Observatory, to begin satellite-tracking geodetic studies.[57]

And "suddenly" the USSR launched *Sputnik.*

It is not as if Indians had been indifferent to rockets and missiles, just that the two "streams" of effort in the country—military and civil—were not integrated from the start. Indians had earlier experience with rockets which they had not forgotten.[58] The first brush with modern rockets appears to have occurred when Krishna Menon got excited about missiles about 1958, and asked Nehru for approval for a missile study group in Defence Research, of which he was the minister. Nehru told him to clear it with Finance Minister Moraji Desai, so Krishna Menon sent Kothari to meet Desai and persuade him to approve the missile group. Kothari was reported by Blackett in 1951 to have been very interested in a French ground-to-ground missile.[59] A more curious encounter than one about rockets and involving Desai, Menon and Kothari can hardly be imagined! As one might expect, Minister Desai disagreed with Kothari's plan in 1958, but Menon persisted in bringing it to the cabinet about September 1958, in a showdown with Desai. Kothari was present to answer questions, all of them coming from Nehru, according to Kothari's associates; when Nehru concluded his questioning, Kothari was told to "keep the foreign exchange expenditure to the minimum." Menon and Kothari decided to develop an antitank missile, and Rs 200,000 was allocated from the Defence budget. Two scientific-military officers were sent to a NATO guided-missile course in Great Britain,

and one was sent to the IISc to use the wind tunnel in Bangalore, for which he had to negotiate in 1960 with a skeptical Satish Dhawan. The missile to be copied was French, and it was to be wire-guided! Then in 1960 a Russian SA2 missile brought down an American U2 plane over Russia; Minister of Defence Krishna Menon determined to build something like the SA2 in India. This intent must have filtered to Moscow, because just prior to his 1962 unmasking as a spy for Great Britain and the United States, the rocket engineer Oleg Penkovsky recorded that he expected to be stationed in India on account of his technical knowledge and Soviet plans to assist India in rocketry and missiles.[60]

In 1960 Kothari retired, and the Defence Research and Development Office was created; physicist S. Bhagavantam became defense advisor; the DRDO then moved from Delhi to Hyderabad in 1962, to the nizam's old army barracks. Though India's Space Research Organization (ISRO) had started building a rocket launch site at Thumba in Kerala, DRDO's antitank missile project was still a long slow ordeal for the DRDO; missile guidance was changing from wires to remote electronics, and the DRDO tests did not work well, but apparently researchers learned experimental lessons about fuels and guidance systems.[61]

Sarabhai continued to pursue his first love, which was upper atmospheric physics, and he became more interested in new techniques and technologies. His lab, the Physical Research Laboratory in Ahmedabad, became an autonomous institute in 1960, with continuous DAE funding. In 1961 a small real-time satellite telemetry station was built at PRL using NASA equipment; this was in addition to the optical tracking station still operated with the US at Nainital. In this year India placed space research under DAE jurisdiction, and in 1962 ISRO established Thumba to study "tropical atmosphere": interaction of ionized and neutral atmosphere, electrojet, ionsphere interaction with geomagnetic field, and the like. In 1962–63 a rocket launch and space research team from ISRO went to Goddard Space Flight Center in the US, for training. Homi Bhabha visited Raytheon Corporation's research facilities in Massachusetts in July 1963 and wrote secretly to the Minister of Economic and Defence Coordination T. T. Krishnamachari to propose establishing a missile production facility like the one he had seen in the US.[62]

The first rocket launch at Thumba used a US Nike Apache rocket carrying a French experiment in November 1963, using orange sodium vapor clouds emitted from the rocket to study prevailing high-altitude winds. French radar and Indian-modified Australian radar systems were installed for tracking the rocket.[63] Radar and microwave technology had just been

given much increased importance by the 1962 China-India conflict. In 1963 Sarabhai and Bhabha said that satellite communication might supersede other modes of intercontinental communication, evidence that a plan for Indian satellites was forming. In 1965, in conjunction with the UN's program Peaceful Uses of Outer Space, Indian studies began of problems of physical energy and momentum transport into the lower atmosphere where climate is produced, with special reference to the monsoon. Serious monsoon failures in eastern India had produced drought and famine in 1964–65. These studies started a focus on altitudes above 100 km, above normal range of balloons but below normal operating range of satellites. This provided the rationale for an investment in sounding rockets rising to the height of 35,000 meters.

The other path that Sarabhai carefully pursued was to bring military resources into the space research field, starting with rocket engines. He was aware of the opportunity to enlarge the small links between the military and his space projects, as other countries had done, and so he adopted a wider view of "building India's technical competence." He had contacts and some influence in defense circles and could maneuver to find the funds to achieve joint projects with them, a direction strongly supported by others. As a small example of his approach to integrating civil and military resources, Sarabhai showed Abdul Kalam, of ISRO, and V. S. Narayanan, of DRDO, a Soviet rocket engine on an Indian military base early in 1968. He proposed "reverse-engineering" (copying) this rocket engine and then copying the SA2 surface-to-air missile, examples of which were already in DRDO's possession. This Soviet rocket engine was capable of boosting a jet fighter aircraft on very short takeoffs, a task that Sarabhai had asked Kalam to prepare for back in 1963, following the conflict with China. This capability was intended for the mountainous environment of eastern India where conflict with Chinese forces was expected, though that conflict never came again.

But how had *this particular* Soviet rocket engine shown to Kalam and Narayanan arrived in India? In 1967, when India was just about to buy a British surface-to-air missile, the USSR offered a gift of a battery of the SA75 missiles—"to guard Delhi."[64] Air Force Group Captain Narayanan was sent to Moscow to evaluate the SA75, and the gift was accepted. As joint director of missiles in 1968, he prepared a plan for "reverse-engineering" the SA75 because the problem of Soviet spare parts "had become acute and expensive." Narayanan's Delhi headquarters dithered over implementing this reverse-engineering project. In 1968 Narayanan was also in charge of the rocket-assisted takeoff project for the MiG (Sukhoi) jet fighter, for high-

altitude runways, so he took one of his technical problems about the rocket takeoff project to Sarabhai, who introduced Narayanan to Kalam. Pointing to the Soviet engine that Narayanan had procured, Sarabhai told Kalam and Narayanan that he could fund their work on the condition that they copy it and make a working prototype within eighteen months. According to Kalam, the two men decided to make the prototype. Narayanan was already an expert on Soviet surface-to-air missiles (SA2), having been put in charge of the missiles for a defense ring around Delhi. Using Rs 7.5 million obtained by Narayanan from the defense budget, the joint project moved quickly at both the DAE's space science and technology center and the Defence Ministry's research and development organization, using a team of twenty DAE and military engineers, according to Kalam. Narayanan promised more defense money if necessary.

Kalam said Sarabhai himself approved a new pattern of management for this project using financial approval through one person only instead of a number of officers in a chain; moving all goods and personnel for the space project by air, thus bucking government rules to use the cheapest rates (rail); and competitive private sector contracting. This pattern was eventually established throughout DAE's space program, and Kalam reported that this rocket-assisted takeoff project saved the Defence Ministry about Rs 40 million, most of it in foreign exchange. Kalam and Narayanan were soon made members of a new Missile Panel in the Ministry of Defence in 1968. Kalam said he briefed Sarabhai following every Missile Panel meeting held at the Defence Ministry.[65]

The public imagination in India was transfixed in July 1969 when the first human stepped on the moon. Despite an anti-American attitude well established in the younger generation of Indians, framed by their rejection of American conduct of the Vietnam War, there was utter public fascination with this American achievement. The invocation of the "step for mankind" attracted them too. Even in West Bengal, where the rejection of things American was stronger than in most other states, the moon landing struck a chord of excitement, awe, and possibility. Indians did not have television at the time, but when those photos of people on the moon's surface were seen, people in the space community understood that this rivalry between the Soviets and the Americans might assist them in pursuing their own goals, riding on public sentiment.[66] Prior to this, they had been working in relative isolation from the public.

At Thumba in January 1969, I observed the launch of a Rohini rocket modeled on a French Centaure design. The nose cone was delivered to the pad on a bicycle, and the rocket flew successfully to a height of 110 km. By

1971 the Menaka and Rohini rockets, based on French models, were sufficiently "indigenized" to be considered of Indian design and manufacture.[67] At the same time India quietly opened a 2,000 km rocket/missile testing range with the coastal launch site just north of Madras, at Sriharikota. This is the preferred type of location for rocket launch sites everywhere, facing east to take advantage of the direction of the earth's rotation and over the sea to allow the booster debris to fall with less risk to human populations. The eastern end of the missile range was near the coast of Burma. E. V. Chitnis and Pramod Kale had begun looking for the east coast launch site in 1968, and in May 1969 Sarabhai went to see it, 80 km north of Madras in the middle of Pulicat Lake.

Someone asked for an American estimate of the cost of this launch site, and ISRO were told that it would be Rs 500 million, half payable in dollars. Vasant Gowarikar said to Sarabhai that it could be built for much less, and Sarabhai, who liked tossing these challenges to his colleagues, agreed to let him try. It was completed at the lower cost.[68]

In July 1970, when Bhagavantam retired from Defence, B. D. Nagchaudhuri left the Planning Commission to become the defense advisor to the minister, and the first scientist to be appointed secretary of Defence while also being head of DRDO. No previous defense advisor had held a secretary position.[69] Here was nuclear physicist Nagchaudhuri, Saha's second student to become defense advisor, D. S. Kothari having been the first. According to Nagchaudhuri, he stepped into an unfamiliar culture: "The number one problem then was that Indian [defense] science wasn't focused. Two it lacked cohesiveness, with everyone going their own way. And three: it lacked objectivity." Regarding the DRDO itself, Nagchaudhuri said, "It was all very experimental. Nor did they completely abandon a project which was floundering."[70] Nagchaudhuri told me that he found the change from the organizational culture of scientists to economic planners and then to the defense establishment "fascinating but trying."[71] He referred to the way that hierarchy blocked communication, the same problems that Ramanna found in DRDO eight years later: "military protocol" in the lab, superiors defended by their secretaries from communication with junior scientists (who thus had to write them letters in order to speak to them), no questions allowed at seminars, and certainly no contradiction of seniors' ideas.[72]

In November 1970 Prime Minister Gandhi asked Nagchaudhuri to initiate a defense feasibility study for long-range ballistic missiles, soon called Project Valiant.[73] A month later, Nagchaudhuri went to the rocket division of the DRDO lab in Hyderabad and asked for an 8,000 km range missile with a 500 kg payload, to be ready in four years! If true, these objectives seem very unusual under the circumstances; one is prompted to ask, "Why build such a long-range missile?" Nagchaudhuri's answer was clear: "Well," he told Raj Chengappa, "Mrs. Gandhi wanted it. And if we were going to have a nuclear explosion then we should prepare for it in such a way that all the pieces fell into line."[74] Two years later he modified the plan profoundly, with the prime minister's agreement, to build a missile with a 1,500 km range carrying a 1,000 kg warhead.[75] While in an officers' course at the National Defence College, Narayanan prepared a classified proposal for the Ministry of Defence on "missile self-sufficiency" in 1971, linking the work on the upgrading of the SLV3 to long-range ballistic missiles.[76] Sarabhai read the proposal, asked Nagchaudhuri to read it, and then introduced Narayanan to Nagchaudhuri in November 1971. Now Sarabhai was dealing on a roughly equal basis with Nagchaudhuri, a physicist who suddenly had a lot of money but had previously been simply the director of one of the DAE's laboratories, the Saha Institute. Nagchaudhuri, with resources and expertise available, was now looking for a new director of the Defence Research and Development Laboratories in Hyderabad. He also felt the SA2 reverse-engineering project "was a good cover for India to build competence in missiles."[77] Narayanan was soon appointed director of the Hyderabad missile laboratory. With the minister of Defence and his scientific advisor Nagchaudhuri watching, Narayanan's prototype engine lifted a Soviet fighter from a runway in October 1972, in 40 percent less distance than was usually required. Had Sarabhai not died ten months earlier, surely he too would have been present.

After the 1969 American landing on the moon, a fresh stimulant to the Indian space program came in the form of the successful launch in April 1970 of a Chinese satellite. This was as powerful an influence among Indian elites as the US moon landing, and the well-informed weekly columnist of the *EPW*, Romesh Thapar, summarized and reflected very widespread elite public attitudes about the difference between China and India after this event: he said the Chinese have shown "a massive coordination of skills and resources," arguing that the Indian prime minister has to compare the Chinese success with the lack of Indian results. (This was rather direct: Thapar knew Gandhi personally, and she had dined at his house.) The DAE's announcement that there would be an Indian satellite in five years could not really be taken seriously; Thapar wrote: "There is no vision of self-reliance, no boldness of sovereign purpose, no desire to take meaningful risks. . . . What of the scientific leadership of this country? We need to shake these non-starters from their roots. If they cannot perform, they must find other jobs." Describing defense research as a "sorry mess" where scientists find nonscientists in charge (according to the just-released Kane Report) and the DAE as full of "lazy, sleeping-partner attitudes," "heavily dependent on the technologies of an advanced nation," meaning Canada, Thapar also said secrecy was coming to the DAE's rescue. In Parliament the debate

> takes the usual course—allegations against this or that personality. The personalities are not important; we have to find the treatment for the paralysis in the laboratories. . . . "Secrecy" is now a major weapon in the hands of the incompetent and corrupt. . . . We have no idea of the monumental waste which is permitted from year to year, in secrecy. And this great job waiting to be done is to carefully tabulate technical and productive capabilities, poll and coordinate the resources available in various sectors, and sketch the perspectives for achievement. Nobody is doing this, despite the proliferation of committees and such bodies.[78]

Because secrecy was fairly effective, harsh criticism like Thapar's was closely related to the lack of meaningful information; few knew what was going on, and reporters were deflected with stories of heroic technical achievements. The Chinese satellite overshadowed discussion in Parliament of the Defence Ministry's budget and again prompted calls on the government to reconsider its decision not to make nuclear weapons. Not to have a satellite, not to test an atom bomb, not to have electricity from a functioning reactor, not to have a high-performance rocket or indigenous jet fighters were all taken, separately and together, to be signs of deficiency and weakness. Although all these lacks arose from different factors in different contexts, they were allied together in the "nationalist" public imagination. The well-funded and constantly enlarging DAE was considered at least partially responsible for these deficiencies. So if an electronics and computer program was not quite sufficient to distract public attention from the problems of reactors, perhaps a satellite program with television might be.

As these difficulties around nuclear reactors became better known, Sarabhai constantly spoke of "Space" as the stimulator of new economic links and technology transfer. His emphasis irritated some and inspired others. Sarabhai's enthusiasm for space was perceived in BARC as distracting attention from the main objectives in reactors, nuclear research, and a bomb test. But for people like Satish Dhawan at ISRO, Sarabhai was laying important groundwork, describing how new R&D companies had sprung up on Route 128 outside Boston, how NASA's new space research and development requirements stimulated innovation and industrial growth (a subject he understood because of his regular visits to MIT). He argued that the space center in Kerala should be able to promote innovations like that, and Sarabhai told me he often poked around in small machine shops and electronics shops in Kerala for evidence of new thinking and new skills spinning off from bigger projects, hoping he could place contracts for parts there.[79] Sarabhai traveled to Taiwan and Japan in 1968, 1969, and 1970, and saw their electronics revolution; on return he asked his colleagues to investigate why those countries were successful, and "why are they making new technologies in their back streets and why aren't we?"[80]

The launch of the first Chinese satellite defined the DAE's plan to launch a satellite within five years using the Indian SLV3 being built in Kerala. The satellite would apply its data-collecting capacity to remote sensing of snowpack in the Himalayas (warning of water problems), monsoon patterns, and television broadcasting. The proposed 1970 budget for the next ten years of the *space program* was Rs 1,650 million, roughly 13 percent of the proposed spending on the atomic energy program.[81] Sarabhai's approach to projects, giving managers maximum autonomy from headquarters and from one another, did not necessarily satisfy those who expected timely results. These people now called for a paramilitary approach, where deadlines are real, saying that delays had already cost money. Said one engineer who eventually worked on industrial contracting for the Space Department, "The space program was a woolly-headed organization run by physicists: Sarabhai had twenty-seven people reporting to him in a flat organization. He approved every purchase order himself! None were being rejected because he didn't have a system of controls. I saw this myself. Would we have got a satellite launched under those conditions? I doubt it."[82] Another observer, who worked in both defense and atomic energy projects, said that work under Sarabhai was not likely to be paramilitary.

> Sarabhai inspired people; he didn't drive them. Even engineers would be inspired by him. When I gave a talk at the meeting to plan the Kalpakkam reactor in November 1971, it touched on two questions Sarabhai was thinking about, namely, surface contamination of tubes, and neutron radiography's role in inspecting irradiated fuel. He had lots of other things to do that day, but he stayed after the talk and asked very sharp questions. He immediately saw the connections between my work and the others. I saw his critics try to prick and poke him, but he was a gentleman and a Gandhian, and he remained very calm in the midst of their rude behavior.[83]

The SLV was to be modeled on the US "Scout" and, being the third of four options available, was called SLV3; Sarabhai divided the project in four parts, each under a different leader—Vasant Gowarikar, Abdul Kalam, A. E. Muthanayagam, M. R. Kurup—but he retained for himself the role of group leader. Kalam, whose nickname was "Busy Bee," was required to find a substitute for expensive American fiberglass nose cones. One of Vikram Sarabhai's methods was "to throw the same challenge to two or more people, encouraging a fierce competitiveness"; for instance, he pitted one group against another to produce solid propellant, and both claimed success in the end. When Satish Dhawan visited Thumba, he "found a rapidly swelling organization of over 3,000 people with no formal systems of administration." Moreover, "the five-member committee Vikram had appointed to oversee the operations apparently bickered bitterly in his absence but was united in its keenness to show him results." And yet Gowarikar recalled, "The insistence on indigenization all the time was a great motivation." Life around Thumba and Trivandrum was not as convenient in those days as it was by the end of the century, and for many reasons, said Dhawan, "despite the rough conditions in the early days, few people quit."[84]

Conclusions

The quest for a high-tech economy thus had a number of attractions: it appealed to a growing, influential, and mobile middle class; it caused sparks in a sector of the economy that could have a considerable multiplier effect in the big industries; it might increase exports of industrial goods for hard currency; and it might appeal to richer sections of the rural uneducated population as a matter of prestige. Even if nuclear reactors were not contributing electricity in 1970, a satellite, a rocket, or television might just turn the tide. But the quest for high tech also had a number of frustrations: the fundamentals of the agrarian economy remained contrary to this quest, because so much of the population was dependent on agriculture. Proceeds of foreign trade continued stronger in the traditional sectors of textiles, tea, leather, and similar products than in high-tech products and services. When geopolitics intervened in India in an unexpected way in 1971, a greater emphasis on self-reliance also looked very attractive, something to work toward and sacrifice for, again. That emphasis on self-reliance, recalling the entire first half of the century, is the subject of chapter 22, but we must first understand the pressure to build and test the bomb, the ultimate sign of the high-tech age but one that was also the proof of self-reliance.

TWENTY-ONE

Nuclear Expectations and Resistance in India's Political Economy

High tech was on many lips in the early 1970s, and everyone meant by this phrase the consumer goods they had seen elsewhere. The envy directed at those few who possessed these goods was linked to complex resentment about their absence or denial in India, where these consumer goods were defined as "not essential." Resentment at government indifference, while privileges were allowed to its diplomats and others, including businessmen, was palpable. As shown in chapter 17, most scientists looked forward to having a refrigerator (among married males a measure of the influence of their wives) although they understood well there was no constant electricity to run that fridge! This chapter concerns the expectations (and conflicts about them) within the nuclear establishment, and internal disagreement and resistance to some of those expectations coming from the prime minister's powerful circle; there is also evidence of resistance in the nuclear establishment to some of the prime minister's expectations concerning its performance.

There was another sign and symbol of high-tech capabilities beyond household and consumer goods, and that was a test of an Indian atomic bomb or, as it was coming to be called, "a nuclear weapon," or particularly for India from 1965 onward, "a nuclear explosion." In this chapter this pressure to test, along with its background, is separated from computers and satellites. This approach risks suggesting that thinking about the bomb test occurred in a different box from thinking about other issues, and I trust that the interweaving of issues in this book persuasively shows that thinking about the bomb was actually deeply embedded with other questions. The bomb should not be thought to have been *just* a gesture among other gestures—it clearly referred to high tech in a very special way. The pressures to build and test a bomb came up in an acute way in 1971, and Sarabhai found it hard to continue his resistance. Before December he had been told that he must choose between managing space and managing atomic energy; if he chose space, he would no longer influence the path to the bomb test, and perhaps he was not influencing it much now.

When India's Supreme Court overturned the government's bank nationalization and invalidated the abolition of the prince's purses in December 1970, Indira Gandhi boldly called an election for March 1971, to focus on the central government and her issues. But in the end only one issue counted and that was Indira's leadership and her slogan *Garibi Hatao* ("Down with poverty") vs. *Indira Hatao* ("down with Indira"), the slogan of the other Congress faction and other parties like Swatantra. Prime Minister Gandhi made public her political friendship with Mohan Kumaramangalam, a brilliant Cambridge-educated lawyer who left the Communist Party in 1966 and coined the phrase "a committed judiciary" and who was the spokesman for the Congress Forum for Socialist Action. She was cultivating the left carefully, playing a role rather like the one during her student days in Oxford and London, 1937–38.

In the March 1971 election Indira's new faction of the Congress Party won 325 seats out of 520 and that pushed the old Congress Party into the background. Gandhi won her seat against opponent Raj Narain; eventually he brought a successful court case against her for using a government servant to manage her election work. Indira captured the female vote. She was elected president of Congress on 17 March, and Congress pushed through new legislation to abolish princely purses and to again nationalize general insurance. Soon she pushed through amendments 24 and 25 to the constitution, permitting changes to the constitution and protecting those same changes to the constitution from judicial review. Kumaramangalam was made minister of Steel and Heavy Engineering and argued that there was no reason to place individual rights before or above the rights of the mass and their "urgent needs." He was a leading proponent of "the commanding heights" strategy in which private business was fine on the right scale as long as the state retained control of the key sectors of coal, steel, electricity, and the like, commanding heights of the economy.

In this context one of the pressures on Sarabhai in 1971 was how the ten-year atomic energy plan would be financed, and he used the question of cost to argue against the test. The dismal picture of foreign exchange reserves in 1965–66 that led to devaluation of the rupee had changed for the better by 1970, partly because of strong controls. In March 1970 foreign exchange reserves stood at $1.09 billion while in 1965 they had been down to $524.3 million. This favorable 1970 position was achieved *even after* paying down about $300 million in debt to the IMF. Since India's foreign reserves had not been this strong since 1958, it made managing big projects in science and technology easier, notwithstanding public charges of "waste." International agreements, including their royalty and concealed transaction costs, all required foreign exchange, yet India entered this decade with more financial confidence than before. This was reflected in a sense that the country could "go it alone," particularly if it could make lower-cost deals with countries like France and the Soviet Union. The cost of petroleum was not a major consideration for India in 1971, and even food supplies were, despite malnutrition, deemed "adequate." So a more self-reliant outlook was borne of confidence about the future, a reason why Gandhi had called an election for March 1971. Then, with little warning, just after the national election in which Indira won heavily, the Bangladesh crisis suddenly brought millions of refugees into India and brought India into another war with Pakistan. By early November 1971 India's attack on East Pakistan was planned and imminent, and the financial picture had changed: the unplanned cost of 10 million refugees was becoming a heavy burden.

The availability of foreign exchange and the rate of exchange between the rupee and dollar had become key factors in the scientific community. As exchange rates with the pound were important at the time of Independence, various currency crises in the relation of the rupee and dollar would dominate the volatile financial situation in the early 1970s. This was exacerbated in August 1971 with the de facto devaluation of the US dollar, and the dismantling of the Sterling Area, of which India was still a member. These were years that nuclear reactors went on line, space projects had to be completed, electronics and computer production facilities were being built. Apart from increasing export earnings, foreign exchange was obtained through grants and loans as part of foreign-funded projects and programs. Because new projects had become the paradigm source of funds for ongoing science and technology programs, continuous approval of new projects was an essential activity to science and technology managers and policymakers.

This helps to explain why more attention went to the negotiation and start of projects than to their completion. And the competition for foreign exchange was fierce; for example, the new bulletproof 1971 Mercedes-Benz for India's President V. V. Giri, which cost $81,000, had to be paid for from the same foreign exchange source from which Vikram Sarabhai and the DAE expected money to buy special heat-resistant metals to build a satellite; bullet-proof and heat-proof metals both cost a lot of money. Planning was predicated on the regular supply of foreign loans and grants (called "the pipeline"), and since this supply increased the national debt, more loans had to be obtained also to repay the interest on that debt. The Indian (im)balance of payments, foreign aid, and foreign reserves were thus of direct relevance to scientists. They needed expensive foreign journals for their everyday work, replacement parts for their everyday equipment, and foreign exchange when they traveled to conferences and work abroad. They were also wrestling with how to substitute their own ingenuity for costly imports and how to develop new technologies instead of buying them abroad. In these conditions lay the origins of the war for self-reliance in science and technology, all in a society that already had established a widespread disdain for Indian-made goods (whether batteries, bicycle tires, syringes, or lightbulbs), a society that overvalued foreign-made goods, a society that had little confidence in Indian-made goods. That war over self-reliance is the subject of the following chapter.

For these reasons, the foreign aid consortium called the Paris Club became more and more important in the science and technology community. The foreign aid budget had become the source of most financing of new projects. The Paris Club meetings in May 1971 approved a record $1,150 billion in grants and loans, noting India's performance in the preceding year of a 5.5 percent growth rate. After favorable grain harvests, the market value of food aid was allowed to decline in 1971 from Rs 1.55 billion to Rs 990 million over the previous two years. Debt rescheduling over the previous three years had made available Rs 2.4 billion of foreign exchange, and servicing India's debt cost Rs 4.120 billion in 1969–70 and Rs 4.35 billion in 1970–71. The total pledged at the 1971 Paris Club included $90 million for debt servicing alone. There was still a large supply of unused funds in the aid pipeline, at Rs 17.17 billion in 1971. But foreign exchange reserves held by the Reserve Bank of India had declined to Rs 885 million, in part because repayments had been made for an IMF loan. These reserves were the crucial uncommitted foreign funds for which project promoters had to compete in Delhi. They had to find intermediaries in foreign funding or loaning agencies who could secure funding that circumvented the limits and priorities of an institution like the Planning Commission or who could themselves influence those planning priorities. This is why Bhabha's relationship with B. K. Nehru, ambassador to Washington and in regular touch with the IMF and World Bank, and Sarabhai's relationship with L. K. Jha, chairman of the Reserve Bank of India after 1967, were crucial: they were important sources of information and influence about foreign reserves for the atomic energy program. B. K. Nehru had already been instrumental in pressing for the 1966 currency devaluation decision.[1]

Two extraordinary things occurred to India's currency in 1971. Due to profound changes in the international currency system, India's rupee experienced great volatility and risk. In August 1971, at the height of the Bangladesh crisis, the rupee was unlinked from the pound and linked to the US dollar, but four months later in December 1971, at the time of the Bangladesh war, the US unlinked itself from gold under the Smithsonian Agreement. India immediately decided to relink the rupee to the pound, again, and achieved this in January 1972, just when Britain decided it would let the pound float freely as a result of the Smithsonian Agreement. Then the other extraordinary though unrelated thing happened: the complete suspension of American aid to India due to the US commitment to Pakistan during and following the December 1971 war. As an illustration of the consequence of this decision, when the pound dramatically lost value in the spring of 1972, the Reserve Bank of India stopped forward trading on sterling for three weeks in June 1972, and therefore all official export and import transactions for India were on hold. The Sterling Area, to which India had been unhappily married for so long, was discontinued. With the dollar and pound floating freely from gold, the rupee was in effect set for a managed float too, pegged to the dollar, but very vulnerable.

Normally one would not include in a scientific and nuclear history an account of a local and regional level conflict like the one over Bangladesh, or of currency relations. But this conflict had a geopolitical dimension, had far-reaching effects among scientists and technologists, and is among the reasons for the decision finally to prepare for and commit to the first nuclear bomb test. It may be one of the reasons why American disapproval (and particularly Henry Kissinger's reaction) of the 1974 bomb test was so muted. The background of the conflict is that a terrible cyclone in November 1970 killed about 400,000 people, and this just preceded the election in December 1970 in which East Pakistanis expressed the view that the government in Rawalpindi had responded to the cyclone feebly and indifferently. In the December 1970 Pakistan general elections, a party advocating regional autonomy gained overwhelming support. This party, the Awami League, won the majority of seats in the Pakistan Parliament because it was swept to success in almost all the constituencies of East Pakistan. It had no presence in West Pakistan. But the government was not prepared to step down and let the Awami League govern, as the rules required in a case of such a clear majority. This anomaly lasted for a few weeks into 1971, until the Awami League leader Sheikh Mujibur Rahman was imprisoned in West Pakistan on suspicion of conspiracy. At this stage news about East Pakistan and Bangladesh was on India's front page.

A surprise massacre of East Pakistani elites in Dacca by the Pakistan Army in March 1971, and the following military crackdown on the Awami League, led to about 20,000 refugees moving into West Bengal each day, many of them temporarily housed in shelters on a salt lake where the Saha Institute was supposed be rebuilt, in the suburbs of Calcutta. The Indian government permitted the Bangladesh government-in-exile, led by imprisoned Sheik Mujibur Rahman, to operate just inside the fence at the India-Pakistan border. By September 1971 there was conflict along the border, and the Pakistan Army was deeply entrenched in guerilla warfare against thousands of Bengali fighters in most districts of East Pakistan, driving hundreds of thousands of refugees into West Bengal. This conflict was discussed by Prime Minister Gandhi when she visited Moscow that same month. Guerilla warfare was beginning to have some negative effects on the Pakistan Army's control. By October, 10 million Bengali refugees had arrived in West Bengal. In November Gandhi was talking about the future of Bangladesh to leaders in Europe and the United States, exercising a confidence founded in her new treaty relationship with the Russians. In particular she asked the US to pressure Pakistan to end its military activities and begin negotiations with the head of the Bangladesh government-in-exile, Mujibur Rahman. She found little official response, despite widespread public sympathy and support in most countries, including the US, for Bangladesh's separation from Pakistan. It was then that she began planning for a dry season war in December.

In November a state of emergency was declared in East Pakistan, with daily eastward troop movement by air via Sri Lanka, because Pakistani aircraft were not permitted to fly across India. On 3 December India declared all-out war, troops entered East Pakistan, and on 14 December came the Pakistani offer of surrender, passed through US embassies to the government of India.[2] During those ten days the US Seventh Fleet aircraft carrier *Enterprise* had moved into the Bay of Bengal, carrying nuclear missiles. India interpreted this movement as a hostile act, although the same ship had been welcomed during the 1962 conflict with China. Although the nuclear missiles were only symbolically important, the helicopters and troops on board were thought by India to be for a kind of elite rescue mission for the Pakistani high command, trapped in Bangladesh, including Generals Yahya Khan and Tikka Khan. These two men had become notorious in India and Bangladesh for their brutality and their contempt for East Pakistan and Bengalis. This floating threat of the aircraft carrier sent all the wrong signals around India. Two days later in Dacca there was a cease-fire declared between the armies and the guerillas. This period was, for Henry Kissinger and Richard Nixon, also a period of secret US negotiation with China, and, among other objectives, this aircraft carrier was sent to show solidarity with China's ally Pakistan, following the US logic of the time. Kissinger had already made a secret visit to Beijing in 1969, something India evidently knew when it signed the treaty with the USSR in that year. An Indo-Soviet axis was forming against a China-Pakistan-US axis: India was looking more closely at becoming a country with a long-range missile, a nuclear bomb, and a satellite. At that stage, the US anticipated a war on India's western front following an Indian attack inside West Pakistan, and the US asked the USSR to intervene with India to stop this western front war. It appears that to the Americans the aircraft carrier *Enterprise* was more a symbolic political gesture of presence than a military maneuver, but that was not the way it was understood in India.[3] Official sources in Delhi encouraged a different, more sinister, even conspiratorial, reading of the situation.

The cease-fire was immediately followed by US refusal to recognize Bangladesh, thus making it difficult to get UN recognition of the country, without which it was difficult for UN agencies to operate except through a proxy. All this occurred in the first months of 1972. Thus was created a polarized environment for trade and diplomatic and scientific relationships between India and the West. Part of the balancing act was that in August 1972, India unlinked the rupee from the pound and relinked it to the US dollar, *in spite of the suspension of US aid.* Following the money and network of apparent influence leads us in two different directions. From early 1972 the road to Moscow appeared smoother than the well-known road to Washington. But the expression of the rupee's value in dollar terms had become a staple of life. The ruble was a currency that did not exchange outside the COMECON zone, and though the Russians were a good source of some other things, they were a wholly uncertain source of commodities in popular Indian terms, and their currency had no meaning for Indians. The American government decided to rebalance the US-India financial relationship in 1973 when it contributed to soft loan increases to India from the World Bank and more or less wrote off $3 billion accumulated as rupees in counterpart (PL480) funds, the joint dispersal of which had become politically contentious. When the Americans finally gave recognition to Bangladesh in March 1972, there was a strange balance of power in South Asia. Nuclear navies of the USSR and USA were in the Bay of Bengal. Though remote, a Washington-Islamabad-Beijing axis faced a Delhi-Moscow axis. Indian elite groups with an interest in the bomb pressed onward, expectantly.

Disagreement and Friction Inside the AEC

In 1970–71 Vikram Sarabhai was a conduit for pressures against the development of bigger reactors, against the continuing marriage of atomic energy and space, and against the elitism of DAE establishments. At the same time he was a conduit for demands for more investment in projects with little relationship to atomic energy, having to defend these investments in terms of the eventual payoff in the capacity for basic science and/or the spin-off to new industries. This was the context in 1970 when AEC's Ten-Year Plan (the Sarabhai profile) for atomic energy received such mixed reception. Even a year before the launch of the Ten-Year Plan, when some of its elements were aired, it was subjected to criticism in both political and scientific circles. Tension between the diverging politics and economics of space and atomic energy grew steadily, and an electronics movement within DAE contributed to that tension. Those with loyalty to atomic energy, who had achieved everything through reactors, felt that space and electronics subtracted resources and inevitably drew attention away from atomic energy. Some influential people thought that Sarabhai was less interested in nuclear power than in rockets and satellites, and thus that he devoted less attention to reactors; others thought Sarabhai's disinterest in nuclear test explosions was an implicit obstacle to the national glory that could result from a nuclear explosion, and thus that he was an indirect obstacle to the implicit payoff to the scientific community. Their reference to Sarabhai's low interest in electricity generation through reactors became a code for the criticism they could not widely express, namely that he was not keen on the bomb. Was the bomb not the ultimate sign of high tech? Did other great powers not have one—well, many more than one?

Of course Sarabhai was ambivalent about the bomb, not only when he was appointed in 1966 or after his unsuccessful and frustrating 1967 tour of the nuclear powers asking for a nuclear security umbrella. On the day before taking office as chairman of the AEC in 1966, he gave a press conference and spoke at length on the *insecurity* of the many countries that were not members of one of the alliances related to the superpowers. Not yet in office, he repeated views he expressed three months earlier in his Laski Lecture in Ahmedabad, when he knew he was being considered for the post of AEC chairman, on the security of developing countries, arguing that the great powers should insure collective security or be faced with countries like India and Israel that are tired of being asked to deny themselves that kind of security through acquisition of nuclear weapons.[4] These countries found themselves at a crossroads after the explosions of atomic bombs by

China, Sarabhai said, because an independent nuclear capability was shown to be possible and tolerated by the superpowers. He referred to the fact that others, India included, "are called upon to exercise self-restraint in denying to themselves the acquisition of an independent nuclear capability. This self-restraint must arise out of an act of faith in the political wisdom and maturity of nations of the world acting collectively."[5]

According to Raja Ramanna, "The first thing Sarabhai did when he came [to BARC] was to close down" the Study of Nuclear Explosions for Peaceful Purposes (SNEPP). When asked about Homi Bhabha's low-cost estimates for a bomb, Sarabhai reportedly said at the time, "You can ask me what is the price of two yards of cloth, but two yards cannot be produced unless you have a loom or a mill or something behind it."[6] By October 1967 "Mrs. Gandhi cleared BARC to begin work again on the theoretical physics aspects of the explosion." Sarabhai, who had gone about shutting down SNEPP, was told to "lay off," according to Ramanna in 1998. So the SNEPP slowly developed under the supervision of BARC's director, Homi Sethna, who himself could take little part. "Sarabhai said right from the beginning that he didn't want it. But it made no impression on BARC and we went right on studying the process. Don't forget you don't need a letter to do all these things. BARC itself has the freedom to initiate these things . . . [but] Sarabhai would say do the calculations but don't do anything more."[7]

A month after Sarabhai and Jha concluded their official tour looking for nuclear security, physicist R. Chidambaram was asked by Ramanna in November 1967 to study shock wave propagation in plutonium. (Chidambaram reports that he was then rather more interested in molecular biology.) Having joined BARC nuclear physics division in 1962, Chidambaram was thirty-one in 1967. Sarabhai and Sethna were kept on their toes in case the bomb issue spilled into the public domain; for example, both had to write letters in May 1968 carefully clarifying the quotations attributed to them by a careless reporter writing in *Himmat Weekly* on the cost of nuclear weapons: Sethna had to explain that "kettle" is slang for a reactor in which plutonium can be produced and that the reactor planned at Kalpakkam was intended for the production of electricity. Sarabhai and Sethna (whose letters are published side-by-side) took pains to assert that India needed economic and social development now more than anything else; Sarabhai said that experience showed approximately $1 billion per year was needed for a serious weapons program and that money could be put to better use.[8] Sethna, on the other hand, had quietly warned Gandhi in 1966 that only a force with about 150 weapons would provide a credible deterrent and said that force would raise the annual operating costs alone to $100 million, exclusive of a delivery system. In Karnad's wishful words, "Indeed Sethna appeared to be bargaining for New Delhi's commitment to a fairly major nuclear force, as much as a third larger than anything Bhabha had ever considered."[9] The opponents, and others, were getting ready for a war of the bomb estimates.

In August 1968 Ramanna added S. K. Sikka to work with Chidambaram, and in January 1970 he added V. S. Ramamurthy to simulate computer models to validate and verify equations of state for the explosion. But, according to Ramanna, the SNEPP work went rather slowly.[10] In March 1970 the Consultative Committee on Foreign Affairs and Atomic Energy asked Sarabhai to study the cost of a peaceful nuclear explosion (PNE), and Sarabhai had already asked N. Seshagiri at TIFR to do a cost-benefit analysis of one bomb test, but a bomb defined as a PNE; he may have hoped that the true cost would show the test to be impractical. In April 1970, R. Chidambaram made a presentation at the International Atomic Energy Agency in Vienna on the importance of PNEs, with Sarabhai present in the meeting (see *Negotiating Nuclear Power*). Ramanna and Chidambaram then prepared a detailed report for Sarabhai on PNEs, and discussed it with him during the flight of all three to Thumba in Kerala in June 1970. At that time Sarabhai introduced Chidambaram to Kalam at Thumba, working on the satellite launch rocket. In June Sarabhai was reported in the *Morning Herald* saying that as a nonsignatory to the nonproliferation treaty India was capable of conducting underground nuclear explosions and had the right to do so. In July, Seshagiri concluded his report for Sarabhai, saying PNEs made some economic sense but nuclear weapons did not.[11] Given this evidence of activity to design but not build a bomb, one question is, did the prime minister ask Sarabhai in April 1970 to prepare for a test? On the basis of this evidence, Chengappa stated that "more than any other year it is 1970 that can be described as the defining period for India's nuclear ambitions."[12] This was about as far from electricity generation as Sarabhai could be expected to go, though in the views of Ramanna and Sethna, Sarabhai's small steps to the first test were being taken too reluctantly.[13] They should have remembered that the DAE was involved in other kinds of administrative and political struggles that overshadowed this little project.

In addition to larger heavy water reactors, the DAE was also looking around for a better model in the next reactor designs, especially one using plutonium, which they were just beginning to reprocess, to breed thorium as a fissile fuel. In January 1968, P. K. Iyengar went to the Soviet nuclear research center at Dubna on a visit arranged by Sarabhai: there he saw a fast-pulsed reactor with plutonium core, and liked it. Three months later, in March, Iyengar called a meeting at BARC to discuss this new reactor, and both Sethna, the head of BARC, and Sarabhai, chairman of the AEC, attended. An official advisory group proposed not to build a fast-pulsed reactor on the Dubna model, but a less expensive plutonium reactor for neutron investigation in multiplying assemblies, named PURNIMA. The ostensible purpose of this reactor was research. Shortly afterward Sarabhai sanctioned Rs 1 million for PURNIMA's design, but it needed 18 kg plutonium before it could be commissioned, and there was not sufficient plutonium in India for this purpose, according to Iyengar. So the project was delayed about eighteen months, waiting for sufficient plutonium to be reprocessed, probably from spent fuel from CIRUS. Nevertheless, by 1970, M. Srinivasan had been named chief physicist of PURNIMA, and it was commissioned in May 1972. At that time the decision to plan a bomb test had been taken. PURNIMA gave Iyengar a lot of experience. According to Chengappa, "M. Srinivasan says the work on the PURNIMA reactor helped the team gain tremendous confidence in their calculations for designing the bomb."[14] Nevertheless, this project did not much help the DAE with its electrical power grid problems, and those were politically urgent.

In fact, the DAE had long been supporting activities unconnected to generating electricity from atomic energy. Bhabha began DAE's support to space in 1961, and in 1963 he began major, and eventually expensive, molecular biology and radio astronomy groups at TIFR, on DAE funding. As the combined functions of the DAE grew in all these different fields, so did its budget. But by the late 1960s space activities were enlarging faster and so, consequently, were their demands on the DAE budget. By 1970 space used about 13 percent of the DAE budget and was growing; beyond the launch site in Kerala, Sarabhai selected in 1969 a full-scale rocket and missile launch station at Sriharikota, north of Madras, and expenditures to develop that site were going to be very big.[15] The AEC had agreed to this strange situation all along the way, and so had three prime ministers, Nehru, Shastri, and Gandhi. But could it last? And could Sarabhai continue to juggle all these projects in the air?

I think it is fair to say that Sarabhai was not *simply* against building and testing nuclear bombs; he was against developing and testing one or two, at least while India had no security apparatus or delivery infrastructure. Except in a few ideologically committed people such as K. Subrahmanyam, there should have been more ambivalence on this question, as a bomb test would put the country at risk. Contrary to Ramanna's view that "for us it was a matter of prestige that would justify our ancient past," I agree with P. K. Iyengar, who said about Sarabhai that he "was a shrewd businessman who knew what it would mean economically to incur the wrath of the US. He knew the consequences, so he was much more careful."[16] Sarabhai was seized with the issue; there is no doubt. In February 1970 Sarabhai chaired the IAEA conference on peaceful uses of nuclear explosions; he even had J. R. D. Tata fly an Indian feast from the kitchens of the Taj Hotel for the delegates in Vienna on Air India and arranged for Indian dancers. R. Chidambaram presented a paper on underground copper mining using PNEs and met Sarabhai after the conference to brief him on the high-speed detonators needed to explode the non-nuclear material in a bomb.

Sometimes Sarabhai acted under the prime minister's instructions; sometimes he expressed his own views, based on his own reading of opinion and policy. Indira Gandhi's close advisors on foreign policy matters, P. N. Haksar and G. Parthasarathi, appear to have been supportive about 1970, noting, however, that Haksar is later characterized as a skeptic or even an opponent of the test in discussions in 1972 and 1974, according to Ramanna.[17] When Sarabhai said India would not build nuclear weapons but kept the option open for PNEs, this was clearly at Indira Gandhi's instruction. Shah says that Vikram's son Kartikeya, who like the family followed this as closely as his father would allow, feels Sarabhai "never quite made up his mind on the subject."[18] From all her conversations about him, Shah concluded that "Vikram was extremely perturbed by the [bomb] debate. Rules of secrecy forbade him from discussing certain matters related to atomic energy with anybody," even his family members.[19] It is therefore likely that a mystery grew up behind which Sarabhai felt he could shield his conscience. In 1970, however, the pressures were mounting, and during his next twelve months he encountered opposition to the space and reactor program (as seen in chap. 20), which made him feel more and more isolated on the question of the bomb. The dramatic and swift change in the geopolitical context only clarified the risk of a nuclear test and enhanced that isolation.

During Sarabhai's time as chairman of the AEC, four opinion polls had been conducted on attitudes to "the nuclear option" in 1968 in Delhi, Bombay, Madras, and Calcutta. Conducted by the Indian Institute of Public Opinion, these polls had shown the not surprising result that some Indians wanted to have the bomb for prestige. When Prime Minister Gandhi said in Parliament on 24 April 1968 that "possession of nuclear weapons gives no military advantage," she indirectly confirmed the value of its prestige, but she also said the bomb "may well endanger our national security by imposing a very heavy economic burden." Another study was done in 1969 by Gerard Braunthal, and in December 1971 again by the Indian Institute of Public Opinion. A more influential survey was conducted by Ashis Nandy during 1969–70, asking questions among "the strategic elite," but the results of this work were curiously not published until 1972.[20] Since this study was carried out by a Delhi insider from a respected institute with a wide policy network, the results were in circulation before its publication. The ambiguity with which top political leaders viewed the bomb was well rooted in the views of their supporting elites: this poll was important because the prime minister was more sensitive on this question to the reaction of elites than to her usual constituencies. Abha Dixit reviewed all these polls and concluded, "Indian public opinion wanted to see the nuclear weapon as part of the country's 'prestige,' but this attitude was tempered by a sense of realism over the potential economic costs of such a policy."[21]

Thinking and talking about bomb tests was going on in *both* the DAE and the Ministry of Defence, as shown by a 1970 visit of the joint secretary of Defence to the United States (see *Negotiating Nuclear Power*). At this time Mrs. Gandhi approved the transfer of physicist B. D. Nagchaudhuri from the Planning Commission, to be the scientific advisor to the minister of Defence and thus the head of the Defence Research and Development Office. From this position he played a major, if less recognized, role in the first bomb test. This would have surprised no one more than his teacher Meghnad Saha. It was a long road from producing a PhD during preparations for the first bomb developments at the University of California (1939–41), through the "unproductive years" around the cyclotron in Calcutta, to being chief defense scientist of India and planning missiles and bombs.

The picture of the DAE in 1970 is one of an expanding and expensive empire involved in electronics, computers, satellites, rockets, and nuclear reactors, as well as basic research in physics, mathematics, and biology. This expansion is contrasted with increasing public questioning about the contributions of the DAE to the economy. Moreover the prime minister was hearing that friction between space and the atomic energy "family" had increased to the point where some people were not cooperating with others. Rumors of tension between members of the AEC and other top DAE officials began to circulate "outside the family." For example, "we heard that Sethna and Ramanna simply would not cooperate by 1970," thus setting the stage for the open gap between these two men following the 1974 bomb test.[22] This conflict has been confirmed by many observers.

This tension was due, in part, to the evolution of two cultures inside DAE, and it was *not just* about the difference between space and nuclear power or *only* between those who were committed to work on the bomb and those who were not. Sethna, an engineer and Ramanna, a physicist, seem to have agreed on the value of an early bomb test.

> There were also two *other* cultures in DAE [besides space and reactors], because in some BARC divisions doing "pure science" really well was the valued goal, at any hour of the day or night. But in other quarters, right next door, just making links to higher officials and doing business, doing what you are told, not asking too many questions, coming on time, and working punctually nine-to-five was the valued goal. You can imagine the relationship between these two cultures.[23]

This cultural tension grew elsewhere too, such as at Kalpakkam, south of Madras:

> The place was full of engineers and administrators, many of whom had been drawn from the state government. A greater cultural shock for a person like me cannot be imagined. . . . I was personally treated well, given due respect and all that, but when it came to physics, few there were prepared to concede that we had a legitimate place. . . . When it came to budget we got a pittance compared to the others; it was declared that our programme was not relevant to fast reactors, though many parts of it like radiation damage studies actually were. It was in matters of promotion that scientists got the rawest deal.[24]

Following the 1971 spring election, Prime Minister Gandhi asked Sarabhai if he would give up one of the jobs—space or atomic energy. She knew what his choice would be and had prepared a cushion, namely that she would create a new Department of Space of which he would be the secretary. Sarabhai is reported to have become depressed about this forced choice between space and atomic energy, thinking that it reflected badly on his abilities, and so lobbied with Gandhi's senior officials, including P. N. Haksar, her private secretary, to try to keep the roles in space and atomic energy united. He may also have resisted the change because he anticipated that space would have a more difficult time getting an adequate budget, knowing that the secure "budget bet" was with DAE.[25]

In addition to the personal preferences of individuals, there was a subtle perception that if DAE were responsible for both nuclear tests and rocket launches, the assertion that India's nuclear program was for peaceful purposes would have even less international acceptance than it did; this separation might improve that. Sarabhai was later told that he would have to accept the separation because the present arrangement was just not working. The prime minister knew, in particular, that Homi Sethna was unhappy with the present arrangement, and she had only recently blocked the solicitation of international bids on new reactors, an offer floated by Sarabhai himself. Clearly the charm in their relationship was fading. Sarabhai was then told that Sethna would take over the DAE when Sarabhai moved to space. Sethna would then be his equal in access to the prime minister and the cabinet, which now he was not. According to Chengappa, "In the last week of November [1971] . . . Mrs. Gandhi called Sarabhai for a meeting. She bluntly told him that she planned to set up an independent space commission to be headed by him and asked him to give up his post as AEC chairman."[26] Friends of the prime minister tried to comfort him because he showed "visible anguish."[27]

The refugee crisis that started in March and ended in war over East Pakistan distracted everyone from the question of a bomb. The effect on the treasury of the unplanned expenditure around this Bangladesh crisis was profound. The new Department of Space was not created during 1971; after all, Sarabhai had created an elaborate structure in India's Space Research Organization himself, and the prime minister's ultimatum had only just been given. Meanwhile the debate about the need for India to conduct nuclear tests continued.[28] Sarabhai still did not favor nuclear tests, and though the prime minister could have overruled him, she probably thought it easier to bifurcate the atomic energy portfolio, create the space department that Sarabhai really wanted, and appoint someone like Sethna or Ramanna in atomic energy who was prepared to conduct nuclear tests. A member of this inner group said later that the serious planning of the tests did not really occur until after Sarabhai's death.[29] So, for reasons of political timing, technical unpreparedness, institutional difficulties, and personal preferences, *planning* for nuclear tests was not done seriously in Sarabhai's administration. But there *was* serious *thinking* about tests in 1970–71, even while Sarabhai and Gandhi were discussing the division of powers between space and atomic energy.

In conclusion, Sarabhai also expected his staff to work very hard, as he did; officials of various commissions and departments would ride trains and sit in airports in order to get enough time to explain a file to Sarabhai and have him sign it.[30] He built up a wide loyalty in the department, considering that he was an outsider to it. Perhaps because he was working about eighteen hours a day (in the late 1960s he slept about five hours a day, he said), he gave the impression to each individual of giving them his undivided attention, the same impression he gave me. In 1969 Sarabhai instituted a rhythm of two-hour sleep breaks, accompanied by a massive monthly crash; but once in 1971 Mrinalini Sarabhai recalls her husband waking up suddenly in the night, sleepwalking, going to the window, and talking incoherently about his work.[31] Indira Gandhi was not unobservant or insensitive to her friend; it is sometimes overlooked that when she told him in 1971, only weeks before his sudden death, that he had to choose between space and atomic energy, she also said, "If you don't slow down, you will die."

TWENTY-TWO

Scientists in India's War over Self-Reliance

Preceding chapters described strains and confrontations in two DAE-funded laboratories and conflict in or between CSIR institutes; they established that a new high-technology sector was seen in the late 1960s as a possible extension, enlargement, or even savior of the Indian economy and showed the pressures within the nuclear establishment. Inevitably scientists and their institutions were drawn into these conflicts, and sometimes experienced personal confrontations. Not unique to India, their intensity is partially explained by a wider systemic disagreement concerning the roles that science and technology were supposed to play in the quest for India's self-reliance. This disagreement was understood up and down the hierarchy, but mid-level and upper-level experts expected their immediate leaders to manage the level of conflict so that it did not interfere with either their institution's reputation or their own professional progress. One way forward through this tension about self-reliance was to point to a different kind of Indian economy, a high-tech economy, and this was, to some extent, what Sarabhai and others were working toward. This was further evidence of the structural disengagement between industry, technology, and science, on the one hand, and the truly vast agricultural socioeconomy, on the other. Nevertheless, some protagonists hoped that this high-technology sector would be self-reliant and so successful that it pulled everything else along with it, including agriculture.

So what exactly were the objectives among the protagonists for self-reliant high technology? A high-tech path would build confidence and indigenous capabilities and thus reduce the constant struggle about self-reliance; in this view the arguments about self-reliance were signs of insecurity. Competence would lead the way; higher qualifications would answer questions. A higher-tech economy would move India's experts and exports into a wider world beyond an industrial base of railway carriage production or the export of tea. The public's appreciation of technical expertise would thus increase dramatically, reducing the effort required at the top to constantly reposition science and laboratories and justify the state's support for scientists and technologists. Yet in spite of these rhetorical efforts, the evidence persisted in a contrary direction, showing that the majority of India's export earnings were firmly rooted in traditional products such as textiles, tea, and leather. Moreover, there was much leaking out of precious foreign exchange earned through trade and aid into the purchase of equipment or expertise that some people thought could be better built or obtained at home, *in* the country and *of* the country (*swadeshi*). They predicted the country could become self-reliant not just in food production, which they hoped was just around the corner, but in most advanced technologies. There was even reference to a "leapfrog effect" that would allow India to bypass long slow periods of research and development on new technologies and reach an operational stage more quickly. This chapter explores the disagreement around this question, a conflict that can be called "the war over self-reliance." Everything began anew with competition for new leadership opportunities in the scientific community in early 1972. As before, it fell to these leaders to articulate the contending positions, absorb the anxiety, and modify the unmet expectations of the scientists and technologists in their institutions. Their biographies and personal relations were no less important in this process than the macroeconomic pressures.

Changing of the Guard

On the last day of 1971, near the rocket site he built at Thumba in Kerala, Vikram Sarabhai unexpectedly died of cardiac shock.[1] Thus 1972 began with a sudden shock for scientists, partly due to the confidence Sarabhai, only fifty-two, inspired in the young generation and partly due to the complex leadership changes which had now to come. Like Bhabha, Sarabhai held a great deal of current strategic information in his head and had built effective relationships overseas, and none of that could not be retrieved now. Like Bhabha's, his sudden death was an unexpected opportunity for change and advancement.

So in a short period of six years, the top Indian nuclear positions were opened again for competition. And ambition was not lacking in the wings. Bhabha and Sarabhai both died while they were very young, far from retirement, thus creating an "unnatural" political succession; none of the successors knew they were playing musical chairs on the day before, so when the music stopped suddenly in 1966 and 1971, they were each judged on what they had been doing at that moment. There was again no time to prepare a strategy and get the predecessor's blessing or even his opponent's blessing.

In early 1972, however, the scientific elite was dealing with a prime minister whose faction of the Congress Party had won a Parliamentary election ten months before, signaling the triumphant defeat of her enemies within the Congress Party, and who had won a war with Pakistan in Bangladesh a few weeks before. Then in April 1972 her faction of the Congress Party swept the West Bengal elections, beating a coalition of left parties in a state usually contrary to the center. She made use of the suspension of 1972 US aid to bolster her self-reliant image. Paul Brass described a "wave" of support for Mrs. Gandhi in 1972–73, during which, he said, there were unprecedented assertions of executive power in the political system.[2] Said one longtime observer of science politics in Delhi, talking about this period, "Mrs. Gandhi became a most skillful player among the elite scientists. She knew Indian history. She did not always give them what they wanted."[3] So Prime Minister Gandhi was in a position and mood to make these choices herself, for the first time. The leading scientists who had worked around Sarabhai—Sethna, Menon, Ramanna, and Dhawan—now came to the fore.

Homi Sethna did indeed become chairman of the Atomic Energy Commission, as he hoped, and this sent shock waves among physicists who believed that an engineer, especially one without a PhD, could neither understand the job nor do it well. The fact is that Sethna had been involved in DAE programs, like the rare earths project, since 1950, and so he knew the organization very well and had a strong Mumbai industrial constituency, besides being well connected in the Parsi community there. Indira Gandhi knew about the disagreements over reactor type and reactor size, and she wanted "a technologist, not a scientist" to manage the AEC. Now, when Sethna moved up to lead the DAE and become chairman of the AEC, nuclear physicist Raja Ramanna took over the direction of BARC and remained there six years, until 1978. Neither Sethna nor Ramanna had significant international reputations. Trained in nuclear physics in London in the late 1940s, Ramanna had been chafing under Sethna's gaze at Trombay since 1966, all the time building his national network. After Bhabha's death in 1966, Ramanna had also coveted the position Sarabhai got and had not thought the choice of "a newcomer in the field" like Sarabhai was a good idea, at first. "I had some doubts as to whether Vikram would be able to get a grip on atomic energy developments here and abroad, and be able to give it any special orientation. I thought that, at best, all he would be able to do was to supervise over atomic energy development in a very general way, to fulfill the plans that Bhabha had so solidly laid and extrapolate this plan with the help of all the expertise available."[4] Now head of BARC, Ramanna could exert his intelligence and known charm on the prime minister directly, as Bhabha and Sarabhai had done before him. With his decisive way of talking, wide cultural experience, and thorough knowledge of nuclear and reactor physics, he remained in the eye of the prime minister's office until Mrs. Gandhi left it in March 1977 and even after that. Deeply involved since 1952 with DAE projects, boards, and committees, and a traveler to Russia and Canada since 1955, Ramanna was curiously not a visitor to the United States until 1969. When he was finally the director of BARC and actively planning the first nuclear test, Ramanna visited Los Alamos in 1972. He was surprised that he was permitted only in the seminar rooms, not the laboratories: "I was rather disappointed that the State Department refused to give me permission to see the labs and have not visited the country since."[5]

Indira Gandhi used this moment to make some other changes too. After Sarabhai's death the program analysis group of DAE was closed because the new chairman of the Atomic Energy Commission, Homi Sethna, thought it was not needed. In June 1972 Gandhi, as president of CSIR, established the Committee on Research and Development for the CSIR, with a mandate to receive advice on innovation and technology transfer from the private sector. This is the group she discussed with Blackett six months earlier (see chap. 19). M. G. K. Menon was to remain as director of Tata Institute and secretary of Electronics, reporting to the prime minister directly. Mrs. Gandhi was carrying the atomic energy portfolio herself and making decisions about its leadership and its nuclear tests; during this period when she was replacing Vikram Sarabhai, she finally decided to approve preparations for a nuclear bomb test, as shown in the following chapter and *Negotiating Nuclear Power*. Physicists and engineers were now secretaries in four departments—Defence, Electronics, Space, and Atomic Energy.

A Department of Space was created four months after Sarabhai's death, and this new department was built around its first secretary, Satish Dhawan. Dhawan had completed his doctorate at Caltech in aeronautics and mathematics in 1946 and was actually teaching again (on sabbatical) at Caltech when Sarabhai died. Just before his death Sarabhai asked Dhawan to assist in preparing the rocket-launching system at the new east coast site at Sriharikota and sent a colleague from India's Space Research Organization to California for a tour and study of the Jet Propulsion Laboratories' launch site in the desert with Dhawan as guide in late 1971. In early 1972 the Indian embassy in Washington phoned him, told him about Sarabhai's death, and invited him to consider heading a new Space Commission and Department of Space. Dhawan replied that he would not disturb his sabbatical research or teaching commitments for another two months, but agreed to accept the offer when he was free. He reminded the embassy that he was responsible for the Indian Institute of Science in Bangalore, where he was director. J. R. D. Tata and the director of the Tata Trusts R. D. Choksi both urged Dhawan to accept the Space position, despite the fact that it would limit his attention to the IISc, the institution for which they were council members. In the end, space research and development moved to Bangalore with Dhawan, the way atomic energy moved to Bombay with Bhabha. The complex projects begun by Sarabhai attracted Dhawan, and their nonmilitary applications appealed to him; he said, "Like all Caltech students I had the opportunity to work at nearby military defense labs, because most of our profs spent a day a week there. I didn't do that. It did not really interest me. So these new civilian space applications were consistent with my old preference for civilian work."[6]

One of Dhawan's first appointments was to make Brahm Prakash the director of the newly named Vikram Sarabhai Space Centre in Kerala. This shrewd choice brought into the new space department a senior DAE scientist, former director of metallurgy in the atomic energy complex at Trombay. Prakash was in charge of fuel rods for the CIRUS reactor from 1957, and this appointment ensured respect from his powerful DAE colleagues. Prakash was an expert in the behavior of special metals at both low and high temperatures, essential for space research. Dhawan did not disturb Abdul Kalam from the work he was already doing on the launch rocket and in a few months decided Kalam should be formally appointed the head of that project, ending the consortium of heads of four or five groups. Dhawan decided that the resources at Thumba, Ahmedabad, and Sriharikota should all be directed to getting the Indian satellite launched; this was certainly Mrs. Gandhi's wish, to see a satellite up soon, and something Sarabhai had been working on consistently.

Just weeks before Dhawan took over at Space, the Defence Ministry decided in February 1972 to develop its own guided missile in a project code-named Devil, with a budget of Rs 50 million, half in foreign exchange (this decision occurred at the very beginning of the "crisis" in aid financing). V. S. Narayanan, now promoted to air commodore, was put in charge of the Defence Research and Development Laboratory at Hyderabad for this purpose. The lab was described as a sleepy snake-infested and dilapidated place, reputed locally to be inhabited by a ghost.[7] But, through the efforts of Nagchaudhuri, it also received a huge Rs 160 million budget to reverse-engineer the SA2 ground-to-air missile, making it one of the richest laboratories in the country, though this was hardly known to most scientists in other laboratories.

Overall the leadership was cosmopolitan: a Mumbai-based Parsi engineer was at Atomic Energy in Mumbai (Sethna), a Mumbai-based Kannada physicist was at BARC (Ramanna), a Bangalore-based Punjabi aeronautics-engineer was at Space in Bangalore (Dhawan), a Mumbai-based Malayali-Rajasthani physicist was at Electronics in Delhi (Menon), a Calcutta-based Bengali physicist was advisor at Defence in Delhi (Nagchaudhuri). Each of them would insist that their provenance, their home districts, and linguistic origins were irrelevant and that they were chosen for these roles on merit alone; all said in one way or another that this cosmopolitanism was essential to the ethic of science. What is clear is that the top positions in the scientific community were in young cosmopolitan worldly hands. A special kind of elite had been formed in a crucible of a small number of research institutes and nuclear laboratories in the 1940s and 1950s and was now firmly in power.

What Was the War over Self-Reliance?

This cosmopolitan new leadership who had known each other for many years were faced with India's version of the "cultural revolution," a war over self-reliance. The reason customarily given since the 1940s for building scientific and technical capacity in all these separate fields was that India had to rely more on its talents and resources and less on others. Building nuclear capability stood at the center of the war over self-reliance, and that center too was ambiguous enough. But first, why call it a war *over* self-reliance? The answer can be found in reviewing four major sectors of the economy: fertilizer, jet fighters, steel, and petroleum. These publicly funded megaprojects impinged directly on the priority of nuclear power because each required large amounts of electricity at widely dispersed sites, and none got sufficient electricity to develop rapidly. Electricity demand now far outstripped supply, and the most important source of the thermal generation of electricity was by a low-grade coal, providing a low-cost alternate in use in 1972; so coal (not the uranium nucleus) was the current key to electricity.

So which four sectors are considered here, and why? Fertilizer was considered to be the most important strategic commodity by the government in order to boost wheat and rice production using fertilizer-responsive varieties. The jet fighter MiG (Sukoi) project was high priority from the mid-1960s, and involved precision machining and construction with advanced heat-resistant materials that would also be used on rockets; an "Indian jet

fighter" had high symbolic value. The steel economy was the backbone of both civil and military construction, and steel had been produced in India since the 1920s. Though it was internationally traded, India's requirements for steel were artificially inhibited in order to "fit" with India's limited capacity, and demand was now completely unsatisfied. Finally, petroleum had been a chronic source of difficulty for India, as every sector of the economy, including agriculture, became increasingly dependent on the internal combustion engine, not to speak of the growing role of plastics.

These four cases have enough similarity to and relationship with the nuclear reactor program to enable us to examine four questions: Were the delays and costs associated with the reactors and achieving continuous flow of electrical power from the reactors to the grid system peculiar to the DAE? Or did other megaprojects resemble the atomic energy situation? Did they have similar difficulties and inefficiencies too? Did the agreements concluded in 1963 between DAE and Canada and the United States, and subsequent agreements with Canada in the late 1960s, resemble agreements in other major projects, particularly with respect to indigenous technology transfer, and reverse engineering? Were the roles of technology transfer and foreign expertise in other major projects similar to the nuclear program? Did these *other* megaprojects meet the tests of self-reliance, as characterized in the language of the war over self-reliance?

These four important projects *all* depended on a constant supply of large amounts of electricity because they were all highly energy-intensive in both their construction and operation. Moreover, in each case, members of the scientific and technological community stood to gain through a *national choice* for self-reliance, and so they each vigorously advocated their own capacity to solve India's problems through these technologies. This is hardly surprising; implicitly or explicitly, they were committed to the mobilization of Indian capital, expertise, and technology because to do so was also a measure and proof of their self-worth. But there the similarities of positions and argument stopped. Many scientists preferred a strong state-managed economy, using state funds, with only a subsidiary and regulated role for private capital. Some wanted private capital mobilized through the state banks and not left for investment exclusively in private hands. Others wanted a balanced or even primary (not subsidiary) role for private capital and industry, in concert with the state where necessary, but meeting the industrial and social objectives established by the state. They accepted the existing "mixed economy" but looked for a much stronger role for the private sector and for Indian expertise, perceiving these as trapped. All scientists were observers of the selective indifference of the Indian state toward

indigenous expertise, and all knew of the offense this indifference caused among "true experts." Finally, none supported the retreat of the state from these key sectors, and none advocated leaving the field to a protected "nationalist" private industrial sector. I did not hear scientists in the late 1960s or early 1970s espouse the radical laissez-faire proposition that the state sector should be dismantled and state enterprises sold to "liberate" or "unleash" private industry (as was done, for example, to the disadvantage of Russians in 1991–94).

In each of these four cases, members of the scientific and technical community stood also to gain through a *foreign choice* of engagement with foreign institutions, foreign skills and research, and foreign capital. That engagement had been the door to their upward mobility since the 1920s, having made it possible to do "good science" and "big science," the kind of science that provided good opportunities for engineers (for example, nuclear, aeronautical, and computer engineers). Foreign choice meant India was getting the best available, and the best allowed India to "catch up"; no one could argue directly against the best available, so the easier route for objections was to point to much higher costs of imported technology. But the costs were not always higher, according to some methods of accounting. The Indian war for self-reliance was shot through with this ambiguity around foreign choice and national choice, permitting and/or forcing individuals and institutions to find a shifting balance point, a place where they would get benefits from both national and foreign strategies. In fact, sometimes the higher cost was to be found in the indigenous technology, not the imported technology (where international companies were willing to lower prices on last year's model to get into this potential market). This shifting point was hard to find, each case presenting its own costs, risks, and opportunities, thus accounting for the intensity of the debate and the sparks of conflict. The debate within the scientific community also resonated with the political economy of national planning policy, and thus was influenced by major interests outside the scientific community.

In practice, of course, each of these interest groups exaggerated the advantage and virtue of their position in the debate, and in practice there was a mixture of the two basic tendencies (national and foreign) in most projects. Many shifted their weight from one position to the other and back again, depending on limitations and opportunities. This shifting therefore became a struggle in both institutional and personal terms, in addition to the struggle to explain the significance of their projects to the often skeptical public. Meghnad Saha's complaint about excessive and unnecessary reliance on foreign capacity in 1950 was still applicable, in the opinion of

many scientists, twenty years later, though now such reliance seemed less inevitable, even less necessary. This perhaps explains the continuing popularity of Saha's views through the 1970s. But underlying this struggle at the time was scientists' unease about the dependence on America, Canada, the Continent, Japan, Russia, and Australia for food, loans, industrial equipment, and large structures (e.g., hydroelectric dams or steel mills). These various groups often coalesced into two or three, and at times the three or four positions were framed in terms of a war. We have seen in earlier chapters the disagreement over the best size of the next generation of nuclear power reactors, where Sarabhai, Sethna, and the supporters of each were opposed. We have also seen how the Planning Commission promoted the importation of large ready-made coal-burning thermal electric power plants instead of the smaller model that had been operating successfully in India and could be built with Indian capabilities. It was a war "*over*" self-reliance, because self-reliance as a mantra was not in doubt; the conflict was over how this sacred objective would be defined and who would achieve it, and thus the calculus was about who would and would not benefit from the strategy leading to it.

Who stood most for self-reliance? To the key players, there were different (though not mutually exclusive) means of reaching this end, and there was constant disagreement over which means and which strategies were authentically "Indian." Clearly in any one year some means prevailed over others, some strategies would be blessed with major commitments of cash, while others would merely be permitted (or licensed, as the saying went) to see if they survived or died, and others would be rejected. Scientific leaders bargained to have it both ways, emphasizing their indigenous character at the right time while proving that they could make the best choice among the foreign factors on offer because they had the foreign training or connections to do so.

This war over self-reliance in the 1970s already had a long history, and it reached up to the households at the pinnacle of power, as the following example shows. So keen was Gandhi to win middle-class approval of her approach to self-reliance that she agreed to an open competition to make small "indigenous" and "inexpensive" cars, which were hardly a high priority for the Indian society and economy, but were very important status markers for the upper middle classes. The reference to "indigenous" meant built using largely local materials in a self-reliant method, a very old value. An indigenous car was being built already, called the Ambassador, and although it was heavy and had poor fuel economy, was costly to purchase and repair, it nevertheless had a long buyers' waiting list. But in November

1969 her own son Sanjay got a license to produce 50,000 Maruti cars from indigenous materials, clearly with Indira's agreement. Criticism ensued, and her weak justification for this favoritism was simply that "young men are needed who innovate, here is one, how will others follow if we do not support them?"[8] In November 1972 Maruti had not produced a single car, though the company "bought" 400 acres of farmland outside Delhi from the chief minister of Haryana, one of his mother's supporters. A year later in 1973 Sanjay Gandhi appointed seventy-five dealers of Maruti cars, took Rs 500,000 deposit from each one of them, and promised them a product in six months. Sanjay then turned to the nationalized banks to lend him matching money, using these deposits as collateral; one banker refused and lost his position. Indira Gandhi heard criticism of this project, even from her trusted advisor P. N. Haksar, and she resented it. There was still no product ten years after the license was issued. That nothing came of this expensive adventure during Sanjay's short life (until his accidental death in June 1980) indicates not simply how difficult Indira Gandhi's relations were with her son, but, more important, how ill-defined some of the self-reliance projects and their permission were, leading to the ironic phrase describing India as the "license-permit-raj."[9] When she returned to power in 1980, Indira revived the Maruti project in 1981 as a memorial to Sanjay after his death: it eventually became a joint venture with Suzuki, hardly a symbol of self-reliance. This is an illustration of how disagreements arose about what was and was not authentic in Indian high-tech research and development.

Small Batteries, Big Ideas

Before examining the four industrial sectors in this war over self-reliance, where large scale dominated, let us consider carefully the war over self-reliance in terms of one small, common, but vital product, far from "indigenous cars." The dry cell battery for use in transistor radios and flashlights was one of the most important items in the daily life of the poor or rural populations, who all lived without electricity.[10] Along with kerosene and candles, the battery was the source of light, especially emergency light, for those millions who did not have electricity or whose electricity supplies failed; in addition to their flashlights their small portable radios were their most important means of receiving information, including foreign shortwave broadcasts. The government also well knew that radios were the means by which government propaganda could reach these people, most of whom did not read. That is why batteries really mattered at a number of political and economic levels. The debate about the appropriate size of the next generation of nuclear reactors should be set in the context of the people's use of electricity. Batteries were currently manufactured in large-scale centralized private-sector factories using about 350 workers and producing 60 million batteries annually. Fourteen licenses had been issued before 1973 to factories of this type, and it was becoming accepted that the battery was a modern "scientific" product that did not belong in a cottage industry economy. People were fed up with weak short-lived unreliable batteries made in India and complained bitterly. Returning Indians were begged to bring foreign batteries home for personal use, and lower-cost Japanese batteries were smuggled in. However, a CSIR institute of electrochemical research in South India created a small-scale battery-making process that could be located much closer to the end user and still quickly deliver the same quality product. This process involved a young London-trained chemist, Amulya Reddy, working at this CSIR lab from 1958 until 1961 before accepting a postdoc at the University of Pennsylvania.[11] After returning to India in 1966, he began to work at the IISc in Bangalore on indigenous development of a magnesium–manganese dioxide battery system and was collaborating with a small company, set up by a classmate.

The new battery unit-cost would be about 15 percent higher than the large-scale production cost, but it would employ not 350 but 1,650 workers to produce the same number of units. If there were thirty-three such small-scale units, argued Amulya Reddy, the required 60 million batteries would be produced with much higher social benefits than through the large-scale centralized model. Moreover, the well-known deterioration of batteries in India's hot climate might be circumvented: "The customer is more likely to get a 'full capacity cell' by bicycle distribution from a local producer than by rail-cum-truck transport [requiring a long time] from a distant factory," said Reddy. "Hence, the cost-benefit analysis for the choice of technology is best based, not on market prices, but on so-called shadow prices which reflect the policy maker's best estimates of real costs of capital, labour, and foreign exchange from the social point of view."[12]

Here was an analyst pointing to a small CSIR lab's alternate technology, but one which the licensing manufacturing authorities ignored. This high-profile attack on the government's licensing, monopolies, and restrictive trade practices found resonance with advocates of "small is beautiful" and appropriate or alternative technology in India. This resonance was truly international; this was precisely the time of the mass communication of the appropriate technology movement. Twenty years later, Amulya Reddy recalled the early reception of his argument at a National Committee on Science and Technology meeting in 1973:

> The real personal "break-through" was achieved at Bangalore. C. Subramaniam was organizing conferences of scientists to get reactions to the NCST document called "An Approach to the Science and Technology Plan." I presented a paper on the "Choice of Alternative Technologies" . . . to my amazement, my presentation was received with thunderous applause. . . . There was an interesting episode during the ensuing discussion—a well-known scientist attacked me with the words: "Reddy is asking us to go backwards!" and C. Subramaniam, who chaired the session, jumped up and said "No! No! He is taking us forward."[13]

Subramanian was possibly the most powerful cabinet minister at the time. But thirty years later Reddy said, just before his death, that his proposal did not get a trial on a scale that might have been be successful, and was eventually sidelined.[14]

Big industry had already declared its interest in the method of industrialization, in J. R. D. Tata's famous May 1972 memorandum to the government, "Suggestions for Accelerated Industrial Growth." At the government's invitation, Tata spelled out the requirement for "a great leap forward" in industrial production; in the view of Tata and his colleagues, this requirement was the removal of the "bias" against large-scale private industries in the Monopoly and Restrictive Trade Practices Act of 1969. The act proscribed private entry into specific industrial fields with large commercial application, attempting to break up their tendency toward vertical integration and market dominance, thus protecting and preserving those markets for state enterprises. This act had been the government's move against the industrialists at the same time as the banks were nationalized in 1969. Attacking these restrictions as misplaced, Tata argued that only large industrial houses had the expertise and capital to succeed quickly in new large-scale industries, meet growing national demand, and compete in international markets. Moreover, these industries needed more energy and electrical power. "Without arguing that Tata became the de facto minister of industrial development, it is evident that most of his main points found a place in Mrs. Gandhi's twenty-point program [introduced in the Emergency period] three years later."[15]

Running through this war over self-reliance was about tactics: between moving quickly or moving gradually, between building medium scale or small scale, between hand-making and automation, between importing turnkey technologies for reverse-engineering or adapting, even developing technologies indigenously. These tactics were variably applied: the idea of moving quickly and catching up was attractive to those for whom India started behind and moved too slowly. And surely India could move quickly if the scale was appropriate. No one argued for very small nuclear reactors although some scientists thought that the nuclear-powered submarine was a useful model for small-scale electricity-producing reactors. But many argued for small and dispersed factories for batteries, or for fertilizer, in order to lower the cost of distribution.[16]

What, it might be asked, do scientists have to do with batteries? These projects and products were defined and communicated publicly as manifestations of science, and the proper work of scientists, mostly then called applied scientists, but scientists nonetheless in terms of public image. Such common commodities as batteries were labeled as "scientific," a code word for foreign, "western" (even if they were from Japan), and modern. "Scientists" made these things (the term was more in circulation then than "technologist") and things were invented by scientists. Commodities like batteries embodied scientific thought, although in the market one could hear the curse that the Indian consumer cast upon batteries made in India. So there was a fundamental tension around self-reliance there too: scientists knew that the Indian public had little confidence in Indian technology and saw that few politicians or senior administrators had much confidence either, as the following cases will demonstrate. The result was *an official duplicity:* rhetorical support for indigenous technologies and science, yet a commitment still to import foreign substitutes for these indigenous Indian things if they seemed to take too long, weren't good enough, or required too much money. Importing was the default position to which people returned time after time. This was not just *a craze for foreign things,* long denied. The disagreement over batteries embodied all the issues of the war over self-reliance, and because batteries were a cheap and common technology, they took on greater symbolic power; in short batteries had currency because they were connected with electricity.

What follows is a sketch of the evolution of four major projects in terms of the war over self-reliance until 1975; comparisons with the DAE/AEC projects thus become possible, notwithstanding the fact that each has its own separate technical and political history.

Fertilizer

The most important variable of Indian agriculture of the early 1970s was not seeds, mechanization, or new skills, but fertilizer. The new varieties of rice and wheat, to which the government and international donors had made massive commitments, were fertilizer-responsive. The new seeds were becoming available, but fertilizer was already in short supply. Moreover, the government was attempting to control its price and to subsidize the price preferentially to smaller farmers. More than irrigation or mechanization, the fertilization of soils and the new "green revolution" plants were judged to be the key to self-reliance in food grain. Also required were pesticides to protect vulnerable new varieties, leading to the creation of chemical plants like the one at Bhopal, which, from 1969 formulated Union Carbide's American-manufactured pesticide stock into a saleable product, until the 1984 disaster that gave that lovely city its notorious industrial reputation. Reducing or eliminating food imports and food gifts was widely perceived as a crucial symbolic and practical achievement, though these calculations ignored millions of people who were unable to purchase grain and who thus presented no "effective demand." Here was the war over self-reliance being waged by opposing interests. Fertilizer had been produced in India since the Sindri fertilizer factory opened in 1948, the very project that Meghnad Saha supported, though he criticized industrialist Sri Ram for interfering in it. With the introduction of fertilizer-responsive varieties of wheat and rice about 1967–68, fertilizer production and cost were national issues. Everyone knew the price per bag of fertilizer; everyone had an opinion on yields with and without fertilizer, an opinion on subsidies and hoarding. Everyone was thus an expert, or knew one! In spite of the presence of a number of competent Indian fertilizer experts, with years of experience in India and abroad, the government agreed to a Japanese project loan for two fertilizer factories in 1972–73, tying the Fertilizer Corporation of India to work on a crash basis with Toyo Engineering of Tokyo. Attached to at least one of these refineries was a heavy water plant.

The reason given for this agreement was that India had to greatly increase fertilizer production, *immediately*. In September 1973, amidst clear signals that these Japanese projects would be postponed because of the formidable rise in oil prices in Japan (causing a consequent decline in availability of Japanese aid money), the government sent a mission abroad both to buy fertilizer and complete contracts to modernize existing Indian fertilizer plants. When an Indian delegation went to Moscow a year later in 1974, their shopping list included 1.3 million tons of urea fertilizer. Critics of the higher-yielding grain strategy said that Indian agriculture had embarked on an irreversible course through which it would never be self-reliant because prerequisites like fertilizer could not be obtained within the country. Chemical fertilizer, they said, drove India into dependence. But in fact other sources of soil fertility were already contributing the maximum possible, in kilocalorie terms. Crop residues contributed more to the fertility of Indian soils than fertilizers, followed by cattle dung, for which there were competing uses, namely, cooking fuel; fertilizer was a weak third contributor. There was no new untapped source of either crop residue or cattle dung.

So not only did India have to import food but also to import fertilizer and pesticide for the fields and diesel and petrol fuel for the irrigation pumps. Fertilizer was being smuggled into India from Bangladesh in 1972 and 1973 because of the higher Indian prices. For these reasons the status of agriculture in the hierarchy of problems that could be addressed by the sciences was now near the top of the cabinet's agenda. The belief at this time, shared in other countries in Asia, was that only a scientific breakthrough could transcend the bind between the rising cost of agricultural inputs and a production slump, particularly in rice, when farmers reached the yield-plateau of new varieties in the fields. At this time rice, not wheat, was the fundamental staple of the majority of the population, though the diet was indeed changing. With new Japanese fertilizer refineries half-built, India could hardly look elsewhere. Yet the demand for fertilizer was so strong that India had to import it from wherever the price was right. Though the refineries were delayed and over budget, they were forced on to completion; though Indian building contractors got a lot of work, the design and flow of the refineries were Japanese. The electricity to run these refineries was exclusively Indian, and it was not available, except in the form of coal thermal-generation and hydropower in a few locations.

MiG Jet Fighters

Under an India-USSR agreement concluded at the time of the war with China in 1962, the Sukhoi aircraft company—famous for its MiG fighters—began joint production of the MiG with Hindustan Aeronautics, Ltd. (HAL) about 1965. The much-awaited Indian jet initiated in 1955, the HF-24, was not on the horizon. "The HF-24 never performed well: it killed a number of test pilots and never achieved expected performance levels. There was supposed to be an engine for it built by the Egyptians (also designed by former members of Messerschmidt team), and that engine never materialized."[17]

A year later the three production divisions still had not completed the huge hangars in the remote hills of Orissa required for the workers, and by 1969 the skilled-labor force at Khoraput was still only one-third of HAL's requirements. The project had been located there because of proximity to a small hydroelectric generating dam. In an article that gained notoriety when published in India, Childs and Kidron concluded:

> By 1967–8, although HAL was able to sell MiGs worth Rs 215 million to the Indian Air Force (IAF), manufacture was still confined to assembly of imported components, using Soviet equipment: in two out of three factories the buildings were still not complete (nearly six years after the programme began), and in none of the three were all the workshops in operation. So the IAF bought new aircraft from abroad: 50 Hunters from the UK, and 150 Sukhoi-7Bs from the USSR. . . . But no HAL-'built' engines were installed in HAL-assembled aircraft before March 1969; and in the following year the output of engines was well below capacity. A realistic appraisal of the programme in 1969 would have put it about three years behind schedule: "Phase 3" had just been fully initiated when "Phase 4" was due to begin.[18]

However, with these facts before them, the Aeronautics Committee chaired by C. Subramaniam decided in 1969 to continue MiG production, because so much had already been sunk into the project. These sunk costs and good relations with Moscow were the persuasive factors. The early MiGs were simply assembled in India from Russian-supplied kits of major components; between 1966 and 1971 about 120 aircraft, from kits, left HAL, or about 30 aircraft per year. This is contrasted with China, which produced 60 aircraft of larger Tupolev-16 bomber type annually. The MiG, it was pointed out, had half the range of American or French jets, was slightly slower, and carried half the number of bombs, half the number of air-to-air missiles, and fewer guns. By 1973, Iran's military buildup with American Phantom jets was cited by the IAF as destabilizing, and the IAF argued that India could counter Iran only if it had Jaguars (a joint British-French jet; one Jaguar cost Rs 50 million). Indian military leaders argued these longer-range jets were necessary to keep the Arabian Gulf open for India's oil movement, in case of Iranian blockades.[19]

Patrick Blackett had been involved in 1955 decisions about an early HAL jet fighter project (HF-24), but in the ensuing eighteen years what had happened to indigenous production? Childs and Kidron concluded in 1973 that "past attempts have met with scant success. The HF-24, designed by a substantial team of German experts led by Kurt Tank, has still not been flown at the supersonic speeds intended for it, while the modified UK engine used in an attempt at improving its performance exploded, destroying the test aircraft. . . . So the MiG programme has probably not added a great deal to HAL's capacity to design its own aircraft." The data and techniques required to work with high-quality steels and alloys were kept secret from HAL by the Russians. So it is not surprising, said Childs and Kidron, that "the Chairman of HAL stated in 1972 that in the field of design and development our current capabilities are somewhat limited, and to some extent, non-existent." HAL's chairman in 1972 confirmed that the foreign exchange content of MiG aircraft then in production was between 30 percent and 55 percent. Childs and Kidron concluded, "It is quite plain that HAL has not yet achieved self-sufficiency in aircraft production, and in fact, were Russia to cease supplying the materials and components that HAL needs, manufacture would very soon have to stop. It is very doubtful whether HAL could rapidly secure alternative sources of supply." HAL sold aircraft only to the Indian Air Force and Indian Airlines Corporation. This was a "closed and secure market" except when the IAF wanted something imported quickly from abroad. So there was an effective government subsidy to HAL between 1965 and 1972 of Rs 900 million, according to Childs and Kidron; MiGs constituted half of Hindustan Aeronautics business by 1973.[20] The adverse terms and conditions of the MiG project became the stuff of the debates in the war for self-reliance. What, experts asked, was the *real* difference between dependence on a Soviet corporation and on any other corporation from any other country?

Oil and Petrol

Foreign oil was the lifeblood in the veins of the economy and the state by 1970. Because India had only minor proven reserves, the import of Mideast crude oil was substantial and increasing, and refineries were running at capacity. Government decree controlled all prices, which were in turn negotiated with foreign oil companies. India had accepted Soviet assistance in 1955–56 for exploration and training, and this was followed by the construction of Rumanian and Soviet refineries; importing kerosene and diesel from the USSR had forced American and British companies to drop their prices slightly, and more efficient "western" refineries and pipelines had forced the Soviets to lower the cost of refinery-building and equipment in the mid- to late 1960s.[21] Becoming more skillful in playing the oil game, India still remained hugely dependent on imports. Although India's support in 1971 for its largely Muslim neighbor Bangladesh did gain it some temporary support among oil producers like Iraq and Saudi Arabia, by the time of the Organization of Petroleum Exporting Countries price and production controls of October 1973, this goodwill had been forgotten. India was totally dependent on the world market petroleum price for its refineries. After October 1973 the world price began to rise steadily and did not stop rising until 1980, draining a great deal of the foreign exchange out of India. By 1974 the national budget for import of oil was estimated at Rs 13 billion, but there were large additional requirements for oil not included in this estimate; for example, the 1974 Defence budget estimate of Rs 17 billion contained a budget of Rs 750 million for oil. The cost of imported oil was projected to consume *50 percent of all Indian export earnings* in 1974, as it was used for transport, plastics, light (in the form of kerosene), and many other types of energy—factors directly affecting both the poor and the rich. The diesel tractor and gasoline irrigation pump had become *essential* in rice and wheat agriculture, wherever the "green revolution" was declared. These were key political commitments to Indira Gandhi's supporters. There was public discussion in 1973 and 1974 of rationing petrol if the trend continued, because crude oil bought by India had effectively doubled in price on world markets.

Though the offshore Bombay High crude oil exploration acquired high political status even before the OPEC crisis, after 1973 it became an imperative. Identified in the 1960s by a joint Indian-Russian team using a Soviet undersea exploration vessel, the field took on the air of high political drama. In early 1973 the minister of Petroleum and Chemicals, H. R. Gokhale, backed by the cabinet, decided to accept proposals for offshore oil exploration on an "Indonesian-style" basis, meaning a "general contracting arrangement" in which the exploring company retains a 40 percent share of what was found (rather than the 15 percent share hitherto offered by India in previous exploration permits). A year later, just before the May 1974 Pokhran nuclear tests, two American firms (Bates Oil and Gas and Carlsberg and Reading) were chosen under this arrangement, based on their offshore drilling experience. The foreign drilling rigs were put in place. This greatly upset the Indian oil exploration community because those experts felt that they now had the capacity to carry out this exploration "indigenously" and at less expense. The tension lay between what was possible in a few years using an India-only approach and satisfying immediate needs. In 1974 the first drilling occurred. For example, when Indian diplomats went to Moscow later in 1974 to negotiate trade and aid, they asked for 1.5 million tons of kerosene for the lanterns of poor people who had no electricity.[22] The government knew the actual flow of oil from these Bombay High fields was a long way off, long after their proven reserves were estimated in 1974–75 and after the first oil flowed at one thousand barrels per day in 1976 from the first well. In this process BARC played a role, too, contributing its senior engineer N. Bhanu Prasad to be the chairman of the Oil and Natural Gas Commission and accelerate Bombay High. Market pressure led to increasing output to 80,000 barrels per day by 1978. Meanwhile, in the absence of significant contribution from atomic energy and in light of the large increase in oil prices, the importance of the large coal resources for thermal power generation of electricity in India was enhanced, notwithstanding its massive contribution to air pollution. Thermal power based on coal was now incontrovertible, as chapter 24 will show.

Steel

Steel was at the commanding heights of the economy, essential for all civil, military, business, and private construction. Steel-making coal and iron ore were very limited. The production of ingot steel in 1952 was 1.3 million tons, in 1964 was 6 million tons, and India planned for production to be doubled to 12 million tons by 1970. In Taya Zinkin's pithy essay concerning her studies of Indian steel history in 1960 and 1964, she described "the bottlenecks of forced growth" and compared the very different approaches taken by foreign collaborators in building "gift" steel mills at Bhilai (USSR), Rourkela (West Germany), and Durgapur (Britain), and the very different technical and social communities they created around steel in the 1950s and 1960s.[23] Though the government nationalized noncoking coal for thermal power generation in 1973, privately mined coking coal was already fully dedicated to steel production and enjoyed good prices and low freight rates. Steel was, in 1972–73, being produced at four state plants, all still with a foreign collaboration: Bhilai and Bokaro with the USSR, Rourkela with West Germany, Durgapur with the Great Britain and Japan. Two smaller steel plants, the private Tata steel mill at Jamshedpur and the Martin Burns plant near Calcutta, had operated for decades in a relatively independent manner. But the dreamed-of 12 million tons was far away; production was not 12 million tons but actually 4 million tons in 1972–73.

A huge open-pit iron mine was being constructed at Bailadila, in Bastar district of Madhya Pradesh, but most of its ore, shipped out by rail through Orissa, was destined for Japan. The growing Indian demand for steel simply could not be met, even with these large mills, even with iron ore available. The common explanation for shortages was that Indian steel mills were inefficient, but despite their reputed inefficiencies in both production and management, these state steel mills were also expected to compete on international contracts—exporting steel if possible. To do this the government decided to consolidate their management and build a conglomerate in the steel sector in 1973 called the Steel Authority of India, a holding company for all state-owned steel mills. Further, during Leonid Brezhnev's visit of late 1973, there was an Indian request to have the Soviets modernize the Russian Bokaro steel plant, built in the early 1960s, and ten months later, in Moscow, an Indian request for 500,000 tons of Soviet rolled steel was presented as part of an aid/trade package.

Parts of the engineering community asked whether such foreign dependence was really necessary? "If Indian engineering and design services are not organized and given responsibility for this task—as has often happened in the past—reliance on foreign agencies for consultancy services will create conditions in which import of machinery and equipment will increase substantially at the cost of fuller utilisation of established capacity."[24] As if to pacify dissatisfied Indian experts and frustrated interests, and continue the integration of the private and public sectors, the government appointed Wahdud Khan as managing director of the new Steel Authority of India, hiring him away from managing Tata Iron and Steel at Jamshedpur; this was evidence that the cabinet had decided that the state steel groups lacked people with sufficient experience and needed private-sector influence. This appointment followed an old Tata pattern of placing its senior people at the disposal of the government, giving guidance to the state, building confidence in the private sector, meanwhile gathering crucial information. This new authority followed an old pattern of managing (and even controlling) a number of steel plants under one roof, previously Hindustan Steel, Ltd., a board reporting in the 1960s through a government secretary to the minister of Steel. It was sometimes said in defense of DAE's difficulties in the 1970s that nuclear reactors were India's first encounter with complex modern technical projects with precision engineering prerequisites. In complexity and size, convergence of scarce skills and resources, and interdependent timing, these multinational steel mill projects are clearly precedents and analogues of the different results obtained in the Tarapur (American) and Rajasthan (Canadian) power reactors. The DAE knew about the difficulties of dealing with so many different steel projects and co-builders at one time and attempted to discipline the reactor-building process. Construction delays, cost overruns, misunderstandings and mistakes, and problems of working together were all seen in both steel mills and nuclear reactors.

Atomic Energy's Response to Self-Reliance

By 1972 the DAE had become a very large and widely dispersed organization of almost 8,000 support staff on payroll, including 2,400 scientists; thousands more technical staff (e.g., machinists) worked on "continuous" casual labor pay. The reactor projects proceeded at high speed, to prove their completion and effectiveness. In May 1972 the PURNIMA reactor went critical at Trombay using plutonium in the fuel assembly, generating lots of technical interest because it was both a research reactor and an additional plutonium breeder like CIRUS. This new reactor was good news for the bomb builders. The next big step was to be a fast breeder reactor, such as the one that Americans were experimenting with at Savannah, Georgia. With only modest reserves of good-quality natural uranium available, the fast breeders and thorium cycle reactors seemed a logical step.[25]

The Tarapur reactor built by GE had begun to deliver electricity, though not continuously, and it regularly tripped the Gujarat and Maharashtra grid systems. Suddenly the US Atomic Energy Commission said it would not reprocess the spent fuel from Tarapur when a fresh charge was put in (according to the "joint determination agreement" signed in 1963).[26] Therefore larger and larger storage tanks had to be built to hold the 20 tons annual production of spent fuel, immersed in water, requiring "a huge annual cost to the DAE." Meanwhile small leakages and radioactive buildup occurred "but the DAE has kept the whole matter a secret," and the Tarapur reactor was thus kept running but at lower power output because it was soon found that it "could not be operated even at 140 MW output though its unit cost calculation was based on 180 MW output."[27]

Finally in August 1972 the Rajasthan reactor went critical, the first pressurized heavy water reactor in India, based on Atomic Energy of Canada Ltd.'s Douglas Point model in Ontario. The year before, because Canadian supplies were so limited, US heavy water was used for the initial fuel pass for the Rajasthan reactor. Within months serious technical flaws and faults became evident in this "premature Canadian technology" caused by cracking due to radiation "embrittlement" of the end shield and leaking pumps in the heavy water circuit causing heavy water to leak onto the reactor floor.[28] Rajasthan grid fluctuations caused tripping and shutdown as in Tarapur, flushing water into the nearby lake, from which the reactor drew its clean water. In early 1973 the DAE admitted only to tripping the grid system, not to leaks and cracks, maintaining the need for secrecy. In February 1973 an official in the prime minister's secretariat received a letter about safety at Rajasthan from inside the DAE safety group; he briefed the prime minister, who ordered it shut down until it was safe. But when P. N. Dhar, the new advisor replacing Haksar, spoke about these instruction to Homi Sethna, Sethna said all this would be cleared up in the next routine maintenance shutdown in a few months, and so the status quo persisted, though the DAE continued to charge the Rajasthan power grid for the same price of electricity from a reactor running half the time! Questions were raised in Parliament about Rajasthan and answered by reference to the newness of the Canadian equipment and extreme variations in the local power supply;

the Rajasthan grid system simply had a very low capacity to absorb additional electrical power.[29]

In May 1973 the NCST asked the DAE to submit their Five-Year Plan, but Homi Sethna wrote back to ask how and why NCST would be competent to assess a power-oriented development program. Sethna asked if the NCST was overreaching its mandate? Sethna reminded the NCST that the prime minister had said it was not to interfere in the working of scientific agencies! The NCST response came from V. Siddhartha and P. K. Bhatnagar, an officer on secondment from DAE, saying that NCST found that one-third of DAE's planned expenditures were non-nuclear, so the NCST was inquiring about those projects! When asked about this, the prime minister said there should be no interference in the DAE plan, "but," she said, "there must be one forum which has the whole picture." A month later Dhar organized a meeting with Sethna and Subramaniam at the prime minister's request; it was decided that DAE would drop some non-nuclear projects, but would keep some in which DAE had exclusive expertise. In Parthasarathi's opinion, "never before and never since did the powerful DAE have to go through such a process."[30]

Sethna and the AEC, having decided against Vikram Sarabhai's plan to build bigger 500 MW reactors, scaled new reactor projects back to half the size, a model familiar to Indian designers and builders. Collaboration was established with France for the Madras fast breeder reactors and a French aerial survey was conducted for radioactive minerals; French, Japanese, and German firms were now building heavy water plants, while India adjusted to the difficulty of getting heavy water from old allies. In November 1973, coinciding with Brezhnev's visit to India, the twelve-year-old scientific cooperation program with the USSR was revived; there had already been a major Soviet BESM-6 computer purchased and installed at BARC, a Soviet extrusion press for the manufacture of zircalloy and seamless tubes for reactor fuel bundles was bought and installed at Hyderabad, and India proposed to purchase nearly 100 tons of Soviet heavy water, to buy time until its own facilities were complete in 1977. Interestingly, this heavy water deal took two more years to negotiate and in the end was consummated in September 1976, just after the final termination of Canadian cooperation; the agreement was now for 200 not 100 tons; 25 percent of the first heavy water to arrive was not formally safeguarded though the Soviets were cooperating with the London Suppliers Group; the price per ton is unknown.[31]

Meanwhile there were projects now maturing but begun long ago. By 1970, the DAE was supporting the variable energy cyclotron (VEC) in Calcutta and in view of that support the Saha's Institute's abandoning the Pagladanga project seemed more justifiable. The experience gained in this model study could, however, have reduced the gestation period for the big project (the VEC of 1977) and many of the hurdles faced in the latter could have been avoided. Though it was difficult for the key actors in this earlier project to complement the larger one under way in the early 1970s through BARC at Salt Lake, a few scientists from SINP did become involved with BARC in the VEC project, even after the tensions in the institute (see chap. 17). Including SINP's director D. N. Kundu, a DAE team visited several accelerator laboratories around the world in 1969 to choose a suitable machine as a model and chose the sector-focused VEC of Berkeley, a replica of which was being constructed at Texas A&M University. Under an agreement with the US Atomic Energy Commission, the design specifications and drawings were obtained of the 88-inch AVF cyclotron at Berkeley, as adopted for construction by Texas A&M University at College Station (and called TAMVEC). Later, members of the VEC project staff were deputed to participate in the installation of the TAMVEC and to work in the Berkeley laboratories. The responsibility for fabrication and installation of the VEC at Calcutta was entrusted by the DAE to BARC, and the actual work was carried out at government and private facilities across India. Given the delicate relations between SINP and BARC from 1969 onward, it took time to accept the relevance of SINP's experience with its cyclotrons, but things changed gradually when the institute accepted in 1968 that it had to abandon its own construction for both security, land, and financial reasons.[32]

So it was that the whole portfolio of the DAE projects now took on a very eclectic network of international support. In one sense the eclecticism appears quite rational: consider the crisis in fertilizer production caused by dependence on Japanese aid and technology alone. But in another sense, fueled by the war over self-reliance, eclecticism resulted in a very inefficient and quarrelsome mixture. The separation of space research and development from atomic energy for electrical generation led to calls in 1973 for the separation from DAE of other activities like cancer research or food sterilization and preservation. Critics argued that some of these projects could find more suitable homes, including universities or CSIR laboratories. And still the level of DAE's foreign collaboration was seen by some experts as inexplicably high, prompting questions whether DAE was *really* contributing to self-reliance, or just saying others should? The DAE certainly enjoyed the lion's share of the R&D budget in the early 1970s. If one compares the official expenditure estimates of the DAE and CSIR in the Third Plan (1966–69) and the Fourth Plan (1969–74), we see a strong contrast in financial strength between the two, with DAE at Rs 560.5 million in the Third Plan and Rs

211.81 million in the Fourth Plan (1969–74), and CSIR at Rs 545.5 million in the Third Plan and Rs 121.12 million in the Fourth Plan. The budget and expenditure reduction of the CSIR are dramatic, if the figures reflect the situation accurately.[33]

This situation drew criticism, even within the expert community favored by DAE's support; again the *EPW* publicly expressed these doubts: "the fact that the DAE claims nearly one-third of the total research expenditure in the country cannot be explained in terms of national economic relevance, but only in the 'personal' relevance of the powers that be."[34] Too many DAE projects were not connected to atomic energy, said critics, and finished reactor costs were too high. As a consequence of the 1973 oil crisis, the DAE budget for 1974–75 was to be cut by 13 percent. This budget would be Rs 960 million (or roughly $83.4 million) and did not account for inflation, so the effective cut was greater.[35] At first there was talk of minimal cuts in atomic energy and space, but in the end the DAE experienced the cut in its budget just at the time the expensive preparations for the first nuclear bomb tests were being concluded.

Rockets and Missiles—Self-Reliant Responses at Space?

Satish Dhawan brought space activities south to Bangalore in 1972 and continued his other role as the director of the IISc, insisting therefore on the same Rs 1 salary that Sarabhai had received. Although he was the secretary of the department, he was also chairman of the Indian Space Commission and the Indian Space Research Organization. Space work was now distributed in Trivandrum, Ahmedabad, and Bangalore. There were inevitable tensions as Ahmedabad, Sarabhai's home city and laboratory, lost its primacy in space activities. A Space Commission was created in April 1972 by the cabinet and announced in May; the commissioners were P. N. Haksar, senior advisor to the prime minister; T. Swaminathan, cabinet secretary; M. G. K. Menon, director of TIFR; Brahm Prakash, then of BARC; and I. G. Patel, secretary of Finance. Their first decisions were to rename the Thumba launch site to be the Sarabhai Space Centre and to push ahead on the development of the satellite and the Sriharikota launch site. Brahm Prakash assumed the directorship of Thumba; Yash Pal, of TIFR, became head of Space Applications; TIFR cosmic ray physicist Devendra Lal was appointed head of Sarabhai's Physical Research Lab in Ahmedabad; and Abdul Kalam remained head of the satellite launch rocket.

But the real tension, in 1974–75, the year before the first satellite launch, was that other agencies wanted a part of the programs and resources of the

Space Department. These agencies, namely, the Ministry of Communications, the Ministry of Information and Broadcasting, the Meteorological Department, the Geological Survey, and so on, all saw the privileged space budget as an opportunity in their time of crisis with budget cuts. The Space Department reported that the expenditure on space increased from Rs 40 million in 1972–73 to Rs 190 million in 1973–74.[36] Because the space program was announced as an applications program in satellite broadcasting, rural education, remote sensing, and monsoon forecasting, other agencies began asking to be more involved and, when rebuffed, wanted to know why. By this time the prime minister had built her confidence in Dhawan's abilities to complete the satellite launch in 1974 and to manage the demands from other agencies for more share of the budget. A senior scientist from TIFR, cosmic ray physicist Yash Pal, long involved in satellite planning, was seconded to chair a national committee to harmonize the competing demands for uses of the satellite.

In May 1973 tenders were let out for a multipurpose satellite for telecommunication, television, and meteorological functions: four bids came, from Philco-Ford, Hughes Aerospace, Fairchild, Messerschmidt-Bolkow-Blohm, but none were satisfactory. The Planning Commission "had serious reservations" about this multipurpose project, and the prime minister did not wish to overrule its members, so there were serious delays. "Dhawan, though deeply disappointed, took the decision stoically." In the end the satellite instructional television experiment proved that a nationwide rural-oriented television system based on direct reception from satellite at the village level "would be next to impossible" and therefore much more work had to be done.[37]

Space expenditures *had increased* dramatically, from Rs 40 million in 1969–70 to Rs 190 million in 1973–74. Though rushing toward a geostationary satellite, the Space Department's budget estimate was cut down 10 percent in late 1973, slightly slowing the work on its satellite rocket launch. The Space Commission and Department officials quietly sought to justify its costly satellite launch rocket on grounds that it would also provide a vehicle for defense purposes. Space explained more publicly that it wanted to use its planned satellite work in important civil projects, like information, broadcasting, meteorology, and geology. Others claimed jurisdiction in these fields, of course, and in the end joint agencies had to be established, most prominently with television. Critics asked why the Department of Space was setting up a remote sensing agency when the Department of Science and Technology was setting one up with the same objectives. Reflecting opinion in other departments regarding Space's ambitions, the *EPW* assessed the new sensing agency: "It should get funded by potential users, rather than by the exchequer directly. It is the bane of many Indian scientific organizations that they do everything 'in-house' and this seems characteristic of Department of Space as well."[38] But this in-house practice had been well established in the DAE itself, for example, going into cancer research in 1973 in direct competition with the Indian Council of Medical Research, even as that council was experiencing a funding cut. Indian leaders clearly believed that the country was so large that it could afford to go beyond a "one of everything" philosophy, and DAE-related agencies were not shy to step on anyone's toes. Some in the Planning Commission thought that "one of everything" was not enough in India and that the lack of competition was also a problem for innovation. Implicitly, and contrary to pronouncement, the whole system thought so at a tacit level.

The competition between civil and military expertise for recognition and resources is well illustrated by the relationships between the Defence Research and Development Office in Hyderabad and India's Space Research Organization in nearby Bangalore. The government had established a new company, Bharat Dynamics Ltd., at Hyderabad in 1970 expressly to build guided missiles in cooperation with France (the Daimant project), but to India's disappointment that project was cancelled by France abruptly about 1971. Both ISRO and DRDO were under new leadership from mid-1972, though one cannot presume a unity of purpose or expertise. Early in 1972 Nagchaudhuri appointed Narayanan as director of DRDL in Hyderabad in order to produce SA2 (SA75) missiles in India within seven years, code-named Project Devil. Initially the lab got a five-year budget of Rs 160 million for this and other projects. In 1971 Narayanan wrote a concept paper about India's self-reliance in missile technology, and a number of influential people read it, including Sarabhai, who showed it favorably to others. Chengappa provides a good description of the kind of work being done in Hyderabad, obstacles in cooperating with other R&D agencies, innovating with insufficient resources, and actually retro-designing the Russian SA2's fuel injection system. The SA2 (named BACA in Soviet terminology) had been designed in the 1950s and had an archaic wiring system 16 meters long.[39] But this was not all the work that DRDO was expected to do; in fact, it was seen as an R&D training program.

In June 1972 Nagchaudhuri proposed to the cabinet a Rs 160 million project to start work on Project Valiant, a long-range missile at DRDL Hyderabad. The cabinet "made a show of turning down the project," but Nagchaudhuri said Prime Minister Gandhi secretly approved setting up parts of the project in different locations; "the expenditures were broken up so that it was not obvious what was going on. . . . I was not happy . . . [because we seemed to be] afraid that nothing could be kept secret in India for long even if only the Cabinet knew." In November 1972 Gandhi and Nagchaudhuri decided to develop a Valiant liquid-fuel missile with a realistic 1,500 km range as a DRDO/DRDL project but a far cry from the 8,000 km objective they discussed in 1971. Expenditures in Project Valiant were broken into Rs 500,000 segments because DRDL itself could authorize these payments without requesting permission from Delhi. Combining the two projects, and two budgets, DRDL grew quickly, and by 1974 had a staff of 2,500, and a Rs 160 million annual budget.[40]

In July 1972 Satish Dhawan pushed ahead with the objective of building the SLV3 for a satellite launch in 1974. Kalam was still in charge of it, and Dhawan fought hard for full autonomy of this project. Instructions from the top were to obtain maximum integration of Space and Defence R&D, but "when Dhawan became Secretary of DoS . . . he even created the impression with the DRDL scientists that the PM's targets did not bind him."[41] Nagchaudhuri at Defence was responsible for both a short-range surface-to-air missile *and* a long-range missile project for nuclear warhead delivery. Work at DRDL paid well and attracted fresh recruits from the very sources that ISRO had previously used almost exclusively. In early meetings between Nagchaudhuri and Dhawan, they tried to cooperate. But Dhawan had publicly stated in Bangalore with Nagchaudhuri present that he thought cooperation was premature and he did not want it. Gandhi heard about this and asked Dhawan, to explain; he said that if foreign state space companies such as French rocket manufacturers heard about Indian Defence-Space cooperation, they would withdraw from negotiations, which were at a delicate stage; "the PM remained silent, indicating acquiescence."[42] A lot of the recruits to the DRDL projects were trained at IISc Bangalore and tension developed between the DRDO and ISRO, particularly as they drew close to their project completion deadlines in 1974.

Can we surmise a parallel path between the weapon missile and satellite launch rocket? Was there a kind of phony separation? There were certainly some design and technology similarities at this stage in 1973–74, notwithstanding the political distance between the two organizations. There had already been suggestions that the ISRO rocket would be converted to the DRDL missile. But there is only one source on this question: Chengappa reports that Dhawan stated, with respect to ISRO's plans for rocket to missile conversion, "I am afraid I was not party to any of that business. Nobody ever asked ISRO to do anything during my time as far as I know."[43] When asked later about the convertibility of the rocket to a missile, Dhawan said,

"The technology of rocketry—the aerodynamics, the structures, the analysis—is identical to missilery with minor variations. What's the damn difference? Only the software! You make a few minor changes and the damn thing goes differently."[44] But Dhawan was astute enough to realize that such conversion through "minor changes" was an intellectual trade-off: "while the base technology may be the same, there must be somebody who could see what can be traded off and what must be done." According to Dhawan, DRDL just did not have that "somebody."[45] So self-reliance did not always mean cooperation and could sometimes mean exclusion; Gandhi had to find a way around this impasse, so, according to an inside observer, "given the barriers to development of aerospace science and technology, the PM had no option but to centralize policy-making for the rapid and *silent* evolution of military missile and civil space launch vehicles."[46]

Enlarging Indigenous Electronics

Pressure mounted for indigenous development of computers and, if not, at least for making computers under license. There was elite disagreement about minicomputers related to differing political obligations and appointments; for example, Electronics Secretary Menon came into conflict with the Electronics Corporation of India, Ltd., in particular, with the chairman of its board Homi Sethna, also the chairman of AEC and secretary of the DAE. The conflict apparently also extended to Menon's relationship with the Electronics Corporation's managing director, A. S. Rao, also a senior BARC electronics expert. The Electronics Corporation's board imagined itself capable of producing India's planned computer on a large scale and was officially expected to do so. But the corporation resented the intrusion into its plans by the Electronics Department headed by Secretary Menon. This was a context in which IBM had withdrawn from India, Russian equipment and software was proliferating, and the cheaper small-scale computers from Taiwan had not yet appeared.

To complicate matters, the political implications of the anti-automation movement were now strong enough to show up in the Automation Committee of the Electronics Commission, where arguments in favor of strictly limiting the application of computers in the workplace found official support. The committee's policy work also revealed the integration of public funds and private enterprise; for example, there was a prolonged workers' struggle against electronic data processing in a large private engineering manufacturing firm, Voltas, Ltd., in 1973. Shares in this firm were held 12 percent by Tata, Ltd., and 30 percent by two state financial institutions—Life Insurance

Corporation and Unit Trust of India. These three organizations had a heavy dependence on routine applications of the computer for their day-to-day business. Of course the government was not going to be deflected from automation but hoped the anti-automation movement could be slowed by sympathetic consideration or discredited with internal politicization. The Public Accounts Committee of Parliament started to subject electronics to scrutiny at this time, and eventually, in April 1976, it concluded that "the use of computers and other sophisticated machines for traditional labor-saving applications may not be desireable or even expedient."[47] This was said in 1976 at the height of the Emergency, when the computer business was finally booming, but the section in chapter 17 on the anti-automation movement and the installation of the Calcutta computer shows that this began in 1968 and was a long struggle.

The Electronics Commission's operational program was dominated by the issue of television set production, complete imports versus assembly versus indigenous production. The policy was supposed to lead to an "entire TV industry being based on indigenous know-how and companies with foreign equity using foreign brand names being completely excluded." In addition to television, computers and defense electronics were promoted. Menon found a stack of frozen license applications in 1972 and decided they should be dealt with as a group, to be tested against government electronics policy. But the policy had to be clarified, and "soon there were rumblings, particularly from the private sector about 'delays' in decision-making."[48] There were also complicated and unhappy "partnerships" in play with government laboratories, as Visvanathan's account of a ferrite project and competition between the CSIR, Indian Telephone Industries, and the Electronics Commission show, pointing to "the utter lack of consensus regarding science policy for developing indigenous technology."[49]

In the context of winding up IBM's operations in India, the government sent a delegation to Moscow to check out computers and all other electronics of COMECON manufacturers. In September 1973 Soviet RIYADH computers were tested by the Department of Electronics under the "Rupee-Rouble Agreement" and found compatible with the popular IBM 360. Nevertheless, there was continuing resistance to the official introduction of this hardware and its software, though this suited the increasing number of Indian electronics engineers trained in the COMECON countries.

C. R. Subramaniam, from the vantage point of his role in the electronics industry, argued that the self-reliant approach in computers could be interpreted as a failure, namely that India got "indigenous" technologies that were not state of the art and paid higher prices than if import strategies had been used. These were classic issues to those people suspicious of self-reliance via the indigenous route, namely that made-in-India technologies took too long, thus cost India more, and therefore kept India "behind": the result was a technology conceived among and for a previous generation. Advocates of self-reliance disagreed, some more and some less, pointing out that developing technologies from scratch in India built up the essential capacity to understand, adapt, or modify state-of-the-art technologies, and then to *innovate* with new ideas with respect to those very technologies in the next generations. The self-reliance advocates pointed out that technology transfer from other countries was costly too, that there were often hidden costs (including wasting indigenous expertise), and that foreign suppliers did not share everything that was important to know about the technology. "Being behind" was as bad as "being innovative" was good, and balanced the equation so that the tension was perfect.[50] No one could win. Just as going forward and going backward were central to the debate around the best way to produce and distribute batteries, so were they central to the debate about computers, rockets, and reactors. There really were losers in this struggle, and thus it is called a "war," although no advocate for either strategy was dying of their wounds.

The Electronics Department did not enjoy protection any more than other agencies, and so its finances became embattled during the OPEC crisis: the Rs 590 million budget for 1973–74 approved in July 1973 was *cut in half* two months later. Still it pushed bravely on: in December 1973 a quarrel between the Department of Electronics and the Home Ministry erupted because of a persistent complaint that Bharat Electronics favored Defence contracts (being under the Ministry of Defence) and thus neglected the national police wireless production requirement.[51] The Electronics Commission's solution was to bring the police business to Electronics Corporation of India, developing and pursuing competition between state enterprises, and this solution was approved by the prime minister, who was unable to discipline Bharat Electronics but hoped market forces would.

Limits to Growth in Science and Technology?

How did the NCST, the elite body of scientists advising the cabinet, respond to the war over self-reliance? The NCST was placed in the middle of this question in 1972 and began to prepare its plan by involving 1,800 scientists in the planning panels for a major conference called the Approach to the Science and Technology Plan, launched in June 1973. Indira Gandhi inserted technology into the popular debate, where her father had spoken more often about science "on its own terms." The 1973 approach emphasized the fullest possible application of Indian scientific capabilities as well as technical know-how. The meeting at which Amulya Reddy spoke about batteries was part of this process of wide consultations.[52] The subtitle was intended to refer to the buzz in India generated in 1972 by the report of the Club of Rome titled "Limits to Growth"; Indian scientists had participated in the Stockholm conference in June 1972 on environment and the limits of the supplies of materials essential to an industrial revolution (chemicals, oil, electricity, surplus food, fertilizer, etc.). Indian media and scientific circles regularly referred to limits to growth in this way.

The functional agencies had their own ongoing projects to contend with, as they had the budget and mandate to do. They too continued with an official duplicity about self-reliance in those projects, urging and supporting indigenous development in science and technology, yet importing when it was expedient or cheaper, or otherwise advantageous. The war over self-reliance was also waged inside these very ministries and agencies, for they were not at all monolithic in practice. Moreover, they also tended to engage in rivalrous development; for example, the Department of Science and Technology and the Department of Space both created their own remote sensing agencies, and the Department of Atomic Energy and the Indian Council of Agricultural Research both created their own programs in nuclear applications in agriculture. In this rivalry they took advantage of the willingness of competing foreign agencies to help fund competing projects and to pay for that help (usually using their foreign experts as consultants), as this advantage could be tallied against their internal rivals. The impact of the OPEC oil crisis meant that any form of self-reliance that substituted for imports became more politically and economically attractive; rivalry intensified when 1973–74 budgets in most science and technology departments were cut back by 10–14 percent, as will be shown below.

Attentive to political details, Gandhi also had the big picture on her mind and on her desk. Seen from the viewpoint of science and technology planners in the prime minister's office in 1972, both the Planning Commission and the Finance Ministry were advocating *economic* self-reliance by pursuing rapid reduction of aid from the World Bank and the donors but were indifferent to *technological* self-reliance by cutting the foreign technology transfers, which came with many strings, because, in the opinion of one planner, "no foreign company transferred its latest technology." The NCST thought that with India's wage rates one-tenth of rates in industrialized countries, India could compete but needed "complete autonomy of decision-making regarding *the make-or-buy option.*" But there was negative reaction from parts of the civil service, particularly in some ministries, and private industry had "deep animosity to that horrible phrase 'self-reliance' and 'technological self-reliance.'" From that viewpoint "the structure of our economy and the nature of the domestic market combined with a set of government policies made foreign technology cheaper than indigenous technology to the entrepreneur."[53]

In September 1972 the draft text of the Approach document, written in part by economist Sukamoy Chakravarty at the Planning Commission, had a character that portrayed indigenous skills and hardware as a burden on economic development, whereas foreign technology was seen as "a liberating input." In Parthasarathi's view this was "mere lip service being paid to indigenous technology," and when he complained to the prime minister, she told him to go back to Chakravarty and get the text rewritten before it came to the cabinet in November 1972. The budget contained a Rs 25 billion outlay for the science and technology plan that she had pushed for personally. She lined up loyal cabinet ministers like D. P. Dhar, of Planning, Mohan Kumaramangalam, who died shortly thereafter in an accident, Jagjivan Ram, Swaran Singh, Y. B. Chavan, C. C. Subramaniam: all pushed hard this time, but the resistance to these changes in practice was massive, and the plan had to wait many months before being passed. When Parthasarathi discussed this deep resistance with Gandhi in January 1973, she said, "I am not surprised."[54]

The quiet preparations for the bomb test ran in parallel with noisy disagreement over science policy and debates about the role of new ministries like Space, Electronics, and Science and Technology. Effects of budget cuts in 1973 and 1974, from which the test makers in the DAE and Defence were shielded, were public and pressing for the other science ministries. The Space Commission and the Electronics Commission each had a member for Finance, just as the AEC did, thus keeping the commission's business largely out of the claws of the Ministry of Finance. That member had the authority from his minister to decide on the commission's budget, and because the member was invariably from Finance, that department knew quite well where cuts could be made, albeit with resistance from other well-placed commissioners.

The new Department of Science and Technology (DST) was empowered to supervise older agencies, some of them much "richer," like the Survey of India—richer because it had a budget for salaries of a much larger staff complement, land, and buildings, being an amalgamation of the geological, archaeological, botanical, and other classical survey departments, founded in the nineteenth century. The political heart of this new department was

Table 5. Budget Allocations for Science and Technology Agencies, 1973–74

	1st est. (January)	2nd est. (September)
NCST	Rs 2.79 m	Rs 1.99 m
NRDC	Rs 2.20	Rs 1.77
National Committee on the Environment	Rs 1.39	Rs 7.95
Survey of India	Rs 96.50	Rs 80.30
Department of Science and Technology	Rs 3.23	Rs 2.15

Source: R. S. Ganapathy, "Science and Technology—A Department in Search of a Role," *Economic and Political Weekly,* 22 June 1974, p. 973. The source for the budget cuts is the DST annual report, 1973–74, Delhi.

the NCST, which was a replacement of the COST that previously advised the cabinet: all commissioners were active scientists, six of the nine were located in Delhi, and none were heads of scientific agencies. This prestigious commission was drawing up a national plan. How else could scientists counteract the intrusion of the Planning Commission, now headed by powerful C. Subramanian, into their new technology projects? If valuable scientific resources were idle, shouldn't scientists make their own plan for mobilization, not someone else? But were agencies capable of implementing the science and technology plan actually listening, and was the money available?

The NCST was studying how government's policies were reshaping the foreign-technology-first environment for the entrepreneur, because it was believed that some government policies "positively discriminate *against* domestic technology." In April 1973, a month after the drafting of the Foreign Exchange Regulation Act, Gandhi told the CSIR directors that experience with foreign technology substitution was still unsatisfactory. In August the prime minister raised concerns about repetitive imports of same foreign stuff (machinery for papermaking, truck tires, inner tubes, etc.). In December 1973 the NCST committee proposed how to use a 1 percent tax on sales of foreign equity companies, consultancies, design engineering, and manufacturing. These organizations would then get a rebate from their tax (cess) contribution equal to the cost of their projects if they were approved by the NCST. This was intended to reduce conflict within industrial policy and direct R&D attention to the core objectives; "companies would know precisely where they were with regard to commercialization of know-how flowing from the NCST-approved R&D projects."[55]

Like other ministries, DST experienced cuts in late 1973. Small agencies like the National Committee on Environmental Planning and Coordination, well focused after the 1972 Stockholm conference on the environment and development, were nearly devastated by these cuts, as was the National

Research and Development Corporation. Even the administration of the whole department itself was cut from Rs 3.2 million to Rs 2.1 million. The cabinet asked DST to take the role of monitoring the implementation of scientific projects and technological plans, but insiders knew that effective monitoring was, in practice, very difficult unless there was commitment to it at the top, plus resources. Expressing a common view of such objectives, the *EPW* said of the DST's overall responsibility that "the failure of the monitoring divisions of the Planning Commission and the Bureau of Public Enterprises should serve as an eye-opener to the DST. It should know that without executive funding power and accountability 'monitoring and evaluation' mean nothing."[56] All NCST members were capable of guiding such monitoring (so long as it was over someone else's project) but none of the commissioners had the time to do it. The NCST, and the secretariat supporting it, believed it should have more power and should exercise that power. But this was a time of greater executive concentration of decision-making in India, and odds were against scientific planning of the sort envisaged.[57] In any case, a plan was afoot that would spring forth suddenly and affect most scientific and technological plans in India for a long time after May 1974.

In the context of this official picture, there was a struggle over the use of Indian capabilities in various aspects of the economy that hinged on different understandings of time and timing: What is being ahead and what is falling behind? One view was that these capabilities were already proven and continued reliance on foreign technology or capital was backward looking. This view then distinguished those who wanted an alternate kind of technology and science serving "the people" from those who wanted the Indian approach to better resemble "the western approach" but to be authentically "Indian." The other view was that Indian capabilities were immature and would take such a long time to deliver the results that time should be bought by the state using quick importation, thus satisfying internal markets and managing people's frustrations. The state would thus buy time for indigenous capacity to mature. A more complex nationalist response was that most of these investments would all pay off eventually, so though the path was long the country should stay on it, only if the state remained in charge. In this debate every voice claimed to be nationalist and there were no technical grounds for disqualification and no deep challenge to the consensus around the shift to a high-technology economy. This was all set in the international debate over the "Limits to Growth," the Club of Rome study published in 1972, based on simulated projections of the relation between human consumption and natural systems. Though it had its critics, it also had subtle effects. Indians, being more intimately acquainted with scarcity than most of those who entered the debate in richer countries, took those limits more seriously whatever they thought about the methodology of the studies.

By the time NCST began the approach study, a new post-Independence generation of alumni of the IITs had taken up positions in various middle-ranking driver's seats in the research agencies and public-sector enterprises. Personal loyalties were replaced by group or professional loyalties looking for a focus. A close observer could perhaps have predicted the emergence of an NCST from the Third National Conference of Scientists, Technologists, and Educationists as a natural outcome of group aspirations. Insider V. Siddhartha suggested the dynamics: "Although the Axes had thus become diffused [referring to powerful deal-making and appointment networks like the Allahabad-Calcutta axis, and the Bombay-Ahmedabad axis], the mandate of the NCST was a result of pressure from below rather than the result of need perceived at the top. It was therefore rather vulnerable to dilution by the Axis powers."[58]

These questions about self-reliance were the same questions asked by advocates of the first nuclear bomb test, saying that the bomb test may not be state of the art, but it was necessary now in order to do anything more sophisticated later. They also pointed out that the nuclear suppliers with whom they dealt were explicitly opposed to sharing any know-how that was *crucial* to atomic bombs. They concluded that the great powers were not likely to permit import of a small atomic bomb for an experiment, noting that every other country had followed precisely the path they advocated: they had all done it themselves. Indian experts and diplomats involved in the discussions at the IAEA in Vienna trumped any other voices and knew that working indigenously—even for an objective some disagreed with—could be criticized but would not be fundamentally opposed. The 1970s were not the birth of confrontations over technical choice, but actors now confronted the effects of earlier choices, including the choice of reactors. Starting with buses in the 1940s (and penicillin, batteries, and tractors, etc.) and sustained by the political economy of the 1970s (like the OPEC oil price increase and foreign exchange crisis), the war over self-reliance was fought in the shadows of the contradictions caused by the fast drive for industrial modernity. Indians in the 1970s had something different from earlier generations who disagreed over these questions—they had the ability to do almost all of it themselves.

TWENTY-THREE

The First Bomb Test: Its Context, Reception, and Consequences in India

During the struggle around indigenous capabilities and the rising economic and political costs of dependency on other countries, Prime Minister Gandhi agreed in early 1972 to a focused discussion about a nuclear test. The difficulties facing other major projects like fertilizers, reactors, and rockets were swirling around her—so, would a bomb test be successful, or could its preparation and even a failure be kept quiet? Would a failure just add to the embarrassment in other fields? How much would a test really cost? Could everything be done indigenously? What would the consequences really be? These were the questions raised in discussions, probably starting as early as 1970, when a group within Indira's Congress Party, led by Krishan Kant (later vice president of India, who had an MSc in physical chemistry), started to push for nuclear tests. In fact, these tests had been contemplated before, but there were always technical and timing obstacles. And although the views of the voting public were quite mixed, it was the view of the prime minister that mattered most in early 1972.

Resistance in the office of the chairman of the AEC ended with the death of Sarabhai, and Indira Gandhi had in her hands the success of Indian military forces in the Bangladesh war, the final arrival of electricity from the Tarapur reactors, and the relatively favorable balance of trade and foreign currency reserves. This situation provided the important precursors to the test decision; the spring 1972 election victory for Gandhi's faction of Congress was probably the deciding factor. Finally, in October 1972, after about nine months of regular discussion, Gandhi gave the go-ahead to scientists to make bomb test preparations and authorized the formation of a steering group composed of senior officials and scientists, namely, P. N. Haksar, the prime minister's principal secretary;[1] P. N. Dhar, the cabinet secretary in the prime minister's office; B. D. Nagchaudhuri, scientific advisor to the minister of Defence in whose laboratories the explosive lenses would be fabricated; Homi Sethna, AEC chairman who would supervise building the trigger device and other electronics; and Raja Ramanna, director of BARC and the bomb project's chief physicist. Ramanna soon involved P. K. Iyengar and R. Chidambaram, later chairman of DAE himself from 1993.[2] This group began to meet in late 1972; according to Ramanna, "In the initial meetings considerable discussion ensued on the economic repercussions and possible fallout of the experiment." Meanwhile, according to an insider at BARC: "Thereafter things began to move fast but so very quietly that hardly anyone (except for a very select few) knew what was going on. I can testify to this, and to all appearances lab life was normal in every conceivable way. But since I was tuned to this explosion business right from Bhabha's time, I could see extremely subtle signals, hardly perceptible, that something was cooking."[3]

Planning for the bomb test was finalized in early 1974 when India was facing a financial crisis and Gandhi's earlier popularity was in doubt. The immediate context made the final decision more clearly one of political symbolism, for people in India and abroad. To use the words of its protagonists, it was "a demonstration" and was intended to have "a demonstration-effect."

Changes in the Economic and Political Context of the Nuclear Test

When the test finally occurred, the government was in difficulty. First there was the 1972 shock of unplanned government expenditure on 10 million Bangladesh refugees accommodated in West Bengal and then the effect of the unplanned expenditure on the war itself. Though popular support for the government and Gandhi continued for a year at least, the fiscal consequences of these activities appeared within twelve months: an effective devaluation of the rupee occurred following the Smithsonian gold-delinking conference in late 1971, and this meant that the value of exports increased 20 percent in rupees (although not in dollars). Looked at carefully, however, these exports were still in "traditional" items—leather, cashew nuts, shrimp, wool, and jute products—not the favored industrial and high-tech projects so heavily invested in by industry and government. True, metals were also exported. For example, exports of manganese ore, so sought after by the Americans in the 1950s, had fallen in value from $30 million in 1961 to $14 million in 1971—not just because of a fall in the world price of manganese due to wider competition but also because Indian high-grade ore had already been exhausted and second-quality manganese was now being exported.

From mid-1973 there were food shortages and rising prices. Rice production was 6–7 percent lower in 1973 than in 1972, and the Ministry of Food had to quickly find $200 million in foreign exchange to import food from Canada, the United States, and Argentina. There had been an expectation that Soviet wheat might fill the gap, but that country's 1972 wheat harvest had been at abnormally low levels, and the USSR too was driving the world price up, buying wheat and paying for it by selling gold. There was also sharper social stress and conflict; for example, in April 1973 there were food riots in Nagpur in the middle of a regional drought, and Gandhi decided to create Operation Grain Trade using the Food Corporation of India to try to break the grain traders' hold on the movement and distribution of rice and wheat. She and her officials saw the traders profiting (as they always had) from a timely manipulation of the market. But these traders had regional economic power and political influence (including in her party) and were angry with the paramilitary interference in their grain-buying and trading business.

At the end of 1973, after a year of spending Rs 1.5 billion on an emergency food production increase program (mostly spent on subsidized inputs like diesel, pumps, and seeds for rice and wheat cultivation), it was admitted officially that overall food production was not up but *actually down* about 5 percent from the previous year. This massive emergency investment had simply not led to production increases, the increase of 15 million tons of new grain production was not going to occur, and the government would now have to spend another similar sum procuring domestic and foreign food for the public grain market (through which it controlled grain prices). The government was now forced to buy from the very private traders it had been trying to control through Operation Grain Trade. Since political stability depended on it, the government was determined to remain massively interventionist. And India was coming to the world market for grain in 1973–74 just when prices were rising there as a function of oil price increases everywhere.

In spite of the US decision to terminate its bilateral aid in early 1972, the Aid-India consortium of the Paris Club did not reduce total aid to India in 1972 or 1973. But most of this new aid money was earmarked for debt repayment, not new projects. The hidden and unexpected problem was that aid decisions were not linked to the rise in the world price of oil that started in 1973, so that the importation of vital supplies (oil, food, paper, etc.) simply cost more than calculated by the Paris Club in June 1973. The fact that only two-thirds of India's trade was realized in dollars or pounds sterling meant the crisis was very real. India's serious (if reluctant) effort to shift to trade with COMECON countries was clearly not going to have a strong effect in a timely fashion.

After introducing the 1973 Foreign Exchange Regulation Act, revenue intelligence controlled by the prime minister's office showed there was still massive smuggling and evasion of the regulation, to be understood as a popular and logical outcome of the history of strong currency regulation and feeble practice. The value of smuggled goods coming into India, in avoidance of customs and excise duty, was about Rs 4 billion in 1973, plus a probable Rs 3 billion in leaked foreign exchange, all in avoidance of Reserve Bank of India and Customs regulations (through a number of methods, one commonly known as false invoicing). So serious was the situation that the government created a committee on foreign exchange leakage, chaired by the Finance secretary. The loss due to leakage of hard currency was later calculated to be equivalent to the entire food import bill for 1973–74.[4] Smuggling of commodities and avoidance of currency controls were also intimately related to the steady expansion of the "black economy," a large volume of unrecorded and untaxed capital (usually held in cash) that was used to finance other activities, including much of the film industry, political party activity, and smuggling itself. The prime minister's control of revenue intelligence was going to be applied to this question two years later when the Emergency procedures were introduced, but an occasional crackdown was widely announced before the Emergency, for instance, in the *New York Times*.[5]

In a situation of scarcity of foreign exchange, people found imaginative methods to circumvent the new currency restrictions. These methods included the smuggling of gold into India, where the demand was insatiable and people were prepared to pay more than the world price. Gold enabled some people to sell it and raise dollars within India.[6] There was also a large unofficial trade in official currency itself, a leakage from official uses of foreign exchange to facilitate trade and industry. This occurred through sophisticated methods of false invoicing, as established by the government's 1971 commission.[7] Thus the unofficial market in India for exchange of the rupee with some foreign currencies was growing larger, not smaller, with the progressive application of controls. Though the state was becoming poorer, a few sections of the private sector were getting rich and benefiting greatly.

Although scientists participated in these methods of circumvention when necessary, I have not heard of their projects benefiting from these techniques except in the importation of equipment and spare parts. Scientists told me of an informal self-help system of purchasing small lab equipment parts and carrying them quietly to India when they traveled abroad. Although cumbersome, official permission was eventually granted to purchase most but not all spare parts for scientific equipment, and the scientists simply sought ways to speed up the process or buy better quality. The real struggle was over costly new equipment, and sometimes the use of foreign exchange was not approved.[8] As in industry, people in the scientific community may have then made use of these unauthorized methods. Though we do not know the extent to which people were prepared to circumvent the limitations imposed by the government, we do know the degree to which the measures adopted were unpopular. Those who were not favored by the official foreign exchange rules resented or envied those institutions that were. Given the demand in the large middle class for foreign-produced goods, foreign education, foreign travel, and the like, the status of those who had access to these valuables was definitely enhanced. Remember that this was well before the popular Indian passion for television and computers. The scientists' projects were touched with glamour because their projects appeared to have unimpeded access to foreign exchange (which was not always true; there were obstacles for many). Although it was certainly not unimpeded, access to dollars was often granted because it was expected that these scientific projects would pay off in the long run.

This is not surprising in light of an intensification of the war over self-reliance described in chapter 22. In a period of increasing aid flows, with attached conditions of foreign consultancies, tied-purchases, royalties, and foreign expertise, the argument for the use of Indian capacity and Indian technologies was strengthened. It was argued that they had lower costs, were not inferior, and helped make the economy stronger. The prime minister herself had a preference for indigenous skills: the mainstream of the Congress had stood for the idea (if not always the practice) of *swadeshi* since the 1930s. And in a situation as complex and large as India, the prime minister was not making all the decisions anyway, hence, the value of making a decision about testing a bomb.

The tantalizing fact that the value of exports was *increasing* spurred some voices to assert more strongly that India's techniques and labor skill in advanced technologies could now be exported. India's large private industrial firms (Mahindra and Mahindra, Larsen and Toubro, Tatas, Birlas, Kirloskars, etc.) were buoyed up by state contracts and a large uncompetitive internal market in which they could operate with privilege. Although they did almost no R&D, they too wished to profit from exports of nontraditional goods, such as electronics, pharmaceuticals, and other products. From 1973 they also wished to do construction and infrastructure projects in the newly rich "near west" or "west Asia" (Middle East) or in Africa. Indians were moving to the gulf to work and were sending considerable remittances home. State corporations were driven by the same imperative to find export markets. For example, in late 1972 the DAE had proposed the sale of Indian-made zircalloy-clad fuel rods for CANDU (Canadian deuterium uranium) reactors *to Canada* and to Canadian General Electric, arguing that they were equal or better quality than Canadian-made fuel rods and because of the difference in the cost of highly skilled labor involved, those rods cost much less in India.[9] This was precisely eleven years after the debate between India and Canada regarding the quality of CANDU fuel rods made in Bombay and Chalk River.

The emphasis on private industrial involvement in state-funded R&D projects and stress on each unit raising foreign exchange for its project inevitably had unintended effects. This war for self-reliance, rhetorically so familiar from the 1930s and 1940s, was now engaging an entirely new generation of scientists and technologists who were discovering it anew. But it occurred in an economy of paradox in 1973–74 because Indian exports were successful in foreign markets, but most of them were not new products but the traditional commodities whose image India wanted to shake off; moreover, real gains in dollar terms were modest. The leakage and loss of dollars through currency control evasion equaled the value of the export earnings, a chilling fact to the minister of Finance and his prime minister.

How had conditions changed at the time of the May 1974 nuclear tests? In early 1974 it was officially accepted that India was experiencing an astonishing 25 percent inflation rate—profoundly affecting the cost of food and energy. The index number of wholesale prices, according to the official economic survey, showed that prices rose 26 points between 1972 and 1973, and 64 points between 1973 and 1974. In March 1974 India borrowed $74.4 million from the IMF for compensatory financing to offset losses in trade and immediately after the nuclear tests began negotiating special oil credits from Iran and Iraq. New money was needed for sophisticated and costly projects like reactors, satellites, and missiles, though the cost of rice and supply of kerosene were issues of even higher national importance. The prime minister thus created and chaired two new special cabinet committees, one on essential commodities and the other on energy, in addition to her special duties in atomic energy, electronics, space, and industrial research. All this added to her near complete control of foreign affairs and to the power of her office over other cabinet ministers. The sudden increase of oil and food import costs, coming together, rapidly and dramatically raised the state's cost of control and lowered its capacity to intervene. Amplified by the shock of the OPEC oil price increase in October 1973 and the tough market for food, the supportive wave around Indira Gandhi was fading: by early 1974 she was facing her enemies once again.

The First Pokhran Test

The events leading up to the 18 May 1974 bomb test have had a variety of public interpretations. The first insider to speak in detail was Raja Ramanna, who died in 2004. In the following years, the books of Abraham, Perkovich, and Chengappa, based on interviews, laid the story out more clearly. Insiders Haksar, Dhar, and Sethna have published nothing about it. Knowledgeable people confirm that the whole project was completed without paper files, something of which Indians were particularly proud. At the moment Raja Ramanna remains the most accessible and descriptive firsthand observer:

> Like all important decision-making processes, the final meeting on Pokharan [in early 1974] was one which involved heated discussions. PN Dhar was vehemently opposed to the explosion as he felt it would damage our economy; PN Haksar took the view that time was not ripe and gave his reasons; my own view was that it was now impossible to postpone the date given the expense, time and the critical stage the experiment had reached. Fortunately for my team Mrs Gandhi decreed that the experiment should be carried out on schedule for the simple reason that India required such a demonstration.[10]

Ramanna gives no clue to Sethna's role in the test preparations, but the two men maintained an uneasy arm's-length relationship throughout. Sethna, an engineer not involved in the technical problems, has maintained a discreet silence about his role and view, except for occasional interviews, but it is known that he resented the manner in which Ramanna sought and appropriated the glory derived from the test after May 1974.[11] It will be noted that there was no military officer named by Ramanna in this group: Nagchaudhuri, a scientific advisor to the minister of Defence, was a physicist, not a military officer.

> Prior to 1972, the BARC and the DRDO had not worked together on any project involving high-level secrecy. This collaboration was thus surprising as the two were culturally opposed, but thanks to Dr. Nagchaudhuri things went smoothly on the Pokharan project. The DRDO's contributions to the Pokharan

> experiment—the development of the lenses and fabrication of the high explosives—were very significant. . . . Dr. Nagchaudhuri was an old friend of mine. With his help the project was not difficult to streamline.[12]

According to a military expert, Nagchaudhuri provided more operational and planning assistance to the test than is commonly recognized and deserves more credit than had been accorded to him; Nagchaudhuri confirmed the two agencies were indeed culturally opposed.[13] But it is significant that Gandhi's two closest advisors, Dhar and Haksar, initially did not support the plan for the nuclear test and were overruled, "fortunately for my team," said Ramanna. Nagchaudhuri was advisor to the minister of Defence but did not advise him about the bomb test plan until cleared to do so ten days before the event.

Filling positions held by Sarabhai, Sethna became chairman of the AEC, and Ramanna the director of BARC in 1972. They had never been close, but distance between them now grew; Ramanna hoped Sethna would leave the management of BARC to him, but Sethna had been there as director and project manager for many years and would not let go entirely. Moreover, the large amount of space work previously attached to the job had gone to Bangalore, and there was thus a greater concentration on fewer issues for the AEC chairman. This continued through to May 1974, after which Sethna felt Ramanna sought the limelight and did not give credit to others, including himself, for their role in the bomb test.[14] There was not a great change in BARC procedures in 1973 and 1974, to the puzzlement and consternation of security experts and inspectors on routine visits to Trombay just weeks before the test. According to Ramanna, inspectors found and criticized the absence of standard security methods like secret logs and safe rooms: "There I was sitting with the biggest secret of my life and they were trying to teach me how to preserve secrets." This secrecy was later confirmed in interviews by Perkovich and Chengappa, and in writing by former AEC chairman M. R. Srinivasan and BARC physicist G. Venkataraman.[15]

The bomb test project team became dramatically larger. Through Nagchaudhuri, Ramanna, and Sethna, fifty-six other specialists were brought in to work on components, such as Anil Kakodkar and S. K. Sikka (system integration). P. R. Roy (radio metallurgy and neutron trigger) had to build an alpha-beryllium source of a certain shape and strength, brought together at a certain precise time. Others had to irradiate bismuth in the Trombay reactor to create polonium for the alpha source (neutron initiator), and "this process proved to be somewhat tardy," according to Ramanna. A conventional implosion in the 1,400 kg device would initiate a reaction of beryllium and polonium (separated by a barrier), and the neutrons produced would trigger a chain reaction in the 6 kg of plutonium in the bomb's core, making it super-critical once everything is under sufficient simultaneous compression. Polonium vaporizes in warm air, and the temperature in the desert would accelerate that loss; apparently BARC had problems getting the polonium to adhere to a metal barrier (like gold) to go into the neutron initiator device.[16] George Verghese was head of the seismological team; T. S. Murthy and Vasudev Iya, who was Bertrand Goldschmidt's physical chemistry student at the University of Paris in the early 1950s, worked on neutron initiation; and P. R. Dastidar made and pulled the bomb's trigger.

The army was in charge of the test site and BARC was in charge of the bomb. "Not a single Army person was to be seen near BARC."[17] Ramanna reported that in 1973 they were choosing and preparing the site, looking around in the Thar desert military test range for a site without water 100 meters down: they did not want the explosion disturbing the underground aquifer nor the bomb's radioactivity contaminating the water. Finally they chose an old abandoned well because it was dry enough. The shaft was re-dug, and the soldiers were told they were actually digging for water. Camels were used to pull the earth from the hole. Soon they hit a lot of water, to the soldier's pleasure and to the surprise and consternation of the bomb testers. So another hole had to be dug quickly, to the puzzlement of the soldiers. At this depth and in this case radioactivity was, according to Srinivasan, "fully contained and posed no hazard."

Gandhi ordered a large press conference on the day of the bomb, in marked contrast with her pattern of one annual end-of-the-year news conference. She gave a long interview to a team of *Newsweek* journalists within a few days of the test, and her office produced a twenty-six-page transcript of the interview, which was distributed to Indian journalists. An observer commented on the singular significance of this interview and its transcript three weeks later, saying that "barring pro-CPI journals, the only interview given to an Indian journalist during the last 12 months was to the biographer of her mother and was largely of a non-political nature."[18] Indian leaders immediately began explaining that this was an explosion for peaceful purposes, in other words, technically, in nuclear jargon, a PNE. This was the explanation soon taken to the IAEA in Vienna. Within a few weeks, the secretary in India's Ministry of External Affairs opened talks with Canada, the only country that immediately suspended its nuclear cooperation with India.

The size and power of this bomb (its yield) was not established with certainty, for curious though interesting reasons. P. K. Iyengar later explained that they did not install elaborate or sensitive equipment to evaluate the test but did not explain why, and so this lack of good data led to years of debate about the yield of the bomb. Eight months after the test, Ramanna and Chidambaram presented their yield estimate of 12 kilotons to the IAEA in Vienna and were asked on the spot about analyzing the molten material formed by the heat of the blast. "We are planning to do this" was their reply, indicating that they had not done so in the preceding eight months! That evidence (and their analysis if any) has not been released.[19] A 1978 report stated that the BARC scientists remained divided on the yield and said there was insufficient good data to draw conclusions; a further 1981 report cited unnamed sources at BARC who stated what had long been suspected, that the yield was lower than 12 kilotons. Interviews conducted later by Perkovich showed that Sethna and Iyengar both conceded that the yield could have been 8 kilotons, and other experts said it could have been even lower.[20] Perkovich reported expert conclusions that inexact simultaneity and asymmetry of the implosion were the cause of the poor yield. "Thus the simplest explanation for not publicly analyzing the Pokhran test results is that the scientists did not want to expose themselves to criticism over technology that they themselves would now try to improve."[21] Because implosion systems could be tested and improved without fissile cores, the BARC and DRDO teams thought they could work on this aspect of the bomb without any further tests, using simulations. In a major 1995 comparative study, Husbeye and Dainty list the blast's official magnitude as 12 kilotons, citing Chidambaram and Ramanna, compared with yields of French tests in 1962 of 52 kilotons and 1965 of 120 kilotons, American tests in 1970 of 220 kilotons, and Soviet tests in 1971 of 500 kilotons.[22] Of greater importance to Indian planners and leaders were China's nuclear tests in the atmosphere: a Chinese thermonuclear test in November 1971 yielded 15 kilotons, and another in January 1972 had a slightly smaller yield. But in June 1974, one month after the Indian test, the Chinese dropped a thermonuclear weapon from a bomber and it exploded in the air with a yield of 200 kilotons.[23] As shown in *Negotiating Nuclear Power,* Indian strategists made a career of pointing to the technical sophistication of Chinese nuclear capabilities and advocating that India catch up.

The Reception and Costs of the Pokhran Test

"In truth compared to other countries it cost us very little and was more a by-product of other activities," said Ramanna triumphantly about the bomb.[24] Bhabha had stated in 1964 on All-India Radio that a 10-kiloton nuclear explosion (equivalent to 10,000 tons of TNT), if conducted, would cost India $350,000 and a 20-kiloton test would cost $600,000.[25] Official estimates of the cost of the test, cited by J. P. Jain, were $400,000 or about Rs 3.2 million. Sethna said immediately afterward that India spent seven times more money on nuclear research in agriculture and medicine than it had on these tests.[26] Even later Sethna explained that because digging the hole for the underground blast "was done by camels[; it] was cheap."[27] These official estimates are consistent but carefully exclude the cost of the necessarily long training of personnel and expensive technical preparations (including plutonium separation) before the decision to go ahead occurred in 1971–72. Official accounting also omitted the cost of medical consequences of the tests and the site cleanup and waste management. It omits the effects and consequences for the atomic energy program and the economy of the withdrawal of all Canadian and most US assistance to existing and planned nuclear power plants. A more complete and realistic estimate by Seshagiri of the probable cost of the bomb project was Rs 1.76 billion between 1969 and 1974, or $220 million in 1974 values, and this figure does not include the post-1974 costs either.[28] But then as this book shows, and others concur, the test was not about cost or economic efficiency.

Putting these costs in comparative context, in 1975 Aqueil Ahmad estimated how much DAE had cost India between the years 1961 and 1974. Total expenditures by DAE were Rs 3.077 billion. In the period 1971–75 the DAE, including the space program in 1971 and 1972, consumed an average of 21 percent of India's R&D budget average each year, followed by an average of 14 percent for CSIR, 13 percent for the Indian Council of Agricultural Research, 12 percent for the DRDO, and 1.6 percent for the Indian Council of Medical Research. Ahmad said budget estimates for the Fifth Five-Year Plan (1975–80) gave space and atomic energy an increased 25 percent of the total; CSIR, 15 percent of the total; agriculture and food, 12 percent of the total; and health came last at 1 percent. Even with this increase in overall share of the R&D budget, Ahmad said that the DAE was producing about 2 percent of the total power generated in the country in 1974–75.

Despite the fact that the plutonium came from a reactor of Canadian design built by India and Canada, a reactor moderated by American heavy water, those who believed in the virtue of indigenous technology development had no difficulty in saying that this first bomb test was indigenous. The preparations had all occurred in India, calculated and thought through by Indian scientists. But for many of the colleagues of these people, a nuclear bomb was not the indigenous technology they had advocated; how, they asked, did it serve the people and how did it strengthen science in India? Conscious of the motivations of the experts who had achieved the successful explosion, Aqueil Ahmad reported eight months later that Indian scientists had indeed "expressed sentiments to the effect that one of the most important pay-offs of the nuclear explosion will accrue by way of a sense of accomplishment and self confidence." But his own analysis was that "a general uplifting of the spirits of Indian scientists and technologists, by this and similar future events, is highly improbable." Echoing a popular question in India, Ahmad asked, "What have we learned from nuclear research that is not already known?"[29] Bomb protagonists argued that years of dependence had rendered most Indian scientists unsure and unable to point to significant accomplishments, and thus confidence and accomplishment were at the heart of the war over self-reliance. Avoiding science that was mere replication in India of what had already been done elsewhere was part of the moving forward or backward dilemma that haunted science planners and activists. To them it did not matter if one was discovering something known about the bomb elsewhere, because the objective was to reach a place where Indians knew as much as experts elsewhere and were then learning in rhythm with everyone else. But people who opposed that view, or opposed the bomb itself, thought that kind of knowledge could also be purchased, or that that kind of knowledge was the wrong knowledge for India, or that Indians should be working on knowledge not known elsewhere: in their view there was little point in doing experiments in India that others had done years before. Although there is no wide survey or assembly of skeptical voices about the first bomb test, there are hints at the differences that emerged. Even the difference between a test bomb and other complex technical apparatus was noted, involving comparisons between large radio astronomy telescopes or large nuclear reactors and the test bomb.

The expressions of wild enthusiasm shown after the May 1998 tests were foreshadowed in May 1974. There was indeed a reception of genuine joy in some quarters, not just among scientists and technologists in BARC and parts of the DAE. The majority of the center-left elite, despite qualified public support, soon articulated an uneasy mixed view of India's new position. This unease was presciently expressed in an editorial written a week after the Pokhran test:

> India is perhaps the only country to have gone nuclear with three-fourths of its population below the poverty line. . . . You do not become a power, and certainly not a nuclear power, when you are on the threshold of economic chaos. . . .
>
> There is no doubt that the Indian elite takes itself very seriously. It must now be hoping that the United States will condemn its "peaceful" explosion in

> harder terms. . . . [Perhaps] the Americans can be trusted to make a silly statement so that the Indian explosion appears a genuine thing. . . . This does not mean however that it will not make an international impact. It will. Strange as it may seem this is the first nuclear explosion by a non-nuclear power which the super powers have chosen to ignore, and which they may over time even applaud. This confirms the alignment on our subcontinent. Attlee's nuclear blast had confirmed Britain's secondary status in the Western alliance system. We shall soon see that something similar has occurred here. . . . Alignment with a super power prevents us from acquiring the privileges of total isolation. It only increases our dependence. India's dependence on the Soviet Union has come to stay.[30]

Although there is no reference here to Pakistan, a country irrelevant to India at this time, the 1974 Pokhran test was conversely the impetus for the Pakistan government's accelerating its nuclear program.[31] Here, by chance, Zulfikar Ali Bhutto was in power again as prime minister, and he remembered his 1965 vow that Pakistanis would eat grass in order to build a testable bomb. Bhutto now quietly proceeded on the path that led to the Pakistani bomb tests of 1998. For unrelated reasons, the Pokhran test was also the prompt for plans to build a French nuclear reactor near Baghdad at Al Tuwaitha in Iraq under Saddam Hussein's supervision and control. And German and American cooperation with Iran accelerated immediately after the Indian test, leading the French to reach an agreement with the shah's government to build a uranium fuel reprocessing plant. Meanwhile French nuclear testing continued in the South Pacific and Chinese testing continued too.

Consequences in India of the First Nuclear Test

Indira Gandhi continued in 1974 her effort to alter the balance of dependence on foreign collaborations, having announced in August 1968 at NPL that she expected CSIR and scientists to reduce "net dependence on foreign assistance" by 50 percent by 1973.[32] The mounting evidence of noncompliance was discouraging to her, and she put her foot down—at least on some agencies, if not all (e.g., on pharmaceuticals and computers, and not DAE or the military). On 22 May 1974, four days after the Pokhran test, the Cabinet Committee on Science and Technology met with the whole NCST, with Gandhi present, to discuss the complete Science and Technology Plan first discussed in 1972. It required Rs 10 billion of committed expenditure plus Rs 10 billion for new plan projects plus Rs 3 billion to be raised through the 1 percent tax discussed for almost two years. Dhar of the Planning Commission and Y. B. Chavan of the Finance Ministry expressed doubt about India's ability to pay for this, but according to a witness, the prime minister said it must be done, and the Science and Technology Plan and tax were approved to the satisfaction of the NCST. During 1974 and 1975 the Committee of Secretaries screened around one thousand foreign collaboration agreements, removing negative technology transfer clauses from 350 of them and putting positive clauses (meaning installing clauses consistent with NCST objectives) in all of them.[33]

Now the NCST was to be transformed when Haksar was rehabilitated and-appointed the deputy chairman of the Planning Commission and the chair of NCST. Haksar drove through appointments to the NCST of 20 brand-new names, including 12 who were heads of agencies and 8 who were not on the first NCST and got the prime minister's approval for these changes. Chief executives of scientific agencies were now happily reinstated as members, along with the president of the Indian National Science Academy and the chairman of the University Grants Commission. The secretary of the NCST was now A. Ramachandran, a former president of IIT Madras, who explained the shift back to something that looked like the old SACC established in 1948: "Our main job is to create an R&D culture in the country, not become big bosses of science and technology." Haksar then called two meetings of the new NCST before December 1974. The committee appeared to be in full bloom. But these meetings were not followed by any more meetings through to the end of Gandhi's regime in March 1977 because Haksar seems to have preferred that the Planning Commission do the planning for science and technology and doubted the utility of the NCST. Gandhi acquiesced in this shift of focus, being preoccupied with her political opponents. The Planning Commission took over, as economists wished. But C. Subramaniam, as minister of Finance until 1977, pushed indigenous science and technology projects, using the plan approved in 1974.[34] Haksar, now with unique power in the prime minister's office, said R&D funding would be the responsibility of individual ministries and not the Department of Science and Technology. This would cut down on overcentralization and produce more user involvement, all of which was supposed to lead to better results. (Critics pointed out that most ministries did not spend much money on R&D, expecting DST to fund it.) The DST now would only coordinate the approach of others. The circles of expertise around the cabinet and the prime minister had thus opened and closed, allowing in voices directly from the institutes and universities and then replacing them with those of the agency heads. Plans formed through direct consultation (like the Approach) were implemented without further engagement with the scientific community. This had happened before, so that when the closing was completed, the pressure built up underneath to force the opening again. Though the prime minister gave the appearance of commanding all her "political scientists," each of them had other pathways to cabinet ministers and could motivate them to present plans originating from outside the inner circle of expertise.

The May 1974 test was a watershed for the careers for senior scientists. Nagchaudhuri left Defence a few weeks after the Pokhran test and soon became the vice-chancellor of the new Nehru University in Delhi. He said he was implored to accept larger responsibilities in government just after the test, though he did not name them, but he ultimately declined these offers.[35] As an indication of the increasing integration of defense with scientists of the atomic energy and space communities, Nagchaudhuri was replaced at DRDO by M. G. K. Menon for the next four years. For a while, Menon remained director of TIFR and was also director general of DRDO, scientific advisor to the Ministry of Defence, and secretary to the government of India for these purposes, with a budget of Rs 250 million. He was still also secretary of the Department of Electronics, a member of the Electronics Commission, and also a member of the Space Commission. Thus electronics and military affairs were truly "united" in Menon's hands. But could one person carry out these roles effectively? Menon soon agreed to give up the position of TIFR director, which he had held proudly since Bhabha's death.[36] TIFR was, after all, a place where he now seldom came, and it was languishing from lack of leadership, despite his new FRS. Menon resigned as chairman of the Electronics Commission and secretary only in 1978, after the fall of the Gandhi government and after "trade lobbies which had wanted free play in the electronics sector" mounted pressure for his replacement.[37]

Sethna held his position as chairman of the AEC, a position that Ramanna hoped to get after the test. However, Ramanna remained director of BARC until he was moved out by Prime Minister Desai to be defense advisor and head of DRDO in 1978, replacing Menon, and he returned to BARC only in 1981. Ramanna did eventually become chairman of the AEC in 1984, when Sethna finally left that position. But the tension between Sethna and Ramanna, so evident in the period before the bomb, continued while Sethna held the chairman's position in the AEC all the way through to 1983. This tension had crystallized in the sudden December 1974 visit by Gandhi to the Pokhran site, accompanied by AEC Chairman Sethna, while BARC Director Ramanna was in Mysore receiving an honorary degree. Ramanna says it was too late to reply to a same-day invitation to attend this bomb site tour with Gandhi, which was interpreted as a kind of pilgrimage to a now "sacred" (national) site. So his glorious moment with Indira Gandhi at the site had passed. The two men each began to tell jokes at the other's expense, trying to cause laughter about what the other had done in the past, sometimes when they had been in the presence of either Nehru or Bhabha.[38] Knowledge of this became so widespread that when Moraji Desai took office in 1977, he heard about the Sethna-Ramanna feud. He asked his cabinet secretary to resolve this tension, and Ramanna was transferred (without warning, according to Ramanna) away from BARC in 1978; he was to be the scientific advisor to the minister of Defence, to replace Menon. Sethna then had, as AEC chairman in 1977, unchallenged power over DAE and BARC, at least for the next four years.[39]

The Satellite, Rockets, and Missiles

When the preparations for the nuclear test in the desert were complete, the first Indian satellite was being constructed for its launch from a Soviet central Asian space station. It had already been decided that an Indian rocket would not be ready, so India's Space Research Organization and Prime Minister Gandhi fixed a launch date scheduled for its satellite, on a Soviet rocket, in December 1974. The prime minister thus planned two technical "coups" in one year: an atomic bomb and a satellite. Given the macroeconomic and political difficulties her government was in, these coups were doubtless seen in her office as a great relief. But since neither the long-awaited long-range missile engine (Valiant) nor the satellite rocket (SLV3) were ready, the decision to accept the Russian offer in Kazakhstan was taken, delaying the launch by four months.[40] This launch was the sign of the increasingly close Soviet-Indian cooperation on science and technology, and how the Indian leadership's use of the Cold War was again paying off inside India, pleasing, as it did, the Communist and other left parties. It also helped to mask the delays in the launch rocket project. Moreover, strategic planners asked if this new bomb would be delivered to an enemy target, and if so, how? Bombers, which India had, were no longer doing that sort of thing, though they were quite capable of it (US bombers had been carpet bombing Laos only four years before). Attention was focused on missiles with nuclear warheads, but those seemed a long way off to Indian defense planners.

The problem was that the Devil and Valiant missile projects were already in some difficulty in 1974 when Nagchaudhuri signaled that he was leaving. DRDL finally tested the Valiant missile jet engine in May 1974, the same month as the Pokhran bomb test. It was not ready yet, though that is hardly surprising. Did the military patrons of this defense laboratory have high confidence in the outcomes? The air force was just then seriously considering a Soviet offer of a more advanced surface-to-air missile called Pechora that could strike at longer range than the SA2 and with greater accuracy. The air force had carefully studied performance of the SA2 in the 1971 war (having used them tactically) and concluded that liquid fuel was going out of fashion and a new solid-fuel missile was inevitable.

All this suggests a lack of consensus on the value of these Indian R&D efforts; to quote a senior officer, "We all regarded defence R&D as a bloody hoax [in the mid-1970s]. They hadn't delivered much and we didn't think in the foreseeable future we were going to get anything anyway. Even the SA2 project was regarded as limited and we knew it would soon be outdated."[41] But V. S. Narayanan was determined to continue because he thought the SA2 would be the catalyst for other new technologies, and so he may even have favored the Devil project at the cost of the long-range missile in the Valiant project. This would be a classic form of DRDO's entrapment in an unsuccessful strategy, seen from the "final product" point of view. But seen in terms of mastery of liquid and solid missile fuels, it was an important learning experience.

In January 1975 Menon took over DRDL and DRDO from Nagchaudhuri, and, knowing of the poor relations between the Departments of Defence and Space, he set up a review committee for the Devil project (SA2 reengineering), chaired by Brahm Prakash of ISRO, with ISRO aeronautic engineer Abdul Kalam as a member. Significant civilian expertise was now judging the future of this military project, and a convergence of military and nonmilitary expertise upon one technology was initiated. The committee did not recommend the termination of the Devil project because they said it had some useful by-products, as Narayanan had thought. In the end both Devil and Valiant were continued by the minister of Defence, though scaled down. Menon later said of his role as scientific advisor, "I tried to tailor the policy to have a commonality of technology so that without too much more effort you could produce an entire family of missiles with the same basic pieces. Like a mechano or [LEGO] set from which you could build all sorts of things."[42] A family of missiles did eventually appear, along with the Indian expertise to build them. As Abdul Kalam later said, "SLVs and missiles can be called first cousins: they are different in concept and purpose but come from the same bloodline of rocketry."[43]

Defence was the largest importer of foreign technology in value and volume. In 1974 it had been decided that all high-value imports of electronics, including Defence requirements, should come under review by the

Department of Electronics, and achieving this agreement involved recalling the participation of Defence secretaries in setting up the DoE. Despite this agreement, some proposals were not shown to DoE, or were shown at the last minute, as in the case of tank simulators and air defense artillery radars. Menon, now responsible for electronics and defense in 1975, called ministries to question on this issue, for example, on the importation of precision approach radars to be imported and manufactured by Hindustan Aeronautics, Ltd., for the Indian Air Force with Tesla of Czechoslovakia. Menon and others emphasized the necessity of developing indigenous radar technology, manufacturing radar spare parts, and preparing for repairs, noting that equipment coming to India would be phased out by the manufacturers. Defence fought back, protesting the interference of DoE, and the prime minister had to intervene, telling Defence to adopt the policy and look inside India. The war over self-reliance continued over other technologies, such as the importation and local fabrication of special metals and alloys for missiles.[44]

A historian of the space program, Dinshaw Mistry, has described the civil and military convergence during the post-test era:

> Soon after the Indian nuclear test in May 1974, ISRO's director at that time, Satish Dhawan, stated in July before an Indian parliamentary committee that India possessed the ability to produce medium-range missiles with locally-developed fuels and guidance systems—he was referring to the SLV-3, which was actually still some five years from its first flight test. The coincidence of increased ISRO activity that signaled its military potential—the launches of Rohini sounding rockets (potential short-range missiles) and the 1974–1975 ground tests of the SLV-3 at Cologne Portz in Germany—with India's nuclear test served the political purpose of demonstrating that India could indigenously build a nuclear arsenal and a credible delivery system.[45]

But another historian, with more evidence available, said that ISRO maintained a separation, even though it did indeed adopt solid-fuel engines, which were available and useful to the Agni missile program later. In fact, just before his death Sarabhai had approved a project to develop liquid fuels delivering higher impulse and greater thrust and power that could have led to a fully cryogenic engine; this very project was overturned by Dhawan when he took over ISRO's leadership, probably because solid fuels looked less expensive, more available, and less complex.[46] In the end the satellite launch rocket was ready at Space in the 1980s, but the missile was ready only a few years after that at Defence. The difference between these two organizations has to be understood before a comparison is made. The DRDO had thirty-five labs with one hundred projects and many users (called "clients" by some planners), and each project required commitment from a military department that would always be comparing it with a comparable item purchased abroad. Whereas ISRO had one set of objectives and one commission to satisfy (its ultimate judge being the prime minister), the DRDO had no specific champion or end-user, and in 1973–74 this further alienated DRDO and Nagchaudhuri from the rest of the Defence Ministry (and Menon later), where some research projects were seen as an unwanted burden. In fact, Nagchaudhuri fought for greater autonomy for his projects (to make them more scientific), further alienating him, and "within a few years of Nagchaudhuri's exit in 1974 the laboratory fractured on civilian-defence service fault lines." Neither DRDO nor ISRO could develop a mature technology for long-range ballistic missiles in the 1970s, and the Valiant jet engine "silently passed into history. A saga of excellence in indigenous technology, it was a victim of inter- and intra-institutional rivalries, conflict of interest, and contradictory perspectives of what constitutes national vision."[47]

Reactors and Results

But what were the effects of the bomb test on the nuclear reactor program itself? They were the very effects Mrs. Gandhi had been warned about in 1972. Although the official relations between the two countries remained distant and cool, India and Canada secretly began to renegotiate their nuclear agreement. Srinivasan, who had helped to negotiate the 1960 reactor with Canada, was among those chosen to go to Ottawa in July 1974 to begin negotiations to restart cooperation (these discussions are examined in *Negotiating Nuclear Power*).[48] Canadian nuclear interests were equally committed to restarting the cooperation program, and Srinivasan felt at home, having completed his PhD at McGill University in Montreal twenty-five years earlier.

Signaling the importance of the relationship with the USSR, Mrs. Gandhi decided to send Dhar, her chief economic advisor, deputy head of the Planning Commission, and previously ambassador to Moscow, back again to Moscow in December 1974, where he died six months later. Soviet officials made no public commentary on India's bomb. The Soviet relationship in trade and diplomacy had become crucial, and the cabinet wanted a person of Dhar's stature in Moscow. His move changed the situation in science and technology planning. The USSR was particularly important to the solution of India's shortage of heavy water for its reactors.

A year after the test, the Canadian cooperation agreement remained suspended, despite regular and improving negotiations between the two countries to revive it. In 1976, however, the agreement was finally terminated because the Canadian cabinet could not agree on reinstating it. Shipments of enriched uranium from the United States to Trombay for the American reactor were under close scrutiny and were described as "delayed" and "under regulatory review," before their termination in 1978. Heavy water was difficult to obtain from alternate sources, whether from India or the USSR.

Nevertheless, in January 1975 a BARC group headed by M. Srinivasan (not M. R. Srinivasan, who was then negotiating a return to cooperation with the Canadians) was formed to make a boosted fission bomb prototype and began accumulating the lithium 6 and tritium required for an improved neutron trigger, according to AEC chairman Homi Sethna.[49] The group also decided to reduce the bulky size and weight of the 1974 bomb by using fewer conventional explosives. As some Canadian officials thought, Indian scientists were working on another bomb for testing while negotiating for the restoration of the nuclear reactor cooperation agreement. But there was clearly no appetite for another test, despite rumors that circulated in 1975: Perkovich records that neither Ramanna nor Sethna received support from the prime minister to continue with plans for further tests.[50] Comparisons were made between projects of different kinds: a key example comes from the engineer who eventually became chairman of the AEC: "It is much more difficult to design, build and operate a nuclear power unit successfully than to set off a nuclear explosion because the power unit has to keep going for twenty, thirty or even forty years, which requires a different approach altogether."[51] It was indeed "much more difficult" technically, but it was much more difficult to explain that to a busy prime minister and have her accept it. In fact, according to Parthasarathi, by 1975 "both the PM and Haksar had become resigned to the failures and poor performance of our nuclear power programme."[52] It was reactors and the electricity they were expected to deliver which haunted the nucleus of DAE scientists and engineers who had been doing so well. Though their projects continued, an impatience was building and becoming louder as the international effects of delay and cancellation were seen in the reactor program itself.

TWENTY-FOUR

The Scientific Community, the State of Emergency, and After, 1975–80

Indira Gandhi's Emergency is one of the more important periods of modern Indian history, but unfortunately one of the less studied. Until 1975 India had been remarkably free of the emergencies and military interventions that characterized the political histories of its neighbors Pakistan and Bangladesh. The 1975–77 Emergency period in India was also the context of crucial negotiations around the future of American and Canadian reactors and around other high-technology decisions. The war over self-reliance took an interesting turn: political interference in the prime minister's plans for indigenous technology first was minimized, and the civil servants had to become more disciplined in their approach to technical choice. To an extent not yet fully understood by outsiders, the scientific and technological community, or its leadership, took advantage of the prime minister's full control of the administration to pursue opportunities created by the absence of party politics. Life in the eighteen-month Emergency was not just business as usual, not even for hardworking scientists, but much effort was expended to make it appear as if it were business as usual.

There were intimations of challenge and unrest before the Emergency, which, in retrospect, might have been the prime minister's grounds for apprehension of an insurrection.[1] The May 1974 editorial in *Economic and Political Weekly* criticizing the bomb test said India was "on the threshold of economic chaos," and a year later Gandhi appears to have believed that both chaos and insurrection were imminent. She said the year after the test there was an organized movement to overthrow the government and pointed to bits of evidence. A showdown occurred in August 1974 over the election of the new president to replace V. V. Giri; her candidate was Fakhruddin Ali Ahmed, an old Congress Party faithful from Assam, but surprisingly the prime minister's move was contested by an unlikely coalition of dissatisfied parties who nominated Tridib Chaudhuri, member of Parliament of the Revolutionary Socialist Party from Baharampur in West Bengal. Here was Meghnad Saha's shadow on the presidential ballot, because Tridib Chaudhuri and Saha were elected together in 1951, and Chaudhuri called him "my political guru" in 1953. Chaudhuri had been reelected to Parliament for the next twenty-three years, but was not elected to this high office this time. In January 1975 the Railway Minister L. N. Mishra was killed by a bomb, an attack commonly seen as retribution for his role in breaking the railway workers' strike nine months earlier.[2] Four weeks later, the imam of the largest mosque in Delhi was arrested in February, with eighty-seven followers, under the notorious Maintenance of Internal Security Act (MISA), and this was followed shortly by an antigovernment riot in a zone of forced land clearance surrounding the Jamma Masjid in Delhi; such riots were sometimes organized, in the capital and elsewhere, by people "known to the authorities."[3] Removal and destruction of all the informal buildings on this state land was part of a larger pattern of elimination of "illegal" neighborhoods, benefiting individuals or groups who controlled them and received revenue from them. She feared a combination of militant and armed left forces, inspired by elements of the militant Naxalite movement that had successfully established itself far beyond the district of Naxalbari in northern West Bengal, where it began in 1967 (in fact, all armed opponents were called Naxalites by 1975). But these young men and women were far away and their threat containable by force and by MISA. Closer to home she certainly had two determined long-term opponents who had ideological disagreements with her as well as personal distaste; one was her old opponent Moraji Desai, who wanted an election in Gujarat and went on fast until death to achieve it, forcing the prime minister eventually to agree to an election there, which she then lost against his faction of the Congress Party. The other was Jayaprakash Narayan, a fellow prisoner with her father in the 1940s who addressed a big rally in Delhi in February 1975 asking the army and police not to obey orders that are illegal or unjust and leading a huge popular march against Gandhi's rule.[4] Because the Emergency affected the internal life of the scientific community and because perceptions of the Emergency affected the views of other governments on nuclear relations with India, we must examine why and how it came about a year after the explosion of the first bomb.

The State of Emergency was declared in late June 1975 when the Allahabad High Court ruled the parliamentary election of Indira Gandhi invalid because of procedural irregularities and corrupt practices in her campaign in her home base of Allahabad. Two weeks later, the prime minister's office ordered the arrest of her principal opponents, including her father's friend and contemporary, socialist Jayaprakash Narayan, former Deputy Prime Minister Desai, and people like L. K. Advani of the Jan Sangh Party, who would sit in the cabinet when she was defeated in 1977. These disparate prisoners formed a loose bond that enabled them to fight and win an election against her, reconnecting to an old Indian touchstone, the leader in a prison experience.[5] The prime minister was courting factions of the middle left and looked for and got support from the Moscow-backed Communist Party of India (CPI); to achieve this she outlawed the party that supported the Naxalites, namely, the "Maoist" Communist Party (Marxist-Leninist); meanwhile she ignored the Communist Party (Marxist) that was powerful in West Bengal. This kind of cultivated support was inconsequential for Gandhi in the coming election, though the suppression was often terrible for the opponents of the CPI in other communist factions and in other opposing parties. The Proclamation of Emergency did not end until eighteen months later, when an election called by Gandhi in early 1977 swung strongly against her and she was swept out of office in March, twenty months after the Emergency period began. By the time of her fall, most of the consequences of the 1974 nuclear tests had been felt, and political benefits (if any) reaped. In fact just before the Proclamation of the Emergency, her Congress Party experienced a massive election defeat in Gujarat, further proof to her that her enemies were working effectively against her. Regionally, Indira Gandhi saw a "foreign hand" behind the assassination of pro-India Bangladesh leader Sheik Mujib by army officers on 15 August 1975, three weeks after her declaration of the Emergency.[6] She imagined that a greater assertion of her authority would win the day.

Indira Gandhi's Big Risk

The ignominious end of her Emergency regime in March 1977 enclosed most of the consequences of the first nuclear test in Pokhran, except one. The new prime minister, her archenemy Moraji Desai, soon declared his objection to bomb-making and tests and then initiated a complete review of the operations and structure of the DAE. The protracted and unsuccessful negotiations with Canada and the United States were over, and those two countries were withdrawing from the Indian nuclear scene. The working relationships with France and Russia in space, reactors, and weapons had to be deepened. But two big and risky demonstrations—the bomb and the satellite—had been successfully achieved just before the Emergency was declared.

In most assessments of politics in India, there is little or no reference to the first nuclear test or its effects.[7] Following the tests there were criticisms, echoed by international comments, but there was also a buoyant self-confidence in large segments of the population for a few months. But when so closely followed by the Emergency, the beneficial political effects of the nuclear tests were lost in the ensuing drama about Gandhi's excessive power and violation of civil rights. The idea that prestigious scientific and technological projects (like bombs and satellites) were an essential part of the political culture of India, and thus influenced its direction (and vice versa), seems to have remained true only on the oral history level but less in the written record. The composition, role, and influence of scientific and technological elites did not change drastically in the Emergency because most of them could adopt a kind of strategic neutrality, silently accepting some of the personal gains and losses that occur in emergencies, appealing to the prime minister's office if possible. It would be misleading to exaggerate the punctuation effect of the Emergency among these elites, but it was not negligible. The nuclear tests did not float long on the surface of political narrative, except when used as a reference in the war over self-reliance. Then they sank into the nation's psyche, mostly out of sight.[8]

It is true that Indira Gandhi had imposed emergencies before, such as during the Indo-Pakistan war in Bangladesh in 1971 (an external emergency was technically still in force). But the 1975–77 period is the only one known as the Emergency, and it had no precedent in the national life of India. For eighteen months civil liberties were suspended, political parties outlawed, the judiciary was manipulated, the constitution altered, and the prime minister maneuvered a law to provide herself with legal immunity. The prime minister's office and its secretariat had been increasing its power relative to other cabinet functions by the time the Pokhran tests occurred in 1974, and the authority it asserted in 1975 with respect to the judiciary was a dramatic extension of that power. Though some people around the prime minister were influential before the declaration of the Emergency, unelected people picked by the prime minister gained even greater influence during these twenty months.[9] More curiously, her son Sanjay and his friends, for example, sought to dislodge her other advisors, who in turn resisted and tried to convince her of the danger of ignoring their drive for power.

Now, in June 1975, the Allahabad High Court ruled that Gandhi was guilty of electoral malpractice by using a government servant to work on her election campaign. Indira first thought she must resign because of this court judgment and discussed her resignation with trusted lawyer Siddharta Shankar Roy, then also chief minister of West Bengal.[10] He urged her to file an appeal instead to provide time, and the Congress Party orchestrated street demonstrations in Delhi in favor of Indira. Chief Justice Krishna Iyer then ruled that Indira could sit in Parliament but not vote pending her appeal.[11] The whole Emergency affair was premised on Article 352 of the Indian Constitution, referring to an "anticipation of internal disturbance" as the country appeared threatened by "armed rebellion." Gandhi described it to the president as "near anarchy": "The purpose of censorship is to restore a climate of trust," said Gandhi on the radio on 26 June 1975, reading a speech she drafted with Roy.[12] She claimed to be besieged from all sides: Maoist Naxalites and Hindu terrorists, Narayan's "fascist" movement (her words), some of it backed by "a foreign hand," by which she usually meant the CIA.[13] In July Parliament convened and both Houses endorsed the Emergency by a large majority provided by her loyalists within the party. But from now on only "urgent and important business" would come to Parliament, routine matters would be dealt with in her secretariat. The Emergency was popular for a few months—prices decreased, scarce commodities became available, buses ran on schedule, police confrontations or strikes did not occur, and so on. Indira plunged confidently into the crowds without security and was unharmed! She said that she intended to continue: "with the re-iteration of my party's confidence in me, and with the Congress continuing to command a clear majority in Parliament, my prime ministership was not in question."[14] Her confidence also stemmed from her sense that she could initiate actions that would be popular among "ordinary people," yet she would not provoke interference, like a coup, from the military chiefs of staff.

The Context for Scientists: Disaffection or Mischief?

In December 1975, six months into the Emergency, the government issued two ordinances prohibiting publication of "objectionable material": this definition included anything that "encourages or incites disaffection" or encourages "any class or community of people to commit mischief." A third ordinance abolished the Press Council (thus saving one of Gandhi's allies, K. K. Birla, then fighting to prevent the council from pronouncing its verdict against his paper *Hindustan Times*). This abolition further reduced protection for journalists. An effect of the censorship practiced during the Emergency, including in the vernacular papers, was that the space allocated to science news shot up over the previous period because science news and news about scientists, being defined as neutral, were not censored.[15] From June 1975, all newspapers had to accommodate censorship officers from the Information and Broadcasting Ministry who would look at all pages every day and would kill all news that was deemed to be influenced by the political factors that had prompted the Emergency. The only news that got through easily was science news, so newspapers used to keep science news as reserve files to fill up space, in case cuts in other matter occurred.

Many scientists remained sensitive to the curtailment of free expression, saying (or thinking) that free expression was essential to science and to the development of a vibrant, vigorous cultural life in which science could flourish. Classifying news about science as apolitical hints at how the scientific community was positioned; it was already nationalized, highly mobile, but without its own capital (its capital was intellectual and social, after all, except for those dependent on equipment). The fact is that thousands of scientists went on about their business in the Emergency period and said little about it. There was a continuing stream of scientists leaving the country for work abroad and returning home. Fear of punishment for criticism had been well established in everyone's mind, as Gandhi and her inner circle had intended and could demonstrate anytime.

But the social environment *had* changed: this deep intrusion of the government's authoritarian politics into many dimensions of ordinary life had so little precedent for the middle classes that even scientists and technologists were disturbed from their protected routines. Though the prime minister called the Emergency a movement for efficiency and unity, and her supporters insisted it was *their kind of efficiency and unity* that reinforced the rational spirit of science, there were skeptics too. Many scientists were attracted to order and regularity because their work invited it, so Gandhi's assertions about the rule of law may have appealed to them at first. But in the end these feelings were contradictory, scientists were divided, and the divisions were not concealed, even among those who supported her earlier efforts. The scent of excess pervaded the air. A year after the Emergency started, there was no evidence on which to base a forecast of its end. Speculation thrived, including rumors of dissatisfaction and unrest among some military officers.

This State of Emergency was simply too much for some scientists, in part because they tended to feel more strongly than others about the freedom of expression, or worked closely with those who did. They were uneasy because the Emergency introduced instability despite the present calm. Some scientists had experienced the excitement of the uprisings in the late 1960s in Europe and North America as graduate students and witnessed the ensuing suppression. In addition, politically active young men and women were held in "preventive detention" without trial across India, and some of these young people came from middle-class families that included scientists or were known to scientists. Counterintelligence penetration was deeper than usual. Conditions on university campuses became more polarized during the Emergency, and the surveillance and polarization were well known even among the privileged groups who worked at DAE installations like BARC or TIFR or SINP.

Coupled with the pro-bomb and anti-bomb difference that was emerging after May 1974, an implicit distance had opened up between people who were carefully observing each other's positions. Foreign intellectuals, civil rights groups, and Indians abroad openly criticized the Emergency, so much that India's prestige was lowered. Scientists with international mobility were more likely to be sensitive to international assessments than others. Her allies began confirming their moral rightness, having committed themselves to Indira Gandhi's Emergency. Not all outsiders were hostile: in 1976 she had favorable visits from Michael Foot, soon to be leader of the British Labour Party, and Margaret Thatcher, leader of the Conservative Party in opposition. Both returned to the UK and made supportive comments about Indira, based on their reading of the positive outcomes of the Emergency. Even people around Mrs. Gandhi were divided; two people who did not support the Emergency were her very old friends Fori Nehru and Pupul Gayakar; Fori Nehru, whose Jewish family escaped the Holocaust during the war in Budapest, broke with her husband Ambassador B. K. Nehru (Indira's cousin) and spoke to caution her that she had gone too far. And, despite concerned warnings from friends that it was useless to caution the prime minister, arts and culture leader Pupul Gayakar confronted Gandhi to no avail.[16] Scientists looked to these prestigious people for some intervention and sought uneasily for modes of resistance of their own. The loyalty of heads of scientific agencies and secretaries of technical ministries to the prime minister (Menon, Sethna, Ramanna, and others) provided an internal discipline that promoted self-control and self-censorship.

The Political Economy of the Emergency and High-Tech Self-Reliance

Gandhi said that she needed the Emergency to address poverty, curtail the activities of those plotting against the state, control basic commodity prices, limit private manipulation of the grain markets by traders, and stop high levels of tax evasion and currency smuggling. In practice the opportunities of the period were used quite variously by her office and other state agencies to crack down on tax evaders, to forcibly clear away slums built on public land in Delhi and other big cities, to increase the procurement of wheat from farmers for sale at fixed prices, to pressure men to have vasectomies in order to meet family planning targets, and so on. Her son Sanjay, his loyal party insiders, and their friends acted with impunity in the shadows. She appears to have calculated that though she would lose the support of the cities, she could gain more support in the countryside, where there were few friends of tax evaders and big smugglers. Her own view of the Emergency is to be found, surprisingly, in an ad she wrote herself and regularly published with her picture after the Emergency was under way: "This is a time for unity and discipline. I am fully confident that with each day the situation will improve and that in this task our people, in towns and villages, will give us their full support, so that the country will be strengthened."

Although "self-reliance" was the battle cry among most sections of the elite, and most particularly scientists and technologists, the government's key agencies were actually practicing a balanced form of dual dependency on the trading blocks dominated by Russia and the United States, a dependency consisting of aid contracts, multilateral loans, and trade for cash in armaments and strategic commodities. The modest effects of the 1974 bomb test can be seen in these sectors, but only briefly. Just before the Emergency was declared, in mid-1975, India's balance of payments deficit was over Rs 9.38 billion (Rs 4.38 billion in non-oil payments, Rs 5 billion in oil imports). The money that India was required to pay annually in interest in order to service its debt (known as "debt service liability") was Rs 6.01 billion. The debt and fiscal situation was defined as "very serious," and compelled a frantic search for new markets and cheaper oil, plus an attempt to stabilize the rupee, meaning "bringing the black market rate and official rate together."

Meanwhile, international aid donors were *actually quite supportive* after the 1974 bomb; a year after the test, the Paris Club decided in June 1975 to allocate about the same amount of aid as the previous year ($1.77 billion) and some countries (Britain, Sweden, the Netherlands) decided to discontinue soft loans and contribute only in the form of grants (a policy not unique to India; they were applying it to all their assistance). From the US and Canada there was also continuing commitment at previous levels (note that this Paris Club decision occurred in June 1975 just weeks before the surprise Emergency declaration). Though it had suspended nuclear cooperation, Canada maintained its full commitment to food aid and agricultural programs. Overall at the Paris Club there was a small 6 percent reduction in commitments in 1975 over the previous year, but the switch from loans to grants certainly made up for that. In net terms, after debt servicing, the aid transfer from the Aid India Consortium nearly doubled from $353 million in 1972–73 to $687 million in 1974–75, a period of poor relations with the United States.[17] The economic situation in 1975 was so serious that India was *importing* wheat, rice, and edible oil—all essential commodities widely produced in India—and these imported commodities were being used to control the price of the same internally grown or produced items. Importing food to manage domestic prices was one of the signs by which the public judged the government's performance; they liked the price stability but noted the vulnerability of the arrangement. Whether in heavy water for its reactors or cooking oil for its kitchens, India had become a high-priced (protected) market unable to produce enough to satisfy its needs.[18]

Just before the Emergency, in April 1975, high-level talks were held in Moscow about rupee-ruble exchange rates. This raised the importance to both countries of Indian expertise in new Russian aid projects, particularly the nuclear program. This coincided with the first Indian satellite being launched that month from a Tashkent base, though we do not know how many "Indian rubles" were charged for the launch. It was widely believed that Indian contributions in Soviet assistance projects in India were systematically undervalued and thus inadequately compensated.[19] As neither the ruble nor rupee was traded internationally, the rate of exchange for the ruble and rupee was fixed by Moscow, but viewed by Delhi (and Indians who understood the issue) as arbitrary. A new trade group directed India's relations with the USSR during the Emergency, namely, T. A. Pai, minister of Industry, P. N. Haksar, deputy chairman of the Planning Commission, and D. P. Chattopadhyay, minister of Commerce. Apart from C. Subramaniam, now minister of Finance, these were the most powerful people around the prime minister. Subramaniam, moreover, was in charge of improving relations with Washington in spite of the vilification of the US in the Indian press: judging from the flow of aid commitments, his efforts were partially successful, although the nuclear negotiations were not.[20]

In mid-1975 the Reserve Bank of India reviewed the rupee's value in hard currency markets. Since it had been pegged to the British pound for three years (June 1972 to June 1975), India was now required to repay its debt in pounds to Japan, the US, France, West Germany, and others, and the value of the pound had slipped 27 percent in relation to the US dollar in those three years, the yardstick currency for those other lender countries. Indian experts pointed out that the "gold content" of the rupee had remained unchanged since the rupee's devaluation in 1966. Debt repayment thus cost India more than when the debt was incurred. This slide of the

British pound also meant an effective devaluation of the rupee, because though the devaluation encouraged an increased purchase of India's exports, it also sharply increased the cost of all imports, among them oil. Britain, not Portugal, was now being called "the sick man of Europe" and India experienced the illness too. The search for foreign exchange by managers of advanced projects now became more and more competitive, and leading importers like the Ministry of Defence were put under even greater scrutiny. The DAE itself was entering a more difficult period.

Science, Technology, and the Energy Paradox

Gandhi and Haksar did not involve the Planning Commission very deeply in their new strategy, and the commission, like the NCST, did not meet regularly in the period just before the Emergency. Haksar, deputy chairman of the Planning Commission and chairman of the NCST, said that it was now essential to capture the value of investments already made in science and technology but that true efficiency depended also on closing down failing programs. The Planning Commission therefore prepared a position paper on the importance of technological self-reliance for the Parliamentary Consultative Committee on Planning, bravely trying to counter the Ministry of Finance's proclamations of success in obtaining more aid at the Paris Club.[21] This policy paper discussed what the Planning Commission thought was the "encouraging progress toward self-reliance" and said that real import substitution was now under way, meaning authentic self-reliance (partially through blocking imports). Talk of closing down failing programs alarmed scientists because the livelihoods of some of them hung in the balance, and there was little precedent for closing programs. To calm their anxiety, Gandhi told the Science Congress in January 1976, "The adjustments in the science and technology plan we are compelled to make will be more tactical than strategic. The core structure of the plan will be protected."[22] Nevertheless, a thorough review was conducted in the public sector, looking for new places to cut, and in the Emergency each of these places and each of these cuts inevitably had a political color.

The Emergency coincided with the arrival of three positive pieces of news for the government: the successful launch of an Indian satellite by the space program in 1975, linked with the start of national television programming with some Indian content, and the announcement of the first offshore oil flowing from Bombay High in 1975. Because these three projects had resulted from government initiatives in science and technology, and seemed to benefit ordinary people, all three were celebrated because they embodied science in its more familiar technological forms and symbols. All three had spectacular appeal to the middle class, and to their leaders. All three embodied precisely what Nehru originally meant when he said, "We must make friends with science." Though the Bombay High offshore oil was not yet petrol in the tank, all three projects had beneficial consequences in the economy, unlike the bomb. As the satellite and television were organized by the DAE and Space, which also spawned the space program, they both helped slightly to deflect the difficulties that the DAE found itself in; equipment fabricated in Canada for the second Rajasthan reactor and for the Madras power reactors was suspended but the firms were still under contract. Heavy water was proving hard to get.

The bomb test briefly enhanced the public status of some scientific experts, but it did not alter the lives of ordinary people who were more interested in batteries for radios and kerosene for night lamps than in nuclear explosions. They looked around and envied those who enjoyed electricity. India's experience with thermal power and electronics suggests that the struggle over self-reliance was by no means suspended during the Emergency. Given the small contribution of atomic energy to the electrical grid, the high price of oil, the intense search for new power sources, and the demand for power from new industries like fertilizers and heavy water, it is hardly surprising that conventional sources of electricity had not lost their attraction. Thus coal-burning "super-thermal power stations" were placed high on the agenda during the Emergency, notwithstanding that they were well-known sources of pollution (so acute was the shortage of electricity that they would also have been considered favorably had there been no Emergency). Meanwhile massive cities like Bombay and Calcutta experienced regular electricity blackouts; though key projects like the big MiG jet fighter project in remote highland Orissa were supplied regularly, ordinary people were not.

The nationalization in 1973 of noncoking coal for thermal power generation was the personal policy of Mohan Kumaramangalam, minister of Steel and Mines.[23] Much of this soft sulphurous coal lay in eastern and southern India, and a large deposit was being mined in open-pit at Neyyvelli in Madras State, where Kumaramangalam had his power base. In fact, the nationalization policy had been initiated in October 1971, when 214 small coal mines (notoriously vulnerable to accident) in West Bengal and Bihar were nationalized, giving the government the upper hand in lowering the price of thermal power generation. But Minister Kumaramangalam died in a plane crash weeks after carrying out the program of nationalization, removing the most effective person to pursue the policy at the political level. Nevertheless, generation of electricity from coal rapidly became the preferred solution for some regions of India, and at least 40 percent of India's rail traffic was carrying coal across the country; for example, the coal-based thermal electric station at Ennore in Madras State was using coal mined on the Bihar-Bengal border, more than 1,500 km away. Railways carried coal westward and southward from eastern India at a favorable rate under a freight equalization policy. No nuclear reactor was planned to be constructed near these coalfields. Electricity shortages were viewed as a most serious obstacle to both industrialization and government business; for example, the electricity consumption for Calcutta was about 600 MW on a "normal day," but by April 1974 actual supply was about 300 MW, even though Calcutta was very close to the coalfields in question. As part of the nationalization and modernization program, just weeks after Premier Brezhnev's 1973 visit to India, the Russian minister of Mines led a delegation of Soviet experts who promised to redevelop three coal mines (Singrauli in Uttar Pradesh, Jhajra in West Bengal, and Kusumunda in Madhya Pradesh). This visit came immediately after a Polish delegation toured the mines, thus extending that country's involvement in two other coal mines. Both Poland and the USSR had large coal deposits themselves and relied heavily on thermal power generation (and on thousands of miners). This foreign activity in Indian coal followed a historical pattern, as the coal industry was one of those most deeply penetrated by British management in the pre-1947 period, so there were strong sentiments about foreign involvement, particularly about payments for foreign expertise. Again, Indian experts questioned whether it was necessary, stating that Indian capabilities had reached the point where they could take much more responsibility. In the end, when nuclear reactors delivered so little electricity to the grid system, low-grade soft coal was massively used to fill the gap.[24] There was so much production and consumption that the safe or even profitable use of coal-ash became a national preoccupation (the production of greenhouse gases was not discussed at the time).

While criticizing the poor performance of nuclear power, the NCST promoted thermal power again in 1974–75. Informed by a team of international consultants, the Ministry of Energy reported in late 1975 that 500 MW coal-burning thermal stations would be more economical than smaller stations, but the larger station would have to be imported. The Ministry of Industry, which owned the firm Bharat Heavy Electricals, Ltd. (BHEL), assured the Energy Ministry, the Planning Commission, and the cabinet that

BHEL could supply thermal stations of 200 MW on an immediate turn key basis, while they agreed a 500 MW station would have to be imported.

Supporters of the Indian 200 MW stations included "eminent scientists and power engineers." Even Homi Sethna and the AEC were reported to support BHEL in its proposal to supply 200 MW stations on a ready-to-go basis. BHEL announced that it could supply 200 MW stations now, and if it could acquire the necessary licenses and collaborative agreements, it would also start to learn to build the larger stations. "This was the position until the World Bank appeared on the scene," wrote an informed observer. The bank, backed by Minister of Finance Subramaniam, who was then in Washington, was immediately forthcoming about financing these stations but pushed harder for the larger imported stations on a turnkey basis. The rationale given for the bank's support was that these larger imported stations would save 20 percent in electrical generating costs, *and* they would be financed by the bank on the stipulated basis that international tenders be invited from suppliers with proven experience. BHEL meanwhile was already building a thermal power station in Iran, worth Rs 1.5 billion, but it looked on helplessly while its domestic role was being bargained away. A more complete illustration of the energy paradox in the war over self-reliance could hardly be found.[25] And the DAE and AEC, which were responsible for nuclear power, knew that there were big new fertilizer plants requiring large amounts of electricity before the cabinet for approval, to enlarge the scope of the green revolution.[26] The AEC knew that their reactors could not supply that electricity now, and they also knew that those fertilizer plants would also produce heavy water for *their* reactors. Though this occurred before the Emergency, these conflicts set the stage for questions during it; with the suspension of Canadian reactor assistance, what were the reactors doing? Where would the electricity come from?

The two Tarapur reactors were online continuously for two hundred days in 1974 before being shut down for refueling, and though refueling took a long time, there was satisfaction with the performance, even if it was lower output than planned. After the first bomb test, however, there were more and more negotiations with the US about Tarapur's spent fuel and spare parts, in which GE and the US Atomic Energy Commission were seen as delaying to cause inconvenience. Although the boiling water reactors built by GE elsewhere were eventually closed, the Tarapur reactors were maintained for years, although at 40 percent lower output than they were designed for. Then, as operations became routine and streamlined, according to a senior DAE engineer, "in due course many Indians who had worked on the construction or operation of Tarapur emigrated to the US to work at nuclear power stations and for the safety and regulatory authorities. Thus the US nuclear community was enriched by the addition of Indian engineers and technicians."[27] These technicians could see that heavy water reactors, not light water reactors, would dominate India's future for the next twenty-five years; they could also see an opportunity in America.

Feeling the pressure on DAE to produce electricity, M. R. Srinivasan, director of BARC's power reactor division for the DAE, predicted in January 1976 that an installed nuclear capacity of roughly 20,000 MW would be ready by the year 2000.[28] (See Srinivasan's surprising conclusion in the next chapter; Sarabhai predicted a nuclear capacity of roughly 2,700 MW by 1980.) But the Canadian nuclear program suspension, the American slowdown in delivery of enriched uranium fuel for Tarapur, the chronic difficulty of getting heavy water, all these were already having cumulative effects in a country where the installed nuclear capacity in 1976 was not yet 1,000 MW. Speaking for many Indians, Ramanna said that this was a good reason to change India's policy. "As a result of the pressures exerted by the West vis-à-vis the [nonproliferation treaty], India could not stick to agreed schedules for the construction of nuclear reactors it had already decided to build. All the troubles we faced were a heaven-sent opportunity for the indigenous industrial sector."[29]

Opportunities for Scientists and Technologists in the Emergency

When Prime Minister Gandhi told scientists in January 1976 that "the core structure of the plan will be protected," she was at the height of her power. Censorship was well established, and political challenges to her were not well organized at this stage. Though some scientists and technologists were among the soft opponents of the Emergency, we have no study of their views or strategies in dealing with it. Moreover, the support for DAE's bomb was undiminished, at least up to the end of the Emergency. According to Dixit, "Whatever the exact motivations for the [first] test, the immediate political result was an increase in support for India's nuclear policy among domestic audiences. Support for the government's nuclear policy became an accepted fact for all political parties."[30] This support did not, however, include the Desai-led Congress faction that opposed her, and they called the bomb a backward step when they were on the verge of taking power from her. This call resonated with Gandhian nonviolence traditions in the opposition coalition and with some leftist party leaders who preferred a focus on socioeconomic struggles instead of dubious technical demonstrations.

Gandhi's core structure of the plan included reactors and all the other DAE projects in physics, space research and development, missiles, the cyclotron, and computers and electronics. These indeed were to be protected: some said overprotected; others said not protected enough, depending on where they stood with respect to the war over self-reliance. The desire to possess and test nuclear bombs was now a subject of open conversation. Ministers and their aides and allies from now on were ranged on a spectrum of support for or opposition to that bomb. If some people knew that there were discreet preparations being made for another test, they understood that there was not yet any commitment to test. Indian officials including AEC experts were presenting the argument at Vienna and in other capitals that the first test was a peaceful nuclear explosion permitted under the IAEA regulations.

The variable energy cyclotron surged ahead to completion during the Emergency. It finally had magnet power supply and other high-current units installed in the shed on the Salt Lake site in Calcutta only in 1972, and thirty-three large power supplies were fabricated during the next three years. Testing of these power supplies posed a great challenge, as the stability of the 2,000–3,000 amp had to be monitored over a number of days. The initial electrolytic tank was found not to attain sufficient stability and thus proved useless, causing further delay. Steel wire resistance banks were later installed and the required precision achieved in 1974. The electrical supply situation in Calcutta was dismal and blackouts were frequent, so when the cyclotron was installed in 1975, the central authorities were forced to install some captive power generators exclusively for the machine. But a dedicated electric line was laid directly, after mid-1977, to the VEC from the coal-burning power plant owned by the Calcutta Electric Supply Corporation at Titagarh. The cyclotron, a DAE national facility, was named the Variable Energy Cyclotron Centre and achieved its first external beam in 1977.[31]

The other important projects rolled along, their "dual use" potential now in full view. Satellite launch rockets helped the missile project, as space rocket builder Satish Dhawan noted in retirement in 1998: "with minor variations—what's the damn difference?"[32] Satellites and remote sensing made groundwater management more effective (or should have) and also made surveillance more effective on the frontiers with Kashmir, Pakistan, and China. Supercomputers helped radio astronomy and also more accurately simulated the trajectories of rockets and missiles. Small electronics satisfied consumers and civil agencies of the state and also helped the military communicate more effectively on the ground. The number of opportunities to wage the war over self-reliance multiplied, and the tension among scientists and technologists over the choices to be made did not diminish during the Emergency.

But great power disapproval of the Indian bomb *was* a major inconvenience to India's energy program, and though official ways were found around that inconvenience with the help of the French and Russians, reactor projects in Rajasthan and Madras were delayed. Hearings and conditions imposed by American officials on the shipment of enriched uranium between 1975 and 1978 certainly delayed the Tarapur reactor's full development and raised its cost. At least since 1974 the US had been looking for the means to relieve itself of the contractual obligation to supply enriched fuel to India (a long contract for the life of the reactor, signed in 1963). The US thus quietly cultivated "the French solution" to its enriched plutonium problems. India and France were already very close in building the fast breeder at Kalpakkam, as Ramanna explained: "We decided to follow the lead of the French with whom we had begun collaborating; they had used mixed oxides of enriched uranium and plutonium and had developed fast reactors as large power breeders."[33]

But the civilian power reactor construction business was slowing down in such a way that it even slowed construction of Dhruva, India's next plutonium-producing research reactor; according to Ramanna, this occurred about 1979 but other sources say the slowdown began earlier, in 1977–78: "The construction of Dhruva had lost its pace after my transfer to Delhi," said Ramanna, noting disagreements between the designers and the builders of Dhruva.[34] In its precritical tests "it had some teething problems and we had to face the criticisms of the newly formed Regulatory Board which had no knowledge of the complexities involved." When he returned to BARC, he faced newspaper reports weekly that he said "were in bad taste"; also "initial small problems were exaggerated, and despite our rejoinders the reports continued to appear and were strangely enough given credence in Delhi."[35] Ramanna was not alone in this irritation about the intrusion of others into the affairs of the nuclear estate, and even his rival Sethna got publicly angry when journalists reported a leak at the Tarapur reactor in March 1980, calling them "irresponsible self-styled experts."[36] This suggests a conclusion that well before Indira Gandhi's return to power in 1980, atomic energy policy and procedure were no longer the exclusive domain of Bombay-based scientists. Media were now less generous or celebratory, and a skeptical and critical "peoples' science movement" called for appropriate technology that did not include nuclear reactors. The antinuclear era had arrived in India. Criticism of "political scientists" began as early as 1970 and sharpened by 1975; there were smart people who thought that some of these political scientists were not unique and that like political doctors or political agricultural experts (modernizers all), they too had tunnel vision and were entrapped by decisions made five and ten years before. They were not, critics thought, more wise nor more capable than others, nor more far-seeing. It was conceded that they knew a great amount about a particular subject (their research, their publications) that was uncommon and difficult for anyone else to know. But for the rest, they had, by the 1970s, become "rather like us."

Heavy Weather for Heavy Water

Well before the 1974 bomb test, the DAE knew that heavy water was in short supply and that Indian production would not be sufficient for years. Bhabha's commitment to the natural uranium reactor now had a growing but unsuspected problem built into it. India's first heavy water plant, built by a German firm beside the Nangal dam and fertilizer plant in the 1950s, from its commissioning in 1962 was producing about 11 tons of heavy water per year in 1971.[37] Promises of heavy water from Canada, promises linked to the CANDU negotiations, were still unfulfilled by 1970 because the Canadians could not achieve constant production at their plant at Glace Bay in Nova Scotia. The aging Nangal heavy water plant was planned to be shut down in 1975. Construction of a new heavy water plant using a BARC-developed hydrogen sulphide–water exchange process began at Kota in 1969, next door to a Japanese fertilizer factory. But delays meant that the new plant was only half built in 1974, and first production was further delayed until 1977, even after huge extra expenditures. Because of this problem with the BARC process at the Kota plant and the age of the Nangal plant, three turnkey ammonia-hydrogen exchange plants were ordered in the early 1970s, beginning with a French project at Baroda, a Japanese project at Tutticorin, and a project at Talcher with West German collaboration. These projects all involved scarce foreign exchange and enriched foreign contractors, yet none was expected to produce heavy water until well after 1977, and, together with BARC's Kota plant, they had already cost Rs 1 billion by 1974. Said one insider, "The cost of Indian-produced heavy water is around Rs 1000/kg whereas the import price is around Rs 800/kg. Even this Indian price is full of hidden subsidies, such as in the form of low interest capital."[38]

Before the first nuclear bomb test, the demand in India for heavy water by the late 1970s was forecast at 1,000 tons annually, but Indian production was clearly not going to meet that demand if the Nangal plant was closed. Importation of small amounts of heavy water had already occurred from USSR through West German broker Alfred Hempel starting in 1973, and more small shipments had been ordered through him. Shipments of small amounts did not trigger IAEA safeguard inquiries. Prime Minister Gandhi herself warned of a shortfall in heavy water supply in September 1974, because of the loss of German equipment for the Talcher heavy water plant in a shipping accident at sea in heavy weather.[39] It is unusual that a prime minister would warn the nation of the result of a shipping accident, but it was her portfolio and she was, four months after the bomb test, drawing political warmth from its radiance, so to speak. Though a serious loss, this accident probably provided cover to pressure the Russians in heavy water negotiations.

According to Ramanna, "Because of the friendly relations that existed between the two countries, the USSR suggested that a modification be made in the agreement which was to allow for the supply of heavy water to the Rajasthan reactors . . . [and] inspection was restricted to the reactors using heavy water from the Soviet Union, and was to be carried out by the IAEA."[40] The initial commitment in September 1976 was for 200 tons, but Russian sources suggest that small quantities of heavy water (100 tons) had been sold to India as early as 1973, also through a West German broker.[41] The formation of the London Club in late 1974, to monitor and restrict nuclear exports, appears not to have greatly affected either India or Pakistan in the three or four years following its formation, though there was indeed an increased sharing of information about India and Pakistan among club members, a purpose which had prompted the formation of the club in the first place.

By the time the Emergency was over, the Canadian contracts in India had been irreversibly cancelled, and the US uranium shipments to Tarapur were suspended while the Americans bargained with the French to reassign their Indian enriched uranium delivery responsibility to France (though that was not the reason given publicly). These consequences had slowed Indian nuclear development down, but this hiatus in foreign cooperation had a stimulating effect internally, both within the DAE and between the DAE and the private industrial-engineering sector.[42]

Scientists Adapt to Post-Emergency Expectations

The eleven-year Indira Gandhi era in Indian politics ended with her sound defeat in elections in March 1977 and the arrival of her archrival eighty-one-year-old Moraji Desai in the prime minister's office. Though Indira's Congress Party obtained 35 percent of the popular vote and still had 153 seats out of the total 540, the new consortium government had obtained 43 percent of the popular vote with 270 seats. The new prime minister also moved into Indira Gandhi's official house, which he forced her to vacate hurriedly and unceremoniously.[43] "Old Congress" Desai soon announced that India would have nothing further to do with a bomb project, although he himself was aware that planning and calculations for another test were indeed proceeding, just in case the cabinet should change its mind in altered circumstances. Knowing that Indira Gandhi had not been so keen on a bomb test during the three preceding years, Prime Minister Desai railed against the bomb test from his first days in office, asking with sharp rhetorical irony what it had produced? This is probably the only case in the world in which a prime minister or president repudiated the work of physicists and engineers on their bomb achievements; India was proving itself rather open-minded on the question of the bomb. Nevertheless, Moraji Desai, an advocate of nonproliferation and opponent of another bomb test, learned soon after his election that an all-party consensus had been established around the nuclear nonproliferation treaty, so, standing before the UN General Assembly in 1978, he confirmed that India would not sign it. Desai also repeated and confirmed his support for "the core structure" in spite of his opposition to weapons development in India; "We ask from others no more than the self-restraint we impose upon ourselves. But our objection to the Treaty is because it is so patently discriminatory." As part of his balancing act, Desai publicly scorned the scientists responsible for the first bomb test, saying it caused trouble and taught India only not to do it again. He sat at the cabinet table with ministers who were committed to the bomb, including the rightist Jan Sangh Party (Vajpayee in External Affairs, Patel in Finance, Advani in Information and Broadcasting) and some members of his own Congress Party. Yet privately he had to review and reconsider his acquaintance with key members of the scientific community and to listen to BARC and other labs, as he himself was also the minister responsible for the DAE. To end dependence in the imported fuel cycle, Desai soon agreed that the DAE should "produce mixed oxide fuel with plutonium from our other reactors, a plan that would have discontinued our reliance on fuel supplies from outside."[44] To repeat Perkovich's lovely image, Indians and their leaders found themselves "in a mirror-lined box."[45]

The new prime minister also made meaningful symbolic changes in the scientific community, not only moving Ramanna out of BARC to the Ministry of Defence in Delhi, but giving seventy-year-old Atma Ram a new postretirement career as the chair of a revived NCST and a member of the AEC. Desai knew DAE finances intimately, having been the minister of

Finance in the 1960s. The same "old-fashioned" characteristics that made the new prime minister a figure of fun sat well with Atma Ram, also a strict nondrinker. Ram had never enjoyed popularity with many scientists in the labs because he was considered "old-fashioned" and did not have a first-class reputation for his own research work as a physical chemist. The idea of his election to the Royal Society had not gone anywhere past a letter to Blackett in 1970. He knew very little about the nucleus or nuclear reactors. Visvanathan, who interviewed him, points out that Ram "began as unpaid panboy in a sugar factory" and "never wrote his autobiography."[46] His importance was as an organizer and institution-builder, starting under Meghnad Saha's gaze in Calcutta from the 1940s: they both understood poverty. Nevertheless, he had the wisdom of an "old hand" at the CSIR and had the new prime minister's confidence—so much so that he was put *in charge* of a review of CSIR's old opponent the DAE; here he skillfully and quickly implicated a nonscientist with great strategic and political influence, K. Subramanyam (see chap. 22).

Atma Ram, now also an AEC commissioner, said that projects like radio astronomy and space research were done more for prestige than for socioeconomic relevance. On the other hand, Ram knew the value of the traditional export commodities (tea, jute, textiles, leather) and promoted applications of science to those unfashionable ends, with a keen eye on the reapplication of traditional skills. So, having long ignored him, *Science Today* devoted four pages to his ideas in early 1978, quoting from his September 1977 lecture given when he received his new appointment as the NCST's chairman. Here he distinguished himself sharply from C. Subramaniam and P. N. Haksar, who had influenced the NCST from its beginning and who had gone with the fall of Mrs. Gandhi's regime: "NCST should not . . . get entangled in considering individual research projects and schemes, reducing itself to a sort of research committee. Nor should it get involved in updating the annual plans, sectoral or otherwise, which the COST and the earlier NCST had done for historical reasons like having a Chairman with official responsibilities of planning."[47]

Illustrating what the new post-Gandhi climate was like, Ramanna did not disappear in Delhi. He was, after all, an important secretary of the government for Defence. He surfaced again as a member of the NCST, of which Atma Ram was chair (two less friendly committee members can hardly be imagined); he also rediscovered the public eye as president of the Indian National Science Academy, the one founded by Saha and then moved to Delhi. Ramanna also sought *Science Today* to present his views on the proper organization of academies and improvement of working conditions of scientists in the labs.[48] Ramanna complained that few of his DAE colleagues came to see him, and given the disapproval of the DAE by the prime minister, one can hardly blame them, because Ramanna was definitely a very "political scientist." The polarization around the bomb and the DAE had settled down to this functional and personal level.

The awaited review of the DAE eventually started: according to nuclear strategist and senior official K. Subrahmanyam, after coming to power in 1977, Desai asked retired CSIR head Atma Ram to conduct a thorough review of the DAE with the objective of reorganizing the department.[49] In an ironic era, a more ironic choice could not have been made, considering the skepticism with which most DAE scientists had viewed Ram and his history with the CSIR. Because of his reputation with DAE scientists (and many CSIR scientists too), Atma Ram was advised to secure help for this review from K. Subrahmanyam, who in turn immediately obtained the services of a retired insider, N. S. Siva, former secretary to AEC chairmen like Sarabhai. Subrahmanyam notes that the prime minister kept himself informed at each step in this review. At some point in this process Ram got himself appointed as a member of the Atomic Energy Commission itself, so better to "diagnose the patient. "While Raja Ramanna was being moved to Defence from Atomic Energy in 1978, according to Subrahmanyam, the DAE and AEC secretariat "made an elaborate and fairly persuasive case of Ramanna's alleged acts of indiscipline and defiance." On the other hand, *some* of the senior scientists working at Trombay, Kalpakkam, and Hyderabad asked reviewers Siva and Subrahmanyam to record their wish to have Ramanna returned to the DAE and BARC. AEC chairman Homi Sethna, however, told Subrahmanyam that he wanted the BARC directorship filled by someone other than Ramanna, and Desai said he "could not stand" Ramanna. Given all these factors, Ramanna did not return from Delhi until 1981, after Indira Gandhi had returned to power, to take up once again his old role as director at BARC.[50] Desai and Ram meanwhile planned the deconstruction of DAE into separate subject-related bodies, created a regulatory commission for atomic energy, and subjected the DAE to the auditor general's scrutiny. But there was insufficient time or perhaps too much infighting in Desai's government to achieve all this.[51] Meanwhile the DAE system kept its head down and carried on with business, counting on a government composed of a coalition of unfriendly political parties, to keep outsiders from interfering too much in its business. There was still a nucleus of "political scientists," from which some retired and to which others were gradually admitted. These people and their institutions continued to draw resources and confidence from their association with the uranium nucleus, but the country's commitment to technical and scientific activities that had nothing to do with uranium and atomic energy was growing, not diminishing. When Indira Gandhi returned to power in 1980, she continued many of the initiatives started during 1977–80 because of the sheer momentum of a large, complex, and growing scientific community that had already absorbed heavy state investment.

But what of the state of play in the growth of nuclear power generation? A former chairman of the AEC, M. R. Srinivasan had forecast in 1976 that the AEC would achieve 2,800 MW output by 1980. But even in 2002 he wrote that "one cannot conclude that nuclear power is securely established in India." Why would he make this surprising statement? He reasoned that in 1947 India had installed electrical capacity of 2,000 MW running through a host of locally owned and controlled grids, beginning with hydropower in Mysore and Madras, then Tata Electric Company in Poona and Bombay, then thermal power near Calcutta. After 1947 states set up electricity boards, leaving the private producers largely alone except for price regulation. Then in the 1960s thermal coal-burning plants became more important, and the electricity boards tended to be directed by thermal power engineers who had a technical understanding of the system. In 1956–59 the DAE studied these grids to forecast the effect of introducing nuclear power into them. On the basis of this study, the DAE subsequently chose western India as the electricity grid best suited for a reactor. Western India was also farthest from the coalfields.

In the 1960s thousands of irrigation pumps required electricity for agriculture, and these pumps were at the heart of the green revolution and therefore very political. By 1970 politicians and civil servants had taken full control of the state electricity boards, and this is partly why most boards gave electricity free, some institutions did not pay for electricity, and states, municipalities, and state agencies and corporations did not pay their electricity bills to the central government, according to Srinivasan. These state electricity boards "were never given autonomy but were run as subsidiary offices of the state governments. Over time they suffered from overmanning and a poor work culture, corruption was rampant, and there was a total absence of professionalism. Given the basic untenability of the tariff structure, they sank into bankruptcy." This is when the reactors began to deliver their first electricity to the grid, but it was not an integrated and uniform national grid at all. It was legendary in the 1970s throughout India to see a tangle of wires on the power poles tapping and delivering "free" unrecorded electricity to houses and businesses.[52] In this context Bhabha had been "supremely confident of taking on a major, first-of-its-kind job, quite unmindful of the realities. . . . [Had he lived longer] he would have realized it would take us considerable time to overcome our lack of industrial preparedness." Reactors were planned and built, but, according to Srinivasan, "the fact that the reactors are small in capacity leads to a modest aggregate power capacity."

Regarding Vikram Sarabhai's passion for the 500 kW heavy water reactor and joining with the French, who were not going to sign the nonproliferation treaty, Srinivasan said, "the power systems in India were not ready" for bigger reactors. In 1980 the DAE set a target of 10,000 MW in 2000 based on nuclear power, but actually DAE realized only 2,800 MW in 2000 in a total supply at the end of the century. This was the very level Srinivasan had forecast in 1976 for achievement in 1980! How could an expert who was later to become the chairman of the AEC have been so wrong in 1976? In 2000 there was a total electricity supply of 100,000 MW, increased fifty times from 2,000 MW in 1947. But, he said, "there is a feeling of inadequacy that, despite all our efforts, we have realized a programme of a mere 3000 MW . . . one cannot conclude that nuclear power is securely established in India. Only when we have a programme of 10,000 MW can we expect it to be self-sustaining."[53] So the difficulty with nuclear electricity lay not just in the low-quality uranium available, the choice of reactor designs, the lack of heavy water, or the costs and delays, but also in the feeble electricity grids themselves, the way they were run, and the policies they followed. This all helps, though perhaps not completely, to explain why after so many years of expensive work, nuclear reactors contributed less than 4 percent of India's total electric supply at the end of the twentieth century.

TWENTY-FIVE

Conclusions

This large story interacts with so many other subjects that my generalizations here may stand only until we see the work of other researchers who will surely follow. Before "the stroke of midnight" in August 1947, good scientists in India experienced many years of competitive individualism. They saw the search for individual and/or local solutions to problems, but in the absence of scientific institutions (particularly national institutions); to them the solutions lay in the establishment of dozens of labs and institutes in various fields, usually supported by local state and private capital, eventually publishing journals in their field. Led by scientists who had established international reputations, particularly through but not limited to the Royal Society, men who enjoyed recognition from other Indian and British scientists, these institutions began journals and articulated their interests through academies and the Science Congress. Awareness of their work increased in Japan, the United States, the Soviet Union, and Europe. Some of these leading scientists like K. S. Krishnan, S. S. Bhatnagar, C. V. Raman even carried this recognition to the privilege of knighthood. Though competitive, these initiatives were held loosely together by a common "nationalist" discipline that articulated a rhetoric of self-reliant science and technology, anchored by a mutual recognition of the others' achievements in international science. Although their FRS status was a convenient signification of such achievements and an important step toward national prominence, they all had to approach the same few sources for financial support. So their individualism was necessarily competitive.

Onto this complex map of various institutional arrangements, at the same moment, two large national schemes were imposed in 1945–47 using big money and full international support. Large laboratories for entire fields were planned and built and, sometimes, directed by foreigners. The break with the UK did not diminish the importance of the Royal Society or British influence, and soon prewar networks through Germany and France were revived too. For the DAE and CSIR, these connections all evolved in parallel through the 1950s and 1960s until they were eclipsed (but not replaced) by American and Canadian networks, and then Soviet networks in the late 1960s. By 1980 India enjoyed—or was weighed down by, depending on who is speaking—a myriad of international agreements, commercial obligations, and sovereign undertakings through its high-technology economy, still buoyed up by very significant flows of international aid, the terms of which tended to benefit donors as much as, occasionally more than, recipients. This part of the economy was supervised closely by scientists employed by the state as managers and offered limited scope for private enterprise. Though the system offered scope for "competitive individualism" (e.g., Sethna vs. Ramanna) in terms of eligibility for the top positions, it was in the 1970s, as it always had been, tested against a mix of political party and regional criteria closely regulated by the prime minister's office. Few heads of state could have been involved in scientific leadership issues and technical choices as much as the first prime ministers of India.

It is important not to overgeneralize and overinterpret the situation of these top "political scientists." Each had their distinctive base in institutions, each with a different ecology. Though competitive individualism was open to everyone in the mid-upper ranks of the scientific community, and beyond "patriotism" there was no single and simple mobilizing ideology in which anyone could claim superiority. Enthusiasm for the nation was acceptable, if it was not overdone, and a vision of globalized capitalism was not yet trendy, even by 1980. International communism was a well-understood ideology, but others already occupied that space, and anti-communism had little support except on the intellectually unattractive far right of the spectrum. Few leaders articulated a clear vision of what became fashionable later, namely, adherence to a Hindu-India nationalism of the variety that flowered in the late 1980s, known to scientists in their youth as Mahasabha or Rashtriya Sywayamsevak Sangh or Arya Samaj, and then Jan Sangh. With a few exceptions that movement toward power was not interested in the conceptual or professional base of the sciences and was interested in technology only to show what Indians had done before others, long ago. So, apart from articulating a world reputation of their work and their institutes, the idea of "national self-reliance" was the mobilizing call for scientists, leading to a drawn-out disagreement and contest about how to achieve it.

International Networks, Diaspora Scientists, and Research Institutions in India

Indian scientists in larger and larger numbers learned from the 1920s that it was not just what you knew that mattered, but who you studied with, where you published, who patronized you, who would recommend your work for publication, and where and when your ideas were recognized. They learned that whether or not there was a Raman effect or Saha equation or bosons or a stellar space probe named *Chandra* depended on very wide appraisals of research, in the most widely circulated media like *Nature,* for example. The production of reputation and prestige turned on social recognition of an individual's hard work; the fascinating and tragic story of the gifted but ignored mathematician Srinivasa Ramanujan[1] showed Indians that individuals could and did accomplish exceptional things, even while working mostly alone. Nevertheless, it was mathematicians, including British mathematics teachers, in India first and then in Cambridge, who opened a door to full recognition of Ramanujan's abilities. From the histories shown here, we conclude that many gifted young people received such recognition, but other scientists remind us that they did not have a supportive environment in families and schools: in short, they *felt* alone. The realization as young adults that they had to create reputations by initiative and effort *and* sustain such recognition raised for young scientists the importance of effective networks. In most cases this required enormous determination, even greater before Independence.

In comparison, we have the report of Pysenson, who compared three colonial scientific traditions, in twentieth-century physics and astronomy—Dutch, German, and French—examining their research contributions and testing them along three dimensions on the ground in their colonial societies: mercantilist, military, and functionary-administrative.[2] Each tradition involving scientists of the three countries did produce good work, which differed in terms of its accommodation to one or two of these imperial dimensions, but each differed little along the research dimension. Pyenson also examined these traditions for the degree of control from the imperial center, where decisions about the employment of scientists usually occurred, whether they worked for companies, bureaucracies, and/or militaries. He found that the scientists reluctantly, "even resentfully," worked for colonial interests, laboring at climate inventory, weather prediction, topography and mineral mapping, time services, and the like as their day jobs. And so what differences emerged, and how "imperial" was their physics and astronomy, often done in parallel with their other work? How much did their employers direct them?

When doing research for peer-reviewed publication in journals they were, Pyenson says, engaged in "cultural imperialism," spreading the influence of their nations, enhancing the reputation of science and scientists, and demonstrating the power of their mode of reasoning in "an empire of reason." Pyenson acknowledged that differences in language and scientific community and diverse styles and competence may have influenced both the choice of researchable questions and the style of work productivity in science. But he says it is virtually impossible to distinguish the physics and astronomy completed by colonial scientists, whether it was done in Hanoi, Bandung, or Beirut. In peer-reviewed publications written in these places, "theories, observations, and data radiated no local cachet" except where and how the data were obtained.

Pyenson's examples are unlike individual scientists in this book in that few characters in this story were essentially colonial scientists even though many of them had employment in agencies and universities that was ultimately regulated and paid for by a colonial government. Indians in this story report benefiting from a world of supporting and enabling institutions that did more teaching than research, and they pointed often to the stimulus or even inspiration they felt from contact with rigorous science and mathematics teachers, some of whom were "colonial scientists" before 1947. These inspirations came from people as different as Belgian botanist and physicist Father Eugene Lafont in the nineteenth century at St. Xaviers College in Calcutta to American physicist Richard P. Riesz at Madurai College in the 1960s, both scientific teachers of a high order. Attendance at convent or mission schools that transmitted a grasp of English and other languages made a big impact, according to those students whom I have interviewed and whose biographies are available. Lafont arrived in India in 1864 at age twenty-seven and lived there until his death in 1908, establishing physics and chemistry labs in the college, an astronomy observatory, and, with others, the Indian Association for the Cultivation of Science and its laboratories.[3] Reisz, lesser known, lived in India from 1962 until 1990.

Saha, Bhatnagar, and Bhabha each went to schools staffed by teachers with religious commitments. Though LaFont and Riesz both worked in a missionary tradition seventy years apart, most of these important enabling scientific institutions and individuals had nothing to do with the missionary enterprise, and Lafont and Riesz moved well beyond the boundaries expected and preferred by their managers and employers.[4] Well-recognized astronomer J. J. Evershed, after whom the Evershed flow effect in the chromosphere (1909) is named, in the Nilgiri Hills of South India was a prime example of a colonial scientist, in Pyenson's terms. It was Evershed, meteorologist more than astronomer, who tried to hire twenty-two-year-old K. S. Krishnan immediately after he graduated from Madras Christian College in 1920. Scientists like Evershed who held state positions were found in the official botanical and geological surveys and many other institutions, and they had a myriad of small influences on young people. And, as we have seen, wholly Indian institutions like the Indian Association for the Cultivation of Science, which welcomed Father Lafont too as a member, were crucial vectors in the inspiration of many scientific careers and indeed provided the laboratory in which the research was done in the 1920s that won the first Nobel Prize in physics. The young man whom Evershed could not hire, K. S. Krishnan, attached himself to the IACS lab in Calcutta and to the slightly older man who ran it, C. V. Raman, age thirty-two, and the rest is history.

Following that logic, young scientists in this story could see that they would have to create or join institutions resembling those of their peers abroad. Living on what they saw as a geographic margin, dependent on the postal system and telegrams plus the occasional, lengthy journey out of the country, scientists in India in the 1930s, 1940s, and 1950s had to work even harder than scientists elsewhere to achieve international reputations and national influence. They looked for and arranged the occasional visit of a scientific star from abroad, no matter how brief. They needed an office, the latest information, and an address in a laboratory, in short, a place to build a recognizable scientific identity. A sound reputation had to be established through recognition abroad and not just at home. Evidence here shows that some influence abroad was necessary, if not sufficient, to the achievement of power in India. Though there were a few exceptions, a foreign reputation was, from 1900, necessary for a favorable appraisal of their science in India. Indian institutions and authorities consistently looked abroad while appraising and valuing their own scientists and so constantly sought the opinions and interventions of visiting foreign scientists. In the colonial period a combined judgment involving Indian, British, and European scientists was a regular feature of scientific life. Later, Englishmen like Blackett and others like Americans John von Neumann and Leo Szilard played a continuing role in some fields, even over many years. For example, whether or not someone could be elected to the Royal Society involved the interplay of international reputations and networks. In January 2005 Prime Minister Manmohan Singh appointed his Scientific Advisory Council, and it included twenty-eight names, of whom four were Fellows of the Royal Society.[5] Throughout the scope of this book, however, from 1920 to 1980 there were recognized, effective, and important scientists who had been trained in India and abroad and who did not win high awards elsewhere. Their competence and capabilities were the foundation of the scientific community, and some of them played a critical leadership role: people like Sarabhai, Nagchaudhuri, and Ramanna are good illustrations. The international networks of all these people contributed to the buoyancy or resilience of the institutions in which they worked, but they were not necessarily "famous scientists" outside India.

For most scientists a national appraisal of individual reputation in India was, though important, incomplete. Gradually the reputation of the Indian scientific institution was factored in, so that if someone worked in the Indian Institute of Science, Tata Institute, or Saha Institute, for example, that made an individual's reputation in the 1950s more buoyant, enabling them to float on the stormy seas of successes and failures in science.[6] This process was intensely social, communicative, and political, which is precisely why scientific institutions in India were advised to exercise their own internal appraisals carefully. Individual scientists had to find or create the institution that would welcome them and had to stay and help build it up by building their reputations there. Finally they had to find the ways and means to sustain the institution that in part sustained their reputation. This all occurred in a society in which scientists knew that almost no one else, other than a few peers, could understand the research that they spent most of their time doing each day.

Some tension arose from the late 1960s on between those who remained in India to rise within the ranks and those who returned having established their reputations abroad. This related particularly to the capacity of those returning to raise money for institution-building, using international reputations, and to obtain land and/or buildings at a very reasonable cost, sometimes on long lease from the state, occasionally from private individuals. For example, astronomer Jayant Narlikar, who arrived home from Britain, and theoretical physicist George Sudarshan, who arrived home from America, eventually founded or revived institutes in the 1980s: Narlikar in astronomy and astrophysics in Pune and Sudarshan in mathematical physics in Chennai. Both already had established reputations abroad when they returned to India. Together they would yield an interesting comparison, but this study awaits its researcher.

The two large systems of science and technology management (DAE and CSIR) responded to these opportunities, including those that came from outside India, and continued a long tradition of competitive institution-building. Senior scientists often told me in 1998 and 2005 that India had too many institutions structured and operating in an incoherent manner, and there was redundancy and unproductive competition between them. This pattern was repeated many times by scientists not as well known as Narlikar or Sudarshan, involving "ordinary" scientists coming to India for a variety of reasons, leaving a growing career abroad, and trying to adapt and fit in. But we have other examples in the previous generation of scientists like Chandrasekhar and Khorana who were begged to return, even by the prime minister, were offered carte blanche, and still declined. Outside the scientific community, by the late 1980s there was a flow of investment and institution-building in all fields by nonresident Indians, and their tax arrangements, social life, and successes and failures were a subject of regular media attention. By the beginning of the twenty-first century, there were webs of Indian kinship and friendship around the world, and Indian communities where special cooking sauces, movie videos, and the sound of Indian languages could be found, making the journey abroad quite unlike the one Bhatnagar and Saha took in 1920. Though there have been Indian communities abroad for centuries, the diaspora accelerated and enlarged in the nineteenth century in East Africa, Burma, the Caribbean and Guyana, Fiji, the Persian Gulf, and North America. Unlike the earlier period, those communities themselves now produce scientists, some of whom will doubtless arrive in India to work professionally.

But what of the reputations, institutions, and networks of the main characters in our story? Saha felt he never regained the reputation that led to his election to the Royal Society in 1927. Knowing that he had been nominated for the Nobel Prize in the 1930s was clearly very satisfying, but his desire to make his mark again in physics never left him. The Nobel nomination was for the work that opened his career, a study of selective radiation pressure and an ionization equation to explain the spectral sequence of the stars on the basis of the differences of conditions prevailing in their atmospheres. This pointed to how to determine the chemical composition of stars, leading to a realization of the common origin of stars, planets, and interstellar medium. The laboratory he built and managed is his lasting legacy. It is curious that Nehru had so much to do with that institute, given that he and Saha were so unlike; Nehru personally backed the institute's financing in 1941–42, in 1947–48, and backed its refinancing in 1955. Saha was disappointed in his leader because he expected socialism with Indian characteristics from Nehru; instead he got socialism with Nehruvian characteristics and too much foreign investment and advice. Cut off by Partition from his East Bengali roots in 1947, he was determined to make his own way, and only in his late years was comfortable. When he was invited, surely with

Nehru's blessing, to run for election in 1948 as a candidate with the Congress Party, Saha declined. He also declined Nehru's personal 1948 invitation to be a member of the Atomic Energy Commission. He bravely and boldly opted for another path, but was bitter when Bhatnagar and Bhabha moved along the smooth one guided by Nehru.

Bhatnagar was not "a person of property" as the term was understood in the Punjab, but when Partition occurred in 1947 he lost everything he had in Lahore, including affection, friendship, and a house. He came from a poor family, had no land, and was determined to make his own way. In the 1930s he achieved surprising success as consultant to an oil company and more or less guaranteed himself a continuing income, in addition to good networks in industrial circles. He could not know Nehru well until Nehru was released from prison in 1945, because Bhatnagar was a government servant and Nehru was in prison as a political prisoner of the government. But from 1946, that distinction was gone, and Bhatnagar was now at the center of Nehru's atomic energy and natural resource planning—without much British interference. Unlike Saha, he was keen on foreign investment, though under the right conditions. He kept his relationship with Saha as balanced as possible, doubtless also hoping Saha would not criticize him in *Science and Culture* or in Parliament. He found the money in 1948 to help build Saha's institute in Calcutta, although Saha had not supported Bhatnagar in the building of the National Physical Laboratory in Delhi, having criticized it and then refused to come to further planning meetings.

Bhabha had no regrets about his life and work, except perhaps not living in London, where he loved the music, theatre, and painting. He believed that had he remained as a physicist in Europe, his reputation might have been larger than just for a theory on the probability of scattering in cosmic ray showers of positrons by electrons in the upper atmosphere. This theory was called for a time "Bhabha scattering." Even if he regretted Nehru's intransigence about not building a bomb, Bhabha might console himself, after the first 1964 Chinese test, that Nehru too, had he lived a year longer, might indeed have agreed to move in that direction. On Nehru's commitment to nonalignment, Bhabha surely realized that he and his government department and institute benefited more than any of its disadvantages entailed. I doubt Bhabha had ever set foot in a jail, and Nehru spent years there, though over tea or whisky, and despite their difference in age and experience, these two men could see eye to eye on many things.

Among the three scientists, Nehru could only have been disappointed in Saha, particularly since he knew about Saha's reputation as an excellent scientist. He genuinely wanted to make friends with scientists but could not achieve this with Saha, whom I think he admired in an unseen corner of his mind. But like Nehru's plan for the economy, Nehru's politics were too mixed for Saha. And being attacked by Saha in Parliament probably hurt Nehru in a way that few other attacks did. Had he not gone to Saha's house in Calcutta to dinner in 1946? Had he not invited Saha to be a member of the AEC in 1948? Had he not intervened three times to support Saha's institute? Had he not brought Saha to his own house in 1954 to discuss atomic energy, informally, to try to reason with him, and to get him to cooperate with Bhabha and Bhatnagar? Had he not tried with these Bengalis?

In Bhatnagar, however, Nehru got an insider he could rely on, a man comfortable with the mixed economy and capable of dealing with foreign investors. Not a patrician in background, Bhatnagar had earned and learned how to be a "Sir," at least in Nehru's terms, and had initiative and spark enough to overcome the weight of Nehru's own bureaucracy. He had a good eye for opportunities and for the story that could be told about them. More important for Nehru, Bhatnagar believed in an efficient well-run state, with slightly British characteristics.

Nehru's relationship with Bhabha is more curious; Bhabha was loyal, but a nongovernment man by outlook; he wanted room to maneuver, and Nehru gave it to him. This was a Tata model of development, officially sanctioned and well supported, but relatively autonomous from the state machinery. Nehru had, after all, embraced the state only late in life and thus may have intuitively understood Bhabha's need for independence for the nuclear program, as well its secrecy. But eventually Bhabha *became the state* with respect to nuclear science and energy.

While young men, Sarabhai, Nagchaudhuri, Ramanna, Menon, and Sethna were touched and motivated by either Bhabha or Saha. When Bhabha died suddenly in 1966, all of them were at the age when political opportunity could be translated into immediate action. Sarabhai's family had long been known to the Nehrus, so Indira Gandhi already had an opinion about Vikram Sarabhai's promise and potential, and he was already the only scientist on the Atomic Energy Commission. Nagchaudhuri, Menon, Sethna, and Ramanna achieved their political and administrative reputations in the few years preceding the first bomb test and brought their international networks with them. Again, the prime minister was attracted to dynamic, ambitious, and smart men who would put India (and her office) on the map. These particular men gave off an effortless superiority, exuding as strong a sense of confidence as the Indian Administrative Service officers with whom they negotiated in meetings. The naturalness of hierarchy surrounded them, and they knew how to play its games. And the shining path to put India on the map, to build science and technology in many directions at once, even into space, was the atomic energy program, through which India could negotiate nuclear power.

Dual Scientific Loyalties in a Divided Landscape

Nonscientists in India in this period thought of scientists as exotic, meaning *not entirely of this place.* Scientists working in India reciprocated by thinking of themselves as exotic, too, but resented the presumption of estrangement with which others viewed them. Scientists and the advocates of science, like Nehru himself, were in the process of cultivating dual loyalties, demonstrating there was an authentic place in India for their international way of thinking. Indians showed they could be practicing scientists on a world scale without losing Indianness and that the apprehensions of the cultural guardians about loss of culture was just the result of the alignment of science with other socioeconomic factors foreign to India. They worked hard to deconstruct and discredit the "science as imperialism" paradigm. Working in a divided landscape, where chronic poverty grew in the shadow of costly projects with fine laboratories set in lovely gardens, scientists were often acutely faced with the question of loyalty, particularly when they had opportunities to move abroad like thousands of others. Even a place like the Saha Institute with its absence of fine laboratories or gardens made these questions of loyalty more acute, not less.

This perception of foreignness suggested an uncertainty about India's place in the world rather than an informed public appraisal of the opportunities and practices of scientists per se. Moreover, given the rise in status of scientists and technologists, moving them closer to power, these apprehensions were also about a shift in the sources of legitimacy and authority in knowledge-making itself. Science and scientists did gain greater authority in the sixty years (1920–80) discussed in this book, and, as the evidence shows, they did so through an uneasy alliance with the projects of an independent nation, first as an imagined community and then as practical working communities in the state, in which these very scientists took key roles. They successfully constructed a laboratory state (chap. 4), in which their laboratories *were sustained* almost entirely by public funds, with notable exceptions of private support, as we see in J. C. Bose, Saha, Bhatnagar, Raman, and Bhabha. They knew they needed support from one or both of two sources: the state apparatus and/or private wealth. Remember that a young and inexperienced British Indian medical researcher working in a very simple laboratory shed in Calcutta, Ronald Ross, was awarded the

Nobel Prize in 1902. Young scientists in the 1920s in Calcutta could see that Jagdish Chandra Bose's private lab received Rs 100,000 from the government of Bengal from 1922 onward, so there was tangible evidence of real opportunity. In an equally simple lab, C. V. Raman and his student K. S. Krishnan did ingenious experiments on light spectra that led to the Nobel Prize in 1930 coming back to Calcutta. But some doors would not open, so in 1949, when the refugee-scientist Har Gobind Khorana from Lahore came "home" from Liverpool to Delhi, he found no opportunity for employment over the next two years, at which point he was permitted by India's rules to accept an offer of a job in Cambridge and returned to Britain. Khorana won the Nobel Prize sixteen years later. Many thousands of talented young people took a similar journey abroad, and thousands stayed. From these patterns we see that scientists' experience in India was extremely varied and sometimes contradictory. They fought to transcend the politics of polarities, which placed them in a double bind, in which their realization that "you are either this, *or* you are that" was too simplistic and "you are *both* this *and* that" was sometimes confusing.

In their loyalty these scientists cultivated whatever importance was available to them through this alliance with the state and as a class grew greatly in intellectual confidence and security of livelihood. Though Indians trained in the sciences in India have been populating the laboratories, universities, technical projects, and corporations of other countries for more than fifty years, the landscape at the end of the twentieth century was divisive far beyond the social conflicts that broke the larger immediate postcolonial India into parts. There was a continuing tension about self-reliance in technology and science that was born in the colonial period and flourished through to the end of the twentieth century. There were economic problems created by the colonial experience that were seen to persist long after 1947 with the "help" of international institutions and interests. Changing strategies of development compelled commitments that tested individual loyalties and strained personal and family relationships. There were evidently good reasons for disagreeing strongly about the paths to be chosen to the sources of industrial energy and military power: political campaigns and party factions were built around these disagreements, and to the end of the century scientists reported feeling this tension at work. In some places there were whispers that caste and community were key variables in social advancement, especially when these factors were active outside the workplace. But they unified around the question, How could the society at large think that there was something not authentically Indian about our scientific work? This question sometimes got resolved by scientists' claiming that they were building a cosmopolitan culture in the scientific community, proudly nurturing two loyalties.

Scientists, Communication, and Urban Culture

During the colonial period, and for a long time after it, the dual loyalty of scientists and their commitment to another kind of reasoning and a worldly cosmopolitan intellectual culture was suspect in skeptics' eyes; perhaps, it was suggested across India, scientists were too oriented to Europe to be *real* Indians in a cultural sense. Perhaps they were really enamored of England or foreign culture and foreign things; perhaps they quietly loved the very source of India's continuing economic problems, which implied Britain and foreign powers like the United States. Perhaps their loyalties were elsewhere? What was often purposefully overlooked in this suspicious appraisal was the vitality of alternate networks running through Germany and France, and to a lesser degree Hungary, Italy, Canada, USSR, Japan, and the United States. What was also overlooked was that many scientists had done very fine research by starting in India. So the nucleus of top scientists unconsciously sensed the need to escape the confines of the India-Britain axis: we observe Saha's learning German at school in Dacca and translating Einstein's papers into English in 1919 in Calcutta; Saha's 1921 sojourn at Nernst's lab in Berlin; Bhatnagar's attempts to learn French in Paris in 1920; Bhabha's sojourn in Germany, where he met Heitler and Heisenberg; Raman's knowledge of German and success in bringing the German physicist-refugee Max Born to Bangalore in 1938. By the 1920s Raman was visiting Caltech and by the mid-1930s Saha was visiting Harvard and Berkeley and meeting Bhabha in Copenhagen. By 1939 Saha's student Nagchaudhuri was studying at Berkeley, a leading nuclear studies center and poised to take a role in the Manhattan Project. Saha's visit to the USSR in 1945 to meet scientists like Kapitza was certainly not the first Indian scientist's visit to Moscow. These alternatives to Britain were pursued all the way after Independence; a French firm was the first to contract rare earth refining with the Indian government and a German firm won the contract to build India's first heavy water plant in the early 1950s. Remember too that foreign scientists came to settle in India for extended periods, from the British chemists who taught Bhatnagar at Lahore in 1917, to physicist Bernard Peters at TIFR in 1951, and to experimental biologist J. B. S. Haldane in 1957.

Well-embedded in Indian society, these top Indian scientists resembled India's politicians, writers, artists, bankers, industrialists, and top military officers—all of them at home in the world. And in the open exciting post-

Independence days, they enjoyed a social and cultural mobility unlike scientists in India before them. They traveled to and lived in international capitals like bankers and diplomats (though living like paupers while students). They ate and worked and traveled and slept with other people, of whom their grandmothers would not have approved. By habit they cast their eyes across the world of science to look for the brightest lights, and they frequently found them outside Britain. Other countries sought out India too, Germans, Italians, and Americans, as well as the French, who had a small colony there, and Swedes and Danes, who had long since given up their small commercial outposts in India. Indians were also rightly curious about and attracted by Japan's rise to prominence after its defeat of the Russian navy in 1905. But English was the language they knew very well and Britain's was the culture they knew well enough, so they drew from these most frequently. It was largely British recognition they sought between the 1920s and 1970s, including the coveted FRS, even as Indian scientists were a key part of the process of disengagement with Britain in the 1950s and 1960s, even as Indian science was asked to show its true nationalist colors. Their international network-building thus was driven by a search for external validation, covering up the fact that very few nonspecialists locally could understand and talk about what scientists were actually doing in their research.[7] This is how the landscape appeared to scientists, with a majority of the population unaware of scientists' work and a minority quite divided about it.

Along with their grasp of new media of communication and assertion of their touchstone values in "the scientific temper," the mark of their surrounding cities is strongly etched on the life of scientists and their institutions in India. It is true there are important scientific organizations and universities in the countryside or small towns, some studying rice, others studying forests, and still others tracking distant stars through radio astronomy. But the big cities of Calcutta, Bombay, Hyderabad, Poona, Ahmedabad, Bangalore, Chennai, and Delhi dominate the story here. Their linguistic compositions and ethnic or religious makeup, their ecological histories and their complex relations with the "outside world," influence very subtly the plans and dreams of scientists. In their self-conception the great majority of scientists aspired to be urbane and not like their country cousins (if they had any). More important, these cities affected scientists' ability to cooperate with one another in productive and creative groups. The styles of governance in their institutions were indeed influenced by the governance of the city and society around them, and the school system became the chief resource for the children of scientists. Their food cultures, book-reading habits, the panoply of city cultural life, the public transportation system, the water and electricity supply, the conflictual or friendly relations among people speaking different languages living in the same street, and the "atmosphere" for loved ones and family—all these factors subtly influenced scientists' lives and outlook.

Cities also were at the heart of the star system. Singing stars, movie stars, and sports stars were all fashioned there. And scientists expected to take part in that urban glow, if lucky. Indeed, the manner in which cities recognized and celebrated their scientific reputations, institutions, and laboratories contributed much to the viability of the scientific life: a good example is the 2005 ad in a Calcutta newspaper providing a list with photos of "unforgettable Calcuttans," paid for by the city's Electric Supply Corporation. In this list of twenty-four best-known individuals, some of them world famous, are four scientists, P. C. Ray, C. V. Raman, J. C. Bose, and Satyen Bose.[8] Even citizens of distant cities can reconsider and remember the sublime (and sordid) histories of Calcutta through the lives of its impressive top scientists. Another example shows loyalty to Calcutta from outside it; in Delhi a project was hatched by a Bengali editor to investigate the "Calcutta psyche," and here too the essay on scientists (referring to P. C. Ray, Meghnad Saha, Satyen Bose) sits easily and logically among the poetry and essays on the city's economic decline and rise, and on the radical political mind of the city.[9] The shift of the national capital during the war from Calcutta to Delhi, the success of the capital markets and industry in Bombay, the departure of its educated citizens—these were the great challenges to Calcutta, contributing to a sense of loss. So to counter nostalgic rosiness about the city (including mine), one should read the sobering reflection on "the age of genius" written on the death of the "unforgettable Calcuttan" physicist Satyen Bose in February 1974 in "Calcutta Diary." This was when Calcutta had little electricity or water, but much lethal political violence in its back streets. The diary ends in this ironic way: "Now the rupture with the age of genius is complete. The Bengalis have entered a period of ribaldry, decadence, and death. This is going to be a long haul. Nothing however can be done about it. History, despite the Boses, cannot be nudged off course. Calamity is around, but with perfect composure and a dry wit, [Satyen] Bose would have asked you to simmer down, maybe he would have invited you to join him in a game of chess."[10]

The age of genius haunts India still. Its scientists were and still are at the center of its great collective life, nudging its directions in history. The cities, which seem to draw scientific genius in and hold it for ransom ("give us more, give us more"), now compete for the best minds in India; these minds do not all stay where they grow up—witness the rivalry between

Bangalore, Chennai, and Hyderabad for computer-literate engineers in the 1990s. These rivalries are also about competition between the large regions and populations that surround them: mobility has been enlarged, and the opportunities appear plentiful elsewhere. Senior scientists in many cities were wringing their hands at the end of the twentieth century wondering where the next genius will come from? Who will be the stars? When Saha left Calcutta in 1923, Chandrasekhar left Madras in 1930, Raman left Calcutta in 1931, Bhabha left Bangalore in 1945, and Khorana left Delhi in 1951, they were doing what scientists have done everywhere, migrating to better conditions of work. But the majority has stayed in one place, looking for good stable institutional environments in which to work and establish their reputations. India's powerful and successful scientific community was started on a network of institutions, some quite fragile, patiently built up by scientists and supported by their farsighted patrons. The inextricable relationship between individual scientists, institutions, and surrounding cities is at the heart of this story. A small nucleus of them, envisaged long ago, has come of age.

But one would be wrong to conclude that everyone inside the DAE or CSIR family simply stayed in the city for the benefits and security of continuous work: two recent examples will illustrate a wider and older process of out-movement, echoing P. C. Ray's founding the Bengal Chemical Company in 1901 and Meghnad Saha standing for election in Parliament in 1952. The first example is M. P. Parameswaran, a nuclear engineer at BARC, with a doctorate in 1965 from the Moscow Power Institute, who established the Bombay unit of the Kerala Sastra Sahitya Parishad (KSSP) among Malayalis in that city in the 1960s. With a leave of absence from the DAE, he acted as the assistant director of Kerala's Institute of Languages between 1969 and 1973, a government-funded institute that was the source of much of the KSSP's science promoting activities.[11] This was the new basis, established first in a small way in Kerala in the late 1950s, to launch what would be called a peoples' science movement—including translating writings of J. B. S. Haldane and J. D. Bernal to Malayalam, teaching science to poor people with little literacy, writing new textbooks, taking adversarial positions on new technological projects in the state, and so on. As Dhruv Raina says, "Popular writing on the sciences in Indian local language celebrating Baconian science dates back to the first half of the nineteenth century. However the peoples' science movements broke out of the self-constructed enclaves of Indian science. Half a century ago [in the 1960s] the movement was restricted to the state of Kerala."[12] This movement was celebrated across the country in the 1980s as a pioneer by similar movements in many states.

In the second illustration of moving out of an established lab, Anil Sadgopal was trained in molecular biology at Caltech and worked in the TIFR group with Obaid Siddiqi until 1974; until then he and others had spent some of their time building up science teaching capacity in poor schools of Bombay and surrounding rural areas. By the mid-1970s he began building a study and action center at Hoshangabad in Madhya Pradesh and succeeded in persuading some TIFR and IIT professors to join him in adapting a science curriculum suitable to the world of nonliterate tribal and nontribal people. He was leaving the institute! Some genuinely helped and responded to his call, a few shook their heads, privately thinking he was wasting a unique opportunity to work in one of the premier biology research groups.[13] No, not everyone remained in the big cities.

Should we conclude that just because they were "modernist" and "urbanist" all DAE projects were the same? No, each of the atomic energy projects had a special culture—Kota for heavy water in Rajasthan had a special town near the lake, Tarapur for reactors in Gujarat had a special town too, Thumba in Kerala for rockets had a special housing project that was a large cosmopolitan community like a village. These projects grew like mushrooms, and though they resembled one another and resembled those towns for steel mills and for jet fighters and hydro dams, they were not everywhere the same, not identical though they all had electricity. In 1967 Taya Zinkin stressed the sociocultural differences between the British, German, and Russian steel project towns (Bhilai, Rourkhela), and the DAE projects experienced the same evolution and differentiation. Each had its special context, its starter's position, its distinctive character, its particular leaders and personalities. In these special DAE places, engineers and technicians enjoyed a status elevation because of their contact with the nucleus, their contact with the high sciences like physics. These DAE projects outside Bombay contributed naturally to the evolution of the different "epistemic communities" in the expanding department—some oriented to construction and operation, others looking ahead to new research and others to practical design in the next generation. Their ways of seeing and thinking gradually diverged and the family relationship, even as metaphor, eventually had less and less meaning. But still institutions referred to themselves as families, as the director of TIFR did in 1968.[14]

Science as Movement, Science as Institution, Science as Temper

From hundreds of conversations and the evidence offered here, I conclude there has been a creative tension running through the Indian scientific community between the idea of *science as a movement* and *science as an institution*. Frustration with institutions, their quarrelsome governance, and the dubious quality of their research produced calls for "the movement" again, appealing to noble values and homage to transcending scientific figures. The existing institutions, once fought for and established, were later described ironically as shrines to old ideas and practices. Ultimately there were calls again for new institutions, to be formed out of the body of the old ones, to carry on the scientific movement. This was a process of *schizmogenesis*, coined and described by anthropologist Gregory Bateson in *Naven* (1958), who applied it to understanding village life in Papua New Guinea. Schizmogenesis was evident in the Saha Institute's emergence from Science College in Calcutta, the Raman Institute's emergence from the IISc in Bangalore, and so on, repeatedly, progressively. Bateson described it as "a process of separation and disconnection" at the levels of both rhetoric and action. Sometimes this movement in the scientific community was about founding new disciplines around unique and costly new equipment such as a radio telescope or an electron microscope, sometimes about personalities and disagreements, sometimes about the prestige and interests of founders and funders, sometimes about pressure from the government and resulting conflict.

These institutions were expressed mainly in buildings, salaries, and labs, but scientists also labored hard to establish the wider community through communication and to maintain disciplinary journals; there was an astonishing growth of Indian scientific journals in almost every discipline between 1930 and 1960, staffed and edited within India but with a broad reach of international editorial advisors. There were originally two, then three news magazines that catered to the scientific community. And the development of *not one but three* Indian scientific academies also suggests a kind of schizmogenesis at the top. "Not too many in a big country," said one critical observer, "but lacking the coherence of the National Academy of Sciences of USA or the Royal Society of Britain." The political scientists understood the importance of improving communication not only within the community but also representing it skillfully to the outside. When Raman founded the radical new magazine *Current Science*, when Saha followed suit with the influential *Science and Culture*, and Bhatnagar took to radio broadcasts in the 1940s and 1950s, which Bhabha also did in the 1960s, they were doing the same work as those in the 1970s who were determined to get science on television for rural audiences as soon as the Indian satellite was launched. Science journalism was almost a profession in India by 1970 though by no means a well-paid one, and *Science Today*, the now vanished biweekly magazine published by the *Times of India*, reached an estimated 900,000 well-educated readers by 1975: professional scientists with busy careers always took the opportunity to write for it because they too needed to reach that audience. *Science and Culture* and *Current Science* both experienced difficulties in the 1970s but were revived spectacularly. The thirst for news about science and scientists seemed unquenched.

The idea of science as a movement was quite in tune with the numerous social movements of the pre-1947 period. These social movements did not die, and the "peoples' science movements" of the 1970s and 1980s and the "technology missions" of the 1980s were their (very different) children.[15] The real disagreement between them was whether scientific thinking and the scientific spirit would spread like wildfire borne by the wind, landing to burn quickly here and there, or would instead move like malaria, through a vector, one infecting another, slowly but steadily. These two metaphors, fire and infection, were not lost on those skeptics who were apprehensive that scientific thinking and temper were going to burn or infect their worlds, as they were based on other forms of transmission and authority. Some critics of science dug in their heels: "broadcasting" science confirmed their concern that the direct master-apprentice model so long practiced in Indian learning (mathematics, music, medicine, for example) might not survive this new more impersonal model. In fact, what has happened is a skillful blend of the older master-apprentice model of direct learning (called *guru-shiksha*) and the new model of science.

The challenges to scientists and their craft, challenges to the movement, particularly from some religious leaders, continued to worry scientific and cultural leaders, so that in October 1980 Raja Ramanna presided over a debate at the Nehru Centre in Coonoor in the Nilgiri Hills on how to express and support the scientific temper more effectively. In his foreword to the text P. N. Haksar said, "There are more than two million scientists and technologists in our country. In addition, we have a large number of economists, historians, sociologists, and anthropologists, lawyers, doctors, administrators, management specialists and teachers who, in one way or another, apply the scientific temper and scientific methodology in pursuit of their respective professions and disciplines." Quoting Nehru at length, Haksar then called for a second renaissance, which would generate "what needed to be done to halt the process of decay of reason and rationality" and thus "catch up with the rest of the world." The statement drawn up in 1980 by prestigious intellectuals gathered in the hills of South India referred to "obscurantism," "deep-rooted structures of an ancient society," "inegalitarian social and economic structures" that "inhibit the spirit of enquiry" and that are associated with "escape into magical beliefs and instant solutions."[16] Here together were two men trained in London, one, Haksar, a lawyer and a skeptic about the first nuclear test, and the other, Ramanna, a physicist and the nuclear test's chief protagonist and actor, collaborating, in their terms, on a national statement that united them in the logic of "scientific values" that transcend strategic differences, in order to catch up with the rest of the world, again.[17] They were addressing 2 million scientists and technologists in India as if they were all in one great epistemic community.

Where did this apprehension of an escape into magical beliefs arise? Was India so profoundly at risk to the smoke and mirrors of religious distraction? I doubt it. The fact that scientists in the nucleus dealt in invisible forces did not alienate them from some large segments of the population who were already convinced there were invisible forces that explained everyday phenomena such as illness, wealth, misfortune, infertility, and the like. The outcome of that special scientific work was unknown to these populations (indeed, most of them still did not know in 1980 of Ramanna's pride in the bomb achievement in May 1974), but the argument that the scientists were doing something tricky with the invisible was not strange to these ordinary people, even though the "secular" reputation of scientists preceded them and was by 1980 taken for granted.[18]

Where did the impetus to organize this costly gathering come from? What really was this "process of decay of reason and rationality"? Indira Gandhi had just returned to power. Ramanna knew he would soon return to Bombay from Delhi to be head of BARC. Haksar was retired but had the prime minister's complete confidence. What deep-rooted structures of an ancient society were being contested? The signatories to the statement are a who's who of filmmakers, artists, novelists, academics, and scientists. And though it is unclear what specific challenge they were responding to, reference to "the second renaissance" reconfirms unambiguously the place of scientific practices and methods of thought in the cultural landscape of India, placing them there along with all others, above all in "a spirit of enquiry."[19] This was to prove that scientists were persons of integrity, and their method could be trusted, even believed, as in "I believe in science." As Visvanathan wryly observed, "we proudly talked about 'the scientific temper' as if it was a vaccine that would immunize us from all forms of superstition."[20] Ashis Nandy said in 1989 that "the Indian state can use the achievements [in modern science and technology] to legitimize itself as a repository of scientific knowledge and *a negation of native irrationalities.*"[21] Was there a link over many years between the incidents noted in this book, the coming of the end of the world during the conjunction of eight planets in 1962, the criticism of physicist Bhagavantam's 1969 religious essay among young scientists, the high-level scientific temper manifesto in 1981, and the criticism of M.G.K. Menon becoming an advisor in 1998 to the "backward looking" BJP? There was continuity in the response of scientists to such phenomena; weren't they all pointing to the scientific temper and the styles and methods of work that made them more reliable and conferred authority on the state and the leaders who supported their work? Their scientific work was transparent, they said, not obscure or esoteric, and certainly not irrational. Moreover, they were no longer solitary seekers of truth, an image well established in Indian traditions, but members of a large community of people, perhaps as many as 2 million, that was disinterested and unbiased in a subtle way that no others were. If they made an error, they said, it could be corrected, making progress continuous with the efforts that went before them but providing that necessary and elusive new edge of change. This was a call to mobilization, but for what? Were they a *whole epistemic community*? By a wave of the hand they all might be, but that wave would pass over a number of distinctions limiting its "wholeness," distinctions evident in the history you have just read. These distinctions enabled some of them to disagree profoundly with others and to fight with them, all the while relying on their "common" scientific temper, scientific methodology, and rhetorical euphemisms like "obscurantism."

Women and Science in India

Though women appear only indirectly in these stories, some of them had influence in the outcomes. It is quite true that through to the end of the twentieth century few women were involved in the elite structures of science described here. Though women became working scientists and successful scientists in larger numbers from the 1940s, they were not engaged in elite structures until the 1980s, and then only in small numbers, varying greatly by discipline. Women were heavily represented in biology, botany, pharmacology, chemistry, and medicine, and had this book focused on those fields, women would have been more prominent. But through the end of the 1970s, there were no women scientists in the committees advising the cabinet on science and technology, and none were directors of major laboratories or projects discussed here.[22]

While true, this picture can be misleading. In a subtle and indirect way, women are present throughout this history, so as illustrations I offer portraits of women who influenced the men in this story. Jawaharlal Nehru was influenced by his bold and bright sister Vijaya and his bold and bright daughter Indira. These women kept reinforcing Nehru's "charm for scientists," and attempting to "correct" his political directions from time to time. On another note, much is made of Nehru's strong affection for Lady Edwina Mountbatten and his influence through her upon Louis Mountbatten and vice versa, but it is not widely known that national medical laboratories were built in Delhi through the efficiency of Nehru's regular contact with Edwina Mountbatten. We know she was asked by Archibald Hill, Bhatnagar, and others to intervene directly with Nehru to get them built.[23] Even less is known about Edwina's influential role in continuously promoting institutions for nurses and paramedics, in India as well as in other countries.

Meghnad Saha's wife, Radharani, was apparently the only woman who influenced him, but although she came from the same village in East Bengal, she was in many ways quite unlike him. She was not educated, married at age fourteen (when he was twenty-six), and never traveled abroad. Her daughter described her as having a "rather submissive and acquiescent devotion" to the Hindu calendar, in contrast to her husband's "explicit atheism." Many relatives came as pilgrims to stay with them in Allahabad during the Magh-mela and Kumbh-mela festivals, to bathe piously among millions at the confluence of the sacred rivers Yamuna and Ganga. Saha's wife patiently cooked and prepared for all these visitors, as her duty. Meghnad welcomed them but did not join them at the river, though he loved to swim. Radharani housed four families of relatives in frustrating conditions in Calcutta, all fleeing the 1947 Partition of East Bengal. And she was Saha's main listener: "As a rule she was uncritical. She was father's confidante and would patiently listen to the day's happenings after he came back home. He needed her and she was an old-fashioned accommodating wife and affectionate mother. She was extremely proud of her husband's achievements, which was rather hard on us, the children."[24]

But in the end Saha did, as he grew older, support the educational and marriage choices of his daughters, and Radharani unreservedly furthered his scientific and political careers, though he made no significant appointments of women scientists. They appeared in institutions he founded, but much later.

Shanti Bhatnagar owed his education to his mother's absolute determination to have her son succeed and then his career support to his wife Lajjawati, a teenage girl with a minimal education who had an arranged marriage to this twenty-one-year-old student in her father's college in Lahore. So Saha and Bhatnagar were more alike than people have thought.

It is unclear whether Lajjawati shared his commitment to a Brahmo spiritual outlook, and we know nothing of her views of his work in science. A greater influence on Bhatnagar was his biographer Norah Richards, his drama teacher and confidante since his late teen years.[25] Richards mentions Bhatnagar's wife only in passing. Bhatnagar's articulate poise, sense of timing, and facility with English on the radio, which bolstered his career from 1940 onward, must surely have been cultivated by Richards, herself an actor who lived in India until her death in 1971; let us say she must have taught him "how to project." Since Bhatnagar was a chemist and helped to build modern chemistry in India, it is probable that he was in contact with many women chemists of the 1940s and 1950s, but he died rather too soon for us to see the extent to which he promoted the cause(s) of prominent women in such institutions as his National Chemical Laboratory. None were directors of CSIR laboratories while he lived. This is not Bhatnagar's "fault," but a feature of the culture of organizational life in science worldwide.[26]

Homi Bhabha grew up in an atmosphere of strong and well-educated women and was particularly under the influence of his maternal aunt. Elite women in Parsi society had both property and privilege, and Bhabha evidently enjoyed their company. We know little of his contact with women at school or at Cambridge, but it must be remembered that women were only just confronting the dominance of men at Cambridge when he was there.[27] Had he gone to the University of Edinburgh he would have seen rather different roles for women, but he went to Cambridge, moreover to the male-dominated engineering department. We do have reports of Bhabha's attractiveness through the 1930s when he was a postdoc in Europe, but it is doubtful that these women influenced him in his scientific thinking or plans.[28] But when he met Pipsi Wadia in the 1940s, a few years older than he, things changed. An untitled pencil drawing of her lovely hands made by Bhabha around 1941 was one of Bhabha's own favorite pictures and became a kind of icon of his skill. Here was someone in Bombay who shared his love of painting and music, cultivated his love of gardens, often saw him off at the airport, went to and gave parties, and supported him as companion. She took a serious interest in the layout, interiors, and paintings of TIFR, and specifically in its gardens and gardeners. Separated from her husband, she met the disapproval of sections of Parsi society, and although not a scientist she valiantly made friends with Bhabha's friends, like Patrick Blackett. I recall a visit Wadia made in 1967 to TIFR, her first since his accidental death. Tall and stately, in a long creamy-white dress with a cream veil, she gave the appearance of a widow visiting a tomb to lay flowers; it was also an official visit in which she was attended by all the dignity and dignitaries TIFR had to offer. She went alone to his office, which was untouched since January 1966.[29] I believe Bhabha might have had friendly relationships with other women too, like Kitty Oppenheimer, Robert Oppenheimer's elegant wife in Princeton, and Constanza Blackett, Patrick Blackett's elegant wife in London. Being unmarried gave him an entirely different air: his brother J. J. Bhabha described him in 1998 as "a gay bachelor," I think in the nineteenth-century meaning of the term.[30]

Vikram Sarabhai grew up close to the nationalist movement in which many strong women played a key role. His mother was Sarladevi, daughter of a relatively poor though intellectually inclined advocate. Vikram's father Ambalal chose Sarladevi as his wife for her independent-mindedness, not her wealth.[31] Vikram's older sister organized unions among textile workers. His father had already somewhat distanced himself from the gossipy circles of the orthodox Jain community. This was a very "nationalist" and very rich family. "But," according to Shah, "there was no suggestion, not in Ahmedabad, not at Cambridge, nor on his return in Bangalore, that the thought of participating in the freedom struggle had even crossed Vikram's mind." He was a twenty-three-year-old physics student in Bangalore when the Quit India movement was launched in 1942. On the other hand, according to Shah, like Ambalal's sister Ansuya, who worked directly with M. K. Gandhi and others among workers, "the Sarabhai women were totally involved in various activities and were frequently arrested. Vikram's aunts Indumati and Nirmalaben organized the city's first *khadi bhandar* [handloom cloth shop]. Sarla and Mridula handled the boycott of foreign cloth and the picketing of liquor shops. . . . The police would come to the house to confiscate a car or some other piece of movable property because a relative was in jail and Ambalal (Vikram's father) in keeping with the spirit of noncooperation, had refused to pay the fine." Shah concludes, "It was as if the women in the Sarabhai household always rebelled against feudalism and patriarchy while the men, however liberal and daring they may have been in social terms, did not disturb or even seek to address class inequality."[32] Sarabhai met, courted, and married his wife on his own terms, a woman originating far from the elite Gujarati industrial and textile circles of his Jain family. His unusual marriage to a young Malayali woman was accepted if not quite approved within the family. Though a beautiful and accomplished dancer, she was not a scientist, and in fact few people in their immediate circles were: science and scientists seem simply to have been considered as a part of a larger cultural whole.[33]

In the late 1950s Sarabhai met Kamla Chowdhry through the new Institute of Management in Ahmedabad. Chowdhry had trained at Harvard in social and educational psychology and soon came to the faculty of the new institute,[34] which was twinned through Sarabhai's effort with the Harvard Business School. She worked to implement his ideas through the 1960s. The intimate relationship with Chowdhry put a real strain on his relations with his wife and children, and often he lived alone in his flat in Bombay: at Sarabhai's sudden death, Chowdhry was included in the funeral ceremony with the family, and she later edited a book of his essays.[35] And one other woman who greatly influenced Sarabhai's civilian and quasi-military projects was Indira Gandhi, who like others reportedly found him very charming and found she was able to move his ideas in her directions, and vice versa perhaps. But this charm was cooled by their confrontation over the ten-year Sarabhai Profile of DAE in 1970 and his reluctance to give up atomic energy while embracing space in 1971.

If any doubt lingered about the influence of women in Indian science, one would need only to look at the interventions of Indira Gandhi after she became prime minister in 1966. She traded on the relationships built up over the years with scientists and technologists, some of them originating with her father; she relied heavily on her skepticism about science in development. But since her academic performance was misunderstood, most Indians in the 1960s and 1970s believed she was a failed person in the professional sense.[36] She was not anything like that and had an important influence on scientists and technologists although she did not complete a university degree and did not have the conceptual fascination for science that her father had. Close observers recalled her astute attention to technical detail, reading and noting questions on long memos, waiting patiently until she understood the answers, and a number of people commented on her exceptional memory.[37] She pushed against the IAS to appoint physicists and engineers as secretaries of ministries, no longer simply as passive advisors but as technocrat policymakers, rising at the same time with economists. She helped to define big strategic projects, rarely in concert with other women, though she did briefly appoint Uma Shankar Dixit to be her junior minister on the atomic energy portfolio. Gandhi was doubtless a role model for many younger Indian women but promoted women's issues in science only in a very indirect way.

Too much is made of the subjugation of women in current Indian science, and I already saw signs by the 1970s that they would be a force to be reckoned with even in male-dominated fields like physics and engineering. Their relative absence in this story really marks the end of an era.

Planning, Unfulfilled Promises, and Scientists

Since the early 1940s science in India has been planned, and scientists were involved in most of the planning: all this time science in India, unlike China, was internationally influenced but not nationally controlled. Institutions grew like mushrooms in the forest after rain, harnessing both private initiative and funds. It was not science's clarity that attracted them to the idea of a nation in the 1930s and 1940s, but its ambiguity and flexibility. Science's internationality and uncontrolled character were its two features that national planners of India disliked and tried to change—but then some planners were themselves beneficiaries of its unplanned character, Mahalanobis being the most obvious example. Supervised by the Indian state, scientific planners would have preferred national controls and priorities that minimized the influence of international factors and maximized local autonomy. In short they would like to pick and choose the fields and persons worthy of investment, attract or retain their citizens who could work at the top in those specific fields, and demonstrate "Indian achievements" on a worldwide stage at a reasonable cost. Not without justification, trained Indian scientists were perceived as a state investment from which India should reap most of the benefits. But the actual experience of planners was otherwise: they picked important fields in response to international economic and political influences and tried to mobilize and direct their best resources toward them. But it didn't often work, and so in this disorderly and reactive process the fields picked did not always get hot or stay hot, and the stars chosen to pursue them sometimes faded early, or stayed too long without producing, or began quarreling. Given this disorderliness there has been no intelligent attack upon the planning of science and technology, other than it was heavy-handed or micromanaging. On the other hand, people with great potential slipped through India on their way to doing greater things elsewhere; the 1968 Khorana story of the Nobel Prize was the paradigm case, but there were others, passed around as oral legends. But there were also choices that produced benefits for India, although over a longer time and requiring more patience than planners intended.

While the nucleus of top political scientists faced the question of how to cultivate their modern science in India, it was more important that they were forced to address what role they and their institutions would have in India's socioeconomic development. But they saw a future quite unlike nonscientists, because it contained their science in it. Imagining their future, they answered affirmatively, "science could do almost anything." So these farsighted planners and schemers planted themselves in the middle of "the dilemmas of technological choice," and few nonscientists could enter that space. While others were choosing which type of public bus to import and whether to build instead of importing buses, these scientists were choosing which type of nuclear reactor to import or build.[38] Their own disagreements over the choice and its political consequences led to a larger recurring pattern I called "the war over self-reliance," in which they stood to benefit from the success of either side or any of the sides in the war because the foreign choice connected them up again with the talents of the outside world and the national choice called again upon them to deliver their capacities to "India," which was both a geographic and conceptual place.

Disorderly too was the competitive interorganizational rivalry among people supported by the same government funds while reporting to different boards and ministers. While the conditions of life and mobility of many thousands of highly trained workers were at stake, these laboratories, projects, institutes, universities, agencies, and the like were themselves the scene of struggles over the best use of resources and best paths of action. Committees met to plan cooperation, often in the private conviction that outcomes would be better than if scientists and their institutions were just left alone. As much lobbying was done for exemption from plans as it was for inclusion in them. There was, however, no government on earth that just left scientists alone: everything else was being planned to improve India, and resources were evidently scarce, so why would science be exempt? Although scientists such as Saha, Bhatnagar, Mahalanobis, Ghosh, Bhabha, Sarabhai, and Nagchaudhuri had been deeply involved in planning from the beginning, what is striking is that the cabinet had good intelligence available through its science and technology committee, yet made so little systematic use of it. It is as if the politicians preferred things a bit disorderly, against the efforts of the Planning Commission, who were supportive of the scientific community and its institutions but opposed to the unplanned use of money by scientists, no matter how creative they or their objectives might be. In this support economists deferred uncomfortably to physicists, admitting an international hierarchy even while trying to control their expensive projects.

At the heart of this book, and of *Negotiating Nuclear Power,* is the decision by Bhabha and Nehru to commit to a specific path toward nuclear electricity and nuclear power, and this included the CANDU model of natural uranium reactor. Relying on natural uranium reactors created inherent difficulties because India did not have high-quality uranium and did not produce enough heavy water. These constraints meant international dependencies and higher costs through the 1970s, particularly after the 1974 nuclear bomb test. Then the natural uranium reactor decision appeared as a form of entrapment: in order to succeed, more and more resources had to be invested in it although there was little evidence (except in DAE minds) it would ever be a successful course of action. This was a classic planning issue because in entrapment it is difficult to develop alternatives concurrently; though it is true there were countries like Britain that had a number of reactor systems on the go, and achieved a bomb from one of them, these cases are in the minority. Though warned about underestimating the opportunity costs of nuclear power (see the early critical signal by economist Ian Little in 1958, and many researchers following him), a great many other energy initiatives could have been pursued with the same money, and it might have been possible to study and develop other sources of energy. Surprisingly there was no sustained push in solar power, which would have provided for India's two big energy uses—night light and cooking—for about eight months of the year. This gap arose because solar power had little value to industry, and India's energy system was an industry-first planning priority; only in the late 1960s did mid-level planners realize that modern agriculture was also going to need energy, and not just from electricity but also from petroleum. It is striking that the installed nuclear capacity announced by the DAE for 1980 was finally reached in the year 2000. One factor that contributed to this unfulfilled promise was that while DAE built reactors, it did not build or manage electrical grid systems, so a holistic planning model combining power generation with distribution took years to overcome the multitude of segmented and often broken grid systems across India.

In this industry-oriented approach the DAE became the conglomerate for the development of the scientific community, first by excluding interference, second by screening for the best young talent, third by supporting "nonessential work," like mathematics, biology, and astronomy. A close observer joked that Sarabhai sometimes called the DAE the "Ministry of Advanced Technology." The approach until about 1970 was science first, with technology following, by which time big DEA engineering and electronics projects were under way, in a parallel track with the industrialization of agriculture. Through to 1970 neither the Planning Commission nor the scientific committees advising the cabinet were able to challenge the plans of the Atomic Energy Commission, not until the confrontation between the prime minister and the chairman of the AEC in 1970–71, and then not again until Moraji Desai's return to power in 1977.

Sometimes valuable long-term DAE projects were questioned but then protected from interference. In the late 1970s, while BARC and the DAE faced critical reviews about its reactors, radio astronomers and molecular biologists at TIFR were doing research and planning that eventually led to the establishment in 2000 of the National Centre for Radio Astrophysics at Poona (and in the countryside far outside Poona) and the National Centre for Biological Sciences outside Bangalore in 1991. These initiatives pioneered brand-new fields of research in India, both of which were initiated by Bhabha at TIFR in 1963, literally while he was negotiating delicate nuclear fuel agreements in North America. The evolution of these centers shows how long and difficult the road to equipment-dependent research is, perfectly illustrated by the case of the great meter radio telescope (GMRT), partially based on the construction of TIFR's half-kilometer-long radio telescope at Ooty in the Nilgiri Hills, itself completed in the 1970s.[39] Outside Poona is built a very powerful telescope operating at low radio frequencies, located in the quiet countryside to be sufficiently north of the geomagnetic equator and have a relatively quiet ionosphere, with access to constant electricity. The biological sciences center was originally to be in the city of Bangalore at IISc, but because land there was scarce it was, after transitional locations, moved 10 km out of the city. Planned in the 1980s, it was formally opened in 1991, with its labs on a reliable constant electricity supply.[40] In both these cases the DAE was the patron for projects that had not much to do with atomic energy directly and supported two farsighted individuals, molecular biologist Obaid Siddiqi and astronomer Govind Swarup, renowned professors at TIFR, who wanted to construct these facilities to very high standards.

When new bomb-making and missile-building projects were well under way in the 1980s, scientists and technologists were given a new opportunity to collaborate in planned "technology missions." First named in 1985–86, the missions were seen as Prime Minister Rajiv Gandhi's opportunity to further his mother's and grandfather's plan to modernize India and change its world reputation in technology and its applications. They followed on the heels of the success of the peoples' science movements, most notably the Kerala movement KSSP. These official missions were thus both a form of thought *and* plan of action for political leaders, officials, and experts, centered around mobilizing ideas and participation just like the science movements were. Physicists of the earlier generation like Yash Pal of TIFR, who had already guided the satellite television project through the 1970s, were involved in this stage. In close relation with the SACC, the prime minister involved business-oriented nonresident Indians like Sam Pitroda in five missions: eradication of illiteracy, improvement of telecommunication, provision of safe drinking water, immunization of children, and higher production of edible oils. There were other candidates for this high-profile marketing initiative flowing through the prime minister's office, but they were set aside for these five big "tractable" problems. There were many dimensions of each mission, but common threads included "demystification of technology" and "stress on local technology, and 'social audits.'" This occurred precisely when the prime minister had appointed another blue-ribbon committee to review the CSIR, whose March 1987 report caused quite a stir ("a storm of protest among the CSIR community," said one response).[41] Almost all scientific institutions of India were doing something that could usefully be applied to these missions, but it was the branding and popularization of missions that made these efforts stand out, refreshing the idea of planning and refreshing the existing plans and action already in place. Again, as in the past, the touch of the prime minister's office in this project was like the wand of the fairy godmother.[42] Still, in the background, the DAE moved steadily toward an achievement of 2,800 MW installed capacity at the end of the century, though twenty years behind its earlier forecasts.

Years after the war over self-reliance, Indian scientists continued to identify risk-taking as a key value, or risk-aversion as a key problem, in research in India. Planners argued that if Indians planned well they would be in the best position to take risks. Echoing B. D. Nagchaudhuri's 1968 exhortation to embrace risk as he moved from being director of the Saha Institute to a member of the Planning Commission, the new director of the Tata Institute from 2003 Shobo Bhattacharya summed up his first two years working in India for the first time by saying that "a kind of Gandhian ethos of self-reliance . . . is still present, but it is not as apparent here at TIFR as it is in places like nuclear research centres. . . . In my childhood something built in India was a source of great pride. Whether it was built well or not was a secondary question. And this idea still lingers on in places. But not so much at TIFR. . . . If we don't re-think we are not going to change and science will leave us behind. It is about taking risks by re-thinking the kind of science we do."[43]

Scientists, India's Self-Reliance, and Industry

The official picture of a gradual but steady development of science and technology covered up a struggle over the use of Indian capabilities in the economy that hinged on different understandings of time and timing: Who is ahead? What is behind? What is this generation? Where are we now? One definition in circulation was that the capabilities sought in India had already been established elsewhere, so that continued reliance on foreign technology or capital was backward looking (meaning neocolonial). This view then subdivided, distinguishing those who wanted an alternate kind of technology and science serving "the people," later called "appropriate technology," from those who wanted the Indian approach to better resemble "the western approach" but to be more authentically "Indian" and to benefit India. By resembling more powerful economies, in this view India could effectively compete with them on their terms, which it was unable to do at the time. It had to transform itself away from an agrarian economy based in a rural population with flat or falling commodity prices; in short, India had to be industrial and to become high tech. The debate raged, almost anyone could engage in it, but institutional positions hardened around their directors. Scientists had one escape clause in the debate because it was acknowledged they could not perform their work properly unless they engaged actively with institutions and individuals far from India.

The other view, both opposing and opposed, was that Indian capabilities were still immature and would take a long time, under Indian circumstances, to deliver the results, and therefore time should be "bought" by using importation. Indeed, most Indians thought that innovation and technical change occurred elsewhere, and later happened in India. This quick importation was an answer to the questions "What is ahead?" and "What is behind?" This quick fix would satisfy growing internal markets, manage people's frustrations, and buy time for indigenous capacity to mature. India could thus catch up on her own terms. In this view India need not (or even should not) resemble more powerful economies and societies but could and should develop independently of those other models, nevertheless using their technologies, as Japan had done. Readers in the twenty-first century should remember that there were few economic models outside North America and Europe that attracted attention in 1970: China's model appealed to very few Indians, and the Soviet Union only to slightly more, but Taiwan and others were not on the screen except to observant visitors. A more complex nationalist response was that these investments would all pay off eventually, so though the path was long and difficult the country should stay on it, as long as the state remained in charge to spread the benefits equitably. Another simpler nationalist view was that "eventually" would take too long, and there had to be, as Saha had said thirty years earlier, a paramilitary state response, with greater discipline, a forced march to industrialization, and the more focused use of capital. As these differences were subtle it was possible to shift positions depending on the challenger at the moment. But in this debate every voice claimed to be nationalist and there were no technical grounds for disqualification. Although people muttered about loyalty to India and neocolonialists taking us backward, and reinventing the wheel, there was no real challenge to the consensus around the shift to an industrialized economy and then to a high-technology economy, always moving away from the agrarian past. These different forms of justification were largely self-fulfilling, and thus the debates were like a dialogue of the deaf.

This debate recapitulated the terms of the late 1930s around what kind of independence was coming and crystallized during the war when the 1942 Cripps Mission signaled Britain's realization that India must have a form of independence. Then too there was agreement about self-reliance yet disagreement about the best path to that destination. Analysts of India as a postcolonial state have perhaps tried too hard to project the 1950s through the 1970s as a radical postcolonial discontinuity, and the critics of neocolonial dependence tendencies in India have perhaps not looked carefully enough at the subtle changes that occurred in different technologies in the same period; there was often too strong a preoccupation with the definition and noisy articulation of what was "authentically Indian" for the various competing voices to hear the sound of these important changes.

Ashok Jain and V. P. Kharbanda described the 1950s and 1960s as an infrastructure-building phase, followed by a reorientation in the late 1960s and 1970s toward protecting the legal, technical, and knowledge environment for "indigenization," partly to save foreign exchange in unequal transactions. This was followed by a period in the 1980s of confidence in and promotion of Indian technologies, but the seeds of that confidence—promotion and success—were sown much earlier. Acknowledging that history does not follow these neat periods, Jain and Kharbanda reviewed the cost of imports and substitution for imports by sector and formal approval of collaborations and exploitation of Indian technologies by industry. They concluded that changes made in the 1970s captured the value of most investments made earlier by the state and laid the structural basis for almost everything that was to follow.[44] I concur with them that the 1970s are thus a watershed period, a reason I have focused attention on that period.

In all countries there is a complex and messy interplay between inventiveness, innovation, playful speculation, state-guided missions, rivalry among entrepreneurs, and financial capital. So too in India the course of the innovation-and-entrepreneurial chain has been neither straight nor smooth. For example, P. C. Ray, with a PhD in chemistry from the University of Edinburgh, while teaching also established an industrial corporation at the beginning of the twentieth century, a move as logical as any chemist of that great chemical age. He created a successful (and explicitly nationalist) business in the Bengal Chemical Company. This company was still making chemicals and exporting toothpaste containing *neem* more than one hundred years after Ray founded it in 1901, but it fell into a difficult phase in the market and had to be rescued. It was bought by the government of India and became a public-sector company in 1980. Again, though Ray's colleague J. C. Bose's research had little to do with industry, he received handsome support from the government in the hope that there would be an industrial payoff. Though the results on Raman's experimental bench of the 1920s were not immediately ready for industrial applications, Raman himself would later invest his Nobel Prize money in an industrial use of beryllium, unable to imagine the profitable flow of his research on optics into industrial applications of the laser late in the twentieth century. By 1990 literally thousands of scientific researchers and technical workers around the world depended on the study or routine application of Raman spectra. That had not been Raman's intention, but by 2000 there had grown up a Raman-spectra industry.

When Bhatnagar was awarded the Steel Laboratory and scholarships in Lahore in 1936 and when Saha reconstructed the 38-inch Berkeley cyclotron at Calcutta in 1942 using Birla and Tata funds, these were not the first instances in which industrialists and industry in India had supported science and scientists. Bhatnagar made explicit comparison between his new laboratories in Punjab University and P. C. Ray's earlier Bengal Chemical Company. But the biggest model of business-knowledge partnership was in Bangalore, where the Tata Trusts took an early lead in 1909 building the IISc on a Johns Hopkins University model; others followed, like Bombay University's Department of Chemical Technology setup in 1932, based largely on a University of Michigan model. The movement to build special institutions for advanced training and research in technical fields started at the beginning of the century: the more recent and celebrated institutions in this tradition were the five IITs, which began life in 1950, and that generation is surely not the last. In these technical institutions the fundamentals of mathematics and sciences have to be transmitted. Most of these people in the technical-industrial community were exposed to undergraduate natural and physical sciences and math, and some did more advanced work. This suggested to the scientific community that less-applied fields are indeed justified through their contribution to industrial development.[45]

Scientists, for their part, trained the bright or orderly minds that industry wanted, and industries, in turn, established scholarships and research programs for the advanced training of top students in chosen fields. This is often as close as industry wished to come to investment in R&D. It was industry's new alliance with science that stood as a lighthouse for the planning committees of the 1940s, acknowledging earlier alliances with independent Indian capital. Modern industry and its products were (and are) both the symbol and instrument of the capabilities of a modern state, notwithstanding the skeptical views of postmodern and postcolonial thinkers about the best uses of those instruments.[46] Stemming the brain drain, retaining trained people, and reattracting the talent of those who had left India dominated the politics of the scientific community. This was done not just for show, but because it was believed that industry would pay off eventually by strengthening private wealth, including the middle class, and thus strengthening the Indian state. The range of joint ventures undertaken by Indian technologists, scientists, and industrialists is astonishing; far from being overly dependent on the British relationship, by 1950 the cosmopolitan networks, diplomatic relations, and thirst of Indian (and foreign) capital for new opportunities had initiated a process linking them to dozens of countries, all at once. Intended by the government to have the appearance of orderliness and rational choice, the internationalization of industry appears more to have been an experiment in applied chaos theory. At the helm the prime minister was giving his approval, for example, when opening the Hindustan Machine Tools factory in 1955, built in partnership with engineering and arms manufacturer Oerlikon of Switzerland; Nehru said, "People who think we are not very anxious to have the cooperation of foreign firms or governments are quite mistaken."[47]

With respect to the atomic energy programs discussed here, there is no doubt that they had become beneficial to private- and public-sector industry by 1970, whether through reactors, rockets, or radio telescopes. There was a unique opportunity to work on projects of the highest standards and to transfer new skill and knowledge contributing to confidence and prestige in the eyes of others. The evidence shows that well-trained young engineers would come to a company that was doing this special work for the DAE. It was not their profitability that drove companies to these projects, but their importance in reputation building. That kind of work had always been the path to mobility, beginning with the construction of the railways and building of trains in the nineteenth century. But for the same reasons that young nuclear engineers trained to operate the Tarapur reactor in 1969–70 were no longer working there in 1974, people who learned to make supercomputers, rockets, and missiles found a greater mobility internationally. We know that many young people trained in the IIT and IIM establishments played a major role in the Silicon Valley and dot-com industries in North America in the 1990s and received extraordinary visas or green cards to come quickly to work in America, right up until the dot-com crash of 2000, when many of these green cards were revoked.

The appearance of Bangalore and then Hyderabad as R&D centers was preceded by individual and sometimes small scientific research and technical training institutions. The DAE and other state agencies assigned contracts to industrial plants there, thus imposing on industry the demands of scientists for high-quality equipment. This new learning had long-lived consequences, even when industrial people did not like them at the time. A good illustration is Bharat Heavy Electricals and the DAE, with respect to the fast breeder test reactor in 1981. BHEL was established at Hyderabad in 1956 and opened for business in 1960. DAE contracts were assigned in the late 1970s on the basis of a lot of previous work done by public-sector BHEL for DAE in the early 1970s, but by 1981—for various reasons like regional politics—big jobs were dispersed around the company's factories, with DAE's main reactor vessel being done in Hyderabad, the turbine in Bhopal, the heat exchangers in Trichinopalli, the generator at Haridwar. An intervening variable was the withdrawal of Canadian cooperation in manufacturing components for reactors from 1974 onward. In 1981 Ramanna and Venkataraman (physicist at Kalpakkam) together visited the Hyderabad reactor vessel at BHEL, and Venkataraman observed: "Now why on earth did BHEL dislike the DAE orders? For one very simple reason—BHEL liked quantity orders whereas DAE wanted just one piece of each item. Next, the fabrication had to be under nuclear-clean conditions and not on the regular shop floor. Finally, the inspection standards were very high, of a level BHEL was not used to. Unable to refuse the order, BHEL took to dragging its feet. Heavy pushing was needed to get every little thing done and progress was excruciatingly slow."[48]

The DAE engineers overseeing the job in the BHEL factory asked Ramanna to put more pressure on BHEL higher up. They had already done so, with little result. The point of this story is that Ramanna was soon able to raise this directly with the prime minister, and she asked why BHEL was "so much against DAE orders?" When Ramanna explained the internal situation, she wrote a letter to BHEL's parent ministry telling them to treat this as a step toward self-reliance, and to hurry up.[49] Ramanna soon became secretary of the DAE and took leading industrialists on a tour of its facilities to impress them with the opportunities in working with atomic energy, as Sarabhai had done in 1966. What is striking in this is that Indira Gandhi's father had written in 1955 about a salt research institute, and here was his daughter writing in 1981 to a state corporation about the slow construction of a piece of a reactor. Presumably prime ministers and presidents all over the world have written letters about such "trivial" details, but it demonstrates how seriously the state had to engage in the scientific and industrial enterprise—prime ministers had to concern themselves with these details, it was often argued in India, "otherwise important things would not get done."

The logic underlying Indian science and industry planning before 1947, and then continuing through the war over self-reliance of the 1970s, was that there would be a connection made between the pieces, whether made by pressure of demand or by contingent accident. The gaps between steel and atomic energy, between microwaves and space research, between salt and steel, would close naturally. This would establish the hopeful preconditions for innovation. Shiv Visvanathan says that the innovation chain "is as central to managerial science as the assembly line." It concerns what Whitehead calls "the invention of invention." Through innovation "the machine from a concrete object is transformed into code," he says, and the experts who animate and operate the innovation chain become an inherent part of it.[50] This combining and transforming was going on all the time between scientific projects and the surrounding industries, a code was established, and the methodology of the transformation was called the "scientific method." It was not just a scientific temper that these pioneers offered to India, but also a (usually) reliable methodology. There was also more of what Atma Ram had long before said was lacking, namely, "machine sense." Natural though it might have originally seemed to scientist-planners that convergence would occur, it took another twenty years even after 1980 to close some of those industrial gaps that appeared to them as just transitional in 1960, in short, almost forty years.

Indian Scientists and the Military

The integration of science and military planning was the objective that advisors, including Blackett, urged Nehru to push for in 1947–48, but this took about thirty years to achieve. Indian Independence occurred when India's armies were very large but exhausted, and, despite its deep recent experience in modern warfare, its strategic capabilities were shallow. At the top of society, there was an aversion to militarization and a desire to see a nonmilitary economy. Though prestige was attached to being *in* the military, there was not wide acceptance for what it could *do*.[51] Although it was the very time the country began importing tanks, guns, ships, and planes, an indigenous defense capacity analysis was established in 1950, and in 1952 Bhatnagar and Bhabha were put on a military-science panel, along with

Saha's student Kothari. When Blackett worked to initiate defense laboratories in the early 1950s, when Kothari and Bhabha used Defence personnel and resources to study the effects of nuclear explosions in 1956, when AEC Chairman Bhabha suggested to Blackett that he would be invited to review Defence research in 1963—these were hints of the perceived need for integration, but there was little need then to share resources. These defense laboratories, when finally built, were at first run by almost-retired service officers (noncombatants) who had little sense of science or scientists. Though the two cultures were still strange to one another, the separation of the military and civilian worlds so evident in the 1950s gradually disappeared through cooperation on major projects conceived in the 1960s and built in the 1970s—the peaceful nuclear explosion, the jet fighter, the satellite, the launch rocket, the guidance computer, and the weapons missile.[52]

With the government's major military-industrial investments of the 1960s, the need to share resources became imperative, and the process of integration accelerated. Scientists accepted the fact that big budgets for projects that had dual purposes could underwrite the expense of some of their own ideas, and they certainly enjoyed those projects that integration made possible. They also enjoyed the relative prestige conferred on them through these big strategic projects. Though scientists sometimes had relatives in the military, many only gradually accepted defense needs and military applications of their work, having little prior experience of the relatively closed military world in India. The memory of the colonial government's internal repression by the military using Indian soldiers was still too fresh for others. But many scientists decided that working on major projects with military support was essential to both national prestige and their own personal well-being.[53] The military and the scientific communities projected quite different organizational personalities, so it was hard to cross through the barriers. Nevertheless, they did have something in common: the military and scientific communities sought to disorganize and subdue the structures of social distinction like caste, language, and economic class. Each of these was powerful outside the scientific and military workplace, but as organizations they tried to minimize such structures by pretending to ignore those characteristics, by moving people around, and by mixing linguistic and cultural groups in work teams. They were also among the early adopters of a professional female employment policy.

Combined civilian and military involvement in the bomb test in 1974 and in the missile project in 1975 shows there was only a partial integration of military and civil applications of science, even in the 1970s. Though it took another twenty-five years, the full integration finally occurred in the making of the 1998 bomb and a delivery system for it. Referring to the early 1970s, Defence advisor B. D. Nagchaudhuri described it as very difficult for a free-thinking scientist to work within the strict limits expected by the military: he was not referring to the requirement of secrecy, which he accepted, but to the absence of lateral thinking, marginalization of speculative thinking, and the rigidity of deference to hierarchy with consequent limits on communication.[54] Defence advisor Ramanna echoed these differences in his 1991 book: he said the situation had not changed much by 1980, although senior military officers were more likely to be interested in and understand some of the science that was being discussed than in the early 1970s but were not more confident in Indian research; he said: "Defence research had not made much of an impact on the armed forces, which was not surprising as several vested interests had ensured that its scope remained restricted. Several users openly commented that defence research in India was essentially a sham, because in any case India would have to continue importing materials either through middlemen or directly from outside."[55]

More of military training contained reference to science, though the skeptical remarks by military experts about "boffins," "eggheads," or "wooly-headed scientists" continued.[56] Such skepticism did not arise fresh in the 1970s, and I expect that D. S. Kothari faced it from the beginning in 1950. Though the sheer size of the military budget guaranteed that its activities would be deeply integrated in the economy, only by 1975 could one begin to speak of an Indian military-industrial complex, in which experts from one side could work among and become accepted as experts for the other side. Despite this gradual integration, and in the middle of the war over self-reliance, India continued to be a major importer of conventional arms like field artillery. Prime Minister Rajiv Gandhi's government was weakened by the scandal surrounding payments to secret foreign bank accounts for contracts related to the import of a conventional Swedish-manufactured Bofors field-gun, a choice based on studies conducted starting about 1980 and finalized in 1986. That such an ordinary gun was not to be manufactured in India haunted some observers, echoing the self-reliance questions of the 1970s. All the scientific advisors to the minister of Defence from 1951 sat on committees that recommended (or didn't) the purchase of equipment from abroad: there they would hear, endlessly repeated, a chorus of doubt about Indian equipment in battle, and why this important item or that one had to be imported.[57]

Scientists who became advisors to the minister of Defence immediately after the first bomb test—M. G. K. Menon and Raja Ramanna—had a very long run of official acceptability, finally being appointed ministers of state (though ironically called "junior ministers" in the their seventieth decade) in the short-lived National Front government of V. P. Singh, elected in January 1990. That government fell within less than a year, and with it fell their opportunity for continued power. Along with strategist K. Subrahmanyam, these two physicists were giving guidance to the cabinet, and the prime minister got a resounding critique for relying on these already overcommitted political scientists in the mass-circulation magazine the *Illustrated Weekly of India*. Ashis Nandy, a writer known for his interest in and respect for scientists and the author of a perceptive biography of mathematician Ramanujan and biophysicist Jagdish Chandra Bose, called Menon and Ramanna "the mimic men"; he reserved for Ramanna the description as "the country's most authentic Dr Strangelove."[58] There had been a hint of antimilitarism before in such appraisals of Menon and Ramanna, suggesting that somehow their enthusiasm for weapons was unbecoming to their generation of scientists, contradicted their earlier philosophies, and/or was a sign of their immaturity. But far too little is known about this list of physicists who were Defence advisors of the period, D. S. Kothari, S. Bhagavantam, B. D. Nagchaudhuri, M. G. K. Menon, and Raja Ramanna. All we can conclude is that the vaunted Indian separation of the military, the scientific community, and civil society gradually faded before 1980. Was there a military-industrial complex? Not yet, but its ingredients were emerging: important and influential forces were pushing for it, calling for a decrease in military imports, an "indigenous" weaponry, and a self-reliant military. Along the way they were slowed but helped by international sanctions against the transfer to India of key dual-use technologies and techniques.

Corruption and the Scientific Community

Moving across thousands of pages of documents and thousands of hours of conversations, one is struck to see almost no reference to corruption among scientists or their institutions. Because corruption is a subject splashed across other aspects of life in India by the 1970s, its absence in science is even more interesting. There is occasional reference to the unreasonable or excessive uses of personal, positional, and family influence by scientists, usually with respect to the opportunities within their sphere, for example, employment opportunities, government housing, international travel, or scholarship funds. Sometimes such influence is explained as self-seeking or as seeking on behalf of others. Sometimes kinsmen received some advantage they might not otherwise expect or did not seem to merit; some scientists were said to exaggerate their own achievements at the cost of others, particularly when speaking to media personnel who were unable to appraise those technical achievements critically. There are cases of scientific fraud in India, as elsewhere, though we have no study to tell us whether it has occurred in a ratio proportionate to the size of the scientific community. But there is little evidence of any public belief that the deeper corruption found in other aspects of life, say in defense contracts or telecommunication projects, was found in the scientific community. This is interesting because Saha, Bhatnagar, Bhabha, and Sarabhai, as well as those around them and those who succeeded them, all handled large contracts, authorized construction, made life appointments, and decided on valuable foreign proposals. They also managed large amounts of foreign exchange and joined in committees that had material power and influence. A recent survey of corruption in India since the 1950s did not discover significant misdeeds among scientists: when questioned the editors said that corrupt practices were probably established in the administrative culture surrounding procurement and supplies for scientific institutions, but that elite scientists in decision-making positions were quite well paid and very well recognized, and "in that era" individuals usually accepted their place in the overall scheme of things.[59] Despite lapses and contradictions, there were reviews of performance and merit. Scientists' expectations were, as one observer said, "focused on recognition and perquisites, not on amassing personal fortunes." There was throughout this period a view of the life of science as a calling and a part of a national movement. Saha and Bhatnagar came from poor families, as did rocket engineer (and president of India) Abdul Kalam. Clearly the wealth of Bhabha and Sarabhai was an exception to a pattern of middle-class or poor family origins for most scientists. Even the very comfortable upbringing of M. G. K. Menon or Raja Ramanna were also atypical for scientists. Nevertheless, this question merits careful study: if this impression is confirmed, then science has not conformed to the widespread assumption held by many Indians of themselves that corruption was *everywhere* and none were immune to its practice.

So what rewards and perquisites might scientists have expected? In the 1950s they desired a telephone and a radio, in the 1960s a telephone, refrigerator, and car (and a driver if possible), and in the 1970s trips abroad and access to foreign exchange for work or for professional travel, a car and television and possibly air-conditioning at work. By the 1970s many scientists had refrigerators and phones, but few had cars; moreover they lived in places with very irregular electricity so a fridge was not always useful. These luxuries were not usually obtained as bribes to scientists. Following the pattern of Bhatnagar and Saha, who had arranged for their children in the 1940s, scientists in the 1960s and 1970s wanted to arrange for their children's study abroad, but usually not before they had negotiated and paid for their daughters' marriages. Nearly all scientists' families I met in the 1960s and 1970s were investing in their children's education, girls and boys alike. Just like administrators of other large public projects, scientific administrators (particularly in the DAE and its sister agencies) may have expected and sought larger rewards, but most scientists seem to have thought that small perquisites and recognition as scientists were satisfactory and sufficient reward. Recognition and status, however, could not really be granted by nonscientists; it was the circle of peers from whom recognition could be sought that counted more. Yet that expression could come only from a limited community in India. That is one reason, in addition to salary and promotion, why the status conflicts in scientific institutions were occasionally so intense. We should not forget the frustration and occasional suicides. Looked at another way, few scientists (except in the important matter of university examination results) stood between the public and something the public desperately wanted, like a passport, ticket, degree, or license; so license and ticket issuers and passport officials had a very different opportunity to accumulate unrecorded incomes. In this sense scientists were a community unto themselves: they had to recognize themselves and reward themselves, and it was understood that they would never be rich. There was nothing to resemble a huge Bofors gun contract opportunity in their world. But their values were in circulation, used in the positions taken in the war over self-reliance and the push for high tech in the 1970s. For example, the dubious and expensive Maruti car project was amenable to the bluff, opportunism, and fraud of Sanjay Gandhi and his friends and prolonged the life of a project that should have been stopped but could not be because of who he was. And it was scientists in the war over self-reliance who had articulated the objectives and values that Indira Gandhi heard when her son spoke to his mother about his peoples' car in 1969, leading to the fateful license to go ahead.

But this does not dismiss the possibility that in the case of some individuals, "corruption" played a key role in getting the work done, in improving, distorting, or delaying scientific work or projects. The classic case of "workers" picking up hammers to bang and look busy at a project when it was visited by Bhabha in 1965 graphically illustrates how the system might have worked. Of course, workers banging hammers to impress the boss is a worldwide phenomenon, and at least in this case the boss was inspecting and there was something to inspect. But that Saha did not have sufficient money to build a house in Calcutta at the end of his career when he was a member of Parliament and that he had to find someone to negotiate a bed for him in hospital illustrates this equally well. His "incorruptibility" was widely noticed and discussed in those terms. This absence of discussion of corrupt practices among scientists, along with the matter-of-fact view that the corruption, if any, was likely to be minor, is interesting and probably very important for Indians and India.

The First Nuclear Bomb Test Seen through Indian Lenses

It is essential to understand that half of India's adult population did not know in 1974 that a nuclear test had occurred, and in 1997, the year before the second tests, when 40 percent of the population was below the age of twenty-five and had grown up after 1974, more than half of India's 1 billion population would still not have known that India's first atomic bomb test occurred in 1974. This subject just did not matter in their young lives, even when they were told about it, as it had no salience and no resonance. It was not for everyone a shining symbol of enhanced prestige; it did not speak to everyone in their language. The scientific institutions, scientists, and technologists who did this work were engaged with ideas and techniques, were focused in cities, worked in close relation to industry and the military, were guided by planners and a few top political leaders while more than half of India's population was structurally and culturally estranged from those worlds. They were "uneducated," rural, and concerned with hard work, neighbors, insecurity, and the weather. I have tried to remain mindful of that large number of unaware people while following the narrative about the bomb, woven into a wider tapestry of individuals and institutions that contributed indirectly to it. I am also mindful of the thousands of scientists whose work and institutions did not impinge on it at all. The focus on the bomb is interesting and important, but one should not think that the threads of this narrative, interrupted in places, stand on their own. It is the complex embeddedness of these individuals, research groups, and institutions in a wider supportive scientific community that lifted their careers and projects to become part of a whole cloth.

But among those who knew about the first test and its outcomes, the perceptions ranged from enthusiasm through skepticism to opposition. Celebrations in scientific and political circles did not match the photo opportunities created in 1998, but Indira Gandhi soon went to the spot and stood near its hot hallowed ground. Those associated with the test knew how difficult it was to conduct and complete successfully, even if the yield was in doubt. Scientists and technologists not involved in the project knew their work would benefit from it now that it was an official achievement of "the system." Confidence in the DAE was sure to rise, they thought, and they accepted the inevitable consequence of international disapproval. How much they believed in the official "peaceful nuclear explosion" explanation, except as a rationale and rhetorical device, we do not know. Their private skepticism was often masked with cheerful acceptance of the inevitable.

But among them were more calculating individuals who expected this test would set off a chain reaction that would be hard to moderate, and they were right. It was immediately clear in 1974 that Pakistan was now accelerating its bomb, and the massive Chinese test a month later confirmed that India was next door to an advanced neighbor, and it was believed necessary to keep up with the neighbors. The contractual difficulties with Canada and the United States, though bravely concealed, began to have their effects, so that other partners were sought to bring new deals, new accommodations and conditions, and new forms of dependency. The skeptical lens was available for those who were disturbed by the 1974 test, but skeptics were criticized for not being committed to self-reliance or to India's destiny of greatness. Arguing that a nuclear test changed little *in India,* they were ignored or dismissed by enthusiasts whose India had *indeed* been changed, though it was *not* the India of the half who knew nothing about the bomb. The bomb enthusiasts, for their worldliness, exhibited a strikingly narrow perspective through their lens: this perspective certainly had made the outcome possible but left more calculating audiences to ask, so what? Opponents of Indian nuclear weapons quietly noted the narrow calculation of advantage by the enthusiasts and pointed to a dozen unmet needs that also limited India's potential greatness and influence in the world. Most of them went home to water they could not drink out of the tap, if there was water in the tap.

When the bomb test was followed with the satellite launch, albeit not in the intended Indian launch rocket, and then with television, and then with an announcement of oil flowing from Bombay High, public approval of the bomb test was embedded in a wider set of good news. Even then the leadership felt under threat, the good news became wrapped up in the declaration of the 1975 Emergency period and some contradictory patterns of opportunism and enforcement that soon emerged. Then in 1977 reappeared a fundamental division in India represented by Moraji Desai. In contrast to Nehru, Shastri, and Indira Gandhi was Prime Minister Desai's deep opposition to the whole bomb project. But although Desai was a Gandhian and the first prime minister to really oppose the bomb, it had been only one reason among others that led to his exclusion from that high office for most of his life. More important, neither his Gandhian spirit nor his opposition to the bomb, both well known before he became prime minister, were strong reasons for which he was brought by others to this powerful office in his old age. But he did not conceal his opposition as a tactical way of smoothing his leadership path.

Subsequently writers began to think that perhaps Nehru had been a less-reluctant bomb project director than previously thought, and people like Bharat Karnad and K. Subrahmanyam began to speak and write about India's early but thwarted ambition to be a big power like others, an ambition compromised by international "allies" who wanted no such thing. Though it would be wrong to limit the list of revisionists to those two names, they participated in calls to weaponize the nuclear bomb and make it deliverable. After 1974 the gloves came off for Bhabha, who, it was said here and there, was a bomb-*wallah* from the very beginning, but who had unfortunately been limited for twenty years by the cautious and naïve indecision of his masters, almost up until his untimely early death in 1966. This indecision or even his indecision, they said, constituted India's vaunted "nuclear opacity." The cultural shift made possible by the first test allowed some experts to describe Bhabha as a bomb advocate all along, instead of describing him as he was—a calculating individual who had to balance a number of scientific, political, and technical objectives in the Indian context. After the first test, the axis of self-reliance converged with the axis of being taken more seriously as a great power, and so the ambiguous record of the past was up for reinterpretation following the bomb tests in 1998 (see chap. 1, n. 6).

The most complete of these "gloves off" books is by Bharat Karnad, who provided a monumental review of the strategic culture of India, based on confidential interviews, including some with DAE scientists, and a review of previously secret documents.[60] He criticizes, in his words, India's deterrence by half measures, a recessed and retarded militarization led by Jawaharlal Nehru and followed blindly by others. Nehru particularly "ignored the compulsions of geopolitics," said Karnad, and India's elite unwisely accepted the message that the US and UK would be India's best friends in the event of a nuclear attack from the 1960s onward. Though he says little about scientists before the 1970s, he writes about Bhabha and implicates a few others in the earlier period. But Karnad is too sophisticated to say "they made us do it," and he analyzes not just foreign pressures but Indian tendencies and priorities too, because India's culture and traditions also contributed to this indifference to or denial of the strategic compulsions. There was too much Indian cautiousness, he says, there was the classic and honored willingness to renounce something powerful that was within reach, and this was mixed with a "dissipative kind of domestic politics" that blurred the distinction between matters of national importance and narrowly partisan issues.

This interpretation, however, is a one-dimensional view of the DAE and AEC's work and India's scientific, technological, and industrial development, fixed upon just one technological item in isolation instead of looking at the whole assembly, which is the subject of my work here. It was Karnad's choice to overlook the deep constraints upon other technical projects with which a bomb was interlocked and to suggest that numerous and important institutional and financial limitations and personal disagreements about this one item could have been submerged or eliminated. He thus thinks that the process could have been speeded up dramatically; instead, he says, it was deliberately and thoughtlessly retarded. His choice to focus on one technical item would not have been mine, but being a useful book it provides enough evidence to draw a different conclusion from the author's, namely that for a number of important reasons and following a different calculus, Indian institutions and leaders (including scientists) were *not ready* to construct and test a bomb nor to think about its symbolic or strategic use until it actually occurred. There was an inertia in the system that delayed this step until 1974, and delayed it again for the next twenty-four years; the surveillance and interference from outside India played a role in the timing of tests, but the evidence he adduces demonstrates that a quite different internal calculus was at work.

Two Final Questions

One only has to mention the role of science in India's overall development to be immediately asked two questions by ordinary people in India: the first is, Why has India's large scientific community produced so few Nobel Prizes? And the second question is, What value has India received from its investment in science and technology? These questions are difficult to answer quickly, but people keep on asking them, and they do not go away.

One might ask the first questioner in return, does she know the number of scientists in India who have been nominated for the Nobel Prize but not chosen, or whose role in work leading to the prize is closely related, and acknowledged in the acceptance speech but not in the prize citation? Or, does she know the number of scientists raised and trained in India who have been nominated for the prize concerning work they have done in other countries? Does the questioner know the number of exceptional scientists and mathematicians in India whose type of work would not really qualify them for the Nobel Prize? Does the questioner understand the difficulties of doing exceptional scientific research at all? Are Indians who ask this question aware that arguably such work is more and more difficult to do everywhere, often requiring more and more money? And finally, have conditions of work changed for prize-winning research? Visvanathan cites two unlikely conversations for an answer, one with Atma Ram and the other with A. Rahman, both with roots in the CSIR. Ram suggested that India's Nobel Prize performance is like almost winning the Olympic bronze medal, over and over again; Indians thought that as GDP rose, he said, perhaps there would be Nobel Prizes. Yes, Nobel Prizes were awarded to Ross in 1902, Tagore in 1913, and Raman in 1930. But GDP has risen and the middle class has benefited, so where are the prizes since 1947? We see Khorana, Chandrasekhar, Sen, and Ramakrishnan, but where are the others, they say? A. Rahman used the analogy of grass hockey at which Indians were once the worlds' best; as the playing surface, the momentum, and the rules of the game changed, India was no longer playing its familiar game. That change has occurred in scientific research, said Rahman, and Indians are no longer working in the right way for a prize.[61] In my view, not disregarding these insights about change and bronze medals, it is important to investigate the cases where nominations are occurring but prizes are not being awarded, to see what patterns if any emerge.

Nevertheless, skeptical Indians asking about the Nobel Prize can reply that even recently the prize has been given for research from the geographic margins, focused on a subject from the conceptual margins of science. The 2005 Nobel Prize in Physiology or Medicine was awarded for research on ulcers to two Australians, one from the "remote" city of Perth in western Australia. Surely, few subjects are less glamorous anywhere than ulcers. Said prize winner Barry Marshall, who was thirty when he began to work with co-winner Robin Warren in Perth on the bacterium *Heliocobacter pylori* in ulcers, "The idea of stress and things like that [causing ulcers] was so entrenched nobody could really believe that it was bacteria," Marshall said. "It had to come from *a weird place* like Perth in western Australia because I think nobody else would have even considered it."[62] How similar is that to a young medical scientist doing research on malaria in a hut in Calcutta, resulting in a prize in 1902? Or to young physicists doing research on a simple workbench in a Calcutta lab in the mid-1920s that led to a prize in 1930—wasn't Calcutta "a weird place" too?

The second question asked frequently is what value has India received from its investment in science and technology? There is one approach to an answer based on forensic accounting of separate projects, tracing the expenditures of all public and private funds for scientists and their institutions, and then estimating their specific benefits. This is the approach taken by science councils, public accounts committees, and auditors general the world over; a monumental comparison of all the enquiries and parliamentary committee reports would begin to yield part of the answer, punctuated by expert disagreement about what is value for money. We do not really know how much has been spent on this total effort in India, because the methodology to track the expenditures was itself doubtful. If accurately traced, however, how do science and technology expenditures compare to expenditures on gold jewelry? Should we ask if India could do without gold jewelry? The result is that it is hard to conceive of India without that gold, as it has been a form of assurance against the future for generations. Now think what India could achieve without the institutions and results of modern science and technology? The evidence here suggests that science and technology are essential to India under almost any conceivable political framework, governed by almost any number of political parties, as a form of assurance against the future—and, though this is not well known, as a link with its intellectual past. "Assurance against" sounds like a vaccination and should be balanced by "commitment to" the future.

But there is another approach, says the counter-questioner; what value does India receive from its dancers, painters, musicians, and writers? "Inestimable" is an archaic term that seems appropriate here because painting, music, and dance are ancient arts and skills that modern Indians perform exceptionally well, to the world's delight. Some of its modern architects, painters, some cinema directors, musicians, and writers are admired across the world, and their value to India is hard to estimate. In contrast the state supported the "nucleus" of the scientific community more handsomely than it supported artists. But should we ask if cultivating modern science is as important to India as cultivating modern painting or classical dance? If the answer is that painting is at least as important as science, though on a different scale, then the frame for an answer is established, and Indians can judge how much more important scientists might be than painters and therefore how much more of India's resources, if any, scientists deserve? But isn't it necessary to rephrase the question: what else in India would not be possible without this investment in science and technology, or to what else is science and technology a means? This line of reasoning gives a hint of why influential nonscientists signed the statement on scientific temper in 1981. The question should be turned around: What was the Indian contribution to other intellectual traditions, across the world, through science and technology? And what might it be in the future? Is there an intrinsic inestimable value of science and technology, and should its costs be underwritten by all states and societies?

Though funds went to nuclear India and its achievements, clean drinking water from a tap remained a privileged luxury at the end of the century, leading to widespread medical problems for half the population, wasted time for women who carried water for hours each day, and so on. Low quality of life, high preventable infant mortality rates, weak infrastructure, all evolved alongside nuclear India. And as for one of the original intents of all the nuclear money, the delivery of very cheap electricity to the economy, how much has that "nucleus" contributed to India, and particularly to its poor people? Of course a thousand national and international connections have been made through the nuclear program, and the atomic energy reactors will eventually contribute 10 percent of India's total electrical supply (or even perhaps 20 percent), but how deeply have poor people tapped into those connections, how much of the electricity can they use? Throughout the period of this book, official approaches to conditions suffered by the poor were incoherent, incomplete, and inadequate.

In the grand sweep of the twentieth century, there is an abiding tension between seeing science as movement and seeing science as institution. There was commitment and excitement in both views and rewards from both. One, the movement, was intended to transform culture and politics through ideas like scientific planning, scientific socialism, scientific culture, even Nehru's scientific temper or the enthusiasm of scientism. The other idea, the institution, was rooted in institutions that were established from the beginning of the century; indeed, some of them like the universities and the IACS were formed in the nineteenth century, and the IISc in Bangalore was conceived in the first years of the century. This institutionalization was incremental and cumulative, but the institutions themselves were open to change and were the sites of ongoing disagreement about the best ways to proceed; in short they were (or their members were) the scenes and sites of movements. Even the resistance of institutions to change generated a form of scientific practice as movement. Here was the open space for the combination of intense personal politics, as we have seen throughout this book, and structural change, resistance, and adaptation. From those struggles new institutions were born and nurtured the purposes and interests of a new generation of men and women in science and technology, refashioning scientific India and its nuclear estate.

CHRONOLOGY OF EVENTS

1893 Shanti Swarup Bhatnagar is born in Shahpur, Punjab. Meghnad Saha is born in Mymensingh, East Bengal.

1909 Homi J. Bhabha is born in Bombay.

1919 Vikram Sarabhai is born in Ahmedabad, Gujarat.

1939 Homi Bhabha visits India on holiday, stays in Bangalore when war starts, and observes history of open confrontation between C. V. Raman, Meghnad Saha, Shanti Bhatnagar, and others. Indian physicists like Meghnad Saha, C. V. Raman, Homi Bhabha, Satyen Bose, and K. S. Krishnan are reading scientific papers on fission of uranium.

1940 Council of Scientific and Industrial Research (CSIR) is created; M. N. Saha, J. C. Ghosh, Nazir Ahmed, S. S. Bhatnagar are scientific members. Prominent industrialists are also appointed. Subhas Bose is elected president of the Indian National Congress Party.

1941 Shanti Bhatnagar is appointed director of the CSIR. Subhas Bose escapes to Germany in order to form an army to liberate India.

1942 Japanese capture Burma; Cripps Mission to India fails; Quit India movement launched; two US missions to India (Johnston & Grady) are sent by Roosevelt. Berkeley cyclotron arrives in Calcutta.

1943 Bengal famine threatens imperial power relations; Japanese troops fight inside Indian border in Assam. Saha builds cyclotron in Calcutta. Subhas Bose moves to Tokyo, to command Indian National Army.

1943 Archibald V. Hill is sent as viceroy's envoy to communicate with "disaffected scientists" and reassure the scientific and industrial community of their importance in war effort; plans for research institutions are discussed as part of independence negotiations.

1944 Roosevelt, Churchill, and McKenzie King meet in July at Quebec; their

Declaration of Trust identifies Indian radioactive materials including thorium as Allied strategic property.

1944–45 Official Indian scientists' tour of UK and US research establishments includes meetings with those in strategic war research; mission is made up of Shanti Bhatnagar, leader, Meghnad Saha, Nazir Ahmed, and others. Scientists Mission meets with King George to discuss royal charter for National Institute of Science; Saha and Bhatnagar are "interviewed" by US intelligence agencies in 1945 after visits near secret Manhattan Project facilities.

1944 Tata Trusts approve funding for Homi Bhabha's nuclear research institute in Bombay; Saha searches for funding for similar lab in Calcutta; Bhatnagar begins to build major research laboratories for physics and chemistry.

1946 Atomic Energy Committee formed, with members Bhatnagar, Bhabha, and Saha. Bhatnagar assumes leadership of CSIR, which finances the Atomic Energy Board; great powers commence secret prospecting for uranium in India, some under guise of archaeology.

1947–48 Independence and Partition are concluded; secret shipment of Canadian uranium is sent to India; Allies compete for use of thorium/beryllium deposits in India. Bhabha becomes chairman of Atomic Energy Commission and begins regular communication with atomic energy agencies in UK, US, France, and Canada; Saha opposes formation of the Atomic Energy Commission and refuses Nehru's offer of membership in it. British physicist Patrick Blackett is invited by Nehru to become his advisor on science and defense.

1948–49 Scientific Advisory Committee to the Cabinet is formed. D. S. Kothari is appointed scientific advisor to the minister of Defence and creates Defence Research Organization within CSIR's National Physical Laboratory. Physicist P. C. Mahalanobis appointed advisor on economic planning and statistics and builds Indian Statistical Institute in Calcutta. Bhabha and Bhatnagar meet regularly with Nehru.

1951 Colombo Plan created by Commonwealth to counterbalance US strategic and commercial activity in South Asia; officials from Canada hold preliminary discussion of nuclear cooperation in this region; France secures exclusive contract to extract and refine thorium in Kerala.

1952 Saha decides to run for election to Lok Sabha and wins seat in North-West Calcutta; Saha publicly criticizes Nehru and Congress, privately criticizes Bhatnagar and Bhabha. Bhatnagar, at height of power, creates 22 new laboratories.

1954 Department of Atomic Energy is created; Bhabha becomes secretary to government and reports exclusively to Nehru. UK decides to help build small reactor in Bombay.

1955 Bhatnagar dies suddenly in Delhi. Bhabha is appointed chairman of Geneva Conference on Peaceful Uses of Atomic Energy; Indo-British nuclear energy cooperation agreement is signed; Canada-India reactor deal is negotiated between Nehru and Saint Laurent. Khrushchev visits India to initiate nuclear cooperation. US agrees to supply heavy water for India's next reactor.

1956 Saha dies suddenly in Delhi. Bhabha is now without political competition and without his ally Bhatnagar; first Indian reactor (APSARA, using a UK design) is operational; construction of the Canada-India reactor begins.

1957 *Sputnik* satellite is launched by USSR. Defence Research and Development Organization is established in Delhi; nuclear fuel negotiation with Canada breaks down.

1957 Parliament passes Science Policy Resolution; first National Conference of Scientists, Technologists and Educationists is organized by cabinet committee.

1959 Nehru confirms that one or two reactor contracts will be awarded to best bidder. US confirms it will loan 10 tons of heavy water for Canada-India reactor, which is called CIRUS eventually. Bhabha states India capable of building atomic bomb.

1960 France, UK, and Canada tender reactor bids, but then US bid is entered through the side door.

1961 First Indian agreement is signed with NASA for cooperation on space research.

1962 India-China border conflict erupts. India accepts US proposal for first nuclear power reactor at Tarapur; Canada and US conclude reactor contracts with Bhabha and Nehru. Tata Institute of Fundamental Research building is opened in Bombay.

1963 India begins to develop national anti-aircraft radar system and other post–China conflict reforms. Bhabha starts work on Electronics Commission; military expenditure almost doubles.

1964 Nehru dies; Chinese test their first nuclear bomb; Bhabha receives approval to plan bomb development; construction of nuclear fuel reprocessing plant begins.

1965 India-Pakistan border conflict breaks out; famine rages in Bihar and east Uttar Pradesh, resulting in larger shipments of US food and increase of US president's influence on agricultural policy; priority for agricultural research and population control shift upward. Bhabha states 18 months required to build a bomb; rocket launch base opens for ionosphere research in Kerala.

1966 Bhabha dies suddenly in plane crash in the Alps. Indira Gandhi becomes prime minister. India's foreign exchange crisis peaks, and rupee is devalued by 57 percent; imports are severely restricted; massive expansion of agricultural projects with research and teaching gets official commitment. Vikram Sarabhai succeeds

Bhabha as chairman of Atomic Energy Commission, does not agree with preparations for bomb, opposes bomb testing.

1967 Canada-India negotiations proceed for second Rajasthan reactor; problem of heavy water becomes acute.

1969 Indian-built rockets are launched successfully on west coast; missile launch base on east coast north of Chennai opens. US and Canadian reactors under construction; second Chinese thermonuclear bomb is tested. Indian banks are nationalized.

1970 Department of Electronics, administratively close to the Department of Atomic Energy, is created.

1971 Conflict over Bangladesh breaks out between India and Pakistan; Chinese satellite is launched; plans are made for creation of new space agency; US (Gujarat) and Canadian (Rajasthan) power reactors open. Sarabhai dies suddenly in Thumba, Kerala.

1972 India creates Department of Space and Department of Science and Technology. US aid is suspended after disagreement with India over liberation/invasion of Bangladesh. Decision to build and test nuclear weapon is made by Prime Minister Gandhi; foreign exchange reserves are pronounced in "good shape."

1973 OPEC production control and rise in oil prices begin; government finance is in crisis. First shipments of heavy water are sent from USSR; US-France-India secretly negotiate on enriched fuel. Crisis over Indian balance of payments begins.

1974 India's first nuclear bomb is tested; official announcement of yield is approximately equivalent to US bomb dropped on Hiroshima. Paris Club increases allocation of aid; US at first denies use of American material in bomb preparations; US approves shipment of enriched uranium to GE reactor in India; London Club of Nuclear Suppliers with USSR-US cooperation is created to track nuclear exports. India persuades Canada to negotiate to reestablish nuclear cooperation.

1975 India launches first satellite on Soviet rocket. Prime Minister Gandhi declares State of Emergency; Canadian and Indian negotiations to restore nuclear cooperation are started; US delays enriched nuclear fuel shipments; probable Pakistan nuclear bomb building program via plutonium separation starts; India promotes debate on Peaceful Nuclear Explosions at Vienna.

1976 Unsubstantiated rumors circulate of second Indian nuclear bomb test; Canadian cabinet votes against resumption of nuclear agreement with India, which remains under Emergency rule; Canadian firms lose contracts and market position. US ships nuclear fuel only after President Carter intervenes; antinuclear movement opposes US enriched uranium fuel exports to India at hearings.

1977 Gandhi and Congress Party lose election; new coalition government repudiates nuclear bomb testing though experts continue to develop capacity to test bombs; full review of Atomic Energy is initiated by Prime Minister Desai; US negotiates withdrawal from its fuel contract for GE reactor, passing supply role to France.

1978 Most non-nuclear forms of Canadian aid resume. US ends enriched nuclear fuel supply.

1974–78 US and Canadian cooperation on nuclear projects in India (except completion of US fuel shipments) ends; USSR and France cooperate more and more.

1980 Indira Gandhi returns to power by election; first successful launch into orbit of Indian satellite carried up by Indian rocket.

1998 Series of nuclear bomb tests in Rajasthan, India, is followed immediately by bomb tests at Chagai, Pakistan; testing of Indian missiles continues.

BIOGRAPHICAL NOTES

Information here complements biographical information in the text. Not all persons important to the text are adequately described here; for example, Homi Sethna, director of Bhabha Atomic Research Centre and chairman of the AEC from 1972 to 1983, has successfully limited what is publicly known about him. Meghnad Saha, Shanti Bhatnagar, and Homi Bhabha are fully described in the text.

NAZIR AHMED studied physics in the Cavendish Laboratory at Cambridge, completed his PhD in 1927, and returned to Lahore to teach physics. Ahmed moved from Lahore to Bombay to work in cotton research, proving the flexibility of elite training at Cambridge, and the mobility offered in the growing industrial research community. A member of the exclusive team of scientists that toured Allied nuclear research installations in 1944–45 with Bhatnagar and Saha, he became involved in planning the CSIR and was thus a major figure in Indian science planning and politics in the late 1940s. He left India suddenly during Partition in 1947 and soon became the first chairman of the Pakistan Atomic Energy Commission.

MAULANA AZAD was born of an Arab mother and Pashtun-Afghan father in Saudi Arabia in 1888; he spoke his father's Pashtu and Urdu and his mother's Arabic as well as English and Hindi. He came to Calcutta in the 1930s and learned Bangla there. A newspaper publisher and supporter of the revolutionary party Juganthar, Azad knew Saha (five years his junior), particularly after the British banned the newspaper he published in Calcutta. He was elected president of the Congress Party in 1940 and was arrested (while actually writing a letter to President Roosevelt) and imprisoned with other Congress leaders in August 1942 until early 1945. Though Shanti Bhatnagar learned about Azad in Calcutta, as a government official Bhatnagar would have to have been very careful in communicating with Azad up until about 1946 because of government surveillance. Azad had no intellectual preparation for his portfolio of natural resources, so he relied on Bhatnagar's considerable knowledge. Both poets, they were drawn to working together in the late 1940s. Though Shanti Bhatnagar could see Nehru alone, whether on Atomic Energy or CSIR business, Azad was Bhatnagar's intermediary to the cabinet if necessary and usually supported Bhatnagar's projects.

SURI BHAGAVANTAM was a physicist trained by C. V. Raman at his lab in Calcutta. He published his first research results at age twenty-four in the prestigious *Physical Review* in 1932; without an advanced degree (like Raman) he did not study abroad but eventually received his DSc from the University of Madras based on his published papers on classical crystallography. There is a suggestion that he was nominated for an FRS by C. V. Raman in 1944, but he was not elected. Originally from Hyderabad, he was scientific liaison officer at age forty at the Indian high commission in London (1947–49), where he met Krishna Menon, who became minister of Defence in 1957. Bhagavantam worked as a physicist at the IISc in Bangalore, where he became director in 1957, succeeding Thacker; soon he was appointed by Minister Menon as scientific advisor on defense and concurrently the first chairman of the board of Bharat Electronics Corporation, a state company created and owned by his ministry.

JAGADISH CHANDRA BOSE was born in Mymensingh, East Bengal, in 1858, the year when the British government took control of the administration of India, succeeding the East India Company. He came to study in Calcutta at age eleven and was trained initially by Jesuits, including the legendary Father Eugene Lafont, who lectured in physics when the IACS opened in 1876. Bose graduated from the University of Calcutta in 1880, went to study medicine at the University of London, and then moved to Cambridge in 1882 to complete the natural science tripos. He was appointed the first Indian professor of physics in the Presidency College in Calcutta in 1885 and began to conduct research in electricity and magnetism in the laboratories of the IACS. However, "at thirty-six [1894] he had no reputation as a scientist. Nor had he the requisite infrastructure" (Subrata Dasgupta, *Jagadish Chandra Bose and the Indian Response to Western Science* [Delhi: Oxford University Press, 1999], p. 47). His research, which began in the 1890s, led to recognition at British scientific meetings in 1896. He built an international reputation for research and instrument building, which produced increasing support from the government of Bengal, and became a legend by developing an "Indian approach to science" in his study of radio waves and plants. He rapidly acquired star status among poets, businessmen, and British and Indian politicians, was awarded a knighthood in 1917, and elected to the Royal Society in 1920. He built his own institute, poured his own fortune into it, and persuaded the government to support it too, well before his death at age ninety-nine in 1957. (See J. Lourdusamy, *Science and National Consciousness in Bengal,* chap. 3, "Re-defining Science" [Delhi: Orient Longman, 2004].)

SATYENDRA NATH BOSE was born the eldest child in 1894 in a middle-class family in Calcutta, where his father, who studied science in college, was an accountant in the Engineering Department of the East Indian Railways. When Bose was eleven years old, Bengal was partitioned by the British in July 1905 by Lord Curzon, and Bose was specifically ordered by his father to stay away from revolutionary activities; though Bose obeyed his father's orders, he sympathized with those who were free to work with the revolutionaries who went underground. This was one major connection with his friend Meghnad Saha, who had identical sympathies. Bose entered a middle school in 1907 where the headmaster and the senior teacher of mathematics encouraged him. He demonstrated a grasp for mathematics by scoring 110 points out of 100 in a school test. Despite his weak eyesight, Bose was an avid reader like his father and showed a talent for languages, especially Sanskrit, French, and German, as well as English. After completing high school in 1909, Bose decided to study science at Presidency College in Calcutta. Some of Bose's classmates were Meghnad Saha, J. C. Ghosh, and his teachers were pioneers like J. C. Bose and E. P. Harrison in physics, P. C. Ray in chemistry, and D. N. Mallik and C. E. Cullis in mathematics. He then went to the University of Calcutta to get his master's in physics and mathematics, despite the fact that a government position was open to him with his performance in his undergraduate degree. Following their master's degree examination in 1915, Bose and Saha continued their studies in physics and applied mathematics at the newly established University Science College at the University of Calcutta. Asutosh Mookerji arranged for the appointment of both Saha and Bose as lecturers in applied mathematics. But neither of them could get along with Professor Ganesh Prasad, the head of the department. Ganesh Prasad was awarded his DSc from Allahabad University and later completed his math tripos at Cambridge, working with Felix Klein and David Hilbert at Göttingen. Prasad, who was from North India, did not have a high estimation of Saha and Bose, an attitude that Bose attributed to their "unsatisfactory" preparation in mathematics at Presidency College, with an added suggestion of Prasad's condescension toward Bengalis. Bose himself remarked, "Disappointed, I came away. I decided to work on my own." So Bose and Saha asked to be transferred to the Department of Physics, where S. K. Mitra was already a lecturer. And Bose did work on his own, completing his PhD in physics in Calcutta in 1920, so that now he was in a position to become a founding member of the physics department in the new Dacca University. He received an offer for a readership in 1921 from its first vice-chancellor. The professor of physics Walter Jenkins recommended that either Bose or Saha be appointed; soon Bose was a professor and gradually built a reputation based on his extraordinary interaction with Albert Einstein before and during a visit to Europe in 1924–26, work which was eventually called the Bose-Einstein statistics (see chap. 2).

SATISH DHAWAN was born the son of a Lahore judge in 1920, studied engineering before doing an apprenticeship at Hindustan Aeronautics, Ltd., in Bangalore repairing Liberator bombers, the planes previously used by Bhabha to fly cosmic ray experiments in 1942–43. He was unusual in having previously completed an MA in literature, before going on to complete a doctorate in aeronautical engineering under Hans Liepmann at Caltech. He joined the IISc in 1950 and set up its first wind tunnel for testing. By

1955 he was head of the department, and in 1962 was the youngest person appointed as director of the IISc. An expert on supersonic fluid dynamics and shock waves, he developed a widely used technique for the measurement of skin friction; he remained active in space research after he ended his role as chairman of the Space Commission in 1979. He was also most proud, when interviewed in 1998, of his work on the flight of birds. The rocket launch base at Sriharikota, north of Chennai, was renamed the Satish Dhawan Space Centre after he died in 2002.

JNAN CHANDRA GHOSH was born in East Bengal, entered Presidency College in 1909, and lived in the same hostel as Meghnad Saha. Ghosh graduated nine years later with a PhD and went to the University of London on a state scholarship in 1919, working in F. G. Donnan's labs. He returned to Bengal by joining the new chemistry department at the University of Dacca and he soon became its head. During this period he published landmark papers on ionic attraction of the strong electrolytes, for which Debye and Huckel gained their reputations. Following the removal of C. V. Raman from the directorship of the IISc in Bangalore, he was appointed to replace Raman as director, work for which he was given a knighthood. He was India's director general in charge of Industries and Supplies in Delhi from 1947 to 1950, at which time he was appointed director of the new Indian Institute of Technology at Kharagpur, just outside Calcutta. There he built a Fischer-Tropsch reactor for coal liquefaction, planning to do research and produce oils and gas from the abundant coal resources that lay nearby, but before he could realize that project he was appointed by Nehru to the Planning Commission in 1955. Thus he remained sensitive to energy questions and alerted his friend Meghnad Saha, now in Parliament, about official plans for nuclear power.

PIARA SINGH GILL was born to a farming household in a village in the Doaba region in the Punjab in 1911, graduated from high school, and promptly left India in 1929 at age eighteen, first to drive a taxi in Panama and then to work in California's fruit orchards and restaurant kitchens. Within two years, while still working, he was studying at a college in Sacramento before winning a tuition scholarship to the University of Southern California. In 1936 he began studying for his PhD with Arthur Compton (Nobel Prize in physics, 1927) at the University of Chicago. Focused on cosmic ray distribution at lower latitudes, his work took him on board a ship from Vancouver to Tasmania fifteen times in 1937–38, confirming that cosmic rays have a minimum incidence at the equator and maximal incidence at 45 degrees north and south. Similar work on board ship had been done ten years previously by Jacob Clay, of Bandung, Java, but Gill's research data were probably more complete. Gill was present at the 1939 international conference on cosmic rays in Chicago, where Clay (now at Amsterdam), one of the discoverers of the latitude effect in cosmic rays, presented five papers. Declining Compton's offer to stay at Chicago, Gill accepted a gift of the lab equipment he had built there and returned with it to Lahore. He passed through the port at Calcutta in 1940, helped by Meghnad Saha to get his lab equipment past skeptical customs officials. Though he was immediately offered a military job, he declined it in favor of a lectureship at the Forman Christian College in Lahore, where Bhatnagar had worked twenty years before, giving a weekly lecture on physics to students at the University of Punjab, among them Har Gobind Khorana and Abdus Salam. He met Bhabha during a brief visit to Bangalore in 1940, at Nehru's suggestion, but he later said that "the prejudice against American degrees coupled with my refusal to work for the Defence Ministry" worked against his getting a university position. Like Ahmed, Bhatnagar, and Sarabhai before him, he too went to high altitudes in the Himalayas in 1945 to study meson production (at 5,000 meters and then flew up to 11,000 meters) for experiments in a small Royal Air Force plane stationed at Lahore, again through Saha's intervention. He was offered a position in TIFR in 1945, which he did not accept. After spending 1946 on secondment working with M. S. Vallarta at MIT on solar flares, he decided to accept a position at TIFR, where he arrived in July 1947. He stayed briefly at TIFR, then left India again and worked during 1948–49 at the National Bureau of Standards in Washington, DC. He finally returned to India for good in 1949 and used his connection with Nehru to get appointed as officer on special duty in the new office of the AEC in Delhi. Looking for a place outside Bhabha's influence, Gill became professor and head of the Department of Physics at Aligarh Muslim University, where he continued high-altitude cosmic ray research at Gulmarg in Kashmir. Relations with Bhabha did not improve, however, and TIFR built a better-equipped cosmic ray lab nearby, at a higher altitude than Gill's lab. Competition in this research field, and interpersonal tensions, led to a 1954 meeting near Delhi between Bhatnagar, Gill, and Compton to try to clarify Gill's role in relation to the AEC and to Bhabha. Gill became head of the new Indian Central Scientific Instruments Organization at Chandigarh in 1963, under the CSIR, with a push from Nehru; this was a position he held until CSIR required him to retire in 1971.

P. N. HAKSAR was born in Allahabad and met Jawaharlal and Indira Nehru in the 1930s, being distantly related to Nehru's wife. Having studied physics with Meghnad Saha (along with B. D. Nagchaudhuri) in Allahabad, Haksar moved to study anthropology at the London School of Economics and then became a lawyer. At the LSE he had occasionally cooked meals for Indira and Feroze Ghandy, her future husband. Called to the Bar at Lincoln's Inn, Haksar was persuaded by Jawaharlal Nehru to join the Indian Foreign Service and so was appointed India's deputy high commissioner in London in 1947, under Krishna Menon, where he stayed until 1955. While working in London he met physicist Raja Ramanna in the late 1940s. He became Prime Minister Gandhi's secretary in October 1967 at age fifty-four and was made a member of the AEC in 1968. He was the second member of the prime minister's office staff to be a commissioner of the AEC, the first having been Cabinet Secretary Dharma Vira in 1964–66. When he came to the prime minister's office, he helped formulate Indira Gandhi's Ten Point Program, set up the Research and Analysis Wing (RAW), and created a revenue intelligence unit inside the prime minister's office in 1968–69. Haksar masterminded the eclipse of the cabinet secretariat in favor of the prime minister's office. By early 1975 P. N. Haksar was deputy chairman of the Planning Commission. (For observations on his role in Indira Gandhi's government, see Katherine Frank, *Indira: The Life of Indira Nehru Gandhi* [London: HarperCollins, 2001], p. 139.)

DAMODAR DHARMANDA KOSAMBI was born in Goa in 1907, accompanied his father to the US, and completed his high school education in Cambridge, Massachusetts, in 1924. He was immediately admitted to Harvard at age seventeen but returned instead to India with his family. When his father returned to Harvard in 1926, Kosambi returned to study mathematics and completed a BA in 1929. He began to teach mathematics in India at the Benares Hindu University in 1930 and at Aligarh Muslim University in 1931 under the direction of André Weil; he eventually left Aligarh and moved to Poona in 1933, where his wealthy in-laws lived. Though a strong teacher and recognized mathematician, publishing in path geometry, statistics of infinite dimensions, and probabilistic number theory, he did not complete any further formal degree. Meanwhile he expanded his interest in archaeology, numismatics, Indology, and Marxist political theory. While in Bangalore he met Homi Bhabha, and in 1945 he accepted Bhabha's invitation to work at TIFR, while commuting between Poona and Bombay. After three years he left on a UNESCO traveling fellowship to study "calculating machines," during which he renewed his friendship with old friend Norbert Wiener at MIT, leader in "calculating machine theory." During this time he taught geometry as visiting professor at the University of Chicago. On return in 1949 he stayed in Bombay during the week, began again to teach and write in TIFR's School of Mathematics, but did not really build a group and school in the way Bhabha hoped. He was troubled by the split between the communist parties of China and Russia, countries he visited more than once. He was superseded at TIFR by younger mathematicians during the 1950s, and his work outside mathematics bloomed, including Indian history and the international peace movement. From 1960 he was placed on annual contracts, and in 1962 when he turned fifty-five his contract was not renewed. Kosambi died in Poona in 1966.

DAULAT SINGH KOTHARI was born in Rajasthan in 1906 into a poor family but did well at school and so won a scholarship and interest-free loan to study physics with Saha in Allahabad. He went on a state scholarship to Cambridge for his PhD in astronomy in 1930–33, where he met people like Blackett and Chandrasekhar. His dissertation on the formation of neutron stars in white dwarfs attracted considerable attention, enabling Saha to get him transferred and appointed at the new University of Delhi in 1934, where he was promoted to professor in 1942. He led the field on ionization of compact stellar objects under high pressure at this time and was nominated for an FRS by Saha, Bhatnagar, and Birbal Sahni, backed by Sir Arthur Eddington, E. A. Milne, R. H. Fowler, and others in 1944. Against his candidature were Hermann Bondi, Fred Hoyle, and Raymond Lyttleton, and it was unsuccessful in votes over the next five years. In 1948 he was appointed scientific advisor to the minister of Defence, where he remained until 1961 (succeeded by Bhagavantam); thereafter he chaired the influential University Grants Commission and participated in numerous other public bodies. He died in 1993.

KARIAMANICKAM S. KRISHNAN studied in American College, Madurai, and Madras Christian College and then was awarded the post of demonstrator in chemistry in Madras Christian College, in the countryside outside the city. In 1923, he went to work as a research scholar with C. V. Raman in the Indian Association for Cultivation of Science (IACS) in Calcutta and joined in the crucial light-scattering experiments, for which Raman was awarded the Nobel Prize in 1930. In December 1928, he moved to Dacca University as the head of the Department of Physics, joining theorist Satyen Bose. In 1933, he came back to Calcutta as the Mahendralal Sircar Professor of Physics at the IACS, inheriting Raman's position. For his light-scattering work he was elected FRS in 1940, and in 1942 he was named professor at the University of Allahabad, inheriting Saha's position. About 1944–45 he was identified (as shown in chap. 9) as a suitable first director of the National Physical Laboratory, yet to be built, was knighted in 1946, and appointed in 1947 director of the NPL in Delhi. It was his specialized knowledge of the behavior of electrons in graphite and of carbon in general that made him most useful to the AEC, plus his broad knowledge of physics and his high international reputation. Not insignificant also was his command of the resources of a very large laboratory, at the center of power. Krishnan died in 1961.

PRASANTA KUMAR MAHALANOBIS was born in 1893 in Calcutta in a family whose roots lay in East Bengal; he entered Presidency College in 1909 and graduated with an honors degree in physics in 1913 at age twenty. Though not from a rich family he left immediately for Cambridge and completed the tripos in 1915 while at Kings College; having stood first in the Part II exam in physics, he was awarded a senior research fellowship. Though his performance in the Part I mathematics exam was just average, this curiously foreshadowed a distinguished career in statistics. In 1913–14 he made contact with mathematician Srinivasa Ramanujan and poet Rabindranath Tagore, who had just won the Nobel Prize in Literature. He returned to Calcutta, participated in the Einstein translation project with Saha and S. N. Bose in 1919, and then worked at the new Visvabharati University founded by Tagore and later, after 1921, at the prestigious Presidency College, where his uncle was a professor of physiology. There he set up a small institute inside the physics department, doing statistical studies of floods in Bengal with Saha in 1923. That institute eventually became the independent Indian Statistical Institute (ISI) in 1931. In times of high political activity, he was appointed acting principal of Presidency College, covering for British principals unable to manage, and sometimes he sided with students in less-confrontational gestures (such as refusing to wear a suit under the convocation gowns in 1934 and replacing suits with dhotis). Having started the statistics journal *Sankhya* in 1933, Mahalanobis published widely and built an international reputation; he was elected as FRS in 1945. By 1950 he was the senior planner-statistician to the government of India, with direct access to Prime Minister Nehru. R. A. Fisher worked at ISI, as did J. B. S. Haldane. At the time of his death in 1972, the ISI had a staff of 2,000 in offices all over India, with an annual budget of Rs 20 million.

MAMBILLIKALATHIL GOVIND KUMAR MENON was born in 1928, the son of the dewan of the princely state of Jodhpur in Rajasthan. Menon achieved his PhD in physics at the University of Bristol in 1953. He specialized in the study of neutrinos in deep underground experiments in the Kolar gold mine in South India and was a leader in the TIFR cosmic ray group that greatly advanced nuclear emulsion techniques flown on high-altitude balloons for x-ray and gamma-ray astronomy (see chap. 16). Following an administrative career in the 1970s as secretary to the government in electronics, defense, and environment (chaps. 20–24), he was appointed director general of the CSIR and then made a member of the Planning Commission in 1982, in addition to a large number of public directorships. He was India's minister of Science and Technology in 1989–90.

SISIR KUMAR MITRA was born in Calcutta in 1890, son of a doctor and teacher. After studying at Presidency College, he joined the new Science College of the University of Calcutta, where he received his first DSc in 1919. His early interest in radio waves was influenced by the teaching and research of J. C. Bose, and he went to work in the labs of Charles Fabry and Marie Curie in Paris, where he achieved a second DSc. When Saha left Calcutta and the Khaira Professorship in Physics was vacant, he immediately sought this very position and returned from Paris in 1923 to teach the propagation of radio waves, though cautioned about expectations of rapid growth by the now retired Asutosh Mookerjee. He focused on studies of the ionosphere and is credited with identifying the role of ultraviolet radiation in layer formation in the ionosphere (after his work the Mitra belt gained its name). When Saha founded *Science and Culture* in 1935, Mitra was a co-editor for a number of years. Mitra, who was three years older, had a slightly rivalrous relationship with Meghnad Saha. They traveled together on the special tour of Allied nuclear research facilities in 1944–45. After establishing ionospheric research and publishing an influential text, he was elected a Fellow of the Royal Society in 1958, five years before his death in 1963.

ASUTOSH MOOKERJEE was born in Calcutta in 1864, pursued two remarkable parallel careers, first through two master's degrees in mathematics and physics and second in law, where he was astonishingly successful. His initial objective of becoming a university professor of mathematics was tested in 1895–96, when he made his appointment conditional on being paid the same salary as comparably trained European members of the "public instruction" civil service. This condition was not (and could not be) met by the government of Bengal under colonial service rules then in place, in spite of the fact that Mookerjee had published original papers and "a whole class of theorems designated as 'Mookerjee's theorems' was included in the curriculum at Cambridge" (J. Lourdusamy, *Science and National Consciousness in Bengal* [Delhi: Orient Longman, 2004], pp. 189–90) and in spite of the fact that he was already a member of the mathematical societies of London, Paris, and New York. He then decided to study law and was soon appointed as a judge on the high court in 1904 at age forty, eventually rising to be chief justice. This legal career was paralleled by his appointment by Bengal's governor as vice-chancellor of Calcutta University from 1906 to 1914 and later in 1921–22. Working from outside the university, and at its top, Mookerjee had direct and sustained contact with physicists C. V. Raman, S. N. Bose, M. N. Saha; biophysicist J. C. Bose; and chemist P. C. Ray. A member of the Sadler Commission in 1917–18, he encouraged Saha to testify on the desperate life of poverty of many students at the university and took a personal interest in the establishment of the departmental labs where empirical research work would take place, thus weaning the university away from its preoccupation with preparation for book-dependent examinations. He retired with illness in 1923 and died a year later, unable to see the success of Raman's and Saha's researches, which he had deliberately nurtured.

BASANTI DULAL NAGCHAUDHURI is seen as responsible for the Calcutta cyclotron and had a long career as a scientist after he left the Saha Institute in 1967. He was born in 1917 in Barodi village outside Dacca, the eldest of five sons of U. C. Nag, professor of English. (It has been suggested that the Nag sons might have attached the title *chaudhuri* to their name for the purpose of upward mobility, as it meant "landowner." The same title is used in Punjab. The practice of adopting this name/title was common in Bengal.) His father was professor first at Dacca and then, at the time of Partition, at Benares Hindu University. Nagchaudhuri's early education in physics was at Allahabad, where he met Haksar, and he said he "tagged along" when Saha moved back to Calcutta in 1938. With Ernest Lawrence's support on top of Saha's recommendation, Nagchaudhuri arrived in Berkeley, California, in late 1938 to do his PhD at age twenty-three, eventually working under Emilio Segré. He succeeded Saha as the director of the Saha Institute for Nuclear Physics in Calcutta and was the first scientist with an American degree to hold an important role like this in India, except for Homi Sethna (at BARC), who was trained in engineering at the University of Michigan. Nagchaudhuri was the member for science at the Planning Commission from 1967, as successor to M. S. Thacker. He met and married Deepali Talukdar, the daughter of a professor at Aligarh Muslim University; she eventually became a renowned singer of classical music with her performer's name of Deepali Nag. This marriage was significant because Deepali Talukdar was the niece of B. C. Ray, physician and powerful Congress politician who became the chief minister of the state of West Bengal in 1954, remaining in office until 1962. Nagchaudhuri, thus connected to the political elite of West Bengal and known to P. N. Haksar, became scientific advisor to the minister of Defence in 1970–74, then vice-chancellor of Nehru University in Delhi from 1974 to 1979.

ATMA RAM was born in village of northern Uttar Pradesh in 1908, son of an unsuccessful cloth merchant; his father was a staunch supporter of Arya Samaj, reformist Hindu movement (e.g., against the dowry). He had no opportunity for an education in English or sciences, so he learned Urdu and Persian, and in1926 though poor got a chance as an undergraduate to study at Benares Hindu University by paying his costs as a tutor. Ram ranked first at the University of Allahabad in MSc in chemistry in 1931, became a student of Nilratan Dhar, and got a small scholarship (he sent 90 percent of it home to help with the family's debts). Then he mingled unpaid work at a sugar factory with paid work as a tutor in order to complete his DSc in 1937 at the University of Allahabad, where he met Meghnad Saha, with whom he found much agreement except about the wearing of *khadi*. In 1941 he began work in Bhatnagar's new CSIR labs in Calcutta in a team that successfully developed a new foam fire suppressant for the military, to be used in petroleum fires. In 1945 Ram began helping Saha to plan and build the Central Glass and Ceramics Research Institute in Jadavpur, was appointed by CSIR as officer in charge of project at Rs 500 per month, and shared a secretary with Saha, who was building a new lab for the IACS nearby. In 1946 a scientific team traveled to Germany and the UK, in part to procure equipment for a new lab and in part to study new techniques; but Ram was injured in a vehicle accident in Germany and lost sight in one eye. After a sojourn in India to recuperate, he toured American glass research and development labs in universities and corporations, where he met N. J. Kreidl, Austrian glass expert now working for Bausch and Lomb at Rochester. In 1949 Ram returned to CGCRI and worked as "joint director" under newly arrived N. J. Kreidl as director of the new lab and eventually himself became director in 1952 on Kreidl's departure, focusing on the development of an optical glass production unit. This brought him into contact with industrialists, cabinet ministers, and defense planners. In 1966 he was appointed director general of CSIR in Delhi but had to wait ten months for his concurrent appointment as secretary to the government of India. He maintained a frugal way of life, guided by his family's Arya Samaj philosophy. In 1971 Atma Ram retired from CSIR at age sixty-five but in 1977 Prime Minister Desai asked him to return from retirement and become chairman of the National Committee on Science and Technology and also to review and propose a restructuring of the DAE. He died in 1983 at age seventy-five.

C. V. RAMAN is the only physicist to have won a Nobel Prize for research done in India. Born in 1888 in South India, he completed his BA in 1904 in Madras with gold medals in English and physics, and completed his MA at age nineteen in 1907. Already married, he had to work and immediately passed the financial civil service examination, enabling him to work in the same all-India financial administration where his uncle worked. Soon transferred to Rangoon and to Nagpur for revenue work, he finally was assigned to work in Calcutta in 1909; this was his opportunity to begin research at the IACS and he deliberately moved into a house next door to that laboratory. When Asutosh Mookerjee offered him the new Palit Chair in Physics in 1917, Raman, age twenty-nine, accepted a salary of half the value of his civil servant's income, but when told that a period of study in Britain was a condition of the position itself, stipulated by the donor, Raman refused to comply. With Mookerjee negotiating, the university governors and the donor yielded on this point, and Raman began teaching at the university while continuing his research at the IACS. This is when his professional contact with the older well-recognized scientists J. C. Bose and P. C. Ray was formalized, and when he first met Meghnad Saha, Satyen Bose, S. K. Mitra, and others (see chaps. 2, 3).

RAJA RAMANNA was born in Mysore in 1925, son of a judge of the court and influenced by a gifted aunt, he studied science and music before graduating from Madras Christian College in time to win a Tata Scholarship to study at the University of London in 1946. While there he was interviewed by Bhabha, whom he had met before in 1944, through their love of European classical music, and was offered a job when he completed his PhD. He joined TIFR in 1949, took a major role in DAE projects, and rose to be the senior physicist of the whole atomic energy establishment at Trombay, after Bhabha's death in 1966. He was also the DAE representative on the Board of the Saha Institute of Nuclear Physics in Calcutta throughout the late 1960s and early 1970s. He played an extensive organizing and intellectual role in the first nuclear test (chap. 23) and its political consequences (chap. 24).

PRAFULLA CHANDRA RAY was born in 1861 and grew up in an East Bengal village before coming to Calcutta to study at Presidency College, where he completed his first degree in 1881. He won a scholarship to the University of Edinburgh, where he studied chemistry with Alexander Crum Brown and was awarded the DSc in 1888. On his return to Bengal he applied often for employment in the Educational Service of India but found that, despite his excellent qualifications, those positions were protected for Britons. Eventually he got a job teaching chemistry at Presidency College, where he stayed until the University Science College opened in 1916. During this time he founded the successful Bengal Chemical and Pharmaceutical Company in 1901, and during the war it supplied essential chemicals for the British war effort. He received a knighthood for these activities. He was unmarried, lived frugally, but toured frequently as a guest abroad, playing a role in deciding on appointments across India (as he did in London in 1921 in Bhatnagar's case). When he was sixty, in 1922, and expected to retire, the university intervened and asked him to remain professor for five more years, which he accepted on condition that his salary be set aside for poor chemistry students of the university. He was then in the middle of organizing a massive flood relief campaign, in cooperation with Meghnad Saha. Ray was appointed as a director of the IISc in Bangalore in 1931 and took part in the launch of Meghnad Saha's *Culture and Science* in 1935 at age seventy-four. Ray wrote the two-volume *History of Hindu Chemistry* (1907, 1925) and *Life and Experiences of a Bengali Chemist* (1932, 1935). He died in June 1944.

VIKRAM SARABHAI was born into a family of merchants belonging to the Dasa Srimali sect of Jains; the family business house had been named after Karamchand Premchand, "who lived in the early 1800s and amassed his fortune by advancing loans to local chieftains and trading in Chinese opium and silk" (Amrita Shah, *Vikram Sarabhai: A Life* [Delhi: Viking/Penguin, 2007], p. 4; all page numbers refer to Shah). Two generations later Vikram's father Ambalal "set up the Sarabhai identity," having inherited two textile mills (Calico and Jubilee) and the whole *pedhi* (family inheritance) at age eighteen; he soon was considered a leader among Dasa Srimali sect. But Ambalal chose to marry outside the sect and married Sarla Devi, daughter of middle-class lawyer from a large poor family with ties to both Jain and Hindu communities; Ambalal financed the liberation of his sister Ansuya from an unhappy child marriage (to the disapproval of the community); she went to England, where she studied medicine and became an active suffragette, an activism she brought back to India. Vikram Sarabhai, born in 1919, was surrounded by strong and educated women. His mother was very active in the community, and one of his sisters was a labor organizer in Ahmedabad's textile mills. Ambalal held both textiles and the Jain community at arm's-length: he was seen as being on the margins by the more conservative elements of that community. He started to move beyond textiles to chemicals (through the link of dyes) in the 1940s, and although Sarabhai Chemicals was small, under his son Vikram's direction it grew steadily in the 1950s, (personal communication, Amrita Shah, 2 November 2007). In the 1920s and 1930s, "the family was particularly close to the Nehrus. Motilal and Jawaharlal both stayed at the Retreat, and when Jawaharlal's daughter Indira moved to Poona for her studies, Ambalal and Sarla Sarabhai looked in on her. These connections were to prove invaluable for Vikram later" (p. 29). Tagore visited Ahmedabad and the Sarabhais and also spent one month as their guest in Shillong in the mountains. Scientists J. C. Bose and C. V. Raman stayed there, along with nationalist leaders Gandhi and Maulana Azad. Though the men of Sarabhai's community supported these leaders, they did not join them in action, unlike the women. In March 1930, Vikram Sarabhai witnessed Gandhi's Salt March at age eleven and went to the Sabarmati Ashram on the evening before the march: Gandhi "never made me feel immature or that I was talking nonsense," said Vikram Sarabhai in the late 1960s (p. 32), and though he was a twenty-three-year-old physics student in Bangalore when the Quit India movement was launched in 1942, there is no evidence that he was tempted to participate. Sarabhai first moved far outside the sphere of influence of Ahmedabad before returning to it: while studying in Bangalore he met and married the dancer Mrinalini Swaminathan from Kerala in 1942, their mothers had both been involved in a social work project together. These unconventional yet comfortable and confident surroundings shaped his approach to scientific institution building and the politics of scientific research.

He and his young wife went together to live in Cambridge for two years in order that he could complete his dissertation in 1947, and he returned home to build up his research lab, businesses, and other institutions in Ahmedabad.

HOMI SETHNA was born in 1923 in Bombay and completed his BSc in the Department of Chemical Engineering at the University of Bombay in 1944. He went to the University of Michigan in 1946 to complete his master's in engineering and worked at Imperial Chemical Industries in Britain during 1948–49. At age twenty-six, in 1949, he joined the Atomic Energy Commission and was, by 1952, in charge of the monazite extraction and refining plant at Alwaye in Kerala. In 1955 he moved to Trombay to build a pure thorium nitrate production facility, and by 1964 he was in charge of building the plutonium plant at Trombay and soon after that the Jadaguda uranium mill in Bihar. Chosen as the director of the new Bhabha Atomic Research Centre (BARC) at Trombay in 1966, he left to become chairman of the Atomic Energy Commission from 1972 to 1983. After retiring at age sixty he ceased to circulate in "atomic energy circles" and has offered no memoirs on his life or work.

K. SUBRAHMANYAM was born in 1929, graduated from the University of Madras with an MSc in chemistry, and joined the Indian Administrative Service in 1951. He specialized in strategic studies and rose in administrative and strategic circles, coming to the attention of Prime Ministers Shastri and Gandhi. As deputy secretary to the Ministry of Defence, he had easy access to government officials, journalists, scientists, and planners (and confidential information). After a fellowship at the London School of Economics (1967–69), he began to influence others in favor of a bomb project, including Pitambar Pant of the Planning Commission and Homi Sethna of the DAE; this path was particularly pursued through his role as director of the Institute of Defence Studies and Analyses, Delhi. He also communicated his views as guest editor of the *Economic Times of India.* Ultimately he was architect of India's security policies in the 1980s and 1990s, advising the prime minister and cabinet.

NOTES

PREFACE

1. Charles P. Snow, *The Two Cultures and the Scientific Revolution*, BBC Reith Lectures (New York: Cambridge University Press, 1959); Thomas Kuhn, *The Structure of Scientific Revolutions* (Chicago: University of Chicago Press, 1962).
2. Edward Shils, *The Intellectual between Tradition and Modernity: The Indian Situation* (The Hague: Mouton, 1961).
3. Edward Shils, *The Torment of Secrecy: The Background and Consequences of American Security Policies* (Glencoe: Free Press, 1956).
4. This was explored, though rather incompletely, in Robert Anderson, "The Social Organization of a Scientific Laboratory" (M.A. thesis, University of Chicago, 1967).
5. A good example of that more typical anthropology can be found in our classmate Paul Rabinow's perceptive book on fieldwork, *Reflections on Field Work in Morocco* (Berkeley: University of California Press, 1977).
6. Gerald Swatez, "Social Organization of a University Laboratory" (diss., Berkeley, University of California, 1966). See also Swatez, "The Social Organization of a University Laboratory," *Minerva* 8, no. 1 (January 1970): 36–58.
7. Robert Anderson, *Building Scientific Institutions in India: Saha and Bhabha* (Montreal: McGill University Centre for Developing Area Studies, 1975).
8. Robert Anderson, "Growing Science in India," *Science Today* (Bombay), October, November, December 1976. This mass-circulation magazine has been discontinued.

CHAPTER ONE

1. Conversations with young scientists in India in 1977 showed a subtle shift from ten years earlier. Midnight's children born in 1947 were now thirty years old, not twenty; they were clearer about India's socioeconomic needs and more articulate about what could have been done with the great talents around them.
2. There are now excellent histories that build on the work of people like S. N. Sen, founder of the *Indian Journal of the History of Science*, and A. Rahman, Roy MacLeod, and others. The organized study of the history of science began in India in 1950, and in concert with UNESCO and the National Commission for the Compilation of History of Sciences of India, Indian historians began to work in all fields. I still have his article given to me by historian S. N. Sen, arising from the 1961 Symposium on History of Sciences in Ancient and Medieval India, "Transmission

of Scientific Ideas between India and Foreign Countries in Ancient and Medieval Times," *Bulletin of the National Institute of Sciences in India,* no. 21 (1962). *The Indian Journal of the History of Science* published its first two issues in 1966. An early example of that focus on the "transmission of ideas and techniques" is found in a Symposium on History of Sciences of India, held in Delhi in October 1968. But for a long time there were few, if any, studies of science in the twentieth century; an early exception of a general nature intended to render scientists and their institutions amendable to a modern management and traditions analysis is Ward Morehouse, ed., *Science and the Human Condition in India and Pakistan* (New York: Rockefeller University Press, 1968). That 1966–68 project was marked by the direct involvement of leading scientists like Abdus Salam, Vikram Sarabhai, Husain Zaheer, M. G. K. Menon, and others. That bleak situation has, however, changed; recent important work includes Deepak Kumar, *Science and the Raj, 1857–1905* (Delhi: Oxford University Press, 1995); Roy MacLeod and Deepak Kumar, *Technology and the Raj: Western Technology and Technical Transfers to India: 1700–1947* (Delhi: Sage Publications, 1995); Zaheer Baber, *The Science of Empire: Scientific Knowledge, Civilisation, and Colonial Rule in India* (Albany: State University of New York Press, 1996); Santimay Chatterjee, M. K. Dasgupta, Amitabha Ghosh, eds., *Studies in the History of Sciences* (Calcutta: Asiatic Society, 1997); Gyan Prakash, *Another Reason: Science and the Imagination of Modern India* (Princeton, NJ: Princeton University Press, 1999); David Arnold, *Science, Technology, and Medicine in Colonial India* (Cambridge: Cambridge University Press, 2000); Dhruv Raina and S. Irfan Habib, "The Missing Picture: The Non-emergence of a Needhamian History of Sciences in India," in Raina and Habib, eds., *Situating the History of Science: Dialogues with Joseph Needham* (Delhi: Oxford University Press, 1999); other essays by Dhruv Raina, *Images and Contexts: The Historiography of Science and Modernity in India* (Delhi: Oxford University Press, 2003). See also J. Lourdusamy, *Science and National Consciousness in Bengal* (Delhi: Orient Longman 2004); a collection of readings edited by S. Irfan Habib and Dhruv Raina, eds., *A Social History of Science: Themes in Indian History* (Delhi: Oxford University Press, 2006). As for the history of technology and industry, the earliest work I am aware of is Henry T. Bernstein on steam engines, steam boats, and inland water transport in the nineteenth century, research completed during 1953–56 as *Steamboats on the Ganges: An Exploration in the History of India's Modernization through Science and Technology* (Calcutta: Orient Longmans, 1960). Much later there were very valuable histories of technology and industry, such as A. J. Qaiser, *The Indian Response to European Technology* (Delhi: Oxford University Press, 1982); Shiv Visvanathan, *Organizing for Science* (Delhi: Oxford University Press, 1984), esp. chaps. 6 and 7 about capacitors and transistors; Nasir Tyabji, *Colonialism and Chemical Technology* (Delhi: Oxford University Press, 1995); Thirthankar Roy, *Traditional Industry in the Economy of Colonial India* (Cambridge: Cambridge University Press, 1999); Nasir Tyabji, *Industrialization and Innovation: The Indian Experience* (Delhi: Sage, 2000); Ashok Parthasarthi, *Technology at the Core: Science and Technology with Indira Gandhi* (Delhi, Longmans, 2007). The unwritten parts of these fascinating stories and the silences between them, however, are vast.

3. The published accounts of this history built upon the interviews conducted by Shiv Visvanathan, George Perkovich, Raj Chengappa, and Bharat Karnad are particularly valuable in an oral world in which so little useful written material is available. Karnad may be the only insider in this group; his father was a DAE employee and engineer at the Tarapur Atomic Power Plant around 1970.

4. The origin of my study in India was a confrontation in February 1962 during which I witnessed scientists attempting unsuccessfully to reassure people across India that the end of the world was not imminent, even though eight planets were in a unique alignment (*astograha*) according to important astrologers. The public's conviction was stronger than assurances from scientists, and so many abandoned their houses and camped out in the fields. See preface.
5. These ideas are more fully explained in the companion to this book, to be called *Negotiating Nuclear Power,* where I enlarge the concept of networks. I first presented these ideas to a colloquium as "Nuclear networks: state secrecy, personal privacy, and professional reputations" at MIT in November 2003, and thank colleagues there for discussion of this subject.
6. One can hardly list all the writing about the Indian nuclear tests and nuclear establishment, as it proliferates. One of the earliest publications was Dhirendra Sharma's *India's Nuclear Estate* (Delhi: Lancers Publishers, 1983), followed by Sharma, ed., *The Indian Atom: Power and Proliferation* (Delhi: Philosophers and Social Action, 1986). A short list includes M. Z. I. Cheema, "Indian Nuclear Strategy, 1947–1991" (PhD diss., University of London, 1991); Robert McMahon, *The Cold War on the Periphery: The United States, India, and Pakistan, 1947–1965* (New York: Columbia University Press, 1994); W. P. S. Siddhu, "The Development of Indian Nuclear Doctrine" (PhD diss., University of Cambridge, 1995); Ian Steer, "Asia's Rival Reactors a Cause for Concern," *Jane's Intelligence Review,* October 1998; Itty Abraham, *The Making of the Indian Atomic Bomb: Science, Secrecy, and the Post-Colonial State* (London: Zed Press, 1998); George Perkovich, *India's Nuclear Bomb: The Impact on Global Proliferation* (Berkeley: University of California Press, 1999); N. Ram, *Riding the Nuclear Tiger* (Delhi: LeftWord Books, 1999); Harold Feiveson, ed., *The Nuclear Turning Point: A Blueprint for Deep Cuts and De-alerting of Nuclear Weapons* (Washington: Brookings Institution Press, 1999); Raj Chengappa, *Weapons of Peace: The Secret Story of India's Quest to Be a Nuclear Power* (Delhi: HarperCollins, 2000). Haider Nizamani, *The Roots of Rhetoric: Politics of Nuclear Weapons in India and Pakistan* (Westport, CT: Praeger, 2000); M. R. Srinivasan, *From Fission to Fusion: The Story of India's Atomic Energy Progamme* (Delhi: Viking, 2002); Bharat Karnad, *Nuclear Weapons and Indian Security: The Realist Foundations of Strategy* (Delhi: Macmillan, 2002). More recently, see also Strobe Talbott, *Engaging India: Diplomacy, Democracy, and the Bomb* (Delhi: Penguin and Washington: Brookings Institution Press, 2004); T. V. Paul, ed., *The India-Pakistan Conflict: An Enduring Rivalry* (New York and Cambridge, England: Cambridge University Press, 2005); Rafiq Dassani and Henry S. Rowen, eds., *Prospects for Peace in South Asia* (Stanford: Stanford University Press, 2005); Kanishkan Sathasivam, *Uneasy Neighbors: India, Pakistan, and US Foreign Policy* (Aldershot, UK: Ashgate, 2005). Itty Abraham, ed., *South Asian Cultures of the Bomb: Atomic Publics and the State in India and Pakistan* (Bloomington: Indiana University Press, 2009) contains a large bibliography focused on the 1998 tests by India and Pakistan, with an excellent comparison of their cultural and social contexts.
7. Editorial, "Wages of Silence," *Economic and Political Weekly,* 16 May 1998.
8. Sunmit Ganguly and Devin Haggerty, *Fearful Symmetry: India-Pakistan Crises in the Shadow of Nuclear Weapons* (Seattle: University of Washington Press, 2005), p. 66.
9. In 1998 I visited India both before the May bomb tests and after them. I visited India in 2005 after the removal from power of the BJP party. By this means I obtained wide-angled snapshots of the state of affairs prevailing at the time, including the relationships of scientists with the technological-industrial system, and with the political

system. I sincerely thank the scientists of the Tata Institute for Fundamental Research, Saha Institute for Nuclear Physics, the Raman Institute, and the National Institute for Science, Technology and Development, the Sarabhai Space Centre, the National Centre for Biological Sciences, and the National Centre for Radio Astrophysics for their courtesies to me during these visits; their time is precious to them, and I am grateful to them for some of it.

10. The chairman of the Atomic Energy Regulatory Board, Dr. P. Rama Rao was our host at the Indian Science Congress. The chairman of the Nuclear Power Corporation, chairman of the Atomic Energy Commission, and former principal secretary to the prime minister were all present in the discussions.
11. O. P. Sabherwal, "FBTR: India's Success Story," *Times of India,* 3 February 1998.
12. Editorial, "Bitter Rice," *Times of India,* 21 February 1998; "Over 100 Indian Plants in US Patent Net," *The Observer,* 21 February 1998.
13. Y. P. Gupta, "Why Do Our Scientists Commit Suicide?" *The Hindu,* 6 January 1998. I do not think that the official record, such as it is, could be complete.
14. "Scientists File PIL against CSIR," *Indian Express,* 24 January 1998.
15. Dhruv Raina, *Images and Contexts: The Historiography of Science and Modernity in India* (Delhi: Oxford University Press, 2003), chap. 5, "Science, Scientists, and the History of Science in India (1966–1994)." A reading of Raina's footnotes alone is valuable; he said he oriented himself to "a realist constructivist theory of the production of knowledge, and a theory of history that recognizes how hitherto drafted histories are premised on a global narrative but discounts the inference that there is no ground for a global history at all. This would imply that while there exist global versions of this history, we do not have a satisfactory version so far" (p. 6).
16. Not everyone finds this kind of work easy, but the rewards are clear, as seen by the result of a pathbreaking 1984 book about the National Physical Laboratory: Shiv Visvanathan, *Organizing for Science: The Making of an Industrial Laboratory* (Delhi: Oxford University Press, 1984). I have explained the necessity and requirements of this kind of study in Robert Anderson, "On the Necessity of Fieldwork in the Study of Scientific Research," in Everett Mendelsohn and Y. Elkana, eds., *Sciences and Cultures* (Dordrecht: D. Reidel Publishing, 1981). I am using some findings here from my own firsthand study in the late 1960s. See also V. S. Mouly and J. K. Sankaran, *Organizational Ethnography: An Illustrative Application in the Study of Indian R&D Settings* (Delhi: Sage, 1995), a study of about 100 scientists in one team in an (unnamed) public sector laboratory.
17. I have spoken and written about the role of this kind of imagination for years, before it became popular, but am grateful for the clear and powerful expression of the role of imagination provided by Benedict Anderson, *Imagined Communities: Reflections on the Origin and Spread of Nationalism* (New York: Verso, 1991).
18. Though the autobiographical tradition among Indian scientists is thin, there are exceptions, such as the recent work of A. P. J. Abdul Kalam, *Wings of Fire: An Autobiography* (Hyderabad: Universities Press, 1999; see also updated 2002 edition); and R. Ramanna, *Years of Pilgrimage* (Delhi: Viking, 1991), and Piara Singh Gill, *Up Against Odds: Autobiography of an Indian Scientist* (Delhi, Allied Publishers, 1992). In 1923 P. C. Ray wrote *Life and Experiences of a Bengali Chemist* (Calcutta: Chuckervertty, Chatterjee), but an autobiographical tradition hardly followed it. We thus mainly rely on biographies such as Ashis Nandy, *Alternative Sciences: Creativity and Authenticity in Two Indian Scientists* (Delhi: Allied Publishers, 1980), focused on mathematician Ramanujan and biophysicist Bose; and the monumental study by G. Venkataraman,

Journey into Light: Life and Science of C. V. Raman (Bangalore: Indian Academy of Sciences, 1988). Venkataraman has also written shorter biographies of S. N. Bose, M. N. Saha, S. Chandrasekhar, and Homi Bhabha in order to encourage, he told me, public and popular understanding of their theories and experiments. See also the major work by Ashok Rudra, *Prasanta Chandra Mahalanobis: A Biography* (Delhi: Oxford University Press, 1996); Subrata Dasgupta, *Jagadis Chandra Bose and the Western Response to Western Science* (Delhi: Oxford University Press, 1999); and a study of mid-nineteenth-century Delhi mathematician Jesudas Ramchandra by Dhruv Raina and Irfan Habib, *Domesticating Modern Science: A Social History of Science and Culture in Colonial India* (Delhi: Tulika Books, 2004); J. Lourdusamy's biographies of science organizers Asutosh Mookerjee and Mahendralal Sircar, pioneering chemist P. C. Ray, and biophysicist J. C. Bose are in his *Science and National Consciousness in Bengal* (Delhi: Orient Longmans, 2004). Abha Sur has written excellent biographical essays on Raman and Saha, all referred to here. Amrita Shah's *Vikram Sarabhai: A Life* (Delhi: Viking/Penguin, 2007) is arguably the best biography of a scientist in India to date: it emphasizes the social and historical context that enabled Sarabhai to succeed, even if it does not examines his scientific work in great detail. Other books and essays are listed in subsequent footnotes; my book would not have been possible without the research of these biographical authors. A striking exception lies in the remarkable monthly magazine *Seminar,* which in 1993 commissioned an entire issue on autobiography and science, opening with Visvanathan's essay "Autobiography as Experiment," a tour of autobiographical experiments from chemist P. C. Ray through M. K. Gandhi to planner-engineer Sir M. Visvesvaraya, to chemist-politician Atma Ram. Then there are short, fascinating autobiographical pieces from people as different as physical chemist Amulya Reddy and cosmologist Jayant Narlikar, ended by reviews of books by or about ornithologist Salim Ali, physicist C. V. Raman, and cosmologists S. Chandrasekhar and Stephen Hawking. See *Seminar,* no. 409, September 1993, "Our Scientists."

19. A visitor to archives in May 2009 stated that he saw no papers of Homi Bhabha, Vikram Sarabhai, Homi Sethna, or Raja Ramanna at the Nehru Memorial Library in Delhi, but he saw the papers of political advisor P. N. Haksar and former Foreign Secretary T. N. Kaul there. There was little of direct relevance to the study of the nuclear and scientific history of modern India, such as official papers of the Atomic Energy Commission or the Department of Atomic Energy, to be found at the National Archive in Delhi. The relevant papers of Jawaharlal Nehru were still not available in 2009, forty-five years after his death. (Ryan Touhey, historian, personal communication, 30 May 2009). Other historians confirmed that some of Meghnad Saha's papers are at the Nehru Library while others are at SINP in Calcutta, and some of the Bhabha papers are at TIFR, Bombay.
20. See the brilliant essay by Norbert Elias, "Scientific Establishments," in N. Elias, H. Martins, and R. Whitley, eds., *Scientific Establishments and Hierarchies* (Dordrecht: D. Reidel Publishing, 1982).
21. India had an antimilitary and antiviolence atmosphere after WW II, in part because hundreds of thousands had been troops in the war and had experienced traumatic fighting and seen violent death, and in part because of the popular nonviolent image of M. K. Gandhi. The violence of Partition further appalled large parts of the population and Gandhi's assassination in 1948 appealed only to a few extremists. This atmosphere survived well into the 1960s, retreating gradually after the wars with China in 1962 and Pakistan in 1965. For a systematic and critical appraisal, lamenting

this unmilitary dimension of Indian political culture, see Bharat Karnad, *Nuclear Weapons and Indian Security* (Delhi: Macmillan, 2002).

22. The key reference is to Ranajit Guha's essay "On Some Aspects of the Historiography of Colonial India," in R. Guha, ed., *Subaltern Studies*, vol. 1 (Delhi: Oxford University Press, 1983); for an example of postcolonial theorizing with reference to science and technology (including India), see Ziauddin Sardar, "Above, Beyond, and at the Center of the Science Wars," in K. M. Ashman and P. S. Baringer, *After the Science Wars* (London: Routledge, 2001); regarding cosmopolitan studies' disinterest in science and technology, see for example Steven Vertovec and Robin Cohen, *Conceiving Cosmopolitanism: Theory, Context and Practice* (New York: Oxford University Press, 2003).
23. A useful summary of the applications of and disagreements about actor-network theory is John Law and Johan Hassard, *Actor Network Theory and After* (Oxford: Blackwell, 1999). The networks in this book are more complex than many others in that some of their individual members operated simultaneously on both a public and secret level in the same domain and in the same action, a feature I develop fully in *Negotiating Nuclear Power*.
24. "On the Annals of the Laboratory State," in Shiv Visvanathan, *A Carnival for Science* (Delhi: Oxford University Press, 1997). I prefer to reserve the term genocide for its justiciable applications, but appreciate that the author uses it partly for its shock value. I completely agree that the matters of displacement and impoverishment he discussed are very serious.

CHAPTER TWO

1. Personal communication, Laura Dawson, Royal Society, London, 17 April 2008.
2. Personal communication, Chitra Saha, Meghnad Saha's daughter, 6 July 1999. The failed "English-educated" elder brother eventually got a job as a jute-trader, supplementing the family income and paying for the children's schooling. Jute trading and traders provided a lucrative and sympathetic network into which Meghnad tapped later, for research funds. A younger brother saved his soldier's wages, bought a small medicine shop and eventually became a physician. The family grocery shop slowly expanded to a prosperous business and was still in Saha hands in 1973. But it was still a small grocery shop in a small village off the highway.
3. Abha Sur, "Egalitarianism in a World of Difference: Identity and Ideology in Meghnad Saha's Physics," unpublished article, p. 6. Abha Sur was given some of this information by space historian David DeVorkin, based on information from Ajit Saha, M. N. Saha's eldest son and former director of the Saha Institute.
4. S. N. Sen, ed., *Professor Meghnad Saha: His Life, Work, and Philosophy* (Calcutta: Meghnad Saha Sixtieth Birthday Committee, 1954). The editor states that "the account of Professor's life and work given in this volume is mostly based on materials and notes received from him." The editor told me in 1969 that when the book was being written, Saha told editor Sen that his (Saha's) philosophy should be clearly and forcefully put, otherwise "what was point to all this work?" Readers should be aware of the book's autobiographical character.
5. In 1973 one of Seortali's elderly inhabitants told me he recalled having read an article in Bengali written by Meghnad Saha on Halley's comet in the magazine of Dacca College about this time.
6. Santimay Chatterjee, "Meghnad Saha, Scientist with a Vision," *Science Today*, February 1968, p. 46. See Leonard Gordon, *Bengal: The Nationalist Movement, 1876–1940*

(New York: Columbia University Press, 1974). Cf. "German Conspiracy: The Organization of the Revolutionaries," pp. 148–58.

7. Sen, *Professor Meghnad Saha*, p. 6.
8. See Benjamin Zachariah, Subhas Chakraborti, and Rajat Ray, "Presidency College: An Unfinished History," in Munshirul Hasan, ed., *Knowledge, Power, and Politics: Educational Institutions in India* (New Delhi: Roli Books, 1998), p. 69. See also Gordon, *Bengal* (1990), pp. 49–50.
9. Sen, *Professor Meghnad Saha*, pp. 5–6.
10. Somaditya Bannerjee, "Satyendranath Bose and the Construction of Theoretical Physics in Colonial India," unpublished paper, May 2008, p. 8. Working at the University of British Columbia, Bannerjee is assembling all known sources on Bose in Bangla and English.
11. Sen, *Professor Meghnad Saha*, pp. 7, 44.
12. Both Saha and Satyen Bose borrowed physics books in German from P. J. Breuhl, an Austrian botanist who taught physics during the war to engineering students at Bengal Engineering College; G. Venkataraman, *Bose and His Statistics* (Hyderabad: Universities Press, 1992), p. 10.
13. Though Einstein had given the publishing rights to Methuen in London (who wanted to stop the distribution of these English translations), Einstein told Methuen and the editor-translators that as long as the book remained in circulation in India, he had no objection. The first printing sold out rapidly; Bannerjee, "Satyendranath Bose," 2008, p. 9.
14. D. M. Bose, "Meghnad Saha Memorial Lecture" (1965), *Proceedings of the National Institute of Sciences of India* 33, A, nos. 3, 4 (Calcutta, 1967): 111–12.
15. Sen, *Professor Meghnad Saha*, p. 9; David DeVorkin, having read Saha's letters to astronomers, concurred with the authorized biography, in "Quantum Physics and the Stars (IV): Meghnad Saha's Fate," *Journal for the History of Astronomy* 25 (1994): 155–88.
16. David DeVorkin, "Quantum Physics and the Stars (IV): Meghnad Saha's Fate," *Journal for the History of Astronomy* 25 (1994): 163.
17. Quoted in D. S. Kothari, "Meghnad Saha," *Biographical Memoirs of the Fellows of the Royal Society* 5 (1959): 222. On the importance of German journals until 1928, see Charles Weiner, "New Site for the Seminar: The Refugees and American Physics in the Thirties," in D. Fleming and B. Bailyn, eds., *The Intellectual Migration* (Cambridge: Harvard University Press, 1968), p. 200. See also N. G. Barrier, *Banned: Controversial Literature and Political Control in British India* (Columbia: University of Missouri Press, 1974).
18. "As far as I can ascertain, and I went over the records again, the British did not ban German scientific literature, at least in the Indian theatre"; N. G. Barrier, personal communication, 25 March 1975. See also Barrier, *Banned.*
19. M. N. Saha to H. H. Plaskett, undated letter, but probably 21 December 1946; dated according to David DeVorkin, based on Saha's papers in the archives. Quoted in David DeVorkin, "Quantum Physics and the Stars (IV)."
20. David DeVorkin, "Quantum Physics and the Stars (IV)," p. 158.
21. M. N. Saha, statement to the Sadler Commission, cited in Abha Sur, "Scientism and Social Justice: Meghnad Saha's Critique of the State of Science in India," *Historical Studies in the Physical and Biological Sciences* 29 (2002): 87–105. See also Sur, "Egalitariansm," p. 22.
22. Norah Richards, *Life and Work of Sir S. S. Bhatnagar* (Delhi: New Book Society, 1948), p. 6. This book has been republished, with a review essay by Rajesh Kochhar, Delhi:

NISTADS, 2004). The book provides a brief, but important sketch of the unusual life of Norah Richards, who as a young actress married an English lecturer coming to teach at Dayal Singh College in Lahore. On his death in 1920 she went to the UK but returned to India four years later and reestablished herself (and a drama school) in the Kangra Valley, on fifteen acres of land given to her by a departing British settler. Here she painted and wrote, and from here she influenced the course of playwriting and drama production in the Punjabi language; she attracted and encouraged actors, painters, sculptors, and other artists, of whom Bhatnagar considered himself one. She lived rather frugally and died in 1971, greatly respected throughout Punjab. Her authorized and intimate biography of Bhatnagar was written mostly in 1944–45 but not published until 1948.

23. Richards, *Life and Work,* p. 3.
24. Personal Records of Dr. S. S. Bhatnagar, ca. 1948, p. 4, Bhatnagar Papers, NISTADS Archives, Delhi.
25. Dayal (or Dyal) Singh was a successful businessman (real estate, jewels) from the noble Shergil lineage whose father had been Maharajah Ranjit Singh's chief of military ordnance. His father drew his wealth from revenue villages (*jagirdar*) in Amritsar, Gurudaspur, and Majitha districts around Lahore. Although the family had previously been in charge of the Amritsar Golden Temple sacred to Sikhs, Dayal Singh's father and mother had become reformist Brahmo Samaj adherents. Born in 1848 and brought up in Benares, Dayal Singh was educated in a missionary school. Before his death he established the Union Academy in Lahore, the nucleus of Dayal Singh College. A poet who wrote in Urdu, he also established the English-language *Tribune* newspaper in 1881 and was in regular contact with wealthy Brahmo patrons in Bengal, confronting Lahore's orthodox customs in his own Punjabi style. His will generously aided the University of the Punjab (founded in 1882) and also established the Dayal Singh College, built after his death in 1898. I am grateful to both Saima Shah, of Palo Alto and Burnaby, and Mohammad Gill, of Lahore, for providing some of this information.
26. Sen, *Professor Meghnad Saha,* p. 12.
27. Ibid., p. 15.
28. Richards, *Life and Work,* p. 61.
29. DeVorkin, "Quantum Physics and the Stars (IV)," p. 161.
30. David DeVorkin, "Saha's Influence in the West," in S. B. Karmohapatro, ed., *Meghnad Saha Birth Centenary Commemoration Volume* (Calcutta: Saha Institute of Nuclear Physics, 1993), p. 171. Saha said in this 20 August 1921 letter to Mookerjee that this equipment would be useful in high-temperature research on glass, ceramics and enamels; twenty-five years later Saha was designing and building a national laboratory dedicated to research on these materials.
31. Subrata Dasgupta, *Jagadis Chandra Bose and the Indian Response to Western Science* (Delhi: Oxford University Press, 1999), esp. chap. 7, "Empire Building." It must be emphasized that this was a large amount of money for any scientist in a private laboratory anywhere at the time, let alone a scientist in India.
32. Deepak Kumar, *Science and the Raj, 1857–1905* (Delhi: Oxford University Press, 1995); also David Arnold, *Science, Technology, and Medicine in Colonial India* (Cambridge: Cambridge University Press, 2000). Ronald Ross left India to study medicine in London at 18, returned with a degree in medicine and did parasitology research in India from 1881 to 1899, working in Bangalore and Hyderabad: at forty he did the research in Calcutta for which he won the Nobel Prize in 1902.

33. Pramathanath Bannerjee, "Reform and Reorganization: 1904–24," in N. R. Ray and P. C. Gupta, eds., *Hundred Years of the University of Calcutta* (Calcutta: University of Calcutta Press, 1957), pp. 277–78.
34. Ibid., p. 273.
35. J. Lourdusamy, *Science and National Consciousness in Bengal* (Delhi: Orient Longman, 2004), pp. 223–24. Sir Taraknath Palit, a Calcutta barrister, had donated about Rs. 3 million to the Science College of the University of Calcutta. A separate laboratory and physics professorship were named after him.
36. Sur, "Egalitariansim," p. 18. Sur's statement is based on correspondence between Saha and Vice-Chancellor Asutosh Mukherjee.
37. DeVorkin, "Quantum Physics and the Stars (IV)," p. 160.
38. One of the first to try to place Bose in his historic context was William Blanpied, "Satyendranath Bose: Co-founder of Quantum Statistics," *American Journal of Physics* 40, no. 9 (September 1972): 1212–20. Also Virendra Singh, "The Discovery of Bose Statistics," and E. C. G. Sudarshan, "A World of Bose Particles," *Science Today* (Bombay), January 1974. Satyen Bose told me in 1969 that he had lost all the correspondence from Einstein. My search for any correspondence involving Bose at the University of Dacca in 1972–74, made with full official university cooperation, showed nothing remained.
39. S. K. Mitra immediately accepted this very position and returned from Paris in 1923 to teach the propagation of radio waves, though cautioned about expectations of rapid growth by a now chastened Asutosh Mookerjee. See J. A. Ratcliffe, "Sisir Kumar Mitra," *Biographical Memoirs of the Fellows of the Royal Society*, vol. 5, 1959, p. 222. (London, Royal Society).
40. Sen, *Professor Meghnad Saha*, p. 17.
41. Ibid., p. 17.
42. D. V. C. Mallik, "KS Krishnan and the Kodaikanal Observatory," *Current Science* 79, no. 5 (September 2000): 665.
43. Kamalesh Ray, *The Life and Work of Meghnad Saha* (Delhi: National Council of Education Research and Training, 1968).
44. This section relies on Somaditya Bannerjee, "Satyendranath Bose and the Construction of Theoretical Physics in Colonial India," unpublished essay, May 2008.
45. According to Somaditya Bannerjee, there is some confusion in the secondary literature over whether Saha visited Bose or not, and when it occurred. Since Saha frequently visited Dacca to see his family, and since raising money for the 1923 floods caused him to travel in East Bengal, I have little doubt that this visit occurred, whether late in 1923 or early in 1924.
46. S. N. Bose "Plancks Gesetz und Lichtquantenhypothese," *Zeitschrift für Physik* 26 (1924): 178.
47. Einstein Archive, Doc. 6–128.
48. DeVorkin, "Quantum Physics and the Stars (IV)," pp. 161–62. A confusion of astrophysicists is found here: Alfred Fowler (b. 1868) at the Royal Society in London and Ralph Fowler (b. 1889) at Cambridge, both astronomers, both known to Saha.
49. Exchange of letters, James Jeans and Holland, May 1925, cited in DeVorkin, "Quantum Physics and the Stars (IV)," p. 163.
50. Based on interview with B. K. Dutta, 7 July 1968, cited in Guatam Chatopadhyay, *Communism and the Bengali Freedom Movement*, vol. 1, *1917–1929* (Delhi: Peoples Publishing House, 1970), pp. 50–60. I am grateful to Dr. Sulagna Roy for bringing this interview to my attention, and to Subhas Chakravarty for meeting with me and

the late Gautam Chattopadhyay in Kolkata in January 2005. Dutta, the source of information about the secret code, was a person whom Gautam Chattopadhyay said he had no reason to doubt.

51. It is unclear what the Royal Society thought was risky about this nomination: Saha's name does not appear in subsequent official publications on communists or terrorists up to 1940, such as *Communism in India* (Calcutta: Government of India Press, 1927); the secret *Politico-criminal Who's Who of Bengal Presidency* (Calcutta: Bengal Intelligence Branch of CID, 1930); *Review of the Terrorist Situation 1939–1940* (Calcutta: Bengal Intelligence Branch of CID, 1940). All these documents were read in the British Library Oriental and India Office Collection, Box L/P and J/12 395 and /669.
52. Note that Alfred Fowler and Walker played key roles in later nominations for the Fellowships of K. S. Krishnan (1940), S. Chandrasekhar (1944), and Prasanta Mahalanobis (1945).
53. A. E. Milne to M. Saha, 7 September 1935, cited in DeVorkin, "Quantum Physics and the Stars (IV)," pp. 171, 186.
54. K. C. Wali, *Chandra: A Biography of S. Chandrasekhar* (Delhi: Penguin, 1992), pp. 259–60.
55. Sen, *Professor Meghnad Saha,* p. 19.
56. Ibid., p. 30.
57. Ibid.
58. Ibid., p. 19.
59. Sur, "Egalitarianism," p. 20.
60. For a fuller account of his thinking while on this journey, see DeVorkin, "Quantum Physics and the Stars (IV)," pp. 168–70.
61. Sen, *Professor Meghnad Saha,* p. 24. See also Rajinder Singh, "Arnold Sommerfeld—The Supporter of Indian Physics in Germany," *Current Science* 81, no. 11 (10 December 2001): 1489–94.
62. S. P. Pandya, "Meghnad Saha: A Pioneer of Nuclear Physics," *The Glittering Spectrum of Meghnad Saha* (Calcutta: Saha Institute of Nuclear Physics, 1994), p. 10.
63. The role of two committees in 1921 was not without precedent. In 1920, just after being elected as a Fellow of the Royal Society, Jagdish Chandra Bose was awarded an annual government grant of Rs 100,000. To satisfy British doubters in Bengal, J. C. Bose proposed "an English Committee" (his words) as the means to ensure continuity and standards at his private institute, and this model was accepted in spite of the fact that a permanent grant had to be approved by vote every year by the Bengal legislature. Composed of distinguished scientists, some of whom had nominated Bose for his FRS, this committee was backed by visits from the most senior education official in the government of India. The annual grant of Rs 100,000 was larger than any other research institute in India at the time, even though it was understood by the donors, at least by 1925, that proclamations that the institute's research would have commercial application were very unlikely to come true. Dasgupta says Bose successfully pursued a dialectical strategy that worked, using British and Indian pressures against one another; see Subrata Dasgupta, *Jagadis Chandra Bose and the Indian Response to Western Science,* pp. 234–40, esp. "Bose and the Raj: Characteristics of the Dialectic."
64. Richards, *Life and Work,* p. 44.
65. I am grateful to Nasir Tyabji for questioning the grounds on which Bhatnagar was chosen for this post (personal communication 7 March 2005). Tyabji reminded me

that Bhatnagar's British superiors may have identified him as a vector for serious cooperation between themselves and Indians in chemistry. My sense is that Bhatnagar was better qualified than his competitor and had the backing of the influential Professor F. G. Donnan in London, thus enhancing the reputation of Lahore and this laboratory in relation to others in India. The two calculations go hand in hand.

66. Bhatnagar saw equipment in this high-altitude project which led him to think about the invention of other research equipment, for physical chemists; his skill was demonstrated in 1928 when he produced (with his old friend from Benares K. N. Mathur) the Bhatnagar-Mathur magnetic interference balance, which was an extremely sensitive instrument for measuring very small variations in magnetic properties. This instrument was manufactured by the British firm Hilger and Company, exhibited in London at the Royal Society in 1931, and Bhatnagar benefited from the royalties for years. This is one origin of his interest in intellectual property.

67. S. S. Bhatnagar, "Personal Records," 1947, p. 12, Bhatnagar Papers, NISTADS Archive, Delhi. This document appears to have been used by Bhatnagar to satisfy requirements for an official government curriculum vitae but includes more information than is normally the case. An earlier version may have been used for his election as Fellow of the Royal Society.

68. "Steels" were established by William Steel in Rangoon in 1870 in order to mill and ship rice; 3 years later he opened an office in London. Using a Scottish network in business recruitment, capital from Rangoon, Glasgow, and London, and new experimental and local technologies, Steels grew rapidly in the rice and timber business, establishing the first weekly market and auction for rice in London. At the same time a Gujerati-speaking merchant in Rangoon, Abdul Karim Jamal had bought leases to prospect for oil in central Burma and then built a small refinery. Almost broke in 1908, Jamal asked William Steel to combine with him in a new company to be called Indo-Burmah Petroleum, shipping kerosene and paraffin from Burma throughout India. Steels is an early version of the multinational corporation, with decisions made in Burma and India as much as in the UK. It was into this world that Bhatnagar stepped in 1934. See H. E. W. Braund, *Calling to Mind: Being Some Account of the First Hundred Years (1870 to 1970) of Steel Brothers and Company Limited* (Oxford: Pergamon Press, 1975).

69. Ibid., p. 65.

70. S. S. Bhatnagar, "Personal Records," 1947, Bhatnagar Papers, NISTADS Archives, Delhi p. 12.

71. Ibid., pp. 12–13.

72. In reference to the School of Mechanical Technology and the School of Chemical Technology (with German links) at Kala Bhavan in Baroda between 1890 and 1910, see Dhruv Raina and Irfan Habib, *Domesticating Modern Science: A Social History of Science and Culture in Colonial India* (Delhi: Tulika Books, 2004), pp. 182–98; also in Bombay University, chemical engineering for textiles was introduced in the 1920s with a major private gift in 1930: see Nasir Tyabji, "Exemplar of Academia-Industry Interchange: The Department of Chemical Technology at Bombay University," *Ambix* (Journal for the Society of the History of Alchemy and Chemistry), 51, pt. 2 (July 2004): pp 149-66. Note that after 1932 the German language was a required subject for training chemical engineers in Bombay.

73. Quoted in T. R. Seshadri, "S. S. Bhatnagar," *Biographical Memoirs of the Fellows of the Royal Society* 8 (1962): 1–17. Attock Oil was formed in 1913 to exploit the Khaur oil field of Punjab and was still registered in Pakistan in the 1970s. Though it was

managed by Steel Brothers, 42 percent of its shares were owned by Indo-Burmah Petroleum. It was economically strong at this time; as Dasgupta says, "towards the end of the 1930s [Attock] made good all losses and paid back all its debts before the end of the 30s." This payback was the money Bhatnagar was awarded; Steel was investing in the future of the Punjab. Biplob Dasgupta, *The Oil Industry in India—Some Economic Aspects* (London: Frank Cass and Co., 1971), p. 62.

74. I saw P. C. Ray's rooms at the Science College in 1969, preserved for 40 years after his death by the authorities of Science College as a kind of shrine to this scientist and entrepreneur. Despite his commercial success, Ray chose to live extremely ascetically and gave away most of his income.
75. Richards, *Life and Work,* p. 121.
76. Ibid., p. 105.
77. Chitra Saha, personal communication, 11 June 1999. A self-described atheist, Saha loved swimming in the river and his devout wife loved the sanctity of the spot. Swimming and walking were among the few things they could do together.
78. Sen, *Professor Meghnad Saha,* p. 21; DeVorkin found current references to the crucial importance of Hund's theory of line spectra in Saha's diary written in Europe; "Quantum Physics and the Stars (IV)," p. 185.
79. P. Dirac to M. N. Saha, 18 May 1936, Churchill College Archives, Cambridge. With one or two exceptions, most of Dirac's letters (in English) to physicists adopt this blunt tone, though to Homi Bhabha he was less severe.
80. A. Compton to M. N. Saha 15 February 1937, Saha Institute Archive. The two scientist brothers (Arthur and Carl) had a sister who was married to the principal of Forman Christian College and lived in Lahore. Hence the Compton visits to Lahore.
81. Dr. D. M. Bose Seventieth Birthday Commemoration Volume, *Transactions of the Bose Research Institute,* vol. 20 (Calcutta, 1955). Regarding the founding of this institute, see S. Dasgupta, *Jagadis Chandra Bose and the Indian Response to Western Science.*

CHAPTER THREE

1. J. W. Wheeler-Bennett, *John Anderson* (London: Macmillan, 1962).
2. D. M. Bose, *Memorial Lecture,* p. 112. Bose is characterizing Saha's views when he writes "I" in his lecture text.
3. Mahendralal Sircar, ed., *The Indian Association for the Cultivation of Science, Proceedings and Annual Reports, 1880–1906* (Calcutta, 1906). On the role of Sircar's friend Father Lafont in the IACS, see A. K. Biswas, *Science and India* (Calcutta: Firma K. L. Mukhopadhyay, 1969). For the relation between Lafont and J. C. Bose, see Ashis Nandy, *Alternative Sciences: Creativity and Authenticy in Two Indian Scientists* Delhi, Oxford University Press, 1995), and Subrata Dasgupta, *Jagadis Chandra Bose and the Indian Response to Western Science* (Delhi: Oxford University Press, 1999).
4. G. Venkataraman, *Journey into Light: Life and Science of C. V. Raman* (Bangalore: Indian Academy of Sciences, 1988), pp. 57–59. Venkataraman records the larger number of Bengal-based researchers attached to the IACS. His is the most nuanced account of the 1931 Raman-Saha confrontation.
5. Ibid., p. 39.
6. K. Wali, *Chandra,* p. 249. In German physics journals it was called the *Smekal-Raman* effect.
7. There were people like Richard Gans at the University of Jena who, independently of Saha, questioned Raman's findings, thinking they were wrong. I am grateful to Somaditya Bannerjee at the University of British Columbia for this observation.

8. Author's conversation with C. V. Raman, Bangalore, February 1968.
9. See annual reports of IACS for 1934 and 1935 in the *Indian Journal of Physics* of the same years. The experimental history leading to the 1930 Nobel Prize has been summarized in G. Venkatraman, *Journey into Light*. See also S. Ramaseshan, "A Conversation with K. S. Krishnan on the Story of the Discovery of the Raman Effect," and Shiv Visvanathan, "The Tragedy of K. S. Krishnan, a Sociological Fable," *Current Science*, 10 December 1998, pp. 1265–75.
10. G. Venkataraman, *Journey into Light*, pp. 57–58.
11. D. M. Bose, personal communication, 8 May 1971.
12. Sen, *Professor Meghnad Saha*, p. 87.
13. M. N. Saha, "The Academy Committee's Report to the Indian Science Congress," *Proceedings of the National Institute of Sciences in India* 1 (1935): 11.
14. Sen, *Professor Meghnad Saha*, p. 41.
15. Ibid., p. 36.
16. L. L. Fermor, "Inaugural Address," *Proceedings of the National Institute of Sciences in India*. 1 (1935): 17.
17. Ibid., p. 19.
18. It would be rewarding to study how much Saha accounted for the different levels of scientific literacy in India, the England, and the United States, and what feedback he received from the Indian audience in terms of understanding his early messages in *Science and Culture*. The printing of the 1936, 1937, and 1939 issues was of high quality on quite good paper for the time, and while the photos were only average in quality, the diagrams were very fine and the chemical and mathematical symbols sharp and clear to read.
19. F. R. Harris, *Jamshedji Nusserwanji Tata—A Chronicle of His Life* (London: Oxford University Press, 1925). Cf. chap. 6.
20. For the history of the IISc and the Raman years, see B. V. Subbarayappa, *In Pursuit of Excellence: A History of the Indian Institute of Science* (Delhi: Tata McGraw-Hill, 1992).
21. Sewell Report, *First Quinquennial Reviewing Committee of the IISc* (Bangalore, 1931), 14.
22. Ibid. p. 19.
23. P. C. Ray, *Life and Experiences of a Bengali Chemist* (Calcutta: Chuckervertty, Chatterjee, 1932), p. 202.
24. M. Born to E. Rutherford, 26 January 1936, Rutherford Papers, University Library, Cambridge. I quote Born's observations at length, not because he was necessarily accurate in his interpretation, but because his views would have carried weight with the council, and Rutherford.
25. Rutherford to M. Born, 19 October 1936, Rutherford Papers, University Library, Cambridge. I looked for evidence that Raman asked Rutherford to intervene but have not yet found any.
26. Raghavan Iyer, *The Moral and Political Writings of Mahatma Gandhi* (Delhi, Oxford University Press, 1986), pp. 310–15. I am grateful to Deepak Kumar for drawing my attention to this earlier visit; personal communication, 23 May 2005.
27. Letter to the editor, from B. V. Thosar to *Times of India*, 4 October 1967, and author's conversation with Thosar, Bombay, November 1967.
28. When I first met Lalitha Chandrasekhar in Chicago in the late 1960s she did not mention her incipient scientific career.
29. It was a suicide of a person on the way to being a scientist. The effect of occasional suicides by scientists reached all the way to the prime minister's office, starting in 1950, as we shall see in chap. 14.

30. Abha Sur, "Dispersed Radiance: Women Scientists in CV Raman's Laboratory," *Meridians: Feminism, Race, Transnationalism* 1, no. 2 (2001): 95–127.
31. There is another interesting example in Asima Chatterjee, who did her M.Sc in 1938 and D.Sc in 1944 in chemistry at the University of Calcutta and later worked with Linus Pauling at CalTech about 1948: did she have any exchanges with Bhatnagar and the CSIR?
32. Irvine Report, *Second Quinquennial Reviewing Committee of the IISc* (Bangalore, 1936). I am grateful to the staff at the India Office Library in London for retrieving these documents (1)5467 for me in 1972.
33. G. Venkataraman, *Journey into Light*, p. 271.
34. He might have won it earlier but the prizes were suspended from 1940 to 1943.
35. M. Born to P. Dirac, 16 October 1936, Churchill Archives, Cambridge. On Raman and the plans of Hungarian scientists in 1935–37, see Rajinder Singh, "Sir C. V. Raman and His Contacts with Hungarian Scientists," *Indian Journal of History of Science* 37, no. 2 (2002). With the full cooperation of their husbands, Dirac's wife along with others similarly well-educated ran a network of communication aimed at informing and aiding eastern European scientists (many of them Jewish, some of them Hungarian) to emigrate or escape. This group, while not "of one mind" on other questions, included Edward Teller, Eugene Wigner, Leo Szilard, George von Hevesy, and John von Neumann. It is possible that B. K. Nehru's wife Fori Friedmann, from a Jewish family in Budapest and once a student at the London School of Economics was part of this network, and thus as he moved into international circles and then to Washington, Bhabha met these people. Bhabha, of course, had direct and earlier access to such networks by virtue of being Dirac's student at Cambridge and traveling with him in North America (e.g., in 1948 to Banff, Alberta). Fori Nehru remained a friend of B. K. Nehru's cousin Indira Gandhi throughout their lives, even in their disagreement during the 1975–77 State of Emergency. She lived until 2005 when she was 95 years old, in the Indian foothills of Kausali.
36. Young physicist Rudolf Peierls, age twenty-seven, a student of Sommerfeld, applied to be an assistant professor at the IISc in 1934, but Raman was delaying for a bigger prize, trying to prepare the council to pay a bigger salary, and when Max Born agreed to come in 1935 the council was obliged to do so. See Rajinder Singh, "Arnold Sommerfeld—The Supporter of Indian Physicists in Germany," *Current Science* 81, no. 10 (10 December 2001): 1492.
37. S. Ramaseshan, *C. V. Raman: A Pictorial Biography* (Bangalore: Indian Academy of Sciences, 1988).
38. Abha Sur, "Aesthetics, Authority, and Control in an Indian Laboratory: The Raman-Born Controversy on Lattice Dynamics," *Isis* 90 (1999): 25–49.
39. The Rutherford Papers at the University Library, Cambridge, contain letters between Max Born and Rutherford. I am grateful to Charles Weiner (formerly of the Niels Bohr Library, New York, and MIT) for enabling me to see copies of these letters, to A. E. B. Owens and M. A. Hoskin of University Library for enabling me to refer to them, and to G. V. R. Born for permission to use them. I reviewed the Rutherford-Born correspondence in Cambridge again in 1998.
40. As Ramaseshan explains, "it was finally shown that Raman's theory was just a small part of the more comprehensive Born theory. If Raman had presented his papers as a simple physical approach for deriving optical modes or obtaining the physical properties of crystals, they would have been considered significant or even important at that time. But Raman would not have it that way. He could not see the missing

elements of his theory. . . . But the greatest sadness and the pity of it was the bitterness this awful controversy created between two old and dear friends"; *C. V. Raman: A Pictorial Biography,* p. 17. According to Raman's biographer G. Venkatramanan, "Raman was locally correct (for second-order spectra for diamonds) but globally wrong. . . . Looking back, one cannot but marvel at Raman's ingenuity . . . [but he] did not appreciate the nuances of theory, while the theorists did not understand the experiment. . . . The irony is that he was not completely wrong but only partially so"; *Journey into Light,* pp. 400–401.

41. Max Born, *My Life and My Views* (New York: Charles Scribner's Sons, 1968).
42. Max Born to Ernest Rutherford, 26 January 1936, Rutherford Papers, University Library, Cambridge.
43. Tata representatives on the council of the IISc were particularly sensitive to colonial government wishes to limit or remove the controversy generated by Raman's term as director. Amiya Bagchi, *Private Investment in India, 1900–1939* (Cambridge: Cambridge University Press, 1972), explains why: "The house of Tata, for example, became very dependent upon government protection in the field of iron and steel. . . . Thus most of the directors had proved their loyalty to the British rulers in the past. Furthermore, the Tata organization embarked on a systematic policy of recruiting former British civil servants." This policy continued until the end of the war in 1945. Cf. "Collaboration and Conflict between European Businessmen," ibid., p. 198.
44. This memorandum, printed but not bound, includes excerpts of council minutes from 1935 to 1937 and was among the papers of Sir P. C. Ray, who was a member of the Court of the IISc. A copy of the official memorandum was typewritten by permission in 1971 for Paul Greenough, and a photocopy of that typed copy was kindly given to me by Professor Greenough.
45. M. Born to E. Rutherford, 22 October 1936, Rutherford Papers, University Library, Cambridge.
46. C. V. Raman to Chairman of the Council, IISc, 1 June 1937, Sir P. C. Ray papers, Science College, Calcutta.
47. G. Venkataraman, *Journey into Light,* p. 280.
48. Rajinder Singh, "Sir C. V. Raman and His Contacts with Hungarian Scientists." After German-born Rudolph Peierls was blocked from returning to Germany in 1935 by new laws, he and his wife took in Hans Bethe as a boarder in Manchester and moved to Cambridge in 1935; he was awarded the D.Sc. by Manchester in 1936. In 1937 he was offered a chair at Birmingham (with Chadwick) as professor of applied mathematics. Peierls became a British citizen in 1940 and one of the key physicists in the British nuclear program. He did not come to India. He participated in the Manhattan Project, was made an FRS in 1945, and was knighted in 1968 (when he was known for his opposition to nuclear weapons).
49. Abha Sur, "Aesthetics, Authority, and Control in an Indian Laboratory," p. 38.
50. "The second-order Raman spectrum is the entire continuum and not just the features which gave early indications of their presence via faint lines. Raman ignored the full spectrum, choosing to retain and explain only the prominent features, and that was his mistake"; Venkataraman, *Journey into Light,* p. 398.
51. Sur, "Aesthetics, Authority, and Control in an Indian Laboratory," p. 41.
52. Author's conversation with A. Rahman, Delhi, November 1998; also G. Venkataraman, personal communication, 1 March 2005.
53. G. Venkataraman, *Journey into Light,* p. 281.
54. Author's conversation with B. V. Thosar, Mumbai, October 1967.

CHAPTER FOUR

1. Nasir Tyabji, personal communication, 20 May 2005.
2. Benjamin Zachariah, *Developing India: An Intellectual and Social and Intellectual History, 1930–1950* (Delhi: Oxford University Press, 2005), p. 131. The best treatment available of this contest of ideas (and who backed them) is Zachariah's chapter 5, "Development: Possible Nations," which explains the roles of Saha, Mahalanobis, and other scientists.
3. Dhruv Raina and S. Irfan Habib, "The Unfolding of an Engagement; the *Dawn* on Science, Technical Education and Industrialization in India, 1896–1912" and "Bhadralok Perceptions or Science, Technology, and Cultural Nationalism," in *Domesticating Modern Science: A Social History of Science and Culture in Colonial India* (Delhi: Tulika Books, 2004).
4. See "Technical Content and Social Context: Locating Technical Institutes—The First Two Decades in the History of the Kala Bhavan, Baroda 1890–1910," in Raina and Habib, *Domesticating Modern Science*, pp. 182–98.
5. Nasir Tyabji, "Examplar of Academia-Industry Interchange: The Department of Chemical Technology at Bombay University," *Ambix* 51, no. 2 (July 2004).
6. Leo Amery to Lord Linlithgow, 27 April 1942, cited in Zachariah, *Developing India*, p. 100.
7. Sen, *Professor Meghnad Saha*, p. 93.
8. Though little is written about Visvesvaraya, one exception is Dhruv Raina, "Visvesvaraya as Engineer-Sociologist and the Evolution of His Techno-Economic Vision," National Institute of Advanced Studies, Bangalore, 2001 (NIAS Lecture L1).
9. Sen, *Professor Meghnad Saha*, p. 93. The "I" in the biography is in fact Saha.
10. Nehru, *A Bunch of Old Letters* (Bombay: Asia Publishing House, 1960), p. 308. I am grateful to Leonard Gordon for reminding me of this evidence of relations between Saha, Bose, Nehru, and Gandhi, and for helping me to correct errors in an earlier draft. See his *Bengal, The Nationalist Movement 1876–1940* (New York: Columbia University Press, 1974). See especially Tagore's speech supporting Bose, pp. 287–88. See also Leonard Gordon, *Brothers Against the Raj: a biography of Indian nationalists Sarat and Subhas Chandra Bose* (New Yrok, Columbia University Press, 1990).
11. A. K. Shaha, *India on Planning: Planning for Liquidation of Unemployment and Illiteracy* (Calcutta, 1948); for a full account of these circumstances see Benjamin Zachariah, "The Uses of Scientific Argument: The Case of 'Development' in India," *Economic and Political Weekly*, 29 September 2001, p. 3699.
12. J. Nehru to S. C. Bose, 11 July 1939, in Netaji Research Bureau, *The Alternative Leadership: Speeches, Articles, Statements, and Letters of Subhas Chandra Bose* (Delhi: Oxford University Press, 1998), vol 10, pp. 156–57.
13. Sen, *Professor Meghnad Saha*, p. 39.
14. Bose's entire speech can be seen in Netaji Research Bureau, *The Alternative Leadership*, pp. 156–57.
15. John Matthai was secretary to the group of industrialists which included many names who figure prominently in this book: J. R. D. Tata, G. D. Birla, Kasturbhai Lalbhai, A. D. Shroff, and "the knights," Sir Ardeshir Dalal, Sir Puroshottamdas Thakurdas, and Sir Sri Ram. The plan was published in two slim volumes. John Matthai's career took off from this point, so that he became India's Minister of Finance until 1951.
16. M. N. Saha's statement in "Acharya Birth Centary Souvenir," Supplement to the *Journal of the Indian Chemical Society*, August 1961.

17. Sen, *Professor Meghnad Saha*, p. 63.
18. Anderson had been not only the Chancellor of the University of Calcutta but also the patron of NISI, of which Saha was president in 1938. Broomfield describes him thus: "This appointment as Governor in 1932 of Sir John Anderson, a senior British Civil Servant with experience in counterinsurgency in Ireland, was followed by an overhaul of the Government's intelligence system and the movement to Bengal of seven battalions of British infantry, who were deployed to hunt down the revolutionary bands and restore British rule in the countryside"; J. H. Broomfield, *Elite Conflict in a Plural Society* (Berkeley: University of California Press, 1968), p. 303. Anderson's term ended in 1937; he was clearly an adversary of the political movement which Saha supported, but the two men were smart enough to collaborate on the NISI.
19. See J. Nehru's letter of 28 September 1939 to K. Kripalani, in Nehru, *A Bunch of Old Letters*.
20. J. W. Wheeler-Bennett, *John Anderson* (London: Macmillan, 1962).
21. This evidence is available in the Archive of the Royal Society, on its website. Gilbert Walker, Birbal Sahni, S. S. Bhatnagar, A. C. Aitken, Major Greenwood, and H. Jeffreys also signed the nomination for election. Interestingly, Saha's name does not appear in this list for Mahalanobis, although they were now neighbors at work.
22. Zachariah, *Developing India*, p.234. Nehru was so worked up about this that he wrote a twelve-page letter to Bose to explain what happened.
23. M. N. Saha to R. Mudaliar, 20 March and 29 March, 1940, Nehru Memorial Library. Saha also pressured Dalal in person.
24. For an entirely different view of Bose than one available in previous writing about him, see Sarmila Bose, "Love in the Time of War: Subhas Chandra Bose's Journeys to Nazi Germany (1941) and towards the Soviet Union (1945), *Economic and Politicial Weekly*, 15 Janaury 2005, pp. 249–56. This fascinating work is based on a very careful reading of Bose's letters to his wife Emilie Schenkl in Austria from 1934 to 1945.
25. J. Nehru to K. Kripalani, 28 September 1939, in Nehru, *A Bunch of Old Letters*, p. 390.
26. Ibid, p. 391.
27. Ibid, p. 394.
28. Meghnad Saha, "Science in Social and International Planning, with Special Reference to India," *Nature*, February 1945, p. 223.
29. Ibid, pp. 222–24.
30. Deepak Kumar, "Emergence of 'Scientocracy': Snippets from Colonial India," *Economic and Political Weekly*, 28 August 2004, pp. 3893–98.
31. Satyen Bose, "Science and Civilization" [1941] reprinted on the occasion of his death, in *Economic and Political Weekly*, 23 March 1974, pp. 467–68.
32. *Hindustan Times*, Delhi, October 1940.
33. Richards, p. 151.
34. Sen, *Professor Meghnad Saha*, p. 23.
35. Arthus I. Miller, *Empire of the Stars: Obsession, Friendship, and Betrayal in the Quest for Black Holes* (New York: Houghton Mifflin, 2005), p. 136.
36. Sen, *Professor Meghnad Saha*, p. 65.
37. Author's conversation with Binayak Dutta-Ray, February 1998, Kolkata.
38. N. N. Dasgupta and M. L. De, "The Electron Microscope—Its Past and Present," *Indian Journal of the History of Science*, May 1968, p. 35. Also discussion with N. N. Dasgupta, Calcutta, March 1969. The concept of the electron microscope was

patented in 1931 by Siemens in Germany, first built in 1938 at the University of Toronto, and first commercially produced by Siemens in 1939.

39. Herbert Childs, *An American Genius* (New York: E. P. Dutton, 1968), pp. 291 and 267.
40. Saha had earlier organized the publicity and the collection of gifts and donations for flood relief in 1923 in North Bengal. The committee was chaired by Saha's teacher Sir P. C. Ray, and Subhas Bose took a group of relief workers to Santahar. Saha's efforts yielded Rs. 2,300,000; Kamalesh Ray, *The Life and Work of Meghnad Saha* (Delhi: National Council of Education Research and Training, 1968), p. 39.
41. Sen, *Professor Meghnad Saha*, p. 65.
42. These figures were compiled from three sources: the authorized Saha biography, the official report of the Institute of Nuclear Physics, Calcutta, 1955, and N. Ray and P. Gupta, eds., *Hundred Years of the University of Calcutta*, University of Calcutta, 1957; Appendix Six: Benefactors. In 1941, Rs. 60,000 from Sir Dorab Tata Trusts; in 1942, Rs. 60,000 from University of Calcutta; in 1943, Rs 60,000 from G. D. Birla (Krishnarpan Charity Trust Fund: Rs 12,000 for 5 years); and in 1944, Rs. 30,000 from Sir Dorab Tata Trusts (Rs 6000 for 5 years).
43. M. N. Saha to J. Nehru, 27 October 1941, Saha Papers, SINP Archives, Kolkata.
44. Sen, *Professor Meghnad Saha*, p. 66. The Birla family's wealth was Calcutta-based and associated with the fortunes of Bengal, especially in the jute industry; however, Birla anticipated the necesssity of moving the family business to Delhi. In contrast, Tata wealth was Bombay-based but also slightly dispersed in Mysore and Bihar.
45. Author's conversation with B. D. Nagchaudhuri, December 1998, Kolkata.

CHAPTER FIVE

1. To understand the context of this community at the time Homi was growing up, see Jesse S. Palsetia, *The Parsis of India* (Leiden: Brill, 2001).
2. Sir John Cockcroft, "Homi Jehangir Bhabha," *Proceedings of the Royal Institution*, vol. 41, pt. 4, no. 191, p. 142.
3. There was a comprehensive exhibit of Bhabha's paintings at the opening of the TIFR auditorium in November 1968.
4. Lord Penny, "Homi Jehangir Bhabha," in *Biographical Memoirs of the Fellows of the Royal Society*, vol. 13, 1967, p. 37. Also R. P. Kulkarni and V. Sharma, *Homi Bhabha—Father of Nuclear Science in India* (Bombay: Popular Prakashan, 1969).
5. Cockcroft, "Bhabha," p. 8.
6. Author's conversation with A. Rahman, November 1998, Delhi. At Cambridge in the early 1930s Homi Bhabha would have met Canadian student Herbert Norman, who acted as recruiter of Commonwealth students to the Communist Party under direction of John Cornford (later killed in the Spanish War in 1937). Herbert Norman, who was born and grew up in Japan, and Homi Bhabha were the same age and members of the same college. This is depicted in the National Film Board of Canada film "The Man Who Might Have Been: An Enquiry into the Life and Death of Herbert Norman" (1998). On the situation in Cambridge at the time, see also Christopher Andrew and Vasili Mitrokhin, *The Mitrokhin Archive: The KGB in Europe and the West* (London: Penguin Books, 2000), chap. 4, "The Magnificent Five."
7. For an inside account of what Bhabha saw in Rome, see Emilio Segre, *Enrico Fermi–Physicist* (Chicago: University of Chicago Press, 1970), chap. 3, "Professor of Rome"; also Gerald Holton, "Striking Gold in Science," *Minerva*, April 1974; and for what

Bhabha encountered in other countries, see Charles Weiner, "New Site for the Seminar: refugees and American physics in the thirties" in D. Fleming and B. Bailyn, *The Intellectual Migration* (Cambridge: Harvard University Press, 1968).

8. Sen, *Professor Meghnad Saha*, p. 51.
9. Werner Heitler, "Bhabha's Work on High Energy Physics," *Science Reporter*, October 1966.
10. Bernard Lovell, "Bristol and Manchester—The Years 1931–9," in Rajkumari Williamson, ed., *The Making of Physicists* (Bristol: Adam Hilger, 1987), pp. 158–59.
11. Source is a tape of an interview of Sir James Chadwick by Charles Weiner, archived in the American Institute of Physics. See Andrew Brown, *The Neutron and the Bomb: A Biography of Sir James Chadwick* (Oxford: Oxford University Press, 1997), p. 164. I record with gratitude the encouragement of Charles Weiner of MIT that I finish this book within his lifetime.
12. M. G. K. Menon, "Homi Jehangir Bhabha," *Proceedings of the Royal Institution*, vol. 41, pt. IV, no. 191, p. 430.
13. I am indebted for corrections to a draft of this section by members of TIFR in personal communications dated 30 September and 16 October 1970 from G. S. Gokhale. This will be referred to as "Gokhale et al., communication 1970."
14. Author's conversation with Sir C. V. Raman, Bangalore, April 1968.
15. Jacob Clay, physicist at the Technische Hoogeschool at Bandung, Java, from 1920 (with his wife Tettje Clay, physicist in her own right), learned to use balloons to study a free source of high-energy radiation, the "penetrating radiation" or "penetrating component," later to be called "cosmic rays." Before returning to Amsterdam in 1929, for nine years Clay did research on latitude and the penetrating radiation, both at high altitude and on board ship in 1927. During that time he built up Bandung's Tropical Institute of Pure Physics. According to Pyenson, who has read the Dutch and German sources in addition to the English, Jacob Clay was among the very first to have "concluded that the intensity of the radiation varied with latitude." His precise data was highly valued. His research using even more precise instrumentation was repeated at Bandung in 1933. Initially a rival, Arthur Compton, at the University of Chicago, became convinced of his originality and nominated him for the Nobel Prize in 1935. In this period both Millikan and Compton conducted penetrating component research in India at high altitudes too. But, despite the nomination from Compton, Jacob Clay (still at Amsterdam) was passed over for the 1936 Nobel Prize awarded for cosmic ray work. Then Compton and Jacob Clay decided to criticize the powerful figure in this field, Robert Millikan at Caltech, and drafted an unpublished analysis of Millikan's work in contrast to others. But Millikan had prevailed in 1935 in nominating the recipients of the 1936 physics prize (his protégé Carl Anderson and Viktor Hess), and now Millikan fought back. Under pressure from Millikan's allies in 1938 Compton withdrew the appraisal manuscript from the publication process at the *Reviews of Modern Physics*. Clay presented five papers at the international conference on cosmic rays, hosted by Compton, at the University of Chicago in 1939 with Millikan in attendance. The Tropical Institute of Pure Physics was closed during Japanese occupation in 1941, and reopened six years later. See Lewis Pyenson, *Empire of Reason: Exact Sciences in Indonesia, 1840–1940* (Leiden: Brill, 1989), p. 146.
16. Pyenson, *Empire of Reason*, pp. 143–57.
17. C. V. Raman to P. Dirac, 20 July 1940 and 19 October 1940, Churchill College Archives, Cambridge.

18. Author's conversation with Sir C. V. Raman, Bangalore, April 1968.
19. H. J. Bhabha to P. Dirac, 8 November 1939, Churchill College Archives, Cambridge.
20. H. J. Bhabha to P. M. S. Blackett, 1941, Royal Society Archives, London. The first page of this letter bearing the date is missing.
21. Bhabha to Dirac, 24 January 1940, Churchill College Archives, Cambridge.
22. Jitendra Singh, "Organization for Science: A Case Study of the Trombay Establishment," paper mimeographed at the Administrative College, Hyderabad, 1970.
23. Bhabha to Dirac, 24 January 1940, Churchill College Archives, Cambridge.
24. Bhabha to Dirac, 30 October 1940, Churchill College Archives, Cambridge.
25. Bhabha to Dirac, 30 October 1940, Churchill College Archives, Cambridge.
26. Bhabha to Blackett, 1941, Royal Society Archives, London.
27. H. J. Bhabha to Sir Sorab Saklatvala, 12 March 1944; quoted in Menon, "Bhabha," and Cockcroft, "Bhabha."
28. D. M. Bose, "Meghnad Saha Memorial Lecture" (1965), *Proceedings of the National Institute of Sciences of India* 33, A, nos. 3, 4 (Calcutta, 1967): pp. 116–17.
29. H. J. Bhabha to J. R. D. Tata, 19 August 1943, quoted in Menon, "Bhabha."
30. A. V. Hill, *The Ethical Dilemma of Science* (New York: Rockefeller Institute Press, 1960), p. 376.
31. A. V. Hill, "Science and the Problems of Development," *Science*, 4 February 1966, p. 545; also Hill, *The Ethical Dilemma.*
32. Bhabha to Saklatvala, 12 March 1944, quoted in Menon, "Bhabha." p. 423.
33. Menon, "Bhabha," p. 424.
34. Author's conversation with B. V. Sreekantan, January 1968, Mumbai.
35. R. M. Lala, *Over the Last Blue Mountain: A Life of JRD Tata* (Delhi: Penguin, 1992). In November 1998 I asked Mr. Lala whether he could reveal whom the Tata family approached to co-finance the institute with them in 1945, but he declined, having been asked not to do so by the Tatas. Other sources suggested that it was probably the Godrej family.

CHAPTER SIX

1. Peter Clarke provides a finely detailed account of the mission's work seen from the British points of view and perceptions of its opponents in *The Cripps Version: The Life of Sir Stafford Cripps* (London: Penguin, 2002), pp. 257–341.
2. Sanjoy Bhattacharya, *Propaganda and Information in Eastern India 1939–45: A Necessary Weapon of War* (London: Curzon, 2001).
3. Richard J. Aldrich, "American Intelligence and the British Raj: The OSS, the SSU and India 1942–47," *Intelligence and National Security* 13, no 1 (1998): 150.
4. A. V. Hill, "Memories and Reflections," unpublished autobiography, 1974, vol. 3, p. 66, Churchill College Archives, Cambridge.
5. Ibid., pp. 636–42.
6. Saha said all this loudly in an editorial in *Science and Culture* in 1946.
7. Shiv Visvanathan, *Organizing for Science* (Delhi: Oxford University Press, 1984), p. 138.
8. Krishnan had left Raman's lab in Calcutta soon after the big discovery in 1928 to join Satyen Bose at Dacca University and returned, in 1933, to the University of Calcutta to a physics chair that Saha himself coveted. He stayed there almost ten years, before becoming professor of physics and department head at the University of Allahabad, where Meghnad Saha had been prior to moving back to Calcutta. It was from Allahabad that Krishnan was moved to become National Physical Laboratory's director,

having already established a brilliant reputation for work on scattering of light in liquids, magnetic properties of crystals in complex structures, and electrical conductivity of metals and alloys at low temperatures. He was an expert in the properties of graphite and, for this reason, became deeply involved, in 1946, in the atomic energy program with Bhabha, Bhatnagar, and Saha. He was soon appointed as third member of the new Atomic Energy Commission with Bhabha and Bhatnagar. See the special issue on K. S. Krishnan, *Current Science,* 10 December 1998; Kathleen Lonsdale and H. J. Bhabha, "Kariamanikam Srinivasa Krishnan," *Biographical Memoirs of the Fellows of the Royal Society,* vol. 13 (1967), pp. 245–55; D. C. V. Mallik, "The Raman Effect and Krishnan's Diary," *Notes and Records of the Royal Society,* vol. 54 (2000); also D. C. V. Mallik, "K S Krishnan and the Kodaikanal Observatory," *Current Science,* September 2000.

9. Winston Churchill to A. V. Hill, 30 October 1943, Churchill College Archives, Cambridge, 1/315 (20/94/221). J. C. Chaudhuri, *Indian Fellows of the Royal Society and Others* (Calcutta: Academic Press, 1992), p. 69.
10. A. V. Hill to P. M. S. Blackett, 30 December 1942, Churchill College Archives, Cambridge. The letter quotes "the Jamshaheb," famous in English cricket circles as "Ranji," a rajah of a small princely state in Punjab called Nawanagar (later Jamnagar). Ranjit Singh was president of India's Board of Control of Cricket in 1937–38. But Ranji had earlier played for England, too, so his opinion carried weight in both places, thus indicating elite Indian approval of Bhatnagar.
11. A. V. Hill papers, Churchill College Archives, Cambridge.
12. A. V. Hill, *The Ethical Dilemma of Science* (New York: Rockefeller Institute Press, 1960), p. 312.
13. S. S. Bhatnagar to A. V. Hill, 4 April 1944, 11 May 1944, 16 June 1944, 26 July 1944; NISTADS Archives, Delhi; A. V. Hill to S. S. Bhatnagar, 24 May 1944, 2 August 1944, Churchill College Archives, Cambridge.
14. Hill to Bhatnagar, 2 August 1944, NISTADS Archives, Delhi.
15. See Mountbatten's account of Bernal, among others, in Brenda Swann and Francis Aprahamian, eds., *J. D. Bernal: A Life in Science and Politics* (London: Verso, 1999), pp. 191–211.
16. Kameswar Wali, *Chandra: The Autobiography of S. Chandrasekhar* (New York: Penguin Books, 1992), p. 196.
17. A. V. Hill Papers, Churchill College Archives, Cambridge.
18. Nothing serious has been written about Nazir Ahmed; the official biography prepared in 1944 by Bhatnagar and Hill for the tour says he completed his PhD under Rutherford at the Cavendish Laboratory in the University of Cambridge in 1925, focused on "radium emanation, gamma ray absorbsion, and induced radio-activity" and published his work in the *Proceedings of the Royal Society*. Ahmed was not the first Indian to complete a degree in experimental physics at Cambridge: B. N. Banerji did so in 1923, and Nazir Ahmed was followed to the Cavendish Laboratories by fellow-Punjabi Rafi Chaudhuri in 1932. Ahmed returned to Lahore in 1925 and organized a new physics laboratory at Islamia College. But in six years he left physics and moved to Bombay to be assistant director of the laboratory of the Indian Cotton Central Committee, becoming director in 1932; he was still director when he participated in the tour in 1944–45. Ahmed joined Bhatnagar on the board of the CSIR and built a special relationship with the Department of Chemical Technology at the University of Bombay. He became Pakistan's first chairman of its Atomic Energy Commission in 1950. A. V. Hill Papers, Churchill College Archives, Cambridge.

19. No one has yet discussed how scientists with British names were categorized when nominated in/from India, noting that in some cases they spent most of their lives in India.
20. Anderson, former governor of Bengal (1932–37) returned to London in 1937 and by 1939 was Churchill's right-hand man in the war cabinet, in charge of British internal security and intelligence. Churchill had attempted to make him viceroy of India. Instead, because of his thorough knowledge of the colonial situation, in 1940 he was put in charge of cabinet matters relating to India and remained so through the war. In this capacity, beginning in the 1930s in Bengal, he was thoroughly familiar with Megnad Saha and his revolutionary sympathies, having opposed them only a few years earlier.
21. M. N. Saha, "Science in Social and International Planning with Special Reference to India," *Nature,* 26 February 1945.
22. DeVorkin, "Quantum Physics and the Stars (IV): Meghnad Saha's Fate," *Journal for the History of Astronomy* 25 (1994): 176.
23. S. S. Bhatnagar, "Science and Industrial Progress in the New India" *Asiatic Review,* February 1945.
24. Hill, "Memories and Reflections," pp. 636–40.
25. Delhi, CSIR, *Report to the Governing Council,* 24 September 1948, NISTADS Archives, Delhi. Unfortunately, little is written on the profitability of war production for Indian industries, although talk of such profit figures in the legendary rise to prominence of certain Indian industrialists.

CHAPTER SEVEN

1. Epigram quotes are from the following sources: M. N. Saha, speech delivered in London in November 1944 and published as "Science in Social and International Planning with Special Reference to India," *Nature,* 26 February 1945; S. S. Bhatnagar to A. V. Hill, 2 August 1944; H. J. Bhabha to Sir Sorab Saklatvala, 12 March 1944.
2. Norah Richards, *Life and Work of Sir S. S. Bhatnagar* (Delhi: New Book Society, 1948), p. 170.
3. A. Vasantha and D. K. Banerjee, "The Indian Science Congress Association: Organization and Development of Its Chemistry Section, 1914–1947," in Deepak Kumar, ed., *Science and Empire: Essays in an Indian Context* (Delhi: Anamika Prakashan, 1991). See also V. V. Krishna, "The Emergence of the Indian Scientific Community," *Sociological Bulletin* 40, nos. 1–2 (March–September 1991); for the wider context see David Arnold, *Science, Technology and Medicine in Colonial India* (Cambridge: Cambridge University Press, 2000), pp. 190 ff.
4. M. N. Saha in *Science and Culture* 6 (1941): 694, cited in Virendra Singh, "Meghnad Saha: His Science and Life," *Current Science* 64, no. 7 (10 April 1993): 530–36. My search for the original 1941 volume in the National Library in Kolkata was unsuccessful in January 2005 when that volume could not be found.
5. Meghnad Saha and B. D. Nagchaudhuri, "The Story of the Atomic Bomb," *Science and Culture* 11, no. 3 (September 1945): 111–13.
6. "The Logic of the Atomic Bomb," *Science and Culture* 11, no. 5 (November 1945): 214.
7. "Utilization of Nuclear Fission Energy," *Science and Culture* 11, no. 6 (December 1945): 294–96. This was largely based on Henry Smyth's report *Atomic Energy for Military Purposes,* August 1945.

8. "The Atom Bomb," *Science and Culture* 12 (June 1946): 4–6 (text of Saha's lecture to the Royal Asiatic Society in London, January 1946.)
9. "Industrial Utilization of Atomic Power in India," *Science and Culture* 13, nos. 14–15 (September–October 1947).
10. "Release of Atomic Energy," *Science and Culture* 13, no. 5 (November 1947): 167.
11. Many of Bhatnagar's republished articles, radio talks, and speeches refer to the power of this cooperation if it could be brought about; see an excellent collection in V. V. Krishna, ed., *S. S. Bhatnagar on Science, Technology, and Development 1938–1954* (Delhi: Wiley Eastern, 1993).
12. For a fascinating and detailed account of this negotiation from various British points of view, see Peter Clarke, *The Cripps Version* (London: Penguin, 2001), pp. 387–458.

CHAPTER EIGHT

1. M. N. Saha to Vice-Chancellor, Calcutta University, 15 February 1947, Saha Papers, Saha Institute Archives, Kolkata. It is not known whether this unused tax money was eventually channeled to research in Bengal or whether this reported Rs 10 million was actually used for the purpose in Ahmedabad, but circumstantial evidence in Gujerat suggests it was.
2. "Another Achievement of the Allahabad-Calcutta Research Axis," Notes and News, *Science and Culture* 33 (February 1967): 49–50. According to S. N. Sen, "Saha himself wrote most of the editorials in his monthly journal." See also Santimoy Chatterjee, "Saha—the Scientist and the Institution Builder," *Indian Journal of the History of Science* 29, no. 1 (1994): 99–110. This reference to an axis was therefore probably his own image.
3. *The Official Report of the Institute of Nuclear Physics* (Science College, University of Calcutta, 1955), p. 3. The sums and sources were Rs 70,000 capital grant (from central government through Nehru); Rs 350,000 operating costs (from central government, Bhatnagar's department); Rs 200,000 for a new building (from University of Calcutta).
4. John Broomfield, *Elite Conflict in a Plural Society* (Berkeley: University of California Press, 1968).
5. Discussion with R. N. Ray, Calcutta, October 1968. From this also follows a pride in students. See, for example, P. C. Ray's comment quoted in Acharya P. C. Ray Birth Centenary Souvenir, *Supplement to the Journal of the Indian Chemical Society*, August 1961, and Ram Nath Chopra, "Problems and Prospects of a Pharmacological Career in India," *Annual Review of Pharmacology* 10 (1965): 26.
6. *The Official Report of the Institute of Nuclear Physics*, p. 3.
7. Saha Papers, Saha Institute Archives, Kolkata.
8. Sen, *Professor Meghnad Saha*, pp. 71–76.
9. Herbert Childs, *An American Genius* (New York: E. P. Dutton, 1968), p. 458.
10. Author's conversation with B. D. Nagchaudhuri, Kolkata, December 1998.
11. Gregg Herken, *Brotherhood of the Bomb: Oppenheimer, Teller, Lawrence* (New York: Holt Publishers, 2002).
12. M. N. Saha to J. Nehru, 27 October 1941, Saha Institute Archives, Kolkata.
13. Author's conversation with B. D. Nagchaudhuri, Kolkata, December 1998.
14. Saha Institute of Nuclear Physics, Annual Report 1967 (Calcutta, 1968), p. 4. My understanding of the history of that cyclotron, though incomplete, relies on conversations with members of the cyclotron group in the late 1960s and with B. D. Nagchaudhuri in 1998.

15. The amount of Rs 200,000 is confirmed in N. R. Ray and P. C. Gupta, eds., *Hundred Years of the University of Calcutta* (Calcutta: University of Calcutta Press, 1957), p. 376, but there is no mention of the source.
16. See photos, Dasgupta and De, "The Electron Microscope: Its Past and Present," *Indian Journal of the History of Science*, May 1968, pp. 33–34; also author's conversation with N. N. Dasgupta, Calcutta, March 1969. Figures from Ray and Gupta, eds., *Hundred Years*, pp. 509–23.
17. Sen, *Professor Meghnad Saha*, p. 81.
18. *The Official Report*.
19. M. N. Saha to H. H. Plaskett, undated letter, probably 21 December 1946, according to David DeVorkin, based on Saha's archives. Quoted in DeVorkin, "Quantum Physics and the Stars (IV): Meghnad Saha's Fate," *Journal for the History of Astronomy* 25 (1994).
20. In addition to DeVorkin's assessment of Saha's influence in astrophysics, see the *Meghnad Saha Birth Centenary Commemoration Volume* (Calcutta: SINP, 1993). The reference to the Nobel Prize is in "M. N. Saha and His Two Chances for the Nobel Prize," *Science and Culture*, May–June 1999; also Rajinder Singh and Falk Reiss, "C. V. Raman, M. N. Saha, and the Nobel Prize for the Year 1930," *Indian Journal of the History of Science* 34, no. 1 (1999). Saha was nominated for the first time by D. M. Bose and S. K. Mitra in 1929, and the second time by Arthur Compton in 1937.
21. Author's conversation with B. M. Bannerjee, Calcutta, November 1968.
22. J. A. Ratcliffe, "Sisir Kumar Mitra," *Biographical Memoirs of the Fellows of the Royal Society*, vol. 5 (London: The Royal Society, 1959), p. 225; D. DeVorkin, personal communication, 10 November 1998.
23. M. K. Dasgupta, "A Day with Professor M. N. Saha," *Physics Alumni Annual* (Calcutta, 1957), p. 5. Two insights into Saha's personality are also recorded. An English colleague of Dasgupta drove them to Jodrell Bank near Manchester. Saha discussed affairs of Calcutta with Dasgupta in Bengali. When asked by Dasgupta if it would not be more appropriate to speak in English, Saha's reply was "why should I disclose all these unsavoury things of our country to a foreigner" (p. 4). Outside a Manchester building then under construction, Saha picked up a brick, smashed it, and examined the pieces. He exclaimed at the purity of the brick and decried the adulteration of bricks in India (p. 5). Saha was deeply involved in procurement for the construction of large laboratories in Calcutta around 1950, and the uncertain quality of bricks that resulted from poor raw materials and incomplete firing, all ways of lowering costs, was a constant concern. The repair of such buildings was a major downstream cost.
24. Author's conversation with J. K. D. Verma, Calcutta, March 1969.
25. A. K. Sen, "Millimeter Waves: Solar Emissions and Various Applications," in Jayanta Basu, ed., *The Glittering Spectrum of Meghnad Saha* (Calcutta: SINP, 1994), pp. 81, 107.
26. Author's conversation with B. M. Bannerjee, Calcutta, November 1968; also author's conversation with Manoj Bannerji, Maryland, December 1970. In 1998 Binayak Dutta Roy reread Saha's papers of this 1940s period and "was struck by how they were not much influenced by quantum mechanics—it seems to have passed him by." Author's conversation with B. Dutta Roy, Calcutta, November 1998.
27. Author's conversation with Ajit Saha, Calcutta, February 1969.
28. Author's conversation with S. Biswas, Bombay, December 1967.
29. Author's conversation with Manoj Bannerji, Maryland, December 1970.

30. Author's conversation with Satyen Bose, Calcutta, April 1969.
31. D. M. Bose, personal communication, 8 May 1971.
32. Ray and Gupta, eds., *Hundred Years*, pp. 411–14.

CHAPTER NINE

1. "Personal Records of Dr. S. S. Bhatnagar," Delhi, no date, probably early 1948, p. 8. The present tense is used and it is probable that Bhatnagar continued to communicate with China through 1948. In 1944 he was to have traveled in the company of Archibald Hill to Chiang Kai-shek's capital at Chungqing in order to meet Chinese scientists and stimulate more cooperation among Allies. But this proved impossible because of the shortage of space on military transport aircraft flying over the mountains from Calcutta to Kuo Min Tang airbases in Yunnan, and on to Chungqing.
2. S. S. Bhatnagar, *Report to the Empire Scientific Conference*, June 1946, Bhatnagar Papers, NISTADS Archives, Delhi.
3. Curiously we have no record of Bhatnagar's interaction with Steels during this fertile war period, nor how Bhatnagar's Steels Scholars fared as their mentor entered high-level government service in 1943.
4. I am grateful to V. V. Krishna, Bhatnagar's biographer, for checking his record of the patents for me; personal communication, 24 May 2006.
5. Minutes of the Governing Body, CSIR, 18 March 1945, in Sarabhai Papers, Nehru Foundation Archive, Ahmedabad.
6. S. S. Bhatnagar to Planning and Development Department, 1 August 1945, Bhatnagar Papers, NISTADS Archives, Delhi. It is curious that in his detailed personal account written in 1948 Bhatnagar does not mention his patents. Patent aversion goes a long way back in India; J. C. Bose refused until 1901 to patent "a detector of electrical disturbances," unrelated to wireless radio processes he had studied earlier. Had he patented his earlier research inventions, they would probably have made him immensely wealthy and he would not have had to ask governments for funds for his Bose Institute. But he rejected that practice, remaining interested only in his noncommercial research, according to Subrata Dasgupta, *Jagadish Chandra Bose* (Delhi: Oxford University Press, 1999), pp. 90–96. There is no record of P. C. Ray's philosophy concerning patents, nor his practice. We know little about patents Raman may have taken out. Bhatnagar was well aware of the practices of others and the Indian attitude that patents had been used against them in colonial economic policies but worried about the lack of Indian protection for what is now called its intellectual property. This subject deserves its historian.
7. R. B. Pai to K. N. Mathur, 20 May 1946, NISTADS Archives, Delhi.
8. S. S. Bhatnagar to A. V. Hill, 7 November 1945, Churchill College Archives, Cambridge.
9. A good description of 1940s student life in London is in R. Ramanna, *Years of Pilgrimage* (Delhi: Viking, 1991). In 1967–69 I interviewed Indian students of science returned from London, and there were some similarities with the earlier period, except for the poverty of the British postwar surroundings.
10. G. S. Tendolkar to S. S. Bhatnagar, 22 April 1947, NISTADS Archives, Delhi.
11. E. Appleton to S. S. Bhatnagar, 12 February 1945, NISTADS Archives, Delhi.
12. Note that it was Saha's academy, not Raman's, that eventually became the official corresponding body between Britain and India and the "establishment" national academy in Delhi.
13. H. R. V. Iyengar to Royal Society, 13 March 1946, NISTADS Archives, Delhi.

14. There was no CSIR lab planned for either Lahore or Karachi, and these cities, like Dacca in East Bengal, were about to be separated from India.
15. V. V. Krishna, *S. S. Bhatnagar on Science, Technology and Development, 1938–1954* (Delhi: Wiley Eastern Publishers, 1993).
16. S. P. Kaushika, "CSIR—Some Reflections" in *CSIR—Looking Back* (Delhi: CSIR Pensioners Welfare Association, 1997), pp. 69–70. How a former student from Lahore would have failed to recognize Bhatnagar's face and tall stature is not explained in the story.
17. "Indian Scientist's Courage," *Times of India,* 21 July 1946.
18. S. S. Bhatnagar to A. V. Hill, 6 September 1948, Churchill College Archives, Cambridge.
19. S. S. Bhatnagar to A. V. Hill, 4 December 1945, NISTADS Archives, Delhi.
20. S. S. Bhatnagar to A. V. Hill, 25 June 1948, NISTADS Archives, Delhi.
21. N. L. Dutta, "Reminiscences of the Formative Years of CSIR," in *CSIR—Looking Back,* p. 37.
22. Zia Mian, "Fevered with Dreams of the Future: The Coming of the Atomic Age to Pakistan," in Itty Abraham, ed., *South Asian Cultures of the Bomb: Atomic Publics and the State in India and Pakistan* (Bloomington: Indiana University Press, 2009), p. 33.
23. N. R. Rajagopal, M. A. Qureshi, and Baldev Singh, *The CSIR Saga: A Concise History of Its Evolution* (Delhi: CSIR, 1991), p. 60.
24. Special Issue on K. S. Krishnan, *Current Science,* 10 December 1998; Kathleen Lonsdale and H. J. Bhabha, "Kariamanikam Srinivasa Krishnan," *Biographical Memoirs of the Fellows of the Royal Society,* vol. 13 (London: The Society, 1967), pp. 245–55; D. C. V. Mallik, "The Raman Effect and Krishnan's Diary," *Notes and Records of the Royal Society,* vol. 54 (London: The Society, 2000); also D. C. V. Mallik, "K S Krishnan and the Kodaikanal Observatory," *Current Science,* September 2000.
25. A. V. Hill, "Memories and Reflections," unpublished autobiography, 1974, vol. 3, p. 637, Churchill College Archives, Cambridge.
26. A. V. Hill to S. S. Bhatnagar, 21 June 1948, Churchill College Archives, Cambridge.
27. P. C. Mahalanobis to S. S. Bhatnagar, 28 August 1946, NISTADS Archives, Delhi.
28. Curiously there is no study of Bhatnagar's relationship to his laboratory staff or to entrepreneurs in small- and medium-scale industry.
29. S. S. Bhatnagar to A. V. Hill, 25 June 1948, NISTADS Archives, Delhi.
30. A. V. Hill to S. S. Bhatnagar, 21 June 1948, Churchill Archives, Cambridge.
31. S. S. Bhatnagar to A. V. Hill, 25 June 1948, Churchill Archives, Cambridge.
32. I am grateful to S. Irfan Habib for information about Maulana Azad; personal communication, 10 November 2003.
33. M. N. Saha to Atma Ram, 31 May 1948 and 6 October 1948, Saha Institute Archives, Kolkata.
34. NISTADS, *A Review of Science and Technology Policies of the Government of India, 1948–1971* (Delhi, 1989), DST Ref No SP/UR/201/87.
35. "Petroleum in India" Third Foundation Lecture, Fuel Research Institute, Dhanbad, 1952.
36. Shiv Visvanathan, *Organizing for Science* (Delhi: Oxford University Press, 1984), pp. 143–44.
37. Ibid., pp. 144–45.
38. A. V. Hill to S. S. Bhatnagar, 11 May 1951, Churchill College Archives, Cambridge.
39. S. S. Bhatnagar to A. V. Hill, 18 May 1951, Churchill College Archives, Cambridge.

CHAPTER TEN

1. I am grateful to Indira Chowdhury, consulting archivist at TIFR, for sharing her archival notes and insight with me (personal communication, 9 June 2006), including reference to H. J. Bhabha, "Historical Note on the Tata Institute of Fundamental Research," prepared for Prime Minister Nehru on the occasion of laying the foundation stone, 1 January 1954; TIFR Archives D-2004–00061, 10 pp. This section depends on a combination of our joint work in separate archives.
2. Minutes of TIFR, 21 September 1945, based on Bhabha's 10-page memo to CSIR (no date).
3. Bhabha, "Science and the Problems of Development," *Science* 4 (February 1966): 546.
4. Gokhale et al., personal communication, September and October 1970.
5. M. G. K. Menon, "Homi Jehangir Bhabha," *Proceedings of the Royal Institution*, vol. 41, pt. IV, no. 191, p. 424.
6. Author's conversations with M. S. Narasimhan, Bombay, February 1968; R. Narasimhan, Chicago, April 1970; and J. Choksi, Montreal, May 1971.
7. C. K. Raju, "Kosambi the Mathematician," *Economic and Political Weekly* 44, no. 20 (16 May 2009): 43. I thank Raju for showing me a translation by Suman Oak from the Marathi biography by Chintamani D. Deshmukh, *Damodar Dharmanand Kosambi, Life and Work* (Mumbai: Granthghar Books, 1993).
8. Raja Ramanna, *Years of Pilgrimage* (Delhi: Viking, 1991).
9. G. Venkataraman, *Bhabha and His Magnificent Obsessions* (Hyderabad: Universities Press, 1994), pp. 115–16.
10. Ibid., pp. 117–18; also author's conversations with G. Venkataraman, February 1969 and November 1998.
11. Minutes, TIFR Council, 28 July 1948. Gill did research during 1948–49 at the Carnegie Institution and the National Bureau of Standards in Washington, DC, but his reputation at TIFR was deteriorating and he sought Nehru's intervention, so he was passed on from Bhabha to Bhatnagar, and moved from Bombay to Delhi as an officer on special duty with the new Atomic Energy Commission of which Bhatnagar was the secretary. But there was no research in that role, so finally in 1949 he was appointed head of the new physics department at Aligarh Muslim University, where he began to establish a high-altitude cosmic ray research station in 1951. See Piara Singh Gill, *Up Against Odds: Autobiography of an Indian Scientist* (Delhi: Allied Publishers, 1992), pp. 40–97.
12. Raja Ramanna, *Years of Pilgrimage* (Delhi: Viking, 1991), p. 53.
13. One should remember that though Bhabha was known best as a physicist, he had studied engineering and felt at home in the company of engineers and with their approach to problems. He himself was constantly designing structures and tinkering with equipment. The tensions that showed up later among some of his younger physics and engineering colleagues were not unknown to him, but after his death those tensions were installed in a large-scale organizational form and became more and more dysfunctional for some of them, including leaders.
14. Gokhale et al., personal communication, September and October 1970.
15. Curtis Peebles, *The Corona Project: America's First Spy Satellites* (Annapolis, MD: Naval Institute Press, 1997), p. 7. Note also that between 1952 and 1954 there was active US planning for *both* satellites and balloons and improvement of photo imagery and interpretation. In June 1953 reconnaissance balloon programs called "Grandson," then "Grayback," finally "Genetrix" were initiated. December 1954 saw the launch

of eight Moby Dick balloons from Scotland that were intended to float on the jet stream across the USSR but instead made "lazy circles over Yugoslavia and North Africa" (p. 31). In January 1955 a second Moby Dick flight from Scotland occurred, but it was a failure, for none of those balloons remained aloft. The CIA, however, had done one successful balloon flight from Scotland to South Korea by that time. The extent to which Peters or others in the TIFR group were aware of the activities in Corona is not known, but they were certainly in regular communication with the researchers at Minnesota, so there is a high probability they knew all but the most classified information. The high-altitude cosmic ray studies group led by Vikram Sarabhai at Physical Research Laboratory, Ahmedabad, was in regular communication with these Minnesota researchers at this time also, and for years to come.

16. Peebles, *The Corona Project*, p. 32.
17. Author's conversation with Bernard Peters, Bombay, November 1968. See also B. Peters, "Homi Bhabha and Cosmic Rays," *Science Reporter*, October 1966, and press interview with B. Peters, "Cosmic Ray Research at the Tata Institute," *Nuclear India*, March 1969.
18. Bernard Peters, personal communication, 21 January 1970.
19. According to Itty Abraham, relying on his personal review of the Ernest O. Lawrence papers at Bancroft Library, University of California, Berkeley, personal communication, April 2000. I am grateful to Itty Abraham for discussion of this issue and for showing me photocopies of Peters's letters.
20. Charles Thorpe, *Oppenheimer, the Tragic Intellect* (Chicago: University of Chicago Press, 2006), p. 206.
21. B. Peters to E. O. Lawrence, 11 June 1949 (Lawrence Papers, Bancroft Library, University of California).
22. B. Peters to E. O. Lawrence, 20 February 1950, with Peter's petition to Secretary of State Dean Acheson, 18 February 1950 (Lawrence Papers, Bancroft Library, University of California) asking for the evidence against him. In this letter Peters refers to Oppenheimer's statement to the House of Representatives Committee "last June," meaning 1949. At this stage Oppenheimer was still chairman of the Advisory Committee to the Atomic Energy Commission and thus had more or less direct access to President Truman.
23. There appears to have been such a ring operated through the office of the USSR consul general in San Francisco; see Gregg Herken, *Brotherhood of the Bomb: The Tangled Lives and Loyalties of Robert Oppenheimer, Ernest Lawrence, and Edward Teller* (New York: Henry Holt & Co., 2002). See also Christopher Andrew and Vasili Mitrokhin, *The Mitrokhin Archive: The KGB in Europe and the West* (London: Penguin Books, 2000), chaps. 6 and 7.
24. Daniel Kevles, *The Physicists: The History of a Scientific Community in Modern America* (New York: Knopf, 1978). For evidence on how a movement grew among disparate groups and individuals against nuclear weapons, see Robbie Lieberman, *The Strangest Dream: Communism, Anticommunism, and the U.S. Peace Movement, 1945–1963* (Syracuse: Syracuse University Press, 2000).
25. Abraham, *The Making of the Indian Atomic Bomb*, p. 66.
26. Indira Chowdhury, personal communication based on TIFR records, 9 June 2006.
27. G. Venkataraman, *Bhabha and His Magnificent Obsessions* (Hyderabad: Universities Press, 1994), p. 150.
28. Ward Morehouse, "Nehru and Science," Special Issue on Science and Government, *Indian Journal of Public Administration* 15, no. 3 (July–September 1969): 494.

29. J. Nehru to Minister Mahavir, 10 November 1953, *Selected Works of Jawaharlal Nehru*, vol. 25 (Delhi: Oxford University Press, 1999), pp. 193–94.
30. T. R. Seshadri, "Shanti Swarup Bhatnagar," *Biographical Memoirs of Fellows of the Royal Society*, 8 (London: The Society, 1962): 8.

CHAPTER ELEVEN

1. David Holloway, *Stalin and the Bomb: The Soviet Union and Atomic Energy, 1939–1956* (New Haven: Yale University Press, 1994), p. 105.
2. Nehru, *A Bunch of Old Letters* (Bombay: Asia Publishing House, 1960), p. 502.
3. Joseph Rotblatt, CBC Radio interview, 20 February 2003.
4. As an intimate member in that network, Bhabha would have known this too, except I have not yet seen evidence about it.
5. Cited in Norah Richards, *Life and Work of Sir S. S. Bhatnagar* (Delhi: New Book Society, 1948), p. 164.
6. All-India Radio, 17 October 1945, cited in Richards, *Life and Work of Sir S. S. Bhatnagar*, p. 195.
7. S. S. Bhatnagar to S. Siddiqui, 31 December 1945, NISTADS Archives, Delhi.
8. S. Parthasarathy to P. Blackett, 28 April 1952, Royal Society Archives. A full report to the government of India with his findings at Hiroshima and Nagasaki, if one was written, has not been found.
9. Author's conversation with W. Bennett Lewis, Vancouver, January 1977.
10. Itty Abraham, *The Making of the Indian Atomic Bomb* (London: Zed Press, 1998), pp. 57–59, and J. E. Helmreich, *Gathering Rare Ores: The Diplomacy of Uranium Acquisition* (Princeton, NJ: Princeton University Press, 1986), pp. 166–72.
11. Author's conversation with Bertrand Goldschmidt, Paris, November 1998.
12. "The Report of the Indian Scientific Mission on their Visit to the United Kingdom, the United States of America, and Canada During 1944–45," 11 February 1946, NISTADS Archives, Delhi.
13. As to why the British authorities, including the viceroy and staff, were totally preoccupied with other questions in 1946–47, see Peter Clarke, *The Cripps Version* (London: Penguin, 2001), Part 5: Cripps vs. Gandhi 1946–47.
14. P. Blackett to S. Cripps, 17 February 1947 and 11 March 1947, Blackett Papers, Royal Society Archives, London.
15. "Note issued to the Press on the proceedings of the first meeting of the Atomic Research Committee," 15 May 1946, Bombay, Bhatnagar Papers, NISTADS Archives, Delhi.
16. H. J. Bhabha to S. S. Bhatnagar, 15 January 1947, NISTADS Archives, Delhi.
17. H. J. Bhabha to M. N. Saha, 25 January 1947 and 12 February 1947, NISTADS Archives, Delhi.
18. S. S. Bhatnagar to M. N. Saha, 19 May 1947 and M. N. Saha to S. S. Bhatnagar, 23 May 1947, NISTADS Archives, Delhi.
19. M. N. Saha to S. S. Bhatnagar, 23 May 1947, NISTADS Archives, Delhi.
20. Also personal communication, D. Page of the Department of External Affairs to B. M. Morrison, 15 October 1974, in author's personal archive.
21. J. Cockcroft, memo on Commonwealth Cooperation, UK Atomic Energy Commission, 6 November 1947, Public Record Office.
22. For an insider's account of the unsuccessful French negotiations to break this syndrome of exclusion in 1946–48, see Bertrand Goldschmidt, *Atomic Rivals: A Candid Memoir of Rivalries among the Allies over the Bomb* (New Brunswick: Rutgers University

Press, 1990). Goldshmidt was one of four English-speaking members of the Scientific Committee of the Commisariat à l'Energie Atomique, all of whom had worked in Chicago or Chalk River and Montreal during the war.

23. See George Perkovich, *India's Nuclear Bomb: The Impact of Global Proliferation* (Berkeley: University of California Press, 1999), pp. 18–19; also Itty Abraham, *The Making of the Indian Bomb*, p. 52. Unfortunately little is known about Krishnamurthy Rao except that he became deputy speaker of the Lok Sabha (1962–67) and after that deputy chairman of the Rajya Sabha. "Doctor" Sitaramayya, on the other hand, was trained in medicine but never practiced as a physician because he joined the Independence movement at an early age. Though he did not support Nehru this time, his credentials were impeccable, having been swept up in the arrest of the Congress Working Committee along with Nehru in August 1942 and having spent years in prison. His criticism of Nehru did not affect his political career: he succeeded in a bid to be president of the Congress in 1948, with Nehru's backing.
24. Abraham, *Making of the Indian Bomb*, p. 51.
25. Sen, *Professor Meghnad Saha*, p. 70.
26. Meghnad Saha to J. Nehru, June 1955, Saha Institute Archives, Kolkata. The statement about the phone call in 1948 is contained in a "supplement" Saha wrote for Nehru and sent with the 1955 letter, though it is not certain on which specific day in June this supplement was written and sent. Nehru's telephone call, however, is substantiated by Saha himself.
27. Sen, *Professor Meghnad Saha*, pp. 69–70.
28. Saha Papers, Saha Institute Archives, Kolkata.
29. M. N. Saha to B. D. Nagchaudhuri, 18 May 1948 and 21 May 1948, Saha Institute Archives, Kolkata.
30. Kathleen Lonsdale and Homi Bhabha, "Kariamanikkam Srinivasa Krishnan" *Biographical Memoirs of the Fellows of the Royal Society*, 13 (London: The Society, 1967): 245–55.
31. Goldschmidt recalled Bhabha's experience in an expensive Paris restaurant about 1950, where he had arrived for dinner just before his French hosts. Asking for their table, Bhabha said it was the dark color of his skin that caused the headwaiter to refuse to seat him, until his embarrassed host arrived. From then on, Goldschmidt said, French hosts called ahead to specify that Bhabha was their guest to avoid such embarrassment. Author's conversation with Bertrand Goldschmidt, Paris, November 1998.
32. B. K. Nehru met Magdolna Friedman when they were students at the London School of Economics in 1930. Known as "Fori," she was a Hungarian woman from a well-established Jewish family in Budapest; he was of the Nehru lineage from Kashmir, settled in Allahabad. She moved to India to marry and live with B. K. Nehru and raised their children there and in Washington, DC, and London. She was a frequent counselor to Indira Gandhi, whom she had known for decades, and even went to see her in 1976 to admonish her about the excesses of the Emergency. Fori Nehru was friends with influential women like the wife of physicist Eugene Wigner, also Hungarian. All these connections were of value to Bhabha.
33. Robert Bothwell, *Nucleus: The History of Atomic Energy of Canada Limited* (Toronto: University of Toronto Press, 1988). Bothwell reports that during the research period for that book (1985–88) no documents were seen relating to this shipment from Canada to India. I think they may have been severed from the files soon after 1974.

34. J. Nehru to C. D. Deshmukh, 23 June 1952, in *Selected Works of Jawaharlal Nehru,* vol. 18 (Delhi: Oxford University Press, 1996), p. 45.
35. J. Nehru, "Fundamental Research," 1 January 1954, in *Selected Works,* vol. 25 (1999), p. 200.
36. Ramanna, *Years of Pilgrimage* (Delhi: Viking, 1991), p. 41.
37. J. Nehru, memo to the cabinet, 1 April 1955, *Selected Works,* vol. 28 (2001), p. 532–34. Nehru does not seem to have contemplated that the leak may have occurred in his own office. M. O. Mathai, Nehru's personal secretary was, it seems, in a position to do almost anything he liked with Nehru's information. More than one source has suggested that Mathai was passing on confidential information, directly or indirectly, to the American embassy, but this is an unconfirmed allegation; we do, however, have insider S. Gopal's view that "it can safely be assumed that from 1946 to 1959, the CIA had access to every paper passing through Nehru's secretariat." See Ben Zachariah, *Nehru* (London: Routledge, 2005), p. 235; see also Katherine Frank, *Indira: The Life of Indira Nehru Gandhi* (New York: HarperCollins, 2001), pp. 205–7.

CHAPTER TWELVE

1. The main sources for this chapter are the Blackett Papers in the Royal Society Archives, London, including the 1967 interview by B. R. Nanda. My first uses of these sources were two articles in consecutive issues of *Notes and Records* of the Royal Society of London beginning in 1999. The supporting citations for the use by others of this material are found in Peter Hore, *Patrick Blackett: Sailor, Scientist, Socialist* (London: Frank Cass, 2003). I am grateful to the Royal Society and to Blackett's heirs and family for permission to publish from these sources. I am also grateful to Andrew Brown and Mary Jo Nye for valuable conversations about Blackett and to many participants at the Blackett conferences organized by the Royal Society in London and the Royal Navy in Cambridge in 1998. Blackett's papers in the Royal Society seem incomplete, so we may not have a thorough record of all of his activities in India. We thus cannot easily calibrate his experience in India with his other activities and relationships in the rest of his large life.
2. One wonders if Jawaharlal's memory was cast back to 1924, when his father Motilal Nehru was, with Mohammed Ali Jinnah, a member of the Skeen committee, the first to try to "Indianize" the military. The question that the future prime minister posed to Blackett, like other questions in India, already had a long history. But this time it was more urgent.
3. D. K. Palit, *Major General A. A. Rudra* (Delhi: Reliance Books, 1997), p. 67. An anonymous reader of this manuscript speculated this remark attributed to Nehru may be apocryphal as it "goes against the spirit" of Nehru's other statements at the time: "but a military man would have an interest in misrepresenting Nehru's position, as it did not offer the armed forces the power that they craved" (personal communication, 2 August 2007).
4. For a wide account of the period, see Christopher Bayly and Tim Harper, *Forgotten Armies: The Fall of British Asia, 1941–45* (London: Penguin, 2004).
5. Jaswant Singh, *Defending India* (Delhi: Macmillan, 1999), p. 33.
6. Omar Khalidi has tried to plot the relations of these groups in the military since 1947; "Ethnic Group Recruitment in the Indian Army: The Contrasting Cases of Sikhs, Muslim, Gurkhas, and Others," *Pacific Affairs* 74, no. 4 (Winter 2001): 529–52.
7. P. M. S. Blackett to S. Cripps, 17 February 1947 and 11 March 1947, Blackett Papers. Cripps had a special interest in India, having taken two British diplomatic

missions there, and was just about to become—in the middle of the 1947 economic crisis—the government's chief supervisor of the British economy, including its vast nationalization program. Cripps sat on the cabinet committee on India and appears to have been an informal "secretary of state for India." See Peter Clark, *The Cripps Version: The Life of Sir Stafford Cripps, 1889–1952* (London: Allen Lane & Penguin, 2002); also Chris Bryant, *Stafford Cripps* (London: Hodder & Stoughton, 1997).

8. Mary Jo Nye, *Blackett: Physics, War, and Politics in the Twentieth Century* (Cambridge, MA: Harvard University Press, 2004), pp. 65–99.
9. Twenty years after he first set foot in that country, Patrick Blackett was thinking about his influence on military and scientific developments in India. In a free-ranging interview with B. R. Nanda in 1967 he selected his influence as military consultant as probably more important than his other roles as scientific intervener; Blackett Papers.
10. On the history of the association in Britain, but without mention of its Indian counterpart, see Gary Wersky, *The Invisible College: A Collective Biography of British Scientists and Socialists of the 1930s*, 2d ed. (London: Free Association Books, 1988).
11. To the best of my knowledge, the ASWI did not have a branch in any DAE-funded or Defence-funded laboratory through 1970, and I do not know the situation after that.
12. Itty Abraham, personal communication, 9 February 1999.
13. Interview, 1967, p. 4, Blackett Papers.
14. Did Nehru grasp this probability? Jaswant Singh quotes Jagat Mehta's personal communication (no date): "Only Maulana Azad, the then President of the Indian National Congress, had some premonition of the disastrous international consequences of the vivisection of India"; Jaswant Singh, *Defending India*, p. 30.
15. Interview, 1967, pp. 4–5, Blackett Papers.
16. M. R. Srinivasan, *From Fission to Fusion: The Story of Indian's Atomic Energy Programme* (Delhi: Viking, 2002), p. 14.
17. Interview, 1967, p. 3, Blackett Papers. Although we know the close friendly relation of Nehru and Blackett, we do not know its frustrations or what value Nehru *privately* placed on Blackett's advice.
18. Nye, *Blackett*, pp. 148–50, points out that Blackett was first nominated for the prize in 1935, by French physicists; see her fascinating analysis of the nomination letters to the Nobel Physics Committee, weighing priority and fairness with respect to Occhialini at Cambridge and others, and assessing how the award of other prizes influenced the decision and timing on this one for Blackett.
19. This combination of talents was fully realized in Blackett's work as military consultant. It was Archibald V. Hill who involved Blackett in military research committees starting in 1936, and Hill and Blackett communicated frequently about new weapons and strategic questions *prior* to Hill's departure for India in 1943. Hill spoke to senior military staff in India in 1943 about the value of science and scientists, gave a lecture on operational research to the chiefs, and reported on his appraisal of the situation in the scientific community to the viceroy. When Lord Mountbatten needed scientific advisors in 1943 in Supreme Allied Command, Hill was instrumental in getting physicist J. D. Bernal and physiologist Solly Zuckerman to work for him. Bernal spent the last months of 1944 in Calcutta, where he also met scientists like Saha, on the Arakan coast of Burma to test the sand on the beaches, and at the military headquarters in Kandy, Sri Lanka. A memoir of Blackett by Lord Mountbatten is found in Brenda Swann and Francis Aprahamian, eds., *J. D. Bernal: A Life in Science*

and Politics (London: Verso, 1999). Blackett, who slipped effortlessly between the naval and airforce "camps" within the British military, already had long experience of interservice competition when he served on defense committees. Despite being classified as a naval expert, he became widely acknowledged as an expert in bombers and bombing strategy. He had access to ultra secret code material from 1943, traveled to Washington to address the Intelligence Seminar on bombing and code-breaking, and knew Paul Rosbaud, the German who supplied Churchill with a crucial assessment of the lack of progress of Werner Heisenberg and his team on an atomic bomb. Zuckerman was warned *not to talk* to Blackett in early 1944 about the plans for the invasion of Normandy. He [Blackett] debriefed the captured German nuclear physics team at Farm Hall, just outside Cambridge, in September 1945, being the person asked for by both Heisenberg and Eric Welsh, the intelligence controller of Paul Rosbaud, the nuclear physicist–spy. According to Arnold Kramish, "It was a measure of confidence between the two men that Blackett was brought to Farm Hall"; personal communication, 19 February and 20 April 1999, Washington, DC. (The "two men" referred to Heisenberg and Blackett) For a more complete biographical account, see Hore, *Patrick Blackett.*

20. Nye, *Blackett.* Nye's valuable work was not available when I wrote my earlier papers about Blackett, so I failed to understand the complexity of his situation in Britain and thought that his official marginalization was due more to American influence than it was. Blackett's name appeared even on a list of "crypto-communists and fellow travelers" given by writer George Orwell to an intelligence unit of the Foreign Office in May 1949.
21. P. M. S. Blackett to S. Cripps, 11 March 1947, Blackett Papers.
22. Interview, 1967, p. 4, Blackett Papers.
23. P. M. S. Blackett to B. R. Nanda, 8 February 1972, Blackett Papers.
24. Interview, 1967, p. 4, Blackett Papers.
25. Itty Abraham, *The Making of the Indian Atomic Bomb: Science, Secrecy, and the Postcolonial State* (London: Zed Books, 1998), p. 56.
26. I am grateful to Andrew Brown for drawing my attention to this point.
27. Vice Admiral Parry to P. M. S. Blackett, 7 September 1948, Blackett Papers.
28. Richard Aldrich and Michael Coleman, "Britain and the Strategic Air Offensive against the Soviet Union: The Question of South Asian Air Bases, 1945–1949," *Journal of the Historical Association* 74 (October 1989): 400–428.
29. Interview, 1967, p. 7, Blackett Papers; note the "us" and "our" syntax positioning Blackett inside Nehru's framework, and inside the cabinet's framework too.
30. Bharat Karnad, *Nuclear Weapons and Indian Security* (Delhi: Macmillan, 2003), p. 213.
31. P. M. S. Blackett to S. Cripps, 11 March 1947, Blackett Papers.
32. J. Nehru to P. M. S. Blackett, 26 September 1948, Blackett Papers.
33. S. S. Bhatnagar to P. M. S. Blackett, 29 December 1950, Blackett Papers.
34. Interview, 1967, p. 8, Blackett Papers.
35. P. M. S. Blackett to Baldev Singh, Minister of Defence, Delhi, 30 September 1948, Blackett Papers.
36. A. R. Verma, Krishnan Lal, L. S. Kothari, "Biographical note on Daulat Singh Kothari," n.d., manuscript, probably 1994. Kothari was from a poor orthodox Jain family; his chance came when the maharana of Merwar awarded him a scholarship to become an undergraduate student at Allahabad in 1926, where he met Saha.
37. Abraham, *The Making of the Indian Atomic Bomb,* p. 61.

38. Evidence in Blackett Papers.
39. See Rajinder Singh, "Why Did Daulat Singh Kothari not Become a Fellow of the Royal Society?" *Current Science* 81, no. 5 (10 September 2001): 607–10.
40. P. M. S. Blackett, "Report to the Minister of Defence," 10 September 1948, Blackett Papers.
41. P. M. S. Blackett to J. Nehru, 31 January 1951, Blackett Papers.
42. H. A. Sargeaunt to P. M. S. Blackett, 9 May 1951, Blackett Papers.
43. P. M. S. Blackett to H. A. Sargeaunt, 29 May 1951, Blackett Papers.
44. Nye, *Blackett*, p. 90. Blackett also said he misjudged the lethal effects of radioactive fallout.
45. M. N. Saha to D. S. Kothari, 21 July 1948, Saha Institute, Calcutta.
46. S. S. Bhatnagar to P. M. S. Blackett, 22 April 1951, P. M. S. Blackett to S. S. Bhatnagar, April 1951, Blackett Papers.
47. On the history of the solar power research at the NPL, see Shiv Visvanathan, *Organising for Science* (Delhi: Oxford University Press, 1985).
48. M. L Ghai, M. L Khanna, J. S. Ahluwalia, S. P. Suri, "Performance of Reflector-type Direct Solar Cooker," *Journal of Scientific and Industrial Research* 12A, no. 12 (1953): 540–51.
49. The best account of the culture of military and strategic decision-making of the 1940–80 period is found in Karnad, *Nuclear Weapons and Indian Security.*
50. *Lok Sabha Debates,* Part II, 8th Session, vol. 9, 21 December 1954, Delhi, pp. 3587–3610. This debate in Parliament and reference to Saha became the subject of an exchange of letters between Saha and Nehru that is discussed in the next chapter.

CHAPTER THIRTEEN

1. S. N. Sen, ed., *Professor Meghnad Saha: His Life, Work, and Philosophy* (Calcutta: Meghnad Saha Sixtieth Birthday Committee, 1954), p. 124.
2. Ibid.
3. Ibid., p. 125.
4. Though his authorized biography described him as an "independent candidate" (p. 125), he sat in the Lok Sabha (parliament) as a member of the Revolutionary Socialist Party (RSP), founded by M. N. Roy after Roy was purged from the Communist Party by Stalin. The Revolutionary Socialist Party was a new entity, formed in the late 1940s. It had no elected members prior to Saha's election in 1951. But it had a precedent, of sorts: M. N. Roy had earlier formed a party in 1940 called the Radical Democratic Party, but the RDP failed to have any candidate elected in the 1946 elections in Bengal, so the new RSP was Roy's next political platform. He was lucky to have Meghnad Saha's old loyalty, formed when they met in Berlin in 1921, to show some results in India's first general election in 1951; only one other member was elected, named Tridib Chaudhuri. But voters for Saha were voting not for the party but for the individual, in this case a distinguished scientist. Moreover, the RSP appears also to have evolved directly out of the Anushilan Samity that Saha had belonged to years before, which had a very East Bengali character. East Bengali refugee voters in Calcutta were numerous in 1951 and Saha's reputation was high among them in that part of the city. Personal communication, Ben Zachariah, 28 November 2005; see also Nicholas Owen, *The British Left and India: Metropolitan Anti-imperialism, 1885–1947* (New York: Oxford University Press, 2007), p. 273.
5. B. K. Dutta interview, 7 July 1968, cited in Guatam Chatopadhyay, *Communism and the Bengali Freedom Movement,* vol. 1: *1917–1929* (Delhi: Peoples Publishing House,

1970), pp. 50–60. I am grateful to Dr. Sulagna Roy for providing this reference, and Dr. Subhas Chakravarty for arranging a meeting in Kolkata with the aging and frail Guatam Chatopadhyay for me in January 2005.

6. M. N. Saha to H. K. Ghosh, Publisher, The Indian Press, Allahabad, 24 November 1951, Saha Papers, Saha Institute Archives, Kolkata. The *Treatise* sold an average of 4400 copies annually after 1948, as a required text, in spite of what Saha considered poor printing, paper, and binding.
7. M. N. Saha to R. K. Das (Barrister), 23 January 1952, Saha Papers, Saha Institute Archives, Kolkata. Note that six months later Saha closely observed another *hartal* in Bengal, writing "the threatened *hartal* fizzled out. It was raining incessantly so the *hartal* could not be successful under this condition. I think it was a rather ill-conceived move"; M. N. Saha to T. K. Chaudhuri, 25 July 1952, Saha Papers, Saha Institute Archives, Kolkata.
8. Santimay Chatterjee and Jyotirmoy Gupta, *Meghnad Saha in Parliament* (Calcutta: Asiatic Society, 1993), p. x.
9. Sen, *Professor Meghnad Saha*, p. 92.
10. M. N. Saha to A. V. Hill, 8 October 1952, Churchill College Archives, Cambridge. An "Irish member" meant that Saha felt disconnected from the normal political ebb and flow: Members of Parliament elected from Northern Ireland to sit in London even after 1922 were separately identified as Ulster Unionists. Though they normally supported the Conservatives, they were detached from normal party lines. I thank Peter Clarke for clarifying this point. Personal communication, 12 October 2005.
11. Tridib K. Chaudhuri to M. N. Saha, 4 March 1954, Saha Papers, Saha Institute Archives, Kolkata.
12. Tridib Chaudhuri, "Socio-economic Role of Meghnad Saha," in Jayant Basu, ed., *The Glittering Spectrum of Meghnad Saha* (Calcutta: Saha Institute of Nuclear Physics, 1994), p. 141.
13. Hirendranath Mukerjee "Reminiscences of Meghnad Saha," in Basu, *The Glittering Spectrum*, pp. 144–45.
14. For a broad account of the project, see Daniel Klingensmith, "Building India's Modern 'Temples': Indians and Americans in the Damodar Valley Corporation, 1945–1960," in K. Sivaramakrishnan and Arun Agarwal, eds., *Regional Modernities: The Cultural Politics of Development in India* (New York: Oxford University Press, 2003).
15. Sri Ram to C. C. Desai, ICS (1951), quoted in Khushwant Singh and Arun Joshi, *Sri Ram: A Biography* (Bombay: Asia Publishing House, 1968), p. 146.
16. Sen, *Professor Meghnad Saha*, p. 106.
17. Ibid, p. 127. Praises for Jay Engineering Works by Nehru and Bhatnagar are also quoted.
18. Ibid., p. 115.
19. Ibid., p. 118.
20. T. R. Seshadri, "Shanti Swarup Bhatnagar," *Biographical Memoirs of Fellows of the Royal Society*, 8 (London: The Society, 1962): 9.
21. M. N. Saha to C. D. Deshmukh, 18 July 1952, Saha Papers, Saha Institute Archives, Kolkata.
22. Edward W. Chester, *U.S. Oil Policy and Diplomacy* (Westport, CT: Greenwood Press, 1983); also Biplob Dasgupta, *The Oil Industry in India—Some Economic Aspects* (London: Frank Cass, 1971), p. 62; R. Vedavalli, *Private Foreign Investment and Economic Development: A Study of Petroleum in India* (New York: Cambridge University Press, 1976).

23. *Lok Sabha Debates* (New Delhi, 1955), vol. 5, pp. 4118, 4781–82, 4814, 4819–20, and 4838; vol. 7, pp. 3, 11, 131, and 1223; vol. 9, p. 341.
24. J. Nehru to S. S. Bhatnagar, 20 November 1953, *Selected Works of Jawaharlal Nehru* (Delhi: Nehru Memorial Fund and Oxford University Press, 1999), 25: 194–95.
25. J. Nehru to M. N. Saha, 3 December 1953, *Selected Works,* 25: 195.
26. *Lok Sabha Debates,* 6th Session, vol. 5, 5–21 May 1954. Nehru defended the AEC and DAE against Saha's attack, as expected. But there was more to come.
27. Saha published his lengthy and thorough presentation to the seminar at Nehru's house in *Science and Culture* 20, no. 5 (November 1954): 208–22.
28. Itty Abraham, *The Making of the Indian Atomic Bomb: Science, Secrecy, and the Postcolonial State* (London: Zed Books, 1998), p. 74.
29. Opening and closing remarks to Conference on the Development of Atomic Energy for Peaceful Purposes in India, J. Nehru, 26–27 November, *Selected Works,* 28: 416–20.
30. Saha published his remarks to this conference in *Science and Culture* 20, no. 5 (November 1954): 208–22.
31. Ramanna, *Years of Pilgrimage* (Delhi: Viking, 1991), pp. 61–62.
32. J. Nehru speech to lay foundation stone of TIFR, 1 January 1954, *Selected Works,* 26 (1999): 198.
33. M. N. Saha to J. Nehru, 31 December 1954, Saha Papers, Saha Institute Archives. Saha wrote to Nehru not as member of the Lok Sabha but from his new office as director of the Indian Association for the Cultivation of Science, in effect the position Raman held until 1932.
34. J. Nehru to M. N. Saha, 2 January 1955, J. Nehru to M. N. Saha, 7 January 1955, Saha Papers, Saha Institute Archives, Kolkata. Nehru maintained an astonishing correspondence schedule, writing short notes and long memos by the dozens every day, and most of them dictated by him: evidence of this lies in the many volumes of *Selected Works.* It simply cannot be said that the Nehru-Saha relationship, though distanced by many factors, was not *personal.* Note also the speed with which letters traveled from Calcutta to Delhi and back.
35. *Selected Works,* appendix VII, Extracts, p. 45.
36. According to Wayne Reynolds, *Australia's Bid for the Atomic Bomb* (Carlton: Melbourne University Press, 2000), Oliphant expressed views remarkably similar to Bhabha's views on the need for independent nuclear development—reactors and weapons—in Australia. As members of the Scientific Advisory Committee to the IAEA, they had frequent meetings in Vienna and New York. Oliphant's knowledge of the Manhattan Project, his good relations with the Canadian and British teams, his knowledge of Australian uranium resources and its strategic ambitions were all valuable to Bhabha. In 1955 both men were keenly interested in circumventing American restrictions on crucial technology and ideas and yet continued to feel left out. Oliphant was born in 1901 in Adelaide and completed his PhD at Cambridge in 1929 at age twenty-eight. Associated with both Trinity and St. Johns Colleges, he worked at the Cavendish Lab from 1930 to 1937 on proton sources. Elected an FRS in 1937, he was appointed at Birmingham as professor. From 1939 to November 1943 Oliphant worked in radar research and tube alloys, and then in November 1943 he moved to the Manhattan Project, in Tennessee. He was approved for the highest US award to a foreigner but the Australian government objected to awards to its citizens by foreign governments, so he did not receive it. He was the key person in the calculation of the electrical power requirements for U235 production at Oak Ridge, Tennessee. He had been a

co-worker with Lawrence at Berkeley, discovered a new hydrogen isotope with the mass of 3, called tritium, key for bomb-making; Oliphant returned to Great Britain in 1945 and was involved in the choice of the Harwell site with AEC. He was a director of research at Australia National University from 1950 to 1963. building reactors and cyclotrons (homopolar generator), but actually furthering Australia's nuclear policy. Then he resigned in 1963 amidst public doubts about his projects and their costs. From 1955 to 1963 he was chairman of the Atomic Energy Committee of Australia and scientist at the IAEA, meeting Bhabha regularly. It was in this phase that he attended the DAE meeting that decided on the financial future of Saha's institute. Finally, while governor of South Australia (1970–76), he described himself as a "belligerent pacificist" and loyal member of the Pugwash movement, opposing French tests in the Pacific in the 1970s.

37. J. Nehru to K. D. Malaviya, 26 April 1954, *Selected Works*, 25: 120–21.
38. M. N. Saha to K. D. Malaviya, 29 May 1954, Saha Papers, Saha Institute Archives, Kolkata.
39. *Lok Sabha Debates*, Part II, 5th Session, vol. 5, 10 May 1954, Delhi, pp. 7006–7054. See also Chatterjee and Gupta, *Meghnad Saha in Parliament.*
40. I am grateful to the librarians of the National Library, Kolkata, for facilitating my access in 2005 to the 1955 and 1956 volumes of *Science and Culture.* The fragile condition of these precious volumes is alarming.
41. M. N. Saha to J. Nehru, draft, July 1955, Saha Papers, Saha Institute Archives, Kolkata.
42. S. S. Bhatnagar to M. N. Saha Birthday Organizing Committee, 27 January 1953. Quoted in Sen, *Professor Meghnad Saha,* p. 165.
43. Brojen Banerjea, "Memories of Other Days," in S. P. Karmohapatro, ed., *Meghnad Saha Birth Centenary Commemoration Volume* (Calcutta: Saha Institute for Nuclear Physics, 1993), p. 91.
44. Jagjit Singh, *Some Eminent Indian Scientists* (New Delhi: Publications Division, Ministry of Information and Broadcasting, 1966), pp. 130–31.
45. D. S. Kothari, "Meghnad Saha," *Biographical Memoirs of the Fellows of the Royal Society,* 5 (London: The Society, 1959): 220. Kothari's authorship is unusual, in that normally memoirs of Fellows are written by other Fellows.
46. M. N. Saha to H. P. Ghosh, 17 December 1954; M. N. Saha to J. C. Ghosh, 3 June 1955, Saha Papers, Saha Institute Archives, Kolkata.
47. It is surprising that a person of Saha's reputation and a sitting member of Parliament would have to resort to negotiating for a bed in a Calcutta hospital.
48. Hirendranath Mukerjee, in Basu, *The Glittering Spectrum,* p. 145.
49. Jyoti Basu, *With the People* (Delhi: UBS Publishers, 1997), p. 163. When I spoke to Jyoti Basu about the Saha Institute in February 1969 he knew about it in great detail. As chief minister he was, in part, responsible for the institute, but considering the number of other things he had to manage, I was surprised by this level of detail.

CHAPTER FOURTEEN

1. Saha's journal *Science and Culture* ran a balanced three-page obituary for Bhatnagar, probably written by Saha, without any personal memories but showing all facets of his official life, in February 1955. But it was not printed on the front page.
2. J. Nehru to all Chief Ministers, 13 January 1955, *Selected Works of Jawaharlal Nehru* (Delhi: Nehru Memorial Fund and Oxford University Press, 1999), 28: 567.
3. J. Nehru to Secretary of CSIR, 15 May 1955, *Selected Works,* 28: 557–58.
4. K. Lalbhai to H. J. Bhabha, 6 January 1955, Nehru Foundation Archive, Ahmedabad.

5. K. D. Malaviya to K. Lalbhai, 10 March 1955, Nehru Foundation Archive, Ahmedabad.
6. D. V. C. Mallik, a biographer of K. S. Krishnan, notes that there is little evidence of his engagement with the Atomic Energy Commission in his papers; personal communication, 11 June 2006.
7. Thacker was born in 1904 in Ahmedabad, studied engineering at Bristol University, and worked at Bristol Electrical Supply for a year before coming to work at the largest electrical system in India, Calcutta Electric Supply, from 1931 to 1947. He then joined the new Department of Power Engineering at the IISc in Bangalore, and two years later, in 1949, was appointed director of that institute in 1949 at age forty-five.
8. I. M. D. Little, "Atomic Bombay: A Commentary on the Need for Atomic Energy in the Developing Countries," *Economic and Political Weekly,* 29 November 1959. Little was, with P. N. Rosenstein, co-author of *Nuclear Power and Italy's Energy Position* (Washington, DC: National Planning Association, 1957). He said "I could not have written the article 'Atomic Bombay' without this research." He also gave a public lecture in Oxford in April 1958 questioning "the current hype that nuclear energy would transform society especially in developing countries." He traveled to India a few months later with an MIT team focused not on power and energy but on the promotion of empirical research at economic institutes. He reports that he did not see Bhabha's own calculations about the cheapness of electricity from nuclear reactors before writing this essay. Despite its unusual nature in India, Little recalled receiving little comment on "Atomic Bombay" essay in 1958. I am grateful to Professor Little for his reply to my inquiry: I. M. D. Little, personal communication, 1 and 6 October 2006.
9. The July 1960 speech was recorded in D. D. Kosambi, *Atomic Energy for India* (Poona: Peoples' Book House, 1960). See also C. K. Raju, "Kosambi the Mathematician," *Economic and Political Weekly* 44, no. 20 (16 May 2009): 33–45. I speculate that in their conversations, at least before disagreements appeared in the late 1950s, Bhabha was inevitably influenced by Kosambi's forceful arguments.
10. J. Nehru to Lok Sabha, 10 August 1960, cited in George Perkovich, *India's Nuclear Bomb: The Impact of Global Proliferation* (Berkeley: University of California Press, 1999), p. 484.
11. Cited in Perkovich, *India's Nuclear Bomb,* p. 38; "India's Drive for Atomic Power," *Hindustan Times,* 20 August 1961; and "What Price Power?" *Times of India,* 23 August 1961.
12. Perkovich assembles very useful evidence of the absence of nuclear news and debate about weapons from the newspapers and scholarly opinion journals through to 1964; *India's Nuclear Bomb,* pp. 39–40.
13. National Institute of Science, *Technology and Development, A Review of the Science and Technology Policies of India 1948–1971,* prepared by J. K. Ahuja, I. George, S. Mahanti, A. Jain (Delhi, November 1989), DST ref SP/UR/201/87; this document will be known below as "SACC Review," even when it refers to the later Committee on Science and Technology (COST). The quotations in this summary cannot be checked against the original minutes, as they were/are secret.
14. J. Nehru to V. T. Krishnamachari (Deputy Director of Planning Commission), 8 January 1955; J. Nehru to V. T. Krishnamachari, 7 December 1955; J. Nehru to M. S. Thacker, 24 February and 27 February 1956; *Selected Works,* 27: 382 and 31: 87, 88.
15. SACC Review, p. 4.
16. Ibid., pp. 63–64.

17. It is curious that the history of institutions so influential in the late twentieth century is so little written about. The first IIT resulted from the report of the Sarkar Committee formed in 1946; at the time of Independence the committee advised four institutes, to be build in the east, north, west, and south. So in 1950 the IIT Kharagpur opened just outside Kolkata on the site of a abandoned British internment camp for political prisoners; it was a bilateral project with the British government (and designed by a Swiss architect). Bengal's influence was very strong in the Sarkar Committee and technical education was advanced in West Bengal. The other three institutes were also supported as bilateral projects: IIT Powaii (near Mumbai) with the USSR in 1958, IIT Kanpur with a consortium of nine American universities in 1959, and IIT Madras with West Germany in 1959. Other institutes followed later, including a fifth one (at Delhi) in 1963. Technology and technical education have, of course, a much older history in India than the Sarkar Committee (not to be confused with Justice Sarkar's Commission formed to review the CSIR in the late 1960s).
18. Shiv Visvanathan, *Organizing for Science: The Making of an Industrial Research Laboratory* (Delhi: Oxford University Press, 1984), p. 197.
19. SACC Review, p. 65–66. Parthasarathy may be the first scientist suicide in India if, as thought, the Sunanda Bai suicide in 1945 in Bangalore was not related to her scientific profession; see chap. 3.
20. SACC Review, pp. 63–64.
21. Ramanna, *Years of Pilgrimage* (Delhi: Viking, 1991), p. 74.
22. J. Nehru to C. D. Deshmukh, 23 June 1952, in *Selected Works*, 26: 45.
23. All evidence for this paragraph is in letters from Bhabha to Blackett or copies from Bhabha in Blackett Papers, Royal Society Archives, London, dated 24 December 1959, 30 March 1960, 5 December 1960, 4 April 1961, 17 June 1961, and 26 June 1961. See also Rajinder Singh, "Why Did Daulat Singh Kothari not Become a Fellow of the Royal Society?" *Current Science*, 10 September 2001.
24. SACC Review, p. 29.
25. Ibid., p. 39.
26. B. Dey, "Scientific and Technical Personnel: Inter-state Flow and Distribution," *Economic and Political Weekly*, 23 August 1969, pp. 1373–81. This is an interpretation of data on 95,000 people collected in 1961, but published in this form eight years later.
27. SACC Review, p. 101. Note that the name of the Defence Science Organization had been changed five years earlier to the Defence Research and Development Organisation, but Zaheer and the minute taker seem to have forgotten this change.
28. SACC Review, pp. 100–101.
29. Two good recent books are Ben Zachariah, *Nehru* (London: Routledge, 2004) and Judith Brown, *Nehru: A Political Life* (New Haven: Yale University Press, 2003). Hundreds of observers, some with direct experience of him, have something to say about Nehru, and yet it is striking that few writers even comment on this aspect of his life and work. Coupled with his extensive personal involvement in nuclear practice and policy shown here, Nehru's views take on a new meaning. Zachariah skillfully shows the interconnection between his international nonaligned movement efforts and domestic politics, suggesting where the science and technology emphasis best fits, as one of Nehru's "wings," so to speak, in his effort to take flight and transcend the *domestic and rooted nature* of his power base.
30. There are three interesting picture books to bear this out, showing Nehru in dozens of labs, openings, technical displays, giving talks to scientists, in meetings, and the

like, all this recorded by official photographers: CSIR, *Nehru: The Architect of Indian Science* (Delhi: NISTADS, 1989); and *Nehru and the CSIR* (Delhi: NISTADS, 1989); see also Bhabha Atomic Research Centre, *Pandit Jawaharlal Nehru on Atomic Energy* (Bombay, November 1989).

31. Perkovich, *India's Nuclear Bomb*, p. 66.
32. Ibid., p. 84; it is worth reading Perkovich's nuanced tour of the landscape of elite opinion (involving journalists, politicians, and a few officials) in the six weeks of debate in late 1964, pp. 66–85.
33. Author's conversation with S. Jagannathan, former member of the AEC, in Chicago, December 1970. Though an official of the Ministry of Finance, his basic degree was in physics.
34. The pace of these projects was such that Bhabha could only survey them quickly: a concrete contractor working at the atomic energy center in Trombay in 1965 was told by the chief engineer to put 20 extra carpenters on the job the next day because Bhabha would be visiting. "Well," he said to me, requesting anonymity, "I only had four carpenters, but I got 15 more hammers out of Stores and had unskilled laborers banging away on old pieces of wood. I could see that both the engineer and Bhabha were very pleased with the progress." Author's conversation with anonymous concrete contractor, March 1968, Bombay.
35. Ashok Parthasarathi, *Technology at the Core: Science and Technology with Indira Gandhi* (Delhi: Pearson Longman, 2007), p. 17.
36. CSIR Survey and Planning of Scientific Research Unit, *A Study of Expenditure in National Laboratories* (Delhi, 1964). As this information was based entirely on a questionnaire filled out by directors and senior staff in the labs themselves, it reveals head office's notoriously weak grasp of the situation on the ground. The data therefore have more of a notional than empirical value.
37. J. B. S. Haldane to P. M. S. Blackett, 25 January 1963, Royal Society Archives, London. On the relationships between Blackett, Bernal, and Haldane, see Gary Wersky, *The Invisible College: A Collective Biography of British Scientists and Socialists of the 1930s*, 2d ed. (London: Free Association Books, 1988).
38. Blackett later explained, "what has gone wrong, I think, in some government research stations is that the principle of scientific freedom has sometimes been misapplied to mission-oriented R and D, where it is largely inapplicable"; Patrick Blackett, 1967 interview with B. R. Nanda, p. 8, Blackett Papers, Royal Society Archives, London.
39. For a historical explanation of this conflictual relationship, see Druv Raina and Ashok Jain, "Big Science and the University in India," in John Krige and Dominique Pestre, eds., *Science in the Twentieth Century* (London: Harwood Academic Publishers, 1997).
40. M. G. K. Menon et al., Report to the Executive Council, National Physical Laboratory, 9 April 1963, Royal Society Archives, London.
41. See Visvanathan, *Organizing for Science*.
42. H. Zaheer to P. M. S. Blackett, 9 December 1964, Royal Society Archives, London.
43. This was not the first time this warning had been sounded by a British physicist. Newspapers throughout India carried the statement of Blackett's friend J. D. Bernal in December 1954 that "China has made more rapid progress in science than India." Speaking in Madras, Bernal said this in context of his support for the Five Nehru-Chou Principles, and his condemnation of the arms race. Nehru, who had, like Bernal, just returned from China in 1954 after optimistic meetings with Mao Tse Tung, met Bernal in Delhi to hear his views about China. Blackett and Bernal had

convinced Nehru in 1948 to be the patron of the Association of Scientific Workers of India.

44. Blackett was now the newly appointed advisor to the British minister of Science in the new Labour government. He had been offered the top post but declined it. A year later, in 1965, he became president of the Royal Society and was even busier. But he didn't lose interest in India, and so the CSIR and DAE continued to cultivate a role for him in 1964.
45. C. G. Wynne, Report on Optical Designing, for CSIR, June 1966, Blackett Papers, Royal Society Archives, London.
46. C. G. Wynne to P. M. S. Blackett, 15 June 1966, Blackett Papers, Royal Society Archives, London.
47. N. R. Rajagopal, M. A. Qureshi, and Baldev Singh, *The CSIR Saga: A Concise History of Its Evolution* (Delhi: CSIR, 1991), 1: 94–95.

CHAPTER FIFTEEN

1. Ashok Parthasarathi, *Technology at the Core: Science and Technology with Indira Gandhi* (Delhi: Pearson Longman, 2007).
2. R. Ramanna, *Years of Pilgrimage* (Delhi: Viking, 1991), p. 75. I speculate that the AEC member most keen on the prime minister's involvement was J. R. D. Tata.
3. Minutes of the Governing Body, CSIR, Delhi, ninth meeting, 21 September 1945, Nehru Foundation Archives.
4. The Raman-Chandrasekhar relationship was complicated, and the 1961 effort to persuade Chandrasekhar to become a national professor was the occasion of a misunderstanding that occurred when Raman made a careless and dismissive remark to his nephew in Bangalore just while he was unwrapping Chandrasekhar's new book *Hydrodynamic and Hydromagnetic Stability* (Oxford: Clarendon Press, 1961); Raman said big books like this did not lead to Nobel Prizes and he had avoided writing them. (Remember that Raman had already proposed Chandrasekhar for the prize in 1948. He was finally awarded the prize in physics in 1983 for work on the structure of white dwarfs, done in his youth. Raman had by then died.) Chandrasekhar thought Raman was dismissing his life's work while unwrapping the book in 1961 and never forgot it, according to Miller, though Raman made efforts to apologize soon afterward; Arthur I. Miller, *Empire of the Stars: Obsession, Friendship and Betrayal in the Quest for Black Holes* (New York: Houghton Mifflin, 2005), p. 137.
5. K. C. Wali, *Chandra: A Biography of S. Chandrasekhar* (Delhi: Penguin Books, 1992), p. 288.
6. A. Parthasarathi, "Brain Drain from Developing Countries" *Nature* 230 (12 March 1971): 87–90. The text and data in this article were first circulated in draft in late 1967, presented at conferences, and the like, and reworked into the final version in 1971, according to the author, personal communication, November 1998.
7. Author's conversations with S. Chandrasekhar, Chicago, 1966–67, and 1969.
8. Based on her conversations with close observers, Shah said this was a very common interpretation at the time; it suggests the tendency to overinterpret and politicize decisions in the scientific community. Clearly the Sarabhai appointment could not have had *nothing* to do with Moraji Desai. Sarabhai had already contested a political plum within Desai's domain and lost. In 1961–62 he fought against the vernacular language exclusion of English medium education in Gujarat and decided to stand for election in the university Senate as vice-chancellor; according to Shah, "caste played a significant role. The university was firmly in the grip of the Congress-clique of

upper-caste Brahmins . . . their leader was Moraji Desai, then known as uncrowned king of the state." Though Sarabhai had the support of respected Kasturbhai Lalbhai, he was labeled "a mill-owners' man" by the Congress Party of Gujarat who bullied the Senate, disciplined their party members, and elected "a nondescript, tired retired school teacher." Amrita Shah, *Vikram Sarabhai: A Life* (Delhi: Viking, 2007), pp. 113–163.

9. Henry Hart, "Indira Gandhi: Determined Not to be Hurt," in H. C. Hart, ed., *Indira Gandhi's India: A Political System Reappraised* (Boulder, CO: Westview Press, 1976).
10. L. K. Jha, in P. K. Joshi, ed., *Vikram Sarabhai* (Ahmedabad: Mapin Publishing, 1992), p. 46.
11. Shah, *Vikram Sarabhai*, p. 154.
12. Author's conversation with anonymous observer, Ahmedabad, November 1998.
13. Shah, *Vikram Sarabhai*, p. 163. This statement is based on Shah's wide-ranging conversations with a number of scientists.
14. A fascinating portrait of the Ahmedabad into which Vikram was born is found in Makrand Mehta, "Gandhi and Ahmedabad, 1915–1920," *Economic and Political Weekly* 22 (January 2005).
15. R. Tagore to Registrar, University of Cambridge, 1 November 1935; Joshi, *Vikram Sarabhai*, p. 19.
16. Vikram Sarabhai, "Cosmic Ray Investigations: Experiments with Gamma Rays" (PhD diss., University of Cambridge; with the supervision of E. S. Shirl, deposited for examination in September 1946).
17. Ibid., p. xi.
18. I am grateful to Blackett's daughter Giovanna Bloor and son Nicolas Blackett for conversations about their father's work in India; March 1998, London.
19. Vikram Sarabhai, "The Method of Shower Anti-coincidences for Measuring the Meson Component of Cosmic Radiation," *Physical Review*, 1 March 1944; A. P. Zhdanov et al., "Anomalous Rate of Nuclear Disintegration Effected by Cosmic Rays," *Physical Review*, 1 March 1944; S. V. Aiya et al., "Slow Mesons in Cosmic Radiation," *Physical Review*, 1 October 1944; M. S. Sinha, "On the Scattering of Slow Mesons," *Physical Review*, 1 November 1945; John Wheeler, "Mechanism of Capture of Slow Mesons," *Physical Review*, 1 March 1947.
20. V. Sarabhai to P. M. S. Blackett, 14 February 1958, Blackett Papers, Royal Society Archives, London.
21. K. Lalbhai in Joshi, *Vikram Sarabhai*, p. 87. There was another Cambridge-trained physicist already established in textile circles, Nazir Ahmed, but he was a Punjabi not a Gujerati and lived primarily in Lahore until coming to work in Bombay, where he also became justice of the peace and honorary magistrate. We have no record of Ahmed's interaction, if any, with Sarabhai, though he had regular contact with Bhabha. Nazir Ahmed was much older than Sarabhai and in 1948 went to Pakistan, becoming chairman of that country's Atomic Energy Commission. It is regrettable that we have no biographical study of a person of such interest.
22. The 1960s address book and diary seen at Sarabhai Archives, Nehru Foundation, Ahmedabad, November 1998.
23. Shah, *Vikram Sarabhai*, p. 139.
24. Joshi, *Vikram Sarabhai*, p. 117.
25. A. P. J. Abdul Kalam, *Wings of Fire: An Autobiography* (Hyderbad: Universities Press, 1999), p. 49.

26. R. Ramachandran, "Delayed Launch," *Frontline*, 6 April 2007, p. 39. Mrinalini Sarabhai told Amrita Shah that she clearly recalled her husband, within a few months of his death, standing up, moving, and talking while asleep, showing clear signs of sleep disturbance; Shah, *Vikram Sarabhai*, p. 206.
27. Indira Gandhi, Foreword, in Joshi, *Vikram Sarabhai*, p. 164.
28. Bharat Karnad, *Nuclear Weapons and Indian Security: The Realist Foundations of Strategy* (Delhi: Macmillan, 2002), p. 309–10. Given Karnad's antipathy to Sarabhai's reluctance to plan for nuclear bombs, I am not sure his report about Trombay should carry great weight: the divisions within Trombay were of a more complicated character, and Sarabhai was only a small part of them. That he had not been a bigger part of Trombay or Rajasthan made his job more difficult and complicated. Being frugal and looking like others goes a long way back in the Jain community and the outlook of the Sarabhai lineage.
29. As for physicists versus engineers, Raja Ramanna seemed to wish to prolong his DAE secretary's role beyond retirement in January 1987, when PhD engineer M. R. Srinivasan took over the department in which he had served for thirty-three years. According to Srinivasan, "Ramanna, it appears, conveyed to the Cabinet Secretary BG Deshmukh that many senior scientists, especially from BARC, would not accept me as the head of the AEC"; this was "a theory that the head of the atomic energy organization should be a physicist from BARC." Subsequently, "The BARC Officers Association and the [Nuclear Power Board] Officers Association passed resolutions criticizing attempts by certain persons to deny an outstanding engineer the opportunity to head the AEC and DAE"; M. R. Srinivasan, *From Fission to Fusion: The Story of India's Atomic Energy Programme* (Delhi: Viking, 2002), p. 170.
30. Author's conversation with Vikram Sarabhai, Bombay, April 1969.

CHAPTER SIXTEEN

1. See an expert's description of a launch in Hyderabad in the spring of 1965, involving TIFR and many other scientists from Berkeley, New York, Minnesota, Rochester, Bristol, Dublin, plus scientists from US Air Force and Naval labs in Serge Korff, "IQSY-EQEX: Cosmic Ray Expedition to India," *Physics Today*, June 1965. This was a year of the quiet sun (IQSY): the stimulating effect on young Indian researchers of such an international gathering for their subject can be imagined. Some of the balloon launchers I was with in 1967 had participated in the 1965 launches with American collaborators. Note that these researchers worked effortlessly in two measurement systems; all experimental calculations were metric and all distance calculations in India were imperial.
2. Three weeks earlier, C. R. Sathya reports that "when the Prime Minister arrived on opening day, 2 February 1969, I was far away integrating a sodium vapor pay-load into the nose cone for the Centaure rocket, in a building 2 km away to the north of rocket range. Prime Minister Gandhi was to press the launch button at around 6:00 p.m. but as early as 4:00 p.m. I could not find a single vehicle to transport the nose cone to the Rocket Assembly Bay for integration before flight. I had been totally forgotten by our administration in the VIP rush for vehicles. Luckily I found an abandoned cycle lying nearby, so I loaded the nose cone on the carrier, and Velappan Nair and I pushed it all the long way to the rocket assembly area. Halfway along the beach road, I saw the famous French photographer Henri Cartier Bresson furiously cycling towards me (the cycle was his official transport too) and when he approached us,

he threw his cycle aside and started clicking. (I am the one to the right of the cycle). We reached the launch range in time and the PM switched the launch switch on at the correct time, thank God!" C. R. Sathya, personal communication, 28 November 2008.

3. This RH-75 rocket flown on 21 February 1969 was fabricated at Thumba (see map) with Indian components and propellant. It was built with commercially available seamless aluminum alloy tubes and used commercial polyester fuel, both produced in South India. It flew that day to a height of 30,000 meters (100,000 ft.); five days later a French Centaure rocket fabricated at BARC and using French fuel rose from the same launch site to a height of 145 km carrying a 31 kg payload. That successful and powerful French rocket had a big influence on the evolution of the Indian space program. See Gopal Raj, *Reach for the Stars: The Evolution of India's Rocket Program* (Delhi: Viking, 2000), pp. 37–38.

CHAPTER SEVENTEEN

1. This chapter is a greatly changed version of my article "The Government of Scientific Institutions: Case Studies of Two Research Laboratories in the Late 1960s," in *Contributions to Indian Sociology* (NS) 2, no. 1 (1977): 137–68. This is not simply a republication of that original text. I am grateful to the editor of *Contributions* at that time, Satish Saberwal, for his advice prior to publication and the thoughtful initiative he took to consult with directors of TIFR and SINP about the article. Part of the TIFR response is reprinted here. SINP declined to reply to the invitation of the editors of the journal. See also results of a study done about the same time as mine, in other unnamed institutions, focusing on age, status, and mobility: S. P. Gupta, K. D. Sharma, and A. Rahman, "Scientists and Their Commitment to Organizational Goals," *Economic and Political Weekly*, 4 December 1971, pp. 2447–49.
2. Administrators of scientific institutions in the 1960s had little writing to stimulate their thinking. There were exceptions, however; for example, discussion in Jitendra Singh, ed., *Management of Scientific Research* (Bombay: Popular Prakashan, 1973), dealt with the relationship in the 1960s between management and the governance of scientific institutions in India. See also the treatment of governance in Sir John Cockcroft, *The Organization of Research Establishments* (Cambridge: Cambridge University Press, 1965). Cockcroft wrote this after years of his life as director of the Atomic Energy Agency at Harwell, Great Britain. Articles on administration of scientific institutions also appeared regularly in the journal *Minerva,* but few administrators in India read them.
3. Between 1967 and 1969 I visited and discussed working conditions informally with scientists at Andhra University, Waltair; Aligarh Muslim University; Benares Hindu University; Osmania University, Hyderabad; Indian Institute of Science, Bangalore; Indian Institute(s) of Technology at Kanpur and Kharagpur Kanpur; University of Bombay; University of Madras; University of Calcutta, National Physical Laboratory, Delhi; Central Food Technology Research Institute, Mysore.
4. Sources: annual reports for TIFR and SINP, 1968–1969. Conventionally the chief minister of West Bengal was a member of the SINP Governing Body but the United Front Government of 1968–69 was suspended by president's rule in March 1969. The member would have been Chief Minister Ajoy Mukherjee.
5. See the *Rules and By-laws,* as amended up to 1965, Tata Institute of Fundamental Research, Bombay. The Constitution of the Saha Institute of Nuclear Physics states, "The Governing Body may delegate as much of its powers to the Director as it may

think will be in the interests of the Institute"; *Official Report*, 1955, Institute of Nuclear Physics, Calcutta, Appendix 1, p. 3.

6. This interdependence was well described by Singh, *Management of Scientific Research* pp. 162–65.
7. Author's conversation with Raja Ramana, Bombay, May 1969.
8. Thomas S. Kuhn, *The Structure of Scientific Revolutions* (Chicago: University of Chicago Press, 1962).
9. The director of TIFR wrote the following comment in 1976 for an earlier version of this chapter: "The microwave activity in TIFR was not taken because of 'military pressure' as suggested by [Anderson]. As a part of the projects on the construction of electron linear accelerator and the Ooty radio telescope, considerable expertise in the area of microwave engineering was developed at the Institute, which naturally provided the necessary background for undertaking the production of strategic items of national relevance"; B. V. Sreekantan to S. Saberwal, 1976, exact date unknown, probably March 1976. One notes however that neither the telescope nor the cyclotron were under construction when the military pressure was brought to bear on TIFR's microwave engineering group.
10. Author's conversation with anonymous TIFR scientist, Bombay, January 1968.
11. The fact is a number of prominent physicists owed much to Bhabha as a teacher at the summer institutes of the late 1940s and early 1950s in Bangalore. This was soon formalized into a regular training school for BARC scientists and engineers. Bhabha and others organized advanced teaching in physics which those junior scientists would not otherwise have received.
12. B. D. Nagchaudhuri, "Strategy for Science in India: Should the Impecunious Be Impetuous?" *Science Today*, Bombay, July 1968.
13. About *three-quarters* of the scientific staff were Bengalis, in contrast to the cultural heterogeneity of TIFR: all the nonscientists at SINP were Bengali speakers.
14. Author's conversation with anonymous TIFR scientist, Bombay, December 1967.
15. Personal communication, J. V. Kotwal, secretary of School of Physics, TIFR, 1 March 1976.
16. A. S. Divatia, "History of Accelerators in India," *Indian Journal of Physics* 62A (1988): 748.
17. I am very grateful to Atri Mukhopadhyay for his thoughtful assistance in re-constructing the history of the Pagladanga project described in the preceding two paragraphs, personal communication, 12-20 June 2009. The interpretation of the situation, however, is mine.
18. M. G. K. Menon, "The Scientific Community in National Development," *Indian Journal of Public Administration* 15 (1969): 516–17.
19. Author's conversation with anonymous TIFR scientist, Bombay, February 1968.
20. A. K. Rajagopal, "Brain Drain?" *Humanist Review*, 1970, p. 26.
21. Everett H. Hafner, "An Institute in North Bengal," *Physics Today*, June 1967. This account was followed four months later by letters analyzing Hafner's views, drawing a vigorous response, positive and negative, from physicists in India and America, including from William Blanpied, who had organized a summer institute in 1967 at Saugar University in Madhya Pradesh. A letter from S. N. Sen, head of the Department of Physics at North Bengal University, where Hafner's institute took place, was also published; see "Comments on Indian Institutes," *Physics Today*, October 1967, pp. 57–65. The North Bengal University institute chose to base its pedagogy over an intense four weeks on the famous Feynman lectures, as had been done at other

institutes. One letter to *Physics Today* reminded readers that "since most of the graduate students at a well known university in the United States could not answer questions from [the Feynman lectures] in their 1965 qualifying exams, how could one expect teachers with BSc's, from remote areas in India, to learn so much in such a short period of time?" (i.e., three to four intensive weeks), June 1967, p. 65.

22. The director of TIFR wrote an explanatory note regarding the source of pressure among the fellows in 1968–70, saying from the point of view of TIFR, the report of strains and issues was "not objectionable though it contains certain criticisms which are attributed to some members of the Institute. We would like to point out that members who have been quoted are those who have spent a very short time, less than a year, at the Institute and as such their views are partially unbalanced"; B. V. Sreekantan to S. Saberwal, editor of *Contributions to Indian Sociology* (exact date unknown, prior to June 1977).
23. B. D. Nagchaudhuri, "Government's Role in Development of Science," *Indian Journal of Public Administration* 15 (1969): 312–13.
24. Personal communication, 24 November 1971.
25. Author's conversation with Raja Ramanna, Bombay, June 1969.
26. Author's conversation with anonymous scientist, Calcutta, May 1969.
27. "Protest against Computer Plan," *Statesman* (Calcutta), 28 January 1969.
28. Padmanabha Dasgupta, "Editorial: A Note on Automation," *Physics Alumni Annual* (Calcutta: Science College, University of Calcutta, 1969), p. 44.
29. The official conversion rate used by scientific institutions for the rupee in 1969–70 was about Rs 7.50 = US$1.00—the black market rate was considerably higher.
30. "Saha Institute Scholars Want Better Deal," *Hindustan Standard* (Calcutta), 5 June 1969; "Students Refuse to Accept Diplomas," *Statesman* (Calcutta), 3 June 1969.
31. The absence of a follow-up study to track where these scientists went over the next ten years points to the importance of longitudinal studies in this field.
32. A referee for a 1977 version of this chapter suggested that frustrating working conditions "may sometimes spur individuals and groups to greater productivity," saying that it might have been appropriate for me to furnish independent judgments regarding the quality of work done at TIFR and SINP also. Though independent judgments are very valuable, this comment raises a crucial misunderstanding of my point of view, suggesting that I was judging the quality of a research laboratory in terms of the access its scientific personnel had to decision-making within it. But this judgment is not mine. It was scientists themselves who regularly judged their laboratories (and other institutes) this way. A key attribute of governance was this level of participation, and scientists described the value of the laboratory to them in terms of the access they have to decision-making. Typically, senior scientists praise the access they enjoy, in terms of its contribution to their productivity and quality of their work, whereas junior scientists, especially those who have been there for ten years, pointed to their frustrations with governance as an inhibition on creativity and productivity. Independent judgments regarding the quality of work done at TIFR and SINP, while very interesting, would not have altered the great importance of this general attitude among scientists of all ranks.
33. Indira Gandhi, Remarks to the Third Conference of Scientists and Educationists, Delhi, November 1970; reference in National Institute of Science, *Technology and Development, A Review of the Science and Technology Policies of India 1948–1971*, prepared by J. K. Ahuja, I. George, S. Mahanti, A. Jain (Delhi, November 1989), DST ref SP/UR/201/87, p. 112.

CHAPTER EIGHTEEN

1. One senior university scientist familiar with both DAE and CSIR institutions vigorously advocated in 1967 that I do a field study of a CSIR lab, if I wanted to "understand the real situation for scientists in India—choose the best one the CSIR can show you," he said in 1967, voice loaded with irony.
2. Present on 24 March 1964, with Nehru, were council members industrialist K. K. Birla, finance minister T. T. Krishnamachari, AEC chairman Homi Bhabha, head of the space program and industrialist Vikram Sarabhai, defense advisor D. S. Kothari, aeronautical engineer and director of the Indian Institute of Science Satish Dhawan, expert prospector for uranium D. N. Wadia, and minister C. Subramaniam, who would hold the most important cabinet portfolios, except for external affairs, in the coming fifteen years. Sending regrets were people equally well placed: lawyer and Tata insider J. J. Ghandy, physicist and parliamentarian Satyen Bose, industrialist and chairman of Air India J. R. D. Tata, industrialist Arvind Mafatlal. This meeting approved the appointment of P. K. Kichlu as director of the National Physical Laboratory. CSIR, Delhi, Minutes of the meeting of the Governing Body, 24 March 1964; see also Minutes of the 6 May 1965 and 19 November 1966 meetings, for evidence of the economic and political influence available to the Governing Body.
3. CSIR Research Survey and Planning Organization, *Opinion Survey of Scientists and Technologists,* Delhi, March 1967. This survey was circulated in late 1966 by lab directors who made replies obligatory; even then only 25 percent of the 9000 scientists completed the questionnaire. "Not a single questionnaire was returned by some of the institutions (including a few under the CSIR) while from many others the return was poor and slow" (p. 4). Despite their limitations, the data are worthy of consideration and do not vary greatly from my observations (including a questionnaire) in DAE institutions in 1967–69. The exception is the unwillingness of DAE scientists to move from their present situations except to a few choice institutions abroad.
4. It appears that the proximate cause for the Sarkar Committee was the confrontation between the director general Atma Ram, who recommended a major reorganization and elimination of technical directorates such as design engineering, and his Governing Body. The Governing Body (notionally chaired by the prime minister) turned Ram's recommendations over to the SACC, which rejected them. An ad hoc committee of SACC recommended a full scale inquiry into the CSIR, hence Justice Sarkar's appointment. One of the less-visible variables was the extent to which CSIR could engage in foreign collaborations. "The CSIR Controversy," *Link,* 10 December 1967, pp. 41–43. In order to survive, Ram went on a counterattack in the following year; see "Research Results to Be Adopted in Practice: Atma Ram," *Amrita Bazar Patrika,* 30 August 1968.
5. It is curious that there have been so many inquiries into CSIR and its labs, and yet the only published research about such a lab, referring to these inquiries, is the classic study of the National Physical Laboratory by Shiv Visvanathan, *Organizing for Science* (Delhi: Oxford University Press, 1984). About 1962 Dr. A. Rahman started a research analysis unit within the CSIR itself and began presciently to investigate issues like retention of foreign-trained scientists, the role of foreign exchange in research, and so on. Rahman trained younger people to do this kind of analysis, though he naturally experienced difficulty in getting an analytic culture started in the CSIR itself. Nevertheless, this unit evolved into the National Institute for Science, Technology, and Development in Delhi.

6. *Report of the Committee of Enquiry,* Council of Scientific and Industrial Relations, Delhi, Part I (1970) and Part II (1971).
7. Atma Ram, "Some Thoughts on Indian Science," *Science Today,* February 1978, p. 21. It curious that Atma Ram provided few autobiographical details and never wrote anything autobiographical in English except this little piece in *Science Today.* Information for the biographical note on Ram is selected from a biography dictated by Ram to Durga Prasad Nautial, translated from Hindi by R. K. Singhal, "A Great Indian Scientist: Dr Atma Ram," October 2007, unpublished manuscript. I am grateful to Anil Rajvanshi for enabling me to see this valuable text.
8. Visvanathan, *Organizing for Science,* p. 262.
9. P. M. S. Blackett to Atma Ram, 29 November 1967, Royal Society Archives, London.
10. I was present at this January 1968 speech at Benares Hindu University and remember vividly the heated discussions among young scientists in the tea tent following Ram's lecture.
11. P. M. S. Blackett to Atma Ram, 4 December 1967, Royal Society Archives, London.
12. P. M. S. Blackett to A. J. Kidwai, 29 November 1967, Royal Society Archives, London.
13. Bharat Bhushan, "Management of a Research Laboratory: Case History of IICT" in D. Bhattacharji, S. N. Ghatak, and S. S. Iyer, eds., *CSIR—Looking Back* (Lucknow: CSIR Pensioners Welfare Association, 1997), pp. 126–146.
14. Hari Narain, "Pleasure and Pains in Establishing a New Research Institute," in Bhattacharji, Ghatak, and Iyer, *CSIR—Looking Back,* pp. 43–53. Hari Narain published a more complete and thoughtful assessment of the Zaheer initiatives and his own experience as a CSIR director in "My Years with the CSIR," *Journal of Indian Geophysical Union* 8, no. 2 (April 2004): 147–56.
15. Ashok Parthasarathi, *Technology at the Core: Science and Technology with Indira Gandhi* (Delhi: Longmans, 2007), p. 10.
16. Patrick Blackett, Report to the Leverhulme Trust, 12 May 1971, p. 7; Royal Society Archives, London.
17. Research on leather was essentially, to most Hindus, a "polluting" activity, and those who worked with leather (at least around 1971, whether the leather-working Hindu castes or Muslim leather workers) were subject to strong social disapproval and segregation. The contradiction lay in the fact that the export of leather goods was a strong foreign exchange earner, and Nayudamma's institute was well known for its effective R&D relationships with this profitable industry. Such effectiveness with industry had always been the unrealized hallmark of the CSIR's objectives for its other labs.
18. Patrick Blackett, Report to the Leverhulme Trust, 12 May 1971, Royal Society Archives, London.
19. P. C. Mahalanobis to P. M. S. Blackett, 29 June 1971, Royal Society Archives, London.
20. E. H. S. Burhop to P. M. S. Blackett, 30 January 1970, Royal Society Archives, London. Burhop was well informed, and when Desai finally became prime minister seven years later his friend Atma Ram was swiftly rehabilitated.
21. P. M. S. Blackett to E. H. S. Burhop, 6 February 1970, Royal Society Archives, London.
22. Much of this section relies on communication with Ram Prasad, personal communications, 22–30 June 2008. Ram Prasad, having arrived at NPL in 1966, became acting general secretary of the ASWI in 1968, and president of the CSIR's SWA in 1970. I am grateful to Ram Prasad for his thoughtful contribution to my understanding of this subject.

23. Gary Wersky, *The Visible College: A Collective Bibliography of British Scientists and Socialists of the 1930s*, 2d ed. (London: Free Association Books, 1988).
24. Shiv Visvanathan, *Organizing for Science*, pp. 197–99.
25. Personal communication, 30 June 2008.
26. Ram Prasad, personal communications , 22–30 June 2008. I also acknowledge my conversations in 1968-69 with ASWI general secretary K. R. Bhattacharya.
27. Visvanathan, *Organizing for Science*, p. 199.
28. For example, one would expect that there would have been litigation within the DAE too, with dozens of laboratories and other projects, thousands of employees, complex contracts and professional appointments being made every day. But such conflict and legal cases were not public as they were inside CSIR, in part because of the secrecy surrounding the DAE and its projects and institutions, in part because of the culture of "separation" cultivated by Bhabha.
29. Parthasarathi, *Technology at the Core*, p. 182; only four persons are named, G. S. Siddhu, M. G. Krishna, G. S. Choudury, and railway engineer M. M. Suri, director of Central Mechanical Engineering Research Institute.
30. Parthasarathi did not identify Gill in his 2007 book as one of the five directors, but I am reliably informed that Gill was investigated and did resign.
31. R. S. Ganapathy, "Science and Technology, a Department in Search of a Role," *Economic and Political Weekly*, 22 June 1974, p. 972.
32. K. R. Bhattacharya, "A Rational System of Recruitment and Promotion for Scientific Personnel," *Economic and Political Weekly*, 6 January 1973. Dr. Bhattacharya, based at the Central Food Technology Research Institute in Mysore, was an indefatigable analyst of the CSIR's social system on behalf of the ASWI and presented this paper to Justice Sarkar's inquiry in 1971.
33. Parthasarathi, *Technology at the Core*, pp. 191–92.

CHAPTER NINETEEN

1. Sumit Ganguly and D. T. Hagerty, *Fearful Symmetry: India-Pakistan Crises in the Shadow of Nuclear Weapons* (Seattle: University of Washington Press, 2005), p. 36.
2. Ibid. .
3. G. Balachandran, *The Reserve Bank of India, 1951–1967* (Delhi: Oxford University Press, 1998); see chap. 17. See also Rahul Mukherji, "India's Aborted Liberalization—1966," *Pacific Affairs* 73, no. 3 (Fall 2000): 375–92.
4. Dennis Kux, *Estranged Democracies: India and the United States, 1941–1991* (Thousand Oaks, CA: Sage Publications, 1994), p. 261.
5. Mukherji, "India's Aborted Liberalization," 375–92. Mukherji's article is supported by his recent conversations with the key people involved in the 1966 devaluation.
6. Alec Cairncross and Barry Eichengreen, *Sterling in Decline: The Devaluations of 1931, 1949, and 1967* (London: Palgrave Macmillan, 2003).
7. Johnson eventually praised Subramaniam's role; see Vernon Ruttan, *United States Development Assistance* (Baltimore: John Hopkins University Press, 1996), pp. 166–67.
8. Subramaniam interview with Dennis Kux, *Estranged Democracies*, p. 259.
9. Ruttan, *United States Development Assistance*, pp. 166–67.
10. With a PhD from Cambridge in 1952, and wide experience in rice genetics, Swaminathan's very active presence in Delhi set the stage for his 1972 promotion to be director general of the Indian Council of Agricultural Research and secretary to the government for agricultural planning purposes. He was the first agricultural scientists to have this level of influence. He eventually became director of the International

Rice Research Institute in the Philippines. See Robert S. Anderson, Edwin Levi, and Barrie M. Morrison, *Rice Science and Development Politics* (Oxford: Clarendon Press; New York: Oxford University Press, 1991).

11. For example, Indira soon disagreed with her father's reluctance to deal harshly with the elected Communist government in Kerala in 1959; she said she did not share Nehru's view that democratic institutions could survive political disorder, even alleging that the Communist government of Kerala were agents of the Chinese! Katherine Frank, *Indira: The Life of Indira Nehru Gandhi* (London: HarperCollins, 2001), pp. 251–52.
12. Ibid., p. 290.
13. Ibid., p. 296.
14. Ibid., p. 297.
15. National Institute of Science, *Technology and Development: A Review of the Science and Technology Policies of India, 1948–1971*, prepared by J. K. Ahuja, I. George, S. Mahanti, and A. Jain (Delhi, November 1989), DST ref SP/UR/201/87, p. 6; cited as SACC Review.
16. CSIR Research Survey and Planning Organization, *A Study on the Conservation of Foreign Exchange by the National Laboratories* (Delhi, February 1966). This work focused on the way in which sections of sixteen laboratories utilized Indian research results that could be applied to industry and thus could conserve foreign exchange. Because the research unit was not funded to travel and work in those labs, it depended for data on laboratories themselves and did not examine the views or evidence of industries themselves. It had no effective way to check what the labs were telling them.
17. CSIR Research Survey and Planning Organization, *Foreign Assistance to Scientific Research in India* (Delhi, November 1966). The authors stress the incompleteness of their data, based on the spotty replies from embassies and foundations and the scattered way in which the Ministry of Finance stored some its records.
18. SACC Review, p. 70.
19. "50 p.c. Cut in Foreign Know-how: Scientists Face Big Challenge," *Amrita Bazar Patrika* 11 August 1968.
20. Suri Bhagavantam was a physicist named by Krishna Menon in 1961 to be scientific advisor to him, as the minister of Defence. He distinguished himself in 1968 by giving a lecture in Bangalore in which he said that all the techniques of decision-making should be applied to the question of the bomb, and "all sorts of cost-effectiveness studies" should be conducted before making a decision. "A Bit of Science in Politics Urged," *Statesman* (Calcutta), 4 September 1968. This was about the time that he reported on his experience of miraculous occurrences in the presence of a swami; see chap. 24.
21. SACC Review, p. 71; from this point the document is a summary of COST proceedings, but for continuity I shall continue to refer to it in footnotes as the SACC Review. We have only a summary of minutes, so the text is not necessarily verbatim but is likely to be closely based on the notes of participants.
22. It should be remembered that concurrently with COST, the cabinet also had its own Standing Committee of the Cabinet on Foreign Affairs. This is relevant because SACC/COST issues always had a very strong international dimension. The standing committee on foreign affairs, however, did not play an influential role in nuclear matters, according to Paranjpe: "The Indian Cabinet has not functioned as a collective decision-making body. The exception was the decision not to sign the NPT [non-

proliferation treaty]"; Srikant Paranjpe, *Parliament and the Making of Foreign Policy: A Study of Nuclear Policy* (Delhi: Radiant, 1997), p. 77.

23. SACC Review, p. 72.
24. I am grateful to Julia Khorana for checking some details of her father's biography; personal communication 7 December 2008.
25. R. P. Rao, "Hargobind Khoranna," *Illustrated Weekly of India,* 17 November 1968, p. 44.
26. "Plea to Revive Plan for Biological Laboratory," *Statesman,* 20 October 1968. The possibility is very low that Khorana would have left his new position at MIT in late 1968 for the task of building a new undeveloped lab in India, prestigious though it might appear to the CSIR, and regardless of what he said to Zaheer in 1965 and 1966.
27. Thirty-seven years later Indian physicists observed the award of the 2005 Nobel Prize in their field to Roy Glauber, of Harvard University, and pointed to the Nobel committee's selective attention, which omitted the 1963 work of E. C. George Sudarshan on the quantum theory of optical coherence (research done at Rochester and Berne). The Nobel Committee is reliably reported to have considered Sudarshan's role as he had been nominated for the prize on this work in the early 1980s, but in this case Sudarshan's pioneering role was not sustained, presumably solely on the grounds of "micro-sequencing"; R. Ramachandran, "Elusive Recognition," *Frontline* (Chennai), 18 November 2005.
28. The analysis was circulated by Sarabhai and his research assistant Ashok Parthasarathi in early 1968 (when Sarabhai asked me to read a copy and think about the brain drain in terms of my dissertation). It was presented in preliminary form at the October 1969 Pugwash Conference. Finally it was revised and published by Parthasarathi as "Brain Drain from Developing Countries," *Nature* 230 (12 March 1971): 87–90. Rather crude evidence from the study was released in the newspapers, e.g., "About 1000 Indians Holding Faculty Positions in USA," *Statesman,* 14 September 1968.
29. A. Parthasarathi, "The Sources of Technological Growth," *Economic and Political Weekly,* 2 December 1967, p. 2092.
30. A. Parthasarathi, "Appearance and Reality in Indian Science Policy," *Nature* 221 (8 March 1969): 909–11.
31. A. Parthasarathi, "Sociology of Science in Developing Countries: The Indian Experience," *Economic and Political Weekly,* 2 August 1969. A full analysis of Kitchlu's ideas as director of NPL, though not an account of his disagreement and dismissal, is found in Shiv Visvanathan, *Organizing for Science* (Delhi, Oxford University Press, 1984), chaps. 6, 7, and 8. Kitchlu had been a student of Meghnad Saha and began to manage an optical glass factory after he abruptly left NPL.
32. Parthsarathi, "Brain Drain from Developing Countries," pp. 87–90.
33. *The Hindu,* 6 May 1967.
34. Thus Rajagopalan Chidambaram, aged thirty-two, began to work on the equation of state. Experimental physicist P. K. Iyengar was brought in to the team in 1968 by Ramanna. Chidamabaram, who ultimately became director of BARC and commissioner of the AEC in 1990 and eventually AEC chairman in the 1990s, wrote about this early stage in "Plutonium and Thorium in the Indian Nuclear Programme," *Current Science,* 10 January 1996; also cited in George Perkovich, *India's Nuclear Bomb: The Impact on Global Proliferation* (Berkeley: University of California Press, 1999), p. 509.
35. Perkovich, *India's Nuclear Bomb,* p. 142.

36. For a full analysis of India's official positions, see Perkovich, *India's Nuclear Bomb.*
37. SACC Review, pp. 110–12. A marginalized observer of the political structure of the scientific community wrote ironically at this time: "Practitioners of science in our country are divided into a certain number of 'maths' or 'churches.' Each 'math' has a powerful 'mahant' who derives his temporal and sometimes even his ecclesiastical power from one of the gods of the Indian scientific pantheon. In the style of the Indian *puranas* (sacred texts), these gods are ever at loggerheads with one another, and the jealousies of their spouses do not in any way help to improve the situation"; A. D. Bhogle, "Challenges to Indian Science" *University News,* January 1969, p. 3.
38. B. M. Udgaonkar, "Implementation of the Scientific Policy Resolution," *Economic and Political Weekly,* 26 December 1970, p. 2093.
39. Ashok Parthasarathi, *Technology at the Core: Science and Technology with Indira Gandhi* (Delhi: Pearson Longman, 2007). p. 34.
40. Ibid., p. 39.
41. "India Curbs Patents: New Law Upsets Some US Companies," *New York Times,* 15 November 1970. Spokespersons for Merck, Sharpe & Dohme, and American Cyanamid are quoted as critical of the royalty and timing provisions of the new law.
42. P. M. S. Blackett, "Personal notes—Atma Ram," no date, probably May 1971, Royal Society Archives, London.
43. According to Frank, *Indira,* in 1937 September Indira sailed to London from Bombay, on a ship full of Indian students, disembarked at Marseille and having met her lover Feroze Ghandy in Paris, went as a new student to Somerville College at Oxford. Based on her investigation, Frank writes "Indira had never been and was not now a brilliant student, but at Somerville she began to apply herself. She always prepared for and performed credibly in her tutorials. She was industrious, intelligent, and could write well" (p. 127). Frank concludes, "it has been generally held that Indira Nehru was a poor student at Oxford, but Somerville College and Oxford University records contradict this, as do . . . the reported judgments of her tutors" (p. 514). It seems from the records that she took the "pass moderation exam" twice without success. She could have written the exam for a third time but left Oxford before the exam because of illness. Most other students passed these obligatory exams set in Latin and Greek for all Oxford students, and only a "pass" is recorded. But she didn't write it the third possible time; she decided on her own not to sit the Latin exam the third time and knew she would thus be sent down. But Indira was already living a double life, going to political meetings in London, and in Oxford too, and changing her style of dress (consistent with her bravely swimming in an orange bathing suit in the pool on the ship coming from Bombay). After failing the first pass mod exam in December 1937 she went to Berlin, where she observed with alarm the prevalence of the Nazis. In August 1938, eleven months after starting at Oxford, Indira was diagnosed with pleurisy (of which her mother died), and so was unable to return to Oxford but decided to return to India in November 1938 at age twenty-one. In 1939 April after five months in India, mostly unhappy, still ill, she insisted on returning to London and to Feroze Ghandy, who was leaving the London School of Economics in order to work more on journalism and Independence politics. Indira began to see a lot of V. K. Krishna Menon, who had spent 28 years in England, was elected a member of Parliament for Labour, and was called to bar in 1934. She also met again P. N. Haksar, who had studied anthropology at the London School with Malinowski and was now studying law in London, and Mohan Kumaramangalam, just graduated from Cambridge. These two men became among her most important advisors from

1967 onward. Not indifferent to her reputation and ability, Somerville College of Oxford proposed to her father that she return to do the diploma on social and public administration (so that no Latin exam was required for a degree); Indira consulted Harley Street doctors, who advised against this and favored going for treatment in the sanatoriums of Switzerland. This decision was final, and she never became a university student again. Moreover, her son Sanjay began a three-year apprenticeship with Rolls Royce in September 1964 instead of going to the university, and her older son Rajiv also failed his Cambridge and Imperial College exams and did not get an engineering degree; eventually he returned to India and began pilot's training about 1967. So Indira Gandhi's academic accomplishments were not on the level of many of the people who worked around her, but how many of her (mostly male) critics in India in the late 1960s could have passed the Latin exam that Indira Gandhi failed at Oxford?

CHAPTER TWENTY

1. I am grateful to colleagues at the University of Cambridge on History and Philosophy of Science discussion list, particularly John Forrester, Susan Gamble, Rachel Jardine, and Samuel Lipoff for suggestions in May 2005 about the origins of the terms "high tech" and "high technology." The *OED* was, as always, most helpful. The relevant historic citations are John Knox, "Social Theory and Industrial Sociology," *Social Forces* 33, no. 3 (1955): "This desire, together with the other complexities of high technology and large scale production, resulted in an intricate division of labor and high specialization of managers" (p. 243). See also Derek J. de Solla Price, *Science Since Babylon* (New Haven: Yale University Press, 1961); Irving Louis Horowitz, *New Sociology*, no. 292 (1964). An early reference in India is by economist Ashok Mehta, former minister of Planning, and then minister of Petroleum and Chemicals and Social Welfare in 1968: "along with the ordered introduction of higher forms of technology (e.g., 'higher technology in agriculture'), we have made an intensive survey of our natural resources"; Mehta, "Science for a Transitional Society," in Ward Morehouse, ed., *Science and the Human Condition in India and Pakistan* (New York: Rockefeller University Press, 1968), p. 9.
2. NISTADS, *A Review of Science and Technology Policies of the Government of India, 1948–1971* (Delhi, 1989) DST Ref No SP/UR/201/87.
3. According to biographer Katherine Frank, *Indira: The Life of Indira Nehru Gandhi* (New York: HarperCollins, 2001), after 1969 Gandhi's "behaviour—both political and personal—becomes increasingly ambiguous and open to interpretation" (p. 311) and "She misconstrued her obligation and responsibility as necessity" (p. 320). The author of the intelligence reorganization and creation of the Research and Analysis Wing in 1968–69 was P. N. Haksar.
4. Christopher Andrew and Vasili Mitrokhin, *The Mitrokhin Archive II: The KGB and the World* (London: Penguin, 2005), chaps. 17 and 18. An officer at the Soviet embassy in Delhi around 1973–74 who became head of the KGB's counterintelligence Directorate K, Oleg Kalugin, said that the KGB was more successful than the CIA partly because of its skill in exploiting corruption, in his opinion, which became endemic under Indira Gandhi's regime (p. 322). Though the KGB was certainly active in attempting to influence national politics (committing forgeries and organizing demonstrations to achieve this, according to Mitrokhin's hand-copied archive), no evidence has yet been seen concerning Russian intelligence on the Indian nuclear program.

5. For example, in 1969 COST discussed a proposal from Schlumberger Corporation of Geneva to establish a "completely foreign-owned subsidiary for the manufacture of electronics instruments." COST questioned how national interests would be safeguarded, because the Schlumberger proposal did not conform to the standard joint-venture proposal necessarily involving 51 percent Indian ownership. How had the proposal got so far without a thorough treatment of the joint-ownership issue? Which minister's intermediary had carried it all this way? The proposal was turned back. NISTADS, *A Review*.
6. Amrita Shah, *Vikram Sarabhai: A Life* (Delhi: Viking/Penguin, 2007), p. 131, n. 11.
7. Ibid., pp. 178–80, plus citations, partly based on February 1976 Sarabhai Memorial Lecture by L. K. Jha.
8. In November 1970 her son Sanjay was awarded a license to produce 50,000 small Maruti cars out of indigenous materials, with Indira Gandhi's knowledge and agreement. In the face of public criticism, Gandhi offered the weak justification that "young men are needed who can innovate." Sanjay had no qualifications for this project other than his fondness for driving fast cars on Delhi roads late at night. The notorious and unproductive Maruti car project was, for the next ten years, a prime illustration of the dilemmas of a prime minister's role in the war over self-reliance. P. N. Haksar chose to join in the criticism, either before or after this 1970 license award, and Mrs. Gandhi punished him three years later by letting his appointment lapse at the time of his normal retirement in February 1973, "on account of differences with her on her support of the 'peoples' car' project of her younger son Sanjay"; Frank, *Indira*, p. 144; also Ashok Parthasarathi, *Technology at the Core: Science and Technology with Indira Gandhi* (Delhi: Longmans, 2007), p. 61.
9. See *Economic and Political Weekly*, 31 May 1969.
10. Ibid.
11. George Perkovich, *India's Nuclear Bomb: The Impact on Global Proliferation* (Berkeley: University of California Press, 1999); pp. 152, 513.
12. Government of India, *Atomic Energy and Space Research—A Profile for the Decade—1970–1980* (Bombay: Atomic Energy Commission, 1970). A draft version of this plan was announced and circulated about six weeks before the release.
13. On this latter subject he was well briefed by his close friend Kamla Chowdhury, eventually the government of India's chief official for wasteland.
14. I have written about this assumption about displacement and labor in India's largest planned forestry project in the 1970s, namely, the Bastar Caribbean pine project; see Robert S. Anderson and Walter Huber, *The Hour of the Fox: Tropical Forests, the World Bank, and Indigenous Peoples in Central India* (Seattle: University of Washington Press, 1988).
15. M. R. Srinivasan, *From Fission to Fusion: The Story of India's Atomic Energy Progamme* (Delhi: Viking, 2002), pp. 107–8.
16. Personal communication from Australian historian of technology, Carroll Pursell, 1 April 2008. It is not necessary to conclude that the idea originated with Weinberg anymore than it did with Sarabhai; this plan was entirely within Sarabhai's conceptual framework.
17. Itty Abraham, *The Making of the Indian Atomic Bomb* (London: Zed Press, 1998), p. 133.
18. Christopher Alexander, "The City Is Not a Tree," 58–62. This article was reprinted twice (in 1966 in *Design* and in 1967 in *Ekistics*) before Sarabhai handed it to me in 1967 and said "let's discuss it soon": I am not sure where he first read it and had it

copied from. Christopher Alexander was one of the most complex thinkers about architecture and design of the late twentieth century, and it is characteristic of Sarabhai that he, a physicist and businessman, had Alexander's paper typed and reproduced for his personal distribution in the era before photocopiers.

19. Parthasarathi, *Technology at the Core,* p. 7.
20. Author's conversation with Ashok Parthasarathi, Delhi, November 1998.
21. Parthasarathi, *Technology at the Core,* p. 18.
22. Ibid., p. 19.
23. Ibid., p. 100.
24. Under Sarabhai's chairmanship, the AEC's voting members were Homi Sethna, head of BARC, J. R. D. Tata, head of the Tata group of companies in Bombay, P. N. Haksar of the prime minister's office in Delhi, Satish Dhawan, director of the IISc in Bangalore, and I. G. Patel of the Finance Ministry (a future governor of the Reserve Bank and the future director of the London School of Economics). The nonvoting secretary of the AEC was T. N. Seshan, who next became the cabinet secretary. These were people of prominence and influence, but they were all closely related: Homi Bhaha was a relative of J. R. D. Tata; Haksar and Patel were close through the London School in the 1930s and 1940s; Sarabhai and Dhawan were both fascinated with space; Dhawan soon succeeded Sarabhai as head of the space program.
25. Parthasarathi, *Technology at the Core,* pp. 100–103.
26. Ibid., pp. 121–22
27. Srinivasan, *From Fission to Fusion,* p. 230.
28. The plant opened in June 1977 to operate for six months (though fitfully because of Baroda's electricity fluctuation), then there was a big explosion of hydrogen that escaped from a pipe in December 1977. It practically demolished the Baroda heavy water plant for three years, but by 1980 the plant was running again, "reasonably satisfactorily" according to M. R. Srinivasan.
29. Srinivasan, *From Fission to Fusion,* p. 222–23.
30. Parthasarathi, *Technology at the Core,* pp. 123–24.
31. Srinivasan, *From Fission to Fusion,* p. 65.
32. Ibid., p. 67.
33. Ibid., p. 72.
34. Parthasarathi, *Technology at the Core,* p. 110.
35. Ibid., p. 112.
36. Ibid., p. 107.
37. Ibid., pp. 105–6.
38. Ibid., pp. 106–7.
39. Calculations from Perkovich, *India's Nuclear Bomb,* p. 154.
40. Homi Sethna, quoted in Abraham, *The Making of the Indian Atomic Bomb,* p. 136.
41. Author's conversation with G. Venkataraman, Puttaparathi, November 1998.
42. Parthasarathi, *Technology at the Core,* p. 21.
43. Report of the Lok Sabha Estimates Committee on the Rajasthan Power Plant (RAPP), July 1970, Vol. 17, pp. 136–38.
44. Electronic Committee members, with Sarabhai as chair, were A. S. Rao, director of electronics at BARC; S. Bhagavantam, scientific advisor to the Ministry of Defence; H. C. Sarin, secretary of Defence Production; Amarjit Singh, director of Central Electronic Engineering Research Institute of CSIR, who wanted Committee funds for his institute. Parthasarathi was its secretary, and in his opinion A. S. Rao was "committed to self-reliance and a 'we can do it at home' philosophy." Bhagavantam "never

took a position on any issue in the 3 years I worked for the Electronics Commission" (1967–70). Harish Sarin, of Defence, tolerated the committee but kept it away from the decision-making of defense electronics. Parthasarathi, *Technology at the Core,* pp. 6–7.

45. This movement is curiously not described in the excellent book by C. R. Subramaniam, *India and the Computer* (Delhi: Oxford University Press, 1992). Much of what follows relies on Subramaniam's chap. 1, "Self-Reliant Development Fails." Note that the anti-automation movement and the need for computers in nuclear physics were central to the political tensions in SINP, as described in chap. 15.
46. I too was among the short-sighted skeptics concerning the future of small computers in 1969, although I have great respect for everything R. Narasimhan taught me, or tried to.
47. Subramaniam, *India and the Computer,* p. 5.
48. Ibid., p. 6; also "the committee was largely ineffective" and "there was growing disenchantment with the working of the Electronics Committee in government, industry, R&D, and academic circles," hence the National Conference in March 1970 at TIFR; Parthasarathi, *Technology at the Core,* p. 8.
49. "The Electronics Game," from a Special Correspondent, *Economic and Political Weekly,* 6 June 1970, pp. 900–902. See also D. B. R. Chaudhuri, "Empire Building in Electronics," *Economic and Political Weekly,* 8 May 1971, p. 953.
50. Subramaniam, *India and the Computer,* pp. 8–26. The situation in electronics and this comment on it explains why the term "scientific community" in this book includes engineers and technologists.
51. G. Venkataraman, *Bhabha and His Magnificent Obsessions* (Hyderabad: Universities Press, 1994), p. 172.
52. It will be interesting to see evidence in the future for the extent of influence brought on Sarabhai by private industrialists, arguing for more open competitive production of consumer electronics. The 1969 Schlumberger proposal for electronics production, brought before COST, is a case in point. An excellent illustration from 1951 lies in Nasir Tyabji, "Gaining Technical Know-how in an Unequal World: Penicillin Manufacture in Nehru's India," *Technology and Culture* 45, no. 2 (April 2004): 331–49.
53. As evidence of this, just consider how many tea shop owners lined up to buy powerful radios and music systems, and then in the 1980s, how many villagers coveted television sets.
54. Gross revenue for 1970–71 was Rs 115 million. The study team was R. Narasimhan, of TIFR, G. U. Menon, of Hindustan Teleprinters, P. D. Kasbekar, of the Ministry of Finance; Parthasarathi, *Technology at the Core,* pp. 81–89.
55. Subramaniam, *India and the Computer,* p. 13.
56. Letters of 26 July, 1 August, and 15 August 1968, Blackett Papers, Royal Society Archives, London.
57. I am grateful to Teasel Muir-Harmony of the University of Notre Dame for a copy of her presentation, "Tracking Diplomacy: The Smithsonian Astrophysical Observatory's Satellite Tracking Station in India," to the History of Science Society conference, Vancouver, November 2006.
58. In this section I rely on my own studies plus the work of B. P. Sanjay, "The Role of Institutional Relationships in Communication Technology Transfer: A Case Study of the Indian National Satellite System INSAT" (PhD diss., Simon Fraser University, 1989); see also Dinshaw Mistry, "India's Emerging Space Program," *Pacific Affairs* 71

(Summer 1998): 151–74; Gopal Raj, *Reach for the Stars: The Evolution of India's Rocket Program* (Delhi: Viking, 2000); Angathevar Baskaran, "Technological Accumulation in the Ground Systems of India's Space Program: The Contribution of Foreign and Indigenous Inputs," *Technology in Society* 23 (2001): 195–216; A. Baskaran, "From Space to Commerce: The Evolution of Space Development Policy and Technology Accumulation in India," *Technology in Society* 27, no. 2 (April 2005): 155–79.

59. The raja of Mysore, Tipu Sultan, successfully held off attacks against his kingdom in the late eighteenth century using rockets with a range of about one kilometer. Indian and British troops, led by officers of the British India Company, were killed and wounded and their ammunition stores exploded as a result of the impact of these rockets. Launched by well-trained rocket teams, based on wheeled mobile launchers and used to initiate infantry attacks, showers of a hundred rockets fell on the enemy, some blowing up like bombs and others attached to metal-tipped bamboo spears. Impressed by this technology and technique, the Royal Arsenal at Woolwich near London began to study and modify these captured rockets in 1801 and successfully used its own version first in the War of 1812 in the United States and then soon afterward in the British attacks upon Rangoon in Burma. The "modern" ground missile and then ground-to-air missile was thus "born" and circulated around many battlefields.
60. Jerrold L. Schecter and P. L. Deriabin, *The Spy Who Saved the World* (London: Brassey's, 1995).
61. Raj Chengappa, *Weapons of Peace: The Secret Story of India's Quest to Be a Nuclear Power* (Delhi: HarperCollins, 2000), pp. 136–40, based on Chengappa's 1998 interviews with Air Marshal Chandrakant Visvant Gole, Air Marshal O. P. Mehra, Major General Brahm Dev Kapur (first chief controller of DRDO), Jagdish Chandra Bhattacharya (Defence Science Labs, 1958–60), A. V. Ranga Rao (Defence Science Labs, 1958–60).
62. Bharat Karnad, *Nuclear Weapons and Indian Security: The Realist Foundations of Strategy* (Delhi: Macmillan, 2002), p. 309 referring to H. J. Bhabha letter to T. T. Krishnamachari, 2 July 1963, T. T. Krishnamachari Papers, Nehru Memorial Museum and Library, Delhi.
63. Raj, *Reach for the Stars*, 2000.
64. Chengappa, *Weapons of Peace*, p. 148.
65. A. P. J. Abdul Kalam, *Wings of Fire: An Autobiography* (Hyderabad: Universities Press, 1999; see also updated 2002 edition), pp. 48–64. A year later, in 1969, Sarabhai appointed Narayanan to the panel searching for a launch site at Sriharikota on India's east coast. Sarabhai, Narayanan, and Kalam discussed the SLV3 space vehicle rocket project in terms of its ground-to-ground missile potential, according to Narayanan. Chengappa, interview with V. S. Narayanan, February 1998, *Weapons of Peace*, p. 149.
66. Mistry, "India's Emerging Space Program," 151–74. The US Information Service put a closed circuit television in its Calcutta window to show the pictures of the moon landing, and crowds of people stood watching in fascination, whereas normally that office was the object of violent physical attacks.
67. See chap. 16, "Launched and Out of Sight: An Evening at the Thumba Equatorial Rocket Launching Station"; the names of these rockets Menaka and Rohini were inspired by Mrinalini Sarabhai, citing the names of *apsaras*, graceful semi-divine flying feminine figures, often seen in Indian painting and sculpture.
68. Shah, *Vikram Sarabhai*, pp. 48 and 195.

69. Nagchaudhuri's wife, a well-known musician, Deepali Nag, was a friend of Mrs. P. N. Haksar, and P. N. Haksar and Nagchaudhuri were also friends. It was Haksar who had called Nagchaudhuri about the job, and he continued to be helpful to Nagchaudhuri while in the prime minister's office.
70. Chengappa, interview with B. D. Nagchaudhuri, April 1998, *Weapons of Peace,* p. 132.
71. Author's conversation with B. D. Nagchaudhuri, Kolkata, November 1998.
72. R. Ramanna, *Years of Pilgrimage* (Delhi: Viking, 1991), p. 89.
73. Chengappa, interview with B. D. Nagchaudhuri, April 1998, *Weapons of Peace,* p. 129.
74. Chengappa, interview with B. D. Nagchaudhuri, April 1998, *Weapons of Peace,* pp. 131–32.
75. Parthasarathi, *Technology at the Core,* p. 172.
76. Chengappa, interview with V. S. Narayanan, February 1998, *Weapons of Peace,* p. 149.
77. Chengappa, interview with B. D. Nagchaudhuri, April 1998, *Weapons of Peace,* p. 150.
78. Romesh Thapar, "Science or Séance?" *Economic and Political Weekly,* 2 May 1970, pp. 725–26. I heard these views in many quarters, expressed over and over again, including from those who did not or would not read this weekly. This column is quoted at length because it put in print what dozens of people said to me privately at that time. Romesh Thapar and his wife had both been friends of Indira Gandhi, and this may have begun the change in their relationship.
79. This spinoff view of innovation was also put forth in Blackett's 1967 Nehru Lecture. Sarabhai was much more at home with market forces than most other people in the SACC or DAE, and he understood them better than Blackett did, particularly Indian markets. I recall that Sarabhai attended Blackett's 1967 lecture, in the company of the prime minister. P. M. S. Blackett, Jawaharlal Nehru Memorial Lecture, Delhi, November 1967.
80. V. Sarabhai, conversation, April 1969. Sarabhai well knew that many important activities and functions in the textile industry occurred in back streets of Ahmedabad.
81. Perkovich, *India's Nuclear Bomb,* p. 15.
82. Author's conversation with a senior official, Delhi, November 1998.
83. Author's conversation with G. Venkataraman, Puttaparathi, November 1998, and subsequent communication 13 March 2005.
84. Based on Shah's reading of Dhawan's memoir interview of 1997, plus author's conversation with author Gopal Raj, Trivandrum, January 2005; Shah, *Vikram Sarabhai,* pp. 146–48.

CHAPTER TWENTY-ONE

1. Rahul Mukherji, "India's Aborted Liberalization—1966," *Pacific Affairs* 73, no. 3 (Fall 2000): 385.
2. On 3 December 1971 the Pakistan Air Force bombed nine Indian air bases in West and North India; on 4 December Indian troops moved into Bangladesh, aided by a full moon; on 6 December, Indira Gandhi announced official recognition of the country of Bangladesh. The USS *Enterprise* entered the Bay of Bengal on 9 December, so she, P. N. Haksar, D. P. Dhar, and General Manekshaw decided so quickly to invade that even the minister of Defence was not fully informed. Dhar flew on 10 December to Moscow to inform the Soviet Union and get Moscow to move the Soviet fleet to the Bay of Bengal; Indian troops moved forward between 11–14 December,

on 16 December, Indira wrote to President Nixon to explain the situation and India's role, and on 16 December, Pakistan surrendered. Indira Gandhi's stock was very high at this moment, she was called by some writers in the Indian press, "goddess of war" or "*shakti*-power."

3. Dennis Kux, *Estranged Democracies: India and the United States, 1941–1991* (Thousand Oaks, CA: Sage Publications, 1994), pp. 279–307.
4. Vikram Sarabhai, "The Security of Developing Countries," Harold Laski Lecture, 26 March 1966, Laski Institute of Political Science, Ahmedabad.
5. Ibid.
6. Raj Chengappa, interview with Ramanna, August 1998, *Weapons of Peace: The Secret Story of India's Quest to Be a Nuclear Power* (Delhi: HarperCollins, 2000), p. 105.
7. Ibid., p. 112. I have not yet seen direct evidence of Sarabhai's views on this subject in the 1970–71 period. Unfortunately, those who were close to him at that time have not written about it.
8. Letters to the Editor: Vikram A. Sarabhai and Homi N. Sethna, *Himmat Weekly*, 24 May 1968.
9. Bharat Karnad, *Nuclear Weapons and Indian Security: The Realist Foundations of Strategy* (Delhi: Macmillan, 2002), p. 268. Sethna's view on this may have been transmitted to Gandhi when she heard about a May 1966 lunch meeting he had attended with S. Gopal, of External Affairs; Pitambar Pant, of the Planning Commission; journalist Romesh Thapar, of the *Economic and Political Weekly*, and B. G. Verghese, editor of the *Hindustan Times*. This lunch occurred one week before Vikram Sarabhai was announced as chairman of the Atomic Energy Commission. When told this group was drafting a concept paper on the bomb, she reportedly asked for a copy; report on an after-dinner speech by B. G. Verghese given 18 September 2004, retrieved from Rediff.com on 1 May 2008. Whatever the case, Sethna's cost estimate was in circulation and was picked up by American intelligence, as Karnad shows.
10. "Sometimes he struggled all on his own, but at times he managed to get a few young people working with him and helping him. At the same time, he never kept anyone with him all the time; if the youngsters felt like moving on to other things, he allowed them to"; G. Venkataraman, "Dr Ramanna as I Knew Him," *Physics News* 35, nos. 2–4 (April–December 2004): 13.
11. See George Perkovich's observation that the study by N. Seshagiri at TIFR and its relationship to the work of the Consultative Committee on Foreign Affairs and Atomic Energy may have been related parts of a whole, or may not; Perkovich, *India's Nuclear Bomb: The Impact on Global Proliferation* (Berkeley: University of California Press, 1999), pp. 150–51.
12. Chengappa, *Weapons of Peace*, p. 123.
13. With slight differences, Perkovich's account based on interviews concurs with what I have written here, where I rely in part on other interviews and in part on Chengappa's account of his interviews. I think any attempt to sustain a linear chain of action and consequence (particularly without documentary corroboration) would ignore the stochastic nature of the first bomb project, that it started, stopped, went sideways, stopped, started again, and so on. There was continuity of personnel, however, through to the formal decision to prepare for an actual test. For an interesting and slightly different account see Karnad, *Nuclear Weapons and Indian Security*.
14. Chengappa, *Weapons of Peace*, pp. 122–25. See reference to Chengappa's interviews with P. K. Iyengar in May 1998, R. Ramanna in August 1998, and M. Srinivasan

in September 1998. On heavy water, see also Perkovich, *India's Nuclear Bomb,* pp. 200–201.

15. A physicist who worked at the Kalpakkam nuclear site reported how the rocket launch station came to be located at Sriharikota, just inside Andhra Pradesh, north of the Tamil Nadu state boundary. When Sarabhai toured the coast in a helicopter north from Pondicherry, looking for sites, he had seen an expanse of "unused" land just 12 km south of Kalpakkam, south of the Palar River. He asked the state government for the land but they refused it. Disappointed, he flew north of the city of Madras to Pulicat Lake and saw remote semidetached islands barely above sea level. "That was when Sarabhai decided to locate the launch station at Sriharikota"; Venkataraman, "Dr Ramanna as I Knew Him," p. 21.
16. Amrita Shah, *Vikram Sarabhai: A Life* (Delhi: Viking/Penguin, 2007), p. 199.
17. Ibid., p. 205.
18. Ibid., p. 194.
19. He lived separately in a flat in Bombay, his wife and children lived in Ahmedabad, people would drop by his flat but "for all practical purposes it was a lonely life"; his son was in Cambridge, his close friend Kamla Chowdhry was in Ahmedabad, he saw his wife and daughter Mallika in Ahmedabad; Saha, *Vikram Sarabhai,* pp. 173–77.
20. Ashis Nandy, "The Bomb, NPT, and Indian Elite," *Economic and Political Weekly,* 21 August 1972.
21. Abha Dixit, "Status Quo: Maintaining Nuclear Ambiguity" in David Cortright and Amitabh Mattoo, eds., *India and the Bomb: Public Opinion and Nuclear Options* (Notre Dame, IN: Notre Dame University Press, 1996), pp. 60–66.
22. Author's conversation with A. Parthasarathi, Delhi, November 1998.
23. Author's conversation with G. Venkataraman, Puttaparathi, November 1998. See also Chengappa, *Weapons of Peace,* pp. 208–12.
24. G. Venkataraman, "Dr Ramanna as I Knew Him," p. 22. The author explained that the DAE merit promotion system of the 1970s was "heavily weighted in favor of engineers, and the interview standards for engineers "were often quite lenient." He said this was officially explained with reference to the important market differential between what physicists expected to earn and what engineers expected to earn outside DAE. "Naturally this created widespread discontent, and as the leader of the Physics group in Kalpakkam I constantly had to face flak from my juniors on this issue."
25. Author's conversation with A. Parthasarathi, Delhi, November 1998.
26. Chengappa, *Weapons of Peace,* undated interview with "high ranking official," probably in 1998, p. 154.
27. Parthasarathi, *Technology at the Core: Science and Technology with Indira Gandhi* (Delhi: Pearson Longman, 2007), p.108; this is confirmed by T. N. Seshan, his close assistant at the AEC.
28. The books of Perkovitch, Karnad, and Chengappa give a complete account of the debate among political parties and other leaders about whether or not to test a bomb, particularly in 1970 and 1971.
29. R. Ramanna, *Years of Pilgrimage* (Delhi: Viking, 1991), p. 88.
30. My observations concerning Sarabhai stem from four lengthy conversations with him in May, June, and December 1968 and April 1969, combined with opinions of dozens of scientists and engineers, plus what very limited documentary evidence is available.
31. Shah, *Vikram Sarabhai,* pp. 202–6.

CHAPTER TWENTY-TWO

1. No autopsy was performed. One French observer mentioned that his death "raised suspicions in India and elsewhere"; Author's conversation with Bertrand Goldschmidt, Paris, October 1998. Amrita Shah, *Vikram Sarabhai: A Life* (Delhi: Viking/Penguin, 2007), p. 216, confirms that there were "murmurs of foul play." Patrick Blackett wrote a perceptive obituary for Sarabhai in 1972, although Sarabhai was not a Fellow of the Royal Society. Blackett wrote that he counted Sarabhai among his closest friends; Blackett Papers, Royal Society Archives, London.
2. Paul R. Brass, *The Politics of India since Independence* (New York: Cambridge University Press, 1994), p. 40.
3. Author's conversation with A. Rahman, Delhi, November 1998.
4. R. Ramanna, *Years of Pilgrimage* (Delhi: Viking, 1991), p. 121.
5. Ibid., p. 85.
6. Author's conversation with Satish Dhawan, Bangalore, November 1998.
7. Abdul Kalam, *Wings of Fire: An Autobiography* (Hyderbad: Universities Press, 1999), pp. 72–75; also Raj Chengappa, based on interviews with early employees, *Weapons of Peace: The Secret Story of India's Quest to Be a Nuclear Power* (Delhi: HarperCollins, 2000), p. 160. Dhawan denied there was competition for staff between the rocket projects at Thumba and at Hyderabad, but Chengappa heard about it in 1998 from early staff at Hyderabad.
8. Sanjay had already been arrested while a mechanic apprentice in Great Britain for speeding in his Jaguar while driving without license. By 1968–69 he was racing his cars at night in the streets of Delhi. In November 1968 Sanjay, then twenty-two, applied for a license to manufacture a small, fuel-efficient indigenous car like a Volkswagen, along with 14 other applications from established carmakers; all were competing for the same license. A year later he was granted that license, in light of his mother's support for the project. No other license was granted; See Katherine Frank, *Indira: The Life of Indira Nehru Gandhi* (New York: HarperCollins, 2001), p. 201.
9. In contrast, according to one close observer Indira's son and future Prime Minister Rajiv Gandhi "did not subscribe to the philosophy of self-reliance, domestic generation of technology, or domestic production that his mother subscribed to and which was the official policy of her government"; Ashok Parthasarthi, *Technology at the Core: Science and Technology with Indira Gandhi* (Delhi: Longmans, 2007), p. 288.
10. Amulya Reddy, "Choice of Alternate Technologies," *Economic and Political Weekly*, 23 June 1973, p. 1112.
11. Amulya Reddy describes the working condition at the Central Electrochemical Research Institute in Karaikudi in Tamil Nadu in "Reflections of a Maverick," *Seminar* (Delhi), September 1993, p. 19. He also describes his dilemma when he learned that the batteries he developed were being tested for high altitude use by the military at Ladakh. He thought this was about conflict with the Chinese, but I think he knew it was also about possible conflict with Pakistan. Note that the IISc responded to the self-reliance question by establishing, with its Director Satish Dhawan's patronage, an applied research center in which institute scientists, Reddy among them, worked on rural development problems.
12. Reddy, "Choice of Alternate Technologies," p. 1112.
13. Reddy, "Reflections of a Maverick," p. 19.
14. Author's conversation with Amulya Reddy, Bangalore, February 2005.

15. Howard Erdman, "The Industrialists" in H. C. Hart, *Indira Gandhi's India* (Boulder: Westview Press, 1976), p. 137.
16. For an example of the politics of this tension see G. S. Arrora and Ward Morehouse, "Dilemmas of Technical Choice: The Case of the Small Tractor," *Economic and Political Weekly*, 25 August 1972, pp. 1633–44. See also the careful study of Swiss-Indian collaboration and comparison of Calcutta and Madras factories in terms of design and production by K. J. Walker, "Technology Transfer to India: The Case of the Integral Coach Factory," *Development and Change* 18 (1987): 99–127. For an excellent review of innovation literature about India, with discussion of cases like the choice of buses for Chennai in the late 1940s, see Nazir Tyabji, *Industrialization and Innovation: The Indian Experience* (Delhi: Sage, 2000).
17. Personal communication, military historian Stephen Cohen, 31 March 2008.
18. Dennis Childs and Michael Kidron, "India, the USSR, and the MiG Project," *Economic and Political Weekly*, 22 September 1973, p. 1724.
19. *Economic and Political Weekly*, 2 August 1973. The authors knew enough about the nuclear plans of the DAE to state that there was no military nuclear capability in sight: "Although India has kept her nuclear options open by carrying on a civil nuclear programme and a small rocket-launching project, there is no certainty that these could be rapidly transformed into a nuclear strike force"; p. 1723.
20. Childs and Kidron, "India, the USSR, and the MiG Project," pp. 1725, 1727, 1728; for HAL's sales up to 1973 see p. 1724; this is an impressive attempt to assemble data from disparate sources. For total "Arms Supplies to India 1948–1973," see list in *Economic and Political Weekly*, 22 March 1975, pp. 499–502 by country of source, by military service, and type of weapon.
21. Stephen Clarkson, *The Soviet Theory of Development* (Toronto: University of Toronto Press, 1978), pp. 173–77.
22. Note that the supply of Russian kerosene dominated the Indian market from the late 1930s when the only other source of supply was Burma; this Soviet market advantage became decisive in the 1939–45 war, though lines of supply were compromised. India was throughout this period one of the world's largest markets for kerosene.
23. Taya Zinkin, "Steel: The British, American and Russian Contributions," *Challenges in India* (New York: Walker & Co., 1967), pp. 87–140; and see particularly "Innocents Abroad: Russians, Germans, and Britons in India," and "Steel Revisited." She concluded, "To industrialize the way Government of India set about it [in steel] is merely to squander the tax payers' money" (p. 140). Zinkin was a British writer married to a senior Indian civil service officer and had excellent access to relevant people in the steel industry. I am grateful to Paul Brass for reminding me of this astute and well-informed essay.
24. B.M., "Soviet Aid: Where to Draw the Line?" *Economic and Political Weekly*, 1 January 1974, p. 11.
25. "Atomic Energy: Dormant Indo-Soviet Collaboration" *Economic and Political Weekly*, 17 November 1973, pp. 2117–18.
26. As shown in *Negotiating Nuclear Power* the United States continued to refuse to do so after 1974 and onward until responsibility for fuel was handed to France in 1982 and then to China in 1992.
27. Parthasarathi, *Technology at the Core*, pp. 113–15.
28. Ibid., p. 116. This is a decidedly Indian official view of the situation; unfortunately we do not have a comparable Canadian perspective.

29. M. R. Srinivasan, *From Fission to Fusion: The Story of India's Atomic Energy Programme* (Delhi: Viking, 2002), pp. 80–81; Parthasarathi, *Technology at the Core*, pp. 113–16.
30. Parthsarathi, *Technology at the Core*, pp. 54–55.
31. George Perkovich, *India's Nuclear Bomb: The Impact of Global Proliferation* (Berkeley: University of California Press, 1999), p. 201.
32. I thank Atri Mukhopadhyay of SINP Kolkata for providing the information on which the preceding paragraph is based, personal communication, 20 June 2009.
33. "Correspondents: Atomic Energy: Technological Come-down," *Economic and Political Weekly* 8, no. 38 (22 September 1973): 1712.
34. Ibid., pp. 1711–12.
35. Perkovich, *India's Nuclear Bomb*, p. 176.
36. Department of Space, *Annual Report 1973–1974* (Delhi, 1974).
37. Parthasarathi, *Technology at the Core*, p. 148. Finally in 1977 under Prime Minister Desai the Space Commission approved a ground-based microwave system (not a satellite TV system), and the same four companies bid on that as in 1973, and Philco-Ford (renamed Ford Aerospace in 1977) won the bid.
38. R. S. Ganapathy, "Space Research: Un-spelled Objectives," *Economic and Political Weekly*, 8 June 1974, p. 896.
39. Chengappa, *Weapons of Peace*, pp. 160–66.
40. Chengappa, interview with Nagchaudhuri, April 1998, *Weapons of Peace*, various pages; also author's conversation with Nagchaudhuri, November 1998. The budget figure includes the cost of salaries.
41. Parthsarathi, *Technology at the Core*, p. 176.
42. Ibid., p. 173.
43. Chengappa, interview with S. Dhawan, February 1998, *Weapons of Peace*, p. 156.
44. Ibid., pp. 157–58.
45. Ibid., p. 174. Dhawan explained the situation to me in similar terms, saying that converting a rocket to a missile required different software guiding systems and required a kind of intelligence that DRDO did not have in 1974–76. He repeated to me his expression of disinterest in such a 'conversion' project, based on his long personal interest in civil projects rather than military. I have not seen an expression of the DRDO's position on this question. Author's conversation with S. Dhawan, Bangalore, January 1998.
46. Parthsarathi, *Technology at the Core*, p. 170.
47. Cited in C. R. Subramaniam, *India and the Computer* (Delhi: Oxford University Press, 1992), p. 15.
48. Parthasarathi, *Technology at the Core*, p. 77.
49. Shiv Visvanathan, *Organizing for Science*, p. 222.
50. A history of this tension is found in Nasir Tyabji, *Industrialization and Innovation* (Delhi: Sage, 2001).
51. For a fascinating account of how the war over self-reliance was waged around new radar systems in 1972–74, both static-based and troposcatter over-the-horizon radar projects involving US suppliers, Indian military, and electronics agencies, see Parthasarathi, *Technology at the Core*, pp. 156–59.
52. Even then its authors at the NTSC appear to have undermined themselves, when reports appeared that "the Approach Paper on science and technology, prepared by a fairly distinguished group of Indian scientists, has been cited as a straight reproduction in part of British and American documents. . . . Whole paragraphs and sentences

have been reproduced in the Indian document." Later, it was confirmed that this "reproduction" without citation "has been indirectly admitted by the Ministry concerned." This refers to the Ministry of Science and Technology and secondarily to the NCST; "Romesh Thapar, "Plagiarists All," *Economic and Political Weekly*, 7 July 1973, pp. 1174–75, and D. C. Kale, "Technology's Aimless Take-offs," 25 August 1973, p. 1543. Parthsarathi (*Technology at the Core*) does not comment on this curious report and its sources, though he certainly knew about it.

53. Parthasarathi, *Technology at the Core*, pp. 43, 45, 58.
54. Ibid., p. 45.
55. Ibid., pp. 58, 202.
56. Ganapathy, "Space Research: Un-spelled Objectives," p. 896.
57. Atomic Energy was the only portfolio the prime minister held onto after the Home Ministry was given over to Uma Shankar Dikshit and after Information and Broadcasting was assigned to Inder Gujral in early 1974. She created the Department of Electronics and was its minister throughout. When these changes were made the power of her office was paramount. At no time in the previous 25 years did the prime minister's office have such executive authority; it is as if she anticipated a challenge to this unprecedented centralization of executive power. For an explanation from an insider, though without any significant reference to the bomb, see P. N. Dhar, *Indira Gandhi, the "Emergency," and Indian Democracy* (Delhi: Oxford University Press, 2000).
58. V. Siddhartha, "Private Science and Public Policy," in Abdul Rahman, K. D. Sharma, and M. A. Qureshi, eds., *Science, Technology, and Development: Essays in Honour of A. Rahman* (Delhi, Sterling Publishers, 1978), p. 77.

CHAPTER TWENTY-THREE

1. In September 1973 P. N. Haksar's appointment was not extended at the time of his legal retirement, and this was widely seen as unusual, considering how much Gandhi depended on him. But he had recently questioned the Maruti automobile deal involving Gandhi's son Sanjay and was being reprimanded by the prime minister. P. N. Dhar was moved up immediately to replace him in her office. A year later, however, Haksar, rehabilitated through the lobbying of his allies, was appointed deputy chairman of the Planning Commission; this influential position hardly indicates a loss of influence for Haksar.
2. P. K. Iyengar was interviewed by Bhabha and began working at TIFR in 1952; he went to Saclay, France, for reactor research training in 1960. R. Chidambaram completed his PhD in physics at the IISc and started to work at Trombay at age twenty-six in 1962.
3. G. Venkataraman, "Dr Ramanna as I Knew Him," *Physics News* 35, nos. 2–4 (April–December 2004): p. 18.
4. B.M., "Foreign Exchange Leakages," *Economic and Political Weekly*, 2 November 1974, p. 1834. Remember that the government had been leaking evidence of evasion of the new law by foreign firms through "irregular remissions and repatriations" of hard currency from their business in India; see "Alleged Funds Transfer: Some Alien Firms Putting up Defence Lines," *Statesman*, 11 July 1973.
5. "New Delhi Cracks Down on Smugglers," *New York Times*, 22 September 1974.
6. In January 1971 it was estimated that in 1970 about 250–300 tons of gold were smuggled into India, and its minimum value was between $308–370 million at $35 per ounce, then the world price. All this gold was paid for in silver or dollars raised in

India, usually at prices higher than the world price. India was becoming the world's largest market for gold.

7. *Report of the Study on Leakage of Foreign Exchange Through Invoice Manipulation* (Delhi: Ministry of Finance, June 1971).
8. The *New York Times* reported in 1974 a scheme to import scientific apparatus and equipment through small importing companies in relatively autonomous Pondicherry, acting on behalf of companies in Bombay. The import licenses for this were reportedly obtained through bribes; "Inquiry Ordered in Indian Scandal: Mrs Gandhi Acts on Issue of Import Licenses," *New York Times*, 15 September 1974.
9. Abraham Friedman to James Ramey, US Atomic Energy Commission, "Discussion with J.C. Shah" in October 1972; 12 January 1973, Natural Resources Defense Council Archive, Washington, DC.
10. R. Ramanna, *Years of Pilgrimage* (Delhi: Viking, 1991), p. 89.
11. Sethna has asserted his interest was fully supportive of the test and the bomb, but eight years before, "in 1966," he is paraphrased in a US secret telegram as doubting the value of "a small nuclear bomb program," saying that "India cannot just detonate one or two devices and stop. Small nuclear bomb program worse than no program at all because would invite pre-emptive Chinese attack. In terms of Sethna's own figures 150 bombs needed for credible deterrent, operating costs soar. Twenty plutonium fission weapons per year would increase annual operating costs alone to $100 m exclusive of delivery system"; quoted in A. G. Noorani, "The Nuclear Guarantee Episode," *Frontline*, 9 (22 June 2001): 41. Knowing as much as he did about technical capacity in 1966, including the limited stockpile of plutonium, Sethna could hardly have imagined India capable of such a bomb program; presumably by 1972 he decided to settle for one or two devices.
12. Ramanna, *Years of Pilgrimage*, pp. 88–90.
13. Conversation with anonymous expert in Delhi, 13 November 1998; also conversation with B. D. Nagchaudhuri in Kolkata, November 1998.
14. Raj Chengappa provides good evidence on this distance and tension between Ramanna and Sethna; see *Weapons of Peace: The Secret Story of India's Quest to Be a Nuclear Power* (Delhi: HarperCollins, 2000), pp. 208–9 and 211. I have seen no evidence of Sethna's views.
15. Cited in Venkataraman, "Dr Ramanna," p. 19. Venkataraman further says "that Ramanna managed to see that nothing was on record on paper. This way, there was no paper that anybody could have stolen. All the secrets were locked up in the minds of an extremely small circle of trusted people, all of whom acted till the very end as if everything was normal. Some day, that story might be told in detail." See also M. R. Srinivasan, *From Fission to Fusion: The Story of India's Atomic Energy Progamme* (Delhi: Viking, 2002), p. 268: "There was not a single piece of paper that sought approval of the project or expenditures relating thereto."
16. Though unstable at room temperature an isotope of polonium-210 seems to be capable of combining with hot liquids, as seen in its use in a teapot in killing a Russian ex-KGB agent Alexander Litvinenko in London November–December 2006.
17. Venkataraman, "Dr Ramanna," p. 19.
18. *Economic and Political Weekly*, 8 June 1974, p. 899.
19. One kiloton equals 1000 ton TNT explosive energy.
20. George Perkovich, *India's Nuclear Bomb: The Impact on Global Proliferation* (Berkeley: University of California Press, 1999), interviews in 1996, pp. 182 ff., including reference to *Hindu*, 20 June 1978, and *Sunday Observer*, 30 August 1981.

21. Perkovich, *India's Nuclear Bomb,* p. 183.
22. E. S. Husbeye and A. M. Dainty, *Monitoring a Comprehensive Test Ban Treaty* (Dordrecht: Kluwer, 1995), pp. 236–38.
23. Robert Norris et al., *British, French and Chinese Nuclear Weapons* (Boulder, CO: Westview Press for Natural Resources Defence Council, 1994), pp. 324–56.
24. Ramanna, *Years of Pilgrimage,* p. 93.
25. Jain, *Nuclear India,* 2: 159.
26. J. P. Jain, *Nuclear India,* 1 (2 vols.; Delhi: Radiant Publishers, 1974): 139.
27. Perkovich interview with Sethna, February 1997, Perkovich, *India's Nuclear Bomb,* p. 522.
28. Perkovich, *India's Nuclear Bomb,* p. 181, citing N. Seshagiri.
29. Aqueil Ahmad, "India's Nuclear Age: Dubious Cost-Benefit," *Economic and Political Weekly,* 8 March 1975.
30. "Cultivating Our Egos with Delight and Terror," *Economic and Political Weekly,* 25 May 1974, p. 808. For the record, the United States did not make an "outraged" statement in the days after the test, and Kissinger avoided doing so, for reasons explained in *Negotiating Nuclear Power.*
31. Ayesha Siddiqi-Agha, *Pakistan's Arms Procurement and Military Buildup, 1979–1999* (London: Palgrave, 2001), chap. 6, "Military Industrial Complex."
32. "50p.c. Cut in Foreign Know-how: Scientists Face Big Challenge," *Amrita Bazar Patrika,* 11 August 1968.
33. Ashok Parthasarthi, *Technology at the Core: Science and Technology with Indira Gandhi* (Delhi: Longmans, 2007), p. 210.
34. Ibid., pp. 60–62.
35. Author's conversation with B. D. Nagchaudhuri, Kolkata, November 1998.
36. "Research Establishments: Unproductive Empires," *Economic and Political Weekly,* 27 July 1974.
37. C. R. Subramaniam, *India and the Computer* (Delhi: Oxford University Press, 1992), p. 25.
38. The 1975 split between Ramanna and Sethna "caused a major rift between scientists and engineers at BARC that deeply affected the ethos of the organization"; leader M. R. Srinivasan is quoted as saying that this conflict at the top led to "conflicting instructions emanating and priorities being set differently"; Chengappa, *Weapons of Peace,* p. 212. This tension is well described in Chengappa's interviews, pp. 160–220, and it is confirmed by Perkovich, *India's Nuclear Bomb,* and Srinivasan, *From Fission to Fusion.*
39. A relatively unrecognized figure, Additional Director S. Fareeduddin, became BARC's acting director in 1978, when Ramanna left, and was there until Ramanna returned, keeping the position "open." He was not promoted to director probably because he lacked a PhD, though he was a recognized expert on heavy water and head of the important heavy water project division.
40. We have no account of the reasoning behind acceptance of the Soviet offer for a launch, as opposed to a French offer—a country with which India had many years of space cooperation.
41. Chengappa, *Weapons of Peace,* interview with Air Chief Marshall Idris Latif (retd.) September 1998, p. 167.
42. Chengappa, *Weapons of Peace,* interview with M. G. K. Menon, January 1998, p. 214.
43. Abdul Kalam with Arun Tiwari, *Wings of Fire: An Autobiography* (Hyderabad: Universities Press, 2002), p. 61.

44. Parthsarathi, *Technology at the Core,* p. 101.
45. Dinshaw Mistry, "India's Emerging Space Program," *Pacific Affairs* 71 (Summer 1998): 162.
46. Gopal Raj, *Reach for the Stars: The Evolution of India's Rocket Programme* (Delhi: Viking, 2000), pp. 278–80; see also R. Ramachandran, "Delayed Thrust," *Frontline,* 6 April 2007, reporting the recollection of key persons in the liquid-solid fuel tension in the 1970s, suggesting that the war over self-reliance found expression there too and that Dhawan was opting for an easier (or less slow) solution using the French Viking rocket engine with a solid fuel engine. "We lacked an engine with greater thrust" said one rocket expert.
47. Parthasarathi, *Technology at the Core,* pp. 179–80.
48. Srinivasan climbed the ladder of power steadily, and near the end of his career was appointed chairman of the AEC between 1987 and 1990. See his own account of his career in *From Fission to Fusion.*
49. Chengappa, *Weapons of Peace,* interview with Homi Sethna, August 1998, p. 207.
50. Perkovich, *India's Nuclear Bomb,* pp. 188–192, based on a conversation with "a former high-ranking official."
51. Srinvasan, *From Fission to Fusion,* p. 120. One theoretical physicist who did not wish to be named, noting the uncertainty about the bomb test's yield and the official efforts to conceal its full costs, said drily in 1998, "it really was like an expensive fire cracker." For another insider's view, see P. K. Iyengar, *Briefings on Nuclear Technology in India* (Delhi: Rupa, 2009), pp 40–45. Accessed on Iyengar's personal website, 25 October 2009.
52. Parthsarathi, *Technology at the Core,* p. 118.

CHAPTER TWENTY-FOUR

1. There is an apparent parallel between Prime Minister Trudeau's apprehension of insurrection in Quebec in 1970 and sudden declaration of the War Measures Act in October 1970 and Prime Minister Gandhi's same apprehension five years later and the declaration of the 1975 Emergency. Trudeau and Gandhi split after the 1974 bomb test but were undoubtedly attracted to one another when they were new to power in the late 1960s.
2. Union leader, and eventual cabinet minister, George Fernandes declared a strike of India's 1.4 million railway workers and was arrested on 2 May 1974. On 7 May the strike began, and 20,000 striking workers were arrested using the Maintenance of Internal Security Act. The strike had been suppressed by about 15–16 May, just prior to the first nuclear bomb test. Fernandes criticized the first bomb test from prison.
3. Paul Brass, *The Production of Hindu-Muslim Violence* (Seattle: University of Washington Press, 2003). The MISA regularized "preventive detention of suspects."
4. Two months before this speech, a close observer of political change recalled that Narayanan's speech to a big rally at Patna in November 1974 "was a clear indicator that [Narayan's] movement was no longer confined to Bihar but had the potential to change the political agenda of the nation"; L. K. Advani, *My Country, My Life* (Delhi: Rupa, 2008), p. 189.
5. For a lucid account of the prison experience during the Emergency by someone who soon sat in the cabinet, see Advani, *My Country, My Life,* p. 190.
6. Some evidence confirming her suspicions has surfaced through the conversations of writer Lawrence Lifschultz with the US ambassador in Dacca in 1975, Eugene Booster; Booster sensed the involvement of CIA staff in the embassy in a plot and

gave instructions that this involvement was to cease. But, said Lifschultz following Booster, it did not and Mujib's entire family was murdered. Gandhi's suspicions, however, would also have had other sources, such as Soviet intelligence in Dacca coming through the CPI, and this story is told here simply to substantiate her perception of a widening conspiracy against her. Lawrence Lipschultz, interview, CBC Radio, Friday 8 June 2007 (*As It Happens*).

7. James Manor, ed., *Nehru to the Nineties: The Changing Office of the Prime Minister of India* (London: Hurst & Co., 1999); Paul Brass, *The Politics of India since Independence* (New York: Cambridge University Press, 1994).
8. There is no published account of the impact of the first bomb test in the social life of the scientific community: what chilly distance and new alliances emerged between enthusiasts and skeptics?
9. For example, "in the mid-1970s, individuals like P. N. Haksar wielded more influence on foreign policy-making than virtually any other person in the government"; Sumit Ganguly, "The Prime Minister and Foreign and Defence Policies," in Manor, *Nehru to the Nineties*, p. 153.
10. The 1975 judgment against her reelection followed an earlier confrontation with the highest courts. In 1973 April India's Supreme Court issued its judgment on Gandhi's 1969 constitutional amendments—saying that though Parliament could amend the Constitution without altering its "essential features," all changes would be subject to judicial review. She was reportedly angry about this. On the next day the chief justice coincidentally had to retire, and Indira passed over three of the most senior judges to promote A. N. Ray as chief justice, and all the three superceded judges promptly resigned. Minister Kumaramangalam defended the move in Lok Sabha, but the judiciary was on guard against her. Backed by her ministers, she began to talk about "a committed judiciary," meaning committed to *her objectives*.
11. The process she followed resembled a military coup; according to interviews with direct observers, intelligence agents reported to Gandhi that they tracked Moraji Desai saying "they" would overthrow Indira and J. P. Narayan saying he planned to ask the Army to turn against the government; Katherine Frank, *Indira: The Life of Indira Nehru Gandhi* (New York: HarperCollins, 2001), pp. 370–80. Whether true or not, this appears to have convinced her of the risk. Her son Sanjay, Bansi Lal (chief minister of Haryana) and Om Mehta (second-in-command at Home Ministry) had already drawn up a list of leaders to be arrested. Indira went to President Ali Fakruddin Ahmed late at night on 25 June 1975 to persuade him to issue a state of emergency, and then she decided *not to consult* her cabinet before the arrests of J. P. Narayan and other opponents at 5:00 a.m. Immediately after those arrests she called the cabinet and asked them to come together at 6:00 a.m. An Emergency Order promulgated in 1971 for the Bangladesh conflict happened to be still in effect but was explicitly intended for external emergencies. So she and others promptly wrote an Emergency Order Proclamation, showed it to cabinet, and it was approved in 30 minutes. Her son Sanjay and friends had already arranged that no electricity could reach newspapers' printing presses overnight so newspaper production was stopped, except for two, the *Statesman* and *Hindustan Times*, that had separate electricity generators of their own. The government had full control of radio and television.
12. Frank, *Indira*, p. 376.
13. During 1975 the Soviet embassy's office of the KGB spent 10.6 million rubles "on active measures in India to strengthen support for Mrs Gandhi and undermine her political opponents," including payments to journalists to write stories in her favor.

This investment reflects accurately the closer trade and strategic ties in place between the two countries; after all, the Indian satellite had just been launched on a Soviet rocket from a Russian space center. But things change: Andrew and Mitrokhin, having read copies of official Soviet files, report that in early 1977 the Soviet embassy and KGB believed, as Gandhi herself did, in their own narratives, expected that she would be reelected, and took steps to supply funds to some Congress candidates and to provide over 3 million rubles to the Communist Party of India for the unsuccessful election campaign in which they too were defeated. See Christopher Andrew and Vasili Mitrokhin, *The World Was Going Our Way: The KGB and the Battle for the Third World* (New York: Basic Books, 2005), pp. 328–29. It is likely than an American agency with a similar role was engaged in similar efforts to bring about an outcome favorable to Indira Gandhi's opponents.

14. Saeed Naqvi, "Mrs Gandhi, it is being said . . ." *Sunday Times* (Delhi), 13 July 1975. She dismissed the remaining non-Congress governments in Gujarat and Tamil Nadu, imposing President's Rule there too. But the key move was the passage of the 39th constitutional amendment, which actually nullified the Allahabad judgment against Indira Gandhi, giving the prime minister immunity from civil and criminal prosecution for offenses committed before or during her term in office.
15. Author's conversation with Padma Prakash, Mumbai, November 1998, and personal communication, 4 May 2008; see Padma Prakash, "Science Reporting in Newspapers: Content Analysis of Two Newspapers in Two Time Periods" (MA thesis in Sociology, University of Bombay, 1978). In the study, two newspapers were selected for analysis: the *Times of India* and *Prajavani* (published in the Kannada language at Bangalore) for two periods, January–March 1975 and August–December 1975. The Kannada paper published in Mysore (Karnataka) was selected to see if science was more or less reported in the regional language papers than in a major English-language paper. The findings were that science features related to agriculture generally got more prominence in the Kannada newspapers over all other science news. The prestigious English paper *The Times of India* covered science only when there was an event of some international importance, for example, space-related developments, Nobel prize-related news, industry-related science breakthroughs, and the like. Agro-related science news was substantial, explaining and interpreting science for the common reader. But comparing the two time periods, science coverage increased substantially (roughly 40–50 percent) during the post-June Emergency months compared with the pre-Emergency period. "Our original vague hypothesis," said Prakash, "was that science news would be much less, like all other news, considering that the Emergency imposed censorship. These two periods were first chosen arbitrarily but when the samples showed such a remarkable and surprising difference in pattern, for both newspapers, we looked more closely and realized that the Emergency itself might be a factor in the patterns seen; personal communication, Padma Prakash, 4 May 2008.
16. Frank, *Indira*, pp. 378–79.
17. *Economic and Political Weekly*, 18 October 1975.
18. Despite the government's financial crisis, in early 1975 India considered the cost of a response to news of a France-Pakistan Mirage jet fighter construction project, one that would lower the cost of Mirage jets to France and make the Pakistan Air Force almost equal in capability to India's. Note that Iraq was involved in a similar arrangement with France at this time. Impatient with the slow progress of the Hindustan Aeronautics MiG-Sukhoi project, the cabinet authorized a study for the purchase of

both American F 104s *and* French Mirage fighters. This is another instance of the war over self-reliance.

19. In January 1975 the Steel Authority of India, Ltd., had given preference to the renovation ("modernization") and reengineering of projects at Bhilai and Bokaro steel mills. Soviet funding was available for this renovation because these were originally Soviet-built mills. But the renovation agreements were linked to services supplied by Gipromez in Moscow, and the engineering subsidiary of Steel Authority of India, Mecon, had to work under (and pay for) the direction provided by Gipromez. All of this was frustrating to those committed to the development of indigenous capacity. This was another instance of the war over self-reliance.
20. C. Subramaniam had come to American attention ten years earlier during his handling of the 1965 food crisis, when President Johnson was directly involved in agricultural development programs and said he was impressed with how Subramaniam, then minister of Agriculture, handled that crisis and initiated the green revolution at the political level. He had previously been minister of steel as well as mines and heavy engineering, and crucially he was president of Congress during Gandhi's bank nationalization and princely purse policy in mid-1969. He stayed with Gandhi during the Emergency but withdrew from her party in 1977, to reemerge briefly in 1979 as the short-lived Janata government's minister of Defence.
21. B.M., "Delusion of Self-Reliance," *Economic and Political Weekly*, 25 September 1975.
22. Indira Gandhi, speech to the Indian Science Congress, Hyderabad, 3 January 1976.
23. In Gandhi's move to involve communists in government, she had reached out to Mohan Kumaramangalam, a prominent lawyer and member of the Communist Party until he broke with them in 1966. According to Frank, she first met Kumaramangalam in London in 1938, after he completed his Cambridge degree, in the company of her family friend P. N. Haksar, who studied anthropology at the London School of Economics with Malinowski; they were both studying law when the war broke out. Kumaramangalam was most impressive: I can still remember the stir created by his presence among students as a visiting speaker at Madras Christian College in November 1961.
24. See the judgment of the Canadian government's own 1966 study of nuclear reactors and Indian coal sources for electricity in *Negotiating Nuclear Power.*
25. "Power Plants: Back to Imports?" *Economic and Political Weekly*, 20 September 1975.
26. There were growing demands for increased private fertilizer production, but the planning and development division of the Fertilizer Corporation of India remained one of the main concentrations of expertise and competence. Factories at Durgapur and Cochin, built by the FCI during the 1960s, took longer to build than planned and had not reached their full production capacity even by 1970. Committed to the fertilizer-responsive varieties the government thus gave precedence to turnkey projects in the control of Toyo Corporation in 1971, using Japanese loans. The main promise of these projects was the speed with which they would be completed by Japanese engineers. When the OPEC crisis hit Japan in 1973, disbursements slowed down markedly for two years, and in the mid-70s government's attention was again turned to the older FCI projects: why did they not perform better? An end-to-end survey of the Durgapur fertilizer factory was conducted in 1974–75 by Technimont of Italy, printed in 14 volumes, and called for improvements worth Rs 70 million, of which Rs 23 million would be in foreign exchange. Yet Indian experts outside FCI had constantly raised doubts about the Durgapur plant and pointed to the large sums required to improve it. Other new plants were now being mentioned—Phulpur

with a quick $109 million World Bank loan, Paradip with French assistance (eventually dropped for lack of Indian contributions), and Panipat and Bhatinda. All these plants required large volumes of electricity. Now that the off-shore oil had been discovered, a contest between coal-based and petroleum-based fertilizers raged; an observer said, "the much vaunted priority for coal-based fertilizers has again lost ground." Was there a struggle between FCI, advocating coal-based systems, and M. S. Pathak, petroleum engineer, member of the Planning Commission, and chairman of the state-owned consulting firm Engineers India, Ltd, advocating oil-based fertilizer? See B.M., "Vendetta against Indigenous Technology?" *Economic and Political Weekly*, 5 April 1975, pp. 571–72.

27. M. R. Srinivasan, *From Fission to Fusion: The Story of India's Atomic Energy Progamme* (Delhi: Viking, 2002), pp. 62–68.
28. George Perkovich, *India's Nuclear Bomb: The Impact on Global Proliferation* (Berkeley: University of California Press, 1999), pp. 192 and 525.
29. R. Ramanna, *Years of Pilgrimage* (Delhi: Viking, 1991), p. 95. Unfortunately we do not yet have a study of the technical, organizational, and intellectual responses of the private industrial sector to this "heaven-sent opportunity."
30. Abha Dixit, "Status Quo: Maintaining Nuclear Ambiguity" in D. Cortright and A. Mattoo, eds., *India and the Bomb: Public Opinion and Nuclear Options* (Notre Dame: Indiana, University of Notre Dame Press, 1996), p. 61.
31. The success of the machine's first operation in 1977 owes much to this gesture of electrical supply, a courtesy of then Chief Minister of West Bengal Jyoti Basu. Atri Mukhopadhyay, personal communication, 20 June 2009. The irony of a poor state government supplying electricity to a DAE facility of the central government cannot be missed.
32. Raj Chengappa, *Weapons of Peace: The Secret Story of India's Quest to Be a Nuclear Power* (Delhi: HarperCollins, 2000), interview with Dhawan, February 1998, pp. 157–58.
33. Ramanna, *Years of Pilgrimage*, p. 112.
34. Ibid., pp. 101–8; when Ramanna returned to Trombay in 1981 he helped restart the Dhruva reactor, and it went critical in 1983.
35. Ibid., p. 110.
36. M. V. Ramanna, "India's Nuclear Enclave and the Practice of Secrecy," in Itty Abraham, ed., *South Asian Cultures of the Bomb: Atomic Publics and the State in India and Pakistan* (Bloomington: Indiana University Press, 2009), p. 50.
37. Because of the greater energy required to manufacture heavy water and because of the availability of hydrogen (essential to D_20), it is often made along with fertilizer. The first North American heavy water plant, built for the Manhattan Project at Warfield near Trail, British Columbia, was constructed in 1942 within a fertilizer plant. Heavy water was also produced in Norway at the same time.
38. "Nuclear Power—Heavy Water Constraint," *Economic and Political Weekly*, 29 August 1974. It is curious indeed that we have no socioeconomic history of the heavy water plants, which must surely have been among the most costly and closely watched industrial projects in India in the 1970s. Perkovich deals with it, though with a minimum of evidence, and Abraham and Chengappa mention it only in passing. I thank writer and journalist G. Balachandran for a conversation about heavy water, in February 2005 in Delhi. Parthasarathi had already made an analysis in 1971 showing that the DAE calculations were 50 percent too low on the cost of heavy water for new reactors, as shown in Ashok Parthasarthi, *Technology at the Core: Science and Technology with Indira Gandhi* (Delhi: Longmans, 2007), pp. 122–24.

39. Parthasarthi, *Technology at the Core*, p. 288.
40. Ramanna, *Years of Pilgrimage*, p. 95.
41. Author's conversation with two anonymous Russian nuclear historians, Moscow and St. Petersburg, June 2001. Attempts to confirm these earlier shipments of heavy water have been unsuccessful; I speculate that these Russian scholars may have concluded that because the early heavy water negotiations occurred in 1973 they thought the shipments occurred that year. Nevertheless, it is possible they did occur because they were small enough not to trigger IAEA surveillance. Shipments for the Rajasthan CANDU reactors were to IAEA safeguarded sites and did not require permits or an intermediary broker. But Soviet heavy water for the unsafeguarded Kalpakkam reactor required this West German intermediary, so it was done that way. Author's conversation with G. Balachandran, Delhi, February 2005.
42. The post–bomb-test sanctions were the cause of delays but also inspired (or compelled) indigenous self-reliant innovation. A good example is zirconium alloy fuel tubes for the two Tarapur reactors. India had zirconia sands on the beaches of Kerala and Tamil Nadu that were good for zirconium alloys. But zirconium was hard to work with and required special furnaces to prevent oxidation. The fashioning of these fuel tubes produced a lot of waste because a fraction of tubes did not conform to the exact standard and so had to be discarded. The expense was great and the unusable tubes were hard to recycle. New workers at BARC took time to acquire the relevant discipline for "unglamorous work in the laboratories and on the shop floor." Then in 1978 the US embargoed zircalloy channel squares to be used with the enriched uranium fuel for Tarapur. "For the next couple of years we were able to import them from Japan, but then Japan sometimes required US clearance to export such items to India. The test applied was that Japan could export items only if its manufacture had been developed in-country [i.e., Japan]. On the other hand, if the process or material initially had come from the US then it was likely to attract embargo provisions." After that Indians made the zircalloy channel squares on their own. Srinivasan, *From Fission to Fusion*, pp. 239–40.
43. *The Economist* put her loss of the election on its front page, using a photo of her torn election poster, entitling it "India tears up its past," 26 March 1977.
44. Ramanna, *Years of Pilgrimage*, p. 94.
45. Perkovich, *India's Nuclear Bomb*, p. 204.
46. Shiv Visvanathan, *A Carnival for Science: Essays on Science, Technology, and Development* (Delhi: Oxford University Press, 1997), p. 5.
47. Atma Ram, "Some Thoughts on Science," *Science Today* (Bombay), February 1978, p. 18. Ram is referring to the role of Haksar as deputy chief of the Planning Commission and head of the NCST until Gandhi's defeat.
48. Raja Ramanna, "On Nurturing Scientific Excellence," *Science Today* (Bombay), January 1978. This was Ramanna's first presidential address to the Indian National Academy of Sciences.
49. K. Subrahmanyam, "Scientists as Friends and Foes" *Chandigarh Tribune*, 2 October 2004. His report relies on the statements of Nirmal Mukherji, cabinet secretary, made at the time when K. Subramanyam was the chairman of the Joint Intelligence Committee.
50. Ibid. The position was held by an acting director, chemical engineer S. Fareeduddin, who set up the DAE's uranium extraction plant in Jaduguda Bihar and became head of the heavy water production group and the Heavy Water Projects Board. He later worked in Vienna for the IAEA.

51. Dhirendra Sharma, *India's Nuclear Estate* (Delhi: Lancers Publishing, 1983); see also "DAE—Organizational Reform," *Hindustan Times*, 20 February 1984. A pioneer in the critical assessment of nuclear India, Sharma also wrote a later essay about Nehru and Bhabha, "India's Lopsided Science," *Bulletin of the Atomic Scientists* 47, no. 4 (May 1991):
52. India would benefit from a long-term study of the socioeconomic history its electrification, perhaps to be inspired by the classic work by Thomas P. Hughes, *Networks of Power: Electrification in Western Society* (Baltimore: Johns Hopkins University Press, 1983).
53. Srinivasan, *From Fission to Fusion*, pp. viii, 59, 108, 285, 310–11.

CHAPTER TWENTY-FIVE

1. Ramanujan was a poor clerk in the Madras Port Trust in 1912 but was known to local mathematicians. Two years later he was brought by G. H. Hardy to Cambridge as a recognized genius, and four years later, at age thirty-one, he was elected a FRS. Within two years he had died of tuberculosis. See Ashis Nandy, *Alternative Sciences: Creativity and Authenticity in Two Indian Scientists* (Delhi: Allied Publishers, 1980), and Robert Kanigel, *The Man Who Knew Infinity* (New York: Simon and Schuster, 1991).
2. Lewis Pyenson, "Cultural Imperialism and Exact Sciences Revisited," *Isis* 84 (1993): 103–8. He wrote, "physics and astronomy in early twentieth-century India seem to follow the pattern in organic chemistry [probably meaning P. C. Ray and his students], showing local initiative, local funding, and intentional isolation of pure research from practical work" (p. 104). See also Zaheer Baber, "Colonizing Nature: Scientific Knowledge, Colonial Power and Incorporation of India into the Modern World-System," *British Journal of Sociology* 52 (March 2001): 37–58.
3. Father Lafont found a perfect institution-building combination in his partnership with Mahendralal Sircar, and together they successfully pursued private financing for the IACS lab, having received but lukewarm support from the viceroy. The role of Alexander Pedlar, professor of chemistry at the nearby Presidency College completed the "culturally balanced and religiously mixed look" of the enterprise. There are numerous writings about Lafont, who monitored meteorological balloon launches and used an Italian connection to develop the astronomical observatory. See A. K. Biswas, *Collected Works of Mahendralal Sircar, Eugene Lafont, and the Science Movement* (Kolkata: Asiatic Society, 2003).
4. Richard Riesz taught at American College Madurai in Tamil Nadu from 1962 until 1990, setting up the graduate program in physics. Before Madurai, he was a successful researcher at the Bell Laboratories in New Jersey for many years and had several patents to his credit when he left for India. According to Srinivasan, Riesz was paid a very modest "Indian salary," although it came directly from his American Mission, which was at the time responsible for the college (it is now autonomous under a trust). "This financial independence enabled him to do many bold things in the physics department; had he been employed by the college management they would not have allowed any of their professors to do such things! Riesz and his family went to the US every 5 years for a period of one year and during the very first home leave, he went back to Bell Labs to work. During later visits, he taught in well-known American colleges. Although it is not recognized, he had a large impact on the Indian physics scene. His best physics students, mostly originating from the neighborhood of Madurai,

became much sought after, not just in India but across the world. Riesz learned Tamil and later established a computer training and processing center operated by disabled young people." G. Srinivasan, personal communication, 10 June 2009.

5. Government of India, "Scientific Advisory Council to the Prime Minister," 27 January 2005; this list of 28 includes both members and "special invitees" who were scientists appointed as secretaries to the government, e.g., R. A. Mashelkar, FRS, who was also director general of the CSIR. I am grateful to a member of the council for showing me a copy of the list.

6. There were, of course, sharp discontinuities; for example, Saha worked alone in Allahabad through the 1930s, and Saha himself was unsatisfied with the work he did there. Arguably good physics was not done there again until the 1980s. But by and large the progress of institutional development was iterative and cumulative in the opinions of close observers of the Indian situation, and conditions for researchers in 1980 were far better than in 1950.

7. For example, Saha, who largely ignored music and frowned about the cinema, was an avid reader of poetry and novels. Like many Bengalis he wrote poetry when young. He learned German and Sanskrit early in life and won prizes for languages. He expected his children to read, and so they often read more widely than even he had: the story is told that he rebuked his eldest son Ajit for reading "a light-looking novel," whereupon his son, who was then studying physics with Max Born, lectured him about the book and its author Thomas Mann; "father quietly and sheepishly listened to his son's lecture about *The Magic Mountain*. So I suppose he was capable of taking in new ideas outside science." In 1946 Saha returned from London and presented Shaw's play *Major Barbara* to his teenage daughter: "I had hardly finished reading it when he summoned me, asking if I could take that as a model and write something in Bengali on the numerous religious and philanthropic institutions that had mushroomed all around? A tall order for a 16 year old"; personal communication, Chitra Saha, 16 June 1999.

8. Advertisement for Calcutta Electric Supply Corporation, *The Statesman*, 27 January 2005. Meghnad Saha is not included in the list as "unforgettable," though Calcutta has named a big street after him.

9. D. Chattopadhyaya, "Defending the Scientific Temper," in Geeti Sen, ed., *The Calcutta Psyche* (Calcutta: Rupa and Co., 1991), pp. 43–57.

10. A.M., "Calcutta Diary," *Economic and Political Weekly*, February 1974 (annual issue), pp. 169–70.

11. Parameswaran decided to resign from BARC in 1975 and return to Kerala to work full time with KSSP while earning his livelihood as manager of Chinta Publishers of Trivandrum, "owned by the Communist Party of India (Marxist)"; Mathew Zachariah and R. Sooryamoorthy, *Science for Social Revolution? Achievement and Dilemmas of a Development Movement: The Kerala Sastra Satiya Parishad* (Delhi: Vistaar/Sage, 1994), pp. 60–64.

12. Dhruv Raina, *Images and Contexts: The Historiography of Science and Modernity in India* (Delhi: Oxford University Press, 2003), p. 41.

13. By 2000 Sadgopal was dean of elementary and social education at the University of Delhi. For a sensitive semi-biographical review of Sadgopal's work, see Vasantha Surya's review of Sadgopal's 2000 book *The Question of Change in Education: From Social Experience to Policy*, in *Frontline* (Chennai), 5–18 August 2000.

14. The declining meaning of the family metaphor ran in parallel with trends of change in family structures among the middle class to which most scientists referred—they

were accepting that to live together as a joint family was more difficult in dense urban situations, having fewer children, accepting more marriages of choice, accepting university education for young women and completion of degrees even after marriage.

15. Prime Minister Rajiv Gandhi adopted this strategy in 1986 responding to a number of influences, among them a physicist, Yash Pal, formerly in the cosmic ray group at TIFR.

16. A heated debate occurred among a group of young Indian scientists in 1969 when I showed them S. Bhagavantam, "Miracle of Longodbhava: Mahasivaratri at Presanti Nilayam," *Bhavan's Journal*, 18 May 1969. *Bhavan's Journal* was a Sanskritic and Hindu-oriented mass circulation magazine of the Bharatiya Vidhya Bhavan, advertising ayurvedic medicines, astrological advisors, and the like, and sold at bus stops and railway stations. Careful to feature articles by or about various kinds of politicians (e.g., socialist Jayaprakash Narayanan, then acceptable to the Congress Party though he was arrested in 1975), the journal stood solidly for old Congress and Gandhi. Bhagavantam, whose career is described in Biographical Notes, was one of the very few senior scientists to publish an essay in this or any similar magazine, and the debate was focused not on whether the top defense-advisor-physicist should be present among the thousands of pilgrims to the ashram of Sai Baba at Puttaparathi in South India, nor on whether he should agree to speak to the huge crowd, which he did in Telegu and English, nor on whether he should be invited into the inner room after Sai Baba performed "miracles" like producing jewels and pearls out of thin air. Though the depth of Bhagavantam's involvement in that ashram and movement was surprising to this group of secular young scientists, and definitely raised their eyebrows, what concerned these young physicists *most* in 1969 was that there was no effort made to explain in physical terms what was happening, no trace of skepticism, no idea of the physics that might underlie these events. Based on the evening at the ashram, the essay was, in their opinions, an adoration of "Him," as Bhagavantam described Sai Baba, as a god-person. The young scientists said it was irresponsible of physicist Bhagavantam to reinforce the superstitions all around him in 1969 when he had just called for "a bit of science in politics" in 1968; see the *Statesman* (Calcutta), 4 September 1968. Suri Bhagavantam was a physicist named by Krishna Menon in 1961 to be scientific advisor to him, as the minister of Defence. He distinguished himself in 1968 by giving a lecture in Bangalore in which he said that all the techniques of decision-making should be applied to the question of the bomb, and "all sorts of cost-effectiveness studies" should be conducted before making a decision. "A Bit of Science in Politics Urged," *Statesman* (Calcutta), 4 September 1968. This was about the time that he reported on his experience of miraculous occurrences in the presence of a swami at Puttaparathi."

17. "A Statement of Scientific Temper," Nehru Centre, Bombay, July 1981 (reproduced by the State of Kerala Committee on Science, Technology, and Environment, Trivandrum, May 1989: copy in the University of Cambridge Library). This was first published as "A Statement on Scientific Temper" in the journal *Mainstream*, 25 July 1981; see also the "Counter-statement on Humanistic Temper" by Ashis Nandy, *Mainstream*, 10 October 1981.

18. Despite their close working relations there is no public trace of an Indian-Soviet dialogue on the spiritual meaning of their respective space programs, though experts in the Soviet program were quite articulate about it, and Indian experts were surrounded by peoples to whom, like the Soviets, there was spiritual meaning in the stars and space.

19. It appears that seven months passed after the meeting in order to obtain signatures of people not present in Coonoor for the meetings.
20. Shiv Visvanathan, *A Carnival for Science* (Delhi: Oxford University Press, 1997), p. 4.
21. Ashis Nandy, "The Political Culture of the Indian State," *Daedalus* 118 (Fall 1989): 10 (italics mine).
22. We do not have a broad historical survey of the roles of women in the wider scientific community, though we have exceptional studies like Abha Sur, "Dispersed Radiance: Women Scientists in C. V. Raman's Laboratory," *Meridians: Feminism, Race, Transnationalism* 1, no. 2 (2001): 95–127. An explanation always given by women career scientists in the 1960s and 1970s for their subdued role in Indian institutions was "that's just how things are." Subrata Dasgupta has written about changes to this subdued role of women scientists in the 1960s and 1970s in his experimental novel *Three Times a Minority* (Kolkata: Writers Workshop, 2003).
23. The different approaches to Nehru's relationship with women are striking: Ben Zachariah scarcely deals with it, whereas Stanley Wolpert credits the relationship with Edwina Mountbatten with real influence; see also Judith Brown, *Nehru: A Political Life* (New Haven: Yale University Press, 2003); and Richard Hough, *Edwina, Countess Mountbatten of Burma* (London: Weidenfeld and Nicolson, 1983).
24. Chitra Saha, personal communication, 16 June 1999.
25. Little has been written about Bhatnagar since 1948 with the depth and feeling of Richard's book about him; it was reprinted for a receptive public, with valuable new photographs, in 2004.
26. One is reminded of the separate chemistry and biology laboratory for women only, standing in beautiful seclusion in the Darwin College gardens at the University of Cambridge. It was in use by female students until well after the WW II.
27. There were important exceptions: see Edward Shils and Carmen Blacker, *Cambridge Women* (Cambridge University Press, 1996).
28. For reports on Bhabha's attractiveness to women in Europe in the 1930s, see Itty Abraham, *The Making of the Indian Atomic Bomb: Science, Secrecy, and the Post-Colonial State* (London: Zed Press, 1998).
29. I thank S. S. Jha, director of TIFR, for enabling me to see Bhabha's office in 1998. Twenty-two years after his death it was still the office that he designed himself in a kind of Bahaus-Indian modernist style and where he had worked, among his personal mementos. I also thank B. Choksi for long thoughtful conversations about her friend Homi Bhabha, particularly in November 1998.
30. Author's conversation with J. J. Bhabha in Mumbai, November 1998; this phrase was used by Homi's brother with a smile; I cannot say if he used it in the late twentieth-century meaning, as Mr. Bhabha appeared not to hear my request for a clarification. Homi Bhabha's night out dancing in Delhi with Jackie Kennedy in 1961 is legendary. Two women who knew Bhabha have mentioned to me that though he was definitely widely attractive to women, these two individuals sometimes felt a sexual ambiguity when with him. He was certainly very well grounded in his relationship with Pipsi Wadia.
31. Amrita Shah, personal communication, 2 November 2006.
32. Amrita Shah, *Vikram Sarabhai: A Life* (Delhi: Viking/Penguin, 2007), pp. 31, 32, 70.
33. Author's conversation with Mrinalini Sarabhai, Ahmedabad, February 1998.
34. Author' conversation with Kamla Chowdhry in Ahmedabad, February 1968; Chowdhry edited two books of Sarabhai's essays published after his death, though she was working on those books before December 1971; see Kamla Chowdhry, ed., *Manage-*

ment for Development (Delhi: Vikas, 1974) and *Science Policy and National Development* (Delhi: Macmillan, 1974).

35. Amrita Shah's biography is the most sensitive treatment of a scientist in an open triangular relationship and even on that dimension alone is worth reading: *Vikram Sarabhai: A Life*.
36. After failing her first "pass-mod exam" in Latin at Oxford in December 1937 Indira Gandhi went to Berlin, and was alarmed by the Nazis; a month later she warned her father about meeting with Lord Lothian (Philip Henry Kerr), a prominent critic of Anthony Eden and supporter of Chamberlain (and probably pro-Hitler). She declined to join her father at a weekend party with Lord Lothian, but she went in the end, to accompany him in a protective way. Nehru met and introduced Indira to Krishna Menon, Palme Dutt, Harold Laski, Sidney and Beatrice Webb. In June 1938 Nehru and Indira stayed with Stafford Cripps at his country house, then *both* went to Lord Lothian's home, where Indira was pained to hear the extremely conservative and pro-Nazi conversation. Harold Laski counseled Indira to follow her own path sometimes and not tag along with her father, who would inevitably dominate; said Laski, "she did not heed [her father's] advice and they parted acrimoniously"; Frank, *Indira*, p. 135. In March 1940 Jawaharlal wrote an article sympathizing with the predicament of Finland, which had been invaded and overrun four months later by USSR. In a letter to her father, Indira disagreed with his article, and "boldly states an ideological stance at variance with her father . . . a position considerably to the left of Nehru." He was fifty and she was twenty-two. She was influenced also by Feroze's friends (and hers) at the LSE like P. N. Haksar and Mohan Kumaramangalam, as well as her father's advisor Krishna Menon. Katherine Frank, *Indira: The Life of Indira Nehru Gandhi* (New York: HarperCollins, 2001), p. 154.
37. Ashok Parthasarthi, *Technology at the Core: Science and Technology with Indira Gandhi* (Delhi: Longmans, 2007), pp. 307–9.
38. Tyabji presents the case of the nationalization of the bus system and the choices facing planners about buses in Chennai and Tamil Nadu around 1950: were they to be diesel or petrol? Italian, German, or British? Repairs and servicing workshops? Training contracts? Should all buses be made in one factory in Mumbai? Should some buses be manufactured instead in Coimbatore, where textile manufacturing machinery was already advanced? What role should large purchasers like police and the military play in the choice? When should innovation with aluminum pistons occur? Are foreign investors serious about local production facilities, or are these just meant to be branch-plant assembly workshops? Every one of these questions had its relevance for the atomic energy program. Nasir Tyabji, *Industrialization and Innovation: The Indian Experience* (Delhi: Sage, 2000), p. 49.
39. The Ooty telescope was built to look at a large number of compact extragalactic sources (at 1 arcsec level) at the outer boundaries of the universe; its most famous result was that occultation studies showed that the number and density of radio sources at an earlier cosmic epoch was higher and the linear sizes of the sources were smaller than commonly thought. This provided independent evidence for the big bang model of the universe. The telescope was also used to discover a few pulsars and did extensive studies on already-known pulsars. The group and visitors also made very significant contributions to the subject of interplanetary scintillation (IPS), which arises between the Sun and the Earth because of the intervening plasma medium called the "solar wind," which is nothing but a continual expansion of the solar corona. IPS and solar wind create "near space weather" extensively studied

by the Ooty telescope: it discovered scintillations through the ionic tails of comets, giving a good idea of their density. S. Ananthakrishan, personal communication, 1 December 2008.

40. For a full account of this group's development, see Indira Chowdhury, "Local Hierarchies and International Networks: Molecular Biology at the Tata Institute of Fundamental Research, 1962–1980," *Archives Internationales d'Histoire des Sciences* 56, nos. 156–57 (2006).
41. "Document: Pandora's Box—excerpts from the CSIR Review Committee Report," *Illustrated Weekly of India*, 29 March 1987.
42. Jairam Ramesh, "Technology Missions: A Preliminary Perspective," Planning Commission memo, Delhi, March 1990. I am grateful to Jairam Ramesh for a conversation about the missions in Delhi in January 1991.
43. Shobo Bhattacharya interview with Jeremy Webb, "We have to take risks" *New Scientist*, 19 February 2005.
44. Ashok Jain and V. P. Kharbanda, "Strengthening Science and Technology Capacities for Indigenization of Technology: The Indian Experience," *International Journal of Services Technology and Management* 4, no. 3 (2003): 234–53.
45. There is a specific study of the role of cyclotrons in nuclear research; see Jahnavi Phalkey, "Science, State-Formation and Development: The Organization of Nuclear Research in India, 1938–1959" (PhD diss., School of History, Technology and Society, Georgia Institute of Technology, 2007). Unfortunately this work was not released before the time of publication of this book.
46. Concerning a capability approach in development theory, see Amartya Sen, *Development as Freedom* (Oxford: Oxford University Press, 1998).
47. J. Nehru, 6 October 1955, *Selected Works of Jawaharlal Nehru* (Delhi: Nehru Memorial Fund and Oxford University Press, 1999) 30: 162.
48. G. Venkataraman, "Dr Ramanna as I Knew Him," *Physics News* 35, nos. 2–4 (April–December 2004): 24.
49. Ibid.
50. Shiv Visvanathan, "Reflections on the Transfer of Technology: Notes on the New Pantopticon," *Lokayan Bulletin* (Delhi), 6, nos. 1/2 (1988): 151. Referring to Bhopal and its famous December 1984 gas accident, he wrote about a complex of technology designed and built in 1970: "the machine here is not the Carbide plant, the concrete set of pipes and steel. It is the grammar of the system and in a secular way it combines and transforms from the individual to the collective level."
51. The prestige was attached in part to the superior quality of life enjoyed (pensions, housing, savings, medical facilities, better schools, upward mobility) but also to its distinguished record in WWII. The disapproval originated in the association with the military's role in suppression of Independence movements, in India and Southeast Asia, and its inability or failure in 1946-47–48 to protect vulnerable populations.
52. R. Ramanna, *Years of Pilgrimage* (Delhi: Viking, 1991), p. 110; author's conversation with G. Venkataraman in Puttaparathi, November 1998.
53. So unimportant have scientists been in the eyes of the most senior military leaders that a major book on the military published after the second Pokhran nuclear tests in 1998 provides not a single detail about the DRDO. See Jaswant Singh, *The Defence of India* (London: Macmillan, 1999), pp. 138–40. We do not have a study of the series of scientists who have advised the ministers of Defence since Blackett set Kothari up as the minister's adviser in 1950.

54. Author's conversation with B. D. Nagchaudhuri, Calcutta, December 1998.
55. Ramanna, *Years of Pilgrimage,* p. 98.
56. Recall the comments critical of Indian military science by Air Chief Marshall Idris Latif, quoted in chap. 23, n. 41.
57. Physicist G. Venkataraman, who worked in both DAE and DRDO, recalled the standard lines in the military chorus to Indian technology developers from memory: "we are dead against this unproved Indian product. Self-reliance is good patriotic talk and Science Departments can use it if they want. Here we are dealing with the lives of our *jawans* (soldiers) for which we are responsible. We refuse to take responsibility if we are not given the imported equipment," in "Dr Ramanna as I Knew Him," p. 21.
58. Ashis Nandy, "The Mimic Men," *Illustrated Weekly of India,* 11 February 1990. In the same essay Nandy also criticized the reappointment of distinguished but aging economist Sukhamoy Chakravarty to the Planning Commission.
59. Shiv Visvanathan and Harsh Sethi, eds., *Foul Play: Chronicles of Corruption, 1947–1997* (Delhi: Banyan Books, 1998). I am grateful for a conversation on this subject with the two editors and with Deepak Kumar, historian of science and technology at Jawaharlal Nehru University in Delhi, all in February 1998.
60. Bharat Karnad, *Nuclear Weapons and Nuclear Security: The Realist Foundations of Strategy* (Delhi: Macmillan, 2002). Although I have not yet been able to harmonize my evidence with his sufficiently to find a synthesis, a careful reading of this vast work, including its valuable footnotes, is important.
61. Rahman quoted by Visvanathan, *A Carnival for Science,* p. 5.
62. Matt Moore, Associated Press, "1982 Ulcer Discovery Brings Nobel Prize to Australians," *Vancouver Sun,* 4 October 2005; emphasis mine.

INDEX OF NAMES

SUBJECT INDEX